Organic Catalysis for Polymerisation

Polymer Chemistry Series

Titles in the series:
1: Renewable Resources for Functional Polymers and Biomaterials
2: Molecular Design and Applications of Photofunctional Polymers and Materials
3: Functional Polymers for Nanomedicine
4: Fundamentals of Controlled/Living Radical Polymerization
5: Healable Polymer Systems
6: Thiol-X Chemistries in Polymer and Materials Science
7: Natural Rubber Materials: Volume 1: Blends and IPNs
8: Natural Rubber Materials: Volume 2: Composites and Nanocomposites
9: Conjugated Polymers: A Practical Guide to Synthesis
10: Polymeric Materials with Antimicrobial Activity: From Synthesis to Applications
11: Phosphorus-Based Polymers: From Synthesis to Applications
12: Poly(lactic acid) Science and Technology: Processing, Properties, Additives and Applications
13: Cationic Polymers in Regenerative Medicine
14: Electrospinning: Principles, Practice and Possibilities
15: Glycopolymer Code: Synthesis of Glycopolymers and their Applications
16: Hyperbranched Polymers: Macromolecules in-between Deterministic Linear Chains and Dendrimer Structures
17: Polymer Photovoltaics: Materials, Physics, and Device Engineering
18: Electrical Memory Materials and Devices
19: Nitroxide Mediated Polymerization: From Fundamentals to Applications in Materials Science
20: Polymers for Personal Care Products and Cosmetics
21: Semiconducting Polymers: Controlled Synthesis and Microstructure
22: Bio-inspired Polymers
23: Fluorinated Polymers: Volume 1: Synthesis, Properties, Processing and Simulation
24: Fluorinated Polymers: Volume 2: Applications

25: Miktoarm Star Polymers: From Basics of Branched Architecture to Synthesis, Self-assembly and Applications
26: Mechanochemistry in Materials
27: Macromolecules Incorporating Transition Metals: Tackling Global Challenges
28: Molecularly Imprinted Polymers for Analytical Chemistry Applications
29: Photopolymerisation Initiating Systems
30: Click Polymerization
31: Organic Catalysis for Polymerisation

How to obtain future titles on publication:
A standing order plan is available for this series. A standing order will bring delivery of each new volume immediately on publication.

For further information please contact:
Book Sales Department, Royal Society of Chemistry, Thomas Graham House, Science Park, Milton Road, Cambridge, CB4 0WF, UK
Telephone: +44 (0)1223 420066, Fax: +44 (0)1223 420247
Email: booksales@rsc.org
Visit our website at www.rsc.org/books

Organic Catalysis for Polymerisation

Edited by

Andrew Dove
University of Birmingham, UK
Email: A.Dove@bham.ac.uk

Haritz Sardon
University of the Basque Country UPV/EHU, Spain
Email: haritz.sardon@ehu.es

and

Stefan Naumann
University of Stuttgart, Germany
Email: stefan.naumann@ipoc.uni-stuttgart.de

ROYAL SOCIETY OF CHEMISTRY

Polymer Chemistry Series No. 31

Print ISBN: 978-1-78801-184-6
PDF ISBN: 978-1-78801-573-8
EPUB ISBN: 978-1-78801-679-7
Print ISSN: 2044-0790
Electronic ISSN: 2044-0804

A catalogue record for this book is available from the British Library

The Royal Society of Chemistry is a charity, registered in England and Wales, Number 207890, and a company incorporated in England by Royal Charter (Registered No. RC000524), registered office: Burlington House, Piccadilly, London W1J 0BA, UK, Telephone: +44 (0) 20 7437 8656.

For further information see our web site at www.rsc.org

Printed in the United Kingdom by CPI Group (UK) Ltd, Croydon, CR0 4YY, UK

Preface

Polymers are ubiquitous in our daily lives. They have revolutionized almost every aspect of modern life, from low cost materials with short lifetimes such as packaging or other single-use products, to those with longer lifetimes such as clothing or construction, through to higher value materials that are components in high value products such as plastic electronics or healthcare applications. As may be expected from the age of petroleum, a majority of polymers that are commonly used are derived from petrochemical resources. Inevitably, with the high current focus on sustainable sourcing of polymers and the need for reducing plastic waste that enters and persists in the environment, when polymers enter the public consciousness, discussions quickly turn to the consideration of their negative points over the many positives in a blessing-and-curse style of argument.

While the benefits of polymers are clearly understood, the major points of criticism focus on the environmental impact of the polymers, from their end-of-life treatments that include limited recycling, burning to produce CO_2 and other harmful greenhouse gasses, disposal in landfill or worse, littering our environment. Furthermore, the leaching of potentially harmful polymer additives such as plasticisers, residual monomers or catalysts all have potentially negative effects on the environment and ultimately people.

The environmental issues, as well as concerns over the depletion of fossil fuel resources from which many polymers are made, have led to a significant and growing interest in 'green' chemistries. This is a broadly defined topic that, with relation to polymer science, is dedicated to increasing the sustainability of polymer synthesis and the connected process technologies, with a strong focus on improving the sustainability of all aspects of plastics technology from removal of harmful solvents to using more sustainably sourced precursors to alleviate environmental concerns, health hazards and resource efficiency. As one of the guiding principles of green chemistry, the

Polymer Chemistry Series No. 31
Organic Catalysis for Polymerisation
Edited by Andrew Dove, Haritz Sardon and Stefan Naumann

Published by the Royal Society of Chemistry, www.rsc.org

application of catalysis is a critical piece of the puzzle. Of all the tools in the chemists' arsenal, catalysis is probably the one able to contribute the most to these goals. In polymer synthesis, catalysis plays an essential role in increasing the rate of reactions and reducing side reactions that occur to improve selectivity thus leading to more predictable outcomes and less waste. Given the excellent performance and vast array of opportunities that metallo-organic chemistry offers through the almost infinite ligand/metal combinations, it is not surprising that transition metals and organometallic catalysts have dominated the field of polymerisation catalysis. However, in part due to concerns over the availability of some widely used metals such as zinc or silver, which risk complete disappearance in the next 100 years, or ruthenium, lithium or copper, which will be seriously threatened in the future if their consumption continues to increase, the use of organic compounds to catalyse polymerisation reactions is gaining increasing interest. In the past two decades, the remarkable ability of small organic molecules to mediate a variety of polymerisation processes has brought about the rapid evolution of metal-free, organocatalytic polymerisation techniques.

While the interest in organic catalysis had been growing for mediating an array of organic transformations, its initial translation into the world of polymer synthesis was in the ring-opening transesterification polymerisation of lactide in order to produce 'soft-etch' polymers for potential microelectronics applications that were readily degradable in mild acidic etching conditions and would not leave any metal residues. While the potential for leaving no metals in the polymer was also quickly recognised in the biomedical field where metal-based impurities can be prohibitively expensive to remove, organocatalysed polymerisation has now progressed far beyond the fact of merely being 'metal-free', presenting ways to lower toxic byproducts in polymers and to address resource efficiency concerns associated with metals. Inspired by nature with the use of enzymes to catalyse biochemical reactions, metal-free polymerisations offer unique polymerisation pathways, access to various macromolecular architectures and novel selectivities, quite often in combination with excellent control over important polymer parameters such as molecular weights, polydispersity, end groups and copolymer constitution. Indeed, the field has progressed to the point that organic catalysts now provide many potential advantages to their metallo-organic counterparts that mean that in many cases they are now preferred (at least in the academic world) for polymerisation. Many organocatalytic species are inexpensive, commercially available molecules that either can be used as received or can be made and purified through a limited number of steps. Indeed, a wide range of the most commonly used species, although not all, are stable to both water and oxygen, which beyond providing an advantage in handling also results in longer shelf lives without the need for storage in inert atmospheres. While some applications may still require removal of air/moisture (*i.e.* ring-opening transesterification polymerisation), this is a process that is only undertaken to ensure high end group fidelity and a good match of molecular parameters to those predicted/desired rather than any need for the organic catalyst to work. Many of the most commonly used organic catalysts are also stable to a large array of reaction conditions, solvents, and monomers, making them highly versatile. Finally, the acidic, basic or ionic nature of a majority of the materials enables their ready removal from polymer mixtures by low cost methodologies such as washing or trapping in resin beads.

The field of organic catalysis for ring opening polymerisation has grown significantly in the past 15 years to the point that the advantages of the approaches are being recognised and organocatalytic approaches are being sought in preference to metal-based catalysts. A much broader range of monomers can now be polymerised, being extended far beyond its original scope to include not only lactones and carbonates, but also epoxides, anhydrides, siloxanes, lactams, acrylic monomers and many others. Alongside the development of innovative catalyst families that are able to be highly selective towards one functionality over another, the field is blossoming. While there are some reviews and viewpoints in the journal literature, the many facets of organopolymerisation and an ever increasing number of publications has raised the demand for a comprehensive review, especially since no comparable collection of information is currently available. As such, the presented edited book on *Organic Catalysis for Polymerisation* is the first of its kind on this research topic. Eminent experts in their respective fields have taken a detailed look at all relevant aspects of metal-free polymerisation approaches. The close interconnections between catalyst development and the investigation of novel polymers and materials are mirrored in the organization of the chapters, where two different viewpoints are taken. The first part of the book presents the fundamental, metal-free catalyst polymerisation principles (nucleophilic, acid- and base-catalysed as well as dual or supramolecular catalysis, Chapters 1–4). Here all relevant types of organocatalysts are detailed, with an emphasis on polymerisation mechanisms and on elucidating the impact of structural changes in the catalyst on the resulting polymers. Together, this delivers an informative profile on the evolution of the field and describes how the different catalyst families are able to polymerise the various monomer classes, highlighting the crucial differences between them to present clearly the opportunities that remain in those areas. In the second part of the book, the focus is turned onto the different classes of monomer, detailing the existing metal-free polymerisation strategies for a given class of compounds (lactones, lactides, carbonates, epoxides, other cyclic and acrylic monomers, Chapters 5–11) to provide a one-stop guide to select the most appropriate catalyst for any given process, as well as inspiration for where the future challenges lie. These chapters will not only provide a comprehensive overview for the polymerisation of conventional monomers such as lactide, epoxides or trimethylene carbonate but also summarize the most relevant results about the use of organocatalysis for the ring opening polymerisation of aziridines or phosphester monomers amongst many others. Taken together, this will facilitate access to this research field both for readers interested in catalyst design and development as well as those focusing on the best methodologies to synthesise a specific type of polymer using organocatalytic routes. Finally, the focus is shifted to the more nascent, yet highly important areas of organocatalysed step-growth reactions and metal-free controlled radical polymerisation (Chapters 12–13). In this third part, the potential of organocatalysis for step-growth polymerisation reaction and the preparation of industrially relevant polymers such as polyurethanes and in the emerging area of organocatalysed controlled polymerisation for the preparation of well-defined specific polymer structures are described. Subsequently, Chapter 14 is focussed on the emerging subject of organocatalysis in polymer recycling/depolymerisation. No doubt this area will grow significantly in importance over the coming years and, given the environmental credentials of organocatalysis, it could play an important role to facilitate the implementation of inexpensive and sustainable chemical recycling processes. Finally, an outlook (Chapter 15) provides a commentary of the most important developments in organic catalysis for polymerisation to date and summarises some of the major challenges that face the field in the coming decade.

We like to conclude with the grateful acknowledgement of all contributors, who with their expertise and diligence were crucial to the success of this project and wish that readers may find *Organic Catalysis for Polymerisation* an informative and thought-provoking inspiration.

Stefan Naumann, Haritz Sardon and Andrew P. Dove

Contents

Polymer Chemistry Series No. 31
Organic Catalysis for Polymerisation
Edited by Andrew Dove, Haritz Sardon and Stefan Naumann

Published by the Royal Society of Chemistry, www.rsc.org

CHAPTER 1

Nucleophilic Catalysts and Organocatalyzed Zwitterionic Ring-opening Polymerization of Heterocyclic Monomers

OLIVIER COULEMBIER

University of Mons, Center of Innovation and Research in Materials and Polymers (CIRMAP), Laboratory of Polymeric and Composite Materials, Place du Parc 23, Mons 7000, Belgium
Email: olivier.coulembier@umons.ac.be

1.1 Introduction

Organocatalysis refers to a form of catalysis whereby the rate of reaction is increased by an organic molecule preferably used in substoichiometric amounts. The use of organic molecules to perform chemical reactions is not a new concept and organocatalytic reactions look back on a respected history. Both cyanohydrin synthesis from quinine alkaloids[1] and proline-catalyzed Robinson annulation[2] belong to the most popular dated examples of organocatalytic reactions. Although organic molecules have been used at the beginning of the chemistry, their narrow scope of reactions has not really stirred scientific interest in the past. Nowadays, thanks to clever and sometimes serendipitous discoveries, the picture is changing and organocatalysis is becoming an indispensable part of organic chemistry, offering a wide diversity of reactions, catalysts and processes.[3–5]

Polymer Chemistry Series No. 31
Organic Catalysis for Polymerisation
Edited by Andrew Dove, Haritz Sardon and Stefan Naumann

Published by the Royal Society of Chemistry, www.rsc.org

While most of the organo-based reactions concern the enantioselective preparation of small molecules, organocatalysis also offers number of prospects in the polymer community and proposes advantages over metal-based and bioorganic methods.[6] In this chapter, special attention is devoted to organocatalyzed ring-opening polymerization (ROP) of cyclic monomers and more especially the zwitterionic ROP (ZROP) from nucleophilic catalysts.

1.2 Definition of ZROP

ZROP is a chain polymerization in which a growing macromolecule bears two ionic chain carriers of opposite signs at its two ends and which usually grows from one of them.[7] The zwitterion propagating species—either obtained from a neutral nucleophilic initiator or a neutral electrophile—is initially poorly active since the electrostatic work needed for parting the opposite charged ions is the largest when they are close together. Stabilization of a zwitterionic species may involve the positive end of one zwitterion propagating chain acting as the counter-ion of the carbanion end of another zwitterion propagating chain.[8] Termination of the reaction may be caused by the presence of a protic nucleophile or by a charge cancellation step of the highly polar dimer leading to linear and cyclic structures, respectively.[9] While neutral electrophiles have already been used in ZROP,[10–12] zwitterionic polymerizations typically employ neutral nucleophiles that react with heterocyclic monomer to *in situ* produce the zwitterionic initiator (Scheme 1.1).

To date, several types of organic molecules have been employed as neutral nucleophiles to initiate ZROPs of cyclic monomers. Similarities with simple acylation reactions are unquestionable and allows pyridine-based molecules, imidazoles, amidines, tertiary amines, phosphines and N-heterocyclic carbenes (NHCs) to be used as initiating agents. Considering Scheme 1.1 (bottom of the scheme) as the general way of polymerization, the latent electrophile present on the cyclic monomer (Z) is most of time a carbonyl function. Strained lactones, thiolactones, *N*-caboxyanhydrides, carbosilanes and cyclic carbonates are then good candidates to undergo nucleophilic ZROP (see Chapter 11).

1.2.1 Pyridine-based Initiation

More than 50 years ago, β-lactones such as β-propiolactone and β-pivalolactone started to be polymerized by pyridine-based nucleophilic initiators.[13–16] For several reasons, the course of such ZROP was very complex, involving chain growth and step growth kinetics as well as elimination reactions regarding the type of initiator and monomer used. When moderate bases[17] such as pyridine, 4-methylpyridine and 4-(*N*,*N*-dimethylamino)pyridine (DMAP) were used for the ZROP of pivalolactone,[16] linear chains having one pyridinium ion and a carboxylate ion as end groups were observed by ^{1}H NMR and MALDI-ToF analyses. The absence of cyclic structures

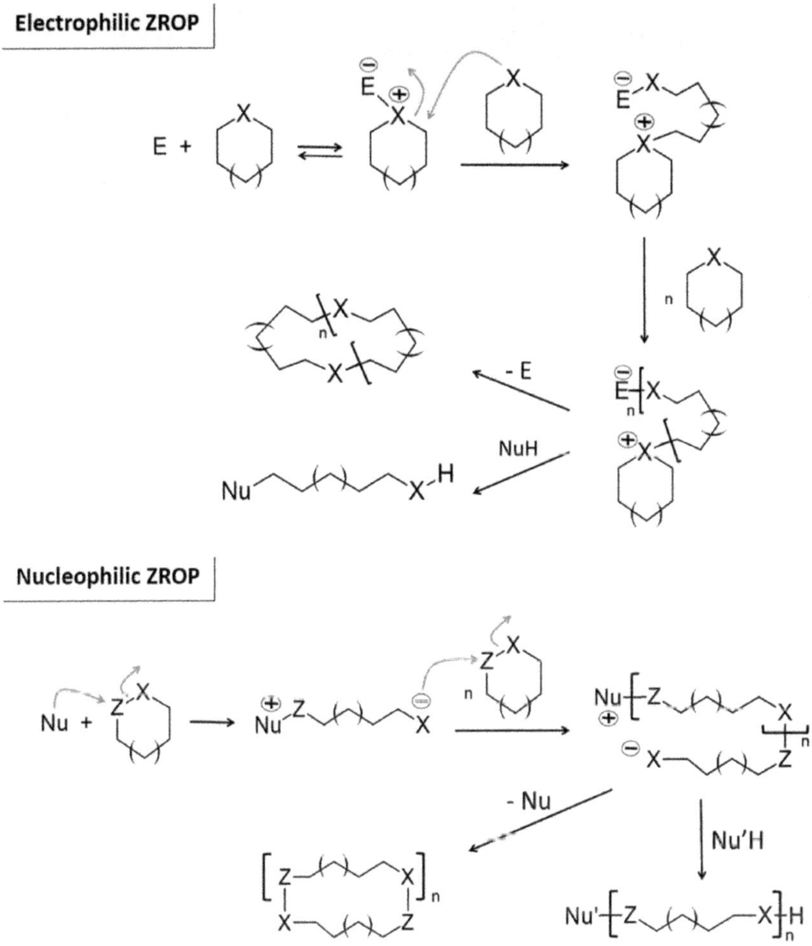

Scheme 1.1 General mechanism of electrophilic (top) and nucleophilic (bottom) zwitterionic ring-opening polymerizations (X and Z represent a heteroatom and an electrophilic site, respectively. Nu and E represent nucleophilic and electrophilic entities, respectively).

suggested that the ZROP proceeded exclusively from the CO_2^- anion (Scheme 1.2, top). In the case of β-propiolactone and β-butyrolactone (BL), complete elimination of the pyridinium ions and formation of acrylate and crotonate end groups, respectively, were observed (Scheme 1.2, bottom).[14,18]

The catalytic potential of donor-substituted pyridines is well established since the report on DMAP by Litvinenko *et al.* in 1967 and by Steglich *et al.* two years later.[19,20] A catalytic improvement was reported in 1978 for 4-(pyrrolidinyl)pyridine (PPY),[21] and in 2003 with 9-azajulolidine.[22] Annelated pyridine derivative is a powerful organocatalyst not only suitable for acylation reactions, but also for other transformations such as the aza-Morita Baylis Hillman reaction.[23]

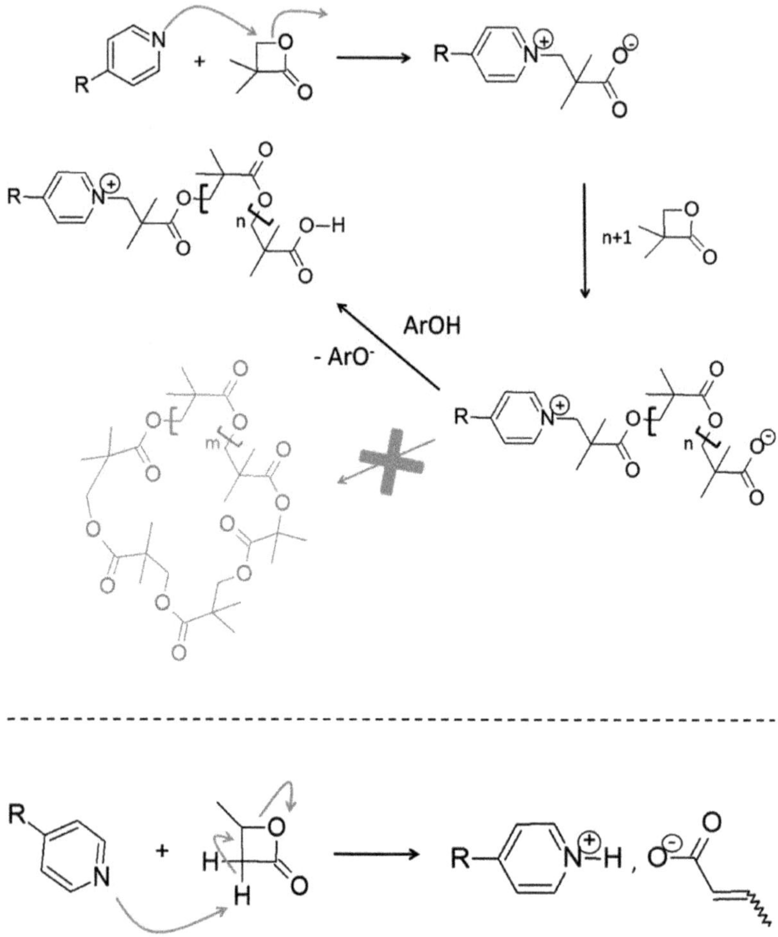

Scheme 1.2 Top: ZROP of pivalolactone by pyridine derivatives (R = H, pyridine; R = Me, 4-methylpyridine; R = -N(CH_3)$_2$, DMAP); bottom: crotonate formation from proton α-elimination of β-butyrolactone induced by pyridine.

The commercially available DMAP catalyst is often the common choice for acylation reactions proceeding by a nucleophilic mechanism involving an acyl pyridinium ion intermediate (Scheme 1.3). The amplified reactivity of aminopyridine derivatives—up to four orders of magnitude higher than pristine pyridine in representative acyl transfer[17,24]—may come from the greater equilibrium concentration of the acyl pyridinium intermediate and its increased electrophilicity because of looser ion pairing.[25–28]

In 2001, Hedrick *et al.* demonstrated that Lewis basic amines such as DMAP and PPY were highly effective for the ROP of lactide (LA) monomers.[29] Although other works on metal-free processes were published earlier,[30–32] his report is considered today as the nucleating agent in the field of the

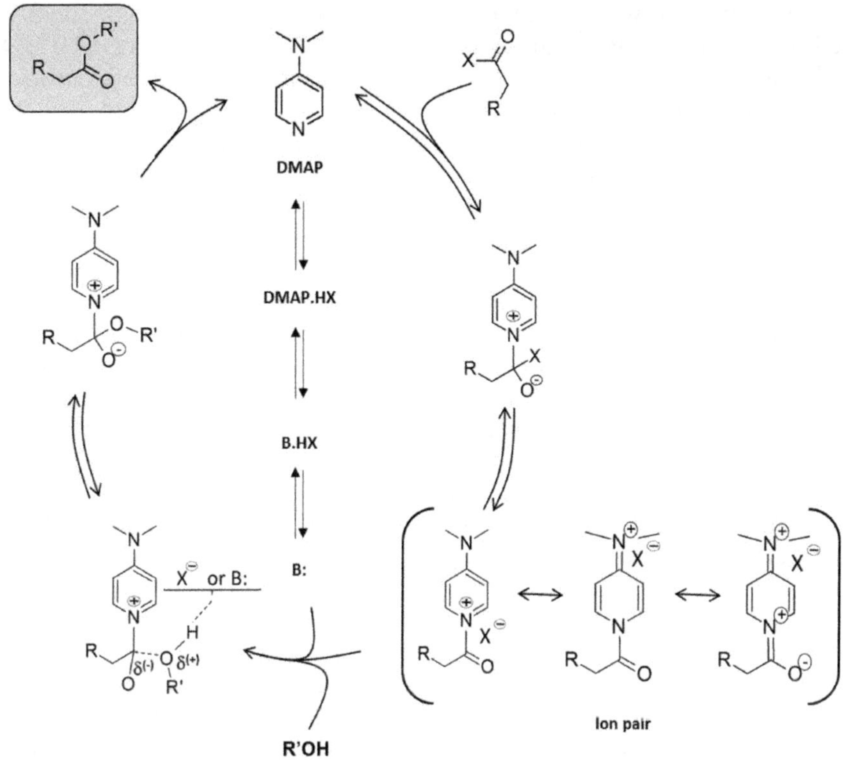

Scheme 1.3 Proposed mechanism for a DMAP-catalyzed acylation reaction. Simplification of the mechanism proposed by Spivey in [25].

organocatalytic approach to the living ROP of lactones. The motivations of such research came from the realization of the potentially dangerousness of the organometallic catalysts used so far in the preparation of polyesters produced for biomedical and electronic applications. The catalytic behavior of both DMAP and PPY in the ROP of LA was studied in dichloromethane at 35 °C using ethanol (EtOH) as initiator in the presence of 0.1 to 4 equivalents of amine as compared to EtOH. Under anhydrous conditions, no polymerizations were observed in the absence of initiating alcohol. Narrowed PLAs were produced for a degree of polymerization (DP) ranging from 30 to 120 in 20 to 96 h. By contrast to most organometallic-promoted polymerizations, the dispersity was kept extremely low well after complete conversion with no noticeable molecular weight modifications. The living character of the polymerization was deduced from the linear evolution of the molecular weight as a function of the conversion, the predictable molar masses and the low dispersities. This living character is the manifestation of the rapid initiation and the weakly nucleophilic propagating species (secondary alcohol) that is active only to the cyclic diester monomer, precluding undesirable transesterification reactions.[33]

Polymerization was originally proposed to occur *via* a "monomer-activated" mechanism through nucleophilic activation of the LA (Scheme 1.4, route A). ^{1}H NMR analyses confirmed that the obtained PLA chains are end-capped in α position by an ester function generated from the initiating alcohol and in ω by a hydroxyl group. This suggests that for such mechanism the alkoxide/acyl pyridinium zwitterion generated by the nucleophilic attack of the DMAP on the LA is subjected to a proton transfer from the initiating/propagating alcohol and an acylation from the resultant alkoxide. However, subsequent computational studies realized by Bourissou *et al.* predicted that a base-catalyzed pathway would be of lower energy (Scheme 1.4, route B).[34] Those simulations were realized in gas phase, implying that in solution, and especially in the presence of alcohol initiators, the nucleophilic monomer activation could be predominant. In 2012, both Hedrick's and Bourissou's uncertain mechanistic concepts were conciliated.[35] Kinetic studies demonstrated that DMAP, generally used in excess as regard to the initiating alcohol, plays a double dealing and is involved in both nucleophilic activation of the LA and the basic activation of the initiating/propagating alcohol. It was demonstrated that ~2 DMAP molecules complete the coordination sphere of the initiating/propagating alcohol and that any excess of DMAP is involved in the nucleophilic activation of the monomer (Scheme 1.5). Interestingly, in the same study, the equimolar ratio of DMAP and *N,N′*-dicyclohexylcarbodiimide (DCC) was demonstrated to better control the ROP of L-LA. As compared to the DMAP alone, a DMAP/DCC mixture was proved to be the only catalytic system totally responding to a livingness criterion.

Ironically, if DMAP represents a very effective catalyst for the metal-free ROP of LA, it is also one of the less active. To circumvent that problem, Péruch *et al.* applied in 2010 the concept of "dual catalysis" to enhance the polymerization activity.[36] That notion, where a Lewis acid supports a nucleophile to gain an increased catalytic effect, broadly falls in the categories of cooperative (synergistic)[37] or cascade catalysis,[38] depending on the proposed polymerization mechanism and is related to the chemistry of frustrated Lewis pairs.[39] In their study, Péruch *et al.* developed an organocatalytic system containing both basic and acidic sites activating cooperatively the alcohol chain end and the LA monomer. To this end, equivalent amounts of DMAP and its protonated form (DMAP.HX) were used as a dual catalytic system for L-LA polymerization initiated by different alcohols (Scheme 1.6). It was shown that the corresponding DMAP/DMAP.HX systems are significantly more active than DMAP alone, and yield well-controlled poly(L-lactide) up to 15 000 g mol^{-1} in DCM at 40 °C after 48 h. While enhancement of the reaction was demonstrated highly dependent on the nature of the counter-anion ($CF_3SO_3^- > CH_3SO_3^- > Cl^-$), transesterification reactions were prevented by fine-tuning the experimental conditions. Note also that this synergetic concept was later applied by Kakushi *et al.* by protonating DMAP with diphenyl phosphate.[40]

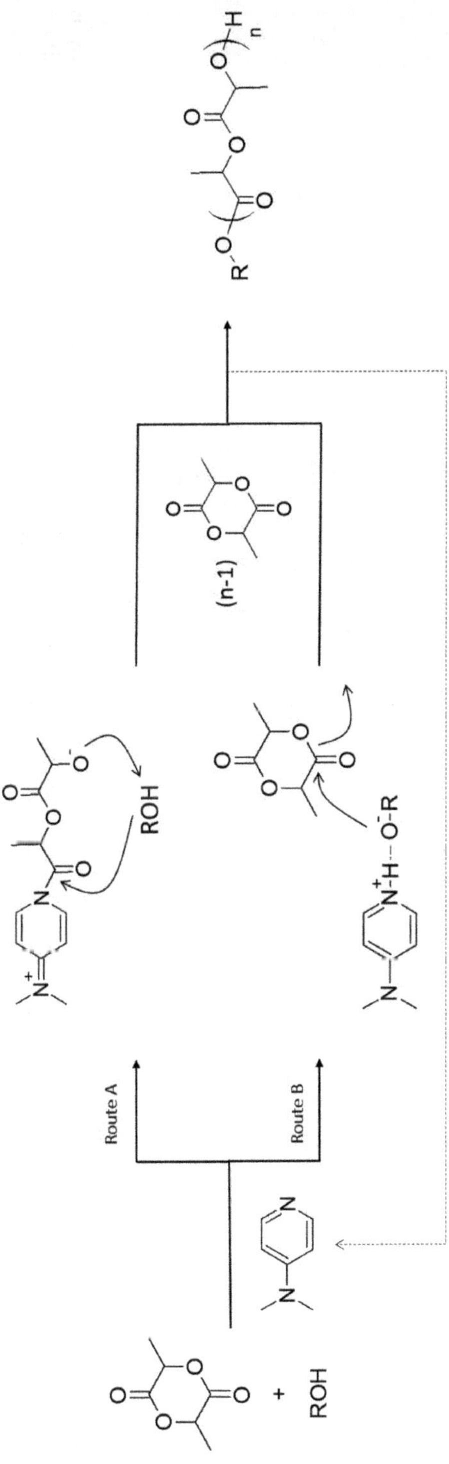

Scheme 1.4 Proposed mechanism for the ROP of LA using DMAP catalyst. Route A: nucleophilic activation; route B: basic activation.

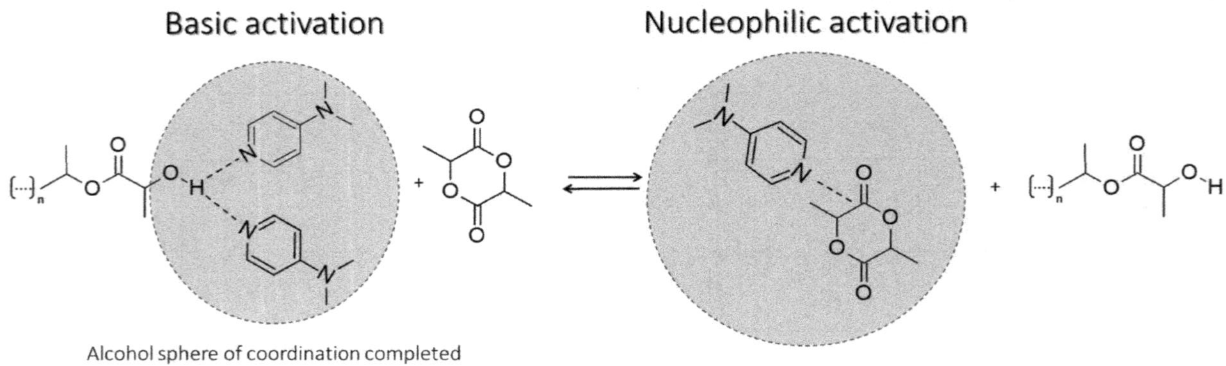

Scheme 1.5 Representation of both basic and nucleophilic activations of the (propagating) alcohol and the LA monomer, respectively. The nucleophilic activation of the LA monomer is here envisioned if free DMAP is present in the reactive medium. (The number of DMAP implied into the coordination of the OH may vary from 1 to 3.)

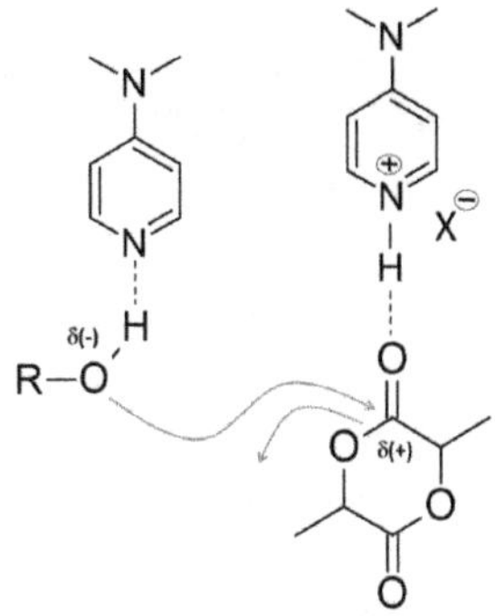

Scheme 1.6 Possible activation mechanism of LA during its polymerization from a DMAP/DMAP.HX mixture.

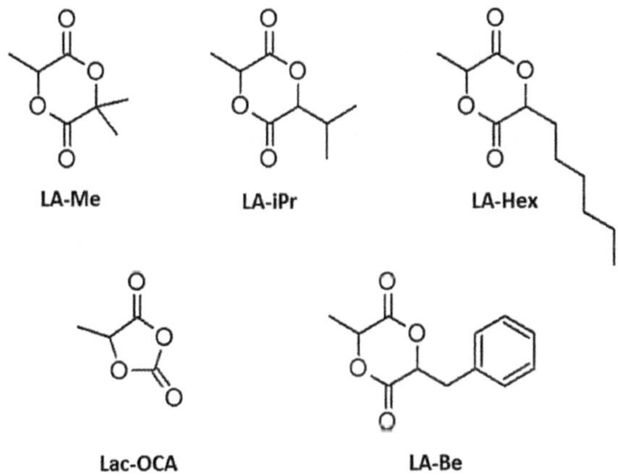

Figure 1.1 Structures of functionalized lactides (LA-Me, LA-*i*Pr, LA-Hex, LA-Be) and lactic *O* carboxyanhydride (Lac-OCA) polymerized by DMAP catalyst.

Next to lactide, a series of other monomers have also been polymerized with pristine DMAP (Figure 1.1). Alkyl-substituted LA were studied by Moeller and coworkers starting in 2004.[41,42] In their study, authors did compare the catalytic efficiency of DMAP to the well-known tin(II) 2-ethylhexanoate ($Sn(Oct)_2$) catalyst. Used in the same concentration (1.5 mol% *vs.* monomer, $[ROH]_0/[DMAP]_0 = 2$), DMAP was demonstrated more active than its metallic homologue. After 24 h in bulk at 110 °C, a 65% conversion was recorded for LA-Me showing good control in terms of molecular weight and dispersity values. By using an excess of DMAP ($[ROH]_0/[DMAP]_0 = 0.5$), polymerizations of LA-Me, LA-*i*Pr, LA-Be and LA-Hex reached 35, 80, 95 and 97% conversion, respectively, in only one hour. Interestingly, this result indicates that DMAP is more efficient in polymerizing steric-hindered lactides (with good control in both M_n and $Đ_M$) than $Sn(Oct)_2$ catalyst.

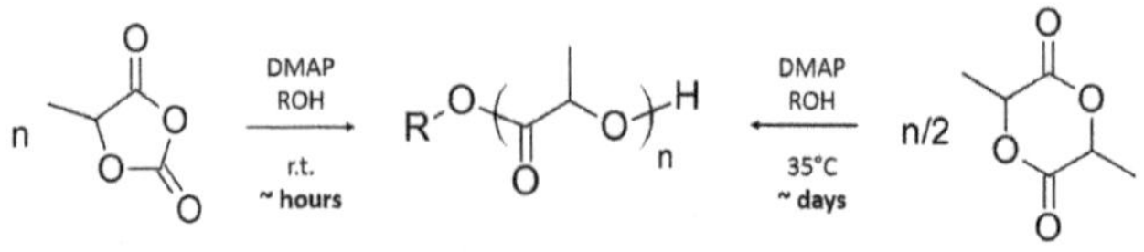

Scheme 1.7 DMAP catalyzed ROP of LA and lac-OCA.

In 2006, Bourissou *et al.* employed DMAP to catalyze the polymerization of the lactic *O*-carboxyanhydride (Lac-OCA, Figure 1.1).[43] This α-lactone equivalent exhibited remarkable reactivity compared to LA in DMAP catalyzed ROP (Scheme 1.7). Depending on the targeted DP (10–600), complete monomer conversions were obtained in minutes to hours (in CH_2Cl_2, 0.75 M, $[ROH]_0/[DMAP]_0 = 1$) while four days were necessary for DMAP to catalyze a DP 10 in LA at 35 °C. Such differences in terms of activity between the two monomers is due to the liberation of carbon dioxide when Lac-OCA is ring-opened and not to the mechanism strictly speaking, which, based on computational investigations, is ascribed to a base-catalyzed route.[44] The optimized intermediates and transition states substantiate the role of multiple hydrogen bonding, evidencing the possibility of the DMAP acting as a bifunctional catalyst. Remarkably, if a basic-catalyzed mechanism may lead to the deprotonation of the α-methine hydrogen of OCA during a ROP process,[45,46] no detectable amount of epimerization of the stereogenic carbon atom of Lac-OCA was observed by homonuclear decoupled ^{1}H NMR spectroscopy.

To face the low propensity of DMAP to polymerize other lactones than LA, *e.g.* ε-caprolactone (CL) and ω-pentadecalactone (PDL),[47–49] Dove *et al.* explored the cooperative effects between Lewis acids and the DMAP organobase (beyond others).[50] While a cocatalysis with YCl_3 and $AlCl_3$ delivered intermediate and no activity, respectively, the combination of DMAP with MgI_2 was revealed to be a very active catalytic duo for both PDL and CL polymerizations. In only two hours, PPDL and PCL samples of 70 000 g mol^{-1} ($Đ_M \approx 1.8$) and 29 000 g mol^{-1} ($Đ_M \approx 1.3$), were prepared in toluene at 110 °C and in THF at 70 °C, respectively.

1.2.2 Imidazole-based Initiation

Next to DMAP, the less toxic imidazole has also been proved to be an efficient catalytic system for transesterification reactions between aliphatic acid esters and alkyl alcohols.[51] In 2003, imidazole was demonstrated as being as efficient as DMAP for a series of reactions between anhydrides and various alcohols under microwave treatment.[52] At the beginning of the 21st century, Kricheldorf and coworkers highlighted the ability of imidazole to promote and lead polymerizations of α-amino acid *N*-carboxyanhydrides (NCA) and L-LA.[53,54] A series of NCAs from sacrosine (Sar), D,L-leucine (Leu), D,L-phenylalanine (Phe) and L-alanine (Ala) was prepared and polymerized in dioxane at 60 °C for two days and various NCA-to-imidazole ratios. As attested by

^{1}H-NMR spectroscopy and MALDI-ToF spectrometry, the authors concluded that polymerizations of Sar, Leu and Phe led majorly to the formation of cyclic oligopeptides obtained from the combination of chain-growth, step-growth and cyclization processes (Scheme 1.8, route B). Comparatively, in the case of poly(Ala), the solubility of the secondary polymer structure induced the rapid precipitation of the growing macromolecule preventing the cyclization step and leading to linear poly(Ala) (Scheme 1.8, route A).

The rare combination of both chain- and step-growth processes has also been observed during the L-LA polymerization initiated/catalyzed by imidazole-based molecules.[54] Bulk polymerization of L-LA (~100 °C) results in complete polymerization within two days. After four hours of reaction, even-numbered PLA cycles were observed and obtained from end-to-end cyclizations. Authors demonstrated that all prepared cycles were amorphous, suggesting a base-catalyzed racemization from reversible deprotonation of the LA α-CH group by the imidazole molecule (Scheme 1.9). Longer reaction times favored the equilibration with odd-numbered PLA cycles and has been observed total after eight hours at 150 °C. If several protic heterocycles are

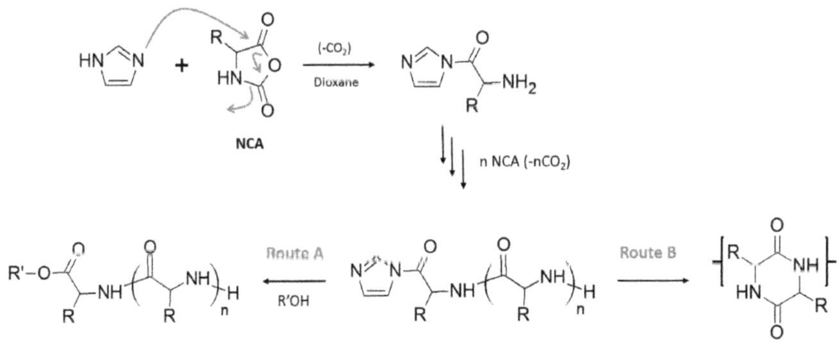

Scheme 1.8 Imidazole initiated ROP of NCAs.

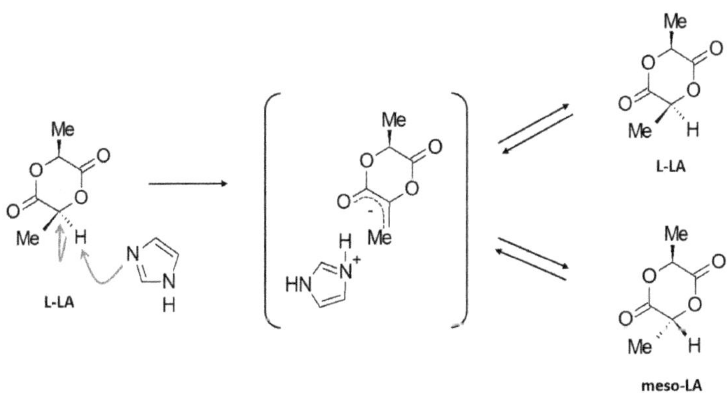

Scheme 1.9 Reversible α-deprotonation of L-LA by imidazole.

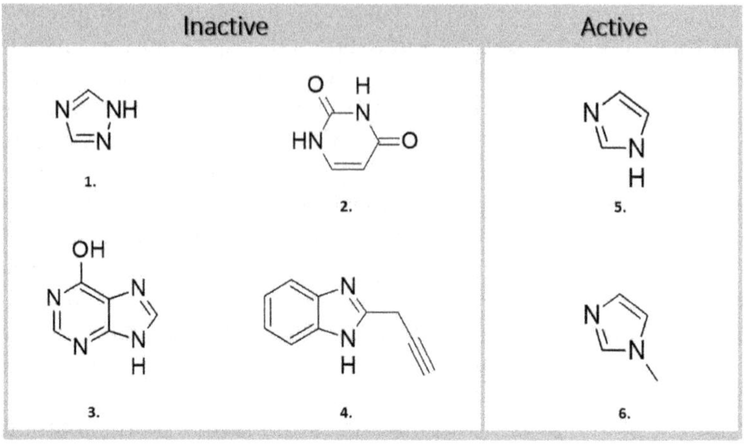

Figure 1.2 Heterocycles tested for the bulk ZROP of L-LA: 1. 1,2,4-triazole; 2. uracil; 3. hypoxanthum; 4. benzylimidazolyl acetonitrile; 5. imidazole and 6. *N*-methyl imidazole.

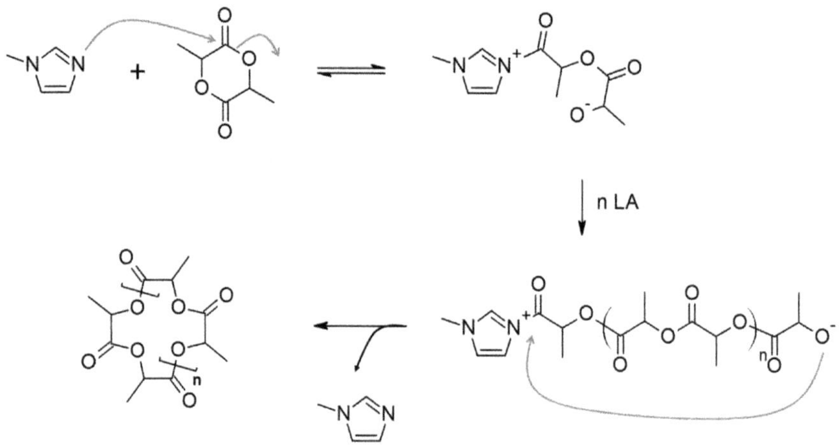

Scheme 1.10 *N*-methylimidazole-catalyzed ZROP of LA.

also tested (Figure 1.2), the *N*-methyl imidazole is the only one active in bulk, leading to cyclic PLA (with several by-products). The formation of PLA macrocycles is explained by a zwitterionic mechanism as outlined in Scheme 1.10.

1.2.3 Amidine/Guanidine-based Initiation

Lewis bases such as DMAP and imidazole-based molecules present the ability to promote acylation processes due to their adequate nucleophilicities.[55] Among potential other candidates, 1,8-diazabicyclo[5.4.0]undec-7-ene (DBU) and 1,5-diazabicyclo[4.3.0]non-5-ene (DBN) have also been demonstrated highly active in representative reactions[56–60] despite the fact they have also been termed as "non-nucleophilic" bases.[61–64] Mayr and coworkers clarified the situation by determining quantitatively their nucleophilicities and their Lewis basic characters.[65,66] As compared to DMAP, both nucleophilicities and Lewis basicities gradually increase in the series DMAP < DBU < DBN.

In 2012, Waymouth *et al.* studied the possibility of using both DBU and DBN to promote the ZROP of lactide in the absence of an exogenous protic initiator (Scheme 1.11).[67] While no reactions were observed in THF at r.t., polymerizations proceeded readily in DCM and THF/DCM mixtures affording PLAs with M_n up to 53 000 g mol^{-1} and $Đ_M < 1.6$. Discordance between theoretical and experimental molar masses is observed, hampering the control of the polymerization process ($M_{n,\text{exp}} > M_{n,\text{th}}$). The rate of polymerization is first order in lactide and demonstrated to be approximately three times higher with DBU than with DBN. MALDI-ToF mass spectrometry and dilute solution viscosity studies revealed that resulting polymers were predominantly cyclic polylactides, along with minor amounts of linear PLA. Theoretical studies suggested that both DBU and DBN may act as nucleophilic initiators leading to the *in situ* generation of a zwitterionic species (Z) in equilibrium with its ring-closed homologue (RC). Addition of lactide to the alkoxide of the zwitterion Z results in chain growth to larger structures (Z_n). Cyclization by attack of the alkoxide end-group to the electrophilic acyl

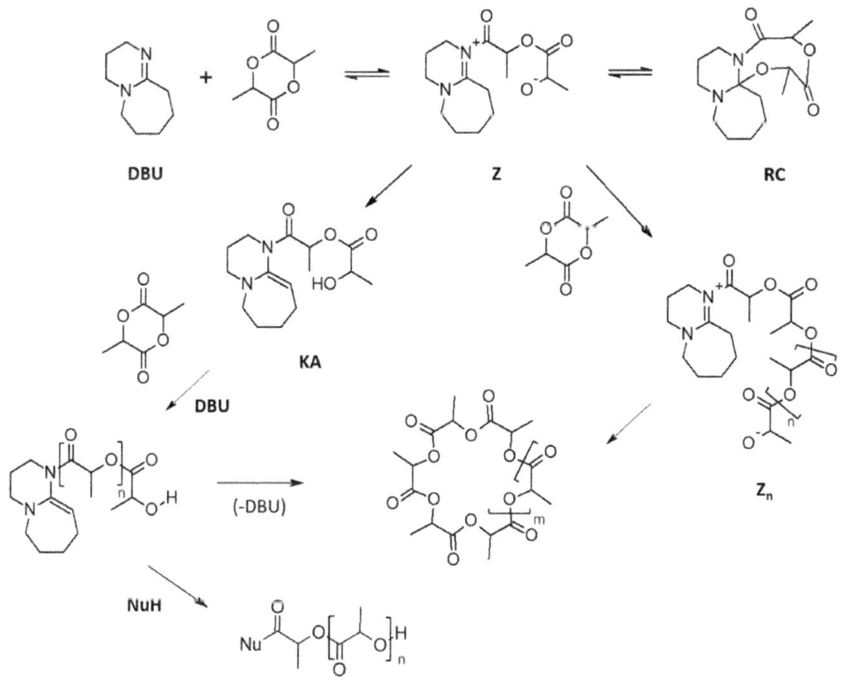

Scheme 1.11 Proposed mechanism for the DBU-mediated ZROP of LA.

amidinium liberates the cyclic PLA and the pristine amidine. Contamination of those cyclic structures by linear species was explained by the energetically accessible ketene-aminal intermediate (KA) obtained by deprotonation of Z. In the presence of excess DBU, KA could undergo a chain growth process generating linear PLA. Under the workup procedure, substitution of the amidine by exogenous nucleophile (NuH) would lead to NuH end-capped chains.

The ability of both amidine and guanidine to initiate the nucleophilic ZROP of β-lactones has also been assessed.[68] Guillaume *et al.* demonstrated the efficiency of both DBU and 1,5,7-triazabicyclo[4.4.0]dec-5-ene (TBD) to act as neat initiators in the ROP of the recalcitrant BL in bulk at 60 °C. Interestingly, the inability of that lactone to be polymerized by using TBD was claimed by Hedrick *et al.* in 2006 when performing the reaction in solution at r.t. (in the presence of an alcohol).[69] The authors demonstrated by ^{1}H and HMBC 2D-NMR that a 1 : 1 mixture of TBD and BL induces the formation of an acyl intermediate stabilized by strong hydrogen bonding of the acidic proton of the ring-opened BL to the adjacent TBD nitrogen atom (Scheme 1.12). Attempts to disrupt this H-bonding at 50 °C led to the generation of oligomers and crotonate byproducts. Theoretical calculations have also proved that this amide-like intermediate is too highly stabilized, presenting an "insurmountable" energy barrier to propagate.[70]

The effective ROP of BL was however demonstrated by Guillaume *et al.* when performing the polymerization reaction in bulk at 60 °C, highlighting the importance of both monomer concentration and temperature. For various monomer-to-catalyst ratios ($[BL]_0/[TBD \text{ or } DBU]_0 = 100–500$), TBD proved to be better than DBU at promoting polymerizations. PBL samples presented experimental molar masses in accord to the theoretical ones (up to 20 000 g mol^{-1}) and proportional to the conversion evolution. All attempts to improve the kinetics by increasing the temperature (up to 90 °C) led to an uncontrolled process characterized by PBL of relatively low molar masses. Interestingly, the introduction of an excess of exogenous alcohol (1 to 10 eq. *vs.* the catalyst) did not affect significantly the polymerization in terms of kinetics and microstructure. Multinuclear NMR spectroscopies and MALDI-ToF spectrometry allowed the authors to speculate on the nature of the PBL end-groups corresponding likely to an amidine (or guanidine) α-chain end and ω-crotonate moieties. The origin of those end-groups was postulated from two possible mechanistic routes (Scheme 1.13) involving either the

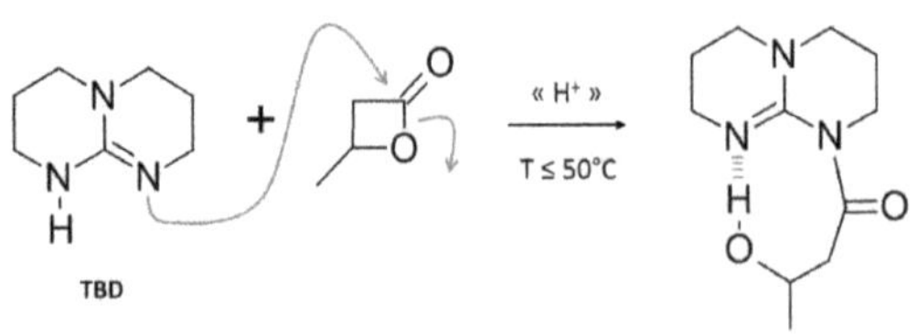

Scheme 1.12 Formation of a stable adduct between TBD and BL (in presence of a protic alcohol, H$^+$).

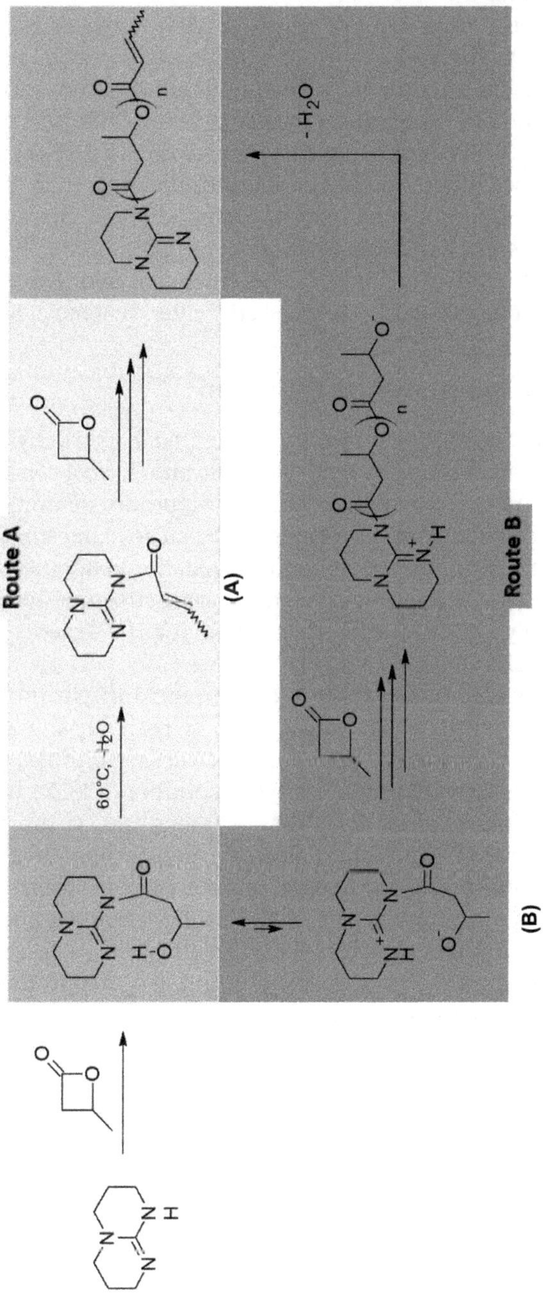

Scheme 1.13 Proposed nucleophilic mechanisms for the BL ROP from TBD in bulk at 60 °C.

in situ generation of an *N*-acyl-α,β-unsaturated species (**A**) (Scheme 1.13, route A) or the generation of a zwitterionic intermediate (**B**) able to ring-open BL monomers (route B).

Quite recently, the bulk ROP of BL from TBD (only) at 60 °C has been reinvestigated.[71] On the contrary to the Guillaume's report, results proved that the process is majorly due to an initial deprotonation of the monomer from the guanidine base generating *in situ* crotonate initiators for which the carboxylate (–C(O)O$^-$) is compensated by the protonated TBD (TBDH$^+$). The association of experimental technics allowed attesting that the previously established nucleophilic mechanism is marginal while the anionic one is dominant. Thanks to ^{1}H/DOSY NMR and MALDI/ESI-MS, TBD was also demonstrated not covalently linked to the PBL chain mainly playing the role of counter-ion in the –C(O)O$^-$, TBDH$^+$ active site (Scheme 1.14).

1.2.4 Tertiary Amine-based Initiation

In the early 1970s, the ability of tertiary amines, such as triethylamine (TEA), to react with carbonyl groups by a nucleophilic attack *via* a "charge transfer" (CT) complex formation was discussed.[72] The Authors demonstrated that a CT complex intermediate, either formed in the steady state concentration or in pre-equilibrium with the reactant, induced the generation of an ionic tetrahedral (IT) structure leading to an ion pair (IP) product. This typical nucleophilic catalysis was peculiarly studied for the reaction of phthalic anhydrides with TEA (Scheme 1.15).[73]

Just like with pyridine (*cf.* Section 1.2.1), Kricheldorf put the nucleophilic behavior of tertiary alkylamines to good use by realizing the ZROP of pivalolactone (PL).[16] To that end, three aliphatic tertiary amines, namely TEA, diazabicyclooctane (DABCO) and 2-ethyloxazolidine (2-EOX) were selected. Polymerizations were realized in NMP at 20 and 100 °C for 48 and 24 h, respectively. No information on molecular masses and dispersities were provided by the authors. Quantitative polymerizations from DABCO were obtained whatever the temperature used. MALDI-ToF mass spectrometry of the as-obtained PVL exclusively displayed mass peaks of linear chains obtained from a chain growth process from both nitrogens. In the case of TEA and 2-EOX, a 100 °C temperature was required to reach, respectively, 98 and 50% conversions. In both cases, the ZROP of PL was the prevailing process but was hampered by side reactions such as initiation from water when TEA was used and formation of cyclic oligolactones from a 2-EOX initiation. The authors concluded that the cyclization process issued from an end-to-end reaction between the nucleophilic carboxylate end-group and the electrophilic methylene group of the oxazolidine moieties (Scheme 1.16).

The polymerization of dithiolane-2,4-dione (DTD) was described by Kricheldorf and coworkers about 45 years ago.[74,75] It was found that tertiary amines were the only useful catalysts to generate high molar mass poly(thioglycolide)s. Due to a lack in characterization technics at that time, the mechanism of the reaction was only elucidated in 2007.[76]

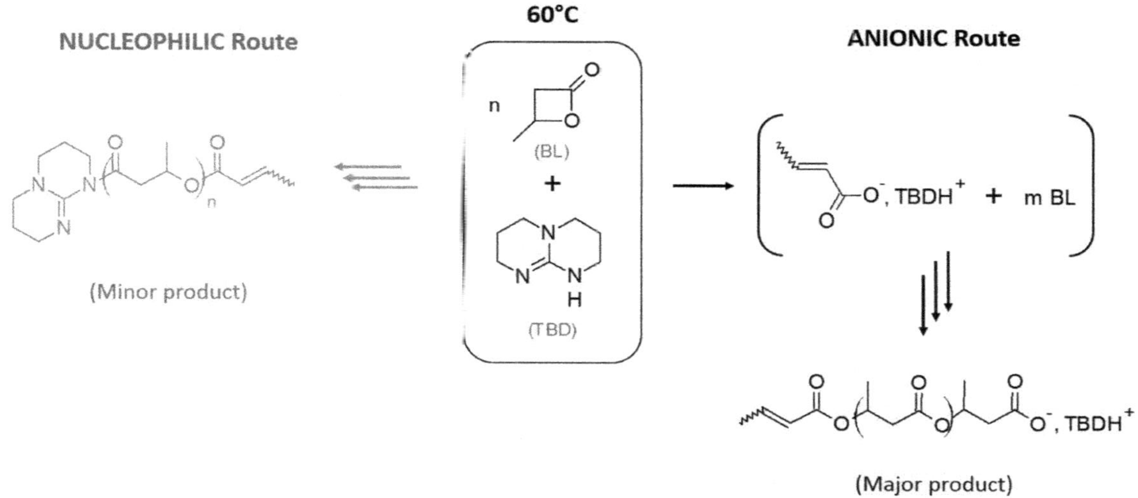

Scheme 1.14 Comparison between nucleophilic and anionic routes leading to PBL chains as obtained from bulk ROP of BL at 60 °C and initiated from TBD.

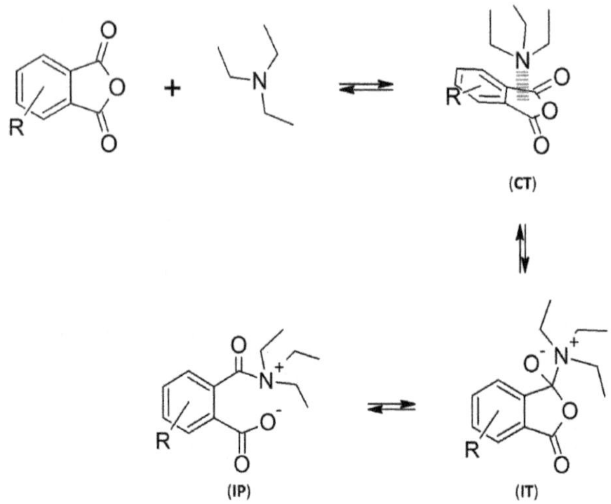

Scheme 1.15 Postulated mechanism of the reaction between phthalic anhydride molecules and triethylamine.

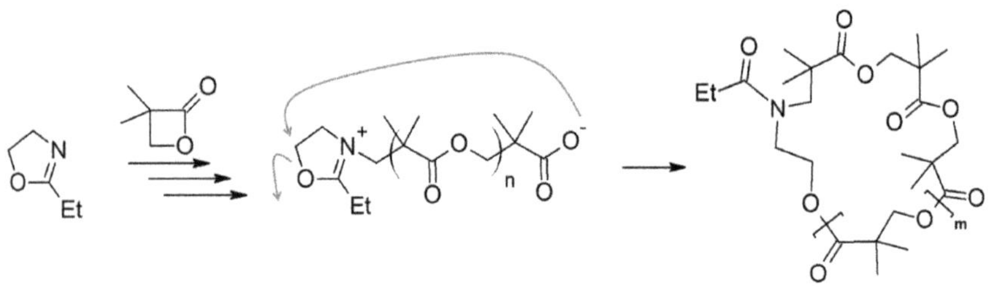

Scheme 1.16 Polymerization of PL from 2-EOX and backbiting cyclization reaction.

Polymerizations were performed in dioxane at 20 °C using TEA as catalyst at a monomer–catalyst ratio of 20. After two days of reaction, polymerizations were worked up with three different NuH non-solvents including water, methanol and ethanol. MALDI-ToF mass spectrometry allowed them to conclude that TEA reacted by formation of an initiating zwitterion leading after precipitation in the NuH non-solvent to a mixture of cyclic and linear chains end-capped by the used NuH. This result contradicted the speculated mechanism of 1973 where the authors assumed that TEA was deprotonating DTD, resulting in a carbanion initiating the ROP process (Scheme 1.17).

As DBU, low and mild basic sparteine and sparteine surrogates have also proven to be highly effective for the ROP of LA monomers by activating the alcohol initiator.[77,78] Inspired by the pioneering works of Kricheldorf on the use of alkali tertiary amines to promote ZROP of various heterocyclic monomers, some of us decided to investigate whether (+)-sparteine (SP) can act as an efficient initiator for the ROP of L-LA without the use of an exogenous protic source.[79] Polymerizations were realized in DCM ($[LA]_0 = 2M$)

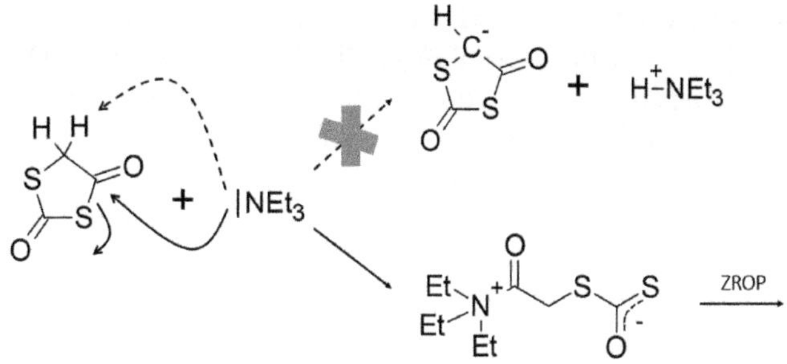

Scheme 1.17 Hypothetical initiation mechanisms and reaction product of TEA-initiated ZROP of dithiolane-2,4-dione.

at r.t. for $[LA]_0/[SP]_0$ ranging from 20 to 300. SEC analyses, ^{1}H-NMR spectroscopy and MALDI-ToF mass spectrometry allowed us to make conclusions on the effective ZROP of L-LA from both nitrogens of the SP initiator. Highly pure (>95%) *cyclo*-PLAs were obtained with dispersities ranging from 1.13 to 1.47. The molecular mass evolution joined to a kinetic study allowed us to make conclusions on a polymerization characterized by a rate of propagation (k_p) much higher than the initiating one (k_i). Therefore, controlled molar masses were only obtained for high LA-to-SP ratios rending possible the generation of controlled *cyclo*-PLAs with M_n up to 13 500 g mol^{-1}. The controlled formation of cyclic polyesters was ascribed to a backbiting process from an *in situ* generated tertiary amine-containing symmetrical binary zwitterion.

1.2.5 Phosphine-based Initiation

In 1993, Vedejs and coworkers reported tributylphosphine (Bu_3P) as a potent catalyst for the acylation of alcohols by acetic acids and anhydrides.[80,81] As compared to DMAP, which is a more versatile catalyst due to its double rôle as nucleophile and base, the authors brought in evidence that Bu_3P is less sensitive and not deactivated by carboxylic acid, does not require basic additives and is appropriate to be used in neutral conditions for acylation reactions proceeding through a nucleophilic activation mechanism. Consequently, Hedrick *et al.* surmised that tertiary phosphines could be another general class of ROP catalyst and studied a series of phosphines as transesterification agents for the LA polymerization.[82] Reactions were realized both in bulk (135 and 180 °C) and in solution (THF, toluene and DCM). When performed in bulk, the catalyst concentration proved to be an important variable and excess of tertiary phosphines as compared to the initiating alcohol must be precluded to limit adverse transesterification reactions. In that condition ($[phosphines]_0/[initiator]_0 = 1$), narrowly dispersed PLA ($M_w/M_n = 1.1$–1.4) with predicted molecular weights (target DPs 30–100) have been produced (in minutes to hours regarding the catalyst). Kinetical studies revealed that phosphine activity is dependent on the nature of their substituents and on the general steric hindrance according the following order: P(n-Bu)$_3$ > P($tert$-Bu)$_3$ > PhPMe$_2$ > Ph$_2$PMe > Ph$_3$P > P(MeO)$_3$ ≈ 0. Alkyl-substituted phosphines are more basic/nucleophilic than phosphines containing aryl ligands promoting then a faster reaction while P($tert$-Bu)$_3$ is less active than P(n-Bu)$_3$ due to the steric constraints. Even by using the most active Bu$_3$P, polymerizations realized in solution were slower and less selective than those catalyzed by DMAP with, for example, a conversion of 60% achieved in one week at 50 °C in THF (target DP 60). ^{1}H-NMR and SEC analyses of the as-obtained PLA, joined to the reactivity order of the various studied phosphines, support a nucleophilic-based LA ROP process (Scheme 1.18).

1.2.6 N-heterocyclic Carbene-based Initiation

The demonstration by Breslow that stabilized singlet carbenes derived from thiamine cofactors are nucleophilic catalysts[83] and the pioneering works of Wanzlick and coworkers[84–88] on the deprotonation of thiazolium and imidazolium salts are to date considered as the initiating steps on the use of NHCs as nucleophilic catalysts.[89] In 2002, independent works of Nolan,[90] and Hedrick and Waymouth[91] on the use of NHCs as performing catalysts for transesterification reactions led to their investigation as catalysts for ROP of lactones.[92]

Initially, NHCs were used in presence of alcohol initiators.[93–96] The first reported example was dealing with the ROP (in THF at 25 °C) of a series of (di)lactones in the presence of the 1,3-bis-(2,4,6-trimethylphenyl)imidazole-2-ylidene (IMes) NHC and various alcohol initiators.[97] Whatever the monomer used, the obtained molecular weights closely tracked the monomer-to-initiator ratios, dispersities were narrow and the process exhibited features of a "living" polymerization. At that time, Hedrick and Waymouth proposed two possible mechanisms: (a) an anionic/basic "chain-end" mechanism where the carbene activates the initiating/propagating alcohol by H-bonding (Scheme 1.19), and (b) a monomer-activated "nucleophilic" mechanism involving a zwitterionic intermediate (Scheme 1.20).

While both mechanisms were conceivable, experimental results suggested that the carbene acts as a nucleophile and not a base. For transesterifications of esters with ethyl alcohol catalyzed by N-alkyl-substituted carbenes, the pK_a of the alcohol[98] is higher than that of the conjugate acid of the carbene (30 *vs.* 22–24 in DMSO).[99,100] Thus, it is unlikely that the carbene catalyzes the ROP process by deprotonating the less acidic alcohol. Additionally, a steric effect observed by Waymouth and Hedrick indicated that the reaction occurs *via* nucleophilic catalysis and not base catalysis.[91] Contradictory results were however postulated in 2005 where DFT calculations suggested that the zwitterionic intermediate (Scheme 1.20) was higher in energy than the H-bonded adduct (Scheme 1.19), implying a more probable "chain-end activation" mechanism when an exogenous alcohol is used.[101] Finally, Arnold *et al.* demonstrated that NHCs in the presence of alcohol can

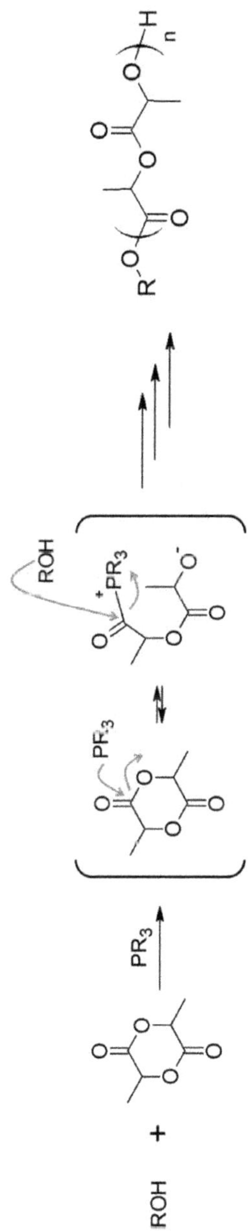

Scheme 1.18 Plausible nucleophilic polymerization of LA from tertiary phosphines.

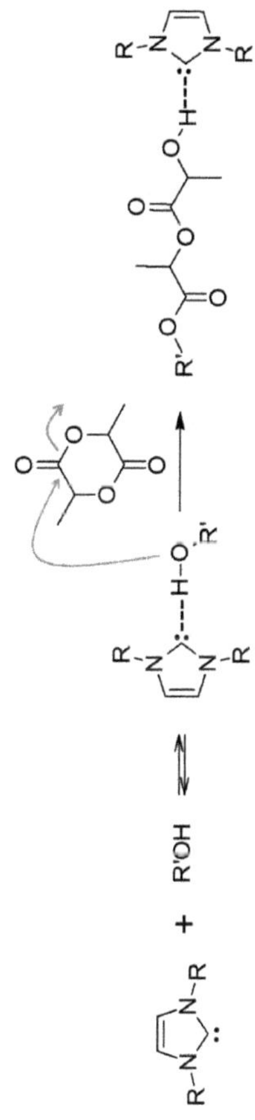

Scheme 1.19 Anionic/basic "chain-end" mechanism for the NHC-mediated ROP of LA in the presence of alcohol.

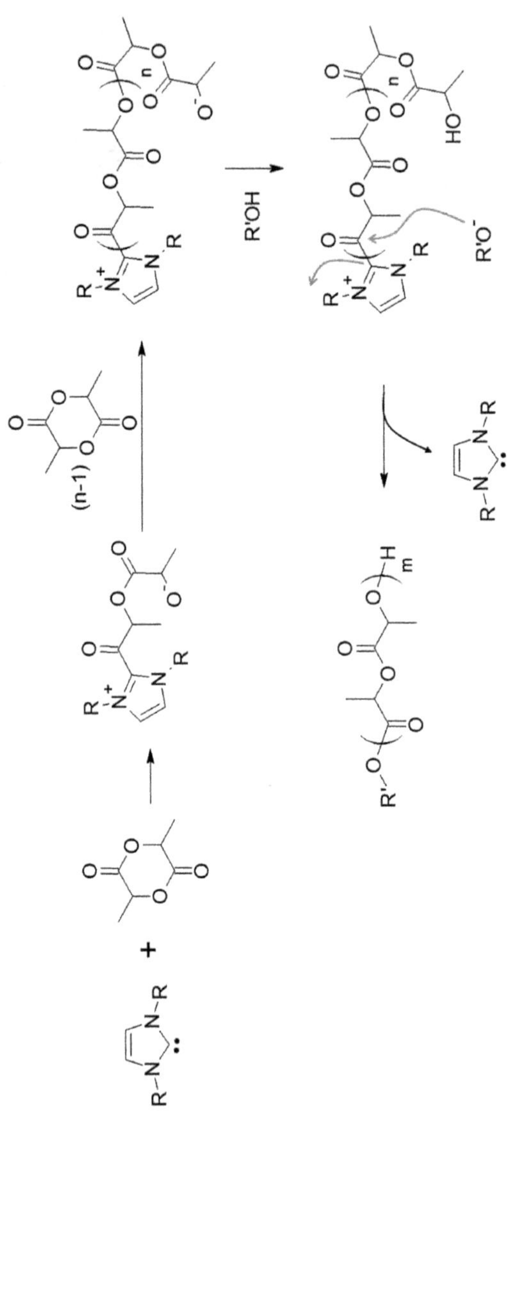

Scheme 1.20 Monomer-activated “nucleophilic” mechanism for the NHC-mediated ROP of LA in the presence of alcohol.

act as bifunctional catalysts, suggesting that both mechanisms may compete.[102] To that end, a reaction between a hydroxylated NHC with one equivalent of D,L-LA in THF, pyridine or toluene afforded two compounds (Scheme 1.21). Those products were identified as being issued from either an alkoxide initiator (path a, Scheme 1.21) or a nucleophilic one (path b, Scheme 1.21).

The zwitterionic intermediate generated *in situ* by the nucleophilic attack of NHC on LA has been indirectly evidenced by Waymouth *et al.* by performing polymerization in the absence of an alcohol initiator.[103,104] Remarkably, the polymerization of *rac*-LA with 1,3-dimesitylimidazol-2-ylidene (IMes) in THF ($[LA]_0 = 0.6$–1.0 M) occurs rapidly (5–900 s) at 25 °C to yield PLAs with molecular weights from 7 to 26 $kg\,mol^{-1}$. The combination of SEC, ^{1}H-NMR spectroscopy and MALDI-ToF analysis indicated that the resulting PLAs were of cyclic nature. As attested by DSC, cyclic PLAs obtained from either L- or D-LA are isotactic. As compared to their linear homologues,[105] the as-obtained semi-crystalline cyclic PLAs present both lower melting point and optical rotation (linear: $T_m = 181$ °C, $\Delta H_f = 85$ $J\,g^{-1}$, $[\alpha]_D = -156°$; cyclic: $T_m = 133$ and 143 °C, $\Delta H_f = 22.7$ $J\,g^{-1}$, $[\alpha]_D = -118°$).[104,106,107] This suggests possible epimerization reactions of either the LA monomer or the polymer by the basic IMes.[103,108] Kinetic investigations revealed a slow initiation step that is second order in monomer rationalizing the reversible formation of the zwitterionic intermediate and a first order propagation, much faster than the initiation step ($k_p \gg k_i$).[109] The observed molecular weight distributions *vs.* monomer conversion are explained by the fast propagation relative to the macrocyclization and the slow initiation at high monomer conversion due to its second-order dependence on lactide. Finally, the cyclization, although slow relative to the propagation ($k_p > k_c$), places limits on the molecular weights that can be achieved liberating the carbene and generating cyclic PLAs (Scheme 1.22).

Cleverly, the possibility of acyl imidazolium being attacked by the alkoxide end group, *i.e.* evidencing the viability of the cyclization step, has been assessed by reaction between sodium methoxide and an IMes benzoyl chloride derivative (Scheme 1.23).[103] To that end, an acyl imidazolium was generated from the reaction of IMes and benzoyl chloride. The as-obtained product was quickly reacted with sodium methoxide yielding the pristine IMes and the expected methyl benzoate.

Stochastic kinetic simulations suggest that the ZROP of lactide from IMes NHC is not a living polymerization, that the initiator efficiency is less than 75% (even at high monomer concentration) and that the carbenes that did not form zwitterions (or those liberated after cyclization) do not initiate chains later in the process. The amount of active zwitterion is not predictable and does not allow the control of PLA molecular weights from the initial $[LA]_0$-to-$[IMes]_0$ ratio.[109] If the kinetic control of the LA ZROP from pristine IMes is certain, however, its feasibility is delicate for a reaction completed in few seconds and a slightly slower process may be of interest. Because the quantity of initiating zwitterion is function of both monomer and NHC

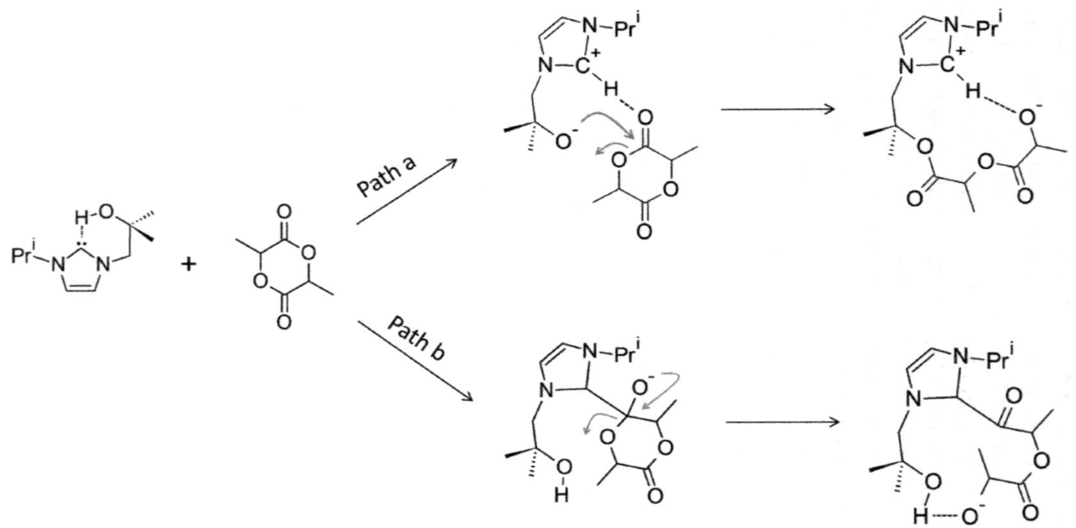

Scheme 1.21 NHC bifunctionality demonstrated on LA; path a: anionic/basic mechanism; path b: nucleophilic mechanism.

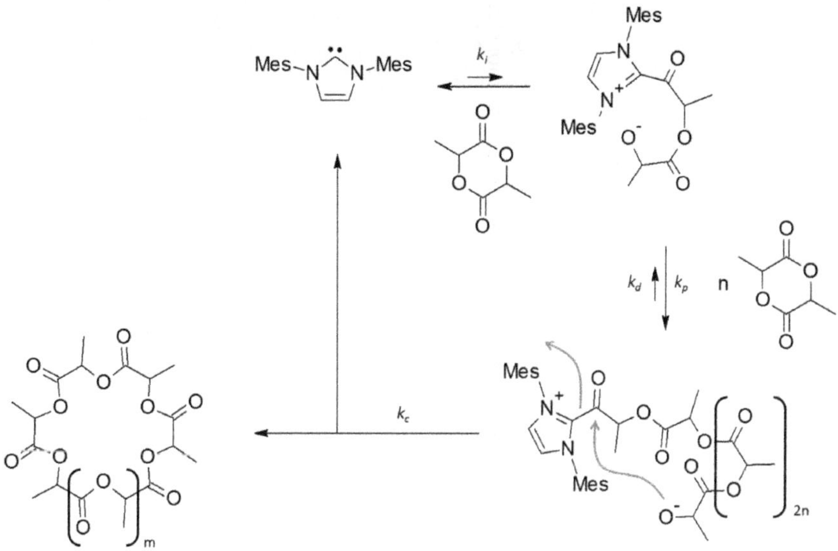

Scheme 1.22 Proposed mechanism for the synthesis of cyclic PLA from 1,3-dimesitylimidazol-2-ylidene (IMes) NHC (with k_i, k_p, k_d and k_c corresponding to rate constants of initiation, propagation, depolymerization and cyclization, respectively).

concentrations, diluting the medium of polymerization would not be of help. To slow down the overall kinetics (while maintaining the same initial $[LA]_0$), the ability that allows the IMes carbene to reversibly form a "dormant" adduct was exploited in 2010.[110] Rather than using an isolated carbene, the authors took advantage of the reversible hydrogen-bonded adduct formed between IMes and *t*-BuOH.[111] The ZROP of LA was then initiated in THF (at r.t.) by using the non-isolated IMes carbene produced *in situ* by reacting its corresponding chloride salt and potassium *tert*-butoxide (*t*-BuOK). As outlined by Scheme 1.24, the generated "dormant" alcohol adduct is in equilibrium with the "active" IMes carbene and free *t*-BuOH unable to initiate the ROP of lactones.[91,112] This allows diminishing the carbene activity by a factor of *ca.* 15 (all other experimental conditions unchanged). Such modification of kinetics allows preparation of "jellyfish" structures based on a PLA macrocyclic inner-core grafted by poly(methyl methacrylate) chains.[110]

Because the aryl substituted IMes carbene is inactive towards larger lactones, such as ε-caprolactone (CL) and δ-valerolactone (VL),[113,114] Waymouth *et al.* used more nucleophilic *N*-alkyl-substituted carbenes[115] such as 1,3-diisopropyl-4,5-dimethylimidazol-2-ylidene (**1**), 1,3-diethyl-4,5-dimethylimidazol-2-ylidene (**2**) and 1,3,4,5-tetramethylimidazol-2 ylidene (**3**) to investigate the ZROP of CL in absence of alcohol (Scheme 1.25).[116]

As compared to the LA ZROP from IMes, the polymerization of CL from carbenes (**1**)–(**3**) generates polymers of ultrahigh molecular weights

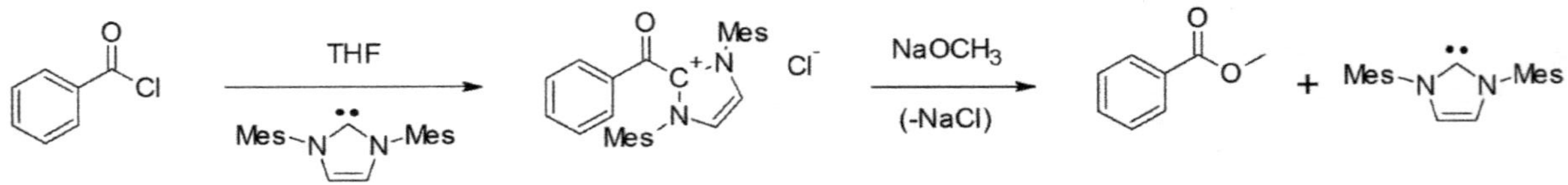

Scheme 1.23 Indirect proof of the cyclization step presented in Scheme 1.21: reaction of NHC acyl-imidazolium with sodium methoxide.

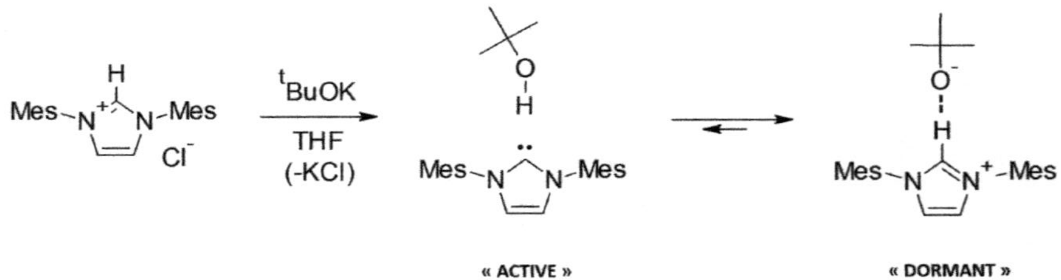

Scheme 1.24 Equilibrium between "active" IMes and its corresponding "dormant" hydrogen-bonded tertiary alcohol adduct.

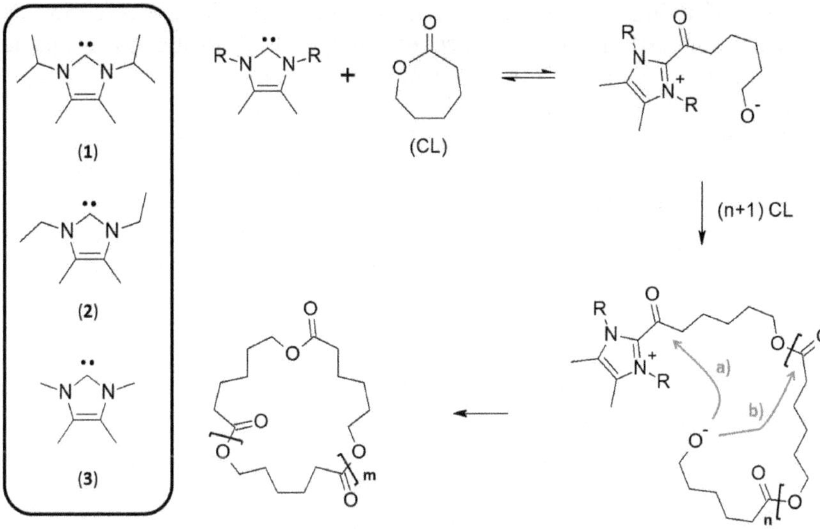

Scheme 1.25 1,3-Diisopropyl-4,5-dimethylimidazol-2-ylidene (**1**), 1,3-diethyl-4,5-dimethylimidazol-2-ylidene (**2**) and 1,3,4,5-tetramethylimidazol-2-ylidene (**3**) alkyl-substituted carbenes used for the ZROP of CL.

($M_n \leq 114.00$ g mol^{-1}) in a few hours. ^{1}H-NMR spectroscopy, MALDI-ToF spectrometry and SEC analyses confirm the cyclic nature of the PCL samples while doubts subsist on the purity of those by linear contaminants. The polymerization, performed in toluene or THF ($[CL]_0 = 1M$), is not controlled, the relatively low efficiency of initiation is dependent on the polarity of the medium and the synthetic strategy is suspected to yield entangled macrocyclic structures. Finally, the high dispersity values ($1.36 \leq Đ_M \leq 2.16$) support the hypothesis of a competition of cyclization between the alkoxide end-group on the acyl imidazolium extremity (path a, Scheme 1.25) and on internal ester group of the macrozwitterion (path b, Scheme 1.25). The ability presented by those alkyl-substituted carbenes to polymerize large lactones has also been taken into account for preparing cyclic gradient polymers by batch copolymerization of CL and VL.[117] To that end, Waymouth *et al.* utilized the unsaturated carbene (**2**) to produce a range of copolymer compositions. Polymerizations were realized in toluene ($[M_{tot}]_0 = 1$ M) at r.t. for reaction times ranging from 7 min to 2 h. The wide difference in reactivity between VL and CL ($r_{VL} = 26.4$; $r_{CL} = 0.38$) with carbene (**2**) gave access to gradient cyclic copolymers presenting melting points that are similar to the ones of homopolymers.

Kinetic studies[109,118] and theoretical simulations[119] brought some light on the complicated sequence of steps involved during the ZROP of lactones by NHCs. The *in situ* generated NHC-monomer complex presents a high energy barrier that account for a slow initiation leading to molecular weight polyesters significantly higher than expected. This is particularly true for the ZROP of VL from NHC;[120] SEC traces of polymer samples revealed the presence of a high molecular weight shoulder all along the conversion evolution suggesting the presence of independent propagating species. Those include ion pairs, solvent-separated ion pairs, free ions and aggregates, all presenting an independent rate of propagation.[121] Because the situation in which the rate exchange between different ion pairs and/or aggregates is slow relative to the rate of propagation favors the appearance of bimodal molecular weight distributions, Waymouth *et al.* proposed a ZROP where the propagating alkoxide anion undergoes ion pairing with the acylimidazolium chain end to explain the as-observed broad and multimodal SEC traces. To evaluate the consequences of ion-pairing, the influence of LiCl to the ZROP of VL from the 1,3-diisopropyl-4,5-dimethylimidazol-2-ylidene carbene (**1**) has been investigated in THF at r.t. (Scheme 1.26). As expected, the addition of LiCl resulted in a more controlled polymerization with a good correlation between experimental and theoretical molar masses. While low distribution samples were obtained from a very efficient initiation, MALDI-ToF analysis revealed the presence of both linear and cyclic products whose proportions depend on the lithium salt initial content.

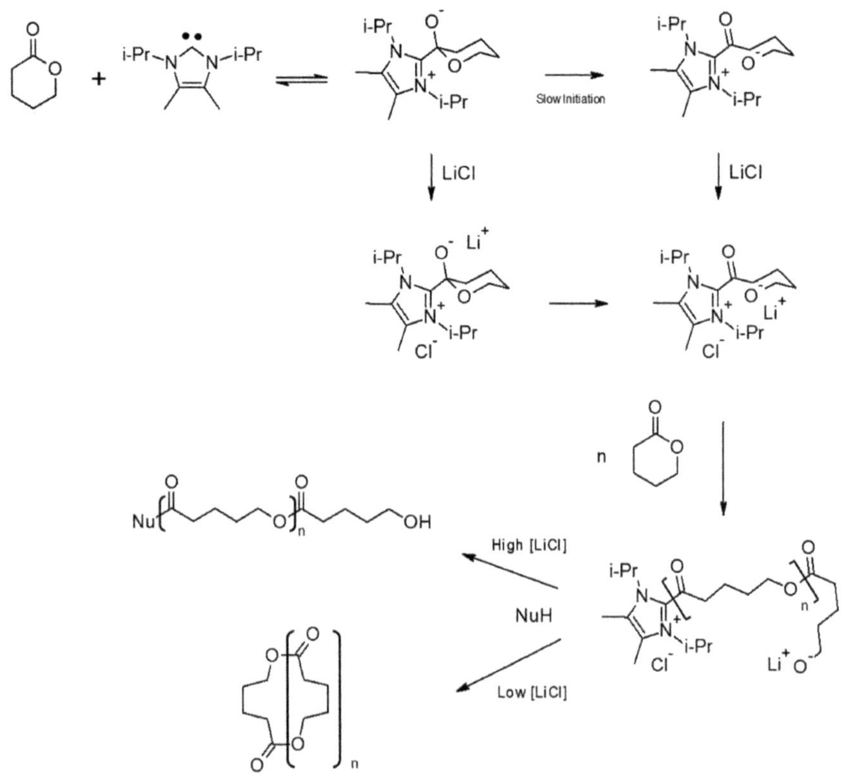

Scheme 1.26 Influence of the LiCl content on the ZROP of VL from 1,3-diisopropyl-4,5-dimethylimidazol-2-ylidene carbene.

The ZROP of β-lactones, such as β-propiolactone and β-butyrolactone (BL), with saturated IMes (SIMes) carbene also leads to polymers of cyclic nature further evidencing the intermediacy of zwitterions in that kind of process.[122] The authors discovered that reacting the SIMes with one equivalent of BL generates a zwitterion, which collapses to an isolable spiro imidazolidine compound (**SpI**) (Scheme 1.27). DFT calculations indicate that this compound is generated due to the release of ring strain from the four-membered lactone to the benefit of a five-membered ring. As compared to the ZROP of LA, CL and VL, the initiation of the polymerization from the spiro (**SpI**) occurs much more rapidly and is quantitative. That difference is due to the slightly higher nucleophilicity of the SIMes carbene relative to IMes[115] and the higher ring strain of the four-membered lactone[123] relative to that of larger lactones. Polymerizations present characteristics of a "living" process with a linear semilog plot evolution and molecular weights tracking the initial $[M]_0/[SIMes]_0$ molar ratios. As compared to the ZROP of LA, CL and VL, the authors proposed a novel mechanism involving a reversible collapse of the zwitterionic species to macrocyclic spirocycles all along the propagation. This mechanism responds to the definition of a ring-expansion polymerization on the same title as the ring-expansion of spirocyclic tin initiators[124] or cyclic Ru carbenes.[125]

If unsaturated carbenes give access to cyclic polyesters, Zhang and coworkers also demonstrated their ability to (co)polymerize *N*-carboxyanhydride monomers (NCA).[126–129] The carbene-initiated ZROP of NCAs provides an elegant strategy for preparing cyclic poly(α-peptoid)s of controlled molecular weight (up to 30 000 g mol^{-1}) and characterized by narrow dispersities.[126] When realized in THF or toluene, the loss of CO_2 from the *in situ* formed zwitterion favors the efficiency of the initiation and explains the perfect agreement between the evolution of molecular weight with both NCA conversion and the initial $[NCA]_0$-to-$[NHC]_0$ ratio. Polymerizations performed in DMF or DMSO result in low oligomers whatever the starting $[NCA]_0/[NHC]_0$.[126] Top-down multidimensional mass spectrometry methods allowed Wesdemiotis *et al.* to demonstrated that NHC-mediated ZROP in THF at 50 °C yields spirocyclic poly(α-peptoid)s, *i.e.* carrying the NHC initiator on the *cyclo* poly(α-peptoid) (Scheme 1.28).[130]

Very interestingly, Gnanou and coworkers also reported that NHCs could trigger the ZROP of oxirane-based monomers, such as ethylene oxide (EO) and propylene oxide (PO).[131–133] By contrast to the ZROP processes applied on (di)lactones or on NCAs, no competitive intra- or intermolecular transfer reactions were observed during the ZROPs of those monomers leading exclusively to linear polyether chains. Cleverly, this strategy allowed the authors to use NuE moieties as terminating agents, leading quantitatively to α-Nu, ω-OE polymers through nucleophilic substitution of the imidazolium moiety by $Nu^{(-)}$ and concomitant reaction of the ω-growing alkoxide chain with $H^{(+)}$ (Scheme 1.29).

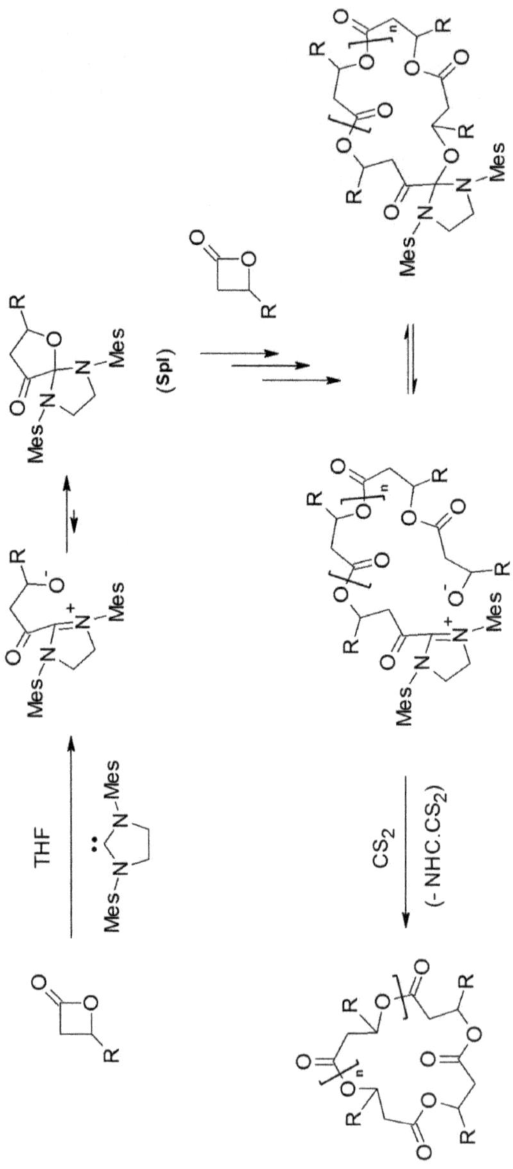

Scheme 1.27 ZROP of β-lactones using SIMes carbene.

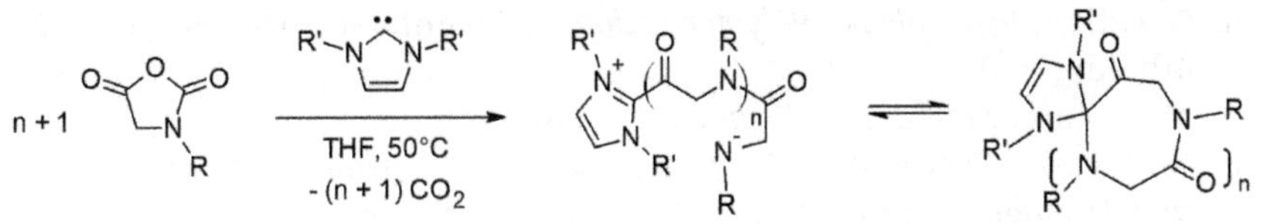

Scheme 1.28 Spirocyclic poly(α-peptoid) generated by NHC-mediated ZROP of NCA.

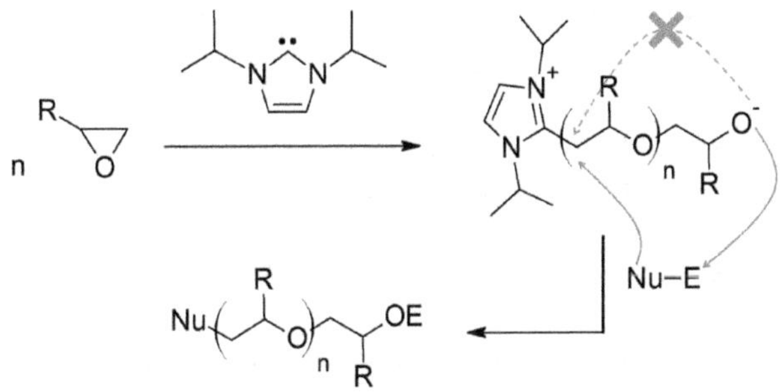

Scheme 1.29 Linear end-functionalized polyether chains as obtained by ZROP of oxirane-based monomers (R = H or CH_3) followed by a NuE quenching step.

Acknowledgements

This work has been supported by the EU Horizon 2020 Research and Innovation Framework Program – European Joint Doctorate – SUSPOL-EJD [grant number 642671], by the European Commission and Région Wallone FEDER program and OPTI^{2}MAT program of excellence, by the Interuniversity Attraction Pole Programme (P7/05) initiated by the Belgian Science office and by the FNRS-FRFC. O.C. is Research Associate for the F.R.S.-FNRS in Belgium.

References

1. G. Bredig and P. S. Fiske, *Biochem. Z.*, 1912, **46**, 7.
2. Z. G. Hajos and D. R. Parrish, *J. Org. Chem.*, 1974, **39**, 1615.
3. A. Berkessel and H. Gröger, *Asymmetric Organocatalysis: From Biomimetic Concepts to Applications in Asymmetric Synthesis*, Wiley-VCH, Weinheim, Germany, 2005.
4. P. I. Dalko, *Enantioselective Organocatalysis*, Wiley-VCH, Weinheim, Germany, 2007.
5. R. A. Sheldon, I. Arends and U. Hanefeld, *Green Chemistry and Catalysis*, Wiley-VCH, Weinheim, Germany, 2007.
6. M. K. Kiesewetter, E. J. Shin, J. L. Hedrick and R. M. Waymouth, *Macromolecules*, 2010, **43**, 2093.
7. S. Penczek, *J. Polym. Sci., Part A: Polym. Chem.*, 2002, **40**, 1665.

8. G. Odian, *Principles of Polymerization*, Wiley-Interscience, Hoboken, NJ, 4th edn, 2004.
9. M. Szwarc, *Makromol. Chem.*, 1960, **35**, 132.
10. M. Suzuki, *Zwitterionic Polymerization, Encyclopedia of Polymer Science and Technology*, John Wiley & Sons, Inc., Hoboken, NJ, 4th edn, 2013.
11. I. Asenjo-Sanz, A. Veloso, J. I. Miranda, J. A. Pomposo and F. Barroso-Bujans, *Polym. Chem.*, 2014, **5**, 6905.
12. I. Asenjo-Sanz, A. Veloso, J. I. Miranda, A. Alegria, J. A. Pomposo and F. Barroso-Bujans, *Macromolecules*, 2015, **48**, 1664.
13. Y. P. Etienne, R. M. Soulas and R. J. Thiebaut, *J. Polym. Sci., Part C*, 1963, **4**, 1061.
14. Y. Yamashitu, K. Ito and F. Nakakita, *Makromol. Chem.*, 1969, **127**, 292.
15. H. R. Kricheldorf and G. Schwarz, *Macromol. Rapid Commun.*, 2003, **24**, 359.
16. H. R. Kricheldorf, M. Garaleh and G. Schwarz, *J. Macromol. Sci., Part A*, 2005, **42**, 139.
17. R. Murugan and E. F. V. Scriven, *Aldrichimica Acta*, 2003, **36**, 21.
18. H. R. Kricheldorf and N. J. Scharnagl, *J. Macromol. Sci., Chem.*, 1989, **A26**(7), 951.
19. L. M. Litvinenko and A. I. Kirichenko, *Dokl. Akad. Nauk SSSR, Ser. Khim.*, 1967, **176**, 97.
20. W. Steglich and G. Höfle, *Angew. Chem.*, 1969, **81**, 1001.
21. A. Hassner, L. R. Krepski and V. Alexanian, *Tetrahedron*, 1978, **34**, 2069.
22. M. R. Heinrich, H. S. Klisa, H. Mayr, W. Steglich and H. Zipse, *Angew. Chem.*, 2003, **115**, 4975; *Angew. Chem., Int. Ed.*, 2003, **42**, 4826.
23. X. Lu and U. Schneider, *Chem. Commun.*, 2016, **52**, 12980.
24. G. Höfle, W. Steglich and H. Vorbrüggen, *Angew. Chem., Int. Ed.*, 1978, **17**, 569.
25. A. C. Spivey and S. Arseniyadis, *Angew. Chem., Int. Ed.*, 2004, **43**, 5436.
26. S. Xu, I. Held, B. Kempf, H. Mayr, W. Steglich and H. Zipse, *Chem. – Eur. J.*, 2005, **11**, 4751.
27. I. Held, A. Villinger and H. Zipse, *Synthesis*, 2005, **9**, 1425.
28. V. Lutz, J. Glatthaar, C. Würtele, M. Serafin, H. Hausmann and P. R. Schreiner, *Chem. – Eur. J.*, 2009, **15**, 8548.
29. F. Nederberg, E. F. Connor, M. Moller, T. Glauser and J. L. Hedrick, *Angew. Chem., Int. Ed.*, 2001, **40**, 2712.
30. A. Kumar and R. A. Gross, *J. Am. Chem. Soc.*, 2000, **122**, 11767.
31. A. Kumar and R. A. Gross, *Biomacromolecules*, 2000, **1**, 133.
32. S. Kobayashi, H. Uyama, S. Namekawa and H. Hayakawa, *Macromolecules*, 1998, **31**, 5655.
33. F. Nederberg, E. F. Connor, T. Glauser and J. L. Hedrick, *Chem. Commun.*, 2001, 2066.
34. C. Bonduelle, B. Martin-Vaca, F. P. Cossio and D. Bourissou, *Chem. – Eur. J.*, 2008, **14**, 5304.
35. O. Coulembier and P. Dubois, *J. Polym. Sci., Part A*, 2012, **50**, 1672.

36. J. Kadota, D. Pavlovic, J. P. Desvergne, B. Bibal, F. Péruch and A. Deffieux, *Macromolecules*, 2010, **43**, 8874.
37. A. E. Allen and D. W. C. MacMillan, *Chem. Sci.*, 2012, **3**, 633.
38. J. Muschiol, C. Peters, N. Oberleitner, M. D. Mihovilovic, U. T. Bornscheuer and F. Rudroff, *Chem. Commun.*, 2015, **51**, 5798.
39. D. W. Stephan and G. Erker, *Angew. Chem., Int. Ed.*, 2015, **54**, 6400.
40. K. Makiguchi, S. Kikuchi, K. Yanai, Y. Ogasawara, S.-I. Sato, T. Satoh and T. Kakuchi, *J. Polym. Sci., Part A: Polym. Chem.*, 2014, **52**(7), 1047.
41. T. Trimaille, M. Moeller and R. Gurny, *J. Polym. Sci., Part A: Polym. Chem.*, 2004, **42**, 4379.
42. T. Trimaille, R. Gurny and M. Moeller, *Chimia*, 2005, **59**, 348.
43. O. Thillaye Du Boullay, E. Marchal, B. Martin-Vaca, F. P. Cossio and D. Bourissou, *J. Am. Chem. Soc.*, 2006, **128**, 16442.
44. C. Bonduelle, B. Martin-Vaca, F. P. Cossio and D. Bourissou, *Chem. – Eur. J.*, 2008, **14**, 5304.
45. Y. Lu, L. Yin, Y. Zhang, Z. Zhang, Y. Xu, R. Tong and J. Cheng, *ACS Macro Lett.*, 2012, **1**, 441.
46. A. Buchard, D. R. Carbery, M. G. Davidson, P. K. Ivanova, B. J. Jeffery, G. I. Kociok-Kohn and J. P. Lowe, *Angew. Chem., Int. Ed.*, 2014, **53**, 13858; *Angew. Chem.*, 2014, **126**, 14078.
47. H. Feng and C.-M. Dong, *J. Polym. Sci., Part A: Polym. Chem.*, 2006, **44**, 5353.
48. M. Oshimura, T. Tang and A. Takasu, *J. Polym. Sci., Part A: Polym. Chem.*, 2011, **49**, 1210.
49. M. Bouyahyi, M. Pepels, P. F. Mark, A. Heise and R. Duchateau, *Macromolecules*, 2012, **45**, 3356.
50. S. Naumann, P. B. V. Scholten, J. A. Wilson and A. P. Dove, *J. Am. Chem. Soc.*, 2015, **137**, 14439.
51. X. Q. Liu, J. Zhao-Xia, W. Yu-Tang and Y. Zuo-Long, *J. Nat. Gas Chem.*, 1999, **8**, 76.
52. T. Hirose, B. G. Kopek, Z.-H. Wang, R. Yusa and B. W. Baldwin, *Tetrahedron Lett.*, 2003, **44**, 1831.
53. H. R. Kricheldorf, C. Von Lossow and G. Schwarz, *J. Polym. Sci., Part A: Polym. Chem.*, 2005, **43**, 5690.
54. H. R. Kricheldorf, N. Lomadze and G. Schwarz, *Macromolecules*, 2008, **41**, 7812.
55. S. E. Denmark and G. L. Beutner, *Angew. Chem., Int. Ed.*, 2008, **47**, 1560.
56. W.-C. Shieh, S. Dell and O. Repic, *J. Org. Chem.*, 2002, **67**, 2188.
57. W. Zhang and M. Shi, *Org. Biomol. Chem.*, 2006, **4**, 1671.
58. K. E. Price, C. Larrivée-Aboussafy, B. M. Lillie, R. W. McLaughlin, J. Mustakis, K. W. Hettenbach, J. M. Hawkins and R. Vaidyanathan, *Org. Lett.*, 2009, **11**, 2003.
59. C. Larrivée-Aboussafy, B. P. Jones, K. E. Price, M. A. Hardink, R. W. McLaughlin, B. Lillie, J. M. Hawkins and R. Vaidyanathan, *Org. Lett.*, 2010, **12**, 324.

60. J. E. Taylor, M. D. Jones, J. M. J. Williams and S. D. Bull, *Org. Lett.*, 2010, **12**, 5740.
61. D. Basavaiah, A. J. Rao and T. Satyanarayana, *Chem. Rev.*, 2003, **103**, 811.
62. N. Ghosh, *Synlett*, 2004, 574.
63. H. Oediger, F. Möller and K. Eiter, *Synthesis*, 1972, 591.
64. I. Hermecz, *Adv. Heterocycl. Chem.*, 1987, **42**, 83.
65. M. Baidya and H. Mayr, *Chem. Commun.*, 2008, 1792.
66. B. Maji, C. Joannesse, T. A. Nigst, A. D. Smith and H. Mayr, *J. Org. Chem.*, 2011, **76**, 5104.
67. H. A. Brown, A. G. De Crisis, J. L. Hedrick and R. M. Waymouth, *ACS Macro Lett.*, 2012, **1**, 1113.
68. C. G. Jaffredo, J. F. Carpentier and S. M. Guillaume, *Macromol. Rapid Commun.*, 2012, **33**, 1938.
69. B. G. G. Lohmeijer, R. C. Pratt, F. Leibfarth, J. W. Loga, D. A. Long, A. P. Dove, F. Nederberg, J. Choi, C. Wade, R. M. Waymouth and J. L. Hedrick, *Macromolecules*, 2006, **39**, 8574.
70. L. Simon and J. M. Goodman, *J. Org. Chem.*, 2007, **72**, 9656.
71. S. Moins, C. Henoumont, J. De Winter, A. Khalil, S. Laurent, S. Cammas-Marion and O. Coulembier, *Polym. Chem.*, 2018, DOI: 10.1039/C8PY00206A.
72. J. B. Nagy, O. B. Nagy and A. Bruylants, *Bull. Soc. Chim. Belg.*, 1973, **82**, 539.
73. O. B. Nagy, J. B. Nagy and A. Bruylants, *Tetrahedron Lett.*, 1969, 4825.
74. H. R. Kricheldorf, K. Bösinger and G. Schwarz, *Makromol. Chem. Phys.*, 1973, **173**, 43.
75. H. R. Kricheldorf and K. Bösinger, *Makromol. Chem. Phys.*, 1973, **173**, 67.
76. H. R. Kricheldorf, N. Lomaze and G. Schwarz, *Macromolecules*, 2007, **40**, 4859.
77. D. J. Coady, A. C. Engler, H. W. Horn, K. M. Bajjuri, K. Fukushima, G. O. Jones, A. Nelson, J. E. Rice and J. L. Hedrick, *ACS Macro Lett.*, 2012, **1**, 19.
78. R. Todd, G. Rubio, D. J. Hall, S. Tempelaar and A. P. Dove, *Chem. Sci.*, 2013, **4**, 1092.
79. O. Coulembier, J. De Winter, T. Josse, L. Mespouille, P. Gerbaux and Ph. Dubois, *Polym. Chem.*, 2014, **5**, 2103.
80. E. Vedejs, S. N. Bennett, L. M. Conn, S. T. Diver, M. Gingras, S. Lin, P. A. Oliver and M. J. Peterson, *J. Org. Chem.*, 1993, **58**, 7286.
81. E. Vedejs and S. T. Diver, *J. Am. Chem. Soc.*, 1993, **115**, 3358.
82. M. Myers, E. F. Connoer, T. Glauser, A. Möck, G. Nyce and J. L. Hedrick, *J. Polym. Sci., Part A: Polym. Chem.*, 2002, **40**, 844.
83. R. Breslow, *J. Am. Chem. Soc.*, 1958, **80**, 3719.
84. H. W. Wanzlick and E. Schikora, *Angew. Chem.*, 1960, **72**, 494.
85. H. W. Wanzlick and H. J. Kleiner, *Angew. Chem.*, 1961, **73**, 493.
86. H. W. Wanzlick, *Angew. Chem., Int. Ed. Engl.*, 1962, **1**, 75.

87. H. W. Wanzlick, F. Esser and H. J. Kleiner, *Chem. Ber.*, 1963, **96**, 1208.
88. H. J. Schoenherr and H. W. Wanzlick, *Liebigs Ann. Chem.*, 1970, **731**, 176.
89. A. P. Dove, R. C. Pratt, B. G. G. Lohmeijer, H. Li, E. C. Hagberg, R. M. Waymouth and J. L. Hedrick, in *N-Heterocyclic Carbenes in Synthesis*, ed. S. P. Nolan, Wiley-VCH Verlag GmbH & Co. KGaA, Weinheim, 2006, ch. 12, p. 275.
90. G. A. Grasa, R. M. Kissling and S. P. Nolan, *Org. Lett.*, 2002, **4**, 3583.
91. G. W. Nyce, J. A. Lamboy, E. F. Connor, R. M. Waymouth and J. L. Hedrick, *Org. Lett.*, 2002, **4**, 3587.
92. A. P. Dove, R. C. Pratt, B. G. G. Lohmeijer, D. A. Culkin, E. C. Hagberg, G. W. Nyce, R. M. Waymouth and J. L. Hedrick, *Polymer*, 2006, **47**, 4018.
93. O. Coulembier, J.-M. Raquez and Ph. Dubois, *Polimery*, 2008, **53**, 255.
94. M. K. Kiesewetter, E. J. Shin, J. L. Hedrick and R. M. Waymouth, *Macromolecules*, 2010, **43**, 2093.
95. N. E. Kamber, W. Jeong, R. M. Waymouth, R. C. Pratt, B. G. G. Lohmeijer and J. L. Hedrick, *Chem. Rev.*, 2007, **107**, 5813.
96. H. A. Brown and R. M. Waymouth, *Acc. Chem. Res.*, 2013, **46**, 2585.
97. E. F. Connor, G. W. Nyce, M. Myers, A. Möck and J. L. Hedrick, *J. Am. Chem. Soc.*, 2002, **124**, 914.
98. W. H. Olmstead, Z. Margolin and F. G. Bordwell, *J. Org. Chem.*, 1980, **45**, 3295.
99. R. W. Alder, P. R. Allen and S. J. Williams, *J. Chem. Soc., Chem. Commun.*, 1995, 1267.
100. Y.-J. Kim and A. Streitweiser, *J. Am. Chem. Soc.*, 2002, **124**, 5757.
101. C.-L. Lai, H. M. Lee and C.-H. Hu, *Tetrahedron Lett.*, 2005, **46**, 6265.
102. D. Patel, S. T. Liddle, S. A. Mungur, M. Rodden, A. J. Blake and P. L. Arnold, *Chem. Commun.*, 2006, 1124.
103. S. Csihony, G. W. Nyce, A. C. Sentman, R. M. Waymouth and J. L. Hedrick, *Polym. Prepr.*, 2004, **45**, 319.
104. D. A. Culkin, W. Jeong, S. Csihony, E. Gomez, N. Balsara, J. L. Hedrick and R. M. Waymouth, *Angew. Chem.*, 2007, **119**, 2681.
105. M. H. Chisholm, S. S. Iyer, D. G. Mccollum, M. Pagel and U. Werner-Zwanziger, *Macromolecules*, 1999, **32**, 963.
106. R. E. Drumright, P. R. Gruber and D. E. Henton, *Adv. Mater.*, 2000, **12**, 1841.
107. J. R. Sarasua, R. E. Prud'homme, M. Wisniewski, A. Le Borgne and N. Spassky, *Macromolecules*, 1998, **31**, 3895.
108. D. Enders, O. Niemeier and A. Henseler, *Chem. Rev.*, 2007, **107**, 5606.
109. W. Jeong, E. J. Shin, D. A. Culkin, J. L. Hedrick and R. M. Waymouth, *J. Am. Chem. Soc.*, 2009, **131**, 4884.
110. O. Coulembier, S. Moins, J. De Winter, P. Gerbaux, P. Leclère, R. Lazzaroni and P. Dubois, *Macromolecules*, 2010, **43**, 575.
111. J. A. Cowman, J. A. Clyburne, M. G. Davidson, R. L. Harris, J. A. K. Howard, P. Küpper, M. A. Leech and S. P. Richards, *Angew. Chem., Int. Ed.*, 2002, **41**, 1432.

112. O. Coulembier, X. Delva, J. L. Hedrick and R. M. Waymouth, *Macromolecules*, 2007, **40**, 8560.
113. G. Nice, T. Glauser, E. F. Connor, A. Mock, R. M. Waymouth and J. L. Hedrick, *J. Am. Chem. Soc.*, 2003, **125**, 3046.
114. N. E. Kamber, W. Jeong, S. Gonzales, J. L. Hedrick and R. M. Waymouth, *Macromolecules*, 2009, **42**, 1634.
115. B. Maji, M. Breugst and H. Mayr, *Angew. Chem., Int. Ed.*, 2011, **50**, 6915.
116. E. J. Shin, W. Jeong, H. A. Brown, B. J. Koo, J. L. Hedrick and R. M. Waymouth, *Macromolecules*, 2011, **44**, 2773.
117. E. J. Shin, H. A. Brown, S. Gonzalez, W. Jeong, J. L. Hedrick and R. M. Waymouth, *Angew. Chem., Int. Ed.*, 2011, **50**, 1.
118. H. A. Brown, S. L. Xiong, G. A. Medvedev, Y. A. Chang, M. M. Abu-Omar, J. M. Caruthers and R. M. Waymouth, *Macromolecules*, 2014, **47**, 2955.
119. A. K. Acharya, Y. A. Chang, G. O. Jones, J. E. Rice, J. L. Hedrick, H. W. Horn and R. M. Waymouth, *J. Phys. Chem. B*, 2014, **118**, 6553.
120. Y. A. Chang and R. M. Waymouth, *Polym. Chem.*, 2015, **6**, 5212.
121. S. Penczek, M. Cypryk, A. Duda, P. Kubisa and S. Slomkowski, *Prog. Polym. Sci.*, 2007, **32**, 247.
122. W. Jeong, J. L. Hedrick and R. M. Waymouth, *J. Am. Chem. Soc.*, 2007, **129**, 8414.
123. A. Duda and A. Kowalski, in *Handbook of Ring-opening Polymerization*, ed. P. Dubois, O. Coulembier and J.-M. Raquez, Wiley-VCH, Weinheim, Germany, 2009, ch. 1, pp. 1.
124. H. R. Kricheldorf, *J. Polym. Sci., Part A: Polym. Chem.*, 2004, **42**, 4723.
125. C. W. Bielawski, D. Benitez and R. H. Grubbs, *Science*, 2002, **297**, 2041.
126. L. Guo, S. H. Lahasky, K. Ghale and D. H. Zhang, *J. Am. Chem. Soc.*, 2012, **134**, 9163.
127. L. Guo and D. Zhang, *J. Am. Chem. Soc.*, 2009, **131**, 18072.
128. C.-U. Lee, T. P. Smart, L. Guo, T. H. Epps and D. Zhang, *Macromolecules*, 2011, **44**, 9063.
129. S. H. Lahasky, W. K. Serem, L. Guo, J. C. Garno and D. Zhang, *Macromolecules*, 2011, **44**, 9574.
130. X. Li, L. Guo, M. Casiano-Maldonaldo, D. H. Zhang and C. Wesdemiotis, *Macromolecules*, 2011, **44**, 4555.
131. J. Raynaud, C. Absalon, Y. Gnanou and D. Taton, *J. Am. Chem. Soc.*, 2009, **131**, 3201.
132. J. Raynaud, C. Absalon, Y. Gnanou and D. Taton, *Macromolecules*, 2010, **43**, 2814.
133. J. Raynaud, W. N. Ottou, Y. Gnanou and D. Taton, *Chem. Commun.*, 2010, **46**, 3203.

CHAPTER 2

Ring-opening Polymerization Promoted by Brønsted Acid Catalysts

BLANCA MARTIN-VACA* AND DIDIER BOURISSOU*

Université de Toulouse, UPS, and CNRS, LHFA UMR5069, 118 route de Narbonne, F-31062 Toulouse, France
*Email: bmv@chimie.ups-tlse.fr; dbouriss@chimie.ups-tlse.fr

2.1 Introduction

Over the last two decades, organocatalysis has been an area of extremely intense investigation and remarkable advances have been achieved in the scope of chemical transformations, as well as in activity and selectivity (including stereoselectivity).[1,2] The application of organocatalysis has also been extended to polymerization processes, and more particularly to the ring-opening polymerization (ROP) of heterocyclic monomers.[3–5] Among the different families of organocatalysts capable of promoting these transformations, Brønsted acids (BAs) occupy a forefront position. The large variety of BA, their ready availability and their functional group compatibility enable to tune the catalytic activity and to address the specificities of the different types of cyclic monomers. Last but not least, chiral scaffolds permit to envision stereoselective ROP of chiral monomers such as lactide.

Brønsted acids are suitable catalysts for the controlled ROP of heterocyclic monomer of adequate basicity, mainly lactones (δ-valarolactone, δ-VL; ε-caprolactone, ε-CL and ω-decalactone, ω-DL) and cyclic carbonates (tetramethylene carbonate, TMC and 1,3-dioxepan-2-one, 7CC) (Figure 2.1).

Polymer Chemistry Series No. 31
Organic Catalysis for Polymerisation
Edited by Andrew Dove, Haritz Sardon and Stefan Naumann

Published by the Royal Society of Chemistry, www.rsc.org

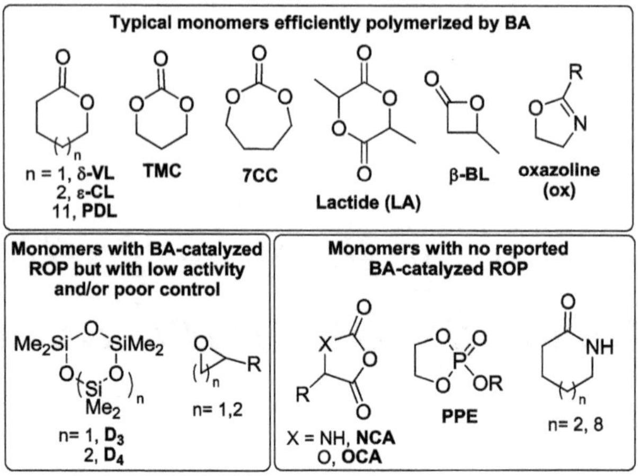

Figure 2.1 Classification of the main heterocyclic monomers according to their behavior towards BA-catalyzed ROP.

The strongest acids perform also nicely with lactide (LA), β-butyrolactone (β-BL) and 2-alkyl-oxazolines (ox). In marked contrast, no example of controlled ROP of *N*-carboxyanhydrides (NCAs), *O*-carboxyanhydrides (OCAs), phosphoesters (PPEs) or lactams have been reported. Cyclic ethers, for which BA-catalyzed ROP has been reported but with low control, will not be included in this chapter.

In this chapter, the different sub-classes of BA shown to efficiently promote the ROP of these heterocycles will be presented, classified according to their nature (sulfonic acids, sulfonamides, phosphoric, carboxylic acids, *etc.*). For each sub-class, the scope and performance in ROP will be discussed, followed by the mechanistic insights obtained into their mode of action. For BA-catalyzed ROP, two main mechanisms are to be considered (Scheme 2.1): (i) the active chain-end (ACE) propagation mechanism, resulting from the nucleophilic attack of a non-activated monomer to an activated one, leading to the formation of an onium species that ensures propagation, and (ii) the activated monomer (AM) propagating mechanism, consisting in the nucleophilic attack of the initiating/propagating group, typically an alcohol, to an acid-activated monomer.[6] The occurrence of these two competitive propagation pathways might affect the control of the polymerization, leading eventually to two different polymer structures. The experimental and computational data shedding light on the operating mechanism will be discussed.

The nice balance between activity and functional group compatibility of BA catalysts makes them suitable for the use of functionalized initiators and/or monomers, and for coupling with other polymerization methods.[7] The contribution of BA catalysts to the preparation of well-defined functionalized polymers and copolymers (including block copolymers) is briefly summarized at the end of the chapter.

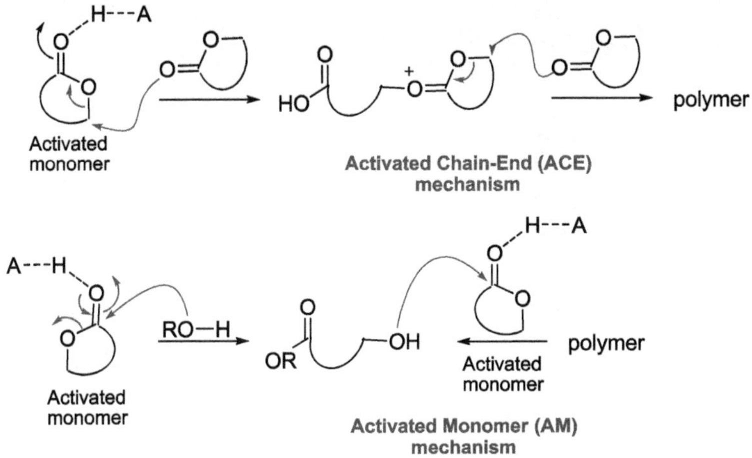

Scheme 2.1 ACE (top) and AM (bottom) propagating mechanisms illustrated for ROP of a lactone.

2.2 Organic Acids

2.2.1 Inorganic Strong Brønsted Acids

Although not organic acids, strong BAs such as chlorhydric acid deserve being mentioned as they are among the first BAs reported to promote controlled ROP of lactones, cyclic carbonates and even 2-alkyl-2-oxazolines when combined with a protic initiator (Figure 2.2).[8–12] More recently, the super acid $[H(Et_2O)_2][Al\{OC(CF_3)_3\}_4]$[13] has been used in this context as well.

In the late 1990s and early 2000s, Endo and Jérôme pioneered the development of acid-promoted ROP of lactones and cyclic carbonates using $HCl.Et_2O$ as catalyst. First investigations by Endo on 7CC ROP showed that while no reaction occurred in the presence of only $HCl.Et_2O$ in CH_2Cl_2 solution, its combination with water as a protic initiator promoted the polymerization under mild conditions (1 mol L^{-1}, 25 °C) to yield polycarbonates of M_n around 10 400 g mol^{-1} and narrow molecular dispersions ($Đ<1.23$) in less than one day.[8,9] The same team reported one year later that benzylic alcohol (BnOH) and 1,4-butane-diol can also be used as initiators of this ROP reaction and demonstrated the incorporation of the initiator as a chain-end on the basis of ^1H NMR spectroscopy.[10] Kinetic studies revealed a nice linear trend of the semi-logarithmic plot for the reactions initiated with an alcohol, indicating first order in monomer, but a slight deviation upwards was observed for reactions initiated with water. However, when considering two –OH propagating chain-ends, linearity of the plot is met. This observation is consistent with a decarboxylation process following ring opening of the first 7CC unit by water, which leads to the formation of a diol that acts as bifunctional initiator and propagates in two directions, most likely by AMM (Scheme 2.2). Endo also reported the polymerization of δ-VL and δ-CL

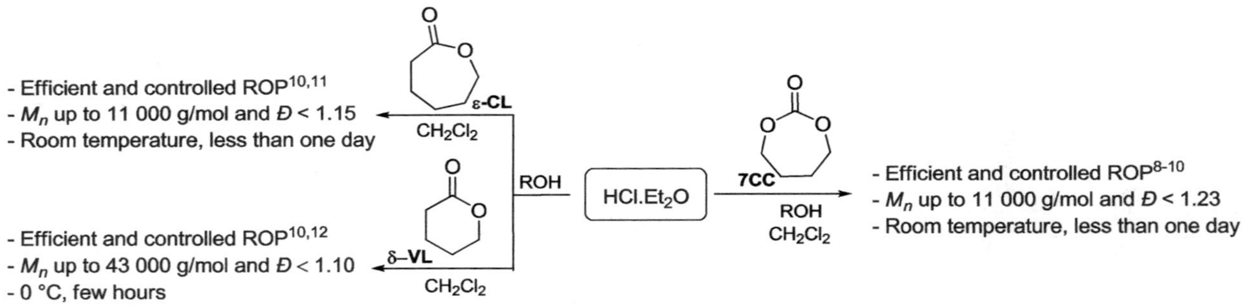

Figure 2.2 Main features of the ROP of lactone and cyclic carbonates promoted by $HCl.Et_2O$.

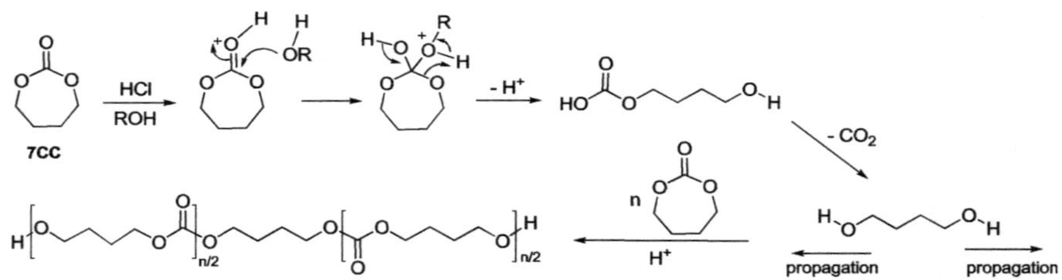

Scheme 2.2

using similar reaction conditions ($HCl.Et_2O$ 5% in 1 mol L^{-1} solutions at 0–25 °C).[10,11] PCL and PVL with M_n in the range of 2700–10 000 g mol^{-1} and narrow molecular distributions (*Đ*<1.17) were prepared in few hours to one day. As a more reactive monomer, δ-VL allowed working at lower temperatures (0 and even −45 °C), which helped to reach high conversions with better control that working at r.t. (*Đ* of 1.10 instead of 1.17 for an M/I ratio of 30). Optimization of the reaction conditions by Jérôme (in particular working at 4 mol L^{-1}) permitted the preparation of PVL with M_n values up to 43 000 g mol^{-1} at 0 °C in less than one day.[12] At this low temperature, the polymerization reaction is well controlled and low dispersities are obtained (*Đ*<1.10). The compatibility of the reaction conditions with functionalized (alkene, ester, brominated alkyl chains) or polyol initiators was also substantiated.[12]

More recently, the super acid $[H(OEt_2)_2][Al\{OC(CF_3)_3\}_4]$ featuring a non-coordinating and non-nucleophilic counter-anion has been applied to the ROP of 2-alkyl-2-oxazolines, a sort of monomer typically polymerized by CROP (cationic ROP) *via* an ACEM using alkyl halides, tosylates or triflates as promoters and for which no example of controlled acid-promoted polymerization existed.[13] Working at relatively high temperatures (80 or 140 °C), polymerization takes place quite rapidly (a few minutes to two hours) in a rather controlled manner to yield polyoxazolines of M_n around 10 800 g mol^{-1} and narrow molecular distributions (*Đ*<1.13). The reaction is first-order in monomer and the M_n of the polymer chains grow linearly with monomer conversion, which is consistent with a controlled process without side transfer or deactivation. The structural characterization of the polymer is consistent with an ACE mechanism, in which the activation of one monomer unit by protonation of the N atom promotes N-nucleophilic attack of a second monomer unit on C_β atom provoking ring opening. Final hydrolysis of the activated chain-end results in a hydroxyl chain-end (Scheme 2.3).

2.2.2 Sulfonic Acids

2.2.2.1 Performances in ROP

Sulfonic acids were the first organic BAs to be used as catalysts for the ROP of lactones. Trifluoromethane sulfonic acid (HOTf) occupies a forefront position. The ability of HOTf to promote the polymerization of lactide in the absence of protic initiator was reported by Kricheldorf in the mid-1980s.[14] Further studies concerning the scope and mechanism of the process in solution[15,16] and in the bulk,[17] were reported in the early 2010s by Kubisa and Basko. The reaction is very slow at room temperature but full conversions can be obtained in several hours at higher temperatures (120–160 °C). Although relatively high M_n can be obtained (up to 25 10^3 g mol^{-1}), the reaction is far from controlled (*Đ*≈1.8–2.5). In 2005, Bourissou and Martin-Vaca reported that the combination of HOTf with a protic initiator,

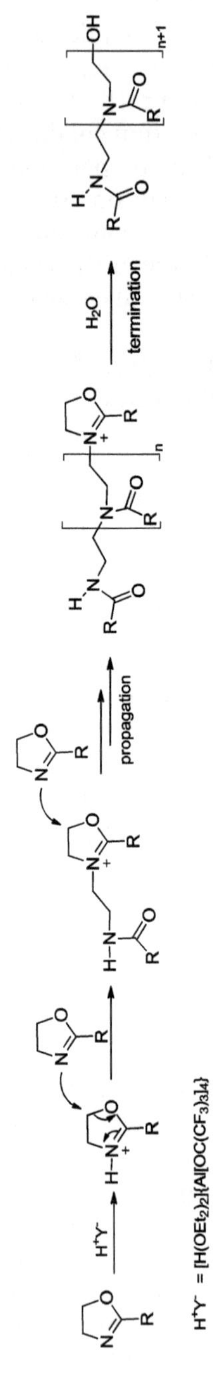

Scheme 2.3 ACE mechanism proposed for the ROP of 2-alkyl-2-oxazolines promoted by $[H(OEt_2)_2][Al\{OC(CF_3)_3\}_4]$.

typically an alcohol, enables the controlled polymerization of lactide under mild conditions (25 °C, 1 mol L^{-1} CH_2Cl_2 solution).[18] PLAs of M_n up to 15 000 g mol^{-1} and $Đ<1.5$ were obtained within hours. Excellent initiation fidelity with different initiators such as primary and secondary alcohols, as well as water was demonstrated by ^1H NMR spectroscopy and MALDI-TOF mass spectrometry. Since then, the activity of HOTf has been evaluated for the ROP of the most common lactones and cyclic carbonates (Figure 2.3).

Particularly relevant is the ROP of β-BL catalyzed by HOTf. The ROP of this monomer faces two main issues: (i) due to its small ring size, competition between *O*-acyl bond cleavage (leading to a propagating hydroxyl chain-end) and *O*-alkyl bond cleavage (leading to a propagating carboxylic acid chain-end); and (ii) propensity to undergo crotonization reactions yielding acrylic moieties (non-propagating chain-end).[19] In a preliminary study, Pol suggested the feasibility of the ROP of β-BL promoted by HOTf in the presence of a protic initiator,[20] but the efficiency and controlled character of the process were not evaluated. Bourissou and Martin-Vaca demonstrated in 2014 that the combination of HOTf with an alcohol promotes the ROP of β-BL with excellent *O*-acyl bond cleavage selectivity and chain-end fidelity without crotonization reactions.[21] This behavior is in marked contrast with that of other organocatalysts (basic or nucleophilic) that act as initiator and promote the dehydration reaction of the terminal hydroxyl group, precluding copolymerization reactions. Using mild conditions, (toluene, 1 mol L^{-1} solutions, 30 °C) polybutyrolactones (PBLs) with M_n values up to 8500 g mol^{-1} and relatively narrow molecular distributions ($Đ<1.25$) were prepared.

HOTf has also been applied to the ROP of ε-CL and TMC, but the performances were less remarkable that those obtained with other milder BAs. ε-CL indeed polymerizes within hours in the presence of HOTf and a protic initiator (1/1 ratio) to yield PCLs of controlled M_n and narrow molecular distributions.[22] However, a deactivation phenomenon was observed for HOTf/ROH ratios higher than 1 affecting the polymerization kinetics.[22] In the case of TMC, although ROP also takes place readily at room temperature, the occurrence of a decarboxylation process leads to polycarbonates featuring ether linkages whose extent increases with the reaction temperature.[23] Better results were therefore obtained for these two monomers using methanesulfonic acid (MSA) as catalyst. The acid strength of MSA is significantly lower than that of HOTf (pK_{aH_2O} of −14 for HOTf *vs.* −1 for MSA).[24] PCL of M_n around 10 100 g mol^{-1} ($Đ<1.15$) could be prepared under mild conditions (1 mol L^{-1} toluene solution at 30 °C).[22] Despite the difference in acidity, the reaction rate was similar with both acids (MSA and HOTf) when using a catalyst/initiator ratio of 1/1. However, for catalyst/initiator >1/1, MSA was surprisingly more active than HOTf, revealing that activity does not directly correlate to acidity. It has to be noted that despite the good performance of MSA with ε-CL, ROP of LA and β-BL with this catalyst are not efficient. Only one monomer unit is opened at r.t. with LA, propagation being probably hampered by the steric hindrance of the monomer and the propagating alcohol. Although better results were obtained with β-BL using a

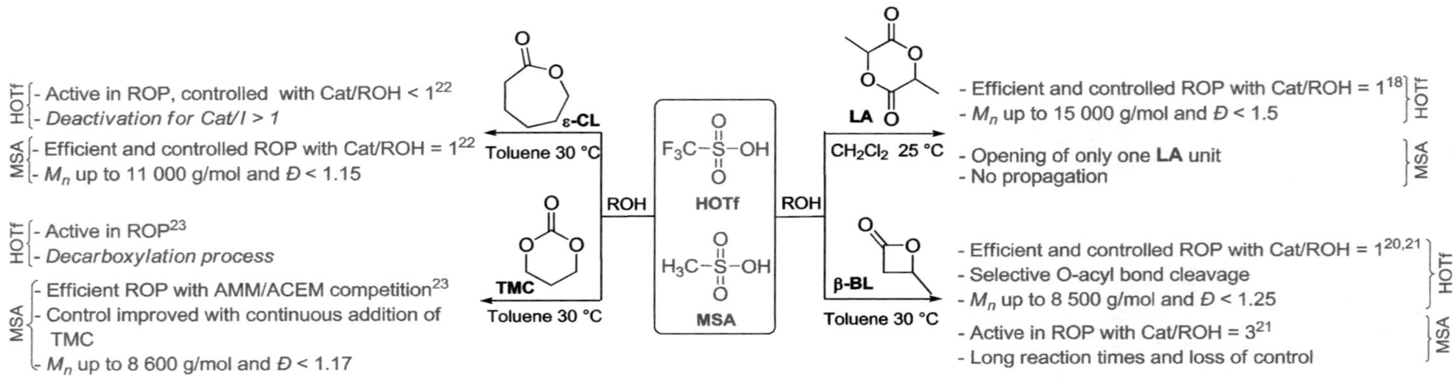

Figure 2.3 Main features of the ROP of lactones and cyclic carbonates promoted by HOTf and MSA.

MSA/ROH ratio of 3, the reaction is still rather slow and far less controlled than with HOTf.[21] Concerning cyclic carbonates, Nakano initially reported the ROP of dimethyltrimethylene carbonate promoted by MSA in toluene at 50 °C.[25] Although spectroscopic analysis revealed complete consumption of the monomer and good incorporation of the initiator, SEC analysis showed bimodal profiles for DP > 30, attesting for the presence of different polymer populations. This was initially explained by the occurrence of transfer reactions during polymerization. Later on, Bourissou and Martin-Vaca studied the ROP of TMC using the combination of MSA with an alcohol at 30 °C in toluene and observed also the formation of two polymer populations.[23] Thanks to detailed mechanistic investigations, competition between AM and ACE mechanisms was suggested to be at the origin of the low polymerization control, and the presence of two polymer populations. The incidence of the last mechanism could be significantly reduced by carrying out the reaction with a continuous addition of the monomer (TMC) into the reaction media, so that its concentration remains low.[26] Thanks to these operational modifications, mono-modal SEC profiles could be obtained for PTMC up to DP 80 ($Đ$ <1.12).

Other sulfonic acids such as camphorsulfonic acid (CSA) and *p*-toluenesulfonic acid (PTSA) have also been applied to the ROP of ε-CL, in particular by Zinck.[27] These sulfonic acids (pK_{aH_2O} 1.2 and −2.8, respectively) were investigated in aqueous media at 25–50 °C using polysaccharides as initiators with the aim of preparing grafted copolymers. The grafted copolymers could be obtained thanks to a combination of ROP of ε-CL initiated by water or by the polysaccharide and polycondensation process between grafted and non-grafted growing chains. The polymerization is most likely favored by the low solubility of the PCL chains in water. Thus, they precipitate and aggregate into hydrophobic domains from where water is expelled, which may favor ROP *versus* monomer hydrolysis and displace the esterification equilibria. Grafted polymers with PCL chains of relatively low M_n are obtained (between 500 and 3000 g mol^{-1}) (Scheme 2.4).

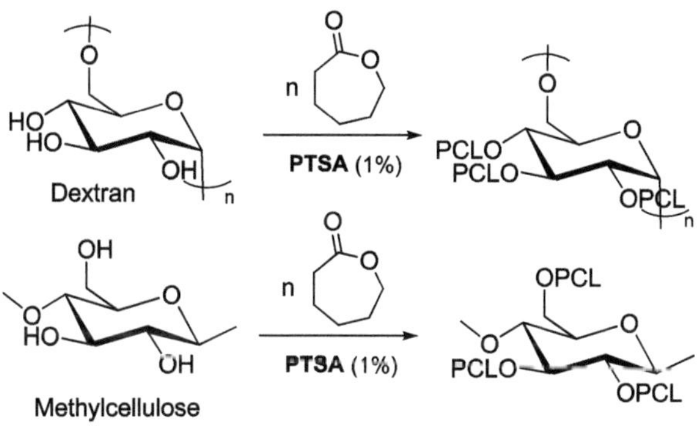

Scheme 2.4 ROP of ε-CL catalyzed by PTSA and initiated by dextran or methylcellulose.

Besides the monomers discussed above, macrolactones have also been polymerized by sulfonic acids, although activity and polymerization control are usually lower (Figure 2.4). Dodecyl-toluenesulfonic acid and HOTf have been reported by Mecerreyes to promote the ROP of pentadecalactone (PDL), ambrettolide (AMB), globalide (GB) and ethylene brassylate (EB) in the bulk at 80 °C using benzylic alcohol as initiator.[28,29] Despite the rather harsh reaction conditions, full conversion requires one to several days for DP ranging from 42 to 136 (M_n up to 9700 g mol^{-1}), confirming the low reactivity of these monomers whose ROP is driven by entropic rather than enthalpic factors (lack of ring strain in these monomers).[30] In addition, the ring size makes the ester function of the polymer chain as reactive as that of the cycle, which favors transfer reactions and leads to rather high $Đ$ values (1.65–2.94).

HOTf has also been used as catalyst for the ROP of cyclic tetra-dimethylsiloxane (D_4) in the preparation of oligodimethylsiloxanes (ODMS). Here, the use of a protic initiator does not permit controlled polymerization,[31] in contrast to tetramethyldisiloxane ($HMe_2Si\text{-}O\text{-}SiMe_2H$) (Scheme 2.5).[32] Interestingly, the terminal Si–H moieties of the siloxane polymer chains allowed for further functionalization of the polymers *via* hydrosilylation reaction.

2.2.2.2 Mechanistic Aspects

Kricheldorf first, and then Basko and Kubisa carried out investigations on the mechanism of the ROP of LA catalyzed by HOTf *in the absence of protic initiator*. They envisioned the occurrence of both the ACE and AM mechanisms (in which triflic acid would also act as an initiator) (Scheme 2.6).[14–16] Experimental characterization of the PLA by SEC revealed that the M_n values of the obtained PLA increase linearly with the LA/HOTf ratio, but are

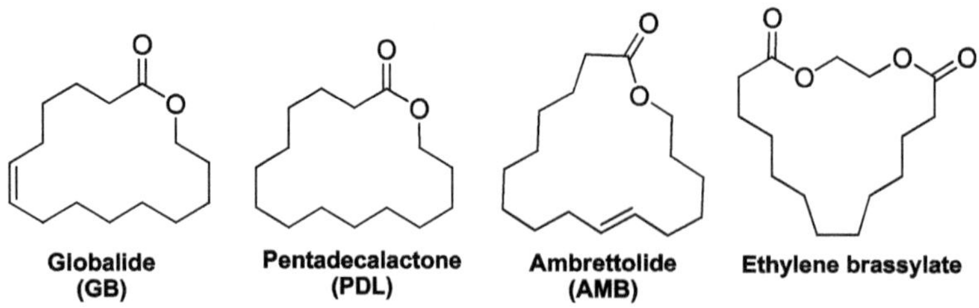

Figure 2.4 Macrolactones with ROP promoted by HOTf and MSA.

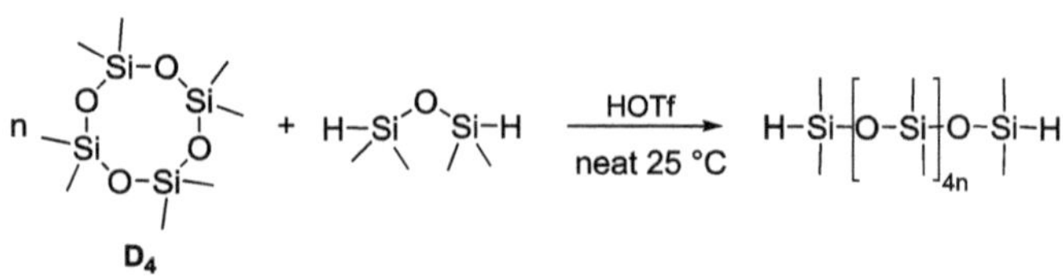

Scheme 2.5 ROP of D_4 catalyzed by HOTf and initiated by $HMe_2Si\text{-}O\text{-}SiMe_2H$.

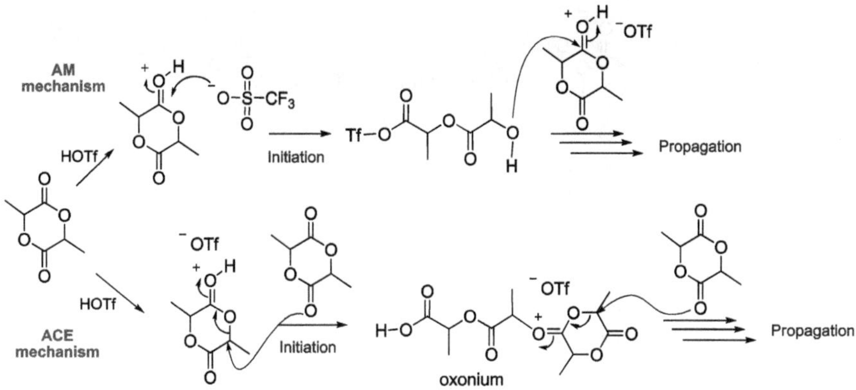

Scheme 2.6 AM and ACE mechanisms for the ROP of lactide promoted by HOTf without initiator.

four-fold those calculated from these ratios.[16] This indicates that the number of growing chains is about one quarter the amount of HOTf. Basko and Kubisa also observed an upwards deviation from linearity of the semi-logarithmic plot of LA conversion. This can be explained by the fact that the ester in the polymer chains are less basic that those on the cyclic monomer (esters are more basic in *anti* than in *cis* conformation).[33] Thus, as the monomer is consumed, the amount of acid available formally increases, resulting in a speed up of the reaction. These two observations are in fact consistent with HOTf playing both the role of catalyst and, to some extent, of protic initiator. Ring-opening of an acid-activated LA unit by nucleophilic addition of a triflate would yield a dilactate bearing a terminal hydroxyl group that can then act as a propagating group. Propagation would thus occur by addition of the hydroxyl group on another acid-activated monomer provoking *O*-acyl bond cleavage, as demonstrated by the retention of configuration when working with L-LA.

Trapping experiments further support the AM *vs.* ACE mechanism.[16,17] Monomer conversion is instantaneously stopped by the addition of a proton sponge onto the reaction media, suggesting the presence of protonated active species. More importantly, upon addition of PPh_3 to the reaction media in the course of polymerization, only the formation of tertiary phosphonium species (Ph_3PH^+) is observed by ^{31}P NMR spectroscopy. No quaternary phosphonium resulting from the nucleophilic attack of the phosphine to the oxonium intermediate was observed (Scheme 2.7).

Concerning the ROP of LA with HOTf *in the presence of a protic initiator*, analysis of the obtained polymers by NMR and MS demonstrated the exclusive incorporation of the initiator as an ester chain-end, as well as the control of the M_n by the initial monomer/initiator feed.[18] Similar observations were made for the PCL prepared by ROP of ε-CL with HOTf or MSA as catalyst and a protic initiator.[22] This is in perfect agreement with an AM mechanism. To get more insight into the mechanism and better understand

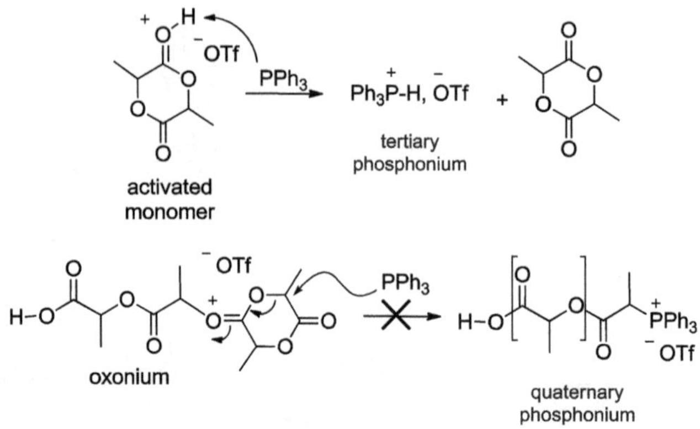

Scheme 2.7 Trapping of propagating species of AM (top) and ACE (bottom) mechanisms with PPh_3.

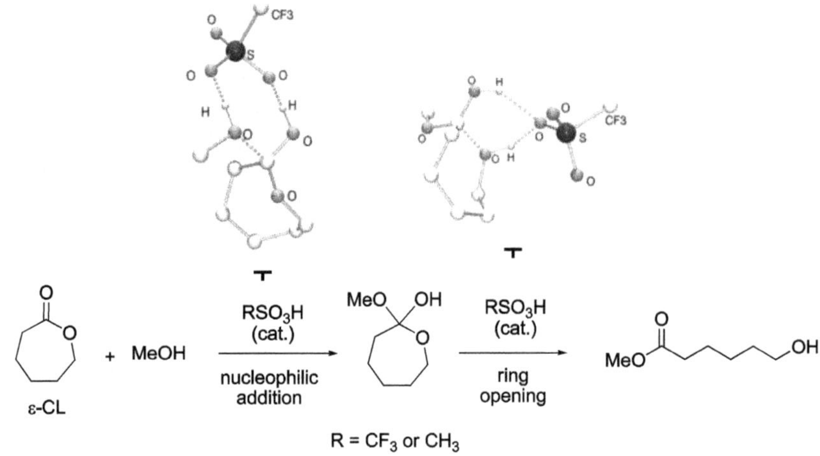

Figure 2.5 DFT-proposed mechanism for the ROP of ε-CL promoted by HOTf and MSA (bottom) and TS proposed for the two steps (top).

the absence of a direct relationship between the acid strength of the catalyst and its activity, a DFT computational study was carried out by Bourissou and Maron on the model reaction of ring-opening of ε-CL with methanol.[34] Both sulfonic acids were in fact predicted to behave as proton shuttles, acting as bifunctional catalysts during the two steps of the reaction, namely the nucleophilic attack of the alcohol leading to the tetrahedral intermediate and the ring-opening (Figure 2.5). In the first step, the monomer is activated by the acid OH group and concomitantly, the alcohol is activated by a basic S=O unit. Activation barriers of *ca.* 16 (respectively, 20) kcal mol^{-1} are predicted for HOTf (respectively, MSA), compared to 38–42 kcal mol^{-1} for the sole activation of the monomer by protonation. The same trend is observed for the ring-opening step, in which the bifunctional pathway (with contribution of the O–H and S=O groups) occurs with barriers of 13 (respectively, 18) kcal mol^{-1}, *vs. ca.* 42 kcal mol^{-1} for the sole contribution of the protic activation. These data are consistent with the reaction taking place at room temperature,[35] and with much higher activity of sulfonic acids *vs.* HCl (pK_{aH_2O}-8)[24] that merely activates the monomer by protonation. Regarding the relative activity of HOTf and MSA, it may be inferred that the higher acidity of HOTf is compensated by the higher basicity of the S=O moieties of MSA, resulting overall in similar activation barriers. The decrease in activity observed for HOTf with cat/initiator higher than 1 may be due to some deactivation of the propagating alcohol by protonation of the OH group, a phenomenon that would have higher impact with this acid than with the weaker acid MSA.

Mechanistic studies on the ROP of lactones promoted by sulfonic acids thus converge on a process occurring *via* an AM rather than an ACE mechanism, whether or not in the presence of a protic initiator.[16–18,22,34] The low nucleophilicity of lactones compared to cyclic ethers probably explains why the AM mechanism prevails over ACE.[36] In the case of LA and β-BL, steric factors also most likely come into play to explain the poor results obtained with MSA with these monomers; propagating with a secondary alcohol and a substituted monomer has a detrimental impact on the kinetics of the reaction. The situation is somewhat different with cyclic carbonates and the AM mechanism is less favored over ACE: the carbon atom of the C=O moiety is less electrophilic than that of lactones and the exocyclic oxygen atom is more basic. As a result, there is a noticeable competition between the two propagating mechanisms, influencing the control of the polymerization:[23,25] (i) the obtained M_n values are higher than those expected from the monomer to initiator ratios, (ii) the integration of the ^{1}H NMR signal associated with the hydroxyl chain-end CH_2OH exceeds that associated with the CH_2OCO moiety of the alkyl ester chain-end, and (iii) two polymer populations are present in the MALDI-TOF spectra. The AM mechanism can explain the main population (AM, Scheme 2.8). The minor population can be explained by the occurrence of the ACE mechanism, leading, after decarboxylation, to the formation of a hydroxyl–oxonium species that favors then bidirectional growth of the polymer *via* a combination of AM and ACE mechanisms. After hydrolysis, the AM–ACE polymer can be formally considered as being initiated by propan-1,3-diol. These two competitive propagation pathways lead here to two different polymer structures that can be differentiated analytically. MALDI-TOF-MS spectrometry allows direct observation of the two PTMC populations. The AM mechanism can be favored by slow addition of the monomer, so as to decrease its proportion relative to the propagating alcohols. The incidence of the undesirable ACE mechanism can be thereby significantly reduced, so that well-controlled polymerization of TMC can be achieved with MSA as catalyst.[23]

Later on, Coady and Hedrick studied by DFT the ROP of TMC catalyzed by HOTf and by MSA *via* the AM mechanism.[37] The results were comparable with those reported for ε-CL. The sulfonic acids were predicted to act as

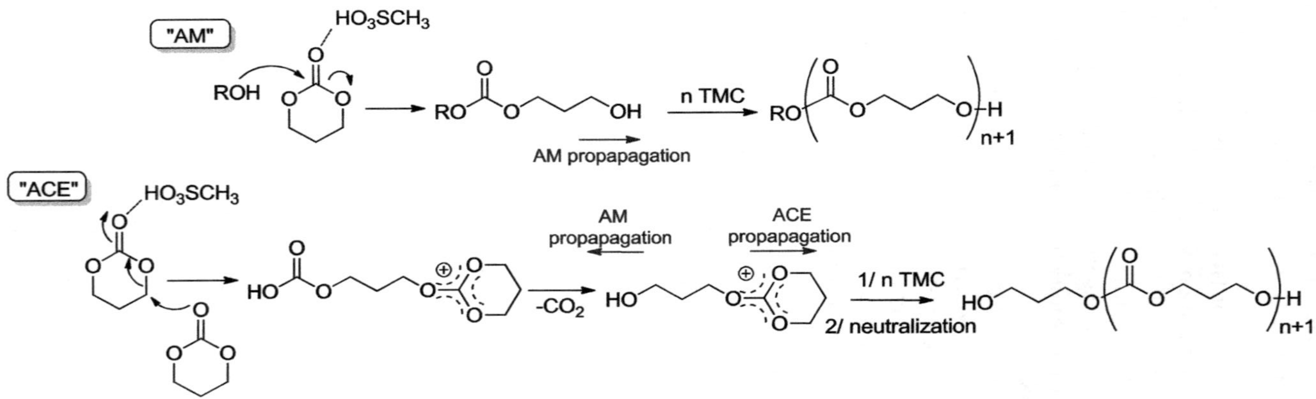

Scheme 2.8 ACE and AM competing mechanisms in the MSA catalyzed ROP of TMC.

proton shuttles in the two steps, namely the nucleophilic addition of the alcohol to the electrophilic C=O and the ring-opening (Figure 2.6). The energy barriers computed for the two acids are very close (17.4 and 16.9 kcal mol^{-1} for HOTf and MSA, respectively) and fall in the same range as that reported earlier for ε-CL.

2.2.3 Sulfonamides

2.2.3.1 Performance in ROP

The structural similarities with sulfonic acids (S=O groups and a polar N–H bond) and their acidic properties make sulfonamides suitable candidates for the ROP of lactones and cyclic carbonates. Perfluorinated bis-sulfonamides $(RfSO_2)_2NH$ with Rf = CF_3 or C_4F_9 (estimated pK_a approximately −12)[38] (RfNH) have indeed been reported to be active in the ROP of δ-VL,[39] ε-CL[40–42] and, to a much lower extent, LA (Figure 2.7).[40] Tf_2NH-catalyzed polymerization of δ-VL was first reported by Kakuchi in 2010.[39] Reactions were carried out in the presence of an alcohol as initiator (3 mol L^{-1} CH_2Cl_2 solutions, 27 °C, Tf_2NH/initiator ratio = 0.1). High monomer concentrations were used to favor full conversion by displacing the polymerization ⇆ depolymerization equilibrium of this six-membered ring monomer.[43] Under these mild conditions, high conversions were attained in several hours (2 to 14 h for M/I from 50 to 130) and PVLs with M_n values up to 12 700 g mol^{-1} and $Đ$ <1.12 were obtained. In 2011, Takasu published the ROP of ε-CL promoted by Tf_2NH or Nf_2NH and an alcohol as initiator.[41] As expected for this less polymerizable monomer, reaction times were longer than for δ-VL. Nf_2NH showed slightly more active than Tf_2NH (8 h, instead of 10 h with Nf_2NH, were required to get full conversion of 40 equiv. of ε-CL in 3 mol L^{-1} toluene solution of at 25 °C with a $(RfSO_2)_2NH$/initiator ratio of 0.1). As with sulfonic acids, toluene was also pointed out as a more suitable solvent than the more polar CH_2Cl_2 or $CHCl_3$, leading to shorter reaction times. In addition, both triflamides afforded shorter reaction times than HOTf (8–10 h with Nf_2NH–Tf_2NH *vs.* 16 h with HOTf). Activation of the ROP of ε-CL by microwave irradiation was also reported by the same authors two years later.[42] Conducting the reaction under MW irradiation at 50–60 °C resulted in the reduction of the reaction times without affecting significantly the control of the polymerization (Figure 2.7).

The good performance displayed towards δ-VL and ε-CL prompted Kakuchi to apply Tf_2NH to the ROP of LA.[40] The reactions were carried out in 1 M CH_2Cl_2 solutions (LA has low solubility in toluene) at room temperature using a cat/initiator ratio of 3. Although ROP indeed took place to yield PLA with M_n values close to the expected ones (6400 g mol^{-1}) and narrow distributions ($Đ$ = 1.17), reaction times were extremely long: more than two weeks were required to convert 50 equiv. of LA. This confirms the negative impact of the LA structure (substitution pattern and lower basicity of the ester moieties) on its ROP catalyzed by acids.

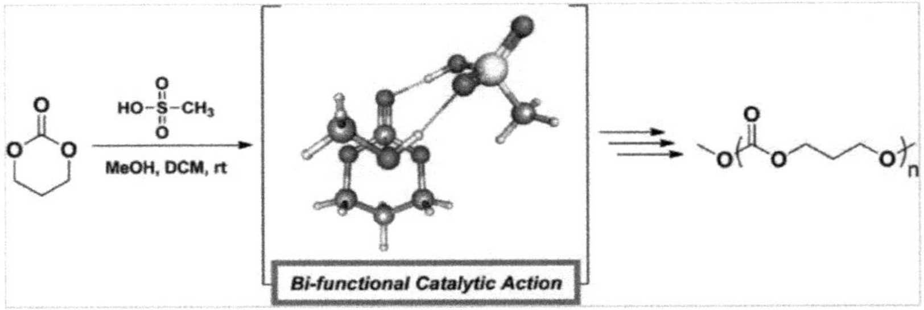

Figure 2.6 Illustration of the bifunctional mode of action of MSA in the ROP of TMC. Reproduced from ref. 37 with permission from American Chemical Society, Copyright 2013.

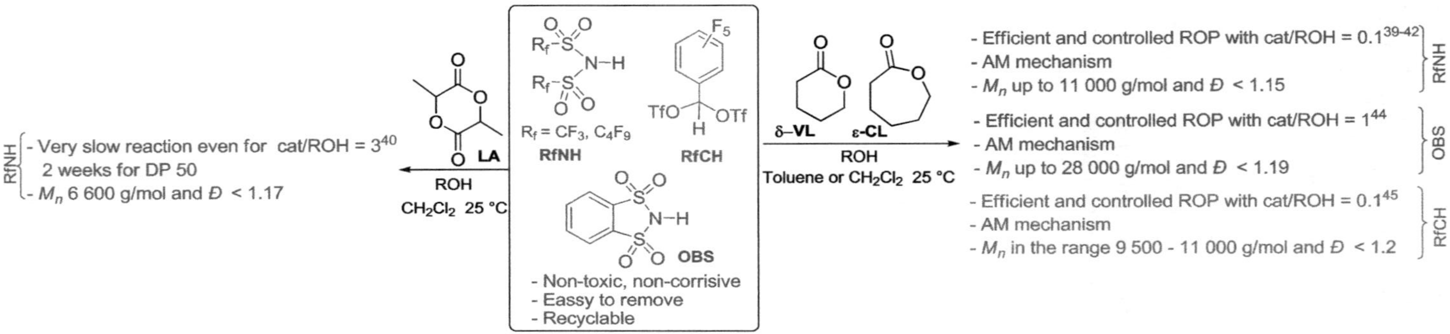

Figure 2.7 Main features of the ROP of lactones and cyclic carbonates promoted by sulfonamides.

O-benzenedisulfonamide (OBS), whose pK_a −4.1 is close to that of Tf_2NH, has also been applied to ROP with the objective of taking advantage of the non-toxic, non-corrosive nature of this compound and of its potential recycling (in marked contrast to the perfluorinated bissulfonamides, OBS is stable in water).[44] In addition, this compatibility with water permits the use of water as initiator, as in the case of MSA. Controlled ROP of δ-VL and ε-CL takes place under mild conditions (3 mol L^{-1} toluene solution, 30–35 °C) leading to polymers of M_n up to 28 000 g mol^{-1} and $Đ<1.16$ in several hours (δ-VL) to less than three days (ε-CL). It is worth mentioning that compared to HOTf, $HCl.Et_2O$ and Tf_2NH, OBS revealed the most active catalyst, evidencing again that with this kind of catalyst, activity does not correlate directly with the pK_a values.

Although pentafluorophenylbis(trifluorometyl)methane ($C_6F_5CHTf_2$, RfCH)[45] is not a sulfonamide, the results obtained in ROP using this compound as catalyst are discussed in this section since its activity results, as for Tf_2NH, from the presence of the two triflate groups on the same atom, which makes the adjacent C–H bond acidic (pK_a in AcOH ≈ 1.5 *vs.* −0.96 for HOTf).[46] $C_6F_5CHTf_2$ was applied in 2011 by Kakuchi to the ROP of δ-VL and ε-CL in 3 mol L^{-1} toluene solution at 25 °C to yield PVLs and PCLs of M_n consistent with the initial monomer/initiator ratios (in the range 9500–10 000 g mol^{-1}) and dispersities $Đ<1.20$. Reaction times with $C_6F_5CHTf_2$ are slightly longer that for Tf_2NH and OBS but still full conversion of 100 equiv. can be achieved in 8 h for δ-VL and 2 d for ε-CL.

2.2.3.2 Mechanistic Considerations

Mechanistic discussions on sulfonamide-promoted ROP are based exclusively on experimental data.[34–39] No theoretical investigations have been carried out to date. NMR and MALDI-TOF MS analyses of the polymers confirm in all cases the efficient incorporation of the protic initiator (alcohol) as an ester chain-end, as well as the formation of a unique polymer population. These facts exclude the occurrence of side reactions such as chain-transfer process, with the exception of traces of water initiation with OBS. Kinetic plots support first-order reactions in monomer. The M_n values increase linearly with monomer conversions and monomer/initiator ratios, indicating a constant number of growing chains along the polymerization process and the absence of termination process. These observations strongly suggest a living character that is further supported by second feed experiments showing nice SEC profiles consistent with a lengthening of the polymer chains. All these data are thus consistent with a controlled living polymerization resulting from an AM mechanism. Activation of the monomer by the acidic proton is in fact substantiated by ^{13}C NMR spectroscopic studies in which the noticeable deshielding of the C=O signal supports the protonation of the O atom.[40,44,45] As an example, Δδ of 4.89 and 1.32 ppm are observed for 1/1 mixtures of Tf_2NH with ε-CL and LA,[40] respectively. The smaller shift observed for LA is consistent with the low activity observed with lactide compared to ε-CL. It can be explained by the ester/acid ratio of 2 (lactide is a diester) and by the lower basicity of each ester unit of LA due to the electro-withdrawing effect of the second ester moiety. Furthermore, the absence of direct relationship between the pK_a value and activity when comparing sulfonamides with HOTf and HCl, strongly suggests that here also the situation is more complicated than the simple protonation of the monomer. The basic S=O groups may contribute by activating the propagating alcohol as for sulfonic acids, and thus, a bi-functional behavior can also be envisioned for sulfonamides.

2.2.4 Phosphoric Acids

The high valence of the P atom in phosphorus-based BAs make them easily and highly tunable, and a great variety of such compounds have found applications in organic catalysis, in asymmetric catalysis in particular.[47,48] These P(v)-based acids have also been applied to promote the ROP of all common cyclic monomers. Modification of the acidic X–H moiety and of the ArO groups permit to tune the acidity (Figure 2.8),[49,50] and thereby the activity in ROP catalysis, but also the steric hindrance, which can strongly impact the selectivity of the process, in particular in the case of a chiral monomer such as LA.

2.2.4.1 *Performance in ROP*

The simplest member of the family, diphenylphosphate (DPP, pK_{aH_2O} 1.1 and pK_{aDMSO} 3.88),[49] is a commercial and non-toxic acid that was first used for ROP in 2011 by Kakuchi, Bourissou and Martin Vaca.[51,52] δ-VL and ε-CL are readily polymerized in 1 mol L^{-1} toluene solution at room temperature (DPP/ROH ratio = 1), leading to the formation of PVL and PCL with M_n values of about 21 600 and 27 500 g mol^{-1}, respectively, and narrow molar distributions ($Đ$ < 1.10) (Figure 2.9). Similar to what was observed for the acidic catalysts discussed above, δ-VL reacted more rapidly than ε-CL: 300 equiv. of δ-VL were consumed in six hours whereas around two days were necessary for ε-CL.[51] The ready availability, low toxicity and relatively weak acidity of DPP makes it particularly suitable to prepare biodegradable polymers under mild and environment friendly conditions and with good functional compatibility. As such, the activity of DPP in ROP has been

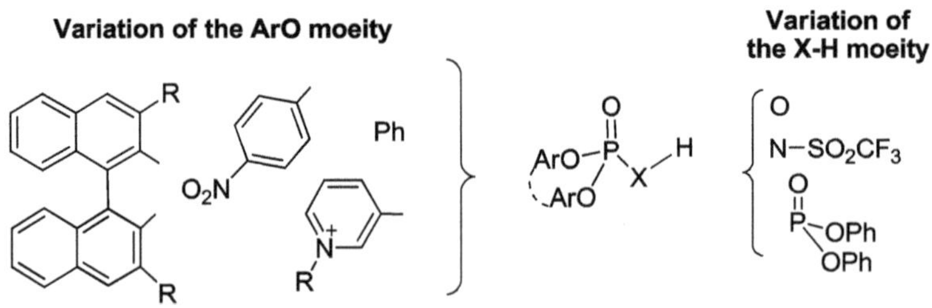

Figure 2.8 Structural modifications undertaken on phosphoric acids.

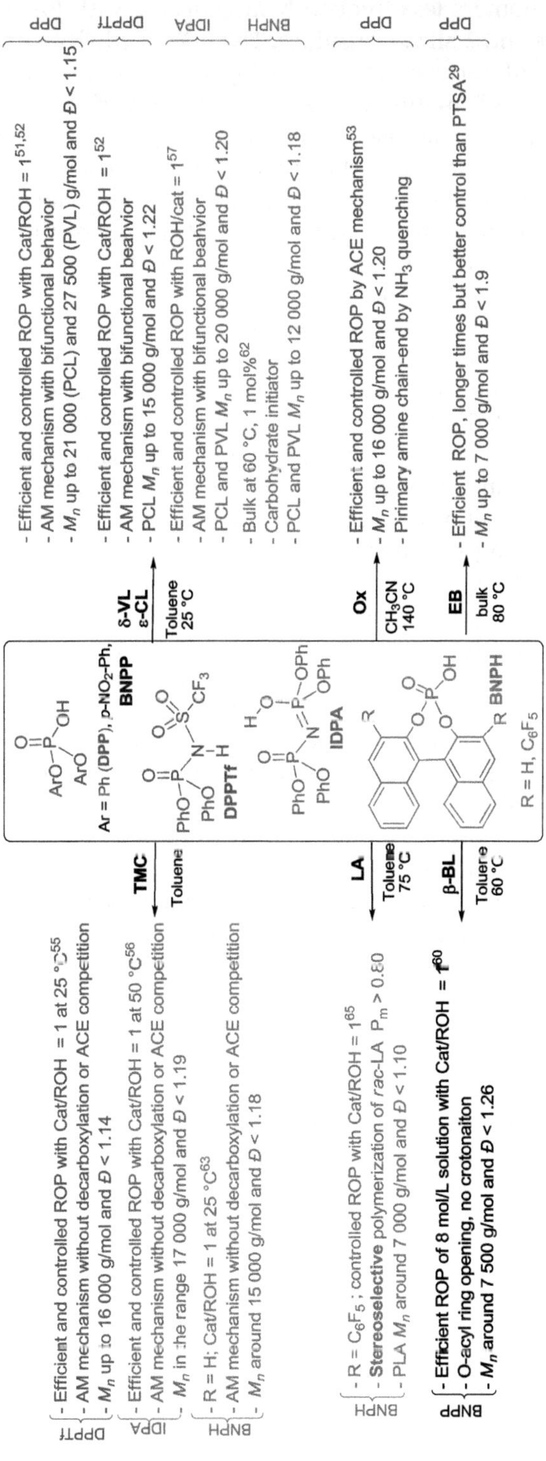

Figure 2.9 Main features of the ROP of lactones and cyclic carbonates promoted by phosphoric acids.

studied with monomers less frequently investigated with BA catalysts such as macrolactones and 2-alkyl-oxazolines. The ROP of ethylene brassylate (EB) was reported by Mecerreyes in 2014 using BnOH as initiator.[29] The low polymerizability of this monomer requires performing the polymerization in the bulk at 80 °C, as in the case of sulfonic acids (PTSA and DBSA). Despite these rather harsh conditions, the reaction is slow compared to the one catalyzed with sulfonic acids (112 h are required to convert 42 equiv., *vs.* 38 h with PTSA) but the obtained polymer has a M_n closer to the targeted one and narrower molar distributions (M_n 7100 g mol^{-1} and $Đ=1.9$ *vs.* M_n 2000 g mol^{-1} and $Đ=2.7$ with PTSA).

In addition to lactones, DPP has been shown by Guo in 2015 to promote the ROP of 2-alkyl-2-oxazolines.[53] With the exception of the example discussed above with the super acid $[H(Et_2O)_2][Al\{OC(CF_3)_3\}_4]$,[13] this is a very rare example of a BA promoter of cationic ROP of oxazolines that competes with the typically used alkyl triflates or tosylates. The reactions were conducted in quite harsh conditions (140 °C, 4 mol L^{-1} CH_3CN solution) in the absence of protic initiator, leading to an ACE propagating mechanism (Scheme 2.3). Varying the monomer to DPP ratio, polyoxazolines with M_n values up to 16 000 g mol^{-1} were obtained in rather controlled manner ($Đ<1.21$). The ACE propagating mechanism allows functionalization of the active chain-end by quenching the reaction with a nucleophilic protic reagent. Typically, a primary amine chain-end was installed using NH_3 (Scheme 2.9).

The first structural modulation reported in the context of ROP was the replacement of the polar OH group of DPP by a NH group. Although the NH bond is less polar, the higher valance of the N atom allows the increase of the acidity of the proton by introducing additional electro-withdrawing substituents. Two different phosphoramides have been reported: DPPTf and IDPA. The activity of DPPTf, carrying a triflic group on the N atom, in the ROP of ε-CL was reported by Bourissou and Martin Vaca in 2011.[52] This acid, which can be recognized as a hybrid between Tf_2NH and DPP, has a pK_{aH_2O} of approximately −3.[54] ROP of ε-CL proceeded in toluene at 30 °C slightly more rapidly with DPPTf than with DPP (1.5 *vs.* 4 h for M/Cat/ROH of 40/1/1), but the properties of the resulting polymer were comparable ($M_n \approx 3000$ g mol^{-1} and $Đ \approx 1.08$). Optimizing the reaction conditions (3 mol L^{-1}, 60 °C) permitted the preparation of a PCL of M_n 15 000 g mol^{-1} and $Đ \approx 1.22$ in only 2 h. In 2013, Guo reported the ROP of TMC catalyzed by DPPTf.[55] Although reaction times were longer than for ε-CL, the obtained PTMC suggest a controlled polymerization (M_n up to 163 000 g mol^{-1} and

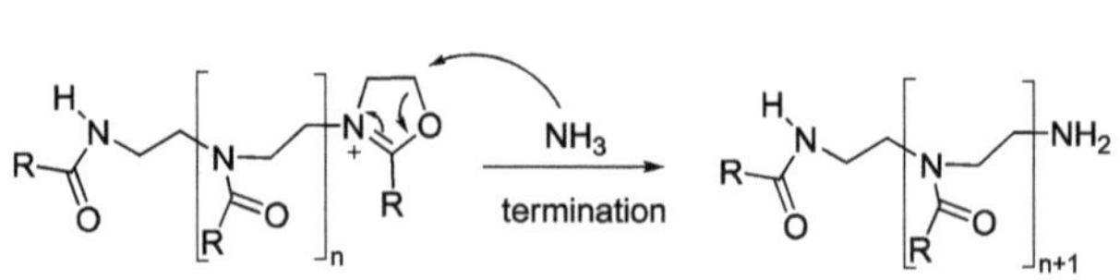

Scheme 2.9 NH_3 quenching of the DPP catalyzed ROP of 2-alkyl-2-oxazolines.

$Đ \approx 1.14$ in less than 40 h) and the absence of decarboxylation or another competing process. This is strongly supported by MALDI-TOF MS analysis of a PTMC of DP 30 depicting a unique polymer population, and monomodal SEC profiles for DP in the range 30–150, precluding the occurrence of an ACE propagating mechanism.

Introduction of a $(PhO)_2P(O)$– group on the N atom results in imidodiphosphric acid (IDPA). The acid proton migrates from N to one of the O atoms, giving rise to a structure ideally pre-organized for bifunctional behavior *via* the acidic OH and the basic P=O groups (Figure 2.10).[56] The activity of this compound in ROP was evaluated by Gou first towards ε-CL and δ-VL and later on towards TMC.[57,58] PCL and PVL with M_n values up to 20 000 g mol^{-1} and narrow molar distributions ($Đ<1.20$) can be prepared in less than 5 h for PVL and one day for PCL under mild conditions (r.t. and 1 mol L^{-1} toluene solution),[57] a performance matching that of DPP. Concerning TMC ROP,[58] as for DPPTf, controlled polymerization without a competing process takes place, and PTMC with M_n up to 17 000 g mol^{-1} and $Đ<1.19$ were obtained in slightly longer times than with DPPTf.

The impact of the variation of the ArO groups has been explored by introducing either aryl groups bearing electron-withdrawing (EW) groups to increase acidity, or a binaphthyl group that, in addition to bearing EW groups, increases the rigidity and the steric hindrance of the catalyst.[47–50] The reaction rate of the ROP of δ-VL was two-fold increased by replacing the Ph groups of DPP by *N*-MePy$^+$ groups, without noticeable impact on the properties of the obtained PVLs.[59] Moreover, the introduction of two *p*-NO_2–PhO groups was foreseen by Kakuchi in order to overcome the poor results obtained in the ROP of the poorly reactive β-BL with DPP.[60] High temperature (100 °C) and catalyst loading (cat/ROH = 5) were indeed required to get full conversion of β-BL with DPP, which resulted in the loss of the polymerization control due to side reactions leading to the formation of macrocycles and oligomers. Recently it has been proposed that the decrease in activity and the loss of polymerization control result most likely from an alcoholysis reaction between the hydroxyl chain-end of the growing polymer and DPP itself.[61] In marked contrast, *p*-NO_2-DPP promoted the ROP of β-BL under milder conditions (cat/ROH = 1), which allowed the preparation of PBLs of M_n up to 7300 g mol^{-1} in a relatively controlled manner ($Đ \approx 1.26$) and acceptable reaction times (9 h). On the basis of the structural characterization of the obtained PBLs, polymerization takes place with a good incorporation of the protic initiator, *via O*-acyl ring

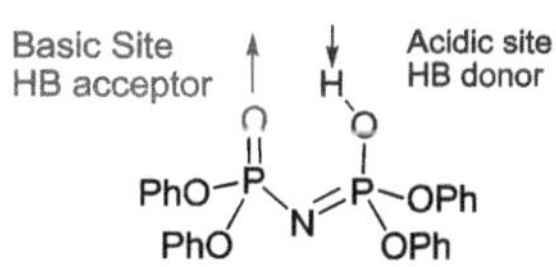

Figure 2.10 Illustration of the bifunctional behavior of IDPA.

opening of the monomer and without a crotonization process, as for HOTf-catalyzed ROP.[21,60]

Phosphoric acids based on the binaphthyl skeleton, which have found numerous applications in stereoselective organic catalysis,[47,48] have also been applied to the ROP of lactones and cyclic carbonates. The simplest one, the 1,1′-binaphthyl-2,2′-diyl-hydrogen phosphate (BNPH, pK_{aDMSO} 3.56), a chiral phosphoric acid commercially available in racemic and enantio-pure forms, was evaluated towards δ-VL and ε-CL by Zinck in 2013.[62] Working in bulk at 60 °C in the presence of BnOH as protic initiator, PVL and PCL with M_n values close to those expected from the monomer to initiator ratios were obtained (M_n 4000–12 000 g mol^{-1}, $Đ$<1.17). These conditions are compatible with the use of monosaccharide methyl α-D-glucopyranoside as protic initiator, but initiation takes place indiscriminately if the primary and secondary hydroxyl groups are not protected. The ROP of TMC with BNPH was also studied by Guo in 2016.[63] PTMC of M_n up to 15 000 g mol^{-1} can be prepared in 1 mol L^{-1} toluene solution at r.t. in less than 3 d with good control ($Đ$<1.18). Similar to what was observed for DPPTf and IDPA, only one polymer population seems to be formed, which is consistent with the absence of a competing process.

The copolymerization of ε-CL and LA with BNPH has also been reported.[64] ε-CL is much more rapidly and efficiently incorporated than LA, despite the harsh reaction conditions (bulk, 120–140 °C). This denotes the low reactivity of LA with this kind of BA catalyst. Better results in the ROP of LA were obtained in 2014 by Terada and Satoh with congeners of BNPH bearing electron-withdrawing groups in *o,o′*-positions.[65] Most importantly, using an enantio-pure version of these highly encumbered catalysts, stereoselective ROP of *rac*-lactide could be achieved. In particular, working in toluene at 75 °C, (*R*)-*o,o′*-C_6F_5-BNPH was able to convert 49% of 50 equiv. of *rac*-lactide in 18 h, with an enantioselectivity of 80% for D-LA (Figure 2.11). A PLA of M_n 3730 g mol^{-1}, $Đ$ = 1.13 and P_m = 0.86 was obtained, despite the high temperature of the reaction. This P_m value is close to the one reported in 2015 by Chen with the BINAM-cinchona alkaloid β-isocupreidine (P_m = 0.94),[66] and to those reported in 2017 by Mecerreyes and Cossío using the association of polysubstituted *N*-methyl prolines and DBU (P_m = 0.90–0.96),[67] two classes of organic catalysts. Similar to the β-ICD-BINAM, but not to the *N*-methyl prolines, the high P_m values are associated to a high *ee* for the monomer conversion, which is more in favor of an enantiomorphic site control (ESC) mechanism.

On the other side, the DBU/N-methyl proline association allows achieving D- or L-lactide selectivity by changing the stereochemistry of the proline (Figure 2.11).

2.2.4.2 Mechanistic Considerations

From an experimental point of view, the data available giving insight into the mechanism are quite similar to those discussed for sulfonic acids and

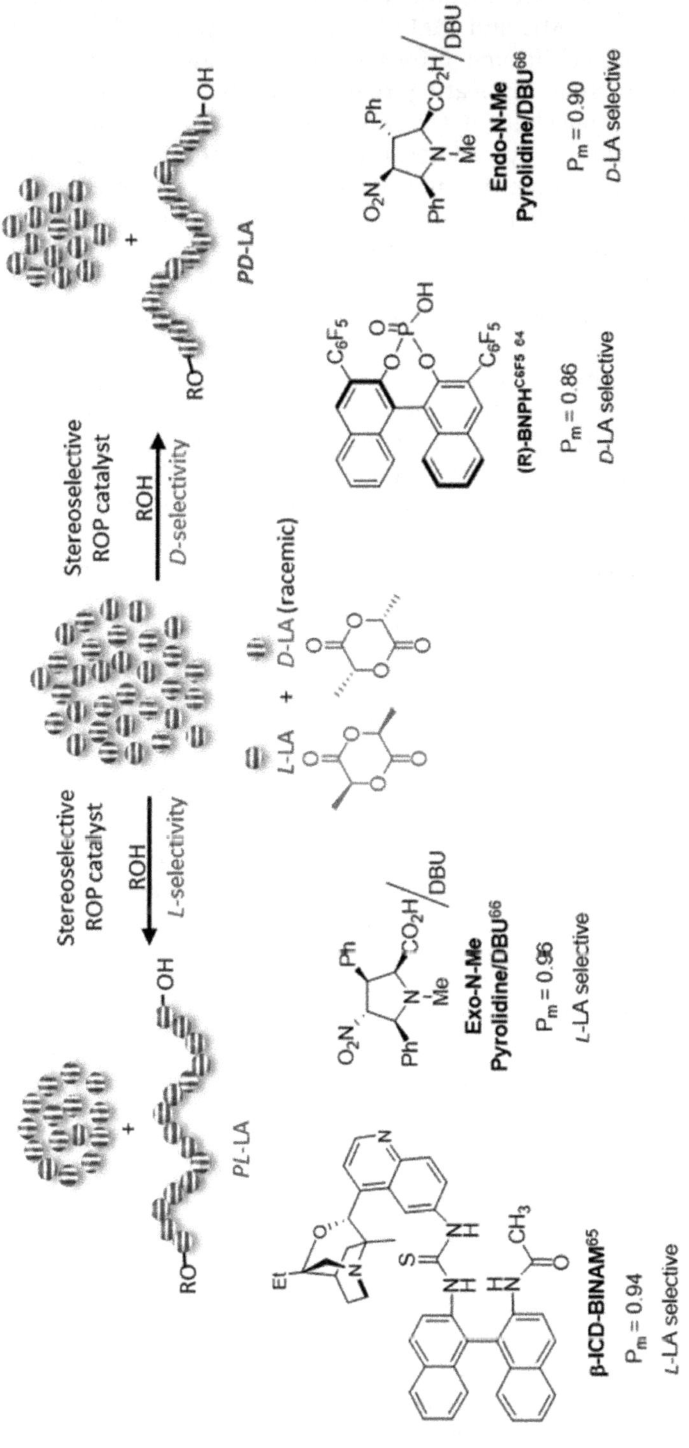

Figure 2.11 Stereoselective ROP of *rac*-LA and results reported with organocatalysts, including (*R*)-BNPH^{C6F5}.

sulfonamides. Exclusive incorporation of the protic initiator is typically supported by ^{1}H NMR and MALDI-TOF SM analyses.[51–54,57–62,65] The last technique gives additional information about the presence of a main polymer population, indicating that the formation of macrocycles by backbiting or of telechelic polymer by ACE (in the case of TMC) occurs to a very small extent, if any. Moreover, controlled and living character of the polymerization with the different P-based acid catalysts are substantiated by second feed experiences, by first-order kinetics in monomer and by the linear growing of the polymer chains with both the monomer conversion and monomer-to-initiator ratio. Only in the case of DPPTf, probably the most acidic catalyst, an upward deviation from linearity of the kinetic plot is observed during ROP of ε-CL.[52] As for HOTf,[22] this may be explained by the lower basicity of the ester moiety of the polymer chains (*syn* conformation) compared to those of the monomer (*anti* conformation), which formally results in an increase of the amount of catalysts available to activate the monomer in the reaction media. A general observation is also the Δδ measured in ^{13}C NMR spectroscopy for the C=O signal of the monomer upon addition of the catalyst (TMC/DPPTf,[55] δ-VL/IDPA,[57] TMC/IDPA,[58] β-BL/BNPP[60]). The observed de-shielding is consistent with an electrophilic activation of the monomer by the acidic proton of the catalyst. Its magnitude depends on the catalyst/monomer couple, on the monomer/acid ratio and on the solvent. All these data are in agreement with an AM mechanism, but again the activities observed for these compounds, generally known as weak to strong acids, cannot be simply explained by their acid strength. The bifunctional behavior of these compounds, in which the proton indeed acts as an acid while the P=O group acts as a base is well recognized in organic synthesis. This scheme is of course reminiscent of the one proposed for sulfonic acids, the acid proton activating the monomer and the P=O activating the protic initiator.[34] Efforts have been devoted to evidence experimentally this basic contribution. The interaction between the alcohol initiator, benzyl alcohol or 3-hydroxybutyrate, and the catalyst (DPPTf, IDA, BNPP) has been evoked on the basis of the shift in the ^{1}H NMR signals corresponding to the CH_2OH protons. The methylene signal is slightly shielded in some cases, while the OH signal is de-shielded. In addition, the $^{3}J_{HH}$ coupling disappears. Note however that these observations may also result from proton exchange between the two polar OH groups (alcohol and acid).

To get further insight into this bifunctional mode of action, the reaction between ε-CL and methanol with DPP and DPPTf catalysts was studied computationally by Maron and Bourissou, similar to what was done for sulfonic acids.[52] The two steps of the monomer ring-opening were investigated, namely the nucleophilic attack of the initiator to ε-CL to form the tetrahedral intermediate and the ring-opening (Figure 2.12). The reaction profile involving bifunctional behavior lies more than 10 kcal mol^{-1} lower in energy than the alternative pathway consisting in the simple protonation of the monomer. This underlines the importance of the cooperativity between

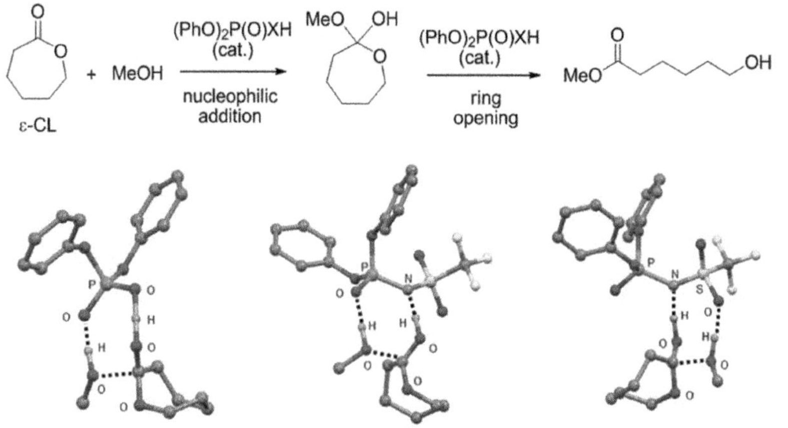

Figure 2.12 DFT-proposed mechanism for the ROP of ε-CL promoted by DPP and DPPTf (top), and TS proposed for the RDS (bottom).

the activation of the monomer by the acid proton and that of the alcohol by the P=O group. For DPP, the formation of the tetrahedral intermediate is predicted to be the rate-determining step with an activation barrier of 17.8 kcal mol^{-1}. For DPPTf, the presence of the SO_2CF_3 group results in two different TS for each step involving either the P=O or the S=O group. For the first step, again the rate-determining one, energy barriers of 20.6 and 24.6 kcal mol^{-1} were predicted for P=O or the S=O groups, respectively. These data are consistent with reactions taking place at room temperature. This bifunctional behavior, with the phosphoric acid acting as a proton shuttle in each one of the two steps, is reminiscent of what was documented for organic reactions catalyzed by phosphoric acids (chiral binaphthyl phosphoric acids are widely used in asymmetric synthesis). This mode of action is likely to also apply to the other P-based catalysts developed and discussed in this section.

2.2.5 Carboxylic Acids

2.2.5.1 Performance in ROP

Although carboxylic acids are far to compete with the other BA discussed above in terms of activity and polymerization control, they are worth mentioning due to their ready availability and environment friendly nature. Among the carboxylic acids that have been reported to promote the ROP of lactones and cyclic carbonates, the results obtained with fumaric acid (FA),[68–71] trifluoroacetic acid (TFA),[68,71,72] lactic acid (LAc)[73–76] and salicylic acid (SA)[77] merit to be commented (Figure 2.13). The best results have been reported recently by Zang using SA, a non-toxic natural product.[77] The ROP of δ-VL and ε-CL initiated by an alcohol was catalyzed by SA in the bulk at 80 °C (SA/ROH ratio = 1/1). Despite the relatively high temperature and the

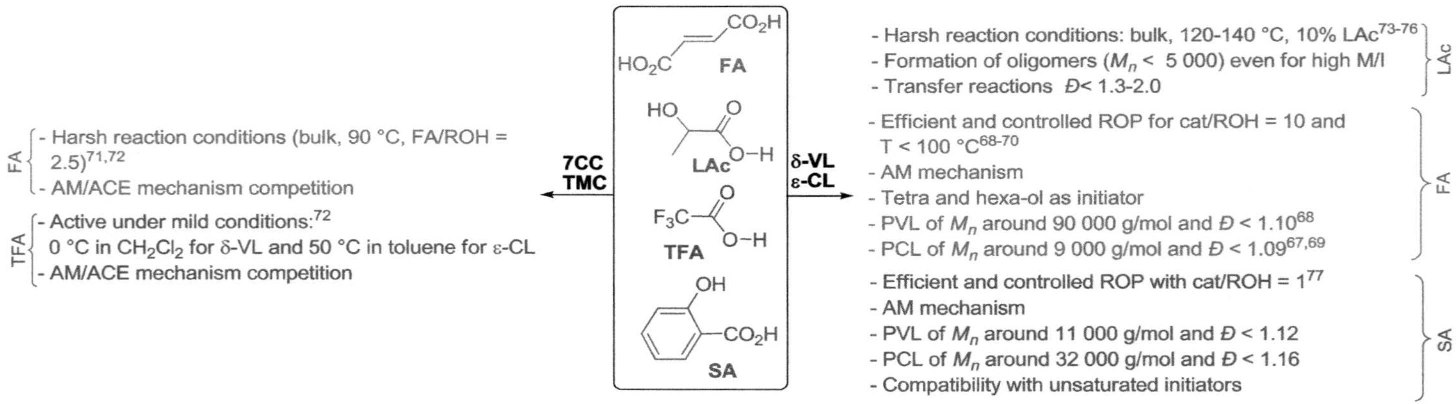

Figure 2.13 Main features of the ROP of lactones and cyclic carbonates promoted by carboxylic acids.

long reaction times (2–3 d and 20–34 h for ε-CL and δ-VL, respectively), well controlled polymers were obtained ($Đ$ <1.16) for relatively high M_n values: 35 000 g mol^{-1} for PCL and 10 000 g mol^{-1} for PVL. TFA has also given interesting results in the polymerization of the cyclic carbonates 7CC and TMC, as reported by Endo in 1998.[72] The rather strong character of this acid (pK_a 0.23, lower than DPP) enables running the reactions under mild conditions, in particular with the more reactive 7CC. Working at 0 °C in 2 mol L^{-1} CH_2Cl_2 solution, high conversions of 7CC were obtained in less than 30 h for 7CC/BnOH ratios up to 100. However, the resulting polycarbonates had lower M_n than expected, in particular for increasing monomer/initiator ratios. Similar observations were made with TMC working in toluene at 50 °C.[72] Recently, a more detailed analysis of the obtained PTMC by de With, in particular by MALDI-TOF MS, evidenced the presence of two different polymer populations, which can be explained by the competition between the two propagating mechanisms (AM and ACE), as already discussed for the ROP catalyzed by sulfonic acids.[71]

The ROP of lactones and cyclic carbonates has also been achieved with FA and LAc. The reaction conditions (bulk, 90–140 °C) are typically harsher than those used with SA and TFA, and in consequence, the polymerization is less controlled. Among the different examples reported, the best performances concern the ROP of δ-VL and ε-CL promoted by FA.[68–70] When working at 90–100 °C with a tetra or hexa-ol initiator (pentaerythritol or dipentaerythritol) and with an acid/ROH ratio of 10, polymers with medium to high M_n values (9500 and 90 000 g mol^{-1} for ε-CL and δ-VL, respectively) and narrow distributions ($Đ$ <1.10) could be prepared. In marked contrast, only oligomers with $Đ$ in the range 1.30–2.00 were obtained using LAc as catalyst due, among other factors, to the high temperatures (120–140 °C).[73–76] FA has also been reported to have some activity for the ROP of TMC,[71,72] but again the reaction conditions (bulk, 90–100 °C) and the occurrence of the competing ACE propagation resulted in low control of the polymerization.

2.2.5.2 Mechanistic Considerations

Very few mechanistic investigations have been performed with carboxylic acid catalysts, and most data concern TFA and SA. Complete and exclusive incorporation of the initiator in the PCL and PVL chains prepared with SA as catalyst is supported by ^{1}H NMR spectroscopy and MALDI-TOF MS.[77] The kinetic of monomer consumption and the evolution of M_n with monomer conversion are consistent with a controlled and living polymerization. Efficient preparation of a PCL-*b*-PVL bloc copolymer by sequential addition of the monomers further supports the living character of the polymerization. Furthermore, a weak de-shielding of the ε-CL signals in ^{1}H NMR is observed for 1/1 ε-CL/SA mixtures, which can result from the interaction between the exocyclic O atom of ε-CL and the acidic proton of SA. All this data are in agreement with a polymerization occurring according to an AM propagation. As already mentioned, in the case of 7CC and TMC polymerization using TFA as catalyst, two polymer populations were evidenced by MALDI-TOF.[71,72] One of them corresponds to the polycarbonate initiated by the added alcohol and propagating *via* AM. The other one, without incorporation of the alcohol, results most likely from the ACE propagation. Activation of both monomers by the acidic proton of TFA was substantiated by ^{13}C and ^{1}H NMR analyses.[72] Notably, Δδ of about 4 ppm were measured for the CO signal, confirming a noticeable interaction between the two units. It appears thus quite clear that carboxylic acids are able to activate lactones and cyclic carbonates. However, in most cases, this activation is not strong enough to enable polymerization under mild conditions so that controlled polymerization can be conducted. Better results can be obtained combining the carboxylic acid moiety with activating factors, as discussed in the following section.

2.2.6 Activated Brønsted Acids

2.2.6.1 Brønsted Acids Activated by a Hydrogen Bond Donor

Increasing the acidity of BAs by hydrogen bonding is a well-known and successful strategy in organic catalysis.[78] It is based on the stabilization of the conjugate-base anion of the BA by a hydrogen bond donor (HB-D), which increases the protic character of the hydrogen atom. This strategy has then been extended to ROP with the aim of increasing the activity of MSA and several carboxylic acids. Two different approaches have been envisioned. One involves intermolecular HB between the BA and the activator, while the other consists in intramolecular HB (Figure 2.14).

2.2.6.2 Performances

(a) Activation of Brønsted Acids via Intermolecular HB

The thiophosphoric triamide (TPTA) was investigated as a tripodal HB-D to increase the activity of MSA. Its geometry is ideally suited to bind the methane sulfate anion as it can form up to three HBs (Figure 2.14).[79] A first proof-of-concept was provided by the ROP of LA, for which MSA alone presents low activity, low control and difficulties to achieve full conversion even for DP as low as 30. Combining MSA with TPTA boosts the activity of MSA and 100 equiv. of LA can be fully converted, although it requires 5 d. Despite the long reaction times, M_n matches the targeted ones (15 000 g mol^{-1}) and the molecular distributions are narrow ($Đ = 1.10$). The increase in activity is also noticeable with other monomers whose ROP can be promoted by MSA alone (ε-CL, δ-VL, TMC).[77]

The distance between the two H atoms in bidentate HB-D such as thiourea and guanidines is suitable for the interaction with the two O atoms of the carboxylate moiety and they have been used for the activation of carboxylic acids, namely TFA and BzA (Figure 2.14). This association is reported to increase the acidity of carboxylic acids *via* two phenomena: the formation of the *transoid* conformation, whose pK_a is known to be up to three units lower

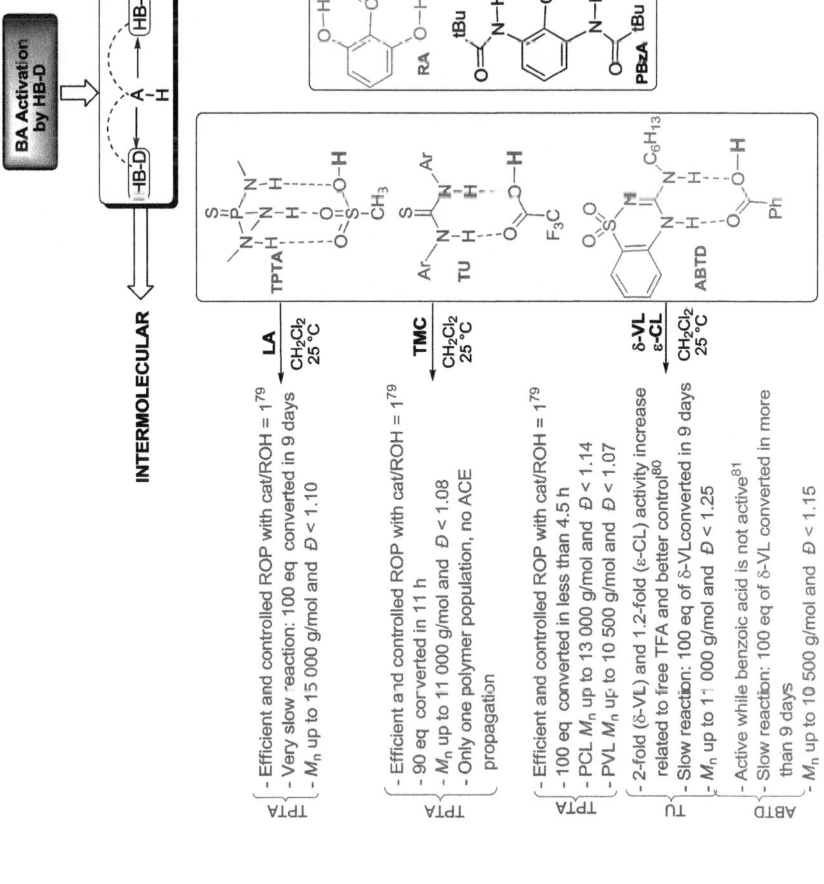

Figure 2.14 Main features of the ROP of lactones and cyclic carbonates promoted by HB-D activated BA catalysts.

than that of the more favored *cisoid* conformation, and at the same time the increase of the polarity of the O–H bond, and thereby the acidity. The activity of TFA in the ROP of δ-VL and ε-CL is increased when combined with 3,5-bis(trifluoromethyl)phenylthiourea (TU).[80] The impact on the reaction rate is more marked for δ-VL than for ε-CL (three-fold *versus* 1.2-fold increase), but with both monomers, polyesters of M_n up to 11 000–12 000 can be obtained in a rather controlled manner ($Đ$ = 1.19–1.28) in 5 and 8 d, respectively.

The sulfonyl guanidine ABTD has been combined with para-trifluoromethyl benzoic acid to promote the ROP of δ-VL (Figure 2.14).[81] Polymerization requires rather extended times (9 d to convert 100 equiv. of δ-VL at 3 mol L^{-1} and r.t.), but the reaction rates are once again around four-fold higher than those of "free" CF_3-BzA. Moreover, despite the long reaction times, the obtained PVL feature M_n close to the expected ones (10 500 g mol^{-1}) and narrow molecular distributions ($Đ$ = 1.14). In marked contrast, very low activity was observed with ε-CL and TMC (conv. < 7% for M/I of 20 after 3 d of reaction). These results confirm the interest of HB for the activation BAs, but the activities achieved with carboxylic acids remain rather modest.

(b) Activation of Brønsted Acids via Intramolecular HB

Intramolecular association of the HB-D to the BA moiety leads to higher activation of the acid and better catalytic performance. This approach was studied by Guo for BzAs flanked by two HB-D groups in the *ortho* position of the carboxylic acid. γ-resorcyclic acid (RA, Figure 2.14),[82] a natural and easily available acid, was first reported to promote the ROP of δ-VL and ε-CL with reaction times that are significantly shorter than those reported for the association of BzA with guanidine, validating thereby the approach.[81] With RA, 100 equiv. of δ-VL were converted in one day at r.t. in 3 mol L^{-1} CH_2Cl_2 solution, while 9 d were necessary with the BzA/ABTD association. However, the activity of RA remains lower than that of DPP or MSA. PVL of M_n up to 13 000 g mol^{-1} and narrow molecular distributions ($Đ$ = 1.08) were prepared under mild conditions, attesting for the controlled character of the polymerization. Interestingly, the activity observed for the monohydroxy-benzoic acid (SA) was significantly lower since only 7% of the 100 equiv. of δ-VL were converted after one day of reaction. This observation validates the approach of activating the acid by double HBs, not only by increasing the polarity of the OH bond but also by enforcing the *cis* conformation. The activity was further enhanced by replacing the two flanking hydroxyls by two amide groups, which in addition opened the door to modulate the activity by varying the electronic properties of the amide substituent O.[83] *o,o′*-bis(pivalamido)benzoic acid (PBzA) features a more rigid structure than RA that strengthens the intramolecular HB interactions leading to two-fold higher activities in ROP of δ-VL, while maintaining similar control of the polymerization (M_n of 10 300 g mol^{-1} and $Đ$ = 1.06). PBA showed also interesting activities with ε-CL and TMC, but no conversion was observed with LA at room temperature.

2.2.6.3 Mechanistic Considerations

Despite the fact than some of the discussed associations lead to rather slow polymerizations, in particular for the intermolecular approach, the analyses carried out on the obtained polymers by NMR spectroscopy and MALDI-TOF MS, the kinetics of monomer consumption, as well as the plots of M_n *versus* conversion strongly support the controlled and living character of the polymerizations promoted by HB-activated BAs.[79–83]

NMR studies seeking to evidence the contribution of HB in the mode of action of these systems have been reported. For example, the ^{1}H NMR signals for the N–H groups and the ^{13}C NMR signal for the C=N moiety of the guanidine ABTD were found to be de-shielded upon addition of increasing amounts of BA.[81] In addition, the impact of the intramolecular HB interactions on the polarity of the OH bond is clearly substantiated by the different pK_a of the species related to BzA (4.0, 2.98 and 1.30 for BzA, SA and RA respectively).[82] The absence of activity of *m,m′*-dihydroxybenzoic acid or *p*-hydroxy-benzoic acid is an additional prove of the determinant impact of HB, indicating that the electronic effect of the OH groups is not the main role. Furthermore, the differences in the variation of the ^{13}C NMR chemical shift of the C=O group of the monomer when mixed with the BA alone or with the HB-D/BA combination is in agreement with a benefic effect of the HB-D.[79–83]

2.2.7 Brønsted Base Activated Brønsted Acids: Ions Pairs as Catalysts

BAs have also been combined with nitrogen-based Brønsted bases (BBs). Two complementary situations have been targeted upon associating BA and BB: (i) increase the activity of weak BAs (as in the case of LA, a monomer towards which most BAs show only low to moderate activity with the exception of HOTf),[18] and (ii) moderate the activity and improve the polymerization control of strong BB catalysts (Figure 2.15).

2.2.7.1 Performances

(a) Combination of BAs with a Weak BB

Peruch was the first to report in 2010 the ROP activity towards LA of mixtures of DMAP (a weak base with low activity) and a BA such as $HCl.Et_2O$, MSA or HOTf in a 2/1 ratio, which corresponds in fact to a 1/1 ratio of "free" DMAP and the corresponding $DMAPH^+,A^-$ ion pair.[84] Working in refluxing CH_2Cl_2 at 1 mol L^{-1} and 5% of the cat loading, all the combinations proved to be more active than DMAP alone, the DMAP/HOTf combination affording the best results. After 24 h of reaction, only 35% conversion was observed for DMAP alone, whereas 72, 77 and 89% of conversion were obtained for the combination of DMAP with $HCl.Et_2O$, MSA and HOTf, respectively. Varying the LA/ROH ratios with DMAP/HOTf

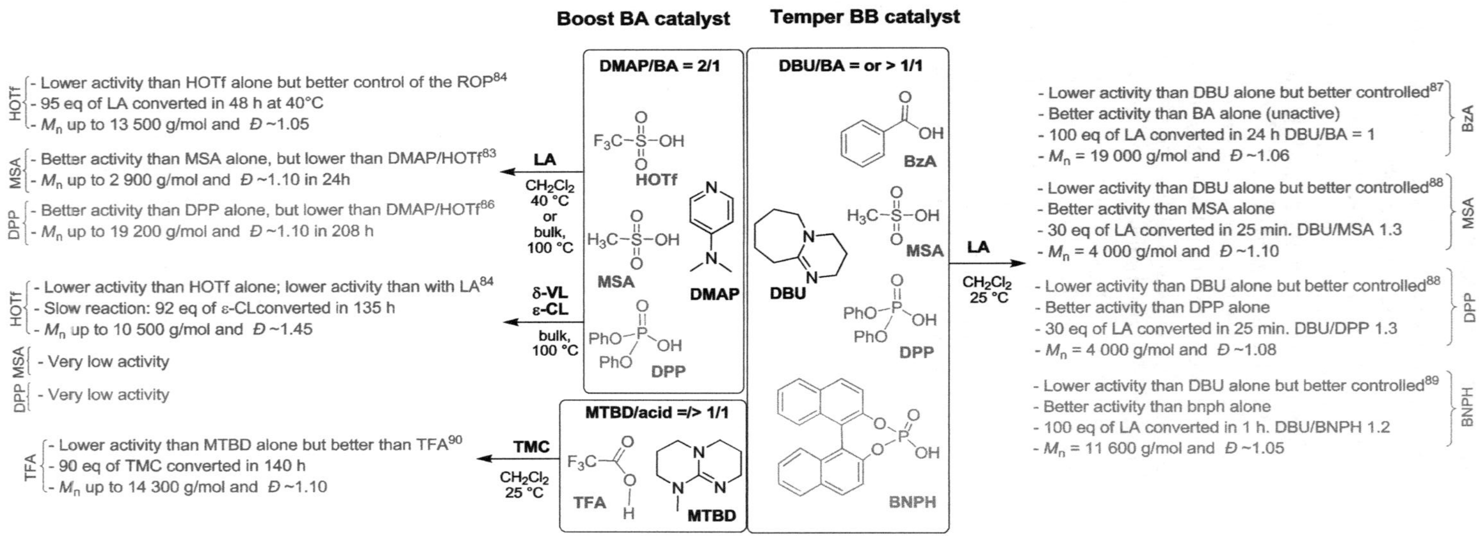

Figure 2.15 Main features of the ROP of lactones and cyclic carbonates promoted by BB-activated BA catalysts.

led to PLA of M_n up to 14 000 g mol^{-1} (LA/ROH of 100) and narrow molar distributions ($Đ \approx 1.05$) in 48 h. It has to be noted that while the combination of DMAP with $HCl.Et_2O$ or MSA results indeed in an improvement of the activity of the two partners separately (the base and the acid), the combination with HOTf results in the decrease of the activity of the acid (100 equiv. of LA are converted in 28 h using 1% of HOTf at r.t.). However, the molar distributions reported are higher ($Đ \approx 1.45$) suggesting that the combination with DMAP leads to a better controlled polymerization of LA with HOTf. The same authors demonstrated four years later that the reaction rates could be significantly increased working in the bulk at 100 °C, without marked impact on the polymerization control.[85] PLA of M_n around 13 400 g mol^{-1} and $Đ \approx 1.16$ were prepared with 5% of the 2/1 DMAP/HOTf combination in only 1 h. MALDI-TOF analyses showed low, if any, impact of transfer reactions up to 90% conversion, but they occurred at higher conversions as indicated by the presence of minor peaks at 72 g mol^{-1} intervals. The combination of DMAP with DPP was studied by Kakuchi in 2014.[86] As for DMAP/MSA, the activity of both partners is improved, but the performance remains inferior to that of the DMAP/HOTf combination (different reaction conditions were used, which makes direct comparison difficult, but the 120 h necessary to convert 90 equiv. of LA in 3 mol L^{-1} CH_2Cl_2 solution at r.t. suggest lower activity). Anyway, comparable M_n and $Đ$ were obtained (13 400 g mol^{-1} and 1.09).

In contrast to what was observed with LA, the combination of BAs with DMAP does not result in an increase of ROP activity for ε-CL and δ-VL, but in deactivation.[85] In fact, only the DMAP/HOTf combination leads to a somewhat efficient polymerization, although with much lower reaction rates that for LA (135 h was necessary to convert 90 equiv. of ε-CL at 100 °C with 5% of the catalytic system *versus* only 1 h for LA). This is a reverse situation than with BAs alone, for which higher activity is observed for ROP of these lactones related to LA.

(b) Combination of BAs with a Strong BB

The second type of combination is the association of a BA with a strong BB with high activity in the ROP of LA, but low control due to transfer reactions (Figure 2.15). In 2011, Hedrick reported that the combination of DBU with a stoichiometric amount of BzA has a marked impact on activity and polymerization control of LA.[87] DBU is a strong BB. It is highly active in the ROP of LA, but the obtained polymers often feature $Đ > 1.40$, indicating the occurrence of transfer reactions, in particular at high monomer conversion. The association of DBU with BzA in a 1/1 ratio led to a dramatic decrease of activity (24 h were necessary to convert 100 equiv. of LA at r.t. whereas only 30 s were necessary with DBU alone), but the polymerization control was significantly improved, $Đ$ around 1.06 being observed. Using a higher amount of BzA resulted in ROP inhibition, as could be expected taking into account that ROP of LA promoted by DBU is typically quenched by addition of an excess of BzA, a BA that is inactive towards LA. The combination of

DBU with DPP and MSA was reported later on by Guo.[88] Here again, the observed activities were higher than those of the acids, but significantly lower than that of DBU (low conversions were observed after 24 h of reaction compared to full conversion with a 1/1 mixture). Interestingly, using 30% excess of DBU (1.3/1 ratio, that is 0.3 equiv. of DBU and one equiv. of the ion pair $DBUH^+,A^-$) resulted in a marked increase of the activity (30 min for full conversion instead of 24 h) while maintaining good control ($Đ<1.10$). Using a 2/1 ratio (as for DMAP) resulted in an even higher gain of activity (10 min for full conversion) but also in some loss of the polymerization control ($Đ=1.34$). Zinck reported in 2015 the combination of DBU with the phosphoric acid BNPH,[89] varying the DBU/BNPH ratio. Once again, the use of a slight excess of DBU (10–20%) was a good compromise in terms of activity and polymerization control, enabling the conversion of 100 equiv. of LA in 1 h to for PLA of M_n 11 000 g mol^{-1} and $Đ=1.05$. Even working with a DBU/BNPH ratio of 2 (1 equiv. of DBU and 1 equiv. of the ion pair $DBUH^+,A^-$) led to a rather controlled polymerization and the formation within 10 min of a PLA of M_n 14 000 g mol^{-1} and only slightly higher molar distribution ($Đ=1.10$). They also showed that DBU could be replaced by the weaker base DMAP in the ion pairs, working with a 1/1/1 mixture of DBU/DMAP/BNPH. High conversions (96%) can be achieved in 30 min to yield a PLA of comparable properties to those obtained with the DBU/BNPH combination ($M_n=12\,800$ g mol^{-1} and $Đ=1.08$).

Similarly to DMAP/BzA, the ROP of ε-CL and δ-VL is dramatically affected when BAs are combined with a strong base such as DBU, leading to rather low activities even for a 1/1 ratio.[88] In marked contrast, Guo reported efficient ROP of TMC promoted by MTBD/TFA.[90] This combination allowed for better polymerization control related to MTBD alone (although with longer reaction times) on the one hand, and for higher activity related to TFA alone (see Section 2.2.5) on the other hand. The best results were obtained with 2/1 MTBD/TFA mixtures that promote the polymerization of 100 equiv. of TMC in 140 h to yield PTMC of $M_n=14\,300$ g mol^{-1} and $Đ=1.10$.

2.2.7.2 Mechanistic Considerations

Characterization of the polymers prepared using BA/BB combinations by NMR spectroscopy and MALDI-TOF MS show good end-group fidelity, indicating efficient initiation with the protic initiator. In addition, the controlled character of the polymerization is strongly supported by linear evolution of M_n *vs.* monomer conversion and *vs.* monomer-to-initiator ratio, and by first kinetic order in monomer.[84,85,87–90]

Investigations of the HB interactions between LA and the ammonium moiety ($DMAPH^+$ or $DBUH^+$)[84,87] and between the protic initiator and the free amine (DMAP and DBU) by ^{1}H and ^{13}C NMR spectroscopy suggest activation of the monomer by the N–H group of the ammonium while the protic initiator is activated by the base when present in excess (base/acid ratio$>$1). Moreover, Hedrick reported than the 1/1 mixture of DBU and

HCl.Et$_2$O was inactive towards LA polymerization, in contrast to the 1/1 DBU/BzA mixture.[87] In addition, a plausible reaction pathway was calculated and the rate-determining step was found to be the nucleophilic attack of the initiator to the monomer ($\Delta G^{\neq} \sim 18$ kcal mol^{-1}). In the corresponding TS, the monomer is activated by the ammonium and the initiating alcohol is activated by the carboxylate (conjugated base of BzA) (Figure 2.16, right).[87] This is consistent with the lack of activity of the DBU/HCl.Et$_2$O combination, since the chloride cannot activate the alcohol. Thus, two different bifunctional mechanisms can be proposed depending on the base/acid ratio. For a base/acid ratio of 1, the ammonium NH$^+$ activates the monomer while the conjugated base of the acid (carboxylate, sulfonate, *etc.*) activates the initiator (Figure 2.16, a). For ratios higher than 1, the amine present in excess activates the protic initiator, while the monomer is activated by the ammonium NH$^+$ (Figure 2.16, b). The conjugate base of the acid has a noticeable influence (different activities are reported for 2/1 combinations of DMAP with MSA, DPP or HOTf), but its exact role remains unclear at this state. It may be envisioned that the conjugate base is involved in competitive HB (for example with the ammonium). This may reduce the activity of the catalyst, all the more so that the conjugate base is more basic. This is consistent with the best performances being reported for the DAMP/HOTf combination.

2.3 Applications of Organocatalyzed ROP Promoted by Brønsted Acids

2.3.1 Functional Group Compatibility

2.3.1.1 Functionalized/Macromolecular Initiators

Thanks to the good functional group compatibility of BA catalysts, initiators bearing an additional functional groups such as those represented in Figure 2.17 have been utilized in the ROP of lactones and TMC. The aim of the additional functional group is to elaborate advanced materials by post-derivatization/polymerization. Block copolymers can be accessed thereby and functional groups can be introduced *via* click chemistry (AHA, PGA, HEMI). The chain-end functionalized polymers can also be used as macromonomers to prepare brush copolymers (HEMA, VBA, NBOH). The functionalized initiators have been applied particularly in the ROP of β-BL with BNPP,[60] and that of δ-VL, ε-CL and TMC with Tf$_2$NH,[39,41]DPP,[51] TfCH,[45] or BNPH.[61,62]

Pre-formed polymers terminated by hydroxyl groups at one or the two chain-ends, usually commercially available, have also been successfully applied as macroinitiators in the ROP of lactones and TMC using DPP, MSA, Tf$_2$NH, SA and even HOTf as catalyst. Block copolymers combining a polyester or polycarbonate block to a block of different nature/chemical structure such as polyether (PEG),[52] polybutadiene (PBD),[52] polystyrene (PS)[91] and

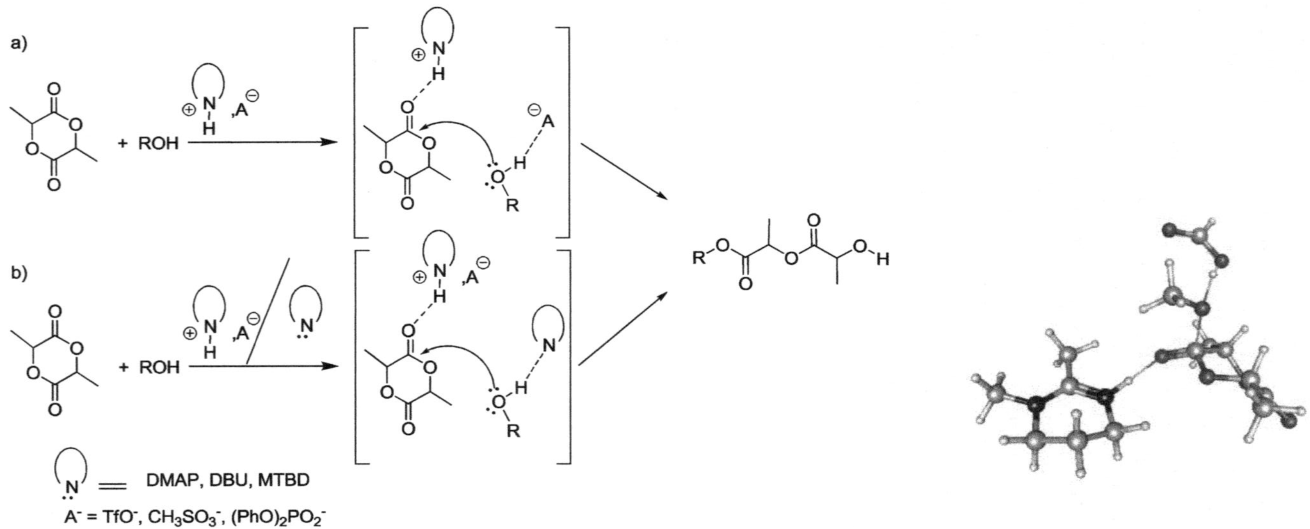

Figure 2.16 Proposed mechanisms for the ROP of LA with BB/BA association depending on the BB/BA ratio (left), and DFT-calculated TS of the RDS for the ROP of LA promoted by BzA/DBU 1/1 (from ref. 87) (right).

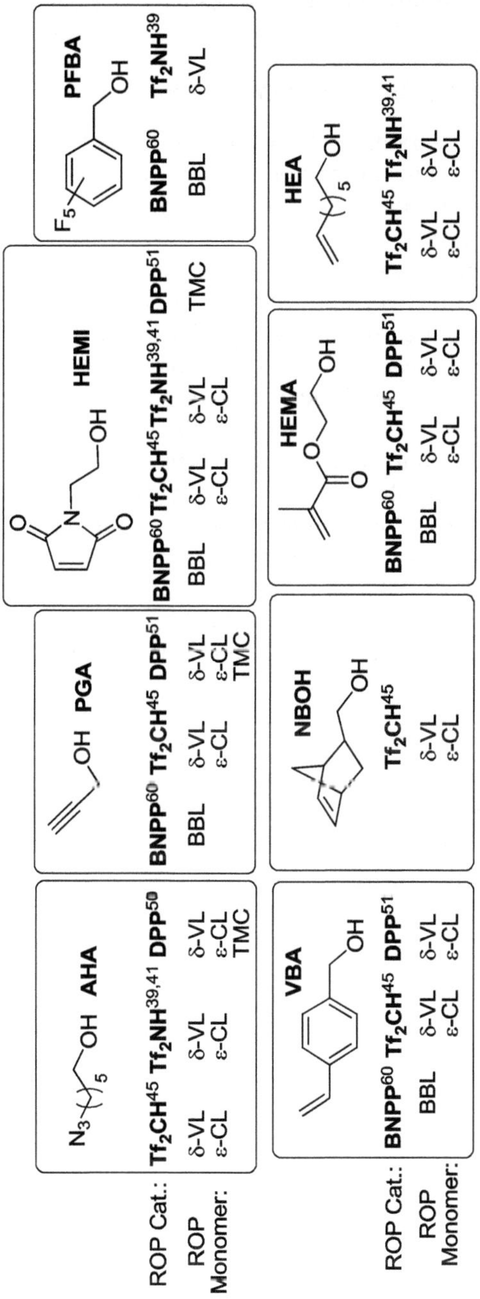

Figure 2.17 Main functionalized initiators used with BA as catalysts.

polyisocyanate (PHIC)[92] have been prepared. The excellent initiation efficiency evidenced for these catalysts ensures selective formation of block copolymers and the absence of free homopolymers.

2.3.1.2 *Using Functionalized Monomers*

Functionalized monomers have also attracted interest with the objective to modify the polyesters and polycarbonates afterwards either by introducing pendant functional groups or by further polymerization leading to graft copolymers (Figure 2.18).

Hedrick and Enders prepared a six-membered carbonate bearing a pendant pentafluorophenyl ester group in the 4-position. It is a common platform for the introduction of a variety of pendant functional groups *via* post-polymerization *trans*-amidation reactions.[93,94] However, the high reactivity of this highly electrophilic ester group makes ROP polymerization with basic or nucleophilic catalysts impossible. By contrast, HOTf is capable of promoting the controlled ROP of this monomer at r.t., and polymers of M_n up to 37 100 g mol^{-1} have been prepared in a rather controlled manner ($Đ = 1.20$). Note that DPP and PTSA are not active enough to promote the polymerization. In marked contrast, DPP was used by Yu to promote the controlled ROP of another six-membered carbonate bearing two iodo-methylene groups in the 4-position, designed to prepare polycarbonates with X-ray opacity properties.[95] Random co-polymerizations with TMC were carried out to produce polycarbonates of modulated radio-opacity.

ε-CLs bearing a bromine atom have also been engaged in BA-catalyzed ROP, aiming at preparing afterwards grafted copolymers by controlled radical polymerization of acrylates (Cu promoted ATRP).[96] α-Br-ε-CL, easily prepared from ε-CL, was simultaneously polymerized with ε-CL using DPP as catalyst. Good control of the copolymerization was evidenced for the two monomers, but ε-CL proved to be significantly more reactive than the brominated monomer (reactivity ratios of 39.0 and 0.016 for ε-CL and α-Br-ε-CL, respectively), which resulted in the formation of block like copolymers of M_n up to 45 000 g mol^{-1} and $Đ = 1.05$. Using these copolymers as macroinitiators for the ATRP polymerization of methyl acrylate

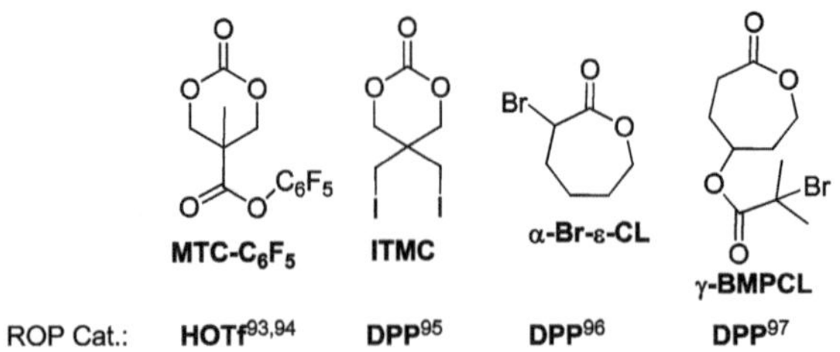

Figure 2.18 Main functionalized monomers used with BA as catalysts.

(MA), P(CL-*co*-α-Br-CL)-*g*-PMA grafted copolymers were obtained. Seeking for a more reactive brominated polymer, Dove introduced the monomer γ-(2-bromo-2-methylpropionyl)-ε-CL (γ-BMPCL) substituted in the 4-position, so that the impact on ROP is reduced.[97] Although γ-BMPCL is less reactive than ε-CL in homopolymerization catalyzed by DPP, close reaction rates were observed during simultaneous polymerization. Copolymers of controlled properties ($M_n = 66\,000$ g mol^{-1}, $Đ < 1.07$) could be obtained, provided monomer conversions are maintained at <60% to avoid the appearance of oligomers. Grafted copolymers were subsequently prepared by surface initiated ATRP with acrylate terminated ethylene glycol oligomers (OEGMA).

2.3.2 Preparation of Copolymers

2.3.2.1 Block and Random Copolymerization Promoted by a Common Catalyst

The most convenient way to prepare copolymers of monomers of similar nature, either random, gradient or block copolymers, is the use of a common polymerization catalyst/initiator. Different structures can be achieved, depending on the monomer feed strategy (simultaneous or sequential), their reactivity ratios and whether transfer reactions take place or not.

When two monomers are polymerized simultaneously, random copolymers are formed provided their reactivity ratios are not too different. Random copolymers are also formed if transfer reactions occur during the polymerization. This is the situation encountered by Basko for the copolymerization of LA and ε-CL catalyzed by HOTf.[98,99] Although LA is consumed faster, NMR characterization of the copolymers shows random distribution of the monomers, and the main presence of secondary LA-OH terminal groups. The absence of primary ε-CL-OH terminal hydroxyl groups is consistent with the occurrence of transfer reactions that most likely take place more rapidly with this less hindered chain-end. Using a diol as initiator, telechelic copolymers of M_n around 10 000 g mol^{-1} and $Đ = 1.30$ were obtained.

Gradient copolymers are formed when the reactivity ratios are different and when transfer reactions do not occur to a noticeable extent. This is the case of the simultaneous copolymerization of TMC and ε-CL catalyzed by MSA at 30–40 °C.[100,101] ε-CL is consumed faster, leading to a copolymer enriched in this monomer at the beginning of the polymer chains and enriched in TMC at the end. Gradient PCL/PTMC copolymers of monomer ratios varying from 4/1 to 1/4 and M_n up to 18 000 g mol^{-1} were prepared in a rather controlled manner ($Đ = 1.20$). In marked contrast, when the TMC/ε-CL copolymerization was catalyzed by DPP at 80 °C, the impact of transfer reactions increased, resulting in random copolymers of $M_n = 40\,000$ g mol^{-1} and $Đ = 1.50$.[102]

Sequential addition of the monomers enables preparation of block copolymers, again provided that transfer reactions are limited. The variety of BA, with acid strengths ranging from strong (HOTf) to weak (SA), that have been developed for the ROP of lactones and TMC have been applied to the block polymerization of several combinations of monomers. Block copolymers combining β-BL with ε-CL and with TMC have been prepared with HOTf as catalyst upon sequential polymerization of the monomers.[52] Using mono and dihydroxy initiators, PBL-*b*-PCL, PBL-*b*-PTMC, PCL-*b*-PBL-*b*-PCL and PCL-*b*-PBL-*b*-PEG-*b*-PBL-*b*-PCL diblock, triblock and pentablock copolymers of M_n in the range 6700–28 700 g mol^{-1} and relatively low Đ values (1.21–1.46) were prepared. The milder catalyst BNPP was also applied to the preparation of PBL-*b*-PTMC leading to copolymers of close characteristics ($M_n = 7200$, $Đ = 1.20$).[60]

LA block copolymers with ε-CL or TMC have been prepared not with HOTf as catalyst (as mentioned before transfer reactions reduced polymerization control) but with the much milder HB system combining TPTA and MSA.[79] Although long reaction times were required for the ROP of LA, the high control of the polymerization enables the preparation of block copolymers with $M_n \approx 10\,000$ g mol^{-1} and $Đ < 1.14$.

The good performance of most BAs towards δ-VL, ε-CL and TMC opens the way to block copolymers of these monomers. Proof-of-concepts have been demonstrated in many cases and copolymers of $M_n \approx 10\,000$ g mol^{-1} and $Đ = 1.05–1.20$ have been prepared.[103,104]

2.3.2.2 *Block Copolymerization Promoted by Different Catalysts: The Switch Catalyst Strategy*

The ideal route for the preparation of block copolymers is the one-pot sequential polymerization of the monomers. However, it remains a challenge when the corresponding monomers polymerize under different polymerization conditions; different catalysts or even different polymerization methods may be needed. In the case of organocatalyzed ROP, the lack of a common catalyst suitable for lactones or cyclic carbonates and for LA hampers the preparation of block copolymers of PLA with other polyesters or polycarbonates. A comparable situation is encountered for the preparation of block polyethers–polylactone copolymers, since controlled ROP of cyclic ethers is promoted by strong basic non-nucleophilic catalysts such as $tBuP_4$, a catalyst showing low control with lactones ($Đ < 1.7$). To overcome these difficulties, the catalyst-switch strategy has been introduced.

The strategy, set up by Hadjichristidis for the preparation of polyether-*b*-polylactone and polyether-*b*-polyTMC copolymers, consists in the ROP of the epoxide (EO, BO, *etc.*) initiated by an alcohol and promoted by $tBuP_4$ (0.2 equiv. related to ROH) in THF or toluene solution, followed by the addition of 1.2 equiv. of DPP, 0.2 equiv. to quench the phosphazene and 1 equiv. to promote the polymerization of the subsequently added lactone or TMC

(Scheme 2.10).[105] Block polyether–polylactone(carbonate) copolymers were thereby prepared combining different epoxides with TMC, δ-VL, ε-CL or δ-alkyl substituted δ-lactones (δ-HL, δ-NL, δ-DL).[106] The M_n of the copolymers are in the range 13 000–15 000 g mol^{-1} and the molecular distributions are rather narrow ($Đ < 1.17$). Using tris(hydroxymethyl)propane as initiator and sequential multiple addition of epoxides (EO, BO) to start with, and TMC/lactone to follow up, enabled the preparation of star-shaped multi-block PBO-PEO-PTMC-PVL(PCL) copolymers of M_n around 46 000 g mol^{-1} and $Đ = 1.16$.[107]

A similar strategy, with the reverse order of addition for the acid and base catalysts, was applied by Guo and Li to the preparation of polylactone–*b*-PLA diblock and polylactone–PTMC-*b*-PLA triblock copolymers.[108] Here, DPP or MSA, known for their efficiency to promote the ROP of ε-CL, δ-VL and TMC, were first used at r.t. in CH_2Cl_2 for the preparation of the polylactone and PTMC blocks (Scheme 2.11). 2 equiv. of DBU was then added, followed by LA. ROP of LA occurred thanks to the 1/1 association of DBU and DBUH$^+$ (Section 2.7.1). Following this strategy, PVL-*b*-PLA and PCL-*b*-PLA diblock copolymers of M_n in the range 12 000–13 000 g mol^{-1} and narrow dispersities ($Đ = 1.16$) were prepared. For triblock copolymers, the order of addition of the monomers has a noticeable impact on the control of the reaction. If TMC is added after LA, although DBU/DBUH$^+$ promotes the ROP of this monomer, the initiation with the terminal secondary alcohol of the PLA block makes the polymerization poorly controlled ($Đ > 1.30$). Much better results are obtained adding TMC before DBU and LA, to afford a central PTMC block. PVL-PTMC-*b*-PLA triblock copolymers of M_n around 13 700 g mol^{-1} were thereby prepared in a nicely controlled manner ($Đ = 1.07$).

2.3.2.3 *Block Copolymerization* via *Two Different Polymerization Methods*

The efficiency of DPP to promote the ROP of lactones and TMC, combined with its functional group compatibility, has enabled the one-pot preparation of block copolymers associating the polyester block with a polyolefin block. Using a bifunctional initiator bearing a reversible addition addition-fragmentation (RAFT) initiation moiety (dithiocarbonate, xanthate, *etc.*) and an alcohol allows the preparation of these block copolymers, either by simultaneous polymerization (lactones, vinyl monomer) or by stepwise addition of the monomers (Figure 2.19). Here, the mild BA DPP is a real asset, since other ROP catalysts such as the strong bases are not compatible with the thiocarbonate group.

The dithiocarbonate 4-cyano-4-(dodecylsulfanylthiocarbonyl)sulfanylpentan-1-ol CPD[109] and the dithioester 4-cyano-1-hydroxypent-4-yl dithiobenzoate ACP[110] have been applied to the preparation of block copolymers incorporating alkylmetacrylates (MMA, BMA, tBuMA) in one block and δ-VL or ε-CL in the other block. Good functional compatibility of all the reactants, including the radical initiator (typically an azo derivative such as AIBN), enables

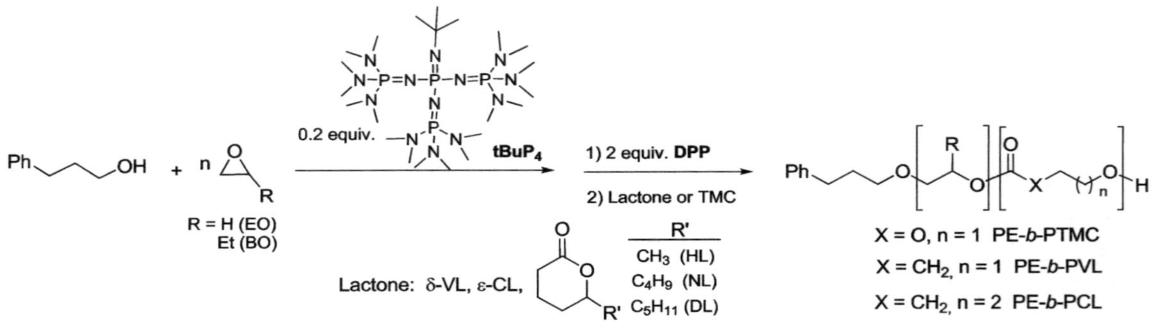

Scheme 2.10 Block copolymerization of cyclic ethers and lactones or TMC by the base/acid catalyst switch method.

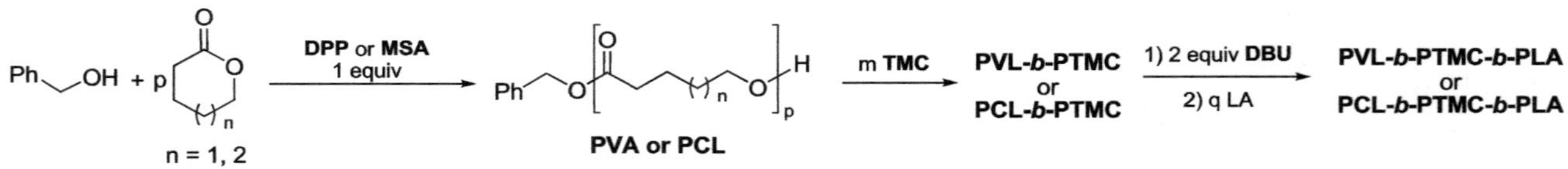

Scheme 2.11 Block copolymerization of lactones, TMC and lactide by the acid/base catalyst switch method.

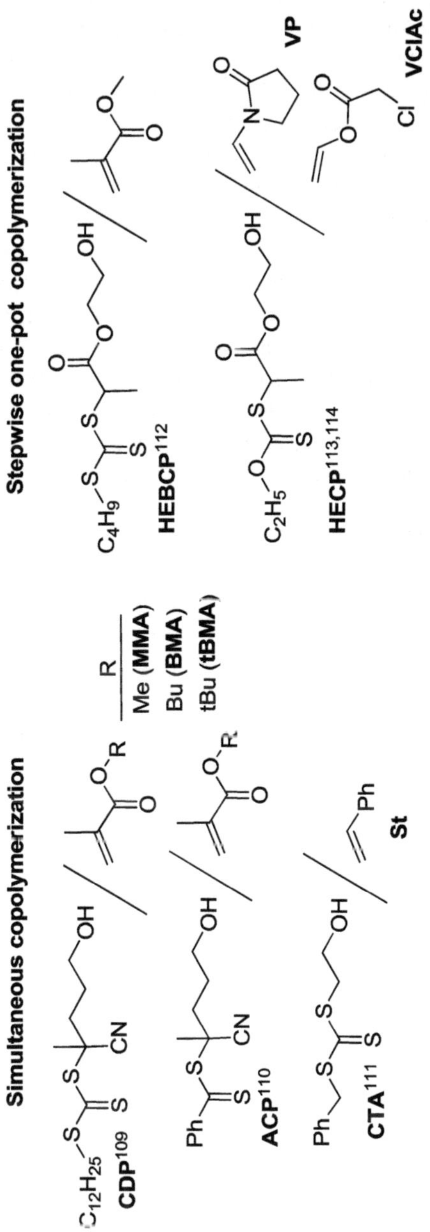

Figure 2.19 RAFT initiator/monomer associations successfully combined with DPP-catalyzed ROP.

simultaneously operating the RAFT and ROP polymerizations without interference. Accordingly, polyolefin–polyester block copolymers of M_n around 30 000 g mol^{-1} and $Đ<1.17$ were prepared in several hours at temperatures in the range 25–60 °C. In all the reported examples, the monomer conversions remain <70%. The dithiocarbonate 2-(benzylsulfanylthiocarbonylsylfonyl)-ethanol CTA has been used as double initiator for the copolymerization of styrene and ε-CL.[111] The higher reaction temperature (100 °C) imposed by the low reactivity of St in RAFT polymerization makes the simultaneous polymerization less favorable due to the occurrence of transfer reactions during ROP. This lack of control is apparent from the relatively broad molar distributions of the copolymers (Đ 1.20–1.57). Photo-activation instead of thermal activation is a useful method to increase the compatibility of the two-step polymerization process, since low temperatures can be used to polymerize the olefin monomer (PET-RAFT). This has been demonstrated by Boyer in the simultaneous copolymerization of MMA and ε-CL at r.t. using HEBCP as bifunctional initiator and $Ir(ppy)_3$ as photo-activator under low intensity visible light radiation ($\lambda_{max}=460$ nm).[112] For reaction times comparable to the previous reactions cited, higher monomer conversions are observed, and the obtained copolymers have M_n in the range of 11 000 to 14 500 g mol^{-1} and $Đ<1.15$.

Combination of RAFT polymerization and DPP-promoted ROP has also been applied to electron-rich monomers, namely vinylpyrrolidone (PV) and vinylchloroacetate (VClAc).[113,114] Here, the RAFT initiator is a xanthate (HECP), which affords better control of the polymerization (Figure 2.19). Here, the simultaneous RAFT/ROP polymerization is hindered by the basic nature of the vinyl monomer that inhibits ROP by interacting with DPP. Notwithstanding, a one-pot two-step polymerization can be carried out, starting by ROP and then adding the electron-rich monomer and the radical initiator (an azo derivative). Block PCP-*b*-PVL, PCL *b*-PVA and PVL-*b*-PVA copolymers of controlled structure ($Đ=1.17$) and M_n up to 23 000 g mol^{-1} were thereby prepared.

Guo reported in 2017 a new strategy combining the RAFT/ROP copolymerization with the A/B catalyst switch enabling preparation of tetrablock copolymers including two polyolefin blocks deriving from electron-poor and electron-rich vinyl monomers, a polylactone and a PLA block.[115] A trifunctional initiator is used, namely 2-hydroxymethyl 2-(methyl(pyridin-4-yl)carbamothioylthio)propanoate (HMCP) (Scheme 2.12). In the presence of 2 equiv. of HOTf or MSA, the pyridine group is protonated, which makes the thiocarbamoyl group suitable for the polymerization of electron-poor vinyl monomers, while the excess of acid promotes the ROP of lactones or TMC. Upon addition of DBU (2 equiv.), the acid is quenched, the thiocarbamoyl group is now able to polymerize electron-rich vinyl monomers and the excess of DBU promotes the ROP of LA. Following this strategy, tetrablock copolymers of $M_n=19 000$ g mol^{-1} and relatively narrow dispersities ($Đ=1.33$) were prepared.

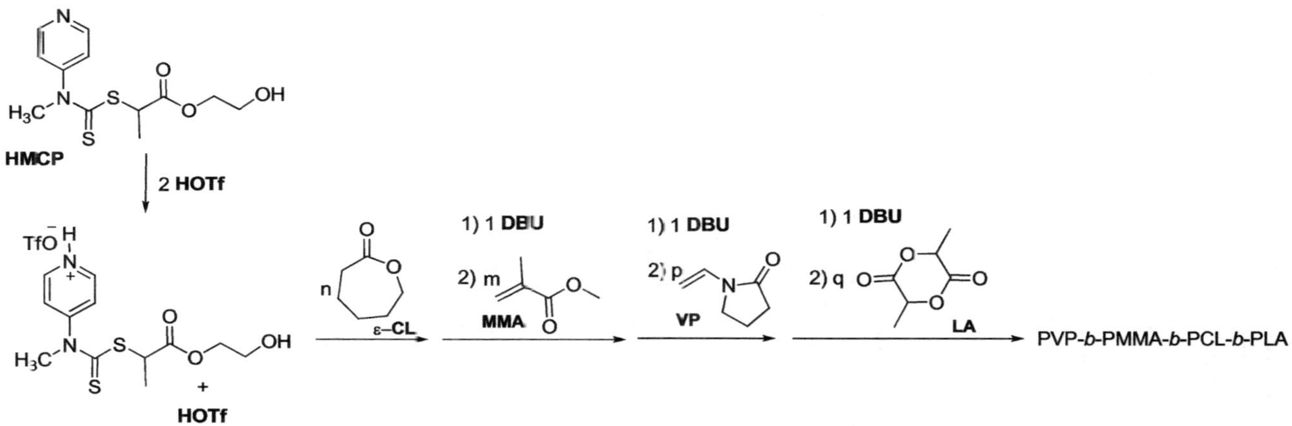

Scheme 2.12 Preparation of tetrablock copolymers by combination of RAFT with base/acid catalyst switch ROP.

2.4 Conclusion

A broad variety of organic Brønsted acids have been shown to be competent catalysts for ring-opening polymerization. Monomers with quite different properties, such as cyclic carbonates, lactones and lactide can be selectively polymerized and copolymerized in a controlled manner. Most of the Brønsted acid catalysts operate under mild conditions and display high functional tolerance, allowing the use of initiators and/or monomers bearing functional groups, or even to couple ROP with of other types of polymerizations. ROP catalysis with Brønsted acids is now mature and stands as a simple yet efficient method to prepare polymers and copolymers of controlled composition and architecture.

References

1. D. W. C. MacMillan, *Nature*, 2008, **455**, 304.
2. T. Akiyama and K. Mori, *Chem. Rev.*, 2015, **115**, 9277.
3. M. K. Kiesewetter, E. J. Shin, J. L. Hedrick and R. M. Waymouth, *Macromolecules*, 2010, **43**, 2093.
4. A. P. Dove, *ACS Macro Lett.*, 2012, **1**, 1409.
5. W. N. Ottou, H. Sardon, D. Mecerreyes, J. Vignolle and D. Taton, *Prog. Polym. Sci.*, 2016, **56**, 64.
6. T. Endo, in *The Handbook of Ring-Opening Polymerization*, ed. P. Dubois, O. Coulembier and J.-M. Raquez, Wiley-VCH, Weinheim, 2009, pp. 53.
7. S. Hu, J. Zhao, G. Zhang and H. Schlaad, *Prog. Polym. Sci.*, 2017, **74**, 34.
8. Y. Shibasaki, F. Sanda and T. Endo, *Macromol. Rapid Commun.*, 1999, **20**, 532.
9. Y. Shibasaki, F. Sanda and T. Endo, *Macromolecules*, 2000, **33**, 3590.
10. Y. Shibasaki, H. Sanada, M. Yokoi, F. Sanda and T. Endo, *Macromolecules*, 2000, **33**, 4316.
11. F. Sanda, H. Sanada, Y. Shibasaki and T. Endo, *Macromolecules*, 2002, **35**, 680.
12. X. Lou, C. Detrembleur and R. Jérôme, *Macromolecules*, 2002, **35**, 1190.
13. T. Rudolph, K. Kempe, S. Crotty, R. M. Paulus, U. S. Schubert, I. Krossing and F. H. Schacher, *Polym. Chem.*, 2013, **4**, 495.
14. H. Kricheldorf and R. Dunsing, *Makromol. Chem.*, 1986, **187**, 1611.
15. M. Baśko and P. Kubisa, *J. Polym. Sci., Part A: Polym. Chem.*, 2008, **46**, 7919.
16. M. Baśko, *Pure Appl. Chem.*, 2012, **84**, 2081.
17. M. Baśko and P. Kubisa, *J. Polym. Sci., Part A: Polym. Chem.*, 2010, **48**, 2650.
18. D. Bourissou, B. Martin-Vaca, A. Dumitrescu, M. Graullier and F. Lacombe, *Macromolecules*, 2005, **38**, 9993.
19. O. Coulembier and P. Dubois, in *The Handbook of Ring-Opening Polymerization*, ed. P. Dubois, O. Coulembier and J.-M. Raquez, Wiley-VCH, Weinheim, 2009, pp. 227.

20. F. A. Jaipuri, B. D. Bower and N. L. Pohl, *Tetrahedron: Asymmetrty*, 2003, **14**, 3249.
21. A. Couffin, B. Martin-Vaca, D. Bourissou and C. Navarro, *Polym. Chem.*, 2014, **5**, 161.
22. D. Gazeau-Bureau, B. Delcroix, D. Martin-Vaca, C. Bourissou, Navarro and S. Magnet, *Macromolecules*, 2008, **41**, 3782.
23. D. Delcroix, B. Martin-Vaca, D. Bourissou and C. Navarro, *Macromolecules*, 2010, **43**, 8828.
24. Estimated by the H_0 method. F. G. Bordwell, *Acc. Chem. Res.*, 1988, **21**, 456.
25. S. Nakano, *Prog. Org. Coat.*, 1999, **35**, 141.
26. M. Bednarek, P. Kubisa and S. Penczek, *Macromolecules*, 2001, **34**, 5112.
27. N. Stanley, G. Bucataru, Y. Miao, A. Favrelle, M. Bria, F. Stoffelbach, P. Woisel and P. Zinck, *J. Polym. Sci., Part A: Polym. Chem.*, 2014, **52**, 2139.
28. A. Pascual, J. R. Leiza and D. Mecerreyes, *Eur. Polym. J.*, 2013, **49**, 1601.
29. A. Pascual, H. Sardon, A. Veloso, F. Ruipérez and D. Mecerreyes, *ACS Macro Lett.*, 2014, **3**, 849.
30. P. Hodge, *Chem. Rev.*, 2014, **114**, 2278.
31. T. Yashiro, H. R. Kricheldorf and G. Schwarz, *Macromol. Chem. Phys.*, 2010, **211**, 1311.
32. M. L. Hawkins, S. S. Schott, B. Grigoryan, M. A. Rufin, B. K. Ngo, L. Vanderwal, S. J. Stafslien and M. A. Grunlan, *Polym. Chem.*, 2017, **8**, 5239.
33. G. Bouchoux, D. Drancourt, D. Leblanc, M. Yanez and O. Mo, *New J. Chem.*, 1995, **19**, 1243.
34. N. Susperregui, D. Delcroix, B. Martin-Vaca, D. Bourissou and L. Maron, *J. Org. Chem.*, 2010, **75**, 6581.
35. When solvents (toluene and CH_2Cl_2) were considered by means of single-point CPCM calculation on the gaz phase optimized structures, the activation barriers were lowered by 5 to 10 kcal mol^{-1}.
36. P. Kubisa and S. Penczek, *Prog. Polym. Sci.*, 1999, **24**, 1409.
37. D. J. Coady, H. W. Horn, G. O. Jones, H. Sardon, A. C. Engler, R. M. Waymouth, J. E. Rice, Y. Y. Yang and J. L. Hedrick, *ACS Macro Lett.*, 2013, **2**, 306.
38. In the same article the pK_a of HOTf is estimated to -11. E. Paenurk, K. Kaupmees, D. Himmel, A. Kutt, I. Kaljurand, I. A. Koppel, I. Krossing and I. Leito, *Chem. Sci.*, 2017, **8**, 6964.
39. R. Kakuchi, Y. Tsuji, K. Chiba, K. Fuchise, R. Sakai, T. Satoh and T. Kakuchi, *Macromolecules*, 2010, **43**, 7090.
40. K. Makiguchi, S. Kikuchi, T. Satoh and T. Kakuchi, *J. Polym. Sci., Part A: Polym. Chem.*, 2013, **51**, 2455.
41. S. Yamada, A. Takasu and K. Kawamura, *J. Polym. Sci., Part A: Polym. Chem.*, 2013, **51**, 3732.
42. M. Oshimura, T. Tang and A. Takasu, *J. Polym. Sci., Part A: Polym. Chem.*, 2011, **49**, 1210.

43. A. Duda and A. Kowalski, in *The Handbook of Ring-Opening Polymerization*, ed. P. Dubois, O. Coulembier and J.-M. Raquez, Wiley-VCH, Weinheim, 2009, pp. 1.
44. Y. Jin, Y. Ji, X. He, S. Kan, H. Xia, B. Liang, J. Chen, H. Wu, K. Guo and Z. Li, *Polym. Chem.*, 2014, **5**, 3098.
45. K. Makiguchi, T. Satoh and T. Kakuchi, *J. Polym. Sci., Part A: Polym. Chem.*, 2011, **49**, 3769.
46. K. Ishihara, A. Hasegawa and H. Yamamoto, *Angew. Chem., Int. Ed.*, 2001, **40**, 4077.
47. M. Terada, *Chem. Commun.*, 2008, 4097.
48. D. Parmar, E. Sugiono, S. Raja and M. Rueping, *Chem. Rev.*, 2014, **114**, 9047.
49. P. Christ, A. G. Lindsay, S. S. Vormittag, J. M. Neudörfl, A. Berkessel and A. C. O'Donoghue, *Chem. – Eur. J.*, 2011, **17**, 8524.
50. K. Kaupmees, N. Tolstoluzhsky, S. Raja, M. Rueping and I. Leito, *Angew. Chem., Int. Ed.*, 2013, **52**, 11569.
51. K. Makiguchi, T. Satoh and T. Kakuchi, *Macromolecules*, 2011, **44**, 1999.
52. D. Delcroix, A. Couffin, N. Susperregui, C. Navarro, L. Maron, B. Martin-Vaca and D. Bourissou, *Polym. Chem.*, 2011, **2**, 2249.
53. W. Wu, S. Cui, Z. Li, J. Liu, H. Wang, X. Wang, Q. Zhang, H. Wu and K. Guo, *Polym. Chem.*, 2015, **6**, 2970.
54. It has been reported that the replacement of the OH by a TfNH group on phosphoric acids results in a 2–4 reduction of pK_a. See ref. 48 and 49.
55. J. Chen, S. Kan, H. Xia, F. Zhou, X. Chen, X. Jiang, K. Guo and Z. Li, *Polymer*, 2013, **54**, 4177.
56. I. Coric and B. List, *Nature*, 2012, **483**, 315.
57. S. Kan, Y. Jin, X. He, J. Chen, H. Wu, P. Ouyang, K. Guo and Z. Li, *Polym. Chem.*, 2013, **4**, 5432.
58. X. He, Y. Ji, Y. Jin, S. Kan, H. Xia, J. Chen, B. Liang, H. Wu, K. Guo and Z. Li, *J. Polym. Sci., Part A: Polym. Chem.*, 2014, **52**, 1009.
59. J. Ma and S. R. Kass, *Org. Lett.*, 2016, **18**, 5812.
60. K. Makiguchi, T. Saito, T. Satoh and T. Kakuchi, *J. Polym. Sci., Part A: Polym. Chem.*, 2014, **52**, 2032.
61. E. K. Macdonald and M. P. Shaver, *Eur. Polym. J.*, 2017, **95**, 702.
62. Y. Miao, Y. Phuphuak, C. Rousseau, T. Bousquet, A. Mortreux, S. Chirachanchai and P. Zinck, *J. Polym. Sci., Part A: Polym. Chem.*, 2013, **51**, 2279.
63. J. Liu, S. Cui, Z. Li, S. Xu, J. Xu, X. Pan, Y. Liu, H. Dong, H. Sun and K. Guo, *Polym. Chem.*, 2016, **7**, 5526.
64. Y. Wang, M. Xia, X. Kong, S. J. Severtson and W. J. Wang, *RSC Adv.*, 2017, **7**, 28661.
65. K. Makiguchi, T. Yamanaka, T. Kakuchi, M. Terada and T. Satoh, *Chem. Commun.*, 2014, **50**, 2883.
66. J.-B. Zhu and E. Y. X. Chen, *J. Am. Chem. Soc.*, 2015, **137**, 12506.
67. A. Sanchez-Sanchez, I. Rivilla, M. Agirre, A. Basterretxea, A. Etxeberria, A. Veloso, H. Sardon, D. Mecerreyes and F. P. Cossío, *J. Am. Chem. Soc.*, 2017, **139**, 4805.

68. F. Sanda, H. Sanada, Y. Shibasaki and T. Endo, *Macromolecules*, 2002, **35**, 680.
69. F. Zeng, H. Lee, M. Chidiac and C. Allen, *Biomacromolecules*, 2005, **6**, 2140.
70. E. Oledzka and S. S. Narine, *J. Appl. Polym. Sci.*, 2011, **119**, 1873.
71. I. Jiménez-Pardo, L. G. J. Ven, R. A. T. M. Benthem, A. C. C. Esteves and G. D. With, *J. Polym. Sci., Part A: Polym. Chem.*, 2017, **55**, 1502.
72. J. Matsuo, S. Nakano, F. Sanda and T. Endo, *J. Polym. Sci., Part A: Polym. Chem.*, 1998, **36**, 2463.
73. J. Casas, P.-V. Persson, T. Iversen and A. Codova, *Adv. Synth. Catal.*, 2004, **346**, 1087.
74. P. V. Persson, J. Schroder, K. Wickholm, E. Hedenström and T. Iversen, *Macromolecules*, 2004, **37**, 5889.
75. P. V. Persson, J. S. Casas, T. Iversen and A. Cordova, *Macromolecules*, 2006, **39**, 2819.
76. R. S. Lee, H. R. Li, J. M. Yang and F. Y. Tsai, *Polymer*, 2005, **46**, 10718.
77. J. Xu, J. Song, S. Pispas and G. Zhang, *J. Polym. Sci., Part A: Polym. Chem.*, 2014, **52**, 1185.
78. A. Borovika, P. I. Tang, S. Klapman and P. Nagorny, *Angew. Chem., Int. Ed.*, 2013, **52**, 13424.
79. X. Li, Q. Zhang, Z. Li, S. Xu, C. Zhao, C. Chen, X. Zhi, H. Wang, N. Zhu and K. Guo, *Polym. Chem.*, 2016, **7**, 1368.
80. X. Li, Q. Zhang, Z. Li, X. Wang, J. Liu, S. Cui, S. Xu, C. Zhao, C. Chen and K. Guo, *Polymer*, 2016, **84**, 293.
81. H. Sun, S. Xu, Z. Li, J. Xu, J. Liu, X. Wang, H. Wang, H. Dong, Y. Liu and K. Guo, *Polym. Chem.*, 2017, **8**, 5570.
82. J. Xu, J. Liu, Z. Li, X. Li, C. Chen, C. Zhao, S. Xu, X. Pan, J. Liu and K. Guo, *Polym. Chem.*, 2016, **7**, 1111.
83. J. Xu, K. Yang, Z. Li, J. Liu, H. Sun, S. Xu, H. Wang, T. Guo, H. Dong and K. Guo, *Polym. Chem.*, 2017, **8**, 6398.
84. J. Kadota, D. Pavlovic, J. P. Desvergne, B. Bibal, F. Peruch and A. Deffieux, *Macromolecules*, 2010, **43**, 8874.
85. J. Kadota, D. Pavlovic, H. Hirano, A. Okada, Y. Agari, B. Bibal, A. Deffieux and F. Peruch, *RSC Adv.*, 2014, **4**, 14725.
86. K. Makiguchi, S. Kikuchi, K. Yanai, Y. Ogasawara, S. I. Sato, T. Satoh and T. Kakuchi, *J. Polym. Sci., Part A: Polym. Chem.*, 2014, **52**, 1047.
87. D. J. Coady, K. Fukushima, H. W. Horn, J. E. Rice and J. L. Hedrick, *Chem. Commun.*, 2011, **47**, 3105.
88. X. Wang, J. Liu, S. Xu, J. Xu, X. Pan, J. Liu, S. Cui, Z. Li and K. Guo, *Polym. Chem.*, 2016, **7**, 6297.
89. Y. Miao, N. Stanley, A. Favrelle, T. Bousquet, M. Bria, A. Mortreux and P. Zinck, *J. Polym. Sci., Part A: Polym. Chem.*, 2015, **53**, 659.
90. X. Wang, S. Cui, Z. Li, S. Kan, Q. Zhang, C. Zhao, H. Wu, J. Liu, W. Wu and K. Guo, *Polym. Chem.*, 2014, **5**, 6051.
91. A. Vora, R. J. Wojtecki, K. Schmidt, A. Chunder, J. Y. Cheng, A. Nelson and D. P. Sanders, *Polym. Chem.*, 2016, **7**, 940.

92. Y. Y. Kim, S. Jung, C. Kim, B. J. Ree, D. Kawato, N. Nishikawa, D. Suemasa, T. Isono, T. Kakuchi, T. Satoh and M. Ree, *Macromolecules*, 2014, **47**, 7510.
93. A. C. Engler, J. M. W. Chan, D. J. Coady, J. M. O'Brien, H. Sardon, A. Nelson, D. P. Sanders, Y. Y. Yang and J. L. Hedrick, *Macromolecules*, 2013, **46**, 1283.
94. J. M. W. Chan, X. Ke, H. Sardon, A. C. Engler, Y. Y. Yang and J. L. Hedrick, *Chem. Sci.*, 2014, **5**, 3294.
95. Q. Ma, K. Lei, J. Ding, L. Yu and J. Ding, *Polym. Chem.*, 2017, **8**, 6665.
96. P. Bexis, A. W. Thomas, C. A. Bell and A. P. Dove, *Polym. Chem.*, 2016, **7**, 7126.
97. F. Ercole, A. E. Rodda, L. Meagher, J. S. Forsythe and A. P. Dove, *Polym. Chem.*, 2014, **5**, 2809.
98. M. Basko and P. Kubisa, *J. Polym. Sci., Part A: Polym. Chem.*, 2006, **44**, 7071.
99. M. Basko and P. Kubisa, *J. Polym. Sci., Part A: Polym. Chem.*, 2007, **45**, 3090.
100. A. Couffin, D. Delcroix, B. Martin-Vaca, D. Bourissou and C. Navarro, *Macromolecules*, 2013, **46**, 4354.
101. J. M. Campos, M. R. Ribeiro, M. F. Ribeiro, A. Deffieux and F. Peruch, *Eur. Polym. J.*, 2013, **49**, 4025.
102. Q. Song, Y. Xia, S. Hu, J. Zhao and G. Zhang, *Polymer*, 2016, **102**, 248.
103. K. Makiguchi, Y. Ogasawara, S. Kikuchi, T. Satoh and T. Kakuchi, *Macromolecules*, 2013, **46**, 1772.
104. T. Saito, Y. Aizawa, K. Tajima, T. Isono and T. Satoh, *Polym. Chem.*, 2015, **6**, 4374.
105. J. Zhao, D. Pahovnik, Y. Gnanou and N. Hadjichristidis, *Macromolecules*, 2014, **47**, 3814.
106. J. Zhao and N. Hadjichristidis, *Polym. Chem.*, 2015, **6**, 2659.
107. J. Zhao, D. Pahovnik, Y. Gnanou and N. Hadjichristidis, *J. Polym. Sci., Part A: Polym. Chem.*, 2015, **53**, 304.
108. X. Wang, J. Liu, S. Xu, J. Xu, X. Pan, J. Liu, S. Cui, Z. Li and K. Guo, *Polym. Chem.*, 2016, **7**, 6297.
109. H. U. Kang, Y. C. Yu, S. J. Shin and J. H. Youk, *J. Polym. Sci., Part A: Polym. Chem.*, 2013, **51**, 774.
110. Y. C. Yu, G. Li, H. U. Kang and J. H. Youk, *Colloid Polym. Sci.*, 2012, **290**, 1707.
111. A. G. O. Freitas, S. G. Trindade, P. I. R. Muraro, V. Schmidt, A. J. Satti, M. A. Villar, A. S. E. Ciolino and C. Giacomelli, *Macromol. Chem. Phys.*, 2013, **214**, 2336.
112. C. Fu, J. Xu, M. Kokotovic and C. Boyer, *ACS Macro Lett.*, 2016, **5**, 444.
113. H. U. Kang, Y. C. Yu, S. J. Shin, J. Kim and J. H. Youk, *Macromolecules*, 2013, **46**, 1291.
114. Y. C. Yu, S. J. Shin, K. D. Ko, W. R. Yu and J. H. Youk, *Polymer*, 2013, **54**, 5595.
115. H. Dong, Y. Zhu, Z. Li, J. Xu, J. Liu, S. Xu, H. Wang, Y. Gao and K. Guo, *Macromolecules*, 2017, **50**, 9295.

CHAPTER 3

Bifunctional and Supramolecular Organocatalysts for Polymerization

KURT V. FASTNACHT, PARTHA P. DATTA AND MATTHEW K. KIESEWETTER*

Department of Chemistry, University of Rhode Island, Kingston, RI 02881, USA
*Email: mkiesewetter@uri.edu

3.1 Introduction

The catalysts in this chapter conduct polymerization *via* non-nucleophilic, H-bond mediated pathways. These catalysts include thiourea or urea plus base, squaramides and protic acid/base pairs—which are unified in a conceptual approach of applying a mild Lewis acid plus Lewis base to effect ring-opening polymerization (ROP)—as well as other supramolecular catalysis. This class of catalyst produces among the best-defined materials available *via* synthetic polymer chemistry through a delicately balanced series of competing chemical reactions by interacting with substrate at an energy of <4 kcal mol^{-1}.[1,2] Indeed, the multitude of simultaneous chemical reactions in a typical supramolecular polymerization is as much awe-inspiring as it is difficult to comprehend, and changing any one factor (H-bond donor, H-bond acceptor, reagent, solvent, temperature, *etc.*) impacts all the interactions in solution. The polymerization catalysis community has been building an understanding of these systems incrementally over the last

Polymer Chemistry Series No. 31
Organic Catalysis for Polymerisation
Edited by Andrew Dove, Haritz Sardon and Stefan Naumann

Published by the Royal Society of Chemistry, www.rsc.org

decade, and our understanding and abilities in rate, selectivity, diversity of polymer architectures available and reaction control continue to evolve.

The purview of the catalysts in this chapter is ring-opening polymerization (ROP), especially of cyclic esters and carbonates. Conceptually, the catalysts in this chapter are ideally suited to perform highly controlled polymerizations. Catalysts for the ROP of lactones and carbonates effect polymerization by (1) activating the chain-end, (2) activating the monomer, or (3) activating both. By separating the roles of monomer and chain-end activation into discrete functions, the dual catalysts can be separately tuned for enchainment *vs.* side reactions. Conceptually, a dual catalyst consists of both a hydrogen bond donor (HBD) (*e.g.* urea or thiourea) for monomer activation and a hydrogen bond acceptor (HBA) (*e.g.* tertiary amines) for chain-end activation. Such dual catalysts may be a single molecule, but in common practice, bimolecular cocatalysts are employed to activate monomer and initiator alcohol/chain ends separately, Scheme 3.1.

The fountainhead of dual catalysis is the 2005 manuscript and its follow-up from Hedrick and Waymouth.[3,4] The roots of organocatalysis reach back more than 100 years to the synthesis of quinine alkaloids,[5] and, in fact, organocatalysts were among the earliest catalysts for the synthesis of polyesters.[6] The renaissance of organocatalysis *ca.* 2000 saw the application of supramolecular catalysts for small molecule synthesis.[7] However, it was the veritable Johnny Appleseeds of organocatalytic polymerization that disclosed supramolecular catalysts for ROP along their continuing journey of discovery and subsequently nurtured the field such that it now encompasses many branches of questioning by several research groups.[4] The first supramolecular catalyst for ROP (the Takemoto catalyst, **1**, Figure 3.1) was adapted from the work of Takemoto, who used chiral H-bonding catalysts for asymmetric Michael reactions.[8] The thiourea/amine base catalyst **1** was introduced into the polymerization community for the organocatalytic ROP of lactide.[4] The inspired (and somewhat miraculous) step of separating the roles of HBD and HBA into discrete cocatalysts facilitated modulation of the individual cocatalysts leading to the ROP of other monomers and launched a field, Figure 3.1.[3,4]

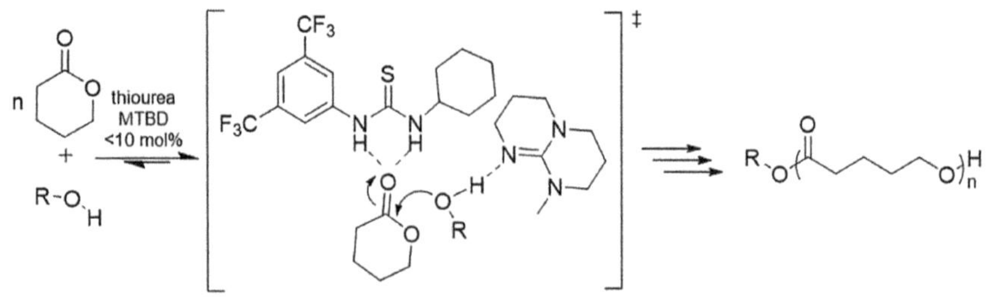

Scheme 3.1 Dual catalyst (bimolecular) mediated ROP of δ-valerolactone. Thiourea and MTBD are exemplary H-bond donors (HBDs) and H-bond acceptors (HBAs), respectively.

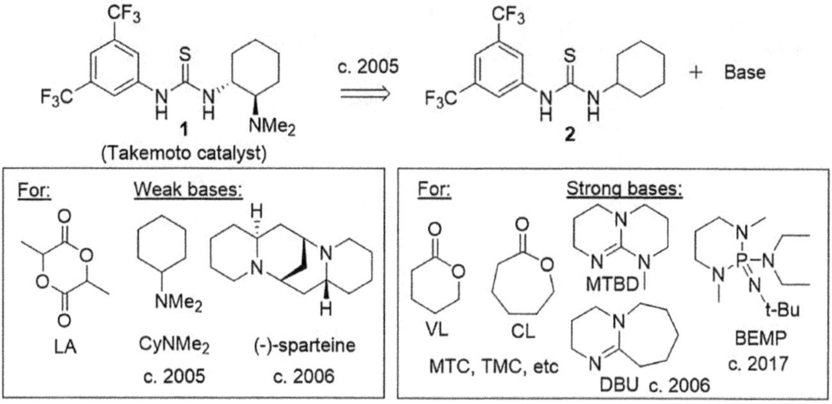

Figure 3.1 The Takemoto catalyst was the inspiration for the popular thiourea plus base catalyst system. Weaker base cocatalysts effect the ROP of lactide, while stronger bases open other monomers.

The class of organic molecules that effects catalysis *via* supramolecular interactions are among the most controlled catalysts available for ROP. Part of this is due to the modular, highly tunable nature of dual catalysts, which results in extremely controlled ROP ($M_w/M_n = Đ = <1.1$) of a host of different cyclic monomers.[9,10] Most of the research in the field of dual catalysis for organic polymerizations has been dedicated to the ROP of cyclic esters and carbonates; however, other monomers will be mentioned. Dual catalysts effect *living polymerizations*, which is a type of *chain growth polymerization* that proceeds without chain transfer or termination.[11,12] This is ultimately a kinetic distinction, and it is often said that a polymerization exhibits the characteristics of a 'living' polymerization: molecular weights (M_n) are predictable from $[M]_o/[I]_o$, linear evolution of M_n with conversion, first order consumption of monomer and narrow weight distributions (M_w/M_n).[11] In practice, these conditions arise when a polymerization has a fast initiation rate relative to propagation rate and few to no side reactions. We shall refrain from pointing out when a catalyst (system) exhibits the characteristics of a 'living' polymerization, and rather point out when it is either especially well controlled or exhibits low levels of control. Several, thorough reviews have been conducted in the wider field,[13–22] but not with quite the level of focus that the current platform provides. Hence, we will attempt to emphasize the virtues and deficits of the various catalysts, especially as they contrast to other organic catalysts for polymerization.

3.2 Dual Catalysts

The dual catalysts for polymerization are a logical mechanistic conclusion of early organocatalysts for ROP, and H-bond mediated (supramolecular) polymerization mechanisms have been implicated for catalysts in a host of architectures.[2,23–25] For example, the pyridine bases 4-(dimethylamino)pyridine

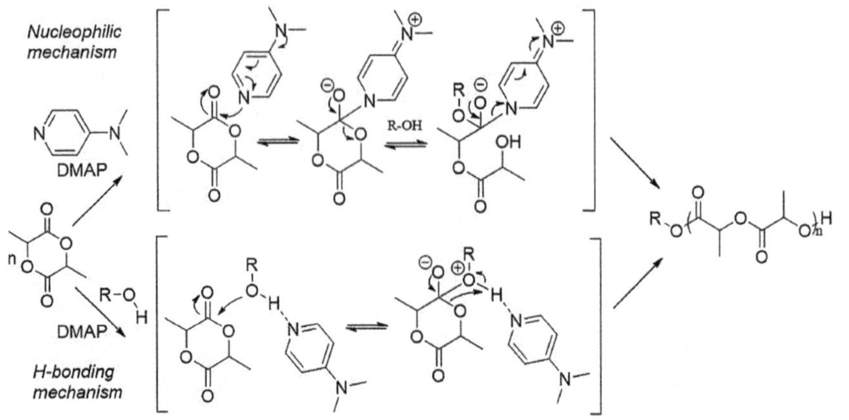

Scheme 3.2 DMAP catalyzed ROP of lactide has been proposed to proceed *via* nucleophilic (upper) and H-bond mediated (lower) pathways.

(DMAP) and 4-pyrrolidinopyridine (PPY) have been proposed to conduct the zwitterionic ROP of lactones (see Chapter 1).[26–30] However, subsequent mechanistic studies suggest that the nucleophilic and H-bonding pathways are both accessible with the hydrogen-bonded pathway being energetically favorable (Scheme 3.2).[31–34] An alcohol-activated mechanism of enchainment has been proposed for the phosphazene bases (*e.g.* P_1-*t*Bu, P_2-*t*Bu, *t*-BuP_4, BEMP), which have been shown to bring about the ROP of lactones in the presence of alcohols.[25,35–38] A similar pathway can be envisaged for the guanidine and amidine bases, 7-methyl-1,5,7-triazabicyclo[4.4.0]dec-5-ene (MTBD) and 1,8-diazabicyclo[5.4.0]undec-7-ene (DBU).[2,24] The dual catalysis conceptual approach of separately activating the monomer and propagating chain end arises from these early organocatalysts, which often suffered from low activity or reaction control.[4,23,24] By separately activating both reactive species, greater specificity and control can be achieved.

3.2.1 Thiourea H-bond Donors

As with many organocatalysts for polymerization, thiourea/base mediated ROP has its roots in small molecule transformations where Jacobsen *et al.* had shown that an array of ureas and thioureas were effective catalysts for Mannich, Strecker, Pictet–Spengler, and hydrophosphonylation reactions,[39–46] among others.[7] Indeed, the parent dual catalyst, **1**, for ROP was used by Takemoto *et al.* for enantioselective aza-Henry and Michael additions.[8,47,48] In the seminal polymerization work, **1** was shown to effect the ROP of lactide with, at the time, remarkably 'living' behavior.[4] Incredibly, failure to quench the reaction after full conversion to polymer did not result in broadening of molecular weight distribution, signifying very minimal transesterification, and minimal racemization.[4] When the HBD and HBA roles of **1** were divided into separate HBD (**2**) and HBA

(*N,N*-dimethylcyclohexylamine) molecules, a field of research was born, Figure 3.1. Polylactide formation was only successful when both **2** and *N,N*-dimethylcyclohexylamine were applied simultaneously, and a range of non-H-bonding solvents were found to facilitate ROP (*e.g.* chloroform, dichloromethane and toluene), while THF and DMF failed.[4] A host of alkylamine cocatalysts (with **2**) has been shown to be effective for the ROP of lactide.[3,49] Strong bases—MTBD, DBU and later BEMP—are effective cocatalysts with **2** for the ROP of other monomers: δ-valerolactone (VL), ε-caprolactone (CL), trimethylene carbonate (TMC), MTC and others, Figures 3.1 and 3.2.[2,50] The stronger bases result in a less-controlled ROP of lactide in the absence of thiourea, but thiourea plus strong base is necessary to open other lactones and carbonates with reasonable rates.[2] The ROP of β-butyrolactone (BL) is not easily performed with most organocatalysts.[2,51] A common red herring in the ROP literature will attribute unexplainable and otherwise 'spooky' observations to *ring strain*. Indeed, it is often observed for organocatalytic ROP that enchainment rates ($k_{LA} > k_{VL} \gg k_{CL} \gg k_{BL}$)[52,53] have no correlation to *ring strain* as measured by equilibrium monomer concentration, $[M]_{eq}$: $[VL]_{eq}$ (low strain) $\gg [CL]_{eq} \approx [LA]_{eq} \gg [BL]_{eq}$ (high strain).[52,53]

The origin of the high selectivity for monomer is thought to arise from selective binding of thiourea to monomer *vs.* polymer. The binding constants of lactones (s-*cis* esters) and open s-*trans* esters to **2** were measured by ^{1}H NMR titration.[2] The s-*trans* ester (ethyl acetate) exhibited minimal binding while binding constants of $K_{eq} \approx 40$ were observed between VL or CL and **2**.[2] Thiourea H-bond donors have subsequently been shown to bind much more strongly to base cocatalyst, where the nature of the cocatalyst binding constant is a better indicator of cocatalytic activity than monomer binding.[50,54–56] The cocatalyst binding can be inhibitory to catalysis under the proper circumstances.[50,54–57] However, the rapid, reversible and promiscuous binding of thiourea to several reagents in solution appears to reduce the overall order of the transformation

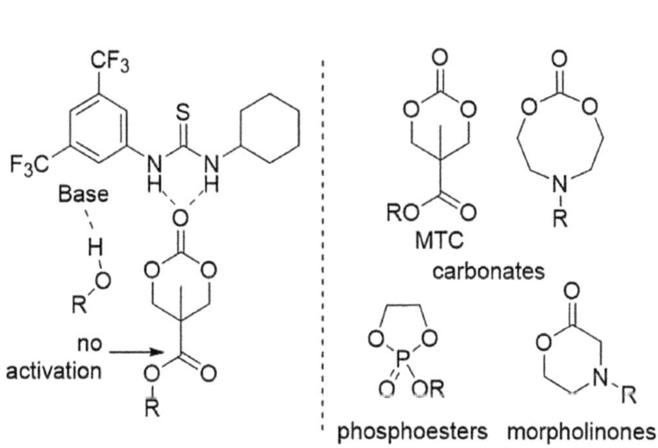

Figure 3.2 Functionalizable monomers that undergo controlled ROP by **2**/base.

(rate = k[M][I]$_o$[cocatalysts]$_o$),[50,55,56] and the notion of thiourea as an entropy trap prior to enchainment has been repeatedly reinforced.[58,59] Indeed, our understanding of the multitude of interrelated interactions that occur during a (thio)urea/base mediated ROP continues to unfold.[60–62] The theme of competitive binding repeats throughout the literature, including the amide and indole H-bond donor catalysts applied to the ROP of LA, which are structurally reminiscent to (thio)ureas.[54,63,64] The major take-away message is that the high selectivity of H-bonding catalysts appears to rise from two sources, (1) selective binding of thiourea to monomer *vs.* polymer, and (2) strong binding (K_{eq} = 100–4200) of thiourea to base cocatalysts, which reduces their relative affinity to other reagents and can become an inhibitory interaction.[50,54] The high selectivity for s-*cis* esters and carbonates has been used to great effect for the generation of classes of functionalizable monomers, Figure 3.2.[65–70]

3.2.2 Thiourea-mediated Stereoselective ROP

The stereoselective ROP of *rac*-lactide is an attractive method for the generation of polylactides (PLAs) with highly regular or novel stereosequences, and the modular scaffold and rich diversity of chiral thiourea H-bond donors has proved an enticing target for several research groups. The ROP of *rac*- or *meso*-lactide to generate highly tactic PLA has been well documented.[71–73] Briefly, stereoselective enchainment of the chiral monomer onto the chiral chain end can occur *via* control rendered by (1) the propagating chain end, (2) a chiral catalyst or (3) a mixed mechanism.[71,74,75] For the ROP of *rac*-LA, a high probability of propagating with retention of stereochemistry (P_m = probability of *meso* enchainment) will result in a highly isotactic PLA.[3,71] Waymouth and Hedrick reported the ***(R,R)*-1** mediated ROP of *rac*-lactide to proceed with modest selectivity (P_m = 0.76); however, **2**/(−)-sparteine catalyzed ROP of *rac*-LA rendered similar selectivity (P_m = 0.77).[3] The polymers did not display a melting point, suggesting low stereoregularity.[3] Exceeding these P_m values has become a benchmark of sorts for the stereoselective ROP of *rac*-lactide by H-bonding catalysts. Despite its successes, (−)-sparteine itself fell out of favor as an organocatalyst when it became scarce *ca.* 2010, but a replacement base, benzyl bispidine, was disclosed that renders similar reaction rates and selectivity in the ROP of *rac*-lactide with **2**, P_m = 0.74.[49,76]

Recent research into photo-responsive azobenzene-based thiourea, **3**, for the ROP of *rac*-lactide suggests a conceptual approach to switchable organocatalysts for ROP.[77,78] Catalysts that are switchable by external stimuli (*i.e.* redox pathways, light, coordination chemistry, *etc.*)[78–96] offer an attractive route to advanced catalyst structures and, presumably, polymer architectures. Thiourea **3** is based on the classic photo-switchable azobenzene moiety, Scheme 3.3. The *trans*-**3** isomer contains an open active site for coordination of lactide by H-bonding whereas *cis*-**3** is blocked by intramolecular H-bonding to the nitro group. The **3**/PMDETA (Scheme 3.3)

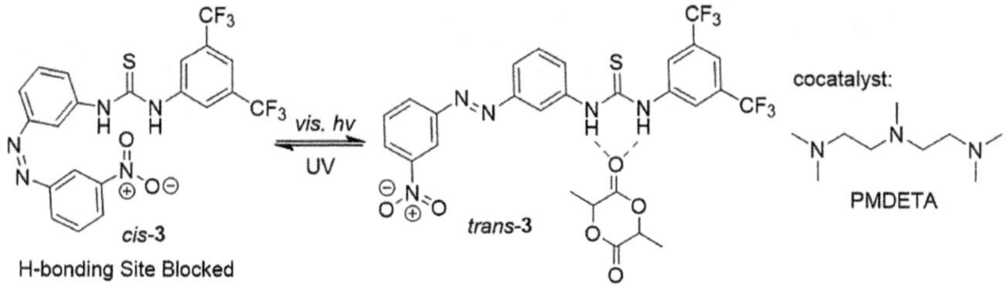

Scheme 3.3 Azobenzene-based switchable thiourea.

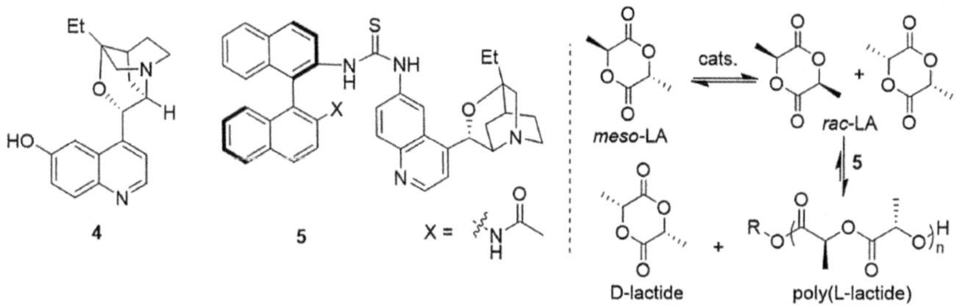

Scheme 3.4 Cinchona alkaloid-based H-bond donors for the stereoselective ROP of *rac*-lactide.

cocatalyzed ROP of *rac*-LA proceeded with moderate isoselectivities ($P_m \approx 0.74$) at room temperature.[77] The ROP was proposed to proceed from the *trans*-isomer, presumably *via* a chain-end control mechanism.[3,77] We make the safe prediction that switchable organic catalysts for ROP will play an important role in the next decade.[78,91]

A thiourea with pendant cinchona alkaloid, **5** in Scheme 3.4, provided the first example of isotactic-rich, stereogradient PLA *via* kinetic resolution polymerization with organocatalysts. The bifunctional **4** (internal nitrogen base) effected the ROP of *rac*-LA to generate isotactic-rich PLA, $P_m = 0.69$.[97] No transesterification was observed in MALDI-TOF MS, and almost no epimerization was observed. Polymerization experiments, isolation of residual monomer and analysis by chiral HPLC suggest that the stereoselectivity in the **4**-catalyzed polymerization of *rac*-LA arises from the kinetic resolution by the catalyst/initiator to produce enantioenriched (stereogradient) PLAs. This motif was later incorporated into a thiourea/BINAM-containing organocatalyst, **5** (Scheme 3.4), for the kinetic resolution ROP of lactide.[98] This stereoselective ROP scheme—arguably the current gold standard—used an epimerization catalyst to transform *meso*- to *rac*-LA, which **5** was able to enchain to isotactic poly(L-lactide) with high selectivity, $k_L/k_D = 53$.[98] Not surprisingly, solvent (and other reaction conditions) dramatically perturb the selectivity.[98] It should also be noted that structurally similar H-bond donors failed to produce ROP with appreciable rates or selectivities,[97,98] which highlights a challenge of stereoselective, organocatalytic ROP. Indeed,

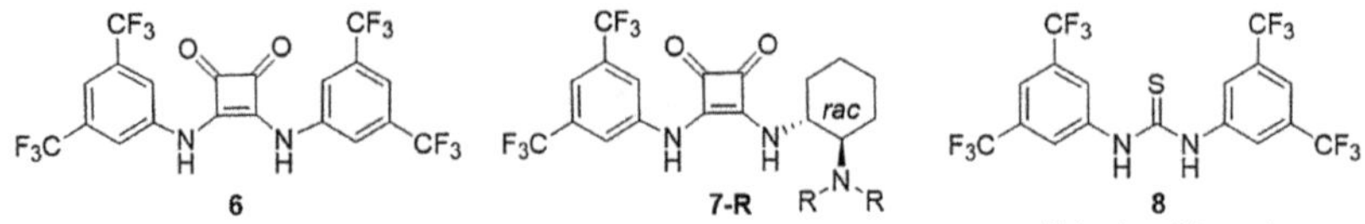

Figure 3.3 Squaramide H-bond donors for ROP of lactide.

a significant amount of inspiring ground work exists upon which to build highly successful stereoselective catalysts for ROP, and the field *could* proceed along this trial and error pathway. However, more fundamental information that might provide a solid mechanistic basis for a path forward may save a tremendous amount of effort.

3.2.3 Squaramides

The squaramide H-bond donor scaffold has been used to great success in small molecule catalysis[99] and may represent an underexplored opportunity for polymer synthesis. Guo *et al.* examined squaramides for the ROP of L-lactide in dichloromethane at room temperature, initiated from benzyl alcohol.[100] Squaramide **6** was unable to effect polymerization alone but was active with tertiary amine, (−)-sparteine, cocatalyst, Figure 3.3. H-bond donor **6** plus sparteine exhibits similar activity for ROP of lactide *vs.* thiourea 2, and squaramides with no electron withdrawing substituents saw less conversion than their electron-deficient counterparts.[100] A slate of bifunctional squaramide catalysts, **7**, was also evaluated for ROP, Figure 3.3.[101–103] The bifunctional catalyst **7-Me** displayed reduced activity *vs.* pentyl groups on the amine motif **7**, which was the only one of the examined structures to achieve full conversion in 24 h.[101] No epimerization was observed during polymerization. A classic H-bond mediated mechanism of enchainment was corroborated by NMR titration studies.[101] The H-bonding ability of squaramides is perturbed *vs.* that of thioureas,[101] but they have approximately the same acidity (Schreiner's thiourea (**8**) $pK_a = 8.5$; **6** $pK_a = 8.4$; both in DMSO).[104,105] The altered structures possessing minimally altered pK_a may have unseen implications for nascent imidate-mediated ROP, see Section 3.3.3 below. The multi-H-bonding scaffold found in squaramides and (thio)ureas are promising for enhanced rates and selectivity and are currently being adopted into other structural motifs.[106]

3.3 Rate-accelerated Dual Catalysis

From the very early days of the field, thiourea/base cocatalysts exhibited remarkably controlled ROP, so remarkable that the poor activity and productivity of the catalysts could be justified. However, with the application of N-heterocyclic carbene (NHC) and TBD organocatalysts to ROP, it became very clear that organocatalysts could possess activity to rival that of metal catalysts.[17,24,51] The dream of combining the rate of NHCs or TBD with the high selectivity of thiourea/base systems became an alluring research goal for several groups. One route that can be envisaged uses internal Lewis acids to stabilize the (thio)urea as it binds to monomer. The challenge became finding synthetically accessible (thio)ureas with Lewis acids that are compatible with ROP.

3.3.1 Internal Lewis Acid Enhanced H-bond Donors

A urea H-bond donating catalyst with an internal boronate ester, **9**, displayed enhanced activity *vs.* its parent urea, **10** (Figure 3.4). HBD **9** was applied with sparteine cocatalyst for the ROP of LA at room temperature ($k_2/k_9 \approx 1$).[107] Importantly, the ROP of LA with **9**/sparteine showed good control and maintained a narrow molecular weight distribution ($M_w/M_n \approx 1.18$) for days after the reaction had finished (initial $M_w/M_n \approx 1.16$), indicating minor transesterification. This motif is an extreme example of the internal H-bond stabilization that is thought to be present in all (thio)ureas bearing electron deficient aryl rings.[108]

3.3.2 Multi (Thio)urea Catalysts

Mechanistic studies on **2**/base cocatalyzed ROP led to the development of highly effective bis- and tris-(thio)urea H-bond donors.[55,109] In general, urea HBDs are more active than thioureas, and tris-donors are more active than bis-, which are more active than mono-; although tris-thiourea (**14**) is markedly inactive, Scheme 3.5.[55,109] These general trends hold for most monomers that have been examined, but the rate accelerations are most dramatic for the slower monomers (*i.e.* CL).[55,109] Just as with **2**, weak alkylamine base cocatalysts are required for the ROP of lactide with **11–15**,[4,49,55] but strong base cocatalysts are required for VL, CL and carbonate monomers.[2,109] For the trisurea (**15**)/BEMP cocatalyzed ROP of CL, a ~500 times increase in rate is observed *vs.* **2**/BEMP, and the reaction is more controlled.[50,109] A typical (thio)urea/base cocatalyzed ROP is run ~2M monomer and displays good control for M_n from $[M]_o/[I]_o \approx 20–500$,[2,55,109] although enhanced (*vs.* **2**) M_n control is observed for **13** and **15** at higher $[M]_o/[I]_o$.[109] The comparisons above are controlled for mol percent (thio)urea *moiety* in

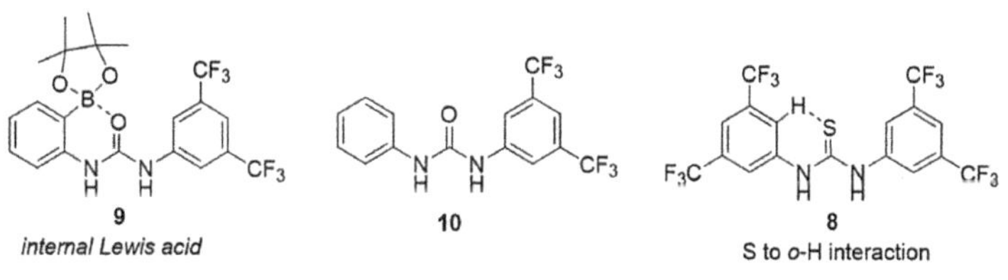

Figure 3.4 Internal Lewis acid stabilized (thio)ureas for ROP.

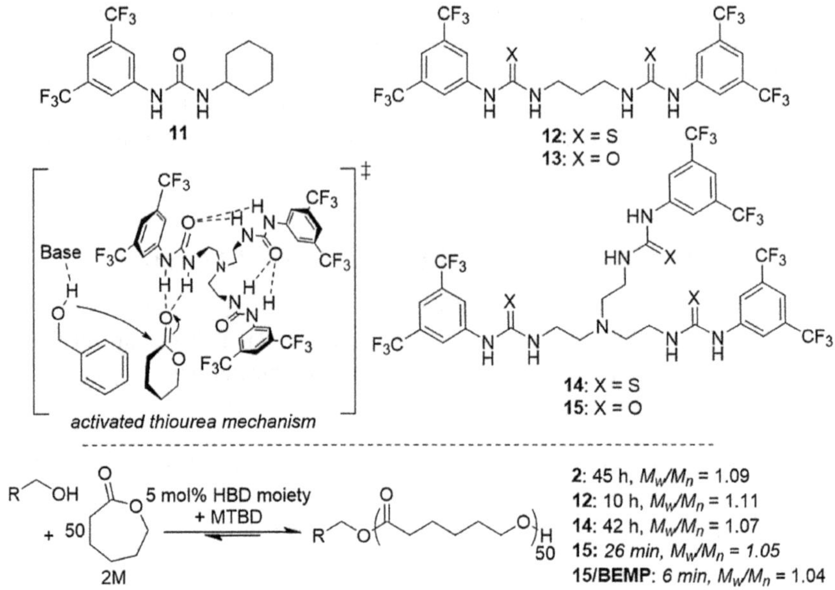

Scheme 3.5 Multi-(thio)urea H-bond donors for ROP.

the ROP; typical catalyst loadings are 5 mol% mono-(thio)urea/base; 2.5 mol% bis-donor/base; 1.67 mol% tris-donor/base.[2,109]

An *activated-(thio)urea* mechanism is proposed for multi-H-bond donor mediated ROP in non-polar solvent, but urea H-bond donors remain highly active in polar solvent. Kinetic studies on the several systems in benzene-d_6 reveal the (thio)urea ROPs to be first order in monomer, initiator, and cocatalysts, suggesting one mono-/bis-/tris-H-bond donor acting at one monomer in the transition state.[50,55,56,109] H-bonds are electrostatic in nature and have low directionality,[110] which allows for the possibility of multi-(thio)ureas directly activating monomer in a multi-activation mechanism. Computational models suggest that tristhiourea **14** is C_3 symmetric (all H-bonded),[109] and an analogue of **15** with *n*-propyl (*vs.* ethyl) linking arms is highly inactive for ROP,[111] suggesting that the (thio)urea moieties prefer to bind to themselves. These experiments, along with computational studies, suggest an *activated-(thio)urea* mechanism is operative in non-polar solvent.[109] Traditional H-bonding catalysts (*e.g.* **2**/base) become very inactive in polar solvent, which limits their utility.[3] The urea HBDs, however, remain highly active in polar solvents (*e.g.* acetone and THF).[109,112] Recent, and still-evolving, studies suggest that a different mechanism involving urea anions is operative in polar solvent.[60–62]

3.3.3 Urea and Thiourea Anions

The deprotonation of urea or thiourea with strong bases (alkoxides or metal hydrides) has been shown to produce the corresponding urea anion or thiourea anion (also: imidate or thioimidate), which are incredibly active for the ROP of lactones.[61,62] An active catalyst system generated by the treatment of urea **17** with an alkoxide base in THF results in the extremely active ROP of L-lactide at room temperature, Scheme 3.6.[61,62] The same ROP with KOMe alone slowed almost 200 times while broadening M_w/M_n (2.22 *vs.* 1.06), and the **17**/KOMe cocatalyst system is ∼25 times more active than the thiourea anion motif.[61,62] Polymerizations with VL and CL were also completed within seconds.[61] An ROP with similar activity can be achieved by a urea (*e.g.* **16**) plus strong organic base (*e.g.* MTBD, DBU, BEMP) cocatalyzed ROP.[112] Urea plus organic base cocatalyzed ROP *may* be more controlled, especially post polymerization.[112] The rates of the two methods appear to be very similar and mark a departure from early H-bond mediated ROP: seconds instead of hours or days! Remarkably, the ROPs remain highly controlled.

The urea/base cocatalyst systems operate by a different mechanism than classic H-bond mediated ROP. For the (thio)urea/alkali base cocatalyzed ROP, the proton transfer to form the 'hyperactive' (thio)imidate is largely irreversible. Hence, more acidic (thio)ureas are thought to generate more basic (thio)imidates, resulting in faster catalysis. Indeed, there is a negative linear correlation between $\ln(k_p)$ against number of CF_3 substituents,[61,112] and Schreiner *et al.* reported a linear reduction in pK_a with number of CF_3 substituents on the diaryl ureas and thioureas in DMSO.[105,113] This mechanism is reminiscent of a bifunctional TBD-mediated ROP of lactones,[24,61] where the imidate can serve as both H-bond donor and acceptor. This same mechanism is believed to be operative for bis- and tris-urea H-bond donors in polar solvent as well.[50,55,109,112]

An antibacterial compound, triclocarban (TCC, Scheme 3.6), was shown to be a very effective H-bond donating catalyst for the ROP of lactones when used with organic base cocatalysts.[112] It was proposed that this compound effects ROP through the same mechanism as other urea/strong base mediated polymerizations, and TCC/BEMP displays the same approximate rate and control behavior as trisurea (**15**)/BEMP, although the trisurea is more active ($k_{15}/k_{TCC} \approx 4$, VL).[109,112] We anticipate that the movement towards readily available reagents will prompt wider adoption of organocatalysts and facilitate new applications; the success of TBD may be due, at least in part, to its commercial availability. To demonstrate this point, TCC/base cocatalyzed ROP was applied to the solvent-free polymerization of several lactones, which was previously limited due to (1) the presumed inactivity of urea HBDs in polar (monomer) solvent, and (2) the large amounts of catalyst required for neat conditions.[60] Solvent-free ROP catalyzed by TCC/base allowed for the one-pot synthesis of di- and tri-block copolymers, and TCC/alkylamines were effective for the solvent-free ROP of LA at 100 °C,[60] a longstanding challenge.[114] The reactions remained highly controlled and 'living' in nature despite solidifying prior to full conversion.

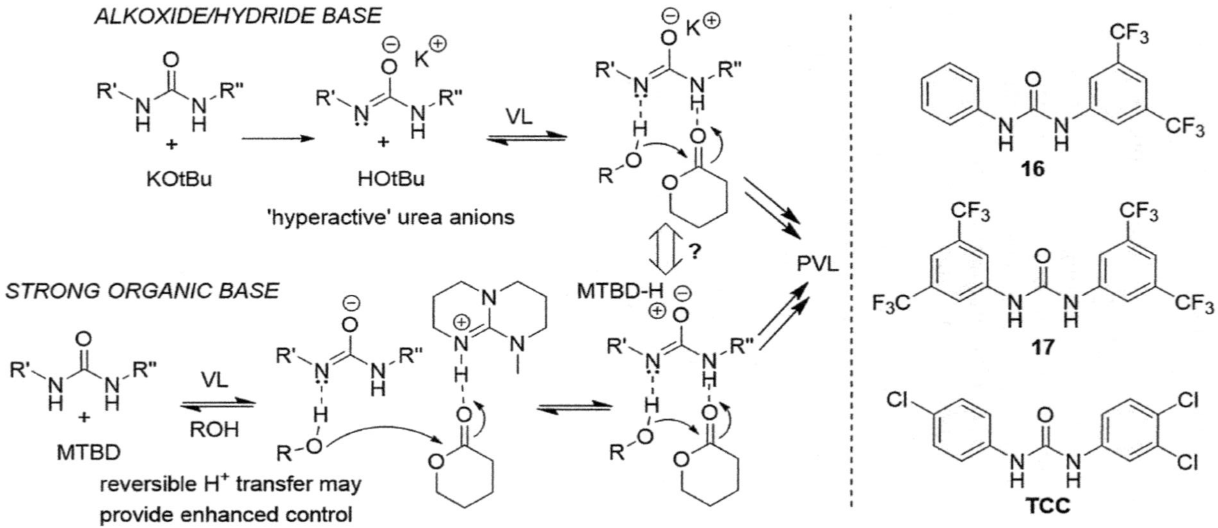

Scheme 3.6 Urea anion mediated ROP.

3.4 Non-(thio)urea Lewis Acid/Base Catalysis

3.4.1 Sulfonamides, Phosphoric and Phosphoramide H-bond Donor/Acceptors

A selection of mono- and bis-sulfonamide HBDs, which have been applied with base cocatalysts for the ROP of LA, are shown in Figure 3.5. The **18**/DMAP cocatalysts produced the most rapid ROP of LA of the HBDs examined, and it was well controlled.[115] Structurally similar catalysts, **19** and **20**, were less active, and no monosulfonamide/base cocatalyzed ROPs of LA have been shown to reach full conversion in 24 h. Neither mono- nor bis-sulfonamides promoted the ring opening of LA in the absence of an amine cocatalyst. For the monosulfonamides, it was suggested that low catalyst activity might arise from reduced H-bond donation *vs.* the bis donors.[115] This account is consistent with observations for the mono-, bis- and tris-(thio)urea H-bond donors.[109]

Phosphoric and phosphoramidic acids, the weak acidity of which contrasts with strong acids used for electrophilic monomer activated ROP,[14] can act as bifunctional organocatalysts for ROP.[116–121] Diphenyl phosphate (**21**), phosphoramidic (**22**) and imidodiphosphoric (**23**) acids were used for the ROP of cyclic esters and carbonates, Figure 3.6. Catalysts **21** and **22** were found to be active towards the ROP of CL, yielding conversion to polymer in 5.5 and 1.5 h, respectively.[116] Catalyst **23** is also active for the ROP of VL, CL and TMC monomers, albeit sluggish.[118–120] The reactions are well controlled ($M_w/M_n < 1.2$). Binding studies between catalyst and monomer or benzyl alcohol (initiator) suggest H-bonding, which have previously been observed with these catalyst motifs (*e.g.* P=O and P–NH).[122] Computational studies on **21** and **22** indicate the possibility of bifunctional activation.[116] Solvent screens performed on **22** and **23** (ROP of TMC) show dramatic slowing of reaction rate in THF (*vs.* CH_2Cl_2 or toluene), corroborating an H-bond mediated mechanism. These systems are part of the vast underpinning of mechanistic studies that have propelled this field forward, and these systems are advantageous in their synthetic modularity and highly controlled nature. This work has roots in the methyl sulfonic acid and triflic acid

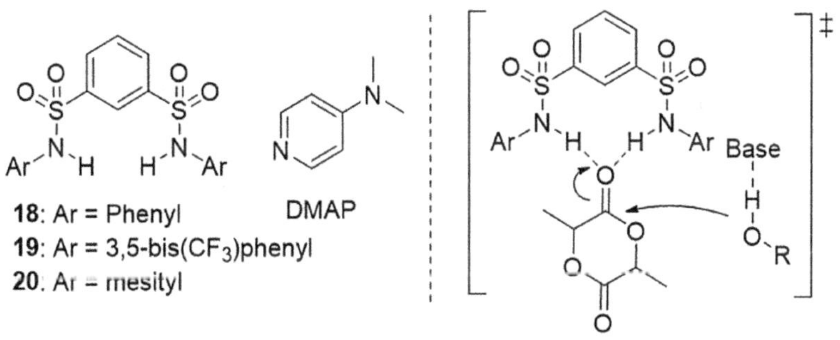

18: Ar = Phenyl
19: Ar = 3,5-bis(CF_3)phenyl
20: Ar = mesityl

Figure 3.5 Sulfonamide H-bonding catalysts.

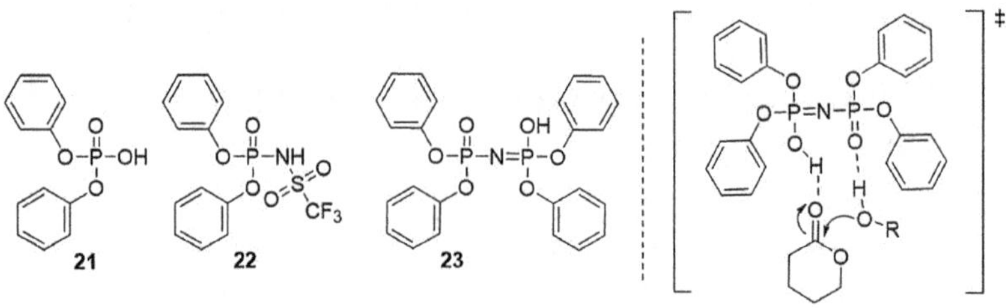

Figure 3.6 Diphenyl phosphate, phosphoramidic and imidodiphosphoric acid catalyzed ROP.

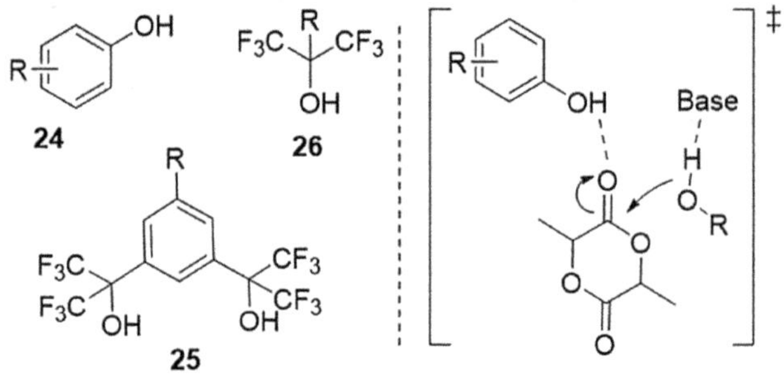

Figure 3.7 Phenol and benzylic alcohol H-bond donors for ROP.

catalyzed ROP of lactones, which have been proposed to operate through both electrophilic monomer activated and bifunctional H-bond activated mechanisms.[117]

3.4.2 Phenol and Benzyl Alcohol H-bond Donors

Considering their efficacy for the ROP of several monomers, electron deficient alcoholic H-bond donors may constitute an underdeveloped class of H-bond donating catalyst. Bibal *et al.* evaluated certain *o*-, *m*-, *p*-substituted phenols **24** for their catalytic activity towards the ROP of LA (Figure 3.7).[123] Full conversion of lactide initiated from 4-biphenylmethanol (a fluorescent alcohol) was observed in 24 h for all phenol/sparteine cocatalyst systems except for *o*- and *p*-OMe-phenol, and the fastest reaction rates were produced from phenols with electron withdrawing groups. MALDI-TOF MS indicated the presence of polymer chains initiated from phenols, an inherent liability with using alcoholic catalysts for organocatalytic ROP of esters and carbonates. Bis-donor catalysts (**24**, *o*-diphenol and *m*-diphenol; Figure 3.7) plus DBU cocatalyst are effective for the ROP of VL from 4-biphenylmethanol.[124] The electron rich diols gave high conversions while the electron poor H-bond donors had lower conversions. Strong binding between cocatalysts has been shown to be inhibitory under some circumstances.[50,54] However, Hedrick *et al.* suggested that steric bulk surrounding the catalytic alcohol would limit initiation from catalyst, producing more controlled reactions (Figure 3.7).[125] The hexafluoroalcohol (**26**, R = H) plus sparteine cocatalyzed ROP of LA initiated from benzyl alcohol resulted in full conversion of monomer in 23 h, but the bulky H-bond donor **26** (R = CF_3) showed no conversion, which may be due to its high acidity (pK_a^{DMSO} $(CF_3)_3COH = 10.7$).[126] In a rare display by H-bond mediated ROP, even β-BL was polymerized by **25** (R = methacryloyl)/sparteine to 71% conversion in 138 h.[125]

Experimental and computational data suggest this H-bond mediated ROP is mechanistically similar to those previously described. Only minimal binding between phenol and VL was observed, but this important observation reinforces early conclusions that weak binding between catalysts and monomer is not vital to catalysis.[50] Rather, a larger picture approach considering all reagent bindings, especially cocatalyst bindings, must be considered.[16,50,54] However, binding measurements on the more effective H-bond donors, **25** (R = methacryloyl) and **26** (R = Me) indicate H-bonding to VL.

Certainly, structural modulation of the established thiourea and urea scaffolds will continue to offer new catalysts—especially if mechanistic advances like the urea anions continue to appear. These changes may occur through the application of these catalysts in new roles. For examples, thioureas have recently been applied as additives in the strong acid mediated ROP of lactones. Guo *et al.* found that thioureas when added to a trifluoroacetic acid (TFA) catalyzed ROP of VL or CL increased the reaction rate by up to three times in an electrophilic monomer activation mechanism; the M_w/M_n was reduced and higher conversions were achieved than with TFA alone.[127,128] However, the drastic departures from the conventional offer a good chance for truly new and exciting developments. The azaphosphatrane (**27**) cocatalyzed (with sparteine) ROP of cyclic esters is the perfect example, Figure 3.8.[129] These structures suggest a new catalytic handle to provide monomer activation with attenuated cocatalyst binding.[129,130] Further, they are highly modular and have multiple sites available for optimization.[129]

3.4.3 Electrostatic Monomer Activation by Cations

H-bonds—a very poor name for the phenomenon—require no orbital overlap and are a type of electrostatic interaction.[110] Bibal *et al.* have demonstrated

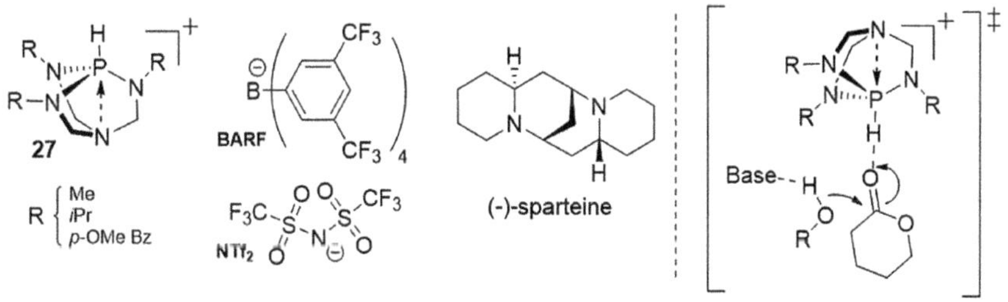

Figure 3.8 Azaphosphatrane H-bond donor.

electrostatic activation of monomer by cationic species along with base cocatalysts to effect the ROP of LA, VL and CL; both tertiary alkyl ammonium salts and alkali metal cations encapsulated in crown ethers have been successfully applied, Figure 3.9.[131] The fastest ROP rates for LA were observed with [15-c-5]Na and sparteine, where full conversion was achieved in 2 h. However, full conversions of LA and VL to polymer were achieved for all cocatalyst systems within 24 h (sparteine for LA; DBU for VL and CL). As usual, the ROP of CL was the slowest, achieving only 53% conversion in 120 h with [15-*c*-5]Na/sparteine. For the ammonium salt mediated ROPs, exchanging NTf_2 for a BARF counterion (Figure 3.9) resulted in a slight increase in reaction rate for all catalytic systems, which is likely attributed to the increased solubility of BARF *vs.* NTf_2.[131] The ammonium species do not polymerize cyclic esters in the absence of a base cocatalyst, which suggests that the native counter-anion is insufficient for alcohol activation. DFT calculations reinforce activation of monomer by the electrophilic portions of the alkylammonium (*i.e.* the methyl groups) and activation of alcohol end group by base cocatalyst, Figure 3.9.[131] Further exploration of this interesting class of catalysts may provide new reactivity and synthetic possibilities.

3.5 Brönsted Acid/Base Pairs

The accepted mechanism for the dual organocatalytic ROP of cyclic esters relies on two factors when promoting polymerization: the activation of monomer and initiator/chain end with a Lewis acid (HBD) and Lewis base (HBA), respectively. One can imagine employing a protic acid in place of a thiourea, for example, which would result in proton transfer to base cocatalyst, generating a new cocatalyst system where the activation of monomer may occur by base-H^+ and activation of chain ends may occur by conjugate base$^-$. Indeed, the previously discussed 'hyperactive' urea anions may operate by this mode when a strong organic base (*e.g.* BEMP) is employed.[60,112] Practically, catalysts of this type are employed by reacting organic bases—many of which are themselves organic catalysts for ROP—with a protic acid to form an acid/base pair. One representative pair, DBU plus benzoic acid (Figure 3.10), was derived serendipitously by incompletely quenching a DBU-catalyzed ROP of lactide.

Benzoic acid, which is widely used to quench organic catalysts by protonating amine bases,[2] forms an active ROP cocatalyst when mixed 1 : 1 with DBU.[132] Hedrick *et al.* found that a 1 : 1 ratio of DBU to benzoic acid produced well-controlled PLA ($M_w/M_n \approx 1.06$) to full conversion in 24 h. When the ratio [benzoic acid]/[DBU] increased to 1.5 and 2, the polymerization rate decreased and stopped, respectively. At lower than 1 equivalence of acid (to DBU), the reaction was faster and less controlled due to free DBU.[2,133] Molecular modeling of the acid/base pair with LA and methanol suggests a catalytic ion pair where DBU-H^+ activates monomer and the benzoate anion (BA^-) activates chain end. The acid/base pairs of DBU with HCl, acetic acid (AcOH) or *p*-toluenesulfonic acid (TsOH) were also evaluated for catalytic

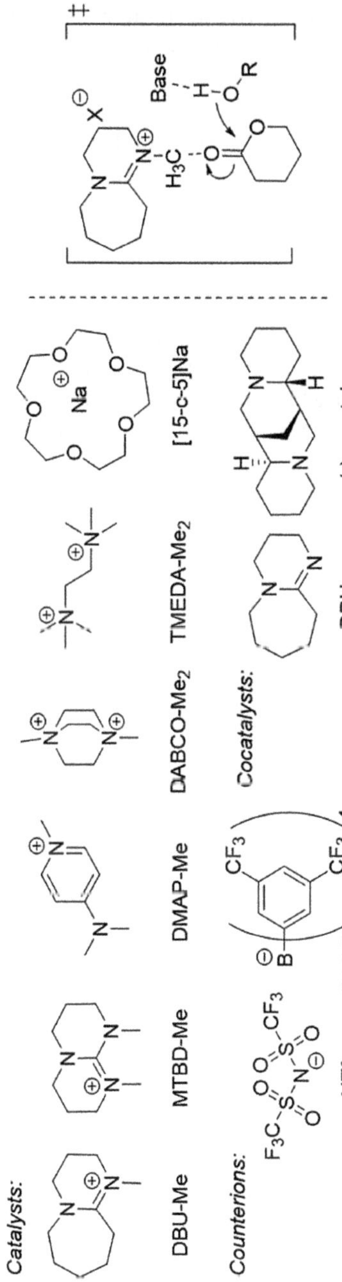

Figure 3.9 Electrophilic monomer activation by stable cations.

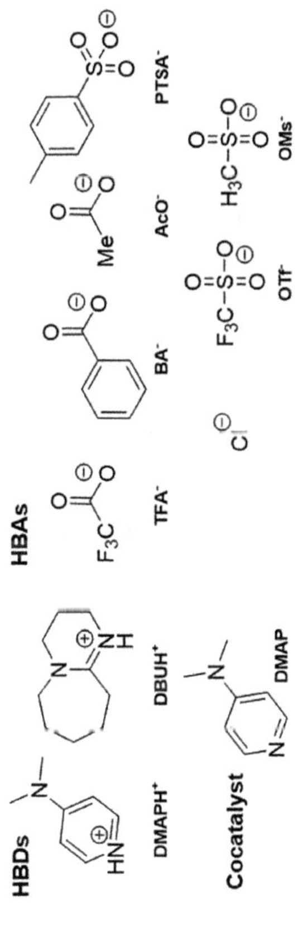

Figure 3.10 Brönsted acid and base cocatalysts for RCP.

activity. No catalytic activity was found after 48 h using HCl. However, the resonance stabilized AcO^- and TsO^- anions both were able to polymerize LA with DBU-H^+ cocatalyst, providing controlled molecular weights and narrow M_w/M_n.[132] On a superficial level, these results provide a clear rationale for using two equivalents of benzoic acid with respect to base to quench an ROP (co)mediated by organic bases.

Several conjugate acid/base pairs have also been applied for organocatalytic ROP.[134] An exemplary pair consisting of 1 eq. DMAP and 1 eq. DMAP · HX (X = Cl, MSA, TfOH) was used as a catalyst for the ROP of LA in solution, and it exhibited augmented rates *vs.* DMAP alone. The conjugate pair with triflate counterion was found to be the most active catalyst, although full conversion to polymer was not achieved in 24 h. The ideal ratio of DMAP to DMAP · HX is 1 : 1. The same group of conjugate acid/base pairs were also evaluated for the ROP of LA, VL and CL in bulk conditions at 100 °C.[135] For LA, the same trend was found in the bulk as was found in solution, with the conjugate pair DMAP/DMAP-H^+/TfO^- system having the highest rate and full conversion in 1 h. DMAP/DMAP-H^+/TfO^- was the only catalyst system effective for the ROP of VL and CL, but full conversions were not achieved within 24 h. VL and CL were not as controlled as LA, giving $M_w/M_n > 1.3$, for reactions with degree of polymerization (DP) ≈ 100. For all ROPs, side reactions that are likely to broaden M_w/M_n often occur at long reactions times. As with many acid mediated ROP, water impurities complicated mechanistic analysis. Several other advancements on this theme have been explored by applying known H-bond acceptors with acids for ROP.[136–141] Conceptually interesting, increased synthetic effort may be able to transition this scheme from concept to practice.

3.6 Supramolecular Catalysts

3.6.1 Betaines

Narrow polydispersity and high molecular weights are possible with ammonium betaine catalysts. Coulembier *et al.* demonstrated that ammonium betaines, used as bifunctional organic catalysts, H-bond with initiating/propagating alcohols at the phenoxide, Figure 3.11.[142] ROP of L-lactide was

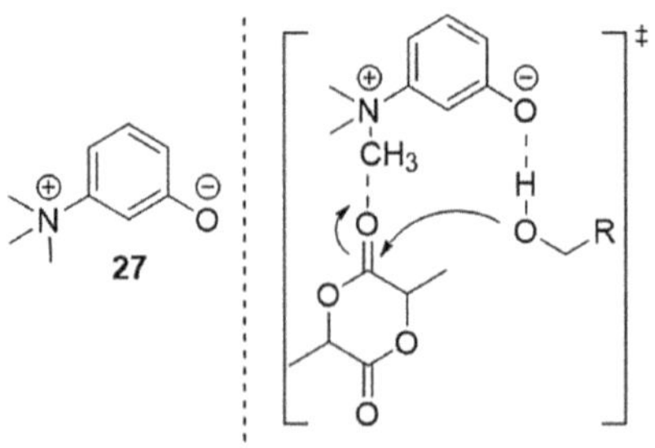

Figure 3.11 Ammonium betaine mediated ROP.

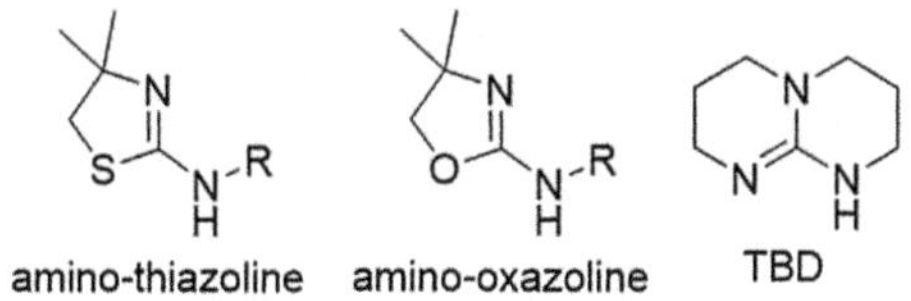

Figure 3.12 Thiazoline and oxazoline bifunctional catalysts.

performed with *m*-(trimethylammonio)phenolate betaine (**27**), producing a controlled polymerization with minimal transesterification and high isotacticity.[142,143] Faster rates are seen in chloroform *vs.* THF, which was taken to suggest that the ionic catalyst acts *via* a H-bonding mechanism.[142] Computational studies suggest that strong interactions are seen between 1-pyrenemethanol and the phenolate anion of *m*-betaine (relative to the other isomers), which is consistent with the rapid ROP with *m*-betaine *vs.* the *p*- and *o*-isomers.[142]

3.6.2 Amino-oxazoline

The structures of amino-oxazolines and thiazolines are analogous to that of TBD. An initial screening of the thiazoline catalyzed ROP of LA determined that thiazolines with electron withdrawing groups resulted in reduced ROP activity and produced atactic PLA.[144] Amino-thiazolines with electron donating alkyl groups are more active, and amino-thiazoline with cyclohexyl groups demonstrated the fastest rates for ROP of LA, Figure 3.12; however, this catalyst is much less active than the 'parent' TBD catalyst.[144] Elevated temperatures indicated little to no rate enhancement, which could arise from weaker supramolecular interactions during the enchainment transition state. ^{1}H NMR binding experiments demonstrate the more electron-deficient compounds have stronger interactions with cyclic esters and conversely have weaker interactions with initiating alcohol. These experiments corroborate the presumption that both the H-bond accepting and donating sites are necessary for effective catalysis.[144] These catalysts are notable because they are mechanistically similar to TBD but far more synthetically modular. With the rising interest in specialized catalyst architectures, these motifs may prove highly useful.

3.6.3 Cyclodextrins

Cyclodextrins (CDs) have garnered interest due to their selective inclusion properties and reactivities,[145–147] and they constitute an example of extremely mild supramolecular catalyst for ROP.[148,149] The ability of CDs to catalyze the hydrolysis of polyesters in water was thought to proceed *via* a polymer inclusion complex with CDs.[146] In the absence of water, CDs catalyze the ROP of lactone monomers.[146] Further, CDs can create selective inclusion complexes with some lactones where the size of a CD can promote or suppress the transesterification of lactones. The inclusion of lactones in the

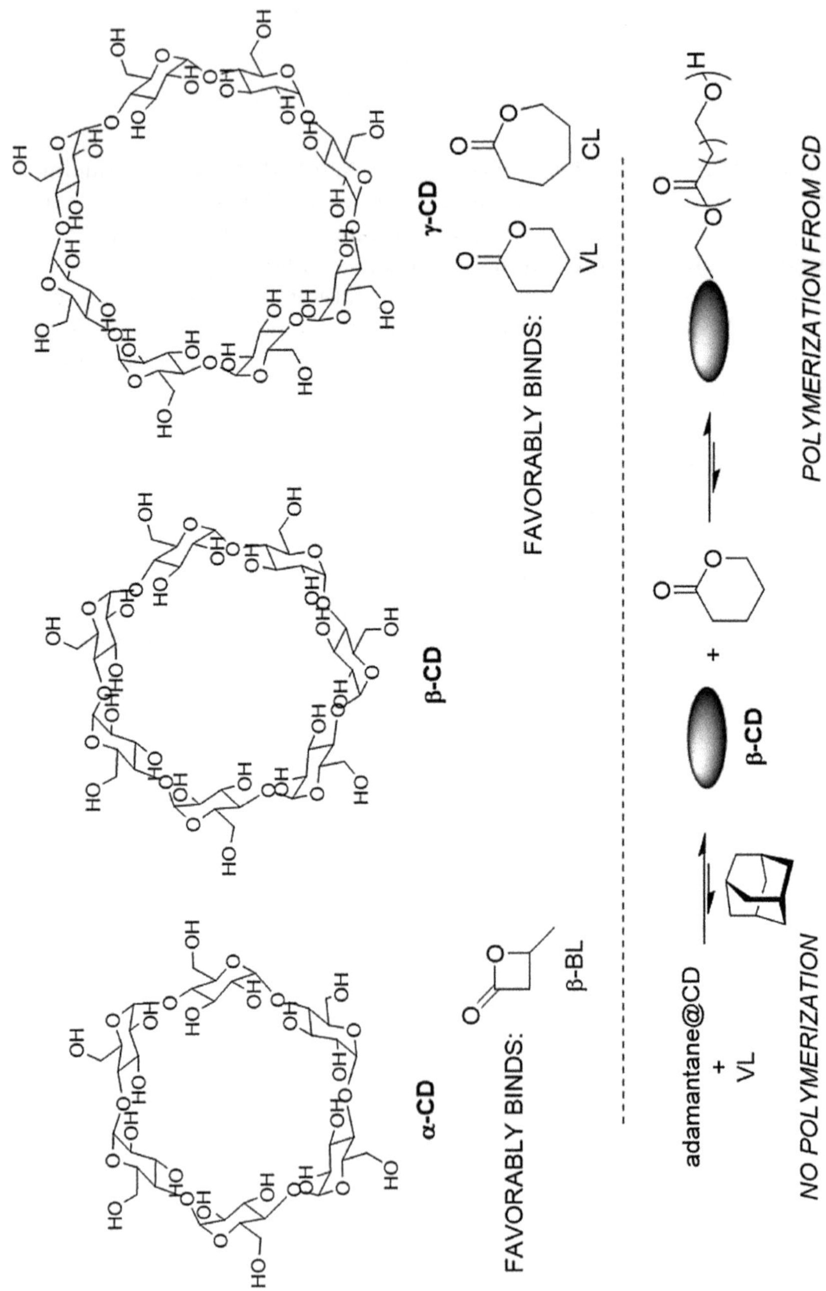

Scheme 3.7 Cyclodextrin promoted ROP of lactones.

hydrophobic CD cavity is believed to be the driving force to yield polyesters,[145] and the existence of hydrophobic, catalytic pockets has been proposed for other organocatalysts for ROP.[58,109,145] Accordingly, the ROPs catalyzed by the CD with a smaller cavity (*i.e.* α-CD and β-CD in Scheme 3.7) produce the fastest conversions of β-butyrolactone (β-BL) to polymer under solvent-free conditions at 100 °C, while β-CD is most effective for the ROP of VL and CL (Scheme 3.7).[145]

Mechanistic studies suggest that ROP is initiated from the CD and that the lactone/CD inclusion complex is vital to catalysis. When ROP is attempted using an acylated CD (no free hydroxyls), no conversion to polylactone is observed, which suggests that CDs are covalently attached to the polylactone chain end in a normal CD-catalyzed ROP.[145] Further, suppression of the ROP of VL was noted with a β-CD/adamantane inclusion complex catalyst system. The adamantane guest is strongly inserted in the β-CD cavity, which excludes VL, suggesting that lactone/CD inclusion complexes are essential for ROP.[145] The mechanistic picture that emerges suggests that, initially, a complex is formed between lactone and CD at a ratio of 1 : 1, and a hydroxyl group at the C_2-position attacks the monomer to begin enchainment. Further development of these or similar extremely mild catalysts for ROP could provide new and exciting methods of ultra-controlled ROP.

3.7 Conclusion

The narrative of this chapter can be summarized by following the circular evolution of dual catalysts away from and back towards the popular organocatalyst, TBD. When the TBD catalyzed ROP of lactones was disclosed in 2006,[24] it was the perfect storm of a successful catalyst. It is easy to use, readily available, highly active, exhibits moderate control ($M_w/M_n \approx 1.2$) and decent selectivity for monomer. While TBD was originally proposed to operate *via* a nucleophilic mechanism of enchainment, an H-bond mediated, bifunctional, mechanism was also envisaged.[24] This mechanism has been much debated,[34,150,151] and, for (thio)urea/base cocatalyzed ROP, it is becoming more clear which mechanism is operative and when.[152,153] Conceptually, a thiourea/base mediated ROP can be viewed as separating the H-bond donating and accepting roles of TBD into separate cocatalyst moieties. This approach, while highly tunable and beneficial for the reasons described above, required sacrificing reaction rate. The various efforts to increase reaction rate without sacrificing control (serendipitously?) brought the community back to an active catalyst which bears a strong structural resemblance to TBD (*i.e.* urea plus strong base mediated ROP). Far from ending up in the same place, the numerous studies that brought us ‘full circle’ have greatly enriched our understanding of how these catalysts operate and have largely mitigated the activity *vs.* selectivity problem of organocatalytic ROP, Scheme 3.8. By no means is this story complete, and as of April 2018 our mechanistic understanding of nascent urea/strong base mediated ROP is still evolving. Indeed, the broader field of organocatalytic

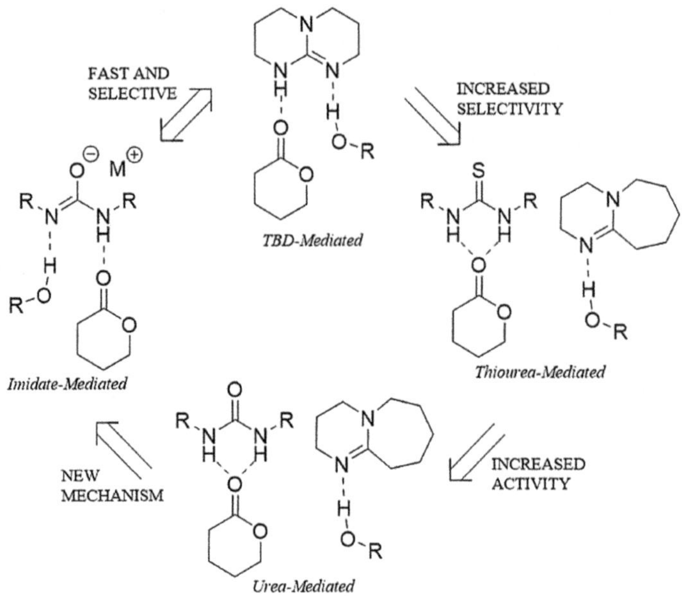

Scheme 3.8 Evolution of dual catalysts for ROP.

polymerization is a bridge between the sometimes disparate worlds of materials chemist (ease of use) and synthetic polymer chemist (mechanistic interest). We assert that the cooperative and collegial nature of our community has facilitated the synergistic evolution of *new mechanism* to *new abilities*—in monomer scope, polymer architecture and level of reaction control. We hope that this will continue.

Acknowledgements

This work has been supported by an NSF CAREER Award (CHE 1554830) and the University of Rhode Island. P.P.D would like to thank the University of Rhode Island Graduate School for a graduate fellowship.

References

1. E. V. Ansyln and D. A. Dougherty, Solutions and Non-Covalent Binding Forces; Binding Forces, in *Modern Physical Organic Chemistry*, 2006, pp. 162–194.
2. B. G. G. Lohmeijer, R. C. Pratt, F. Leibfarth, J. W. Logan, D. A. Long, A. P. Dove, F. Nederberg, J. Choi, C. Wade and R. M. Waymouth, *et al.*, Guanidine and Amidine Organocatalysts for Ring-Opening Polymerization of Cyclic Esters, *Macromolecules*, 2006, **39**(25), 8574–8583.
3. R. C. Pratt, B. G. G. Lohmeijer, D. A. Long, P. N. P. Lundberg, A. P. Dove, H. Li, C. G. Wade, R. M. Waymouth and J. L. Hedrick, Exploration, Optimization, and Application of Supramolecular Thiourea - Amine

Catalysts for the Synthesis of Lactide (Co)Polymers, *Macromolecules*, 2006, **39**(23), 7863–7871.

4. A. P. Dove, R. C. Pratt, B. G. G. Lohmeijer, R. M. Waymouth and J. L. Hedrick, Thiourea-Based Bifunctional Organocatalysis: Supramolecular Recognition for Living Polymerization, *J. Am. Chem. Soc.*, 2005, **127**(40), 13798–13799.
5. G. Bredig and P. S. Fiske, Durch Katalysatoren bewirkte asymmetrische Synthese, *Biochem. Z.*, 1912, **46**, 7–23.
6. P. J. Flory, A Comparison of Esterification and Ester Interchange Kinetics, *J. Am. Chem. Soc.*, 1940, **62**(9), 2261–2264.
7. A. G. Doyle and E. N. Jacobsen, Small-Molecule H-Bond Donors in Asymmetric Catalysis, *Chem. Rev.*, 2007, **107**(12), 5713–5743.
8. T. Okino, Y. Hoashi and Y. Takemoto, Enantioselective Michael Reaction of Malonates to Nitroolefins Catalyzed by Bifunctional Organocatalysts, *J. Am. Chem. Soc.*, 2003, **125**(42), 12672–12673.
9. For reasons of being accessible to the novice, we prefer to use directly M_w/M_n ($= D_m = Đ =$ PDI, polydispersity index).
10. R. F. T. Stepto, Dispersity in Polymer Science (IUPAC Recommendations 2009), *Pure Appl. Chem.*, 2009, **81**, 351–353.
11. G. Odian, Ring-Opening Polymerization, in *Principles of Polymerization*, New York, 1991, pp. 532–603.
12. S. Penczek, Terminology of Kinetics, Thermodynamics, and Mechanisms of Polymerization, *J. Polym. Sci., Part A: Polym. Chem.*, 2002, **40**(11), 1665–1676.
13. M. K. Kiesewetter, E. J. Shin, J. L. Hedrick and R. M. Waymouth, Organocatalysis: Opportunities and Challenges for Polymer Synthesis, *Macromolecules*, 2010, **43**(5), 2093–2107.
14. N. E. Kamber, W. Jeong, R. M. Waymouth, R. C. Pratt, B. G. G. Lohmeijer and J. L. Hedrick, Organocatalytic Ring-Opening Polymerization, *Chem. Rev.*, 2007, **107**(12), 5813–5840.
15. A. P. Dove, Metal-Free Catalysis in Ring-Opening Polymerization, in *Handbook of Ring-opening Polymerization*, ed. P. Dubois, O. Coulembier, and J.-M. Raquez, Wiley-VCH, Federal Republic of Germany, 2009, pp. 357–378.
16. C. Thomas and B. Bibal, Hydrogen-Bonding Organocatalysts for Ring-Opening Polymerization, *Green Chem.*, 2014, **16**(4), 1687–1699.
17. A. P. Dove, Organic Catalysis for Ring-Opening Polymerization, *ACS Macro Lett.*, 2012, **1**(12), 1409–1412.
18. X. Zhang, M. Fevre, G. O. Jones and R. M. Waymouth, Catalysis as an Enabling Science for Sustainable Polymers, *Chem. Rev.*, 2017, **118**, 839–885.
19. J. A. Byers, A. B. Biernesser, K. R. Delle Chiaie, A. Kaur, and J. A. Kehl, Catalytic Systems for the Production of Poly(lactic Acid), in *Synthesis, Structure and Properties of Poly(lactic Acid)*, ed. M. L. Di Lorenzo and R. Androsch, Springer Nature, Cham, Switzerland, 2018, pp. 67–118.

20. R. B. Grubbs and R. H. Grubbs, 50th Anniversary Perspective: Living Polymerization – Emphasizing the Molecule in Macromolecules, *Macromolecules*, 2017, **50**(18), 6979–6997.
21. S. Hu, J. Zhao, G. Zhang and H. Schlaad, Macromolecular Architectures through Organocatalysis, *Prog. Polym. Sci.*, 2017, **74**, 34–77.
22. Q. L. Song, S. Y. Hu, J. P. Zhao and G. Z. Zhang, Organocatalytic Copolymerization of Mixed Type Monomers, *Chin. J. Polym. Sci.*, 2017, **35**, 581–601.
23. W. Jeong, E. J. Shin, D. A. Culkin, J. L. Hedrick and R. M. Waymouth, Zwitterionic Polymerization: A Kinetic Strategy for the Controlled Synthesis of Cyclic Polylactide, *J. Am. Chem. Soc.*, 2009, **131**(13), 4884–4891.
24. R. C. Pratt, B. G. G. Lohmeijer, D. A. Long, R. M. Waymouth and J. L. Hedrick, Triazabicyclodecene: A Simple Bifunctional Organocatalyst for Acyl Transfer and Ring-Opening Polymerization of Cyclic Esters, *J. Am. Chem. Soc.*, 2006, **128**(14), 4556–4557.
25. L. Zhang, F. Nederberg, R. C. Pratt, R. M. Waymouth, J. L. Hedrick and C. G. Wade, Phosphazene Bases: A New Category of Organocatalysts for the Living Ring-Opening Polymerization of Cyclic Esters, *Macromolecules*, 2007, **40**, 4154–4158.
26. F. Nederberg, E. F. Connor, T. Glausser and J. L. Hedrick, Organocatalytic Chain Scission of Poly(lactides): A General Route to Controlled Molecular Weight, Functionality and Macromolecular Architecture, *Chem. Commun.*, 2001, **20**, 2066–2067.
27. F. Nederberg, E. F. Connor, M. Möller, T. Glauser and J. L. Hedrick, New Paradigms for Organic Catalysts: The First Organocatalytic Living Polymerization, *Angew. Chem., Int. Ed. Engl.*, 2001, **40**, 2712–2715.
28. H. R. Kricheldorf, N. Lomadze and G. Schwarz, Cyclic Polylactides by Imidazole-Catalyzed Polymerization of L-Lactide, *Macromolecules*, 2008, **41**(21), 7812–7816.
29. H. R. Kricheldorf, C. Von Lossow and G. Schwarz, Tertiary Amine Catalyzed Polymerizations of Amino Acid N-carboxyanhydrides: The Role of Cyclization, *J. Polym. Sci., Part A: Polym. Chem.*, 2006, **44**(15), 4680–4695.
30. O. Coulembier and P. Dubois, 4-Dimethylaminopyridine-Based Organoactivation: From Simple Esterification to Lactide Ring-Opening "Living" polymerization, *J. Polym. Sci., Part A: Polym. Chem.*, 2012, **50**(9), 1672–1680.
31. C. Bonduelle, B. Martín-Vaca, F. P. Cossío and D. Bourissou, Monomer versus Alcohol Activation in the 4-Dimethylaminopyridine-Catalyzed Ring-Opening Polymerization of Lactide and Lactic O-Carboxylic Anhydride, *Chem. – Eur. J.*, 2008, **14**(17), 5304–5312.
32. A. Chuma, H. W. Horn, W. C. Swope, R. C. Pratt, L. Zhang, B. G. G. Lohmeijer, C. G. Wade, R. M. Waymouth, J. L. Hedrick and J. E. Rice, The Reaction Mechanism for the Organocatalytic Ring-Opening Polymerization of L-Lactide Using a Guanidine-Based

Catalyst: Hydrogen-Bonded or Covalently Bound?, *J. Am. Chem. Soc.*, 2008, **130**(21), 6749–6754.

33. C. L. Lai, H. M. Lee and C. H. Hu, Theoretical Study on the Mechanism of N-Heterocyclic Carbene Catalyzed Transesterification Reactions, *Tetrahedron Lett.*, 2005, **46**(37), 6265–6270.
34. L. Simón and J. M. Goodman, The Mechanism of TBD-Catalyzed Ring-Opening Polymerization of Cyclic Esters, *J. Org. Chem.*, 2007, **72**(25), 9656–9662.
35. R. Schwesinger and H. Schlemper, Peralkylated Polyaminophosphazenes—Extremely Strong, Neutral Nitrogen Bases, *Angew. Chem., Int. Ed. Engl.*, 1987, **26**(11), 1167–1169.
36. A. Molenberg and M. Möller, A Fast Catalyst System for the Ring-Opening Polymerization of Cyclosiloxanes, *Macromol. Rapid Commun.*, 1995, **16**(6), 449–453.
37. I. Kaljurand, A. Kütt, L. Sooväli, T. Rodima, V. Mäemets, I. Leito and I. A. Koppel, Extension of the Self-Consistent Spectrophotometric Basicity Scale in Acetonitrile to a Full Span of 28 pKa Units: Unification of Different Basicity Scales, *J. Org. Chem.*, 2005, **70**(3), 1019–1028.
38. H. Schlaad, H. Kukula, J. Rudloff and I. Below, Synthesis of Hetero-bifunctional Poly(ethylene Glycol)s by Metal-Free Anionic Ring-Opening Polymerization, *Macromolecules*, 2001, **34**(13), 4302–4304.
39. M. S. Sigman, P. Vachal and E. N. Jacobsen, A General Catalyst for the Asymmetric Strecker Reaction, *Angew. Chem., Int. Ed.*, 2000, **39**(7), 1279–1281.
40. P. Vachal and E. N. Jacobsen, Enantioselective Catalytic Addition of HCN to Ketoimines. Catalytic Synthesis of Quaternary Amino Acids, *Org. Lett.*, 2000, **2**(6), 867–870.
41. P. Vachal and E. N. Jacobsen, Structure-Based Analysis and Optimization of a Highly Enantioselective Catalyst for the Strecker Reaction, *J. Am. Chem. Soc.*, 2002, **124**(34), 10012–10014.
42. A. G. Wenzel and E. N. Jacobsen, Asymmetric Catalytic Mannich Reactions Catalyzed by Urea Derivatives: Enantioselective Synthesis of β-Aryl-β-Amino Acids, *J. Am. Chem. Soc.*, 2002, **124**(44), 12964–12965.
43. M. S. Taylor and E. N. Jacobsen, Highly Enantioselective Catalytic Acyl-Pictet-Spengler Reactions, *J. Am. Chem. Soc.*, 2004, **126**(34), 10558–10559.
44. T. P. Yoon and E. N. Jacobsen, Highly Enantioselective Thiourea-Catalyzed Nitro-Mannich Reactions, *Angew. Chem., Int. Ed.*, 2005, **44**(3), 466–468.
45. G. D. Joly and E. N. Jacobsen, Thiourea-Catalyzed Enantioselective Hydrophosphonylation of Imines: Practical Access to Enantiomerically Enriched α-Amino Phosphonic Acids, *J. Am. Chem. Soc.*, 2004, **126**(13), 4102–4103.
46. D. E. Fuerst and E. N. Jacobsen, Thiourea-Catalyzed Enantioselective Cyanosilylation of Ketones, *J. Am. Chem. Soc.*, 2005, **127**(25), 8964–8965.

47. T. Okino, S. Nakamura, T. Furukawa and Y. Takemoto, Enantioselective Aza-Henry Reaction Catalyzed by a Bifunctional Organocatalyst, *Org. Lett.*, 2004, **6**(4), 625–627.
48. Y. Hoashi, T. Okino and Y. Takemoto, Enantioselective Michael Addition to Unsaturated Imides Catalyzed by a Bifunctional Organocatalyst, *Angew. Chem., Int. Ed.*, 2005, **44**(26), 4032–4035.
49. D. J. Coady, A. C. Engler, H. W. Horn, K. M. Bajjuri, K. Fukushima, G. O. Jones, A. Nelson, J. E. Rice and J. L. Hedrick, Catalyst Chelation Effects in Organocatalyzed Ring-Opening Polymerization of Lactide, *ACS Macro Lett.*, 2012, **1**(1), 19–22.
50. O. I. Kazakov, P. P. Datta, M. Isajani, E. T. Kiesewetter and M. K. Kiesewetter, Cooperative Hydrogen-Bond Pairing in Organocatalytic Ring-Opening Polymerization, *Macromolecules*, 2014, **47**(21), 7463–7468.
51. W. Jeong, J. L. Hedrick and R. M. Waymouth, Organic Spirocyclic Initiators for the Ring-Expansion Polymerization of β-Lactones, *J. Am. Chem. Soc.*, 2007, **129**(27), 8414–8415.
52. A. Duda and A. Kowalski, Thermodynamics and Kinetics of Ring-Opening Polymerization, in *Handbook of Ring-opening Polymerization*, ed. P. Dubois, O. Coulembier, and J.-M. Raquez, Wiley-VCH Verlag GmbH & Co. KGaA, 2009, pp. 1–52.
53. P. Olsén, K. Odelius and A. C. Albertsson, Thermodynamic Presynthetic Considerations for Ring-Opening Polymerization, *Biomacromolecules*, 2016, **17**(3), 699–709.
54. S. Koeller, J. Kadota, F. Peruch, A. Deffieux, N. Pinaud, I. Pianet, S. Massip, J.-M. Léger, J.-P. Desvergne and B. Bibal, (Thio)Amidoindoles and (Thio)amidobenzimidazoles: An Investigation of Their Hydrogen-Bonding and Organocatalytic Properties in the Ring-Opening Polymerization of Lactide, *Chem. – Eur. J.*, 2010, **16**(14), 4196–4205.
55. S. S. Spink, O. I. Kazakov, E. T. Kiesewetter and M. K. Kiesewetter, Rate Accelerated Organocatalytic Ring-Opening Polymerization of L-Lactide via the Application of a Bis(thiourea) H-Bond Donating Cocatalyst, *Macromolecules*, 2015, **48**(17), 6127–6131.
56. O. I. Kazakov and M. K. Kiesewetter, Cocatalyst Binding Effects in Organocatalytic Ring-Opening Polymerization of L-Lactide, *Macromolecules*, 2015, **48**(17), 6121–6126.
57. C. Thomas, F. Peruch and B. Bibal, Ring-Opening Polymerization of Lactones Using Supramolecular Organocatalysts under Simple Conditions, *RSC Adv.*, 2012, **2**(33), 12851–12856.
58. P. P. Datta, J. U. Pothupitiya, E. T. Kiesewetter and M. K. Kiesewetter, Coupled Equilibria in H-Bond Donating Ring-Opening Polymerization: The Effective Catalyst-Determined Shift of a Polymerization Equilibrium, *Eur. Polym. J.*, 2017, **95**, 671–677.
59. K. M. Lippert, K. Hof, D. Gerbig, D. Ley, H. Hausmann, S. Guenther and P. R. Schreiner, Hydrogen-Bonding Thiourea Organocatalysts: The Privileged 3,5-Bis(trifluoromethyl)phenyl Group, *Eur. J. Org. Chem.*, 2012, 5919–5927.

60. J. U. Pothupitiya, N. U. Dharmaratne, T. M. M. Jouaneh, K. V. Fastnacht, D. N. Coderre and M. K. Kiesewetter, H-Bonding Organocatalysts for the Living, Solvent-Free Ring-Opening Polymerization of Lactones: Toward an All-Lactones, All-Conditions Approach, *Macromolecules*, 2017, **50**, 8948–8954.
61. B. Lin and R. M. Waymouth, Urea Anions: Simple, Fast, and Selective Catalysts for Ring-Opening Polymerizations, *J. Am. Chem. Soc.*, 2017, **139**, 1645–1652.
62. X. Zhang, G. O. Jones, J. L. Hedrick and R. M. Waymouth, Fast and Selective Ring-Opening Polymerizations by Alkoxides and Thioureas, *Nat. Chem.*, 2016, **8**, 1047–1053.
63. S. Koeller, C. Thomas, F. Peruch, A. Deffieux, S. Massip, J. M. Léger, J. P. Desvergne, A. Milet and B. Bibal, α-Halogenoacetanilides as Hydrogen-Bonding Organocatalysts That Activate Carbonyl Bonds: Fluorine versus Chlorine and Bromine, *Chemistry*, 2014, **20**(10), 2849–2859.
64. S. Koeller, J. Kadota, A. Deffieux, F. Peruch, S. Massip, J. M. Léger, J. P. Desvergne and B. Bibal, Ring-Opening Polymerization of L-Lactide Efficiently Triggered by an Amido-Indole. X-Ray Structure of a Complex between L-Lactide and the Hydrogen-Bonding Organocatalyst, *J. Am. Chem. Soc.*, 2009, **131**(42), 15088–15089.
65. T. R. Blake and R. M. Waymouth, Organocatalytic Ring-Opening Polymerization of Morpholinones: New Strategies to Functionalized Polyesters, *J. Am. Chem. Soc.*, 2014, **136**(26), 9252–9255.
66. A. C. Engler, J. M. W. Chan, D. J. Coady, J. M. O'Brien, H. Sardon, A. Nelson, D. P. Sanders, Y. Y. Yang and J. L. Hedrick, Accessing New Materials through Polymerization and Modification of a Polycarbonate with a Pendant Activated Ester, *Macromolecules*, 2013, **46**(4), 1283–1290.
67. Y. A. Chang, A. E. Rudenko and R. M. Waymouth, Zwitterionic Ring-Opening Polymerization of N-Substituted Eight-Membered Cyclic Carbonates to Generate Cyclic Poly(carbonate)s, *ACS Macro Lett.*, 2016, **5**(10), 1162–1166.
68. S. Venkataraman, V. W. L. Ng, D. J. Coady, H. W. Horn, G. O. Jones, T. S. Fung, H. Sardon, R. M. Waymouth, J. L. Hedrick and Y. Y. Yang, A Simple and Facile Approach to Aliphatic N-Substituted Functional Eight-Membered Cyclic Carbonates and Their Organocatalytic Polymerization, *J. Am. Chem. Soc.*, 2015, **137**(43), 13851–13860.
69. C. J. McKinlay, R. M. Waymouth and P. A. Wender, Cell-Penetrating, Guanidinium-Rich Oligophosphoesters: Effective and Versatile Molecular Transporters for Drug and Probe Delivery, *J. Am. Chem. Soc.*, 2016, **138**(10), 3510–3517.
70. D. P. Sanders, K. Fukushima, D. J. Coady, A. Nelson, M. Fujiwara, M. Yasumoto and J. L. Hedrick, A Simple and Efficient Synthesis of Functionalized Cyclic Carbonate Monomers Using a Versatile Pentafluorophenyl Ester Intermediate, *J. Am. Chem. Soc.*, 2010, **132**(42), 14724–14726.

71. T. M. Ovitt and G. W. Coates, Stereochemistry of Lactide Polymerization with Chiral Catalysts: New Opportunities for Stereocontrol Using Polymer Exchange Mechanisms, *J. Am. Chem. Soc.*, 2002, **124**(7), 1316–1326.
72. B. M. Chamberlain, M. Cheng, D. R. Moore, T. M. Ovitt, E. B. Lobkovsky and G. W. Coates, Polymerization of Lactide with Zinc and Magnesium Diiminate Complexes: Stereocontrol and Mechanism, *J. Am. Chem. Soc.*, 2001, **123**(14), 3229–3238.
73. C. Bakewell, T. P. A. Cao, N. Long, X. F. Le Goff, A. Auffrant and C. K. Williams, Yttrium Phosphasalen Initiators for Rac -Lactide Polymerization: Excellent Rates and High Iso-Selectivities, *J. Am. Chem. Soc.*, 2012, **134**(51), 20577–20580.
74. M. T. Zell, B. E. Padden, A. J. Paterick, K. A. M. Thakur, R. T. Kean, M. A. Hillmyer and E. J. Munson, Unambiguous Determination of the ^{13}C and ^{1}H NMR Stereosequence Assignments of Polylactide Using High-Resolution Solution NMR Spectroscopy, *Macromolecules*, 2002, **35**, 7700–7707.
75. K. A. M. Thakur, R. T. Kean, E. S. Hall, J. J. Kolstad, T. A. Lindgren, M. A. Doscotch, J. I. Siepmann and E. J. Munson, High-Resolution ^{13}C and ^{1}H Solution NMR Study of Poly(lactide), *Macromolecules*, 1997, **30**(8), 2422–2428.
76. R. Todd, G. Rubio, D. Hall, S. Tempelaar and A. P. Dove, Benzyl Bispidine as an Efficient Replacement for (−)-Sparteine in Ring Opening Polymerisation, *Chem. Sci.*, 2013, **4**, 1092–1097.
77. Z. Dai, Y. Cui, C. Chen and J. Wu, Photoswitchable Ring-Opening Polymerization of Lactide Catalyzed by Azobenzene-Based Thiourea, *Chem. Commun.*, 2016, **52**(57), 8826–8829.
78. S. M. Guillaume, E. Kirillov, Y. Sarazin and J. F. Carpentier, Beyond Stereoselectivity, Switchable Catalysis: Some of the Last Frontier Challenges in Ring-Opening Polymerization of Cyclic Esters, *Chem. – Eur. J.*, 2015, **21**(22), 7988–8003.
79. R. Cacciapaglia, S. Di Stefano and L. Mandolini, The Bis-Barium Complex of a Butterfly Crown Ether as a Phototunable Supramolecular Catalyst, *J. Am. Chem. Soc.*, 2003, **125**(8), 2224–2227.
80. J. Beswick, V. Blanco, G. De Bo, D. A. Leigh, U. Lewandowska, B. Lewandowski and K. Mishiro, Selecting Reactions and Reactants Using a Switchable Rotaxane Organocatalyst with Two Different Active Sites, *Chem. Sci.*, 2015, **6**(1), 140–143.
81. J. Wang and B. L. Feringa, Dynamic Control of Chiral Space in a Catalytic Asymmetric Reaction Using a Molecular Motor, *Science*, 2011, **331**, 1429–1432.
82. D. Sud, T. B. Norsten and N. R. Branda, Photoswitching of Stereoselectivity in Catalysis Using a Copper Dithienylethene Complex, *Angew. Chem., Int. Ed.*, 2005, **44**(13), 2019–2021.
83. B. M. Neilson and C. W. Bielawski, Photoswitchable Organocatalysis: Using Light To Modulate the Catalytic Activities of

N-Heterocyclic Carbenes, *J. Am. Chem. Soc.*, 2012, **134**(30), 12693–12699.

84. M. Samanta, V. Siva Rama Krishna and S. Bandyopadhyay, A Photoresponsive Glycosidase Mimic, *Chem. Commun.*, 2014, **50**(73), 10577–10579.
85. T. Imahori, R. Yamaguchi and S. Kurihara, Azobenzene-Tethered Bis(Trityl Alcohol) as a Photoswitchable Cooperative Acid Catalyst for Morita–Baylis–Hillman Reactions, *Chem. - Eur. J.*, 2012, **18**(35), 10802–10807.
86. C. K. A. Gregson, V. C. Gibson, N. J. Long, E. L. Marshall, P. J. Oxford and A. J. P. White, Redox Control within Single-Site Polymerization Catalysts, *J. Am. Chem. Soc.*, 2006, **128**(23), 7410–7411.
87. E. M. Broderick, N. Guo, T. Wu, C. S. Vogel, C. Xu, J. Sutter, J. T. Miller, K. Meyer, T. Cantat and P. L. Diaconescu, Redox Control of a Polymerization Catalyst by Changing the Oxidation State of the Metal Center, *Chem. Commun.*, 2011, **47**(35), 9897–9899.
88. E. M. Broderick, P. S. Thuy-Boun, N. Guo, C. S. Vogel, J. Sutter, J. T. Miller, K. Meyer and P. L. Diaconescu, Synthesis and Characterization of Cerium and Yttrium Alkoxide Complexes Supported by Ferrocene-Based Chelating Ligands, *Inorg. Chem.*, 2011, **50**(7), 2870–2877.
89. H. J. Yoon, J. Kuwabara, J.-H. Kim and C. A. Mirkin, Allosteric Supramolecular Triple-Layer Catalysts, *Science*, 2010, **330**, 66–69.
90. E. M. Broderick, N. Guo, C. S. Vogel, C. Xu, J. Sutter, J. T. Miller, K. Meyer, P. Mehrkhodavandi and P. L. Diaconescu, Redox Control of a Ring-Opening Polymerization Catalyst, *J. Am. Chem. Soc.*, 2011, **133**(24), 9278–9281.
91. C. Romain and C. K. Williams, Chemoselective Polymerization Control: From Mixed-Monomer Feedstock to Copolymers, *Angew. Chem., Int. Ed.*, 2014, **53**(6), 1607–1610.
92. Y. Sohtome, S. Tanaka, K. Takada, T. Yamaguchi and K. Nagasawa, Solvent-Dependent Enantiodivergent Mannich-Type Reaction: Utilizing a Conformationally Flexible Guanidine/Bisthiourea Organocatalyst, *Angew. Chem., Int. Ed.*, 2010, **49**(48), 9254–9257.
93. A. Piermattei, S. Karthikeyan and R. P. Sijbesma, Activating Catalysts with Mechanical Force, *Nat. Chem.*, 2009, **1**, 133–137.
94. O. Coulembier, S. Moins, R. Todd and P. Dubois, External and Reversible CO_2 Regulation of Ring-Opening Polymerizations Based on a Primary Alcohol Propagating Species, *Macromolecules*, 2014, **47**(2), 486–491.
95. X. Tian, C. Cassani, Y. Liu, A. Moran, A. Urakawa, P. Galzerano, E. Arceo and P. Melchiorre, Diastereodivergent Asymmetric Sulfa-Michael Additions of α-Branched Enones Using a Single Chiral Organic Catalyst, *J. Am. Chem. Soc.*, 2011, **133**(44), 17934–17941.
96. V. Blanco, D. A. Leigh, V. Marcos, J. A. Morales-Serna and A. L. Nussbaumer, A Switchable [2]Rotaxane Asymmetric Organocatalyst

That Utilizes an Acyclic Chiral Secondary Amine, *J. Am. Chem. Soc.*, 2014, **136**(13), 4905–4908.

97. G. M. Miyake and E. Y. X. Chen, Cinchona Alkaloids as Stereoselective Organocatalysts for the Partial Kinetic Resolution Polymerization of Rac-Lactide, *Macromolecules*, 2011, **44**(11), 4116–4124.
98. J. B. Zhu and E. Y. X. Chen, From Meso-Lactide to Isotactic Polylactide: Epimerization by B/N Lewis Pairs and Kinetic Resolution by Organic Catalysts, *J. Am. Chem. Soc.*, 2015, **137**(39), 12506–12509.
99. S. M. Banik, A. Levina, A. M. Hyde and E. N. Jacobsen, Lewis Acid Enhancement by Hydrogen-Bond Donors for Asymmetric Catalysis, *Science*, 2017, **358**, 761–764.
100. J. Liu, C. Chen, Z. Li, W. Wu, X. Zhi, Q. Zhang, H. Wu, X. Wang, S. Cui and K. Guo, A Squaramide and Tertiary Amine: An Excellent Hydrogen-Bonding Pair Organocatalyst for Living Polymerization, *Polym. Chem.*, 2015, **6**(20), 3754–3757.
101. A. Rostami, E. Sadeh and S. Ahmadi, Exploration of Tertiary Aminosquaramide Bifunctional Organocatalyst in Controlled/living Ring-Opening Polymerization of L-Lactide, *J. Polym. Sci., Part A: Polym. Chem.*, 2017, **55**, 2483–2493.
102. J. Liu, J. Xu, Z. Li, S. Xu, X. Wang, H. Wang, T. Guo, Y. Gao, L. Zhang and K. Guo, Squaramide and Amine Binary H-Bond Organocatalysis in Polymerizations of Cyclic Carbonates, Lactones, and Lactides, *Polym. Chem.*, 2017, **8**, 7054–7068.
103. D. Specklin, F. Hild, L. Chen, L. Thévenin, M. Munch, F. Dumas, F. Le Bideau and S. Dagorne, Bifunctional Squaramides as Organocatalysts for Lactide Polymerization: Catalytic Performance and Comparison with Monofunctional Analogues, *ChemCatChem*, 2017, **9**(15), 3041–3046.
104. X. Ni, X. Li, Z. Wang and J. P. Cheng, Squaramide Equilibrium Acidities in DMSO, *Org. Lett.*, 2014, **16**(6), 1786–1789.
105. G. Jakab, C. Tancon, Z. Zhang, K. M. Lippert and P. R. Schreiner, (Thio)urea Organocatalyst Equilibrium Acidities in DMSO, *Org. Lett.*, 2012, **14**(7), 1724–1727.
106. J. Xu, J. Liu, Z. Li, S. Xu, H. Wang, T. Guo, Y. Gao, L. Zhang, C. Zhang and K. Guo, Opposite-Charge Repulsive Cation and Anion Pair Cooperative Organocatalysis in Ring-Opening Polymerization, *Polym. Chem.*, 2018, **9**, 2183–2192.
107. S. Xu, H. Sun, J. Liu, J. Xu, X. Pan, H. Dong, Y. Liu, Z. Li and K. Guo, Internal Lewis Pair Enhanced H-Bond Donor: Boronate-Urea and Tertiary Amine Co-Catalysis in Ring-Opening Polymerization, *Polym. Chem.*, 2016, **7**(44), 6843–6853.
108. A. Wittkopp and P. R. Schreiner, Metal-Free, Noncovalent Catalysis of Diels - Alder Reactions by Neutral Hydrogen Bond Donors in Organic Solvents and in Water, *Chem. – Eur. J.*, 2003, **9**(2), 407–414.
109. K. V. Fastnacht, S. S. Spink, N. U. Dharmaratne, J. U. Pothupitiya, P. P. Datta, E. T. Kiesewetter and M. K. Kiesewetter, Bis- and

Tris-Urea H-Bond Donors for Ring-Opening Polymerization: Unprecedented Activity and Control from an Organocatalyst, *ACS Macro Lett.*, 2016, **5**(8), 982–986.
110. E. V. Anslyn and D. A. Dougherty, Solutions and Non-Covalent Binding Forces; Thermodynamic Analyses of Binding Phenomena, in *Modern Physical Organic Chemistry*, 2006, pp. 145–222.
111. K. V. Fastnacht, N. U. Dharmaratne, and M. K. Kiesewetter, unpublished results.
112. N. U. Dharmaratne, J. U. Pothupitiya, T. J. Bannin, O. I. Kazakov and M. K. Kiesewetter, Triclocarban: Commercial Antibacterial and Highly Effective H-Bond Donating Catalyst for Ring-Opening Polymerization, *ACS Macro Lett.*, 2017, **6**(4), 421–425.
113. M. Blain, H. Yau, L. Jean-Gérard, R. Auvergne, D. Benazet, P. R. Schreiner, S. Caillol and B. Andrioletti, Urea- and Thiourea-Catalyzed Aminolysis of Carbonates, *ChemSusChem*, 2016, **9**(16), 2269–2272.
114. L. Mezzasalma, A. P. Dove and O. Coulembier, Organocatalytic Ring-Opening Polymerization of L -Lactide in Bulk: A Long Standing Challenge, *Eur. Polym. J.*, 2017, **95**, 628–634.
115. A. Alba, A. Schopp, A.-P. De Sousa Delgado, R. Cherif-Cheikh, B. Martín-Vaca and D. Bourissou, Controlled Ring-Opening Polymerization of Lactide by Bis-Sulfonamide/amine Associations: Cooperative Hydrogen-Bonding Catalysis, *J. Polym. Sci., Part A: Polym. Chem.*, 2010, **48**(4), 959–965.
116. D. Delcroix, A. Couffin, N. Susperregui, C. Navarro, L. Maron, B. Martín-Vaca and D. Bourissou, Phosphoric and Phosphoramidic Acids as Bifunctional Catalysts for the Ring-Opening Polymerization of ε-Caprolactone: A Combined Experimental and Theoretical Study, *Polym. Chem.*, 2011, **2**(10), 2249–2256.
117. N. Susperregui, D. Delcroix, B. Martín-Vaca, D. Bourissou and L. Maron, Ring-Opening Polymerization of ε-Caprolactone Catalyzed by Sulfonic Acids: Computational Evidence for Bifunctional Activation, *J. Org. Chem.*, 2010, **75**(19), 6581–6587.
118. S. Kan, Y. Jin, X. He, J. Chen, H. Wu, P. Ouyang, K. Guo and Z. Li, Imidodiphosphoric Acid as a Bifunctional Catalyst for the Controlled Ring-Opening Polymerization of δ-Valerolactone and ε-Caprolactone, *Polym. Chem.*, 2013, **4**(21), 5432–5439.
119. X. He, Y. Ji, Y. Jin, S. Kan, H. Xia, J. Chen, B. Liang, H. Wu, K. Guo and Z. Li, Bifunctional Imidodiphosphoric Acid-Catalyzed Controlled/living Ring-Opening Polymerization of Trimethylene Carbonate Resulting Block, Dihydroxy Telechelic, and Star-Shaped Polycarbonates, *J. Polym. Sci., Part A: Polym. Chem.*, 2014, **52**(7), 1009–1019.
120. J. Chen, S. Kan, H. Xia, F. Zhou, X. Chen, X. Jiang, K. Guo and Z. Li, Phosphoramidic Acid Catalyzed Controlled/living Ring-Opening Polymerization of Trimethylene Carbonate, *Polymer*, 2013, **54**(16), 4177–4182.

121. E. K. Macdonald and M. P. Shaver, Understanding the Phosphoric Acid Catalysed Ring Opening Polymerisation of β-Butyrolactone and Other Cyclic Esters, *Eur. Polym. J.*, 2017, **95**, 702–710.
122. T. Akiyama, Stronger Brønsted Acids, *Chem. Rev.*, 2007, **107**(12), 5744–5758.
123. C. Thomas, F. Peruch, A. Deffieux, A. Milet, J. P. Desvergne and B. Bibal, Phenols and Tertiary Amines: An Amazingly Simple Hydrogen-Bonding Organocatalytic System Promoting Ring Opening Polymerization, *Adv. Synth. Catal.*, 2011, **353**(7), 1049–1054.
124. C. Thomas, F. Peruch and B. Bibal, Ring Opening Polymerization of D,l-Lactide Using a Supramolecular Sulfonamide-Tertiary Amine Organocatalytic System, *All Results J.: Chem.*, 2012, **3**, 7–11.
125. O. Coulembier, D. P. Sanders, A. Nelson, A. N. Hollenbeck, H. W. Horn, J. E. Rice, M. Fujiwara, P. Dubois and J. L. Hedrick, Hydrogen-Bonding Catalysts Based on Fluorinated Alcohol Derivatives for Living Polymerization, *Angew. Chem., Int. Ed.*, 2009, **48**(28), 5170–5173.
126. R. W. Taft and F. G. Bordwell, Structural and Solvent Effects Evaluated from Acidities Measured in Dimethyl Sulfoxide and in the Gas Phase, *Acc. Chem. Res.*, 1988, **21**(12), 463–469.
127. X. Li, Q. Zhang, Z. Li, X. Wang, J. Liu, S. Cui, S. Xu, C. Zhao, C. Chen and K. Guo, Thiourea Binding with Carboxylic Acid Promoted Cationic Ring-Opening Polymerization, *Polymer*, 2016, **84**, 293–303.
128. X. Li, Q. Zhang, Z. Li, S. Xu, C. Zhao, C. Chen, X. Zhi, H. Wang, N. Zhu and K. Guo, Tripodal Hydrogen Bond Donor Binding with Sulfonic Acid Enables Ring-Opening Polymerization, *Polym. Chem.*, 2016, **7**(7), 1368–1374.
129. D. Zhang, D. Jardel, F. Peruch, N. Calin, V. Dufaud, J. P. Dutasta, A. Martinez and B. Bibal, Azaphosphatranes as Hydrogen-Bonding Organocatalysts for the Activation of Carbonyl Groups: Investigation of Lactide Ring-Opening Polymerization, *Eur. J. Org. Chem.*, 2016, **2016**(8), 1619–1624.
130. H. Sun, S. Xu, Z. Li, J. Xu, J. Liu, X. Wang, H. Wang, H. Dong, Y. Liu and K. Guo, Preorganized Dual H-Bond Donor Promotes Benzoic Acid Active in Polymerization of δ-Valerolactone, *Polym. Chem.*, 2017, **1**, 5570–5579.
131. C. Thomas, A. Milet, F. Peruch and B. Bibal, Activation of Carbonyl Bonds by Quaternary Ammoniums and a (Na^+:crown-Ether) Complex: Investigation of the Ring-Opening Polymerization of Cyclic Esters, *Polym. Chem.*, 2013, **4**(12), 3491–3498.
132. D. J. Coady, K. Fukushima, H. W. Horn, J. E. Rice and J. L. Hedrick, Catalytic Insights into Acid/base Conjugates: Highly Selective Bifunctional Catalysts for the Ring-Opening Polymerization of Lactide, *Chem. Commun.*, 2011, **47**(47), 3105–3107.
133. H. A. Brown, A. G. De Crisci, J. L. Hedrick and R. M. Waymouth, Amidine-Mediated Zwitterionic Polymerization of Lactide, *ACS Macro Lett.*, 2012, **1**(9), 1113–1115.

134. J. Kadota, D. Pavlović, J. P. Desvergne, B. Bibal, F. Peruch and A. Deffieux, Ring-Opening Polymerization of L-Lactide Catalyzed by an Organocatalytic System Combining Acidic and Basic Sites, *Macromolecules*, 2010, **43**(21), 8874–8879.
135. J. Kadota, D. Pavlović, H. Hirano, A. Okada, Y. Agari, B. Bibal, A. Deffieux and F. Peruch, Controlled Bulk Polymerization of L-Lactide and Lactones by Dual Activation with Organo-Catalytic Systems, *RSC Adv.*, 2014, **4**(28), 14725–14732.
136. G. Gontard, A. Amgoune and D. Bourissou, Ring-Opening Polymerization of ε-Caprolactone Catalyzed by Ionic Hydrogen Bond Activation with Bis-Pyridiniums, *J. Polym. Sci., Part A: Polym. Chem.*, 2016, **54**(20), 3253–3256.
137. X. Wang, S. Cui, Z. Li, S. Kan, Q. Zhang, C. Zhao, H. Wu, J. Liu, W. Wu and K. Guo, A Base–conjugate-Acid Pair for Living/controlled Ring-Opening Polymerization of Trimethylene Carbonate through Hydrogen-Bonding Bifunctional Synergistic Catalysis, *Polym. Chem.*, 2014, **5**(20), 6051–6059.
138. O. Coulembier, T. Josse, B. Guillerm, P. Gerbaux and P. Dubois, An Imidazole-Based Organocatalyst Designed for Bulk Polymerization of Lactide Isomers: Inspiration from Nature, *Chem. Commun.*, 2012, **48**(95), 11695–11697.
139. X. Zhi, J. Liu, Z. Li, H. Wang, X. Wang, S. Cui, C. Chen, C. Zhao, X. Li and K. Guo, Ionic Hydrogen Bond Donor Organocatalyst for Fast Living Ring-Opening Polymerization, *Polym. Chem.*, 2016, **7**(2), 339–349.
140. S. Chen, Y. Liu, Z. Li, X. Wang, H. Dong, H. Sun, K. Yang, H. Gebru and K. Guo, H-Bonding Binary Organocatalysis Promoted Amine-Initiated Ring-Opening Polymerizations of Lactide from Polysarcosine to Diblock Copolymers, *Eur. Polym. J.*, 2017, **97**, 389–396.
141. K. Makiguchi, S. Kikuchi, K. Yanai, Y. Ogasawara, S. I. Sato, T. Satoh and T. Kakuchi, Diphenyl phosphate/4-Dimethylaminopyridine as an Efficient Binary Organocatalyst System for Controlled/living Ring-Opening Polymerization of L -Lactide Leading to Diblock and End-Functionalized poly(L-Lactide)s, *J. Polym. Sci., Part A: Polym. Chem.*, 2014, **52**(7), 1047–1054.
142. B. Guillerm, V. Lemaur, J. Cornil, R. Lazzaroni, P. Dubois and O. Coulembier, Ammonium Betaines: Efficient Ionic Nucleophilic Catalysts for the Ring-Opening Polymerization of L-Lactide and Cyclic Carbonates, *Chem. Commun.*, 2014, **50**, 10098–10101.
143. Y. Tsutsumi, K. Yamakawa, M. Yoshida, T. Ema and T. Sakai, Bifunctional Organocatalyst for Activation of Carbon Dioxide and Epoxide To Produce Cyclic Carbonate: Betaine as a New Catalytic Motif, *Org. Lett.*, 2010, **12**(24), 5728–5731.
144. J. M. Becker, S. Tempelaar, M. J. Stanford, R. J. Pounder, J. A. Covington and A. P. Dove, Development of Amino-Oxazoline and Amino-Thiazoline Organic Catalysts for the Ring-Opening Polymerisation of Lactide, *Chem. – Eur. J.*, 2010, **16**(20), 6099–6105.

145. A. Harada, M. Osaki, Y. Takashima and H. Yamaguchi, Ring-Opening Polymerization of Cyclic Esters by Cyclodextrins, *Acc. Chem. Res.*, 2008, **41**(9), 1143–1152.
146. Y. Takashima, M. Osaki and A. Harada, Cyclodextrin-Initiated Polymerization of Cyclic Esters in Bulk: Formation of Polyester-Tethered Cyclodextrins, *J. Am. Chem. Soc.*, 2004, **126**(42), 13588–13589.
147. M. Osaki, Y. Takashima, H. Yamaguchi and A. Harada, Polymerization of Lactones Initiated by Cyclodextrins: Effects of Cyclodextrins on the Initiation and Propagation Reactions, *Macromolecules*, 2007, **40**(9), 3154–3158.
148. P. Glöckner and H. Ritter, Cyclodextrins in Polymer Synthesis: Free Radical Polymerization of Methylated Beta-Cyclodextrin Complexes of Methyl Methacrylate and Styrene, Controlled by Dodecanethiol as the Chain-Transfer Agent in Aqueous Medium, *Macromol. Chem. Phys.*, 2000, **201**(17), 2455–2457.
149. M. Takayanagi, S. Ito, K. Matsumoto and M. Nagaoka, Formation of Reactant Complex Structure for Initiation Reaction of Lactone Ring-Opening Polymerization by Cooperation of Multiple Cyclodextrin, *J. Phys. Chem. B*, 2016, **120**(29), 7174–7181.
150. M. K. Kiesewetter, M. D. Scholten, N. Kirn, R. L. Weber, J. L. Hedrick and R. M. Waymouth, Cyclic Guanidine Organic Catalysts: What Is Magic about Triazabicyclodecene?, *J. Org. Chem.*, 2009, **74**(24), 9490–9496.
151. H. W. Horn, G. O. Jones, D. S. Wei, K. Fukushima, J. M. Lecuyer, D. J. Coady, J. L. Hedrick and J. E. Rice, Mechanisms of Organocatalytic Amidation and Transesterification of Aromatic Esters As a Model for the Depolymerization of Poly(ethylene) Terephthalate, *J. Phys. Chem. A*, 2012, **116**, 12389–12398.
152. J. U. Pothupitiya, R. S. Hewawasam and M. K. Kiesewetter, Urea and Thiourea H-bond Donating Catalysts for Ring-opening Polymerization: Mechanistic Insights via (Non)Linear Free Energy Relationships, *Macromolecules*, 2018, **51**, 3203–3211.
153. B. Lin and R. M. Waymouth, Organic Ring-Opening Polymerization Catalysts: Reactivity Control by Balancing Acidity, *Macromolecules*, 2018, **51**, 2932–2938.

CHAPTER 4

Base Catalysts for Organopolymerization

STEFAN NAUMANN

University of Stuttgart, Department of Chemistry, Pfaffenwaldring 55, Stuttgart 70569, Germany
Email: stefan.naumann@ipoc.uni-stuttgart.de

4.1 Introduction

Complementing previous chapters, which have detailed the application of acidic and nucleophilic organocatalysts, now the subject of base catalysis for organopolymerization shall be considered. Since the number of compounds eligible to act in such as manner—that is, able to fully deprotonate substrates or "activate" them by interaction with R–H-moieties—is potentially very large, and the underlying mechanisms quite diverse, this chapter is by intention not limited to "superbases"[1,2] or truly anionic mechanisms.[3] Rather, several families of neutral Lewis- or Brønsted-basic organocatalysts, which have successfully been used to synthesize polymer materials, shall be presented. For each group a concise description of physical properties and synthetic accessibility will be given, including a tabular listing of pK_a-values. How the corresponding organocatalysts have been employed in polymerization reactions, and especially the way they can be structurally adapted to meet specific challenges in polymer chemistry, will form the main part of each discussion. In this manner, amidines and guanidines will be explored (Section 4.2), followed by the eminently important class of phosphazene bases (Section 4.3) and by N-heterocyclic carbenes (NHCs)/N-heterocyclic olefins (NHOs, Section 4.4). In Section 4.5 other catalyst types are briefly

Polymer Chemistry Series No. 31
Organic Catalysis for Polymerisation
Edited by Andrew Dove, Haritz Sardon and Stefan Naumann

Published by the Royal Society of Chemistry, www.rsc.org

discussed. Organobase catalysis is then discussed in a brief summary (Section 4.6), followed by some concluding remarks.

When contemplating organic base catalysts (OBCs) for use in polymerization, information about the corresponding pK_a-values is necessary, but hardly sufficient. Apart from being strongly dependent on matrix (type of solvent, gas phase), temperature and measuring methods (conventional titration, NMR experiments, level of DFT calculations), pK_a-data is not trivial to determine and does not represent the full spectrum of decisive factors. These crucially also include kinetic properties (how fast is deprotonation? Is the proton confined or in fast exchange with other OBCs, or the substrate?), but also the complex interdependency of nucleophilicity and basicity. While a large, sterically hindered phosphazene may be considered "non-nucleophilic", for many compounds such a clear statement is much harder, and probably wrong, to make. A prime example for this behaviour is provided by N-heterocyclic carbenes, which have been discussed previously in Chapter 1. Depending on chemical structure, monomer and reaction conditions, NHCs may either act as organobases or nucleophiles; in some cases it even may be possible that both mechanisms apply simultaneously.[4–6] This mechanistic dualism (Figure 4.1) complicates matters, but also offers access to novel polymerization setups. Methods to tailor NHCs for one or the other

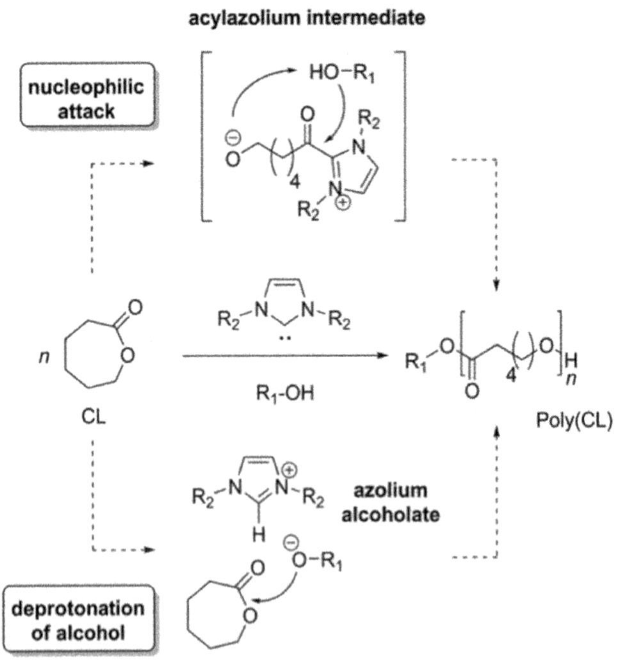

Figure 4.1 Mechanistic dualism as found for NHC-mediated lactone polymerization. In this instance, both pathways result in the identical polymer structure (R_1–O/H–terminated).

pathway do exist, as will be presented in the following. However, beyond NHCs, it is important to be aware of a potentially increasing nucleophilicity as basicity rises.

Several general options exist to increase the basicity of organic compounds. A synthetically very simple method to achieve that is to strengthen the electron density of the active centre, for example by exchanging aryl- for alkyl-substituents. The same effect can be realized by modulation of the ring-structure for cyclic OBCs, most notably for NHCs (Figure 4.2).[7] Geometry can also be employed to advantage when the protonated structure is less strained than the parent OBC, or when bidentate hydrogen bonds can be formed (as in "proton sponges").[8,9] Verkade's base (a cycloazaphosphine) is stabilized by a *trans*-annular N–P bond, which forms after protonation.[10,11] Aromatization upon protonation is also a formidable driving force to generate a strong base, as is of course the enlargement of conjugated systems to delocalize the positive charge. An example for the former is found in NHO chemistry, where aromatization makes the difference between absent polymerization activity and high performance,[12] while Schwesinger's "battery cell-style" extension of phosphazenes[13,14] or the construction of multicyclic amidines[15] made use of the latter principle.

Mechanistically, the way an OBC initiates or catalyses a polymerization reaction falls in one of two fundamental categories: either (i) a direct deprotonation or activation of the monomer takes place, or (ii) an auxiliary substrate—usually a separate initiator—is deprotonated. Polymerization by direct deprotonation is realizable if the monomer possesses a suitably acidic site and if the proton abstraction produces a species reactive enough to initiate the polymerization. A well-documented, large-scale example is the anionic polymerization of lactams to synthesize polyamides (Figure 4.3). Deprotonation of the amide functionality requires a strong base, and consequently only very few instances of metal-free initiation by neutral OBCs have been described.[16–19] Cyclopropenimines have been employed in another example for this type of catalyst activity, where the initiating species resulted from deprotonation of the lactide monomer, generating a reactive enolate.[20]

If the base catalyst is intended to work in combination with a specifically added initiator, for the latter almost always alcohols are used. Since in this case deprotonation/activation is more readily achieved, the range of OBCs is larger. Representative applications include polymerization of lactones, lactide, carbonates or epoxides in the presence of an R–OH-type initiator. Together, this area is responsible for the bulk of reports on organobase-mediated polymerizations. However, deprotonated solvent molecules can also act as initiators.[21]

The vast majority of OBC-mediated polymerizations concern chain-growth mechanisms. In contrast, step-growth is rarely reported on.[22]

Even discounting simple amines and proton sponges, the number of OBCs has markedly increased in the past three decades; a monograph edited by Ishikawa covers the situation until 2009 for organic superbases and their

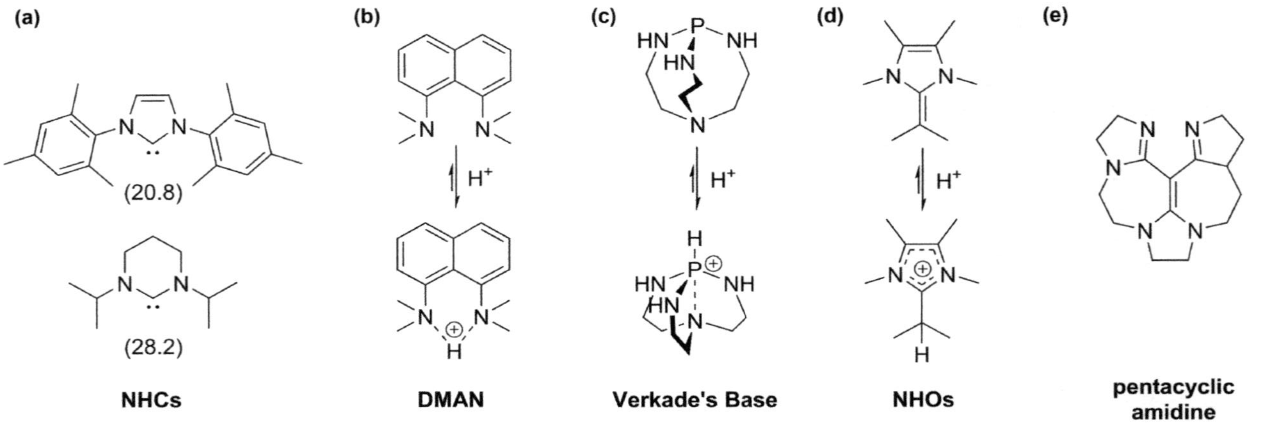

Figure 4.2 (a) Two different NHCs and the corresponding change of the pK_a (brackets, H_2O, 25 °C);[7] (b) 1,8-bis(dimethylamino)naphthalene (DMAN), where prior to protonation an unfavourable *N*/*N*-lone pair interaction exists; (c) Verkade's base with P(III) bonded to amino groups. After protonation, both bridgeheads form a *trans*-annular bond; (d) NHO with the formation of an imidazolium moiety after protonation; (e) amidine with enlarged conjugated system.

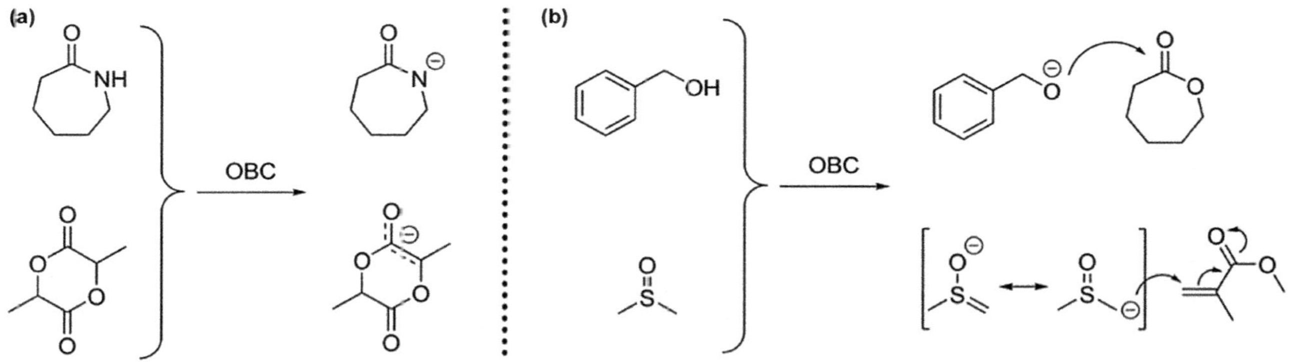

Figure 4.3 (a) Direct deprotonation of monomers, exemplified for ε-caprolactam and lactide; (b) deprotonation of an R-OH-type initiator (benzyl alcohol) and of a solvent molecule (DMSO).

application in organic chemistry.[2] Out of this plethora of compounds only a minority has ever been tested for (organo)polymerization activity, the subject of the following considerations. Certain aspects of organobase catalysis have been reviewed.[5,6,23]

4.2 Amidines and Guanidines

4.2.1 Amidines—Synthesis and Properties

Amidines form a class of readily adaptable and well accessible OBCs, with a general basicity on the lower range of the superbase scale.[2,24] The amidine motif as such is not only of interest as organobase, but also frequently found in biochemistry.[25,26] The simplest representatives, formamidines (**1**), contain an amino- and an imino-functionality; these are in conjugation (n–π), providing the underlying mechanism for the basicity of amidines (Figure 4.4). This conjugation also makes itself noticeable in molecule geometry: typically, amidines are flat (with small deviations from full planarity) and the C=N-bond is elongated, while the C–N-bond is somewhat shortened compared to non-conjugated systems.[27] Acetamidine (**2**), for example, displays corresponding bond lengths of 128 pm and 134 pm, respectively.[28] Both bonds become fully equivalent after protonation. The three tuning sites—substituents on the imino group, the amine and the central carbon—can be used to influence the pK_a of the compounds. For *acyclic* amidines, the substituent effects are considerable (Table 4.1) and strongest for the imino-moiety,[29] whereby only compounds with electron donating groups (alkyl) can be considered as strong bases.

Embedded in cyclic structures, amidines have evolved to increased basicity and popular reagents in organic chemistry and polymerization catalysis. Compounds like 1,5-diazabicyclo[4.3.0]non-5-ene (DBN) and 1,8-diazabicyclo[5.4.0]undec-7-ene (DBU) are commercially available and display pK_a-values of 23.8 and 24.3 in acetonitrile, respectively (Table 4.2). The somewhat increased basicity of DBU relative to DBN can be attributed to a difference in ring tension.[30] The application of both compounds, however, can be

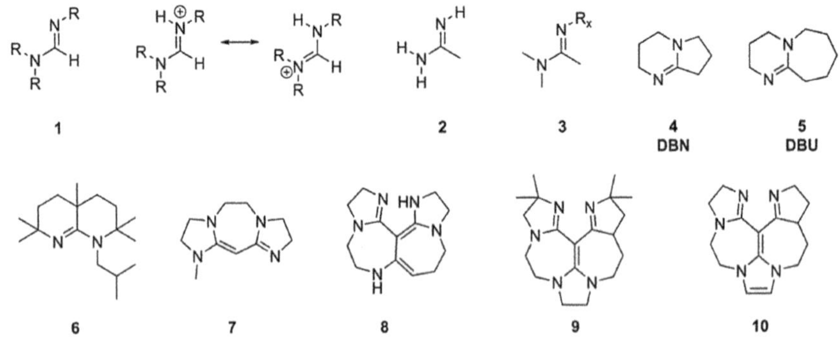

Figure 4.4 Scope of the amidines, acyclic and cyclic structures.

Table 4.1 Several pK_a-values as found for amidine **3** with different substituents (95.6% EtOH, 25 °C).[29]

R_x	p-C_6H_4–NO_2	C_6H_5	*i*-butyl	Cyclohexyl	*i*-Pr
pK_a	5.7	8.3	12.3	12.6	12.6

Table 4.2 pK_a-values for several amidines and guanidines. Data in brackets from extrapolation/calculation.

Compound	pK_a (MeCN) (ref.)	Compound	pK_a (H_2O) (ref.)	pK_a (MeCN) (ref.)
4	23.8 (31)	**11**	13.6 (39)	
5	24.3 (31)	**13**	13.9 (39)	
7	27.0 (31)	**14**	13.8 (39)	
8	25.2 (30)	**16**	(13.8) (31)	(23.2) (14)
9	30.0 (31)	**17**		26.0 (31)/26.0 (50)
10	32.9 (31)	**18**		25.4 (31)/25.5 (50)

complicated by a certain *N*- and *C*-nucleophilicity.[31] A similar, yet sterically more hindered amidine (**6**) has been demonstrated to be much less nucleophilic, resisting alkylation with methyl iodide.[32] Combining amidines into larger, more conjugated polycyclic systems leads to a further increase in basicity. A structure like **7**, containing the "vinamidine" (diazapentadiene) motif, was found to have a pK_a of 27.0 in acetonitrile; a further fusion of these subunits to generate compounds **8–9** pushed the corresponding values up even further (Table 4.2).[15] Introduction of an additional double bond resulted in the most basic vinamidine reported to date (**10**, pK_a = 32.9). However, the concept of enlarged conjugated systems to stabilize the positive charge, thus rendering the compound more basic, is limited by a growing sensitivity towards oxidation; organobase **10** has been shown to be thermally labile and readily oxidized.[15,31]

Synthetically, amidines are well accessible by a number of routes. In 1968, Weintraub and co-workers described a method based on amides as versatile precursors for the generation of acyclic amidines, complementing the previously dominant pathway that had been developed by Pinner.[33] The latter synthesis employs the acid-catalysed reaction between an anhydrous alcohol and a nitrile to form an imidate that is subsequently treated with an amine to afford the amidines (Figure 4.5a and b). The somewhat limited availability of nitriles may be seen as a disadvantage; the alternative strategy starting from amides was shown to deliver excellent yields and allow for a range of different substituents. In this method, the imidate is made by application of triethyloxonium tetrafluoroborate as a strong alkylating agent. Already in 1966 Oediger and co-workers presented a facile route to cyclic amidines, starting from lactams.[34] The reaction sequence involves a base-catalysed Michael addition on acrylonitrile, followed by a reduction of the nitrile group and acidic cyclization. Thereby, DBN can be prepared from pyrrolidone, for example. Inspiration for amidine synthesis can also be taken from the plethora of methods suitable for the creation of NHC precursors, since

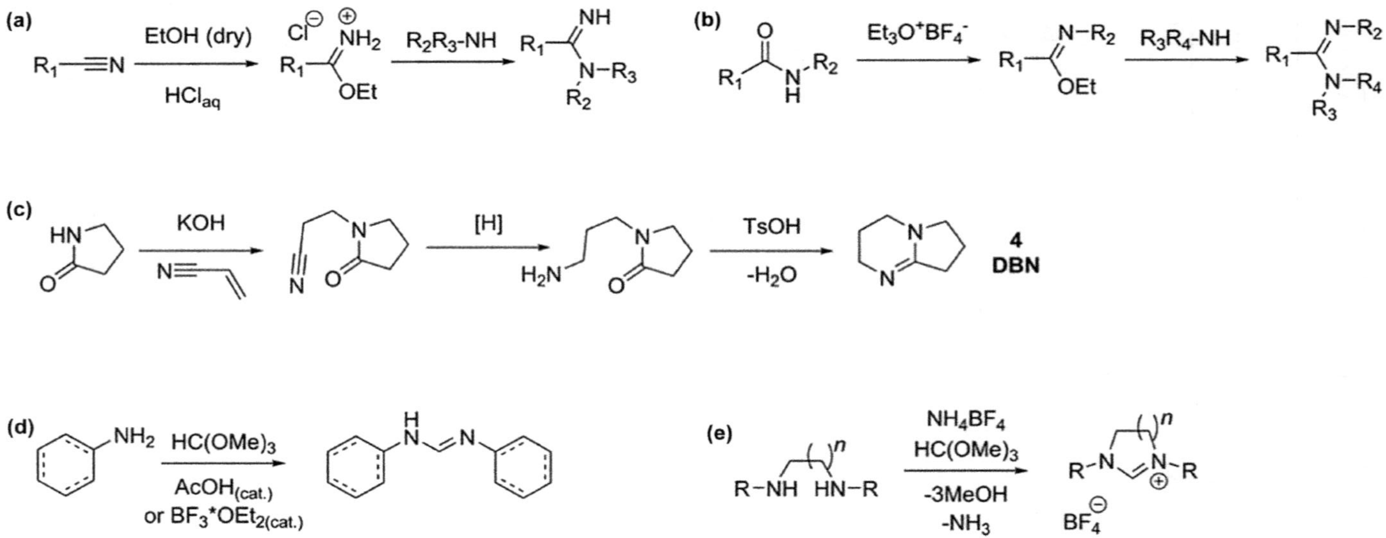

Figure 4.5 Synthetic access to amidines following the route by (a) Pinner, (b) Weintraub, (c) Oediger and *via* amines and orthoesters to generate (d) acyclic and (e) cyclic structures.

many of those involve amidine moieties (Figure 4.5d and e).[35] A very convenient method to prepare symmetric formamidines or amidines with simple *C*-substituents, is the acid-catalysed condensation of amines (aryl or alkyl) with orthoesters.[36,37] A similar principle has also been employed to gain a facile access to cyclic amidinium salts.[38]

4.2.2 Guanidines—Synthesis and Properties

Guanidines are closely related to amidines and can be understood as imides of urea; also, the guanidine motif itself is biochemically important, as is evident for example by its incorporation in the amino acid arginine or in creatine. The parent structure (Figure 4.6), unsubstituted guanidine (**11**), is already on its own a powerful organobase with a determined pK_a of 13.6 (water, corresponding to a value of approximately 23–24 in MeCN, see also Table 4.2) and thus comparable to the hydroxide ion.[39] The guanidine moiety is characterized by one imino and two amino functionalities linked to the same carbon atom, whereby a cross-conjugated hetero allylic system with six π-electrons is formed. As described for amidines, guanidines are protonated at the imino group and only accept one proton, even under strongly acidic conditions.[40] Four mesomeric structures can be drawn as a consequence of conjugation for both **11** and the protonated guanidinium cation **12**; for the latter, this seems favourable as three of those forms are fully equivalent, and the fourth places the positive charge on the central carbon

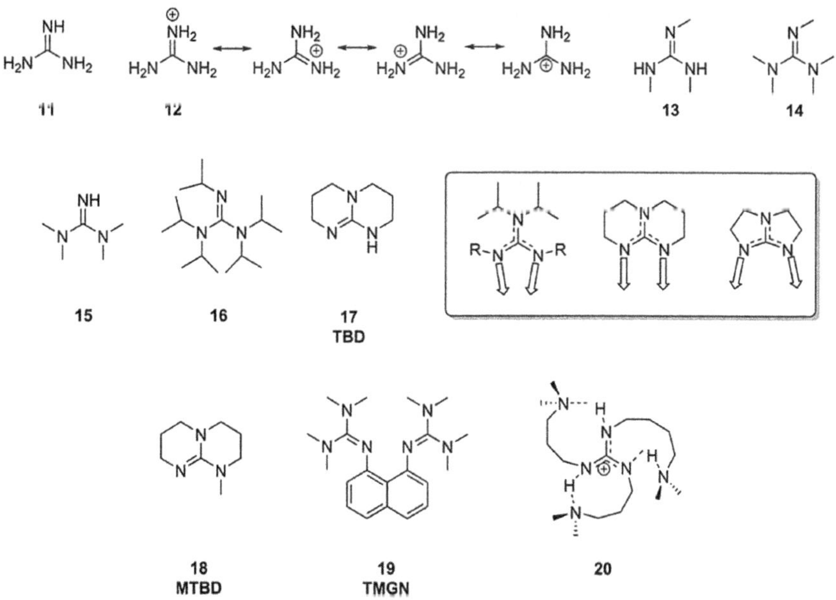

Figure 4.6 Structures of various guanidines as discussed in this chapter. Insert: Orbital projection and its change upon incorporation of the functionality in rigid bicycles.[51]

atom in this highly symmetrical structure. The exceptional basicity of guanidine has been related to this conjugation, which was also considered especially strong as an example for "Y-aromaticity".[41] Fittingly, crystal structures of guanidinium salts have shown planar cations with equal C–N bond lengths.[42] However, theory suggests that resonance stabilization alone does not account for the abnormal basicity observed. Rather, the ability of guanidinium to strongly interact with the solvent *via* H-bonding (note the high number of potential interactions for **12**) and dipoles seem to be chiefly responsible for this property, and a geometry deviating from full planarity (solution, gas phase) is proposed.[43,44] Coherently, the gas phase basicity is lower than the corresponding value in solution.[45] Hence, the electronic and geometric situation in guanidines is labile and readily influenced by a number of factors, which renders this class of OBC a powerful tool for chemists. Methylation of the parent structure **11** to yield compounds like *N*, *N′*, *N″*-trimethylguanidine (**13**) or pentamethylguanidine (PMG, **14**) has only a limited influence on pK_a-values of the corresponding conjugated acid (Table 4.2).[39] *N*, *N*, *N′*, *N′*-tetramethylguanidine (Barton's base, TMG, **15**) is one of the most frequently employed acyclic guanidines with applications in organic synthesis.[46] In these cases an interplay between increased electron density (potentially increasing basicity) and the decreasing number of hydrogens in the structure (and thus less capability to form hydrogen bonds) might be responsible for the weak correlation observed. Moreover, it must be considered that increasing the steric bulk—as found for example in **16**[47–49]—will impede conjugation in the cation, hence lowering basicity. This disadvantage is circumvented in cyclic systems, which are the most heavily employed guanidine OBC species. The iconic representative is 1,5,7-triazabicyclo[4.4.0]-5-decene (TBD, **17**), which has a basicity in acetonitrile of 26.[31,50] The main consequences from incorporating the guanidine functionality in a rigid, bicyclic skeleton arise from (a) an enforced, optimal conjugation (planarity) of the *N*, *N′*, *N″*- plane, (b) blocking of any isomerism (locking for example the N–H group in TBD in an E_{anti} conformation), and (c) changing and fixing the projection of the *N*-lone pair orbitals of the amidine subunit (bite angle, Figure 4.6 insert).[51]

This versatile structure can also be further modified by alkylation, resulting in MTBD (**18**, Figure 4.6) as another frequently used OBC. Compared to DBU, these cyclic guanidines are stronger bases by two orders of magnitude (Table 4.2). The combination of guanidine structures with "proton sponge" moieties has delivered further promising results. 1,8-bis(tetramethylguanidino)naphthalene (TMGN, **19**) was found to be about 10^7 times stronger a base ($pK_{a(MeCN)} = 25$) than the prototypical DMAN (Figure 4.2) and almost equal to cyclic guanidines such as MTBD.[52] Resonance stabilization imparted by the guanidine moieties (which are only marginally conjugated to the naphthalene ring) and a strong intramolecular H-bond upon protonation (resolving the unfavourable *N*-lone pair interaction) combine to provide this compound with its exceptionally high basicity.[53] Interestingly, TMGN was also demonstrated to be less nucleophilic than MTBD, while at the same time being more kinetically active than DMAN, as demonstrated by ^{1}H NMR observation of proton self-exchange rates.[52] Although often ignored, the propensity of an OBC for proton exchange can be expected to be an important parameter in catalytic transformations and might determine whether the base acts rather initiator-like or as a true catalyst. Additional methods to manipulate the base strength of guanidines have been investigated theoretically. Strengthening the intramolecular charge stabilization by H-bonding seems a good strategy to support basicity. For guanidinium **20** a $pK_{a(MeCN)} = 29.4$ was calculated.[54] Likewise, high proton affinities are predicted for guanidino-cyclopropenimines.[55]

Importantly, guanidines can act in a bifunctional manner, which has extended the range of catalytic transformations accessible with this class of organocatalyst (see also Chapter 3).[56,57]

Depending on the complexity of the desired guanidine, a number of preparative routes can be chosen. The important bicyclic structures can be conveniently accessed *via* thiourea intermediates, starting from triamines. Reaction with CS_2 (1 : 1) in xylene affords the target compounds by a sequence of ring-closing steps whereby H_2S is lost, driving the conversion (Figure 4.7).[58]

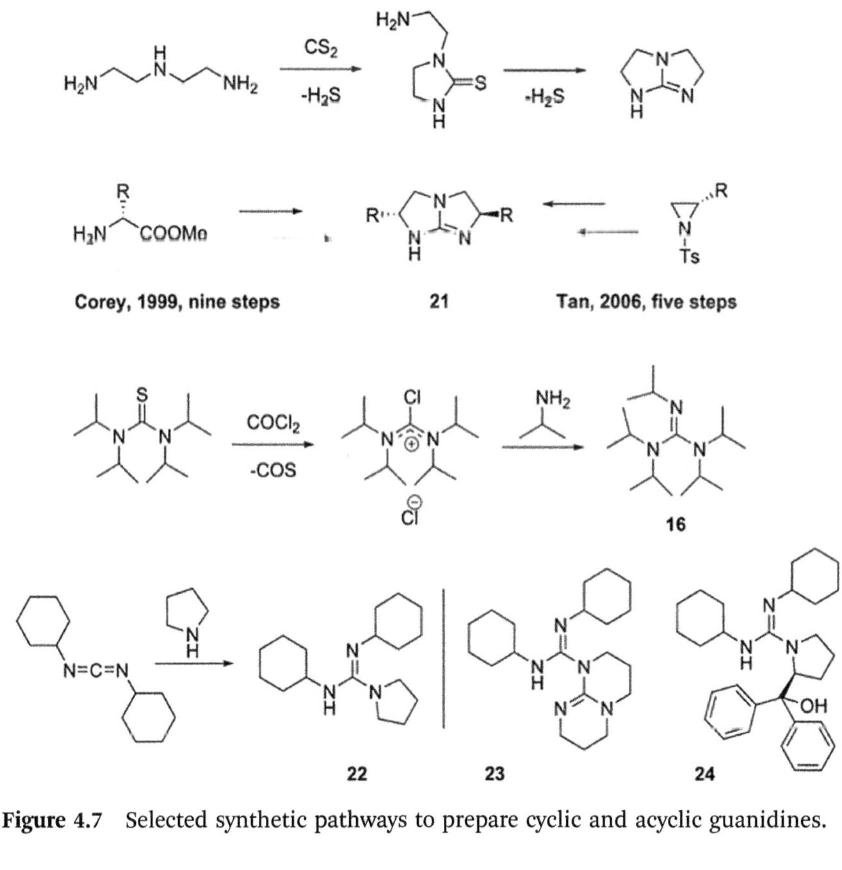

Figure 4.7 Selected synthetic pathways to prepare cyclic and acyclic guanidines.

Similarly, seven-membered motifs can be constructed.[59] Substituted derivatives require more synthetic effort; Corey and co-workers described a nine-step sequence to generate a chiral, bicyclic guanidine of the type **21** starting from a chiral amino acid.[60] Later, this pathway was shortened to five steps by the group of Tan, involving azirdines.[61] Acyclic guanidines are well accessible and can be generated by numerous methods; for a more complete overview excellent literature is available.[46] For the simpler structures, as are those who have been employed in polymerization catalysis, urea- or thiourea-derivatives are readily available starting compounds as reported by Barton and Simchen.[47–49] Carbodiimides can also serve as convenient precursors for simple guanidines. Hedrick and co-workers used a reaction with secondary amines to prepare **22–24**, compounds that were subsequently employed for polymerization catalysis.[62]

4.2.3 Amidines and Guanidines as Base Catalysts for Polymerizations

Applications of amidines and guanidines in polymerization catalysis will be discussed together, since many studies simultaneously investigate both types. In spite of the wealth of compounds described in the literature and the near unlimited range of conceivable structures, only a relatively small number of amidines and guanidines have been employed for organopolymerization. Among those, commercially available, cyclic derivatives (DBU, TBD, MTBD) clearly dominate. Nonetheless, even with a limited set of catalysts a rich polymerization chemistry has developed, with cyclic esters as well as carbonates and cyclic phosphates forming the majority of monomers reported to be polymerized by amidines or guanidines. Much of the following discussion will focus on the influence of the catalyst structure on polymerization behaviour. This not only concerns the catalytic mechanism, especially the extent to which bifunctional processes play a role, but also other factors that directly affect the polymerization performance. These include tuning of the reaction kinetics, the suppression of side reactions or the tailoring of polymer architectures.

While acyclic amidines are notably absent from the portfolio of successful polymerization catalysts, probably a consequence of their comparatively low basicity, acyclic guanidines have been employed to synthesize poly(lactide). Hedrick and co-workers employed **22–24** (Figure 4.7) for the ROP of L-Lactide (L-LA), in the presence of pyrenebutanol as initiator (OBC/OH/Lactide = 1 : 1 : 100).[62] With **22**, after 40 min quantitative conversion was achieved, while dispersity ($Đ_M = 1.06$) and polymerization behaviour suggested a controlled polymerization with living characteristics. Interestingly, **23** was found to be considerably more active (full conversion after 20 min); protonation of the TBD subunit in the molecule can easily be imagined to result in a favourable structure. However, this improved activity was accompanied by a loss of control, most probably a result of frequent transesterification. **24** proved to be more active and more selective than **22**, but stereoselectivity (*rac*-lactide) was rather low, probably because the spatial distance to the stereogenic centre is large. Additionally, DFT calculations clearly showed **22** to be a weaker base than TBD. This was attributed to the out-of-plane rotation of the pyrollidine moiety in the molecule, which diminishes conjugation. Overall, such behaviour is typical for acyclic guanidines and one of the main reasons for their decreased basicity when compared to cyclic derivatives (see Table 4.2). Hecht later showed that the inherently versatile structure of acyclic guanidines enables the facile adaption of their polymerization rates for lactide polymerization by introduction of aryl substituents with suitably tailored electron richness.[63]

However, cyclic derivatives have received much more attention, and indeed the first ever reported example of the application of guanidines in organopolymerization was published in 2006, employing TBD for the polymerization of L-LA, ε-caprolactone (CL) and δ-valerolactone (VL).[64] For all three monomers polymerization succeeded, whereby lactide was consumed rapidly and efficiently with good control over the polymerization (see Table 4.3 for examples). VL was converted more slowly (>90% after 30 min, 0.5 mol% TBD loading), but good polydispersity was achieved when the reaction was stopped early enough. Polymerization of CL was more problematic with a slow progress of the polymerization and a considerable degree of transesterification (Table 4.3). Nonetheless, it was possible to obtain well-defined polyester from this monomer too, if reactions were quenched at intermediate conversions. Interestingly, the polymerization of β-butyrolactone (BL) failed with TBD under these conditions; at elevated

Table 4.3 Polymerization results for cyclic ester monomers, catalysed by TBD, MTBD and DBU (no co-catalyst, with initiator).[64,65]

OBC	Monomer	Cat.[a] [%]	M/I[b]	Solvent	Time (h)	Conv. (%)	M_n (kg mol^{-1})	$Đ_M$
TBD (**17**)	L-LA	0.1	100	CH_2Cl_2	20 s	99	24	1.19
	L-LA	0.1	500	CH_2Cl_2	1 min	95	63	1.11
	VL	0.5	100	C_6D_6	0.5	91	15	1.09
	VL	0.3	200	C_6D_6	0.5	77	17	1.12
	CL	0.5	100	C_6D_6	8.0	72	17	1.16
	CL	0.5	200	C_6D_6	8.0	52	21	1.16
MTBD (**18**)	L-LA	1	100	$CDCl_3$	0.5	92	18	1.05
	L-LA	0.5	500	CD_2Cl_2	0.5	99	55	1.10
	VL	5	100	C_6D_6	72	0	—	—
	CL	5	100	C_6D_6	72	0	—	—
	BL	5	100	C_6D_6	72	0	—	—
DBU (**5**)	L-LA	1	100	$CDCl_3$	1	99	21	1.05
	L-LA	1	500	$CDCl_3$	2	98	85	1.08
	VL	5	100	C_6D_6	72	0	—	
	BL	5	100	C_6D_6	72	0	—	—

[a]Catalyst loading relative to monomer.
[b]Initial ratio of monomer to initiator (pyrenebutanol).

temperatures oligomers could be obtained. Overall, these pioneering findings rendered TBD one of the most broadly applicable organopolymerization catalysts at the time, and while the tendency towards transesterification still persevered, it was less marked than for NHCs. This prompted further investigations into related OBCs, and application of MTBD and DBU to the same set of cyclic ester monomers revealed pronounced differences to TBD.[65] Both organocatalysts effected a slower conversion of lactide and required higher catalysts loadings, but this was balanced by excellent control over the polymerization. Up to 500 repeating units could be targeted and even at high conversion the molecular weights remained controlled ($Đ_M < 1.10$). However, and in stark contrast to TBD, neither MTBD nor DBU were able to deliver polymer from VL or CL, even after prolonged reaction times (Table 4.3). This inactivity could only be mitigated by addition of a thiourea co-catalyst, which additionally activated the monomer.[65] Again, BL proved resistant to polymerization efforts.

While the inherent basicity of the catalysts was crucial to achieve polymerization, the models that were developed to explain these findings did not consider TBD an exclusive organobase. Instead, bifunctional behaviour was put forward. Based on model acyl transfer reactions, Hedrick and Waymouth proposed the formation of *N*-acyl species formed by nucleophilic action of the imine nitrogen of TBD; the neighbouring N–H moiety was thought to meanwhile facilitate proton transfer (Figure 4.8b). These considerations were supported by a number of findings. An exemplary transesterification reaction was successfully catalysed by TBD, where crucially the acylated TBD (**25**) was isolated (Figure 4.8a). Polymerizations in more polar solvents were observed to be not faster than reactions in less polar media, a result that was interpreted to contradict ionic processes. Also fittingly, MTBD and DBU cannot polymerize VL or CL in an identical manner, which was related to their obvious inability to act in the same bifunctional way. In contrast, for the readily polymerizable LA, a (pseudo-)anionic polymerization is sufficient and here the pK_a-values (Table 4.2) mirror the polymerization rates, which are TBD > MTBD ≈ DBU.[65] Later, it was even shown that a catalyst like DBU can form zwitterions to polymerize lactide in the absence of initiator (see discussion in Section 1.2.3).[66]

Calculations helped to refine this model. Independently, Goodman and Rice found that an "acid-base" mechanism is favoured for TBD-catalysed ROP, while direct nucleophilic attack is disfavoured by a high activation barrier (Figure 4.8).[67,68] In this case, the basic imido group of TBD activates the initiating alcohol by H-bonding (partial deprotonation), increasing the electron density on the oxygen and correspondingly its propensity for nucleophilic attack on the lactone carbonyl. The monomer itself is coordinated to the organocatalyst by H-bonding *via* the neighbouring N–H-moiety. Hence, close proximity, dual activation and the strong basicity of TBD combine to enable an effective polymerization; the difference in Gibbs free energy is 7 kcal mol^{-1}, relative to the nucleophilic pathway.[67] It should also be noted that this mechanism allows for a good stabilization of the

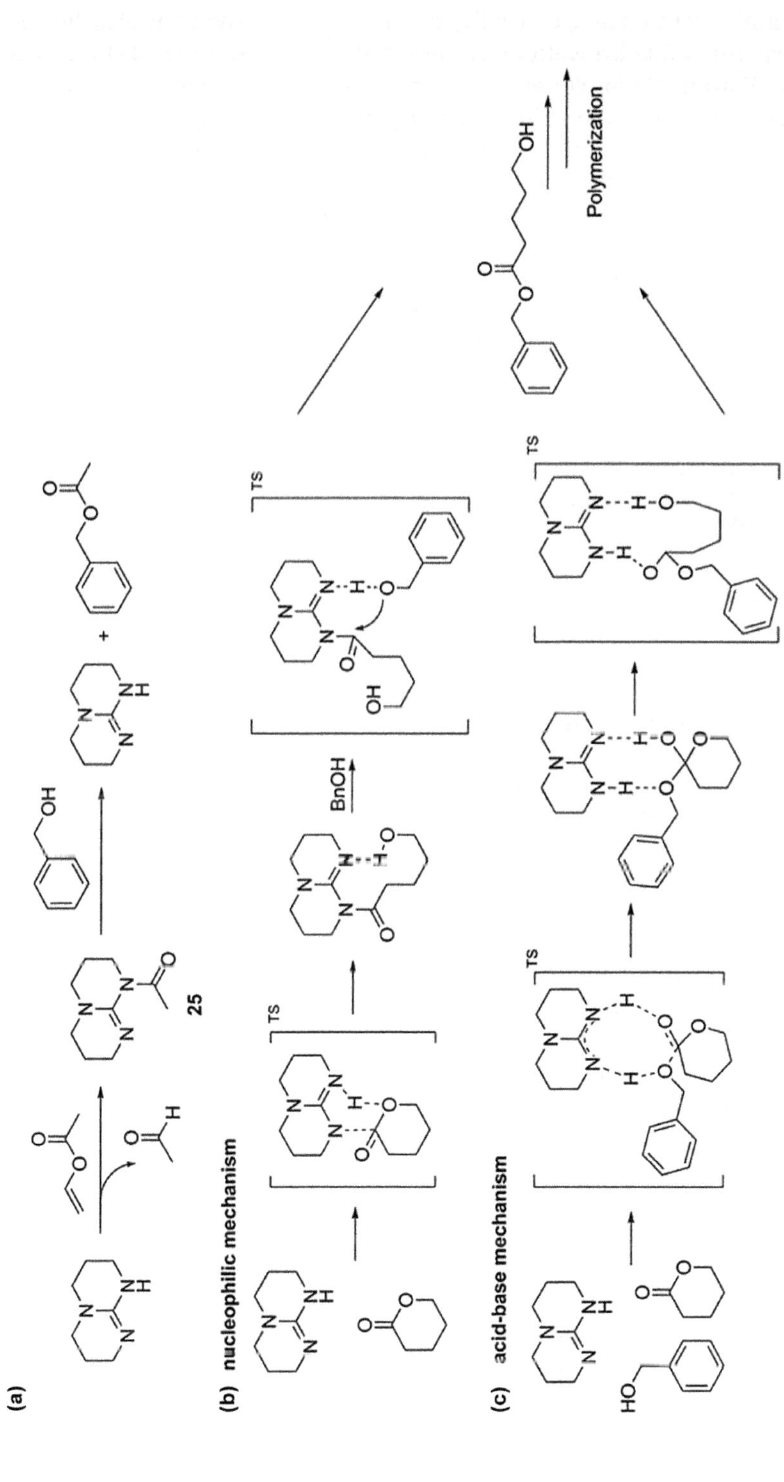

Figure 4.8 (a) Model reaction for transesterification catalysed by TBD;[64] (b) nucleophilic mechanism for ROP of lactones by TBD; (c) acid-base mechanism.[67,68]

tetrahedral intermediate. Crucially, the acid-base mechanism also fits the other experimental observations discussed above: for example, MTBD has no N–H-motif available for H-bonding, hence it should be inactive according to this mechanism. Also, in spite of making use of the strong basicity of TBD, no full charge separation is proposed and thus no acceleration of the reaction can be expected when polar solvents (THF, DMF) are employed. The prevalence of acylation in the model reaction (Figure 4.8c) was convincingly related to the high concentration of BnOH. Under polymerization conditions, [BnOH] is smaller by a factor of more than 200, disfavouring the nucleophilic pathway relative to the acid-base analogue.[68] For reaction with BL, it was shown that the catalyst is trapped in a stable amide-like intermediate.[67] Interestingly, Carpentier and Guillaume later showed that at increased reaction temperatures (60–80 °C) this species breaks up and polymerization is possible with both TBD and DBU.[69] In bulk, BL was polymerized to $M_n > 20$ kg mol^{-1} ($Đ_M = 1.10–1.30$, TBD) and $M_n > 9$ kg mol^{-1} ($Đ_M = 1.15–1.5$, DBU). End-group analysis supports the proposed polymerization mechanisms, where polymer with crotonate- and TBD-derived termini was observed (see also Section 1.2.3).

Problem-free availability and broad applicability—at least in the case of TBD—have prompted many other investigations, including more unusual and functionalized lactones. Duchateau and co-workers showed in a detailed study that the ability to polymerize lactones for TBD also extends to ω-pentadecalactone (PDL), a sixteen-membered macrolactone that is essentially strain-free and notoriously difficult for metal-free catalysts.[70] Thus, reactions in bulk and in solution (toluene) at 100 °C in the presence of initiator (BnOH) delivered PPDL in excellent conversion (up to 99%) and high molecular weights ($M_n > 25$ kg mol^{-1}). Polydispersity, which always tends to be less well-controlled for PDL compared to medium-sized lactones, was found to be in the range $Đ_M = 1.3–2.0$. In spite of a rapidly growing viscosity, it was possible to achieve a linear correlation of M_n and conversion, suggesting living characteristics. One-pot copolymerization with CL was also successfully demonstrated, resulting in fully randomized copolymer with depressed melting points relative to the one for pure PDL. Notably, while the TBD/BnOH setup might be less active than some metal-based catalyst systems, it is one of very few successful organocatalytic approaches for PPDL synthesis (Figure 4.9). In the same paper, it was shown that neither DBU nor MTBD or other acyclic guanidines were able to polymerize PDL; similarly, NHCs, DMAP or imidazoles were found to be inactive.[70] A similar mechanism as suggested for CL or VL (see above) was assumed, namely no "simple" anionic pathway but a base mechanism with dual activation of the monomer.

TBD and DBU were both successfully employed to catalyse a copolymerization of 3,4-dihydrocoumarin (DHC) and styrene oxide (SO).[71] The process was shown to be almost perfectly alternating with barely any SO–SO ether linkages found, helped by the fact that the lactone is non-homopolymerizable. Reaction occurred both in the absence and presence of an initiator (Figure 4.9), suggesting that again zwitterionic growth is

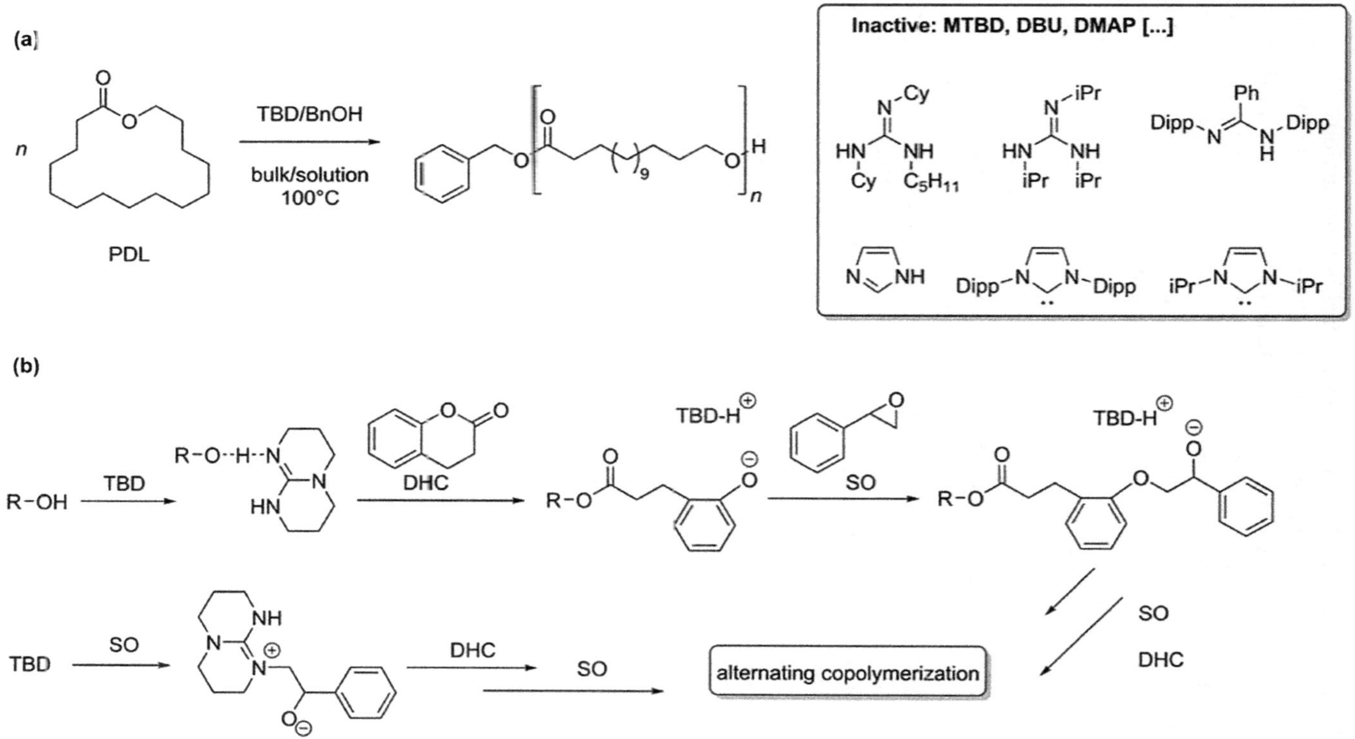

Figure 4.9 (a) Polymerization of PDL by TBD/BnOH and examples of inactive catalysts for this monomer;[70] (b) basic (top) and nucleophilic (bottom) mechanism for alternating copolymerization of DHC and SO.[71] Dipp = 2,6-diisopropylphenyl.

possible (no initiator). A chain-end-activation mode was proposed when TBD/DBU were used in conjunction with an alcohol initiator. Interestingly, macrocycles are generated in both cases by backbiting reactions.

The introduction of functional groups on lactones extends the range of applications and properties for the resulting polyester. However, this might also render the polymerization of monomers with increased complexity more difficult. In this context, Waymouth and co-workers were able synthesize and then polymerize differently substituted 3-mercaptovalerolactones (Figure 4.10), using TBD/BnOH in solution or in bulk (melt).[72] The presence of the β-thioether moiety clearly reduced the propensity for polymerization. In toluene, yields between 10% and 60% were achieved with catalyst loadings of 5% (1–24 h, $M_n = 0.5–5.0\ \mathrm{kg\,mol^{-1}}$, $Đ_M = 1.30–1.60$). In the melt (25–45 °C), higher conversion was observed (60–70%) and TBD was overall more effective than $Sn(Oct)_2$, however, partial degradation of the monomer was also found. Copolymerizations with CL went more smoothly, whereby random copolymers were synthesized with the copolymer microstructure reflecting the monomer feed ratio. In general, the much reduced polymerizability of these substituted lactones was attributed to less favourable thermodynamic conditions. The Thorpe–Ingold effect was identified as a major contributor to this behaviour.

The polymerization of lactide monomers bearing poly(ethyleneglycol) (PEG) units with variable length was met with somewhat less difficulty.[73] Conversion failed with $Sn(Oct)_2$ but succeed upon application of TBD. Molecular weights exceeding 10 $\mathrm{kg\,mol^{-1}}$ were realized ($Đ_M = 1.60–2.10$). Longer PEG substituents caused "internalization" of the lactide moiety, thus aggravating polymerization. The chemistry for successful functionalization

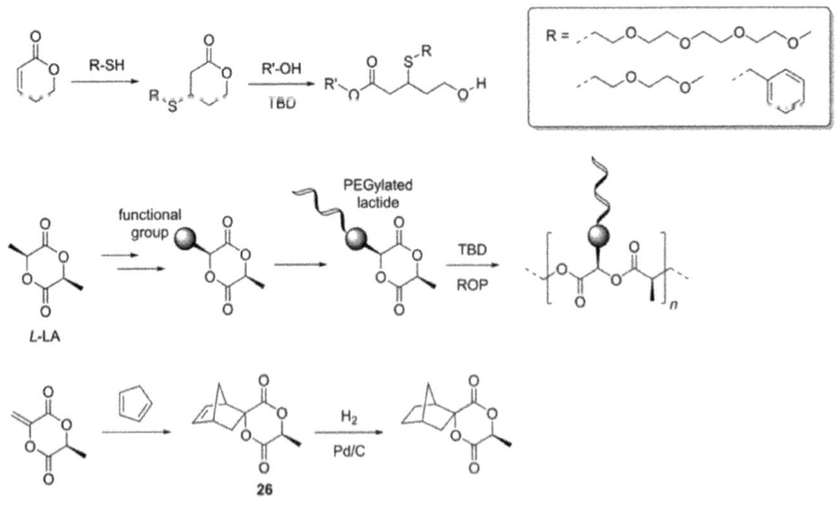

Figure 4.10 Top: Polymerization of mercaptovalerolactone;[72] middle: TBD-mediated polymerization of PEG-bearing lactide moner;[73] bottom: example for Hillmyer's functional lactide-derivatives.[74,75]

of lactide, without ring-opening it had been developed by Hillmyer and co-workers, who used exomethylene-lactide (Figure 4.10) as a platform molecule to synthesize various lactides with functional groups and who applied TBD to copolymerize these compounds with non-modified lactide.[74,75] A monomer such as norbornene-lactide (**26**) can be subjected to ROP as well as ROMP (ring-opening metathesis polymerization), thus opening access to intriguing materials. A hydrogenated version of **26** can be used to tailor thermal degradation and glass transition temperatures of pure PLA.

Amidine/guanidine organocatalysis has also been applied for the preparation of advanced polymer architectures. For example, Coulembier and co-workers published the synthesis of amphiphilic PEO-*b*-PCL-*b*-PLA, using a one-pot, two-step procedure.[76] Notably, this study tried to eliminate hazardous substances in the preparation as far as possible, which was realized by the metal-free catalyst system and the employment of bulk conditions. On PEO as macroinitator a block of caprolactone was grown by application of TBD. As discussed above, this catalyst is prone to engage in transesterification for lactide, so TBD was quenched by addition of acid; then, PLA was introduced as third block by DMAP/acid mediation. The resulting amphiphilic material had molecular weights $M_n > 10$ kg mol^{-1} with a relatively well controlled dispersity ($Đ_M = 1.40$). Interestingly, the influence of the initiator on the molecular weight distribution for TBD-catalysed CL polymerization was also investigated, an aspect that has not yet received much attention. This was prompted by the finding that PEO macroinitiator entailed a better controlled polymerization than analogous reactions using pyrene butanol. Modelling suggested that this might result from the overall time TBD spends in the vicinity of the initiator. For the less flexible, slowly diffusing pyrene butanol this is much longer than for the agile PEO–OH chain ends. The latter could thus result in slower, but more "averaged" polymerization of CL, engendering a more controlled polydispersity of molecular weights. Finally, in the same paper several acyclic guanidines were also tested regarding their activity for lactide polymerization. Conversions were found to be <40%, in accordance with their generally lower reactivity compared to cyclic guanidines. Nonetheless, the best performance was displayed by 1,1,3,3-teramethylguanidine (**15**), with which molecular weights of up to 12 kg mol^{-1} were accessible. Phenyl-substituted analogues, more electron poor compared to **15**, were much less active.

Sly, Miller and co-workers employed TBD-based catalysis to synthesize biodegradable star polymers (Figure 4.11).[77] This was achieved by polymerization of VL (in the presence of initiator, toluene), followed by addition of a cross-linker (**27**). Observation by GPC proved that this reaction went well, with barely any "arms" left free in solution. In line with the previous findings discussed above, the polymerization of VL by this catalyst delivered well-defined linear PVL arms ($M_n = 3$ kg mol^{-1}, $Đ_M = 1.06$). The cross-linked star-like product still had a controlled molecular weight distribution ($Đ_M = 1.20$–1.30). In this context, again the superiority of the organocatalytic approach compared to traditional $Sn(Oct)_2$-mediated polymerizations was

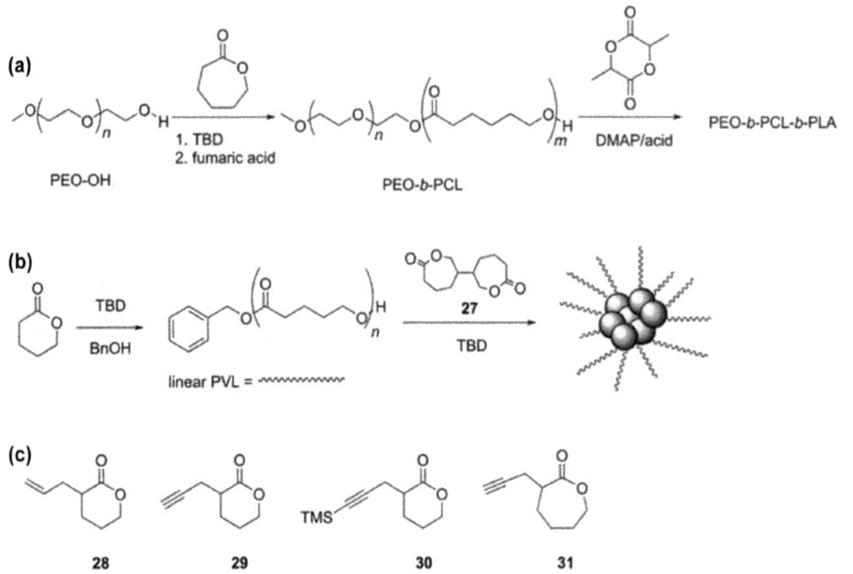

Figure 4.11 (a) Amphiphilic triblock copolymer with the central PCL unit synthesized by application of TBD;[76] (b) preparation of biodegradable nanogel star polymers;[77] (c) alkene and alkyne functionalization of CL and VL.[78]

impressively demonstrated. Not only were the reactions faster and much better controlled with TBD, the residual amounts of Sn(II) in the star polymer were exceptionally high (1800 ppm). Removal of tin species from such a surrounding would be especially demanding, since the corresponding complexation agents would face competition by the highly functionalized polymer. Hence, in this case organocatalysis enables sensitive applications that would be difficult to realize otherwise.

The polymerization of allyl- and propargyl-substituted VL and CL (**28–31**) using TBD was also reported, with the overall aim of nanoparticle preparation.[78] When reactions were stopped at 70–85% conversion, homopolymer with molecular weights of 12–15 kg mol^{-1} was received, with well-defined dispersity ($Đ_M$ = 1.10–1.20). Importantly the authors also determined the apparent rate constants (k_{App}) for unsubstituted VL and CL and the modified monomer ($[M]_0$ = 2 M in toluene, $[M]_0/[I]$ = 140, 2 mol% catalysts loading relative to monomer). In line with the findings by Waymouth and Hedrick,[65] VL was consumed much faster than CL by a factor of 40 (k_{App} = 0.2580 *vs.* 0.0064). Propargyl-bearing CL was somewhat more prone to polymerize (k_{App} = 0.0105), while for the substituted VL the corresponding parameters ranged from 0.0374 to 0.2740, underlining that alkene/alkyne functionalization is not detrimental at all for TBD-mediated organopolymerization of lactones.

Unlike with cyclic ester monomers, where MTBD or DBU (without further additives) are restricted to lactide while TDB can polymerize a whole range of other lactones, the important class of carbonate monomers is more prone to conversion by a number of different OBCs.[79] Inspiring work by Waymouth and Hedrick demonstrated that organocatalytic synthesis of polycarbonates from trimethylene carbonate (TMC) suppresses decarboxylation as a major side reaction and also enables the preparation of polymer in excellent yields with high control over end groups and molecular weights.[80] Again, TBD, MTBD and DBU (among others) have been applied (Table 4.4). While TBD shows the highest activity even for high ratios of catalyst to monomer, the polymerization is much less controlled than found for DBU. In the latter case, an extraordinarily low polydispersity is observed, which importantly does not broaden at high conversion or prolonged stirring (3d). DBU retains these positive characteristics in bulk polymerization (65 °C), where the dispersity increases somewhat ($Đ_M = 1.09–1.15$), but end-group fidelity and molecular weight control is still given. Under these conditions, chain extension experiments were also successful. Mechanistically, the authors were able to observe the H-bonded complex between DBU and the initiator (**32**, Figure 4.12), resulting in a significant low-field shift of the Bn–O–*H̲* signal ($\delta = 6.09$ ppm, C_6D_6). Notably, from the viewpoint of degradation properties, a random copolymerization of TMC and CL was demonstrated in the same paper. The superiority of DBU relative to other amines (trimethylamine, pyridine, aniline, quinuclidine) had already been identified by Endo and co-workers, who used DBU for the polymerization of a functionalized carbonate.[81] Since no initiator was present in this latter case, again also a nucleophilic mechanism seems a reasonable alternative for DBU-catalysed carbonate polymerization; based on that, it was also proposed that a double activation might explain the exceptional polymerization ability of DBU (Figure 4.12).[79]

The group of Guillaume had a closer look at TBD for carbonate polymerizations. Conditions were chosen to be very practicable, that is reactions were conducted in the melt and with non-purified TMC.[82] While control over the polymerization was not as good as found for DBU, the high activity of TBD allowed for some impressive results. It was shown that technical TMC could be converted to polycarbonate with $M_n > 25$ kg mol^{-1} ($Đ_M = 1.54$, TOF = 6000 h^{-1}, 91% conversion after 15 h at 150 °C), notably at an extremely low catalyst loading (TBD/BnOH/TMC = 1 : 100 : 100 000). TBD tolerated impurities and the high alcohol content much better than Zn-based

Table 4.4 Polymerization results of TMC using cyclic amidine/guanidine catalysts.[80 a]

Catalyst	Target DP	Time (min)	Conv. (%)	Reached DP	$Đ_M$
TBD	50	15	>99	50	1.31
TBD	250	60	>99	240	1.32
TBD	500	360	>99	420	1.31
MTBD	50	180	>99	48	1.28
DBU	50	480	>99	51	1.04

[a]DP = degree of polymerization. Conditions: 1 mol% catalyst loading relative to monomer, CH_2Cl_2, $[M]_0 = 2$ M.

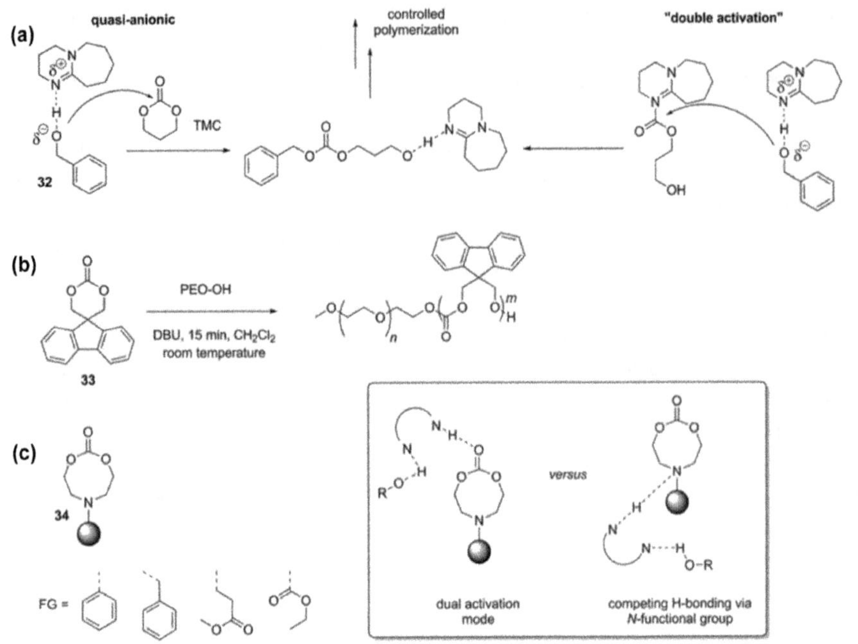

Figure 4.12 (a) DBU-catalysed polymerization of TMC in the presence of initiator, engaging in an anionic-like process (left) or combined with additional nucleophilic activation of the monomer (right); (b) preparation of diblock copolymer based on PEO and fluorine-bearing carbonate;[83] (c) eight-membered carbonate modified with *N*-functional groups and calculated competing reactant complexes.[84]

organometallic catalysts. The polymerization of functionalized (benzyloxy or trimethoxy groups in 3-position) TMC also succeeded.

Yang and Hedrick showed that fluorene-functionalized TMC (**33**) can be polymerized by DBU/BnOH.[83] Albeit this reaction is less controlled than for pristine TMC, it was still possible to obtain conversions of ~90% with acceptable polydispersity. When the reactions were allowed to proceed further, a marked deviation from the linear correlation of M_n and conversion occurred, suggesting side reactions. Interestingly, this process could be used to prepare amphiphilic diblock copolymers, using PEO–OH as initiator (PEO with 2–10 kg mol^{-1}, DP(**33**) = 20, $Đ_M$ (diblock) = 1.15–1.33).

In 2015, Yang, Hedrick and co-workers investigated the DBU/TBD catalysed polymerization of *N*-substituted functional eight-membered carbonates (**34**, Figure 4.12).[84] (Co)polymers of these readily accessible monomers are seen as highly promising for medical applications. DBU and TBD were both effective in this case, with TBD expectedly displaying faster polymerization. However, it was found that the nature of the *N*-substituent (alkyl, aryl, carbamate) had a major influence on the polymerization kinetics. This was elucidated by calculations: interestingly, it was found that competitive H-bonding can occur between the nitrogen atom in the monomer and the N–H moiety of TBD. This disrupts the dual nature of TBD catalysis. In spite of the energy differences being relatively small for the different reactant complexes, this was seen as influential enough to slow down the polymerization. Coherently, the effect should be strongest for *N*-alkyl bearing monomer, which is nicely mirrored by experimental results. Under optimized conditions, the eight membered carbonates could be converted to polymer in high conversion in less than 2 h, with a generally good molecular weight distribution ($Đ_M < 1.30$).

Dove and co-workers have recently demonstrated that TDB can be employed for selective ring opening of carbonate moieties in the presence of epoxy functional groups on the same monomer (**35**, Figure 4.13).[85] With 1 mol% TBD, **35** was converted to polymer with $M_n > 4$ kg mol^{-1} (DP ≈ 20) and $Đ_M < 1.20$ in less than 20 min. The pendant epoxy groups of the polycarbonate were subsequently functionalized by reaction with amines or thiols.

Albertsson and Olsén have studied the copolymerization of CL with an allyl-functionalized, six-membered carbonate (**36**).[86] Using TBD, they had previously shown that this particular organocatalyst polymerizes **36** readily and with good control.[87] For the copolymerization, the group succeeded in the preparation of multiblock material by using a simple "temperature switch". CL is polymerized efficiently only at higher temperature, while **36** is readily polymerized at −40 °C. This enabled a sequential preparation of a well-defined nine-block copolymer in only 2.5 h, rendering this thermodynamics-based approach[88] a useful extension to the toolbox for polymer chemists. Overall, TBD was more suited than a similarly tested phosphazene base, on account of impracticable fast polymerization and the occurrence of side reactions in case of the latter OBC.

Copolymerization of TMC with lactide was also focused on in several studies. Coulembier was able to show that from a 1 : 1 (wt.%) mixture of TMC and L-LA, which melts at 21 °C (eutectic), DBU/BnOH forms homo-PLA.[89] Addition of CH_2Cl_2 as solvent then allows for the still active chain ends to incorporate TMC, resulting in a copolymer with a strong gradient. Interestingly, Carpentier and Guillaume further demonstrated that the microstructure of L-LA/TMC copolymers can be tuned by choice of catalysts.[90] While two metal-based systems resulted in the formation of gradient-copolymers (whereby preference for TMC or lactide was each achieved),

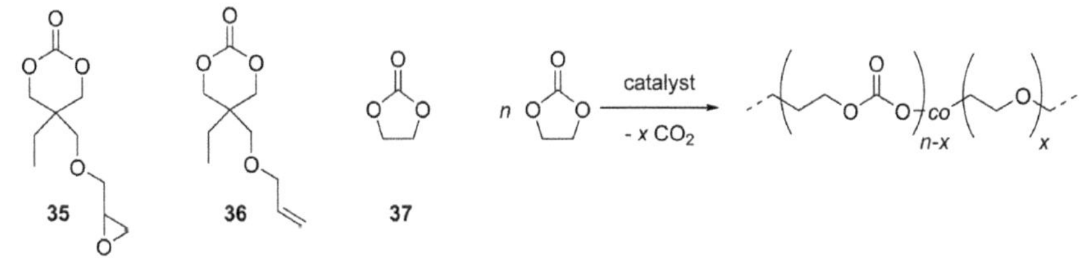

Figure 4.13 Several carbonate monomers and general polymerization scheme for ethylene carbonate.

TBD/BnOH generated fully random copolymers. In toluene at 100 °C, after 1.75 h, a polymer with $M_n = 10$ kg mol^{-1} ($Đ_M = 1.40$) was synthesized, whereby the catalyst loading was quite low (TBD/BnOH/TMC/L-LA = 1 : 5 : 250 : 250). ^{13}C NMR analysis supported the formation of a random structure.

As a final example for carbonate organopolymerization by OBC-type compounds, a recent work by Möller and Keul on ethylene carbonate (**37**) shall be considered.[91] Much in contrast to TMC, this monomer is more resistant to polymerization as a consequence of its stable, non-strained structure. An important driving force is decarboxylation. The partial loss of CO_2 results in polymer with both ether and carbonate linkages (Figure 4.13). This motif is very interesting from a number of perspectives, among them the degradability of poly(ether-carbonate)s. Using TBD/initiator, under demanding conditions (180 °C, 3 h, microwave) high conversions were achieved (>98%), whereby the resulting polymer had molecular weights of 2–3 kg mol^{-1} ($Đ_M$ <1.60–1.70). The carbonate content in the material was determined to be 15–20%. More complex polymer structures were also realized (polyester-*b*-poly(ether-carbonate)-*b*-polyester) and side reactions investigated. Among them, the TBD-catalysed reactions displayed transcarbonylation, a consequence of the higher reactivity of carbonate repeating units compared to the ethylene carbonate monomer. Overall this application again highlights the robustness and versatility of TBD-based organopolymerizations.

Cyclic amidines and guanidines were also found to be excellent organocatalysts for the ROP of cyclic phosphoesters, such as 2-isopropoxy-2-oxo-1,3,2-dioxaphospholane (iPP, **38**, Figure 4.14). Poly(phosphoester)s are considered as very promising materials for a range of applications,[92,93] perhaps best obtained from ROP processes. In 2010, Iwasaki published the first organopolymerization of **38**, employing DBU or TBD in conjunction with an initiating alcohol.[94] Compared to metal-based systems, the organocatalytic approach was described as much less difficult; in general, narrow polydispersity and

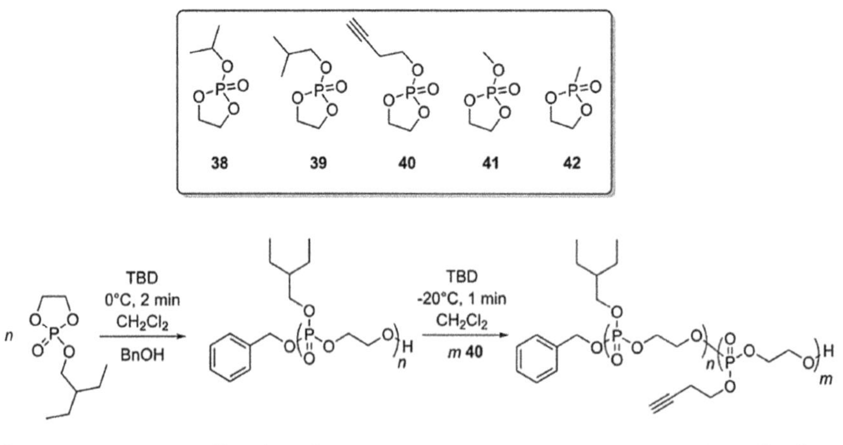

Figure 4.14 Top: cyclic phosphoester monomers. Bottom: preparation of a functional diblock copolymer as described by Wooley.[97]

substantial molecular weights were obtained. For DBU, M_n and conversion were linearly correlated up to 60% conversion. TBD was again observed to have a heightened activity. Supported by NMR experiments, this was again linked by the authors to the mode of activation, which is "quasi-anionic" for DBU and "dual" for TBD. In a typical reaction (TBD/OH/iPP = 2 : 1 : 200), after 20 min a polymer with $M_n = 29$ kg mol^{-1} ($Đ_M = 1.05$) was delivered.

Jérôme and co-workers assessed DBU and TBD for differently substituted phosphoesters, with a special emphasis on transesterification mechanisms.[95] Employing a similar monomer (**39**) to Iwasaki, the authors showed that with DBU, at higher conversion, significant transesterification occurs, broadening the molecular weight distribution from $Đ_M = 1.10$ to 1.40. GPC traces displayed a tail towards lower molecular weight. TBD again performed well and showed very fast polymerization kinetics for **39**, with a DP of 100 reached already after two minutes (DBU took 11 h for the same DP, under identical conditions). However, GPC showed bimodal molecular weight distributions with TBD. Interestingly, the high molecular weight peak still increased after full conversion at the cost of the lower molecular weight one, suggesting a different type of transesterification from DBU. In the same paper it was also shown that addition of thiourea as cocatalyst massively improved results (see Section 3.2.1).

The group of Wooley used poly(phosphoester)s to construct more complex structures. For example, copolymerization of **40** and **41** enabled subsequent click-type reactions with azides or thiols.[96] It was found that for copolymerization, DBU was not active enough. Hence, TBD was applied, but at 0 °C and for only very short polymerization times (20 s), to minimize side reactions. This was the base for further work, where functional diblock copolymers were synthesized from two different phosphoesters in an analogous manner.[97] After further modification this enabled the generation of non-ionic, anionic, cationic and zwitterionic amphiphilic diblock copolymers for various self-assembly processes in water. The same group later also demonstrated that DBU can be used in a one-pot process to prepare poly(phosphoester)-*b*-poly(L-LA) material ($Đ_M = 1.20$, sequential monomer addition).[98]

Wurm and co-workers investigated the organocatalytic preparation of poly(phosphonate)s, a subclass of poly(phosphoester)s where in the monomer one P–O moiety is substituted by a P–C bond (**42**).[99] In contrast to polymer from compounds like **38** or **39**, the polymerization of **42** does not suffer from much transesterification when DBU is employed as catalyst. This may be connected to the fact that fewer modes of transesterification are accessible for the phosphonates.[95] Notably, polymer in the range of 1.5–17 kg mol^{-1} was synthesized with $Đ_M < 1.10$. The linear correlation of M_n and conversion remained valid up to conversions of more than 90% (less than 2 h, 0 °C). Overall, this type of polymer—and especially when polymerized by metal-free catalysts—must be considered to be very promising for biomedical applications as it is water-soluble, potentially biodegradable and well-defined.

An instructive example for amidine and guanidine organopolymerization was reported by Landais and Cramail.[100] This study details the organopolymerization of polyols (PEG-600) and isophorone diisocyanate (IPDI) to the corresponding polyurethanes (Figure 4.15). An extensive range of different organobases was surveyed, including DBN, DBU, TBD and MTBD but also DMAP, DABCO (1,4-diazabicyclo[2.2.2]octane) and acyclic guanidines such as **15** or **19**. Their performance was directly compared to dibutyltin dilaurate (DBTDL), a frequently employed Sn(IV) catalyst. In the bulk (60 °C) conversion was monitored by FT-IR to reveal amidines and guanidines as superior catalysts. While all catalysts delivered high molecular weight PU ($M_n = 20$–35 kg mol^{-1}), the time required for quantitative conversion was very different. DBU, MTBD and **15** were more active than DBTDL, reaching completion after only 15 min. In contrast, DMAP and DABCO required almost 5 h to do the same. TBD, initially quite unexpectedly, performed poorly. A general base mechanism could be ruled out, since if so the activity should scale with the pK_a values; this was found not to be true. However, it was demonstrated that TBD readily reacts with isocyanates to form a urea derivative (**43**). If this derivative was then employed as catalyst, it polymerized at the same slow rate that was found for TBD, suggesting that **43** was the real catalyst—and obviously much less basic and nucleophilic compared to TBD, explaining the low activity in this case. Interestingly, MTB also reacts with isocyanates, but this adduct (**44**) releases MTBD again when warmed up. This reversibility ensures that the catalyst remains active; in a later publication the same group exploited this effect to realize a thermally latent polymerization.[101]

Finally, it should be noted that cyclic amidine/guanidine OBCs have not only been used for polymerization reactions, but also for depolymerization of poly(ethylene terephthalate) (PET).[102] Strikingly, the frequently met differences between the monofunctional amidines DBU and DBN and the

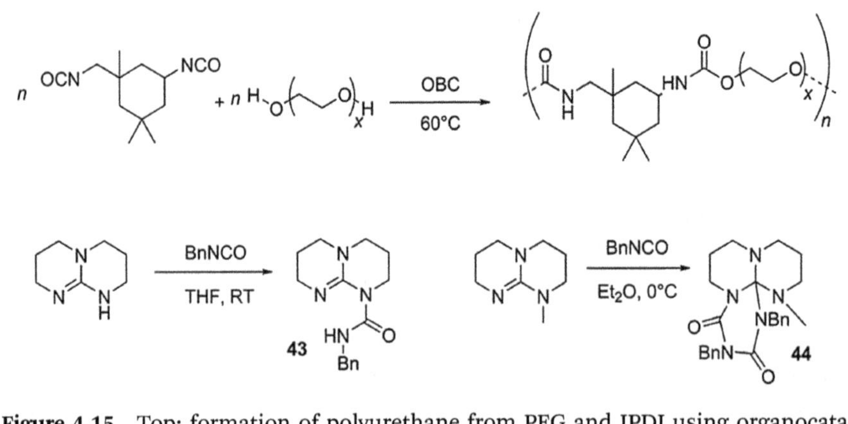

Figure 4.15 Top: formation of polyurethane from PEG and IPDI using organocatalysis. Bottom: adducts formed by reaction of TBD and MTBD with isocyanates.[100]

bifunctional guanidine TBD again surfaced in this context, too. Underlining that pK_a values alone are not a sufficient guideline to judge the activity of an OBC, it was found that the degradation of PET was much faster with DBU and DBN than with TBD, but only if reactions were conducted in excess of short chain alcohols. Once longer-chain alcohols were applied, this tendency was reversed and TBD became the most active depolymerization catalyst (even performing better than a likewise tested phosphazene). The authors attributed this effect to the different modes of activation. Once the concentration of –OH functionalities decreases or the alcohols are sterically more hindered, dual activation is required for efficient catalysis and TBD is best suited. An excess of readily accessible alcohols, on the other hand, allows DBU to compensate for its lack of bifunctionality.

4.3 Phosphazenes

4.3.1 Synthesis and Properties

Phosphazenes (iminophosphoranes) represent a highly successful, versatile class of uncharged, largely non-nucleophilic Brønsted bases that are constituted of at least one P(v)-atom bonded to four nitrogen moieties (three amino, one imino, Figures 4.16 and 4.17). Starting from the literature-known structure **45**,[103–106] it was Schwesinger who developed phosphazenes into the versatile tool they are today by conducting detailed studies into their physical and chemical properties and who managed to also massively increase their basicity (Table 4.5).[13–15,31,107] Nowadays, these compounds are frequently employed in organic chemistry[108] and many examples exist for their advantageous application in polymer chemistry.[23,109–111]

Schwesinger showed that homologization of iminophosphoranes increases the basicity more than expected as a consequence of (among other factors) an improved delocalization of the positive charge. This "basicity battery cell"-style increase of the corresponding pK_a values plateaus at P_5–P_7, that is, a phosphazene structure with five to seven P=N-motifs. For systems with more than two phosphors, isomers can be obtained (Figure 4.17). Initially, it was not clear whether linear or branched phosphazene structures would turn out to be more basic, but for most series of compounds this can now be answered in favour of the branched derivatives. The pK_a data for several phosphazenes, including the most popular in polymer chemistry, are given in Table 4.5; as solvent, acetonitrile has been chosen, but the basicity of phosphazenes has also been determined under other conditions.[112–116]

By structural modification, the basicity of phosphazenes can be varied over a very wide range of about 20 pK_a units. The most significant influence is exerted *via* the number of P=N-groups. For similarly substituted compounds, from P_1 over P_2, P_3, P_4 to P_5 phosphazenes, the corresponding pK_a values stepwise increase from 27 over 33, 36, 43 to 46 (CH_3CN). A related P_7-base was however shown to be not more basic than P_5 derivatives, demonstrating that this strategy cannot be extended without limits. Nonetheless,

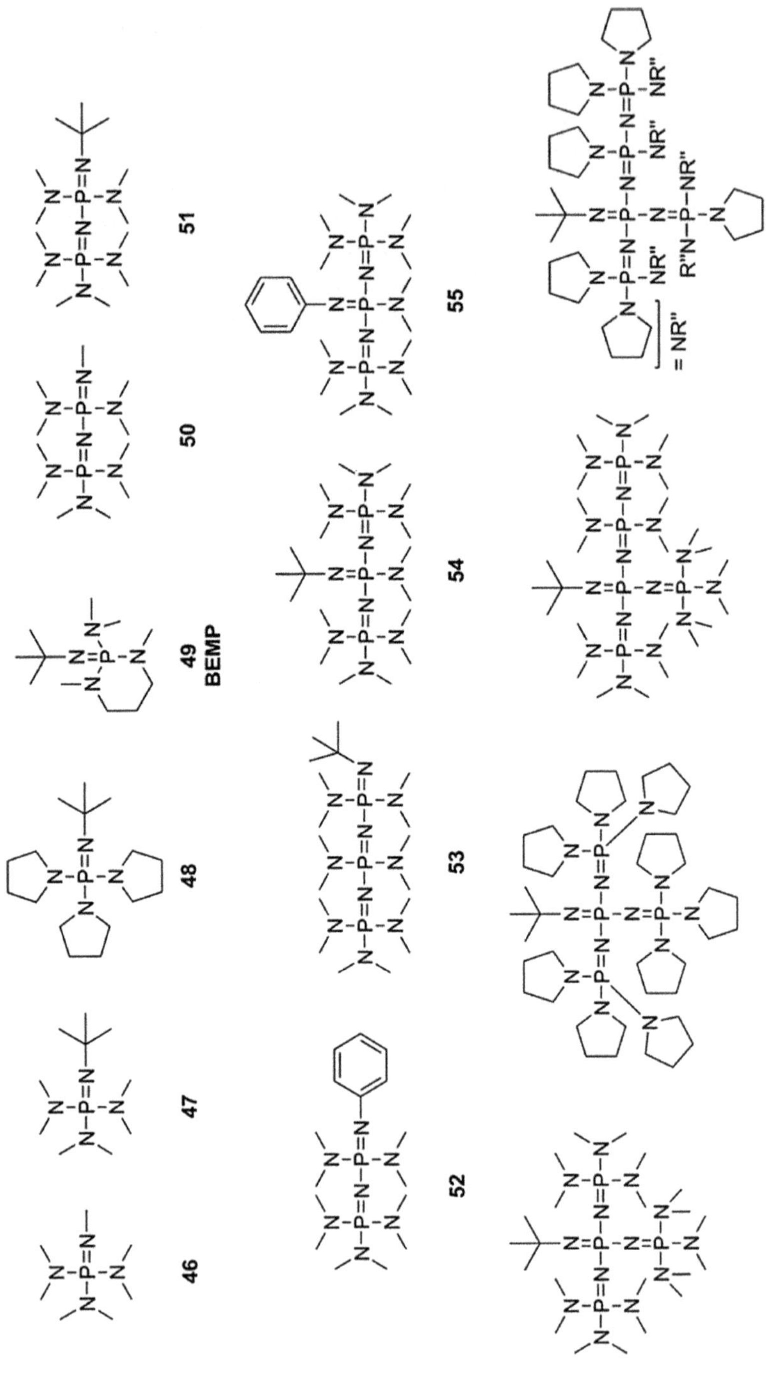

Figure 4.16 Structures of several P_1–P_5 phosphazene bases.

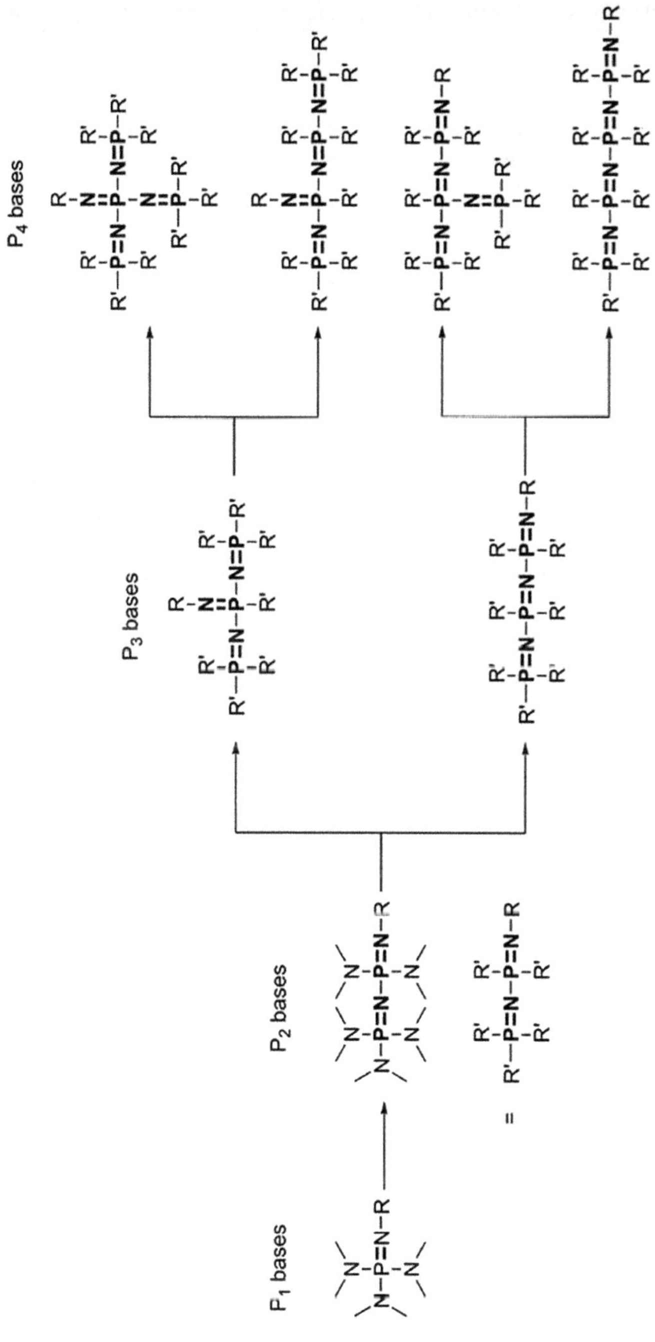

Figure 4.17 Homologization and structural isomerism as established by Schwesinger.[13]

Table 4.5 The pK_a data for various phosphazenes, determined in acetonitrile and DMSO.

Number of P-atoms	Compound	pK_a (MeCN)	pK_a (DMSO)	References
P1	**46**	27.6		13
	47	26.9	15.7	107
	48	28.9		107
	49 (BEMP)	27.6		107
P2	**50**	32.7		13
	51	33.5		13
	52	26.5		50
P3	**53**	36.6		13
	54	38.6	26.2	13
	55	31.5		50
P4	**56**	42.7	30.2	13
	57	44.0		13
P5	**58**	45.3	>32	13
	59	46.9	>32	13

the reported high pK_a values of phosphazenes are remarkable, and the P_4- and P_5-bases are strong enough to deprotonate a wide range of organic molecules; this also shows in the multiple applications found for them in polymerization catalysis (Section 4.3.2). Apart from the number of P=N-moieties, basicity can additionally be modified by several other strategies. For example, the substitution of the imino functionality alone has a significant impact in some cases. Compound **51**, a P_2-base bearing a *tert*-butyl group, is more basic than **50** (methyl group) and much more than **52** (phenyl substation). If, instead of $N(CH_3)_2$, pyrrolidine groups are employed, the pK_a values also increase (compare **56**/**57** and **58**/**59**, Table 4.5). It is also possible to incorporate the phosphazene motif in a cyclic structure; BEMP (**49**, 2-*tert*-butylimino-2-diethylamino-1,3-dimethylperhydro-1,3,2-diazaphosphorane) is the most frequently employed and a commercially available cyclic derivative.

The interplay of steric congestion and conjugation and the resulting effects on basicity are not always clear for phosphazene bases. In some cases, a limited decrease of basicity is found and can be correlated to steric demand, but this trend is not general.[107] Likewise, the higher basicity of branched P_3 species relative to the linear analogues has been attributed to the respective stabilities of the free bases; after protonation, resonance should be comparable for both cations and consequently should not be the root for the considerable differences in basicity. In contrast, for the P_5 series, where also the branched structures display higher pK_a values than the linear ones, differences in cation stabilization were suggested to be dominant.[13] Crystal structure X-ray analysis of a P_1 base shows that the P=N double bond is shorter than the P–N single bonds in the free base (by about 14 pm).[107] Upon protonation (Figure 4.18), the bond lengths roughly equalize. Crystal data of protonated oligophosphazenes (phosphazenium salts) display a tetrahedral configuration of the four nitrogen atoms around phosphorous and the

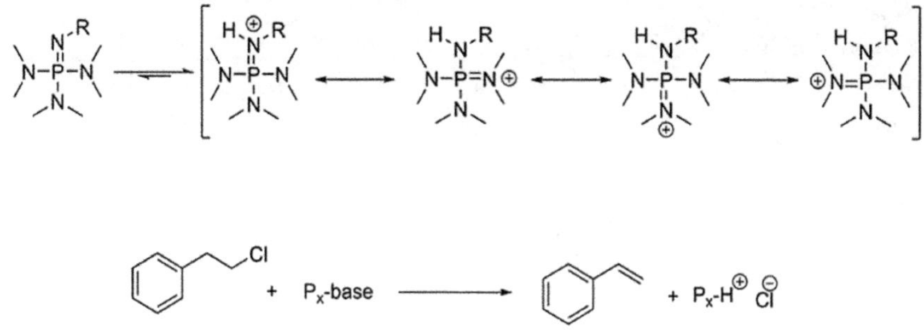

Figure 4.18 Mesomeric stabilization of a phosphazenium ion (top) and dehydrohalogenation reaction (bottom).

P–N–P bridges are bent (138–158° for a typical P_5-base).[117] The P–N–P–N units are not planar. It was suggested that overall, steric congestion controls the conformation of phosphazenium ions rather than electronic preferences. In solution several conformations of comparable energy most likely coexist.

Apart from the scalable basicity, phosphazenes also possess a number of other beneficial properties. Solubility is generally very good, including nonpolar and polar solvents; with the latter care has to be taken as deprotonation can occur (DMSO, acetone, *etc.*). Albeit being hygroscopic in most cases, phosphazenes are unusually stable towards water (hydrolysis only at 160 °C, $t_{1/2} \approx 40$ h)[14] and also against electrophilic attack: control reactions with alkyl halides demonstrate a propensity for dehydrohalogenation while alkylation of the phosphazene is largely suppressed. Test reactions with *iso*-propyl bromide underlined that most phosphazenes are more resistant against alkylation than DBN or DBU.[107] If the steric shielding is decreased (for example by using pyrrolidine substituents instead of $N(CH_3)_2$) nucleophilicity will grow, however. Many of the lower-molecular weight phosphazenes are liquids and distillable. Interestingly, even the sterically most hindered phosphazenes are kinetically active, as evidenced by 1-alkene isomerization.[13] Also, nowadays not only BEMP but also a number of other phosphazenes are commercially available. These include **47** (*t*Bu-P_1), **48, 51** (*t*Bu-P_2) and **56** (*t*Bu-P_4), with the latter probably being the most popular OBC of this class.

The synthesis of these potent base catalysts can be demanding, depending on the desired substituents and the size of the phosphazene system. A "block-like" construction usually gives good results (Figure 4.19), based on building blocks like compound **60.**[13] As an example, coupling of this structure with imino trichloride **61** results in a precursor compound. From this, the P_2-base **51** is liberated by a treatment with KOMe, followed by distillation. Likewise, the P_3-base **54** is accessible in a related manner (Figure 4.19). The linear analogue is prepared by reaction of a P_2-builidng block with the corresponding monochloride **62**. Among the P_4-bases it is *t*Bu-P_4 (**56**) that is synthesized in the easiest manner, employing a one-pot

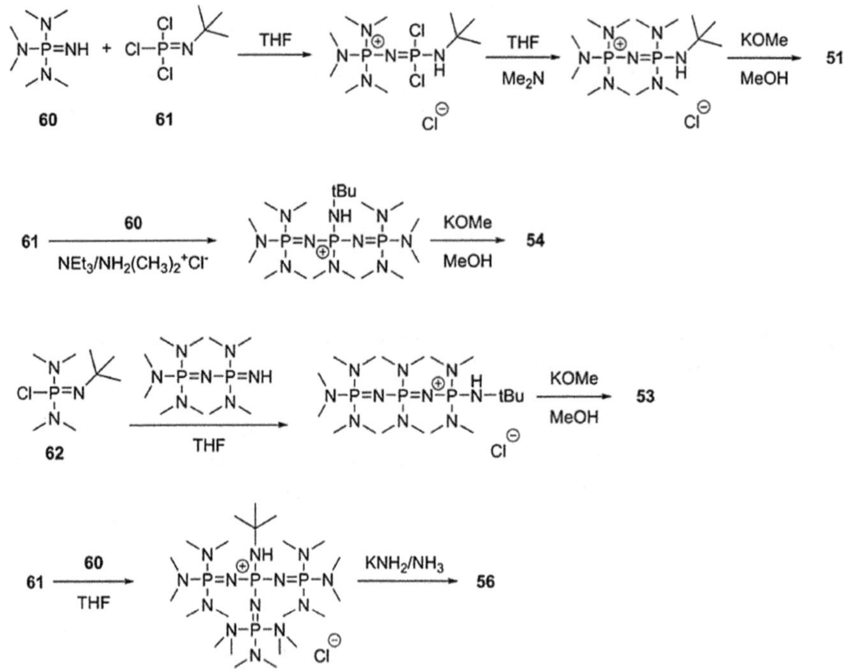

Figure 4.19 Examples for the block-like construction of phosphazene bases developed by Schwesinger.

reaction of **60** and **61**. In this special case, the free phosphazene can only be generated by application of KNH_2 in liquid ammonia, followed by extraction with hexane.

4.3.2 Phosphazenes as Base Catalysts for Polymerizations

The adaptability of phosphazenes with regard to their basicity and the wide range of accessible pK_a values is mirrored in a broader applicability when compared to amidines or guanidines. With these OBCs, cyclic esters and carbonates, but also epoxides, lactams, cyclosiloxanes and even acrylic monomers can be polymerized. On the other hand, the mechanistic ambiguity encountered with amidines and guanidines, and also NHCs and NHOs, is much less pronounced for phosphazenes. Although labelling phosphazenes as "non-nucleophilic" is still not true in every case, the overwhelming majority of proposed mechanisms assumes base catalysis and/or (quasi-)anionic processes. However, in this context it should also be noted that protonated phosphazenes can activate carbonyl groups.[118]

The employment of phosphazenes for organopolymerization of lactones was pioneered by Wade and co-workers, who used BEMP and **47** (*t*Bu-P_1) for the polymerization of VL and lactide (1-pyrenebutanol as initiator).[119] The authors were able to show that a polymerization with living characteristics ensues. The activity of BEMP for lactide polymerization was moderate (with 1 mol% catalyst loading a conversion of 76% resulted after 23 h of polymerization time, $M_n = 13\ \text{kg mol}^{-1}$, $Đ_M = 1.08$), comparable to DBU/MTBD but inferior to the performance of TBD (Section 4.2.3). Lactide epimerization was observed to be minimal, with *rac*-LA an isotactic-enriched polymer was generated (probability of isotactic placement (P_i) = 0.70). Under identical conditions, polymerization of VL proceeded slower (69% conversion after 73 h, $M_n = 9\ \text{kg mol}^{-1}$, $Đ_M = 1.12$). Compound **47** was found to be somewhat less active than BEMP. Interestingly, ^{1}H NMR experiments showed that the –OH signal of benzyl alcohol is shifted from $\delta = 1.13$ ppm in toluene-*d8* to 2.50 ppm once BEMP was added. In contrast, no direct interaction of this phosphazene with the monomers VL or lactide was found, suggesting a mechanisms that operates by activation of the –OH functionality of initiator or propagating chain end (**63**, Figure 4.20). In a follow-up study, the same group later demonstrated that, by shifting to a P_2-base (**51**), *rac*-LA can be polymerized in a stereoselective manner to yield highly isotactic polymer ($P_i = 0.95$).[120] At the time, this was a performance superior to any other organocatalyst and on par with the best metal-based catalysts available. One of the factors favouring this high selectivity was that **51** retains its high activity down to −75 °C, another was proposed to be the considerable steric bulk of the phosphazene. Coherently, the stronger basic P_2-base entailed a stronger shift of the –OH signal of BnOH to $\delta = 7.66$ ppm (toluene-*d8*).

BEMP or **47** delivered an impractically slow conversion of CL, which prompted further studies to illuminate the impact of reaction conditions and of other, stronger phosphazene bases. Hadjichristidis and co-workers screened a range of different initiators and solvents, using **51** as OBC for CL polymerization.[121] With a relatively high monomer concentration of $3\ \text{mol L}^{-1}$ it was shown that the polymerization is fastest in DCM (DCM ≫ toluene > 1,4-dioxane ≈ THF), with almost quantitative conversion in less than 1 h reaction time. However, under these conditions the polydispersity is relatively high ($Đ_M = 1.42$), a problem that still perseveres at lower starting concentrations of CL. Reactions in THF were slow, but more controlled. The fact that the reactions proceeded more rapidly in toluene than in more polar THF was explained by competitive complexation of the propagating chain end. With regard to initiators it was successfully demonstrated that also secondary alcohols can be employed, as well as benzamide, while a thiophenol or a carboxylic acid were not found suitable in combination with **51**.

Notably, also the more challenging lactone monomers PDL, BL and γ-butyrolactone (GBL) can be polymerized by phosphazene bases. BL has attracted a lot of attention, because, as discussed in the previous section, many *N*-based heterocyclic organocatalysts struggle with this monomer. In 2010, Coulembier studied the polymerization of a substituted, four-membered lactone (**64**) in an extensive publication.[122] BEMP, *t*Bu-P_1 (**47**), *t*Bu-P_2 (**51**) and *t*Bu-P_4 (**56**) were employed in combination with a PEG-based carboxylic acid as initiator. It was found that the polymerization kinetics

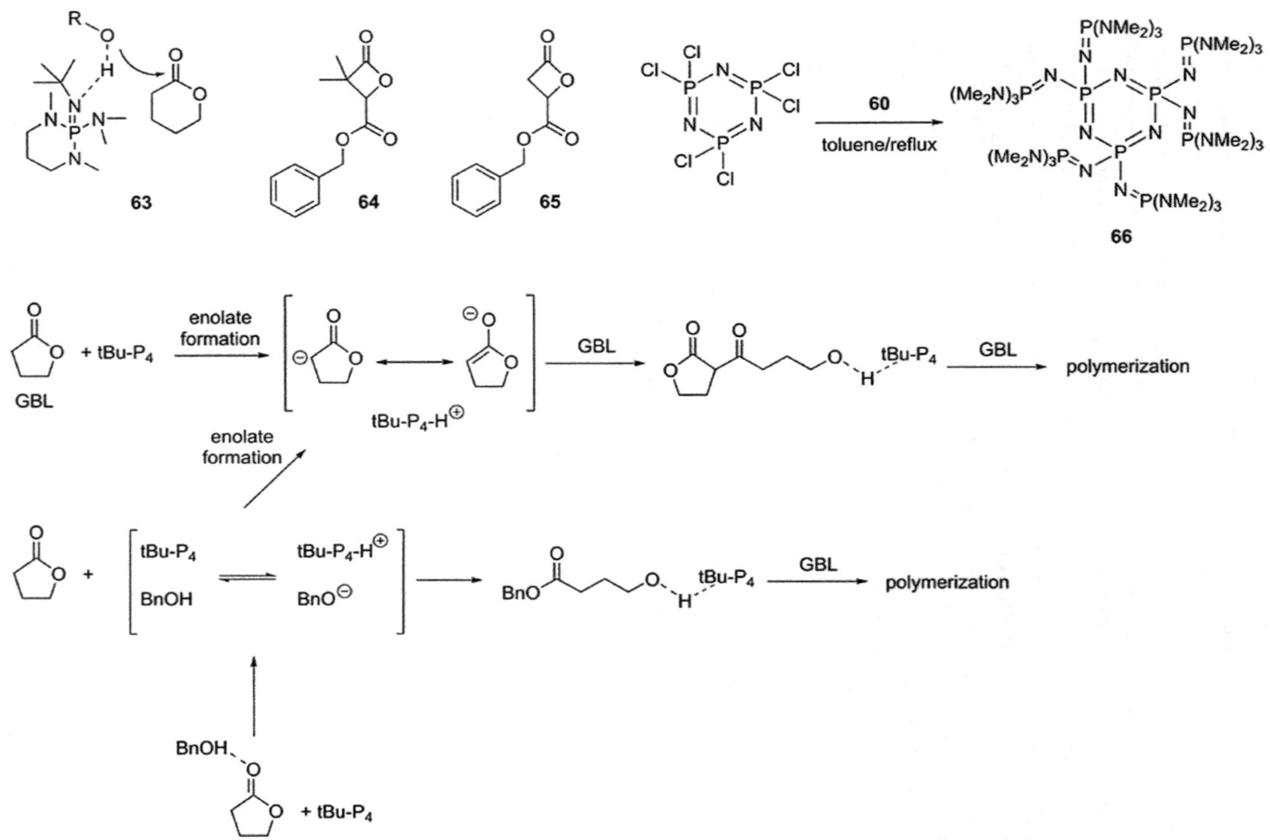

Figure 4.20 Structures as discussed for organopolymerization of lactones using phosphazenes (top) and metal-free polymerization mechanism for GBL using *t*Bu-P_4 as proposed by Chen.[127]

were clearly correlated with the basicity of the phosphazene; the fastest conversion was achieved with *t*Bu-P_4, the strongest base in the inspected series (forming also the largest counter ion). Furthermore, only *O*-acyl scission took place, as evidenced by MALDI-ToF MS analysis, while transfer reactions were absent (helped by 3,3-substitution of the monomer which prevents α-proton abstraction). All phosphazene bases enabled a relatively controlled polymerization and very high molecular weights could be obtained (up to 1.5×10^6 g mol^{-1}). With the same set of OBCs it was shown that the purification method for BL decisively influences the outcome of the polymerization.[123] As briefly mentioned above, Guillaume and Carpentier highlighted that polymerization of BL works in the bulk at elevated temperature with DBU or TBD, and the same is true for application of BEMP in this regard.[69] In that manner, PBL with $M_n > 25$ kg mol^{-1} ($Đ_M = 1.05$–1.23) was obtained. Interestingly, evidence was produced for a direct attachment of the phosphazene at one terminus of the polymer (no initiator present), suggesting that BEMP can initiate polymerization in a nucleophilic manner, too; however, these results have recently been disputed, especially regarding TBD.[124] The same groups also succeeded in the copolymerization of BL with benzyl β-malolactone (**65**),[125] whereby the microstructure of the copolymer was determined by catalytic tuning: BEMP, in contrast to other metal-based catalysts and organocatalysts, effected the formation of a block-copolymer even when both monomers were simultaneously present (bulk, 60 °C, no initiator).

The polymerization of the (almost) strain-free GBL was recently reported by the group of Chen,[126] and interestingly phosphazenes were shown to enable an efficient conversion of this monomer.[127] With *t*Bu-P_4, detailed NMR and MALDI-ToF MS experiments revealed that the phosphazene base can polymerize *via* different mechanisms (Figure 4.20). Polymerization also occurs in the absence of initiator; rather than by a nucleophilic pathway, this finding was explained by direct deprotonation of the monomer, forming an enolate, which is reactive enough to initiate polymerization. ^{1}H NMR clearly showed that *t*Bu-P_4 can abstract a proton from GBL, and mass spectra of the polymer found acylated lactone/H termini. However, the polymerization proceeded much faster in the presence of BnOH, where notably it made a difference whether the phosphazene was pre-mixed with BnOH followed by addition of monomer, or whether GBL and BnOH were mixed with subsequent addition of *t*Bu-P_4. In the former case, conversion was faster, but under the latter conditions more controlled. This was rationalized by assuming that in the first case initiation occurs by both the enolate and the BnO^- anion, while premixing of GBL and BnOH disfavours the enolate pathway. Notably, under suitable conditions, it was possible to achieve conversions of more than 90% ($M_n = 27$ kg mol^{-1}, $Đ_M = 2.0$, -40 °C).

In this setup, both macrocyclic and linear polymer architectures are found simultaneously, with the relative ratio depending on the absence or presence of initiator and the mixing sequence. Very recently, Liu and Li presented a potent cyclic, trimeric phosphazene base (**66**), with which the selective preparation of linear poly(GBL) is possible.[128] For **66**, a pK_a of 33.3 in CH_3CN was determined, rendering it much less basic than *t*Bu-P_4 (42.7). This may account for the fact that not only no macrocycles were found (no backbiting), but also that no enolate-initiated polymer was generated. The same catalyst **66** was also applied to prepare poly(GBL)-*b*-poly(L-LA) copolymers in a one-pot setup with sequential monomer addition.[129] GBL was polymerized first, then a solution of L-LA/THF was added (−50 °C). Full GBL conversion was not possible, and competing transesterification (including macrocyclization) was cited as a major obstacle. Molecular weights of up to 20 kg mol^{-1} with moderate control ($Đ_M = 1.5–2.0$) were prepared in this way.

As a final example for phosphazene-mediated polymerization of lactones, a report on PDL polymerization is considered.[130] While this macrolactone is a challenge for many organocatalysts, with strong phosphazene bases the polymerization runs relatively well, with 91% conversion after 1 h polymerization time at 80 °C ($M_n = 26$ kg mol^{-1}, $Đ_M = 1.63$, toluene, *t*Bu-P_4). With a weaker P_2-base the reaction is much slower. Interestingly, with higher catalyst loading and higher monomer concentration the reaction also proceeds at room temperature, albeit with a very broad molecular weight distribution resulting ($Đ_M = 3.8$). A bicyclic phosphazene was found to be inactive for PDL polymerization.[70]

The polymerization of unpurified, technical grade TMC succeeds with BEMP.[82] With optimized reaction conditions, it was possible to consume up to 100 000 equivalents of the carbonate (10 ppm BEMP-loading, M_n up to 45 kg mol^{-1}, $Đ_M = 1.20–1.60$) in polymerization times of 15–26 h. Sequential monomer addition also allowed for the preparation of block-copolymers from TMC and lactide.[131] With the same phosphazene base, the resulting polymer was found to be well defined ($M_n = 36$ kg mol^{-1}, $Đ_M = 1.32$) and received in high yield (>90% overall conversion, L-LA second monomer). Interestingly, the polymerization of ethylene carbonate, as briefly discussed above with the application of TBD, was also investigated in combination with phosphazenes, specifically catalysis by the strong P_4-base **56**.[132] The authors found a strong dependence on reaction temperature. While bulk polymerizations at 100 °C or 120 °C proceeded sluggishly, with 150 °C or 180 °C the monomer was quantitatively consumed after 8 h and 2.5 h, respectively. The resulting poly(ether-carbonate) displayed a carbonate content of 20–25% ($M_n =$ up to 14 kg mol^{-1}, $Đ_M \approx 1.7$).

The polymerization of commercially important epoxides such as ethylene oxide (EO) or propylene oxide (PO) is a formidable challenge for metal-free polymerization catalysts; no successful incident has been reported using guanidines or amidines, for example. Phosphazenes, in contrast, have attracted attention early on, initially because they could be used for complexation of Li^+ from alkyllithium initiators.[133–137] Hence, as an additive, phosphazenes enjoyed some success in epoxide polymerization, but these examples will not be discussed here, on account of not representing truly metal-free systems. The next step in this evolution was the employment of pre-formed alkoxides with phosphazenium-based counter ions for EO and

PO polymerization.[138,139] The phosphazenium alkoxides were thought to be of interest because the cation—the phosphazenium ion—is large and soft with little propensity for aggregation, thereby supporting the nucleophilicity of the propagating oxyanionic chain end. Indeed, Möller reported the preparation of well-defined PEO this way ($M_n = 1.7–6.5$ kg mol^{-1}, $Đ_M = 1.07–1.13$).[138] For PO polymerization, which is additionally complicated by transfer to monomer as side reaction (Figure 4.21), Mülhaupt observed that the polymerization rates neatly correlated with the size of the phosphazenium cation, with *t*Bu-P_4-H^+ effecting the most rapid monomer consumption.[139] However, the weaker coordination is simultaneously accompanied by an increase in transfer to monomer as observable by the appearance of C=C double bonds. Consequently, the polydispersity was somewhat broadened ($Đ_M = 1.11–1.15$) and the molecular weight limited. Schlaad prepared α,ω-heterobifunctional PEO using *t*Bu-P_4 in combination with either *p*-cresol or α-methylbenzyl cyanide as respective initiators (Figure 4.21).[140] A molecular weight of 2.5 kg mol^{-1} was targeted, which was achieved after 20 h at 45 °C (>90% conversion, $Đ_M = 1.04–1.09$). With a simple potassium alkoxide, the same degree of polymerization requires days of reaction time.

In the following time, much effort was put into the preparation of more complex polymer structures based on the phosphazene/epoxide system. For example, it was shown that carboxylic acids can act as initiators for polymerization of EO in the presence of *t*Bu-P_4;[141] this works in spite of the initiator being obviously much more acidic than the propagating species. NMR and MALDI-ToF MS analysis show that indeed the carboxylate starts the polymerization. The expected product (terminated by one ester moiety and one –OH group) forms the main distribution in mass spectrometry, while two more species were identified as diester- and dihydroxyl-terminated. Alternatively, terpenes as well as poly(isobutylene)–OH have been used as (macro-)initiators for *t*Bu-P_4-mediated polymerization of EO.[142,143]

Hadjichristidis also highlighted that poly(acrylamide) is a suitable initiator *via* its pendant –NH_2-groups.[144] Activation with *t*Bu-P_4 and reaction with EO affords macromolecular combs. Interestingly, the complexity can be increased and it is possible to prepare also double-graft polymer and block- as well as random-copolymer side chains (Figure 4.21). Pispas and co-workers also prepared a graft copolymer by phosphazene catalysis.[145] The group employed poly(*p*-hydroxystyrene) as backbone polymer. Application of *t*Bu-P_4 and a PO/EO feed afforded thermoresponsive brush copolymers. While the material was well-defined ($Đ_M = 1.16$) and the copolyether "arms" were found to be essentially random, it was also observed that the fraction of incorporated PO was lower than expected. At the same time, side reactions occurred (transfer to PO monomer). This again supports the finding that the large, soft cations of a sterically demanding phosphazenium salt favour higher reaction rates but also some side reaction by only slightly coordinating the oxyanionic chain end. Tri- and pentablock copolymers based on EO, CL and L-LA have been realized by application of the P_2-base **51**.[146]

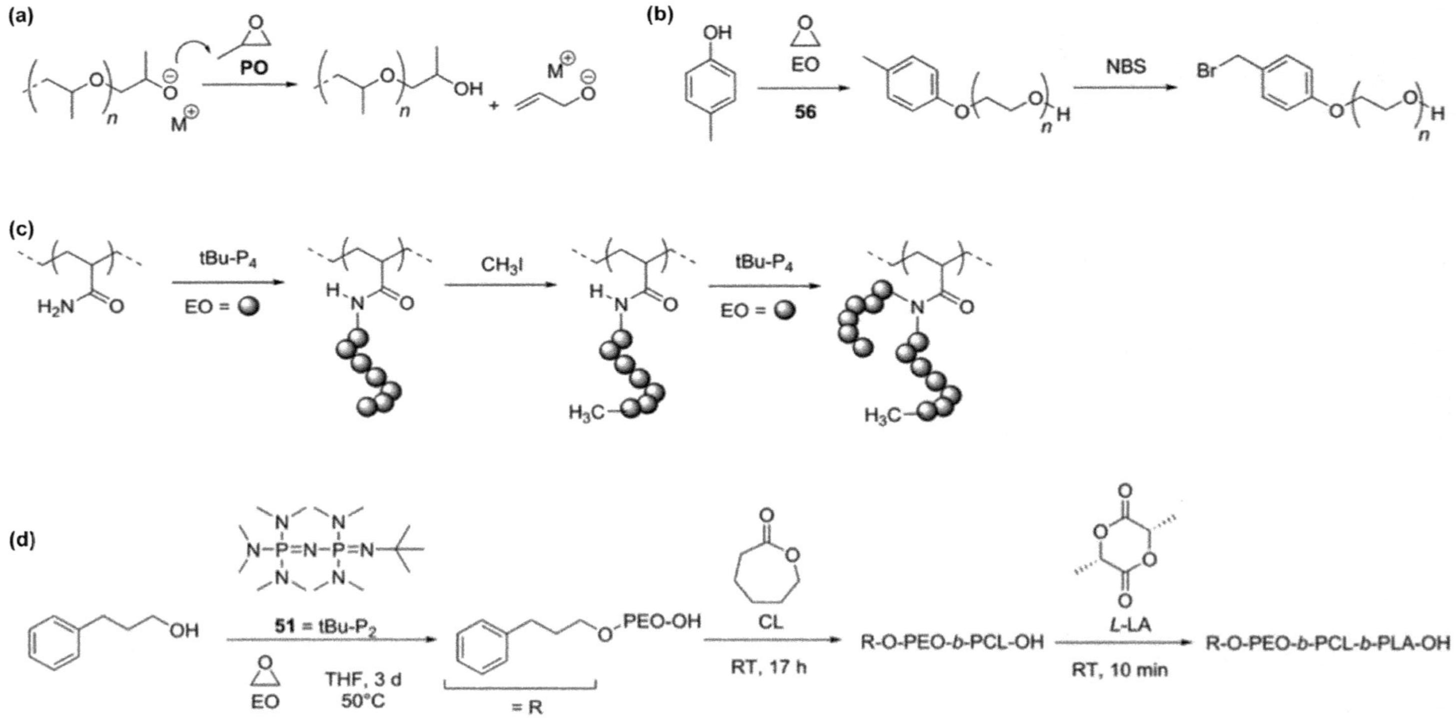

Figure 4.21 (a) Transfer to monomer as complication for anionic PO polymerization (M^+ = metal or non-metal counter ion); (b) hetero-bifunctional PEO as described by Schlaad;[140] (c) and (d) comb-like and multiblock copolymer prepared by application of phosphazene bases.[144,146]

In a one-pot reaction with the same catalyst (sequential addition, PEO first, CL must be polymerized before lactide in this system), well-defined copolymer was received. Importantly, it should be noticed that successful polymerization of cyclic ester monomers requires a milder phosphazene base, hence in this case *t*Bu-P_2 is used in preference to *t*Bu-P_4. The downside is that with this OBC the polymerization of EO takes comparatively long (Figure 4.21).

In an effort to overcome these disadvantages, Hadjichristidis and coworkers developed a "catalyst switch" approach.[147] In this method, the strongly basic *t*Bu-P_4 is employed for the cyclic ether monomers, which not only speeds up the polymerization but also extends the monomer scope to substituted epoxides (such as 1,2-butylene oxide, BO). The polymerization of the epoxide in the first stage was followed by the addition of an excess of diphenyl phosphate (DPP) to neutralize the phosphazenium alkoxide and to polymerize lactone or carbonate comonomers in a second stage. In this manner structures like PEO-*b*-PVL or PBO-*b*-PEO-*b*-TMC could be realized with well-defined properties ($M_n = 3$–17 kg mol^{-1}, $Đ_M = 1.04$–1.17). Contrastingly, a consecutive polymerization of EO and CL using *t*Bu-P_4 throughout resulted in ill-defined copolymer.

The importance of choosing the right phosphazene base for polymerization catalysis was recently also underlined by the work of Zhao, focusing on the copolymerization of DHC, the non-homopolymerizable phenolic lactone mentioned briefly above (Section 4.2.3) with several epoxide monomers.[148] Using *t*Bu-P_4, the polymerization works in principle with EO, PO, BO and styrene oxide (SO) to result in a perfectly alternating copolymer. However, the reactions are slow, conversion is never near quantitative consumption (25–63% DHC conversion after 72 h) and macrocycles appear (backbiting) in considerable amounts as impurities relative to the linear major product. Differently, upon application of the much milder P_1- and P_2-bases **47** and **51** conversion (up to 99%) and molar masses rise considerably, and less cyclic polymer is formed in side reactions.[71] This behaviour was attributed to a lively exchange of protons on the propagating –OH termini, the existence of which is only enabled by the much lower pK_a values of a base like **47** relative to **56** (27 *vs.* 43 in CH_3CN, Table 4.5). Since –OH is also less prone to side reactions compared to $–O^-$ the polymerizations are overall more controlled.

The same group also published the strictly alternating copolymerization of phthalic anhydride (PA) and EO (Figure 4.22).[149] Again, it was found to be advantageous to employ *t*Bu-P_1 in preference to *t*Bu-P_4. The former enables an elegant, self-buffering process: on account of its relatively weak basicity it is able to deprotonate the propagating species resulting after addition of PA (benzoic carboxyl group), while it cannot do the same for the –OH terminus after addition of an EO moiety. The proton thereby shuttles between chain end and phosphazene catalyst, a mechanism that is thought to be responsible for the very well-defined copolymers (M_n up to 50 kg mol^{-1}, $Đ_M < 1.10$).

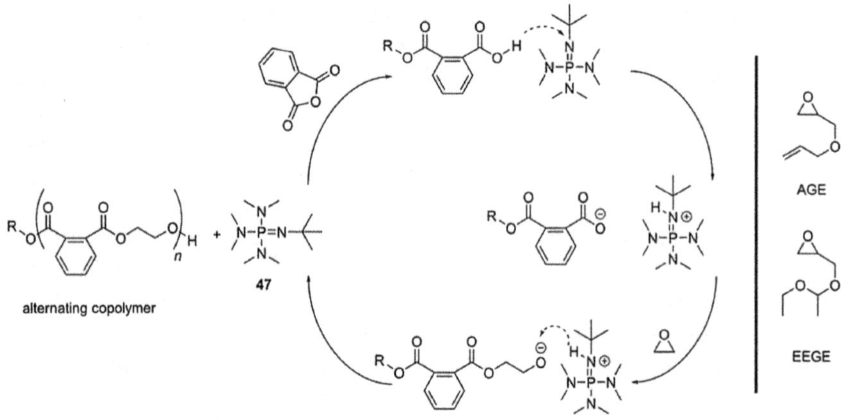

Figure 4.22 Left: mechanism for copolymerization of phthalic anhydride and EO as proposed by Zhao for the mild base *t*Bu-P_1.[149] Right: functionalized epoxide monomers as discussed in the text.

Homopolymerization of SO catalysed by *t*Bu-P_4 proceeds nicely if an alcohol initiator is present, displaying some characteristics of living polymerizations.[150] At room temperature already, high conversion and molecular weights ($M_n > 20$ kg mol^{-1}) result, whereby polydispersity remains controlled ($Đ_M < 1.15$). Propagation occurs mainly through β-scission (attack from the less sterically hindered side), which seems not to be negatively influenced by the strong basicity of the phosphazene (>93% β-selective).

Finally, also allyl glycidyl ether (AGE, Figure 4.22) and ethoxyethyl glycidyl ether (EEGE) have been copolymerized using phosphazene bases.[151] While both monomers were successfully homopolymerized using *t*Bu-P_4, EEGE showed a small propensity for transfer to monomer. Nonetheless, polydispersity remained controlled in both cases ($Đ_M < 1.15$). These beneficial properties also allowed for the preparation of diblock copolymers; the pendant double bonds rendered the material well suited for further modification and phase separation was also found to result in interesting nanoscale patterns.

Other ROP-type applications of phosphazene bases include cyclic siloxane monomers, lactams and cyclopropanes. Octamethylcyclotetrasiloxane (D_4, Figure 4.23) was described to be polymerizable by *t*Bu-P_4/MeOH as early as 1995 by Möller.[152] It was shown that, at room temperature, high molecular weight poly(dimethylsiloxane) (PDMS) is received (M_n up to 440 kg mol^{-1}, $Đ_M$ 1.7–1.9), whereby the phosphazene catalysis compared favourably to other base systems with regard to speed of conversion and to being homogeneous. This was mainly attributed to the size of the phosphazenium cation and the delocalization of the positive charge. The same group later also employed a P_2-base (however in combination with a organolithium compound) to polymerize six-membered cyclosiloxane monomers.[153]

In a detailed study dating from 1996, Memeger and co-workers (DuPont) investigated P_1, P_2 and P_4 bases for the anionic polymerization of lactams, mainly ε-caprolactam (CLA).[16] The traditional mechanism relies on deprotonation of the amide functionality; this requires a potent initiating base and consequently phosphazenes are among the few neutral, fully organic compounds able to polymerize these commercially important monomers. Indeed, it was found that activity clearly scaled with the basicity of the phosphazene, with *t*Bu-P_1 (**47**) being practically inactive while the P_2-base was active but inferior to *t*Bu-P_4. This gives an estimation for the pK_a value required for successful deprotonation of the monomer, similar to what was found for NHC-based lactam polymerization (see Section 4.4.4). The phosphazenes also showed typical characteristics for polyamide (PA) preparation from lactams, specifically the well-known induction period. Intriguingly, the authors also showed that limited amounts of water, otherwise to be strictly avoided, were tolerated by the system. Reaction temperature should not exceed 180 °C. Above, notable degradation of the phosphazene was observed. The molecular weight distribution as observed by GPC (hexafluoro-2-propanol as solvent) was found to be broad ($Đ_M = 2.5–8.0$)—itself not surprising for the non-living anionic polymerization of lactams—with molecular weights up to 50 kg mol^{-1} (M_n). To prove that a proton transfer actually takes place (Figure 4.23), solid state NMR measurements were conducted (^{31}P and ^{13}C MAS). For the non-protonated phosphazene (**56**, *t*Bu-P_4), ^{31}P NMR found *four* signals. This contradicts the seeming symmetry of the molecule (compare Figure 4.16), but fits to crystal structure data reported by Schwesinger, who found asymmetric configurations (see above). In combination with CLA, a protonation is clearly observed, the degree of which rises with temperature. ^{13}C CP-MAS NMR finally confirmed that the proton was abstracted from the monomer. Thus, it was concluded that phosphazenes act in a manner that is fully comparable to standard strong bases for anionic PA 6 preparation. Zhang and co-workers later extended this principle to β-lactams in order to prepare the corresponding PA 3, albeit in the presence of LiCl.[154]

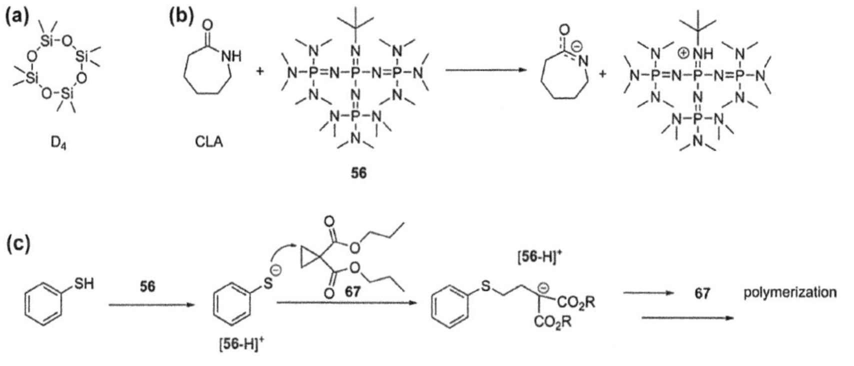

Figure 4.23 (a) Chemical structure of cyclosiloxane monomer D_4; (b) deprotonation of a lactam by a phosphazene base;[16] (c) cyclopropane polymerization as described by Barbier and Penelle.[155]

Substituted cyclopropanes have also successfully been polymerized. Barbier and Penelle reported on the application of *t*Bu-P_4 in combination with thiophenol as initiator.[155–157] Polymerization of di-*n*-propyl cyclopropane-1,1-dicarboxylate (**67**) is achieved in THF (60 °C) to result in well-defined polymer with narrow polydispersity ($Đ_M$<1.10). Molecular weights of up to 30–40 kg mol^{-1} were obtained.[158] As a mechanism, it was plausibly proposed that the phosphazene deprotonated the thiophenol, which in turn ring-opens the cyclopropane. Propagation occurs *via* the malonate carbanion (Figure 4.23). Living characteristics and enhanced polymerization rates render this phosphazene-mediated process superior to conventional approaches. Optimization of this system recommended toluene as solvent and underlined that several other initiators (phenol, carbazole, malonate) can be used, thus enabling end-group control.[156] Strikingly, other organocatalysts than *t*Bu-P4 were found to be completely inactive, including Verkade's base (Figure 4.2), DBN, DBU, 1,1,3,3-tetramethylguanidine (**15**) and *t*Bu-P_2.[158]

Leaving ROP, one of the more unusual applications found for phosphazene bases is their use in polymerization of acrylic monomers. In 1993, Seebach reported the polymerization of methyl methacrylate (MMA).[159] The initiating system consisted of *t*Bu-P_4 and ethyl acetate, the formation of the corresponding enolate was thought to initiate the reaction (Figure 4.24). Polymer properties, typical for the sensitive anionic polymerization of acrylics, heavily depend on the polymerization conditions. The best results were obtained in THF at 60 °C (M_n=16–26 kg mol^{-1}, $Đ_M$=1.1–1.2). Lower temperatures entailed broad molecular weight distribution and higher than expected molecular weights, suggesting incomplete initiation and faster propagation relative to initiation. Next, Heitz published on the phosphazene-mediated polymerization of butyl acrylate.[160] Two initiator systems were used, *t*Bu-P_4/ethyl acetate and *t*Bu-P_4/isobutyric acid methyl ester. Although quantitative conversion was achieved, the polymerizations proceeded uncontrolled. This was attributed to insufficient generation of active species (3–5% of isobutyric acid methyl ester were deprotonated by the phosphazene at −50 °C in THF, and 8–11% of ethyl acetate, base/ester = 1 : 1.1) and also to side reactions between the monomer and *t*Bu-P_4. Curiously, MALDI-ToF MS analysis found that the majority of chains were initiated by butanolate, supposedly liberated from butyl acrylate by interaction with the phosphazene.

A cyclic analogue of MMA, γ-methyl-α-methylene-γ-butyrolactone (MMBL, Figure 4.24), which contains a central GBL unit but polymerizes exclusively through its Michael-acceptor system (no ring-opening), was found by Chen to be readily polymerizable by *t*Bu-P_4.[161] In contrast to earlier reports of organopolymerization of the same monomer employing NHCs (see Section 11.4.1), the strong phosphazene base seems to start the polymerization by direct proton abstraction from MMBL. The reactions proceed with high speed: even at very low catalyst loading (0.02 mol%) quantitative conversion is achieved in less than one minute. The molecular weight distribution

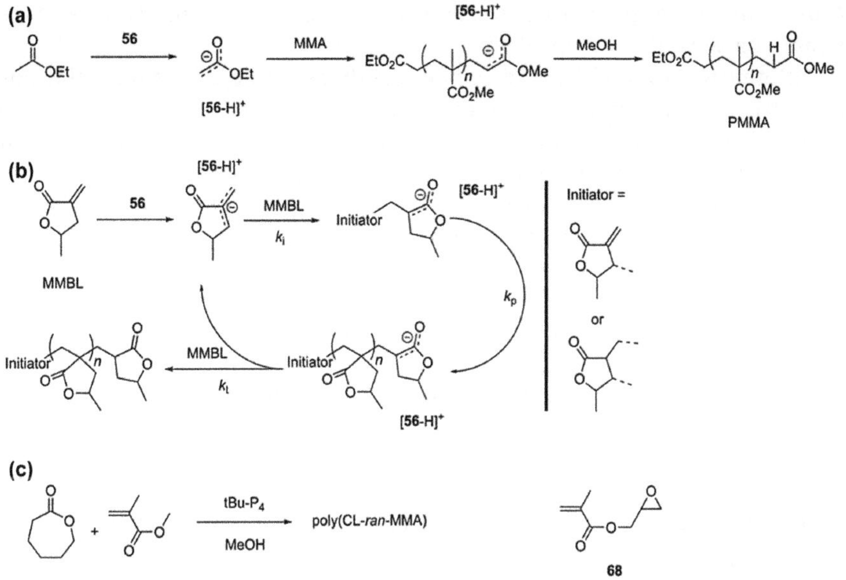

Figure 4.24 (a) Polymerization of MMA initiated by *t*Bu-P_4/ethyl acetate;[159] (b) high-speed polymerization of MMBL as described by Chen;[161] (c) copolymerization of CL and MMA using *t*Bu-P_4/alcohol initiator.[162]

remained broad ($Đ_M > 1.9$), however, as a consequence of a number of side reactions. Most importantly, it was found that more than one polymer chain per phosphazene initiator was produced, indicative of chain transfer processes. This was corroborated by the addition of chain transfer agents, which indeed enabled control over the molecular weights and the number of chains formed. The generally high molecular weights were attributed to the energetic difference between termination processes and essentially barrier-free propagation, as found by DFT calculations.

Remarkably, Zhang and Pispas showed that *t*Bu-P_4 is able to copolymerize MMA and CL.[162] The reactions result in random copolymers, suggesting that during polymerization a cross-over from propagating oxyanionic species (from CL) to enolates (from MMA) must take place. Albeit only moderately controlled material results ($Đ_M = 1.5–2.9$), this kind of hybrid polymerization enables access to an interesting class of polymers. The degradation temperature of poly(CL-*co*-MMA) was shown to be higher than for the respective homopolymers. Also, this approach was extended to *tert*-butyl methacrylate, whereby copolymerization with CL using *t*Bu-P_4 again resulted in random copolymers.[163] The incorporation of the acrylate in the polymer was found to mirror the composition of the monomer feed ($Đ_M = 1.3–3.3$, depending on CL/acrylate ratio). The hydrolysis of the *tert*-butyl groups resulted in the formation of charged, degradable polymer. The same methodology was employed to generate hyperbranched PCL when bifunctional glycidyl methacrylate (**68**) was used as cross-linker (Figure 4.24).[164] Overall, exactly why

tBu-P_4 can bring about this unusual polymerization is not clear yet, since simple base-catalysis is unlikely.

Differently, Coulembier and co-workers demonstrated that phosphazenes can also cyclopolymerize *ortho*-phthalaldehyde.[165] While a strongly basic P_2-and P_4-bases induce side reactions and insufficient control, tBu-P_1 allows for a well-behaved polymerization. With this OBC, high molecular weights (up to 200 kg mol^{-1}), low dispersity ($Đ_M = 1.1–1.3$) and also the formation of block-copolymers were realized (15 min, −85 °C).

As a concluding example for phosphazene-mediated polymerization, a step-growth process shall be considered. The previously mentioned study by Cramail and Landais on poly(urethane) formation also investigated the catalytic performance of BEMP.[100] Although less active than DBU or MTBD, the cyclic phosphazene enabled a faster polymerization than found for TBD, DBN or guanidine **19**.

4.4 N-heterocyclic Carbenes and N-heterocyclic Olefins

Following amidines, guanidines and phosphazenes, now N-heterocyclic carbenes (NHCs) and N-heterocyclic olefins (NHOs) will be discussed in the same section. Both types of compounds share some similarities, mainly regarding their modular synthesis, but also with reference to their polymerization mechanisms: the (co)existence of nucleophilic and Brønsted-basic pathways is characteristic for both. On the other hand also significant differences remain. This distinction is most important for the reactive centres—the carbene-carbon for NHCs and the exocyclic carbon with its high electron density for NHOs—enabling different polymerization reactivity for the two compound families.

Obviously, the following account specializes on the generation of basicity and the application of these compounds as base catalysts. For their employment in the organopolymerization of acrylates or the use of NHCs in ZROP (see Sections 11.4.1 and 1.2.6), information on the polymerization activity of NHCs is also available from recent review articles.[5,6,23,166]

4.4.1 Properties of N-heterocyclic Carbenes

The structure of typical NHCs (Figure 4.25) essentially reflects efforts to stabilize the carbene functionality and harness its reactivity in a practical manner. Carbenes are defined as divalent species with electron sextet. This electronic configuration was long seen as invariably connected to short-lived, transient species in organic synthesis, without much prospective application on their own. Also, early efforts to stabilize carbenes by their incorporation in an N-heterocyclic scaffold enabled the preparation of organometallic complexes with NHC-ligands, but still the isolation of free carbenes remained elusive, compounding this view.[167–169] Change only

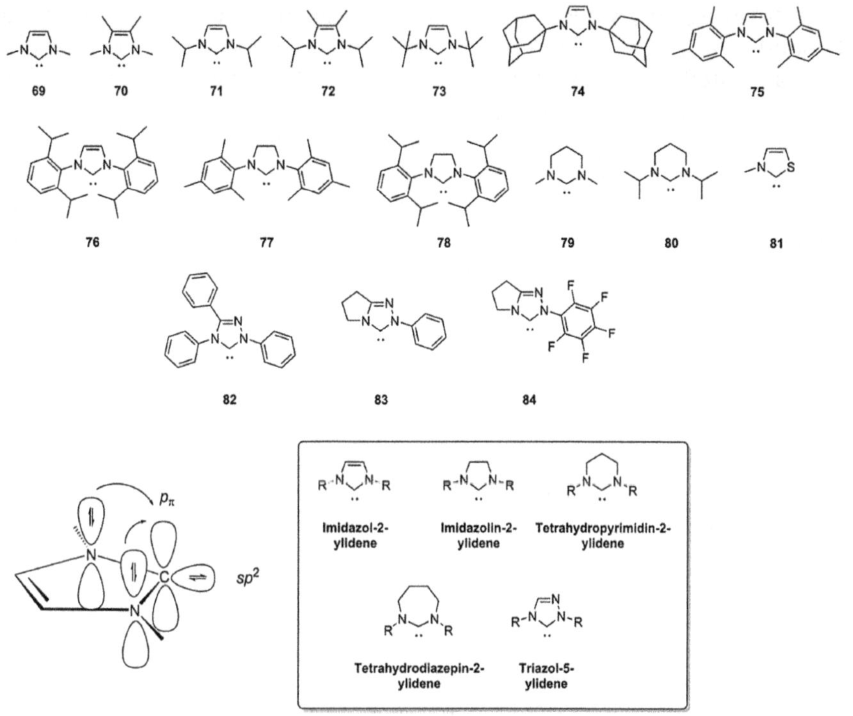

Figure 4.25 Structures of several NHCs frequently employed in organopolymerization and their electronic configuration.

occurred around 1990, based on work by Bertrand[170] and finally Arduengo,[171] who reported the isolation and characterization of **74** (1,3-diadamantyl-imidazol-2-ylidene). This compound could be recrystallized from toluene and was stable in the absence of water or oxygen.

The neighbouring nitrogen atoms stabilize the carbene by donation of π-electrons and by their electron withdrawal *via* the σ-bonds.[172,173] Crucially, this modulates the electronic situation at the carbon in a way that the gap between HOMO and LUMO increases and spin pairing occurs to form a singlet state (unlike the triplet-configured classical carbenes). This is further supported by the cyclic structure, which encourages a bent, sp^2-like hybridization. The HOMO and LUMO orbitals are thus best understood as sp^2-lone pair on the carbon and empty *p*-orbital, perpendicular to the N–C–N plane (Figure 4.25), respectively. The relative electron-richness and the filled sp^2-lone pair explain why NHCs are strongly electron donating, as well as excellent ligands, nucleophiles and also Brønsted bases. However, it should be noted that NHCs are not pure σ-donors. Backbonding does occur and seems to make a significant contribution to the overall NHC–metal bond strength in organometallic complexes.[174]

The most relevant properties of NHCs can be described by the parameters of buried volume ($\%V_{Bur}$), Tolman electronic parameter (TEP) and the corresponding pK_a values. The %V_{Bur} can be used to assess the steric demand of any NHC, from crystal data or by theoretical considerations.[175] It is calculated by looking at a sphere with a radius of a given length (300–350 pm). In this sphere, the distance from the centre to the NHC is defined as 200 pm, mimicking the situation in a typical NHC-metal complex. The percentage of the sphere covered by the NHC is equal to %V_{Bur}, hence high values correspond to a sterically demanding NHC. Obviously, a series of compounds can only be compared in a meaningful manner if the parameters for calculations are exactly the same. In the context of this chapter, steric congestion can provide a way to disadvantage any nucleophilic action of the NHC relative to operation *via* its basicity.

TEP is a measure for the ability of the NHC to donate electron density.[176] This is usually determined by IR measurements of $Ni(CO)_3(NHC)$ (or increasingly of less toxic alternatives) and assesses the stretching frequencies of the carbonyl ligand *trans* to the NHC. Lower values indicate a stronger electron donation. If comparing TEPs, all measurements must have been conducted in the same solvent, as differences in wave numbers are small. The electronic properties of NHCs have been studied in great detail.[177]

Fortunately, pK_a values have been determined for a number of different NHCs (Figure 4.25, Table 4.6).[178–182] Some general tendencies can be identified. For the simplest imidazole-based NHC, **69**, a pK_a of 22 has been determined in DMSO, as well as a corresponding value of 23.8 in water. For CH_3CN, calculations have put the basicity at 32.4. Overall, this means that this NHC is comparable to a P_2-base such as **50**. The introduction of further substituents on the backbone (**70**, **72**) or of more donating alkyl groups on the nitrogen atoms (**71**, **73**, **74**) further raises the pK_a up to levels of 34–36 in acetonitrile, almost on a par with P_3-bases. In contrast, the introduction of aryl-substituents markedly decreases basicity. Frequently employed NHCs **75** ("IMes") and **76** are therefore more comparable to a phosphazene like **48**. The saturated counterparts, imidazolin-2-ylidenes like **77** ("SIMes") or **78**, are somewhat stronger than their non-saturated direct analogues, but the difference is not too large. A much stronger influence is found when so-called ring-expanded NHCs are employed,[183] like the six-membered tetrahydropyrimidin-2-ylidenes or the seven-membered tetrahydrodiazepin-2-ylidenes. In contrast to the five-membered NHCs these structures are not planar, and a compound like **80** is as strong a base as *t*Bu-P_3 (**54**), highlighting manipulation of the ring-structure as one of the most effective ways to tune the basicity of NHCs. In clear contrast, thiazol-2-ylidines and the frequently used triazole-5-ylidenes (**82-84**) are relatively weak bases.

Thus, the pK_a range covered by NHCs is not quite as extensive as the one by phosphazenes, but still considerable. Furthermore, it must be stressed that the compounds discussed above represent but a fraction of the variety of NHC- or NHC-like structures.[184] Other species, with higher basicity in part, are for example so-called abnormal or remote NHCs,[185] as well as cyclic alkyl amino carbenes (CAACs)[186] or diamidocarbenes (DACs).[187] However, the NHCs depicted in Figure 4.25 sum up the organocatalysts employed in

Table 4.6 The pK_a data for the conjugated acids of various NHCs and NHOs, determined in different solvents. Brackets refer to calculated values (as opposed to experimental results).

NHC/NHO	pK_a (H_2O)	pK_a (DMSO)	pK_a (MeCN)	References
69	23.8	22.0/(21.1)	(32.4)	178, 180, 181
70	—	23.4/(23.7)	(34.9)	180, 181
71	—	(22.0)	(33.3)	181
72	—	(24.5)	(35.8)	181
73	25.2	23.2/(22.6)	(33.9)	179–181
74	25.4	—	—	179
75	20.8	—	—	179
76	21.1	—	—	179
77	21.3	—	—	179
78	21.5	—	—	179
79		(27.1)	(38.4)	181
80	28.2	—	—	179
81	19.5	(14.5)	(25.6)	178, 181
82	16.8	—	—	182
83	17.6	—	—	182
84	16.5	—	—	182
85	—	—	26.6	226
86	—	—	26.8	226
87	—	—	28.2	226

polymer chemistry. As noted before for amidines, guanidines and phosphazenes, the structural versatility of organobases has not yet been applied to its full extent for polymerization catalysis.

The physical appearance of NHCs can be oily in cases, depending on substituents. In the absence of oxygen, water or other protic compounds (NMR spectra should be taken in benzene-*d6*), they are stable, yet for longer-term storage low temperatures are recommended. Some NHCs can be purified by distillation or sublimation.[188,189]

Good accessibility, facile tuning, excellent ligand properties and versatile catalytic performance have catapulted NHCs to a central position in several fields of chemistry,[190] of which organopolymerization catalysis is only one—and not the most intensively researched—aspect. For more in-depth reading, excellent literature is available.[172,173,177,185,186,190–198]

4.4.2 Properties of N-heterocyclic Olefins

N-heterocyclic olefins (NHOs) can be seen as alkylidene-derivatives of NHCs, but comparison to NHCs is only justified within limits. The synthetic access is similar (or even directly includes NHCs, see below) and NHOs are structurally identical with the so-called deoxy-Breslow intermediates known from NHC catalysis.[199] On the other hand, the typical carbene reactivity is different from the behaviour of the exocyclic carbon in NHOs which is more comparable to a (tamed) carbanion. As a consequence the polymerization catalysis mediated by NHCs or NHOs can be quite different.

Structurally, NHOs are cyclic derivatives of ketene aminals (ene-1,1-diamines) and possess a very electron-rich, highly polarized double bond (Figure 4.26). This polarization is caused by the tendency of the N-heterocyclic ring to accommodate a positive (partial) charge and can be described in mesomeric formulas as full charge separation. The high electron density on the exocyclic carbon renders it basic and nucleophilic and is also expressed in NMR analysis; the olefinic protons of **92** appear at $\delta = 2.77$ ppm (benzene-*d6*),[200] clearly shifted strongly towards high field from where olefins are usually expected.

Although they have only recently entered the ranks of organopolymerization catalysts, as a class of compounds NHOs attracted interest decades ago. Originally, attention peaked as they showed activity in unusual, reverse Diels–Alder reactions.[201] Derivatives with the exocyclic carbon substituted by a nitro group were patented in the 1970s as insecticides.[202] In 1979, Kaska reported the formation of platinum complexes (using the Zeise dimer) with an NHO.[203] He already observed one of the consequences arising from the strong polarization of the double bond: NHOs coordinate end-on rather than side-on to metal complexes. This has until now been confirmed in numerous cases, including examples of NHOs with lanthanide metals,[204] molybdenum and tungsten carbonyls,[205] Rh(I) complexes[206,207] or Ga(III)-compounds.[208] The electronic flexibility of these ligands was noted,[209] as well as the fact that simple NHOs confer more electron density onto a metal centre than comparable NHCs, as shown by Fürstner.[210] This latter finding is remarkable, since NHCs are generally considered to be among the most electron-donating ligands. Rivard was able to show that, while NHOs are strong σ-donors, almost no π-backbonding occurs.[207] Overall this has the

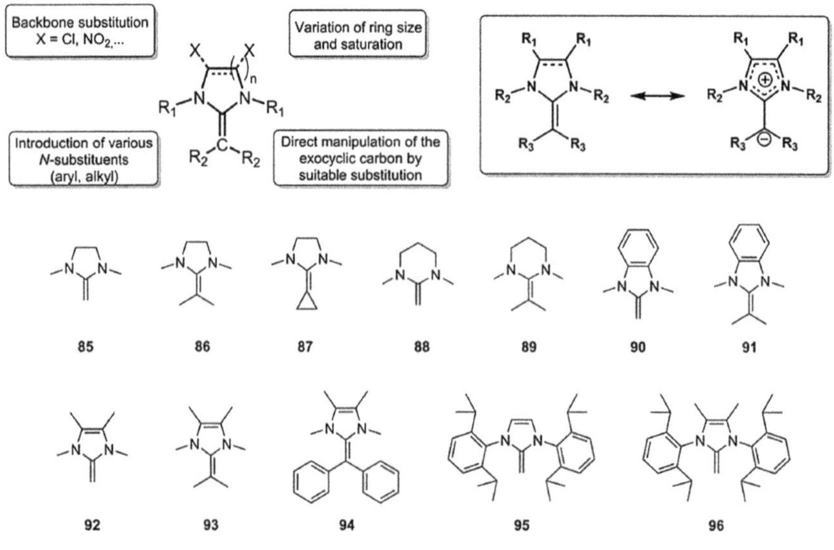

Figure 4.26 Examples for NHOs and their structural variability.

intriguing effect that TEP-values are lower for Rh(I)-NHO complexes than for the corresponding NHC ligands, while at the same time the *bonding strength* to the metal is lower for NHOs than for NHCs.

Likewise, the considerable nucleophilicity of NHOs is obvious from reactions with CS_2, with which stable zwitterions are formed,[211] as well as from the ready manner in which NHOs react with isocyanates,[212] for example. Perhaps the most attention was devoted to the elegant application of NHOs for the sequestration of carbon dioxide as reported by Lu.[213] NHO–CO_2-adducts and the follow-up catalysis have been investigated in detail.[214–220] Recent examples underline that organic chemistry and catalysis in general will certainly profit from this type of organocatalyst.[221–223] For a more detailed reading, competent reviews can be recommended.[224,225]

The structural diversity of NHOs is potentially at least as extensive as found for NHCs, but much less mapped out as yet. Some groups have pioneered the field. In 1987, Heuschmann synthesized >15 different 2-alkylidene imidazolidines, that is, saturated, five-membered NHOs and determined their physical and spectroscopic properties.[226] Kuhn reported the synthesis of the first imidazole-based NHO—1,3,4,5-tetramethyl-2-methyleneimidazoline (**92**)—and characterized the compound, including by crystal data.[200] In the 1990s, Quast prepared a number of different NHOs, including benzimidazole- and tetrazole-derivatives.[227–229] Pittman first reported 1,3-dimethyl-2-methylenehexahydropyrimidine (**88**), a six-membered saturated NHO.[230] More recently, Tamm and Rivard extended the range of known NHOs by several interesting compounds.[206,225] Incomplete as the current knowledge is, from these studies and the following observations in polymerization catalysis several important tendencies can be identified. Generally, the alkylidene imidazolines (**92–93**) are much more reactive than their saturated counterparts. This can be understood when the mesomeric structures are considered: in their case, charge separation results in aromatization, forming a favoured imidazolium moiety. This optimal stabilization of the positive charge localizes the electron density firmly on the exocyclic carbon. Hence, tailoring of the polarization/charge separation is the key to manipulate the catalytic performance of NHOs. Methylation of the backbone (as present in **92** and **93**) further stabilizes the positive charge and minimizes the risk of forming an abnormal NHC by deprotonation.[231] Importantly, with NHOs the catalytically active centre can be influenced directly. Alkylation of the exocyclic carbon (**93**) is readily done and will increase the excess of electron density. Conversely, in a compound like **94**, a mitigated reactivity might be expected. An additional way to tailor the NHO performance will be the adaptation of the *N*-substituent. While systematic studies seem to be absent so far, aryl substituents can be expected to lower the reactivity to some degree, while electron-donating groups will increase this property. The influence of steric congestion is not fully clear yet; calculations by Falivene and Cavallo suggest that for **93** a torsional angle of 25° is necessary because of *N*-methyl/*C*-methyl interactions.[232] It is well conceivable that protonation of the NHO (and the resulting change of hybridization from sp^2 to sp^3) relieves some of that strain, thus providing an additional driving force for base-like activity. In contrast, **92** is fully planar.

A comprehensive study on pK_a values is missing so far. For several saturated five-membered derivatives, the corresponding information was reported to lie in the range 26–29 (acetonitrile, Table 4.6).[226] For **92**, Nguyen found a probable pK_a range of 24–28, suggesting stronger basicity than that of DBU and comparability to weaker phosphazene bases.[221]

While even the simplest imidazolin-based NHOs are solid at room temperature, several of the lower-molecular weight representatives of saturated ring systems are liquids.[12,226] Like NHCs, NHOs will readily react with protic compounds or humidity; in the case of **92**/**93** this reaction is strongly exothermic and fast. With the more reactive compounds not only chloroform, but also dichloromethane can be an unsuitable solvent.[226] Nonetheless, NHOs are not more sensitive than compounds of comparable reactivity. Storage at −36 °C is often recommended, but reports of actual decomposition at room temperature seem to be limited to the most active structures.[206] Solubility is excellent; pentane or diethylether can be readily employed.

4.4.3 Synthesis of NHOs and NHCs

The synthetic chemistry of NHOs[225,226] and especially NHCs[35] has grown rapidly in the last two decades to nowadays include a plethora of strategies, a description of which would be way out of scope for this context. However, in this regard the modular variability shall be highlighted because this hands polymer chemists a tool to systematically alter the organocatalysts' properties in an operationally simple manner.

A typical example is the Radziszewski reaction and modifications thereof (Figure 4.27). The construction of substituted imidazoles is readily achieved with this approach; depending on the aldehyde, either a NHC or a NHO precursor is prepared. Likewise, saturated ring systems can be accessed in an adaptable manner by starting from diamines or amidines in combination with an orthoester. The perhaps highest flexibility is achieved by the amidine route, because it employs primary amines and those are commercially available in manifold variations. In the latter two cases, the application of a suitably substituted orthoester determines whether NHC– or NHO precursors are created. From the precursor salts as depicted in Figure 4.27, the free catalysts are generated by application of strong bases, typically KOtBu or KHMDS for NHCs and KH for NHOs. With increasing frequency reports appear where NHCs are liberated from thermally labile progenitors.[233]

Interestingly, Rivard has described several methods to generate NHOs whereby the synthesis involves free NHCs.[207] This approach seems to be especially useful for an NHO like **95**; the deprotonation of the corresponding (commercial) NHC precursor salt with a strong base such as butyllithium followed by treatment with methyl iodide as source of the $=CH_2$ methylene subunit delivers the target molecules in good yields.

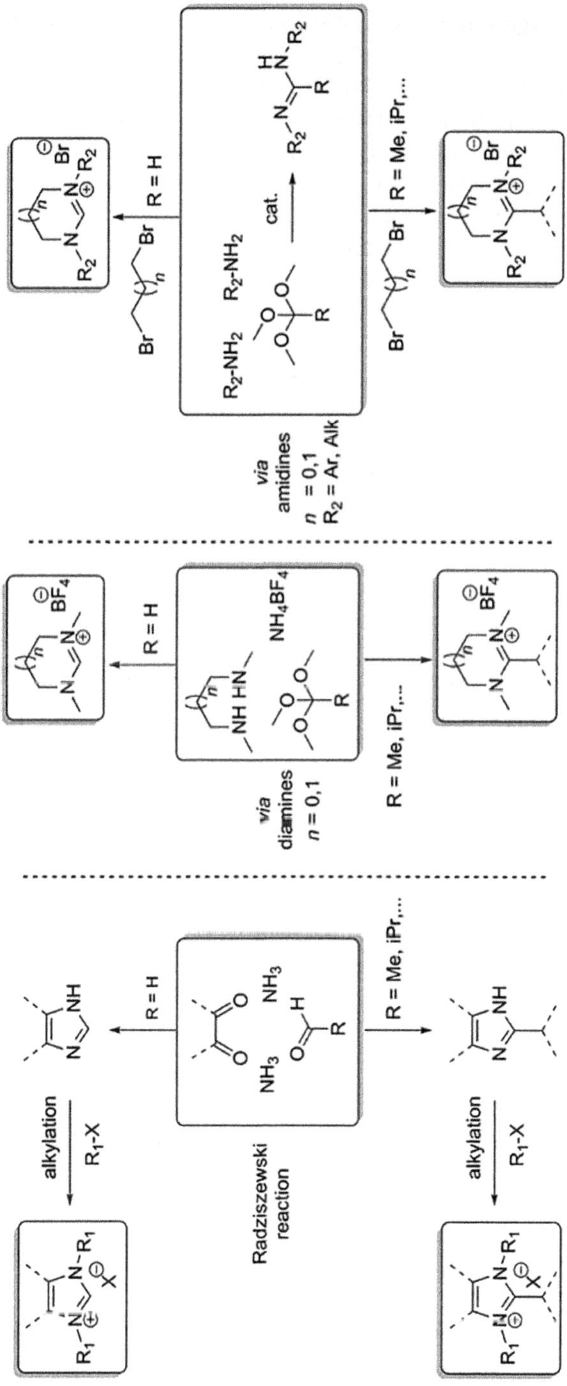

Figure 4.27 Some synthetic approaches for a modular NHC-/NHO-precursor synthesis.

4.4.4 NHCs as Base Catalysts for Polymerizations

The activity as organobase is a side issue in the theatre of NHC-mediated polymerizations. Quite often, this aspect is inexorably intertwined with *nucleophilic* action of the NHC and in many cases results in the same polymer structure as the *basic* pathways. An iconic example was touched on in the introduction: the NHC-catalysed polymerization of cyclic esters can, in principle, proceed *via* these two competing mechanisms (compare Figure 4.1). Crucially, it has been revealed that the chemical nature of the NHC seems to direct which mode of action operates. Although theory has favoured action as base,[234] support for nucleophilic catalysis and the formation of acylazolium intermediates has been found for five-membered NHCs in a number of cases. Waymouth and Hedrick found that smaller *N*-substituents on the organocatalyst entailed more rapid polymerizations;[235] independently, Buchmeiser and co-workers reported a successive decrease in activity upon growing steric congestion of the NHC (going from compound **71** over **73** to **74** lowered conversion (CL) under identical conditions from 97% to 41% and 26%).[236] Because the basicity of the NHC grows in that direction, the performance clearly does not correlate with pK_a values. The mechanism in Figure 4.28 (left) was therefore plausibly proposed[237] and notably sterically non-hindered NHCs can polymerize lactones in the absence of any initiator (ZROP)[238]—clearly that would not be possible if direct attack of the NHC on the carbonyl carbon could not occur. Similar dual mechanisms have been suggested for ROP of epoxides[189] and cyclosiloxanes.[239] More detailed information on nucleophilic NHC catalysis can be found in Chapter 1.

Importantly, for six-membered NHCs a switch towards base catalysis seems to happen.[236] Polymerization of CL with a catalyst like **97** fails in the absence of an alcohol initiator. At the same time, this compound can be expected to have a strong basicity, comparable to that of a P_3-base (Table 4.6). Also, an increased steric demand was not observed to hinder polymerization; inversely, the activity of **97** was higher than that of **80**. Polymerization of CL in bulk at 70 °C, starting from the carbon dioxide adduct of **97**, proceeded to 60–85% yield ($M_n = 5–8$ kg mol^{-1}, $Đ_M = 1.40–1.70$) in several hours reaction time. Together, these findings suggest that the change in the chemical nature of the NHCs favours the base-catalysed mechanism, as shown in Figure 4.28 (right). It should be noted that both pathways result in polymer with the same end groups. However, it is important to understand these finer points, as they might for example widen the range of suitable types of initiators for polyester synthesis. In this context it should also be pointed out that a rare example of polymerization catalysis mediated by an abnormal NHC (*a*NHC) has found that it readily polymerizes lactide, VL and CL.[240] Overall, well-defined polymer was received in the presence of BnOH, especially for CL the *a*NHC was found to be superior with regard to the more frequently employed NHCs ($M_n = 11.3$ kg mol^{-1}, $Đ_M = 1.07$, 99% conversion after 1 h, 1% catalyst loading). This beneficial behaviour was attributed to

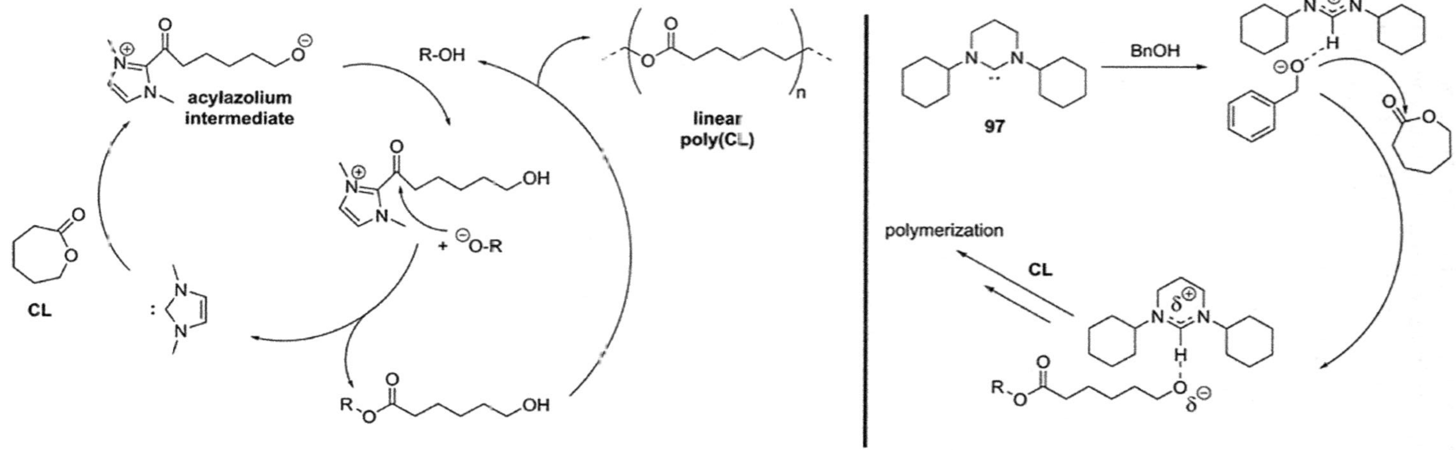

Figure 4.28 NHC-mediated polymerization of lactones, exemplified for CL. The left, "nucleophilic" pathway involves direct attack by the organocatalyst on the carbonyl carbon, while the "basic" mechanism deprotonates/activates the initiator to result in a (pseudo-)anionic process.

the higher basicity of *a*NHCs[241] and coherently it was shown that the OBC interacts with the initiator in such a manner that the –OH signal is shifted strongly towards high field, indicating significant hydrogen bonding.

The high basicity of **97** also resulted in other surprising reactions. Applied for the polymerization of MMA, it was found that the NHC is rather unsuitable for the polymerization of this acrylate monomer by direct addition.[21] While **73** will add to the Michael acceptor system of the monomer as shown by Chen,[242] forming a zwitterionic enolate that is able to initiate polymerization (Section 11.4.1), the analogous reaction with **97** is sluggish and low-performing. This only changes when the polymerization is conducted in DMSO, whereby polymerization rates and polymer properties (M_n, polydispersity) were found to be very sensitive to the concentration of the solvent.[21] It was proposed that the strongly basic NHC deprotonates small amounts of DMSO, which in turn starts the polymerization. At elevated temperatures and with long reaction times, conversions of up to 80% were obtained. However, the initiation process was obviously ill-controlled and the resulting PMMA displayed a broad molecular weight distribution ($M_n = 20–30$ kg mol^{-1}, $Đ_M = 3.2$).

The polymerization of lactams for polyamide synthesis was also investigated employing NHCs. In a paper reported by Buchmeiser and co-workers, the free NHCs were generated from thermally labile CO_2-adducts. This way, the pre-catalyst and the monomer could be stored together over extended periods of time (under exclusion of humidity) to compose readily polymerizable one-component mixtures that could be triggered by heating. This latent, metal-free setup was able to convert CLA into PA 6 with the expected strong dependence of NHC-activity on the basicity. The performance decreased in the order of **97** ≫ **73** > **71** > **77** = 0. This neatly correlates with pK_a values (Table 4.6) and indicates a threshold basicity that has to be surpassed in order to gain polymerization activity for the OBC, as found for phosphazenes.[16] Both accounts suggest that from a p$K_a \geq 33–34$ (CH_3CN) a sufficient reactivity is given. For truly efficient conversion at elevated temperatures (140–200 °C) a stronger basicity is recommended, as present with **97** or *t*Bu-P_4. Thus, with **97** at loadings <1 mol%, PA 6 is received in 70–85% yield (isolated) when applied in bulk at $T = 180$ °C (45 min). The characteristic induction step was observed and a classical anionic polymerization mechanism was proposed (Figure 4.29). Likewise, the less polar laurolactam (LL), raw material for production of PA 12, can be homopolymerized by sufficiently strong NHC bases.[19] With **97**, polyamide was generated in quantitative yield at 180 °C (bulk, 45 min, 0.33 mol% catalyst loading). The resulting material had molecular weights >10 kg mol^{-1} (determined from *m*-cresol GPC) and $Đ_M = 2.3–2.4$. The activity of the NHCs for polymerization again followed the expected pK_a-dependent trend and the fastest catalysts showed more than 90% conversion after 400 s at 200 °C. Interestingly, a copolymerization of CLA and LL was also possible, resulting in gradient copolymers, with the more strained CLA being consumed preferentially at the onset of the polymerization. Molecular weights

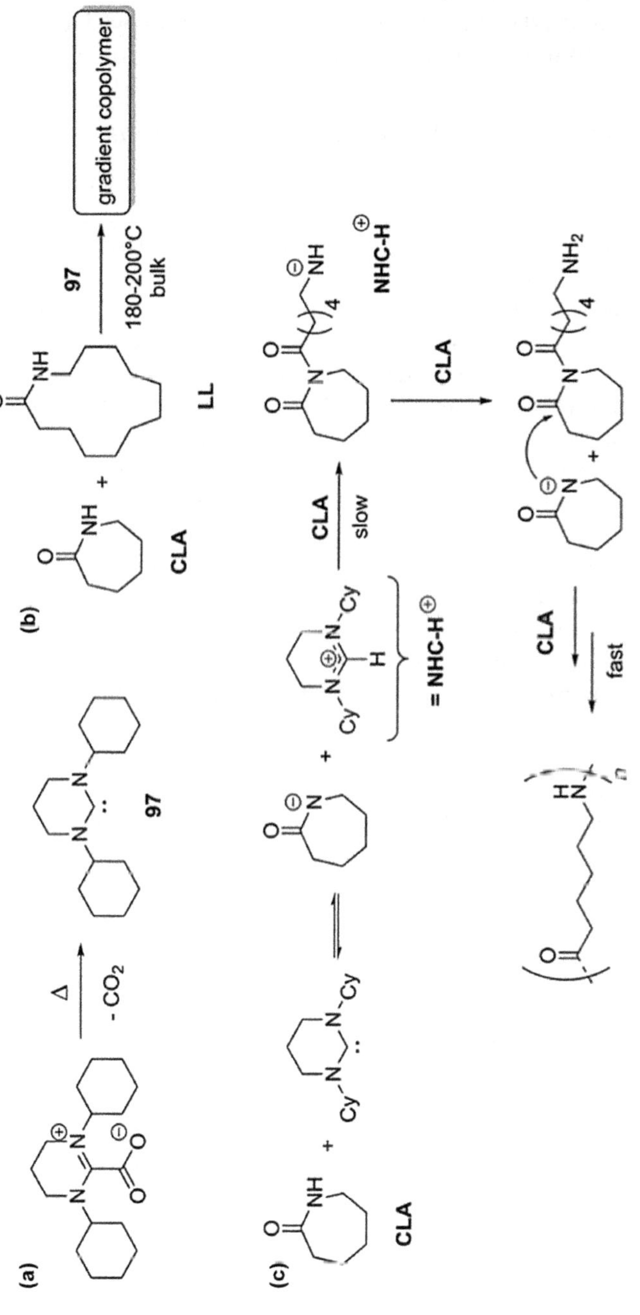

Figure 4.29 Polymerization of lactams by NHCs, including (a) thermal release of the active organocatalyst, (b) the preparation of PA 6/12 copolymers and (c) the anionic mechanism with the NHC as OBC.

of up to 13 kg mol^{-1} ($Đ_M = 2.1–2.4$) were achieved this way, with the copolymers displaying the typical diminished melting points (down to 120 °C) compared to the corresponding homopolymer PA 6 and PA 12. Meanwhile, this thermally latent NHC-based process has been employed to fabricate polyamide fibres from one-component setups by melt spinning.[243]

4.4.5 NHOs as Base Catalysts for Polymerizations

As indicated before, the application of NHOs as (organo-)polymerization catalysts is still in an early stage. Nonetheless, several intriguing examples for their abilities have recently been published, and most of those suggest their action as base to be the major reason for their polymerization activity. Operation purely *via* nucleophilicity seems to be rare for NHOs so far; however, **92** and **93** have lately been shown to enable the organopolymerization of different acrylic monomers (methyl methacrylate, *N*,*N*-dimethylacrylamide) by direct addition to the Michael acceptor system of the monomer.[244] Another case where NHO catalysts make use of their nucleophilicity is the homopolymerization of PO, where a mechanistic dualism with the base-style pathway has been found, not dissimilar to NHC-mediated polymerizations.[12] As outlined above, anionic polymerization of PO is aggravated by several problems, including transfer-to-monomer (Figure 4.21a), which limits molecular weights and impacts control over end groups. Also, the generally slow monomer conversion in metal-free systems is problematic. In this context, the application of NHOs proved to be a significant step forward: when imidazole-derived organocatalysts were employed (**92, 93**) a lively polymerization activity was observed, while for saturated NHOs (**85, 88**) no conversion occurred. Under suitable conditions (**93**/BnOH/PO = 1 : 10 : 1000, 50 °C) it was thus possible to achieve high conversion (up to 96%) while maintaining good control over the polymerization ($Đ_M \leq 1.09$), including molecular weights and end groups. Transfer-to-monomer could be largely suppressed. With very low catalyst loadings (down to 0.01 mol%) and extended polymerization times a TON > 2000 was realized. Of mechanistic importance, it was found that NHO **92** swiftly delivers PPO, but GPC analysis showed that a small impurity of relatively high molecular weight was present, apart from the major product.[12] This observation was related to competing zwitterionic initiation (Figure 4.30), a notion that was supported by control reactions and NMR experiments. Recently, Falivene and co-workers independently corroborated this proposed mechanistic scenario and highlighted spirocyclic resting states as crucial intermediates.[232] It should be noted that in contrast to NHCs, once an NHO has directly ring-opened the epoxide monomer, an interconversion of the zwitterionic species back to the "anionic" pathway is highly doubtful; hence, a clean, monomodal distribution can only be achieved if one of the two mechanisms is supressed. As reported by Naumann and Dove,[12] the application of **93** (instead of **92**) ensures just that: the NHO is too sterically encumbered to directly ring-open PO, while at the same time its basicity is increased, both factors favouring

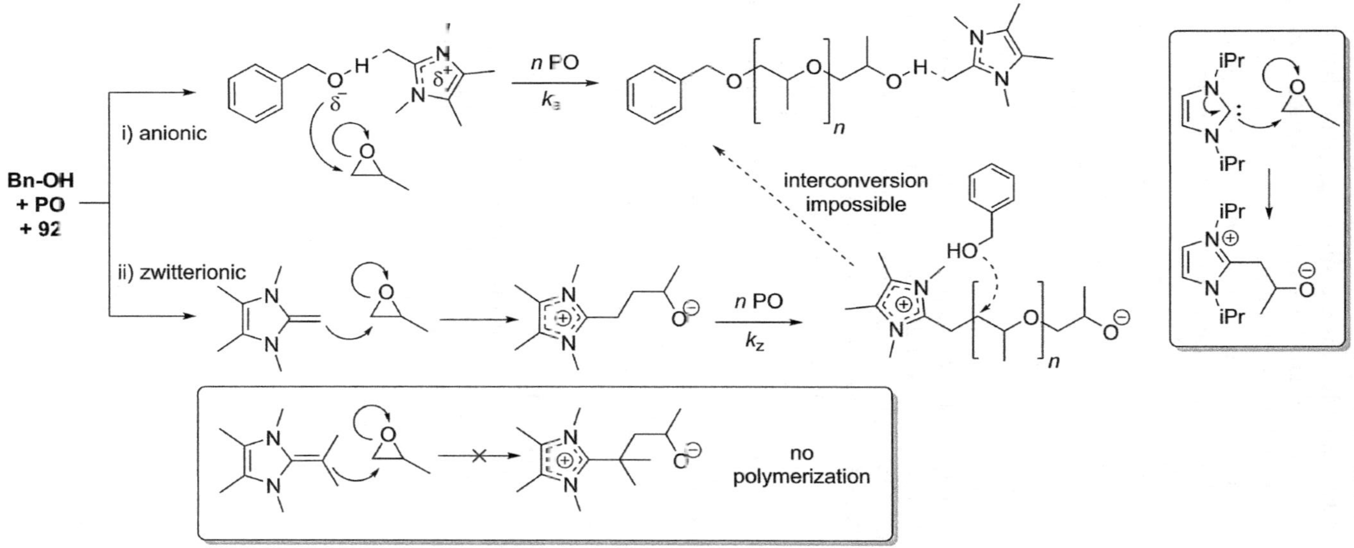

Figure 4.30 NHO-mediated polymerization of PO using NHO **92**.[12] Monomer conversion can take place *via* an anionic mechanism or by zwitterionic polymerization, originating from direct ring opening of the epoxide by the NHO. Application of **93** renders the zwitterionic pathway unfavourable (bottom). The analogous situation for NHCs has been described to allow for nucleophilic substitution of the NHC-moiety (interconversion of both pathways).[188,189]

the anionic polymerization mechanism. Exceptionally well-defined polyether is received this way, in a solvent- and metal-free manner. The fact that transfer-to-monomer was largely absent also suggests that the NHO-mediated polymerization of PO can be used to construct more complex polyethers, especially block-like structures. This was recently realized, using α,ω-dihydroxylated PEO as macroinitiator for the preparation of PPO-PEO-PPO amphiphilic triblockcopolymer.[245] The operationally simple setup allowed for the highly controlled synthesis of a series of polymers with stepwise increased PPO-blocks, which in turn enabled a quick screening of the corresponding polyethers regarding their structure-directing properties in self-assembly processes. Molecular weights of >25 $kg\,mol^{-1}$ (GPC) were achieved in this manner, with $Đ_M \leq 1.03$ in all cases. Interestingly, it was also shown that butylene oxide can be employed in the same manner, albeit at reduced polymerization rates compared to PO.[245]

NHOs have also been applied for the polymerization of lactones. Again, the chemical nature of the organocatalyst strongly influences the outcome. In the first report on this subject it was revealed that $=CH_2$–substituted NHOs (such as **85**, **88** or **92**) suffer from non-living, non-quantitative conversion, both in the presence and the absence of BnOH as initiator when applied for VL or LA.[246] This observation was attributed to an undesired proton transfer, forming enamines (deoxy-Breslow intermediates). The side reaction is enabled by the considerable acidification of the $-CH_2-$ moiety which is placed between a positively charged imidazolium group and a carbonyl functionality (Figure 4.31). Intra- or intermolecular proton transfer (involving oxanionic chain ends or free NHO) can potentially occur and quench the polymerization activity. It should be noted that this reactivity again requires the NHO to directly ring-open the lactone monomer. In sharp contrast, when $=C(CH_3)_2$-type NHOs were employed (bearing exocyclic methyl groups), high conversion and partly unexpected reactivity was observed. Using **93**—as discussed before a compound with a very strong polarization of the olefinic bond—it was revealed that LA, VL, CL or TMC could be polymerized. Even for PDL a moderate reactivity was found (47% conversion after 4 h at 110 °C, 1 mol% NHO loading). Unfortunately, this broad applicability was somewhat offset by a low degree of control over the polymerization for most monomers. Especially for VL **93** possesses a turbulent reactivity: within seconds to few minutes near quantitative conversion can be achieved, down to 0.2 mol% catalyst loading. Interestingly, monomer consumption also works in the absence of BnOH and consequently a mechanism for direct polymerization of VL by this NHO must exist. ^1H NMR experiments (C_6D_6) highlighted that **93** is protonated in the presence of dry VL, strongly suggesting that an enolate is formed (Figure 4.31), acting as the initiator for the polymerization. Polymer spectra showed that the NHO signals were fully lost during work-up, further corroborating that the organocatalyst is not attached to the polymer chain. Overall, this might explain the lack of control over the polymerization process. Based on these insights, NHO **86** was developed.[246] This compound was designed to possess the

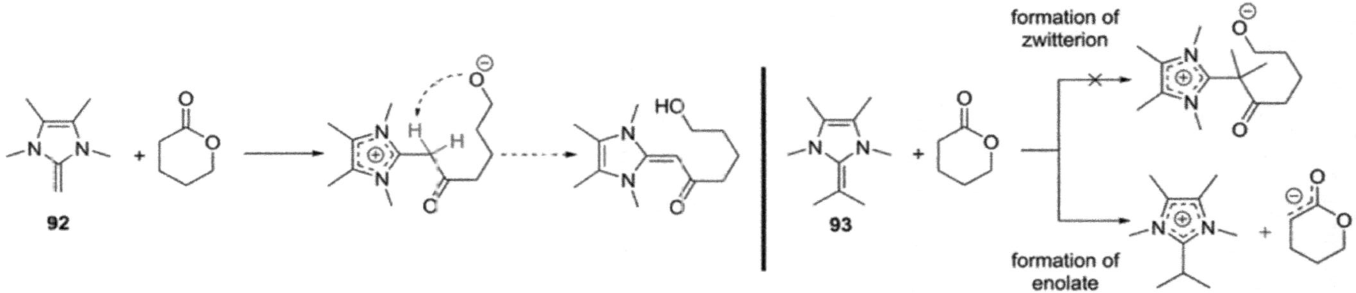

Figure 4.31 NHOs **92** and **93** behave differently towards VL. While **92** can form a zwitterionic species that is prone to deactivation to form an enamine (left), **93** does not ring-open the monomer but seems to generate an enolate as initiating species (right).[246]

exocyclic methyl groups to prohibit any possibility of undesired proton transfer while its backbone was saturated and therefore less prone to accommodate a positive charge. Fortunately, this weakened polarization of the double bond translated into much more controlled polymerization properties: TMC and L-LA were gently polymerized at room temperature to reach high or quantitative conversion with a reasonable control over the molecular weight distribution ($Đ_M = 1.1–1.3$). Nonetheless, for common lactone monomers like CL or VL **86** failed; also, Nguyen reported that NHO **96** is able to polymerize CL and lactide, albeit harsh reaction conditions must be applied (110 °C in toluene, 17–25 h) to result in moderate control ($Đ_M = 1.6–1.7$).[221] Both findings highlight that NHO organocatalysis still requires future catalyst optimization. Further revelations might well originate from indirect sources, since it has been shown in a number of cases that NHOs are excellent catalysts or initiators for a broad range of lactones[247–249] or acrylic monomers[244,250,251] when employed with co-catalysing Lewis acids (simple salts like $MgCl_2$ and LiCl or more complex structures such as $Al(C_6F_5)_3$).

4.5 Other Types of Organic Base Catalysts

There are other classes of OBCs or OBC-like catalysts that will not be discussed here in detail because they are either weaker Brønsted bases than amidines and might thus act *via* other mechanisms, are eclipsed in their performance by most competing systems or are only described in few (but promising) publications for organopolymerization purposes.

The foremost example in this respect is 4-dimethylaminopyridine (DMAP), simultaneously the catalyst employed in the pioneering first studies investigating (living) polymerization by organocatalysts.[252,253] As discussed for a number of other types of OBC, DMAP is neither a pure organobase nor does it act solely *via* its nucleophilicity (monomer-activated mechanism).[254] A detailed discussion is found in Chapter 1; likewise, the polymerization activity of sparteine and derivatives is described there.[255,256]

Bredereck's reagent (**98**, Figure 4.32) originally attracted attention because its structure is similar to alcohol-protected NHCs;[257] those compose one of the main strategies to render NHCs more robust in handling and storage.[233] In a mechanistic study, Waymouth and Hedrick assessed the capability of **98** for polymerization of L-LA.[258] Conversion with relatively good control over the molecular weight distribution was achieved ($Đ_M < 1.2$, $T = 70$ °C) and importantly the reaction works both in the absence and presence of additional alcohol initiator. This is the case because upon (thermally induced) dissociation each molecule of the catalysts delivers one equivalent of alcohol/alcoholate. The evidence provided by the authors cannot fully exclude the intermediate formation of carbene species, but overall favours an anionic mechanism (Figure 4.32). The polymerization is most likely started by the formation of a formamidinium alcoholate, generated from **98**; coherently, the end groups of the PLA were found to originate exclusively from the alcohol moiety released from the organocatalyst precursor. Since

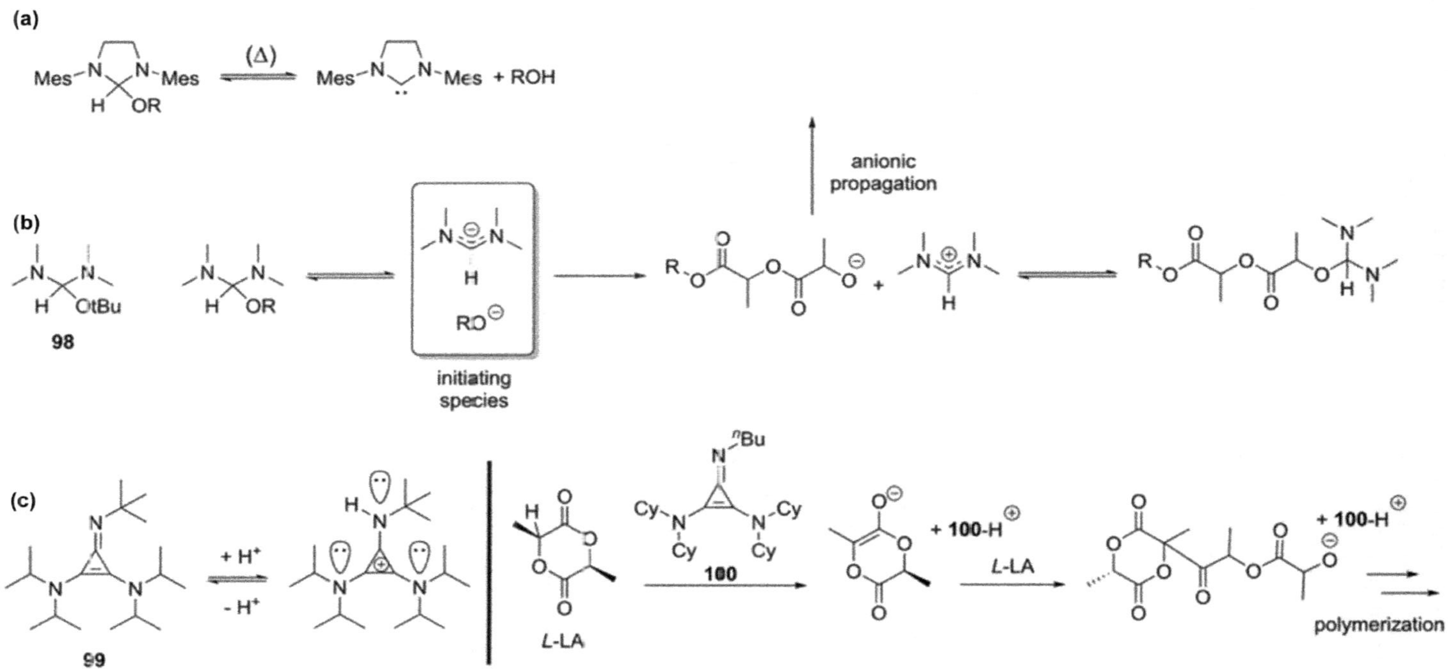

Figure 4.32 (a) NHC-alcohol adduct and its thermally induced dissociation; (b) application of Bredereck's reagent for the metal-free, anionic polymerization of lactide *via* the *in situ* generated formamidinium alcoholate;[258] (c) cyclopropenimine for lactide polymerization *via* enolate formation.[20]

amidines possess a relatively muted basicity, the well-behaved polymerization properties were tentatively attributed to a reversible capture of the counter ion by the propagating oxyanionic chain end.

A recent promising development is found in the application of 2,3-bis(dialkylamino)cyclopropenimines as organobase-type catalysts for polymerizations. For a typical representative such as **99** (Figure 4.32), Lambert has shown that in acetonitrile a pK_a-value of 26.9 ensues.[259] This is perfectly comparable to **47**, a typical P_1-base (Table 4.5). The high basicity of cyclopropenimines is generated by aromatization upon protonation, forming the smallest possible Hückel system (2π electrons). The lone pairs of the three surrounding nitrogen atoms can conjugate, increasing charge delocalization. For the organopolymerization of lactide, these properties have been demonstrated to result in alternative initiation mechanisms.[20] The polymerization using compound **100** proceeds both in the absence and in the presence of alcohol initiator; in the latter case the kinetics were observed to be only marginally slower. With initiator (1-pyrenebutanol, M_n up to 13 kg mol^{-1}, $Đ_M = 1.2–1.4$) the molecular weight distribution broadens over time. Also, MALDI-ToF MS found even- and odd-numbered lactic acid repeating units, confirming that transesterification occurs. Interestingly, with no initiator present polymerization is started by formation of an enolate (as in contrast to zwitterionic ring-opening as described for NHCs). This finding is supported by MALDI-ToF MS data, NMR analysis and viscosimetry (to exclude the formation of macrocyclic polymer, which would display the same mass spectra but lower intrinsic viscosity for the same molecular weight). In CH_2Cl_2, high to quantitative conversion was achieved in a few minutes (**100**/LA = 1 : 100, $Đ_M = 1.3–1.6$) under these conditions. In accordance with enolate formation as initiation step, lack of acidic protons prevents polymerization and indeed it was shown that conversion remains negligible when TMC was employed as monomer, which of course cannot be enolized.

4.6 Summary and Comparison

As evident from above considerations, organobase-mediated polymerizations have evolved into a field with rich and versatile chemistry, a multitude of polymerization pathways and manifold applications. This adaptability is advantageous, but also raises the question whether guidelines can be developed to simplify the selection of a suitable catalyst for a given monomer. It also requires organobase polymerizations to be put into perspective, especially with regard to alternatives such as acid catalysis or the use of metal-containing bases.

4.6.1 Why Use Organobase Polymerization Catalysis?

Compounds like NHCs, phosphazenes or guanidines seem more complex than simple metal-containing bases such as alkyl lithium compounds, alkaline metal hydrides or metal alcoholates. The latter can certainly bring about anionic processes. A number of reasons, however, limit the applicability of metal bases and highlight the advantages that can be gained by the corresponding organocatalysis.

Firstly, of course, the physical presence of metal ions can be problematic, especially if they remain in the polymer. Toxicity issues should not be overstated; for one, if such a process is highly efficient the amount of metal species in the polymer will be very small, and secondly something like a general toxicity for a given element does not exist.[260] Rather, toxicity depends on the specific chemical form (oxide, halide, organic ligands) and compounds like phosphazenes for example can also display a considerable cytotoxicity.[261] Overall however, most organocatalysts seem indeed to be less problematic than many metal species frequently found in polymerizations (such as tin).[262] Furthermore, metal residues in the material can remain catalytically active as Lewis acids, posing a potential danger for the long-term integrity of polymers. And of course there are sensitive applications where the complete absence of metals is required and facilitates clearance for medical applications (sutures, drug release), forming an important area where organocatalysis is clearly favoured.

Solubility is also a constant issue for many metal bases; compared to the neutral organobases, simple hydrides, hydroxides or alcoholates are harder to dissolve. And if they do, the formation and type of ion pair will be decisive for any reaction to occur.

Safety issues are also to be considered. This is usually not problematic for organobases. Alkyl lithium compounds on the other hand are more challenging to handle, especially on larger scale. Likewise, the more reactive hydrides pose technical problems for obvious reasons.

Perhaps most importantly, organocatalysis performs well when a controlled polymerization of heterocyclic monomers is required. Simple metal bases on the other hand would be too reactive (lithium organyls), deliver too slow initiation relative to propagation (solubility, ion pairs) or lack the necessary basicity (hydroxides, some alcoholates). Conversely, the anionic polymerization of weakly activated, double-bond containing monomers is a topic where currently organobases cannot compete. Styrene is still out of scope for even the most basic/nucleophilic organobases, while the polymerization *via* alkyl lithium is well established, of course. A higher degree of activation for double bonds, as present in acrylic monomers, has on the other hand already led to successful polymerization by NHCs or NHOs,[21,242,244,263,264] suggesting that the organopolymerization of activated olefins will be a frontier for future development.[6]

There is no need to be dogmatic about the presence of metal species in a polymerization setup; "hybrid" approaches are emerging. The cooperative application of typical organocatalysts (DBU, NHCs, NHOs) in combination with simple metal halides has recently received increasing interest as a form of dual catalysis,[247,249,265] whereby organobase and Lewis acid support each other in their reactivity, resulting in increased polymerization rates and novel selectivities, underlining that metal-free and metal-based polymerization catalysis should be seen as complementary, well positioned to inspire each other.

At least for some monomers, acid catalysis forms an interesting alternative to organobase polymerizations; the allure of cationic, acid-mediated processes is certainly that this type of catalysis tolerates a different set of functional groups on monomers,[266,267] using robust and often commercially available catalysts. The monomer scope is currently however limited to some lactones and carbonates, whereby frequently relatively slow conversion is observed (with excellent polymerization control in several cases). In Chapter 2 a detailed description of (metal-free) acid catalysis can be found.

4.6.2 Selecting Organobases

As should have become evident, the choice of organocatalyst is pivotal for any given base-catalysed polymerization, routinely impacting not only the more obvious parameters such as polymerization rate and achievable conversion but also selectivity, end groups, tolerable functionalities and applicable reaction conditions. In view of the plethora of already investigated compound families, the question for the best suited organocatalyst for a given monomer might arise. To tentatively answer this, four points should be considered:

1. The ideal organocatalyst, even for well-investigated monomers, has yet to be found. That is to say, catalyst design and modification should not be neglected over the short-term benefits of using well-available commercial compounds. As was pointed out for several instances above, the number of catalysts investigated for each catalyst family is still small, strongly suggesting that future breakthroughs will also depend on synthesis.
2. Is the polymer chain significantly less reactive than the corresponding monomer? If yes, chain-transfer/backbiting will be less of a problem and less coordinating, stronger bases can be employed. If no, a more selective catalyst should be chosen; reversible resting states and/or bifunctional or dual activation might be beneficial.
3. Does the monomer have acidic sites (and what is the approximate pK_a of those)? If yes, then deprotonation of the monomer can be used intentionally (direct polymerization *via* enolates, lactam anions *etc.*), which requires a strong base. If non-desired, transfer-to-monomer and competing end groups can be problematic, suggesting the application of weaker bases in the presence of initiator, typically alcohols.
4. How important is selectivity, especially with regard to molecular weight (distribution)? While in academia additives, solvents and long reaction times are acceptable to obtain a well-defined polymer, larger scale application demands rapid reactions in the bulk, using well-available organocatalysts. Highly defined polymer may not always be the ultimate aim.

A few examples shall be considered. The most obvious tuning site to increase the performance of organobases seems to be an increase of basicity; however, a simple relationship where activity and basicity scale in a reasonable manner is not observed that often and frequently the faster polymerizations come at the cost of increased side reactions. A case where "the stronger the better" seems to be true of organocatalytic polymerization is for example found with the anionic polymerization of ε-carpolactam (CLA) to prepare PA 6. Both for NHCs and phosphazenes a clear correlation of performance and pK_a was observed.[16,18,19] Albeit polyamide chains are reasonably robust, a degree of transamidation will inherently occur and in such a case it is recommendable to speed the polymerization up rather than accepting long reaction times—thus in this case selecting the right organobase is clear and limited to a small group of compounds, namely highly basic phosphazenes (tBu-P_4) and NHCs (**80**, **97**). The situation becomes more complex when moving to polyethers. There, the macromolecular chain is highly resistant to side reactions, with *transetherification* being a rare occurrence (in contrast to *transesterification*) and molecular weight distributions excellently controlled ($Đ_M$<1.10). While for ethylene oxide (EO) strong bases with "non-coordinating" conjugated cationic species can be employed (phosphazenes: tBu-P_4),[138] the same approach would deliver poly(propylene oxide) with an increased number of allylic chain ends resulting from transfer-to-monomer.[139] For PO, reactions work well using some NHCs[268] and NHOs[12] with the latter constituting the currently most active, metal-free polymerization setup for PO, where the right balance between too low reactivity (amidines and guanidines do not polymerize) and excessive basicity (tBu-P_4) seems to be represented by compounds such as **70**, **71** and **93**.

The highest number of potential catalysts is to be considered when lactones are targeted. Here all points 2–4 must be carefully considered. Transesterification (2) is a constant challenge, most of all for those monomers with little ring-tension (GBL, PDL); lactones are also enolizable (3) and many catalyst setups require a balancing of reactivity and selectivity (4).

The requirements for a good organocatalyst are hard to define for lactones in general; indeed, from the numerous literature examples discussed above concerning lactones it should be clear that each species should be considered distinct. Nonetheless, some tendencies can be identified. Perhaps most importantly among those is the finding that a simple base mechanism, best embodied by the "non-nucleophilic" phosphazenes, is problematic. It frequently leads to increased transesterification, which broadens the molecular weight distribution and might even cause competing initiation by enolates. Consequently, strongly basic phosphazenes compete best performance-wise only in cases where transesterification is inherently always occurring (GBL and PDL).[127,130] Using weaker bases does not necessarily help either: DBU or MTBD (alone) are not suited for polymerization of GBL,[127] VL,[65] CL[65] or PDL[70] (while being highly selective for lactide) because obviously in their case the required reactivity/basicity is below the necessary threshold. Thus, for lactones like VL and CL, where in principle very narrow molecular weight distributions can be obtained, a promising approach is *activation of the monomer* rather than increasing basicity of the organocatalyst. The prime example is TBD with its bifunctional character.[65] Likewise, after addition of a cocatalyst coordinating to the monomer *via* H-bonding (thioureas) also DBU and MTBD effectively convert CL and VL.[65] In a similar manner, NHOs readily polymerize several lactones after Lewis acid-activation of the monomer.[247,249] A different but related organocatalytic approach that successfully delivers well-defined polyester relies on zwitterionic ring-opening; in this area in particular NHCs have excelled, where it is important to balance the high reactivity of the NHC by moderate steric shielding.[237,238]

4.7 Outlook

Without doubt, organobase catalysis is one of the most important facets of organopolymerization. The range of suitable compounds is large and growing rapidly, while also the scope of relevant monomers is expanding. Although the field certainly develops on several very different frontiers—the decoupling of activity and selectivity for example, asymmetric organopolymerization or (co)polymerization of double bond-containing monomers, to name but a few—some general tendencies can still be identified.

One major development throughout polymerization catalysis is the need to achieve the highest possible control over the polymerization process itself; a "precise" construction of macromolecular material, regarding molecular weight, end groups, tacticity and topology, is central to improvements in sensitive high tech fields. For the design of organobases this means that a simple adaption of pK_a-values to the task is only a first (albeit important) step. Many of the examples discussed in the previous chapter have highlighted the importance of secondary interactions as well as resting states or dormant species. Quite regularly, these massively improve polymerization control and suppress side reactions. This clearly implies that "the stronger the better" is only true for a limited set of organobase polymerizations and true success is rather determined by manipulating the catalyst in other ways.

At the same time, to be relevant in the competitive field of polymerization catalysis, OBCs have to be robust, easily accessible and in many cases still need to improve their efficiency. This will most likely remain a formidable challenge in the near future. However, in growing numbers reports on highly efficient (*i.e.*, high TON) yet simple organocatalytic polymerization setups occur. This is further supported by the fact that being "metal-free" is nowadays only one of the benefits associated with organopolymerization techniques. Perhaps equally important, intensive research has shown that also different architectures (such as macrocyclic polymer), alternative initiation mechanisms and different selectivities compared to organometallic catalysts can be realized, adding another factor as to why organ(base)polymerization should remain a thriving venture.

References

1. P. Caubère, *Chem. Rev.*, 1993, **93**, 2317.
2. T. Ishikawa, *Superbases for Organic Synthesis*, John Wiley & Sons, Ltd, Chichester, UK, 2009.
3. N. Hadjichristidis and A. Hirao, *Anionic Polymerization*, Springer Japan, Tokyo, 2015.
4. B. Maji, M. Breugst and H. Mayr, *Angew. Chem., Int. Ed.*, 2011, **50**, 6915.
5. M. Fèvre, J. Pinaud, Y. Gnanou, J. Vignolle and D. Taton, *Chem. Soc. Rev.*, 2013, **42**, 2142.
6. S. Naumann and A. P. Dove, *Polym. Chem.*, 2015, **6**, 3185.
7. E. M. Higgins, J. A. Sherwood, A. G. Lindsay, J. Armstrong, R. S. Massey, R. W. Alder and A. C. O'Donoghue, *Chem. Commun.*, 2011, **47**, 1559.
8. R. W. Alder, P. S. Bowman, W. R. S. Steele and D. R. Winterman, *Chem. Commun.*, 1968, 723.
9. H. Staab and T. Saupe, *Angew. Chem., Int. Ed.*, 1988, **27**, 865–879.
10. P. B. Kisanga, J. G. Verkade and R. Schwesinger, *J. Org. Chem.*, 2000, **65**, 5431.
11. J. G. Verkade and P. B. Kisanga, *Tetrahedron*, 2003, **59**, 7819.
12. S. Naumann, A. W. Thomas and A. P. Dove, *Angew. Chem., Int. Ed.*, 2015, **54**, 9550.
13. R. Schwesinger, H. Schlemper, C. Hasenfratz, J. Willaredt, T. Dambacher, T. Breuer, C. Ottaway, M. Fletschinger, J. Boele, H. Fritz, D. Putzas, H. W. Rotter, F. G. Bordwell, A. V. Satish, G.-Z. Ji, E.-M. Peters, K. Peters, H. G. von Schnering and L. Walz, *Liebigs Ann./Recl.*, 1996, **1996**, 1055.
14. R. Schwesinger and H. Schlemper, *Angew. Chem.*, 1987, **99**, 1212.
15. R. Schwesinger, M. Mißfeldt, K. Peters and H. G. von Schnering, *Angew. Chem., Int. Ed.*, 1987, **26**, 1165.
16. W. Memeger, G. C. Campbell and F. Davidson, *Macromolecules*, 1996, **29**, 6475.
17. H. Yang, J. Zhao, M. Yan, S. Pispas and G. Zhang, *Polym. Chem.*, 2011, **2**, 2888.
18. S. Naumann, S. Epple, C. Bonten and M. R. Buchmeiser, *ACS Macro Lett.*, 2013, **2**, 609.
19. S. Naumann, F. G. Schmidt, M. Speiser, M. Böhl, S. Epple, C. Bonten and M. R. Buchmeiser, *Macromolecules*, 2013, **46**, 8426.
20. T. S. Stukenbroeker, J. S. Bandar, X. Zhang, T. H. Lambert and R. M. Waymouth, *ACS Macro Lett.*, 2015, **4**, 853.
21. S. Naumann, F. G. Schmidt, R. Schowner, W. Frey and M. R. Buchmeiser, *Polym. Chem.*, 2013, **4**, 2731.
22. S. Marrot, F. Bonnette, T. Kato, L. Saint-Jalmes, E. Fleury and A. Baceiredo, *J. Organomet. Chem.*, 2008, **693**, 1729.
23. W. N. Ottou, H. Sardon, D. Mecerreyes, J. Vignolle and D. Taton, *Prog. Polym. Sci.*, 2016, **56**, 64.

24. *The Chemistry of Amidines and Imidates*, ed. S. Patai and Z. Rappoport, Wiley, 1991.
25. F. Z. Dörwald, *Lead Optimization for Medicinal Chemists. Pharmacokinetic Properties of Functional Groups and Organic Compounds*, Wiley-VCH, Weinheim, 2012.
26. J. V. Greenhill and P. Lue, in *Progress in Medicinal Chemistry, 30*, ed. G. P. Ellis and D. K. Luscombe, Elsevier, Burlington, 1993, ch. 5, p. 203–326.
27. T. M. Krygowski and K. Woźniak in *The Chemistry of Amidines and Imidates*, ed. S. Patai and Z. Rappoport, Wiley, 1991, ch. 2, p. 101–145.
28. R. Norrestam, S. Mertz and I. Crossland, *Acta Crystallogr.*, 1983, **C39**, 1554.
29. J. Oszczapowicz in *The Chemistry of Amidines and Imidates*, ed. S. Patai and Z. Rappoport, Wiley, 1991, ch. 12, p. 623–688.
30. D. Margetic, in *Superbases for Organic Synthesis*, ed. T. Ishikawa, John Wiley & Sons, Ltd, Chichester, UK, 2009, ch. 2, p. 9–48.
31. R. Schwesinger, *Nachr. Chem., Tech. Lab.*, 1990, **38**, 1214.
32. F. Heinzer, M. Soukup and A. Eschenmoser, *Helv. Chim. Acta*, 1978, **61**, 2851.
33. L. Weintraub, S. R. Oles and N. Kalish, *J. Org. Chem.*, 1968, **33**, 1679.
34. H. Oediger, H.-J. Kabbe, F. Möller and K. Eiter, *Chem. Ber.*, 1966, **99**, 2012.
35. L. Benhamou, E. Chardon, G. Lavigne, S. Bellemin-Laponnaz and V. César, *Chem. Rev.*, 2011, **111**, 2705.
36. E. C. Taylor and W. A. Ehrhart, *J. Org. Chem.*, 1963, **28**, 1108.
37. G. Lehmann, H. Seefluth and G. Hilgetag, *Chem. Ber.*, 1964, **97**, 299.
38. S. Saba, A. Brescia and M. K. Kaloustian, *Tetrahedron Lett.*, 1991, **32**, 5031–5034.
39. Y. Yamamoto and S. Kojima, in *The Chemistry of Amidines and Imidates*, ed. S. Patai and Z. Rappoport, Wiley, 1991, ch. 10, p. 485–526.
40. G. A. Olah, A. Burrichter, G. Rasul, M. Hachoumy and G. K. Surya Prakash, *J. Am. Chem. Soc.*, 1997, **119**, 12929.
41. P. Gund, *J. Chem. Educ.*, 1972, **49**, 100.
42. D. J. Haas, D. R. Harris and H. H. Mills, *Acta Crystallogr.*, 1965, **19**, 676.
43. K. B. Wiberg, *J. Am. Chem. Soc.*, 1990, **112**, 4177.
44. A. Gobbi and G. Frenking, *J. Am. Chem. Soc.*, 1993, **115**, 2362.
45. E. D. Raczynska, M. K. Cyranski, M. Gutowski, J. Rak, J.-F. Gal, P.-C. Maria, M. Darowska and K. Duczmal, *J. Phys. Org. Chem.*, 2003, **16**, 91.
46. T. Ishikawa, in *Superbases for Organic Synthesis*, ed. T. Ishikawa, John Wiley & Sons, Ltd, Chichester, UK, 2009, ch. 4, p. 93–143.
47. D. H. R. Barton, J. D. Elliott and S. D. Géro, *J. Chem. Soc., Chem. Commun.*, 1981, 1136.
48. D. H. R. Barton, J. D. Elliott and S. D. Géro, *J. Chem. Soc., Perkin Trans. I*, 1985, 2085.
49. G. Wieland and G. Simchen, *Liebigs Ann. Chem.*, 1985, 2178.
50. I. Kaljurand, A. Kütt, L. Sooväli, T. Rodima, V. Mäemets, I. Leito and I. A. Koppel, *J. Org. Chem.*, 2005, **70**, 1019.

51. M. P. Coles, *Chem. Commun.*, 2009, 3659.
52. V. Raab, J. Kipke, R. M. Gschwind and J. Sundermeyer, *Chem. – Eur. J.*, 2002, **8**, 1682.
53. B. Kovačević and Z. B. Maksić, *Chem. – Eur. J.*, 2002, **8**, 1694.
54. B. Kovačević, Z. Glasovac and Z. B. Maksic, *J. Phys. Org. Chem.*, 2002, **15**, 765.
55. B. Kovačević, Z. B. Maksić and R. Vianello, *J. Chem. Soc., Perkin Trans. 2*, 2001, 886.
56. P. Selig, *Synthesis*, 2015, **45**, 703.
57. D. Leow and C.-H. Tan, *Chem. – Asian J.*, 2009, **4**, 488.
58. F. A. Cotton, C. A. Murillo, X. Wang and C. C. Wilkinson, *Inorg. Chem.*, 2006, **45**, 5493.
59. F. A. Cotton, C. A. Murillo, X. Wang and C. C. Wilkinson, *Dalton Trans.*, 2006, 4623.
60. E. J. Corey and M. J. Grogan, *Org. Lett.*, 1999, **1**, 157.
61. W. Ye, D. Leow, S. L. M. Goh, C.-T. Tan, C.-H. Chian and C.-H. Tan, *Tetrahedron Lett.*, 2006, **47**, 1007.
62. L. Zhang, R. C. Pratt, F. Nederberg, H. W. Horn, J. E. Rice, R. M. Waymouth, C. G. Wade and J. L. Hedrick, *Macromolecules*, 2010, **43**, 1660.
63. F. Eisenreich, P. Viehmann, F. Müller and S. Hecht, *Macromolecules*, 2015, **48**, 8729.
64. R. C. Pratt, B. G. G. Lohmeijer, D. A. Long, R. M. Waymouth and J. L. Hedrick, *J. Am. Chem. Soc.*, 2006, **128**, 4556.
65. B. G. G. Lohmeijer, R. C. Pratt, F. Leibfarth, J. W. Logan, D. A. Long, A. P. Dove, F. Nederberg, J. Choi, C. Wade, R. M. Waymouth and J. L. Hedrick, *Macromolecules*, 2006, **39**, 8574.
66. H. A. Brown, A. G. de Crisci, J. L. Hedrick and R. M. Waymouth, *ACS Macro Lett.*, 2012, **1**, 1113.
67. L. Simón and J. M. Goodman, *J. Org. Chem.*, 2007, **72**, 9656.
68. A. Chuma, H. W. Horn, W. C. Swope, R. C. Pratt, L. Zhang, B. G. G. Lohmeijer, C. G. Wade, R. M. Waymouth, J. L. Hedrick and J. E. Rice, *J. Am. Chem. Soc.*, 2008, **130**, 6749.
69. C. G. Jaffredo, J.-F. Carpentier and S. M. Guillaume, *Macromol. Rapid Commun.*, 2012, **33**, 1938.
70. M. Bouyahyi, M. P. F. Pepels, A. Heise and R. Duchateau, *Macromolecules*, 2012, **45**, 3356.
71. H. Zhang, S. Hu, J. Zhao and G. Zhang, *Eur. Polym. J.*, 2017, **95**, 693.
72. H. Kim, J. V. Olsson, J. L. Hedrick and R. M. Waymouth, *ACS Macro Lett.*, 2012, **1**, 845.
73. J. A. Castillo, D. E. Borchmann, A. Y. Cheng, Y. Wang, C. Hu, A. J. García and M. Weck, *Macromolecules*, 2012, **45**, 62.
74. F. Jing and M. A. Hillmyer, *J. Am. Chem. Soc.*, 2008, **130**, 13826.
75. G. L. Fiore, F. Jing, V. G. Young, Jr., C. J. Cramer and M. A. Hillmyer, *Polym. Chem.*, 2010, **1**, 870.

76. B. Guillerm, V. Lemaur, B. Ernould, J. Cornil, R. Lazzaroni, J.-F. Gohy, P. Dubois and O. Coulembier, *RSC Adv.*, 2014, **4**, 10028.
77. E. A. Appel, V. Y. Lee, T. T. Nguyen, M. McNeil, F. Nederberg, J. L. Hedrick, W. C. Swope, J. E. Rice, R. D. Miller and J. Sly, *Chem. Commun.*, 2012, **48**, 6163.
78. A. L. Silvers, C.-C. Chang and T. Emrick, *J. Polym. Sci., Part A: Polym. Chem.*, 2012, **50**, 3517.
79. L. Mespouille, O. Coulembier, M. Kawalec, A. P. Dove and P. Dubois, *Prog. Polym. Sci.*, 2014, **39**, 1144.
80. F. Nederberg, B. G. G. Lohmeijer, F. Leibfarth, R. C. Pratt, J. Choi, A. P. Dove, R. M. Waymouth and J. L. Hedrick, *Biomacromolecules*, 2007, **8**, 153.
81. M. Murayama, F. Sanda and T. Endo, *Macromolecules*, 1998, **31**, 919.
82. M. Helou, O. Miserque, J.-M. Brusson, J.-F. Carpentier and S. M. Guillaume, *Chem. – Eur. J.*, 2010, **16**, 13805.
83. S. Venkataraman, J. L. Hedrick and Y. Y. Yang, *Polym. Chem.*, 2014, **5**, 2035.
84. S. Venkataraman, V. W. L. Ng, D. J. Coady, H. W. Horn, G. O. Jones, T. S. Fung, H. Sardon, R. M. Waymouth, J. L. Hedrick and Y. Y. Yang, *J. Am. Chem. Soc.*, 2015, **137**, 13851.
85. P. K. Kuroishi, M. J. Bennison and A. P. Dove, *Polym. Chem.*, 2016, **7**, 7108.
86. P. Olsén, K. Odelius, H. Keul and A.-C. Albertsson, *Macromolecules*, 2015, **48**, 1703.
87. P. Olsén, K. Odelius and A.-C. Albertsson, *Macromolecules*, 2014, **47**, 6189.
88. P. Olsén, K. Odelius and A.-C. Albertsson, *Biomacromolecules*, 2016, **17**, 699.
89. O. Coulembier, V. Lemaur, T. Josse, A. Minoia, J. Cornil and P. Dubois, *Chem. Sci.*, 2012, **3**, 723.
90. W. Guerin, M. Helou, M. Slawinski, J.-M. Brusson, S. M. Guillaume and J.-F. Carpentier, *Polym. Chem.*, 2013, **4**, 3686.
91. G. Kapiti, H. Keul and M. Möller, *Mater. Today Commun.*, 2015, **5**, 1.
92. T. Steinbach and F. R. Wurm, *Angew. Chem., Int. Ed.*, 2015, **54**, 6098.
93. K. N. Bauer, H. T. Tee, M. M. Velencoso and F. R. Wurm, *Prog. Polym. Sci.*, 2017, **73**, 61.
94. Y. Iwasaki and E. Yamaguchi, *Macromolecules*, 2010, **43**, 2664.
95. B. Clément, B. Grignard, L. Koole, C. Jérôme and P. Lecomte, *Macromolecules*, 2012, **45**, 4476.
96. S. Zhang, A. Li, J. Zou, L. Y. Lin and K. L. Wooley, *ACS Macro Lett.*, 2012, **1**, 328.
97. S. Zhang, J. Zou, F. Zhang, M. Elsabahy, S. E. Felder, J. Zhu, D. J. Pochan and K. L. Wooley, *J. Am. Chem. Soc.*, 2012, **134**, 18467.
98. Y. H. Lim, G. S. Heo, S. Cho and K. L. Wooley, *ACS Macro Lett.*, 2013, **2**, 785.
99. T. Steinbach, S. Ritz and F. R. Wurm, *ACS Macro Lett.*, 2014, **3**, 244.

100. J. Alsarraf, Y. A. Ammar, F. Robert, E. Cloutet, H. Cramail and Y. Landais, *Macromolecules*, 2012, **45**, 2249.
101. J. Alsarraf, F. Robert, H. Cramail and Y. Landais, *Polym. Chem.*, 2013, **4**, 904.
102. K. Fukushima, D. J. Coady, G. O. Jones, H. A. Almegren, A. M. Alabdulrahman, F. D. Alsewailem, H. W. Horn, J. E. Rice and J. L. Hedrick, *J. Polym. Sci., Part A: Polym. Chem.*, 2013, **51**, 1606.
103. K. Issleib and M. Lischewski, *Synth. React. Inorg. Met.-Org. Chem.*, 1973, **3**, 255.
104. P. Haasemann and J. Goubeau, *Z. Anorg. Allg. Chem.*, 1974, **408**, 293–303.
105. H. Goldwhite, P. Gysegem, S. Schow and C. Swyke, *J. Chem. Soc., Dalton Trans.*, 1975, 12.
106. R. Appel and M. Halstenberg, *Angew. Chem., Int. Ed. Engl.*, 1977, **16**, 263.
107. R. Schwesinger, J. Willaredt, H. Schlemper, M. Keller, D. Schmitt and H. Fritz, *Chem. Ber.*, 1994, **127**, 2435.
108. See for example: Y. Kondo, in *Superbases for Organic Synthesis*, ed. T. Ishikawa, John Wiley & Sons, Ltd, Chichester, UK, 2009, ch. 5, p. 145–185.
109. S. Boileau and N. Illy, *Prog. Polym. Sci.*, 2011, **36**, 1132.
110. J. Zhao, N. Hadjichristidis and H. Schlaad in *Anionic Polymerization*, ed. N. Hadjichristidis and A. Hirao, Springer Japan, Tokyo, 2015, Part I, p. 429–449.
111. J. Zhao, N. Hadjichristidis and Y. Gnanou, *Polimery*, 2014, **59**, 49.
112. I. A. Koppel, R. Schwesinger, T. Breuer, P. Burk, K. Herodes, I. Koppel, I. Leito and M. Mishima, *J. Phys. Chem. A*, 2001, **105**, 9575.
113. T. Rodima, I. Kaljurand, A. Pihl, V. Mäemets, I. Leito and I. A. Koppel, *J. Org. Chem.*, 2002, **67**, 1873.
114. I. Kaljurand, T. Rodima, A. Pihl, V. Mäemets, I. Leito, I. A. Koppel and M. Mishima, *J. Org. Chem.*, 2003, **68**, 9988.
115. L. Sooväli, T. Rodima, I. Kaljurand, A. Kütt, I. A. Koppel and I. Leito, *Org. Biomol. Chem.*, 2006, **4**, 2100.
116. B. Kovacević and Z. B. Maksić, *Chem. Commun.*, 2006, 1524–1526.
117. R. Schwesinger, C. Hasenfratz, H. Schlemper, L. Walz, E.-M. Peters, K. Peters and H. G. von Schnering, *Angew. Chem., Int. Ed.*, 1993, **32**, 1361.
118. D. Jardel, C. Davies, F. Peruch, S. Massip and B. Bibal, *Adv. Synth. Catal.*, 2016, **358**, 1110.
119. L. Zhang, F. Nederberg, R. C. Pratt, R. M. Waymouth, J. L. Hedrick and C. G. Wade, *Macromolecules*, 2007, **40**, 4154.
120. L. Zhang, F. Nederberg, J. M. Messman, R. C. Pratt, J. L. Hedrick and C. G. Wade, *J. Am. Chem. Soc.*, 2007, **129**, 12610.
121. H. Alamri, J. Zhao, D. Pahovnik and N. Hadjichristidis, *Polym. Chem.*, 2014, **5**, 5471.
122. J. de Winter, O. Coulembier, P. Gerbaux and P. Dubois, *Macromolecules*, 2010, **43**, 10291.

123. M. Kawalec, O. Coulembier, P. Gerbaux, M. Sobota, J. de Winter, P. Dubois, M. Kowalczuk and P. Kurcok, *React. Funct. Polym.*, 2012, **72**, 509.
124. S. Moins, C. Henoumont, J. de Winter, A. Khalil, S. Laurent, S. Cammas-Marion and O. Coulembier, *Polym. Chem.*, 2018, **9**, 1840–1847.
125. C. G. Jaffredo, J.-F. Carpentier and S. M. Guillaume, *Macromolecules*, 2013, **46**, 6765.
126. M. Hong and E. Y.-X. Chen, *Nat. Chem.*, 2016, **8**, 42.
127. M. Hong and E. Y.-X. Chen, *Angew. Chem., Int. Ed.*, 2016, **55**, 4188.
128. N. Zhao, C. Ren, H. Li, Y. Li, S. Liu and Z. Li, *Angew. Chem., Int. Ed.*, 2017, **56**, 12987.
129. Y. Shen, J. Zhang, N. Zhao, F. Liu and Z. Li, *Polym. Chem.*, 2018, **7**, 5096.
130. V. Ladelta, P. Bilalis, Y. Gnanou and N. Hadjichristidis, *Polym. Chem.*, 2017, **8**, 511.
131. W. Guerin, M. Helou, J.-F. Carpentier, M. Slawinski, J.-M. Brusson and S. M. Guillaume, *Polym. Chem.*, 2013, **4**, 1095.
132. H. Yang, M. Yan, S. Pispas and G. Zhang, *Macromol. Chem. Phys.*, 2011, **212**, 2589.
133. B. Esswein and M. Möller, *Angew. Chem., Int. Ed.*, 1996, **35**, 623.
134. S. Förster and E. Krämer, *Macromolecules*, 1999, **32**, 2783.
135. H. Schmalz, M. G. Lanzendörfer, V. Abetz and A. H. E. Müller, *Macromol. Chem. Phys.*, 2003, **204**, 1056.
136. R. Joso, S. Reinicke, A. Walther, H. Schmalz, A. H. E. Müller and L. Barner, *Macromol. Rapid Commun.*, 2009, **30**, 1009.
137. F. Schacher, M. Müllner, H. Schmalz and A. H. E. Müller, *Macromol. Chem. Phys.*, 2009, **210**, 256.
138. B. Esswein, N. M. Steidl and M. Möller, *Macromol. Rapid Commun.*, 1996, **17**, 143.
139. O. Rexin and R. Mülhaupt, *J. Polym. Sci., Part A: Polym. Chem.*, 2002, **40**, 864.
140. H. Schlaad, H. Kukula, J. Rudloff and I. Below, *Macromolecules*, 2001, **34**, 4302.
141. J. Zhao, D. Pahovnik, Y. Gnanou and N. Hadjichristidis, *Macromolecules*, 2014, **47**, 1693.
142. M. Groenewolt, T. Brezesinski, H. Schlaad, M. Antonietti, P. W. Groh and B. Iván, *Adv. Mater.*, 2005, **17**, 1158.
143. J. Zhao, H. Schlaad, S. Weidner and M. Antonietti, *Polym. Chem.*, 2012, **3**, 1763.
144. J. Zhao, H. Alamri and N. Hadjichristidis, *Chem. Commun.*, 2013, **49**, 7079.
145. J. Zhao, G. Zhang and S. Pispas, *J. Polym. Sci., Part A: Polym. Chem.*, 2010, **48**, 2320.
146. J. Zhao, D. Pahovnik, Y. Gnanou and N. Hadjichristidis, *Polym. Chem.*, 2014, **5**, 3750.

147. J. Zhao, D. Pahovnik, Y. Gnanou and N. Hadjichristidis, *Macromolecules*, 2014, **47**, 3814.
148. S. Hu, G. Dai, J. Zhao and G. Zhang, *Macromolecules*, 2016, **49**, 4462.
149. H. Li, J. Zhao and G. Zhang, *ACS Macro Lett.*, 2017, 1094.
150. H. Misaka, R. Sakai, T. Satoh and T. Kakuchi, *Macromolecules*, 2011, **44**, 9099.
151. W. Kwon, Y. Rho, K. Kamoshida, K. H. Kwon, Y. C. Jeong, J. Kim, H. Misaka, T. J. Shin, J. Kim, K.-W. Kim, K. S. Jin, T. Chang, H. Kim, T. Satoh, T. Kakuchi and M. Ree, *Adv. Funct. Mater.*, 2012, **22**, 5194.
152. A. Molenberg and M. Möller, *Macromol. Rapid Commun.*, 1995, **16**, 449.
153. A. Molenberg and M. Möller, *Macromol. Chem. Phys.*, 1997, **198**, 717.
154. H. Yang, J. Zhao, M. Yan, S. Pispas and G. Zhang, *Polym. Chem.*, 2011, **2**, 2888.
155. N. Illy, S. Boileau, J. Penelle and V. Barbier, *Macromol. Rapid Commun.*, 2009, **30**, 1731.
156. N. Illy, S. Boileau, W. Buchmann, J. Penelle and V. Barbier, *Macromolecules*, 2010, **43**, 8782.
157. N. Illy, S. Boileau, M. A. Winnik, J. Penelle and V. Barbier, *Polymer*, 2012, **53**, 903.
158. A. Benlahouès, B. Brissault, S. Boileau and J. Penelle, *Macromol. Chem. Phys.*, 2018, **219**, 1700463.
159. T. Pietzonka and D. Seebach, *Angew. Chem., Int. Ed.*, 1993, **32**, 716.
160. H. G. Börner and W. Heitz, *Macromol. Chem. Phys.*, 1998, **199**, 1815.
161. M. Schmitt, L. Falivene, L. Caporaso, L. Cavallo and E. Y.-X. Chen, *Polym. Chem.*, 2014, **5**, 3261.
162. H. Yang, J. Xu, S. Pispas and G. Zhang, *Macromolecules*, 2012, **45**, 3312.
163. J. Xu, H. Yang and G. Zhang, *Macromol. Chem. Phys.*, 2013, **214**, 378.
164. H. Yang, J. Xu, S. Pispas and G. Zhang, *RSC Adv.*, 2013, **3**, 6853.
165. J. de Winter, A. P. Dove, A. Knoll, P. Gerbaux, P. Dubois and O. Coulembier, *Polym. Chem.*, 2014, **5**, 706.
166. S. Naumann and A. P. Dove, *Polym. Int.*, 2016, **65**, 16.
167. H. W. Wanzlick, *Angew. Chem., Int. Ed.*, 1962, **1**, 75.
168. H.-W. Wanzlick and H.-J. Schönherr, *Angew. Chem., Int. Ed.*, 1968, **7**, 141.
169. K. Öfele, *J. Organomet. Chem.*, 1968, **12**, P42.
170. A. Igau, H. Grutzmacher, A. Baceiredo and G. Bertrand, *J. Am. Chem. Soc.*, 1988, **110**, 6463.
171. A. J. Arduengo, R. L. Harlow and M. Kline, *J. Am. Chem. Soc.*, 1991, **113**, 361.
172. T. Dröge and F. Glorius, *Angew. Chem., Int. Ed.*, 2010, **49**, 6940.
173. F. E. Hahn and M. C. Jahnke, *Angew. Chem., Int. Ed.*, 2008, **47**, 3122.
174. H. Jacobsen, A. Correa, C. Costabile and L. Cavallo, *J. Organomet. Chem.*, 2006, **691**, 4350.
175. A. C. Hillier, W. J. Sommer, B. S. Yong, J. L. Petersen, L. Cavallo and S. P. Nolan, *Organometallics*, 2003, **22**, 4322.
176. C. A. Tolman, *Chem. Rev.*, 1977, **77**, 313.

177. D. J. Nelson and S. P. Nolan, *Chem. Soc. Rev.*, 2013, **42**, 6723.
178. T. L. Amyes, S. T. Diver, J. P. Richard, F. M. Rivas and K. Toth, *J. Am. Chem. Soc.*, 2004, **126**, 4366.
179. E. M. Higgins, J. A. Sherwood, A. G. Lindsay, J. Armstrong, R. S. Massey, R. W. Alder and A. C. O'Donoghue, *Chem. Commun.*, 2011, **47**, 1559.
180. Y. Chu, H. Deng and J.-P. Cheng, *J. Org. Chem.*, 2007, **72**, 7790.
181. A. M. Magill, K. J. Cavell and B. F. Yates, *J. Am. Chem. Soc.*, 2004, **126**, 8717.
182. R. S. Massey, C. J. Collett, A. G. Lindsay, A. D. Smith and A. C. O'Donoghue, *J. Am. Chem. Soc.*, 2012, **134**, 20421.
183. M. Iglesias, D. J. Beetstra, J. C. Knight, L.-L. Ooi, A. Stasch, S. Coles, L. Male, M. B. Hursthouse, K. J. Cavell, A. Dervisi and I. A. Fallis, *Organometallics*, 2008, **27**, 3279.
184. M. Melaimi, M. Soleilhavoup and G. Bertrand, *Angew. Chem., Int. Ed.*, 2010, **49**, 8810.
185. R. H. Crabtree, *Coord. Chem. Rev.*, 2013, **257**, 755.
186. V. Lavallo, Y. Canac, C. Präsang, B. Donnadieu and G. Bertrand, *Angew. Chem., Int. Ed.*, 2005, **44**, 5705.
187. T. W. Hudnall and C. W. Bielawski, *J. Am. Chem. Soc.*, 2009, **131**, 16039.
188. J. Raynaud, C. Absalon, Y. Gnanou and D. Taton, *J. Am. Chem. Soc.*, 2009, **131**, 3201.
189. J. Raynaud, C. Absalon, Y. Gnanou and D. Taton, *Macromolecules*, 2010, **43**, 2814.
190. M. N. Hopkinson, C. Richter, M. Schedler and F. Glorius, *Nature*, 2014, **510**, 485.
191. *N-Heterocyclic Carbenes. Effective Tools for Organometallic Synthesis*, ed. S. P. Nolan, Wiley-VCH, Weinheim an der Bergstrasse, Germany, 2014.
192. O. Coulembier, J.-M. Raquez and P. Dubois, *Polimery*, 2008, **53**, 255.
193. D. Martin, M. Melaimi, M. Soleilhavoup and G. Bertrand, *Organometallics*, 2011, **30**, 5304.
194. D. Enders, O. Niemeier and A. Henseler, *Chem. Rev.*, 2007, **107**, 5606.
195. N. Marion, S. Díez-González and S. P. Nolan, *Angew. Chem., Int. Ed.*, 2007, **46**, 2988.
196. D. M. Flanigan, F. Romanov-Michailidis, N. A. White and T. Rovis, *Chem. Rev.*, 2015, **115**, 9307.
197. K. M. Hindi, M. J. Panzner, C. A. Tessier, C. L. Cannon and W. J. Youngs, *Chem. Rev.*, 2009, **109**, 3859.
198. *N-Heterocyclic Carbenes in Transition Metal Catalysis and Organocatalysis*, ed. C. S. Cazin, Springer Science + Business Media B.V, Dordrecht, 2011, vol. 32.
199. B. Maji, M. Horn and H. Mayr, *Angew. Chem., Int. Ed.*, 2012, **51**, 6231.
200. N. Kuhn, H. Bohnen, J. Kreutzberg, D. Bläser and R. Boese, *J. Chem. Soc., Chem. Commun.*, 1993, 1136.
201. U. Gruseck and M. Heuschmann, *Tetrahedron Lett.*, 1987, **28**, 6027.
202. C. H. Tiemann, W. D. Kollmeyer, *U. S. Pat.*, 3948934, 1976.

203. P. P. Ponti, J. C. Baldwin and W. C. Kaska, *Inorg. Chem.*, 1979, **18**, 873.
204. H. Schumann, M. Glanz, J. Winterfeld, H. Hemling, N. Kuhn, H. Bohnen, D. Bläser and R. Boese, *J. Organomet. Chem.*, 1995, **493**, C14.
205. N. Kuhn, H. Bohnen and R. Boese, *Chem. Ber.*, 1994, **127**, 1405.
206. S. Kronig, P. G. Jones and M. Tamm, *Eur. J. Inorg. Chem.*, 2013, **2013**, 2301.
207. K. Powers, C. Hering-Junghans, R. McDonald, M. J. Ferguson and E. Rivard, *Polyhedron*, 2016, **108**, 8.
208. A. El-Hellani, J. Monot, R. Guillot, C. Bour and V. Gandon, *Inorg. Chem.*, 2013, **52**, 506.
209. M. Iglesias, A. Iturmendi, P. J. Sanz Miguel, V. Polo, J. J. Pérez-Torrente and L. A. Oro, *Chem. Commun.*, 2015, **51**, 12431.
210. A. Fürstner, M. Alcarazo, R. Goddard and C. W. Lehmann, *Angew. Chem., Int. Ed.*, 2008, **47**, 3210.
211. J. Nakayama, K. Akimoto and Y. Sugihara, *Tetrahedron Lett.*, 1998, **39**, 5587.
212. G. Ye, S. Chatterjee, M. Li, A. Zhou, Y. Song, B. L. Barker, C. Chen, D. J. Beard, W. P. Henry and C. U. Pittman, *Tetrahedron*, 2010, **66**, 2919.
213. Y.-B. Wang, Y.-M. Wang, W.-Z. Zhang and X.-B. Lu, *J. Am. Chem. Soc.*, 2013, **135**, 11996.
214. L. Dong, J. Wen and W. Li, *Org. Biomol. Chem.*, 2015, **13**, 8533.
215. Y.-B. Wang, D.-S. Sun, H. Zhou, W.-Z. Zhang and X.-B. Lu, *Green Chem.*, 2015, **17**, 4009.
216. W. Li, N. Yang and Y. Lyu, *J. Phys. Org. Chem.*, 2016, **81**, 5303.
217. V. B. Saptal and B. M. Bhanage, *ChemSusChem*, 2016, **9**, 1980.
218. W. Li, D. Huang and Y. Lyu, *Org. Biomol. Chem.*, 2016, **14**, 10875.
219. L. H. Finger, J. Guschlbauer, K. Harms and J. Sundermeyer, *Chem. – Eur. J.*, 2016, **22**, 16292.
220. A. P. de Lima, Batista, A. G. S. de Oliveira-Filho and S. E. Galembeck, *ACS Omega*, 2017, **2**, 299.
221. M. Blümel, J.-M. Noy, D. Enders, M. H. Stenzel and T. V. Nguyen, *Org. Lett.*, 2016, **18**, 2208.
222. M. Blümel, R. D. Crocker, J. B. Harper, D. Enders and T. V. Nguyen, *Chem. Commun.*, 2016, **52**, 7958.
223. U. Kaya, U. P. N. Tran, D. Enders, J. Ho and T. V. Nguyen, *Org. Lett.*, 2017, **19**, 1398.
224. R. D. Crocker and T. V. Nguyen, *Chem. – Eur. J.*, 2016, **22**, 2208.
225. M. M. D. Roy and E. Rivard, *Acc. Chem. Res.*, 2017, **50**, 2017.
226. U. Gruseck and M. Heuschmann, *Chem. Ber.*, 1987, **120**, 2053.
227. H. Quast, M. Ach, M. K. Kindermann, P. Rademacher and M. Schindler, *Chem. Ber.*, 1993, **126**, 503.
228. H. Quast, M. Ach, S. Ivanova, E.-M. Peters, K. Peters and H. G. von Schnering, *Liebigs Ann. Chem.*, 1996, 1551.
229. H. Quast, A. Fuss and W. Nüdling, *Eur. J. Org. Chem.*, 1998, 317.
230. G. Ye, W. P. Henry, C. Chen, A. Zhou and C. U. Pittman, *Tetrahedron Lett.*, 2009, **50**, 2135.

231. C. Hering-Junghans and A. Schumann, *Eur. J. Inorg. Chem.*, 2018, **2018**, 2584.
232. M. Al Ghamdi, L. Cavallo and L. Falivene, *J. Phys. Chem. C*, 2017, **121**, 2730.
233. S. Naumann and M. R. Buchmeiser, *Catal. Sci. Technol.*, 2014, **4**, 2466.
234. C.-L. Lai, H. M. Lee and C.-H. Hu, *Tetrahedron Lett.*, 2005, **46**, 6265.
235. G. W. Nyce, T. Glauser, E. F. Connor, A. Möck, R. M. Waymouth and J. L. Hedrick, *J. Am. Chem. Soc.*, 2003, **125**, 3046.
236. S. Naumann, F. G. Schmidt, W. Frey and M. R. Buchmeiser, *Polym. Chem.*, 2013, **4**, 4172.
237. N. E. Kamber, W. Jeong, S. Gonzalez, J. L. Hedrick and R. M. Waymouth, *Macromolecules*, 2009, **42**, 1634.
238. H. A. Brown and R. M. Waymouth, *Acc. Chem. Res.*, 2013, **46**, 2585.
239. M. Rodriguez, S. Marrot, T. Kato, S. Stérin, E. Fleury and A. Baceiredo, *J. Organomet. Chem.*, 2007, **692**, 705.
240. T. K. Sen, S. C. Sau, A. Mukherjee, A. Modak, S. K. Mandal and D. Koley, *Chem. Commun.*, 2011, **47**, 11972.
241. R. Tonner, G. Heydenrych and G. Frenking, *Chem. – Asian J.*, 2007, **2**, 1555.
242. Y. Zhang and E. Y.-X. Chen, *Angew. Chem., Int. Ed.*, 2012, **51**, 2465.
243. K. Jovic, J. Unold, S. Naumann, M. Ullrich, F. G. Schmidt and M. R. Buchmeiser, *Macromol. Mater. Eng.*, 2016, **301**, 423.
244. S. Naumann, K. Mundsinger, L. Cavallo and L. Falivene, *Polym. Chem.*, 2017, **8**, 5803.
245. A. Balint, M. Papendick, M. Clauss, C. Müller, F. Giesselmann and S. Naumann, *Chem. Commun.*, 2018, **54**, 2220.
246. S. Naumann, A. W. Thomas and A. P. Dove, *ACS Macro Lett.*, 2016, **5**, 134.
247. S. Naumann and D. Wang, *Macromolecules*, 2016, **49**, 8869.
248. Q. Wang, W. Zhao, J. He, Y. Zhang and E. Y.-X. Chen, *Macromolecules*, 2017, **50**, 123.
249. P. Walther and S. Naumann, *Macromolecules*, 2017, **50**, 8406.
250. Y.-B. Jia, Y.-B. Wang, W.-M. Ren, T. Xu, J. Wang and X.-B. Lu, *Macromolecules*, 2014, **47**, 1966.
251. Y.-B. Jia, W.-M. Ren, S.-J. Liu, T. Xu, Y.-B. Wang and X.-B. Lu, *ACS Macro Lett.*, 2014, **3**, 896.
252. F. Nederberg, E. F. Connor, M. Möller, T. Glauser and J. L. Hedrick, *Angew. Chem., Int. Ed.*, 2001, **40**, 2712.
253. F. Nederberg, E. F. Connor, T. Glausser and J. L. Hedrick, *Chem. Commun.*, 2001, 2066.
254. C. Bonduelle, B. Martín-Vaca, F. P. Cossío and D. Bourissou, *Chem. – Eur. J.*, 2008, **14**, 5304.
255. D. J. Coady, A. C. Engler, H. W. Horn, K. M. Bajjuri, K. Fukushima, G. O. Jones, A. Nelson, J. E. Rice and J. L. Hedrick, *ACS Macro Lett.*, 2012, **1**, 19.

256. R. Todd, G. Rubio, D. J. Hall, S. Tempelaar and A. P. Dove, *Chem. Sci.*, 2013, **4**, 1092.
257. S. Csihony, D. A. Culkin, A. C. Sentman, A. P. Dove, R. M. Waymouth and J. L. Hedrick, *J. Am. Chem. Soc.*, 2005, **127**, 9079.
258. S. Csihony, T. T. Beaudette, A. C. Sentman, G. W. Nyce, R. M. Waymouth and J. L. Hedrick, *Adv. Synth. Catal.*, 2004, **346**, 1081.
259. J. S. Bandar and T. H. Lambert, *J. Am. Chem. Soc.*, 2012, **134**, 5552.
260. K. S. Egorova and V. P. Ananikov, *Organometallics*, 2017, **36**, 4071.
261. Y. Xia, J. Shen, H. Alamri, N. Hadjichristidis, J. Zhao, Y. Wang and G. Zhang, *Biomacromolecules*, 2017, **18**, 3233.
262. A. Nachtergael, O. Coulembier, P. Dubois, M. Helvenstein, P. Duez, B. Blankert and L. Mespouille, *Biomacromolecules*, 2015, **16**, 507.
263. J. Raynaud, Y. Gnanou and D. Taton, *Macromolecules*, 2009, **42**, 5996.
264. M. D. Scholten, J. L. Hedrick and R. M. Waymouth, *Macromolecules*, 2008, **41**, 7399.
265. S. Naumann, P. B. V. Scholten, J. A. Wilson and A. P. Dove, *J. Am. Chem. Soc.*, 2015, **137**, 14439.
266. A. C. Engler, J. M. W. Chan, D. J. Coady, J. M. O'Brien, H. Sardon, A. Nelson, D. P. Sanders, Y. Y. Yang and J. L. Hedrick, *Macromolecules*, 2013, **46**, 1283.
267. Q. Ma, K. Lei, J. Ding, L. Yu and J. Ding, *Polym. Chem.*, 2017, **8**, 6665.
268. J. Raynaud, W. N. Ottou, Y. Gnanou and D. Taton, *Chem. Commun.*, 2010, **46**, 3203.

CHAPTER 5

Ring-opening Polymerization of Lactones

PHILLIPE LECOMTE* AND CHRISTINE JÉRÔME

Liège University, Center for Education and Research on Macromolecules (CERM), CESAM-RU, Building B6a, Sart-Tilman, Liège, B-4000, Belgium
*Email: Philippe.Lecomte@uliege.be

5.1 Introduction

The ring-opening polymerization of cyclic monoesters was discovered in the 1930s by Wallace H. Carothers, who achieved the thermal polymerization of δ-valerolactone[1] and ε-caprolactone.[2] Since then, a steadily increasing interest has been paid to polyesters prepared by ring opening, mainly because of their (bio)degradability and even for some of them their bio-origin.

Ring-opening polymerization can roughly take place by two main mechanisms (Figure 5.1). The first possible one is based on the hydrolysis of the cyclic monoester into an hydroxyacid followed by its step-growth polymerization. The second one is based on a chain growth-mechanism made up of initiation, propagation, transfer and termination steps. The initiation implies the reaction of a nucleophilic compound such as an alcohol and an electrophilic cyclic ester to afford a new alcohol, which is the initiating species. Propagation takes then place by the repetition of the mechanism involved in initiation. Nevertheless, the nucleophilicity of alcohols is not high enough to assure an efficient reaction with the cyclic monoester. A possible strategy to increase the nucleophilicity of the initiator relies on the use alkoxides rather than alcohols. Another possibility relies on the use of catalysts to activate either the alcohol or the monomer. Intra- or

Polymer Chemistry Series No. 31
Organic Catalysis for Polymerisation
Edited by Andrew Dove, Haritz Sardon and Stefan Naumann

Published by the Royal Society of Chemistry, www.rsc.org

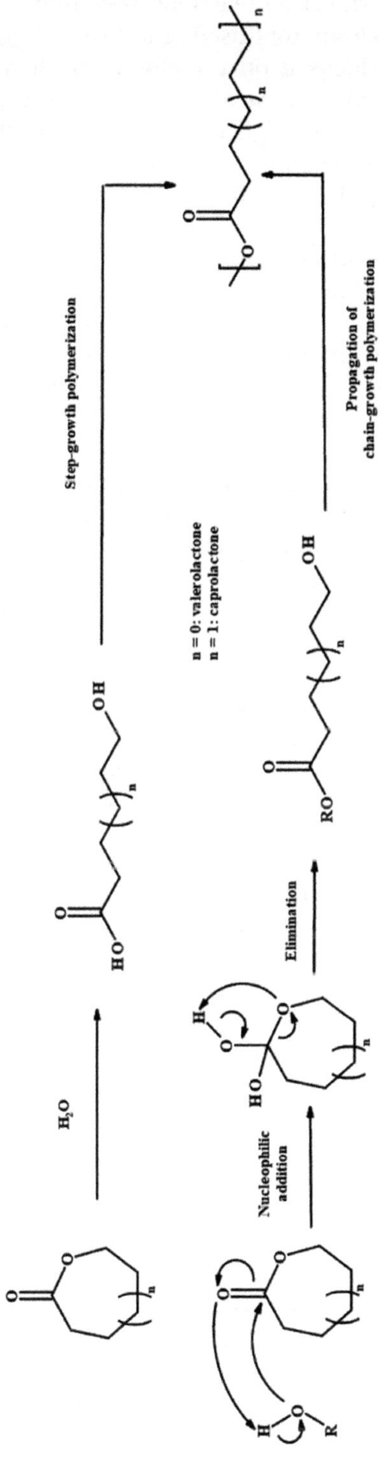

Figure 5.1 Possible theoretical mechanisms of ring-opening polymerization of cyclic monoesters.

intermolecular transesterification reactions are observed, depending on the conditions and catalysts/initiators used, and bring about a loss of control. Many researchers have focused on the chain-growth mechanism because the discovery of active and selective catalysts opens the way to the synthesis of polyesters with predictable high molar mass as well as low dispersity. Indeed when transesterification reactions and other side reactions are as limited as possible, the polymerization meets the criteria of living polymerization.[3]

Since the pioneering works of Carothers, many studies have been devoted by many teams all around the World to develop active and selective catalysts and initiators.[4–8] Cyclic esters have the scarce capability to be polymerized by different routes such as anionic, cationic, coordination enzymatic and organocatalytic routes. The first investigations based on ionic polymerizations show clear limits in terms of control.[4] The control was deeply improved by implementing coordination polymerization and using suitable catalysts based on transition metals. Coordination polymerization is still nowadays implemented for the synthesis of polyesters at the industrial scale, the typical example being poly(ε-caprolactone) synthesized by using tin(II) 2-ethylhexanoate as a catalyst.[9] Nevertheless, coordination polymerization has its limitations, and among them, let us point out (1) the toxicity of many complexes based on transition metals, and especially the most popular ones (Sn, Al...), (2) the high sensitivity of some complexes towards water (for instance the tin and aluminum alkoxides), (3) the safety hazards of some complexes or their precursors, for instance triethylaluminium, a precursor of aluminum alkoxides initiators, ignites in the presence of oxygen.

Polymer chemists decided to investigate non-metallic catalysts for the ring-opening of cyclic monoesters. In the beginning of the 1990s, the groups of Kobayashi[10] and Gutman[11] showed independently that lipases catalyze the ring-opening polymerization of cyclic esters. Biochemists identified within the catalytic site a triad made up of serine, histidine and aspartic acid.[12] In the active site of the enzyme the aspartate interacts though H-bonds of histidine to increase its pK_a (7 to 12). Histidine is then able to interact with the alcohol function of serine to increase its nucleophilicity and catalyze its reaction with the cyclic monoester (Figure 5.2). In addition, the monomer is in the enzyme site in a pocket made up of an oxyanion hole, capable of activating the monomer through H-bonds. The mechanism is thus based on a dual catalysis based on the activation of both the OH group of serine and the monomer. Researchers paid great attention to enzymatic polymerization due to (1) the low toxicity of enzymes, (2) stereroselectivity of polymerization by chiral enzymes, (2) polymerizability of macrocyclic monoesters, less strained than medium size cyclic monoesters.[13,14]

Despite the advantages of enzymatic polymerization, the control remains limited and it is difficult to target high molar masses. Nevertheless, enzymatic catalysts have raised the question whether it should be possible to catalyze the ring-opening polymerization of cyclic esters by metal-free organocatalysts with similar interactions and activation mechanisms

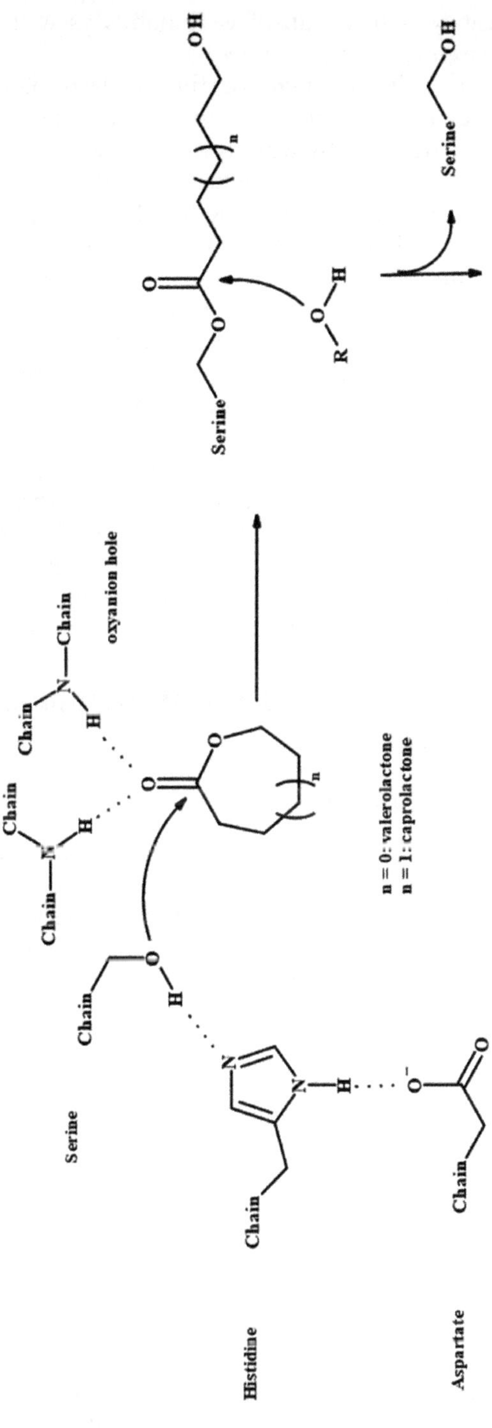

Figure 5.2 Mechanism of the polymerization of cyclic monoesters by lipases.

implemented by enzymes, and more particularly lipases. Organocatalysts are commonly defined as catalysts made up of C–H bonds as well as other heteroatom (O, S, N...) enzymes being excluded.

Organocatalysts have been investigated for the catalysis of the ring-opening polymerization of cyclic monoesters for a very long time, and even before the advent of enzymatic polymerization, typically for cationic polymerization and zwitterionic polymerizations.[4] Nevertheless, for a very long time, these catalysts have remained less popular than transition-metal-based catalysts because of a lack of selectivity. Quite recently, in the 2000s, this field witnessed a rejuvenation under the impulse of the works of Hedrick,[7] who succeeded in implementing very selective organocatalysts for the ring-opening polymerization of cyclic monoesters. Since then, a plethora of organocatalysts have been discovered and, nowadays, are more and more systematically used for the polymerization of cyclic monoesters.

Firstly, this review will focus on organocatalysts able to polymerize medium size cyclic monoesters, and mainly δ-valerolactone and ε-caprolactone. The polymerization of small size cyclic monoesters or macrocyclic monoesters will thereafter be discussed to assess the importance of the ring size. Finally, we show that a wide range of organocatalysts are available for the polymerization of cyclic monoesters, substituted and functionalized, for copolymerization and for the synthesis of different architectures.

5.2 Polymerization of Six- and Seven-membered Medium Size Monoesters

5.2.1 Polymerization Catalyzed by Carboxylic Acids

The ring-opening polymerization of cyclic monoesters relies on the esterification reaction, which is very common in organic chemistry. This esterification reaction catalyzed by organic and inorganic acids is reported in the state-of-the art as the Fischer esterification reaction. As far as a cyclic monoester is polymerized in the presence of water, it is always possible that the hydrolysis of the cyclic monoester releases an hydrocyacid capable of catalyzing the esterification reaction.[15] Researchers investigated this possibility by carrying out ring-opening polymerization in the presence of carboxylic acids, purposely added in the reactor, where the polymerization takes place. Ideally, these studies should be carried out in the absence of water to prevent any hydrolysis of the cyclic ester, and thus avoiding any release of carboxylic acids capable of interfering. ε-caprolactone is reported to be polymerized by alcohols as initiators in the presence of a catalytic amount of carboxylic acids such propanoic acid ($pK_a = 4.87$ in H_2O),[16,17] hexanoic acid ($pK_a = 4.88$ in H_2O),[16,17] citric acid ($pK_{a1} = 3.13$ in H_2O),[17] tartric acid ($pK_{a1} = 3.04$ in H_2O)[17] and lactic acid ($pK_a = 3.86$ in H_2O).[16,17]

It is generally accepted by most authors that catalysis by carboxylic acids takes place through a cationic mechanism. In the absence of any alcohol or other nucleophilic additive, the only remaining nucleophilic species is the monomer itself and the cationic polymerization then takes place through an activated chain-end mechanism (Figure 5.3, path b). Examples of polymerization through the chain-end-activated mechanism can be found in the state-of-the-art but this route is not popular because of the lack of control of the polymerization. As soon as an alcohol is present in the polymerization medium, for instance by ring-opening of the cyclic monoester into an hydroxyalcohol, or by direct addition of an alcohol in the reactor on purpose, the alcohol is by far more nucleophilic than the cyclic monoester and the polymerization does not take place any more by the chain-end-activated mechanism but by a new mechanism reported in the state-of-the-art as the activated monomer mechanism (Figure 5.3, path a). The polymerization involved then the cleavage of the C_{acyl}–O bond (Figure 5.3). The alcohol is then converted into an ester at the α-chain end of the polyester, which is confirmed experimentally, for instance, by NMR techniques. This mechanism suggests that pK_a is a key parameter to quantify the catalytical activity of the organic acid. The screening of different carboxylic acids suggests that

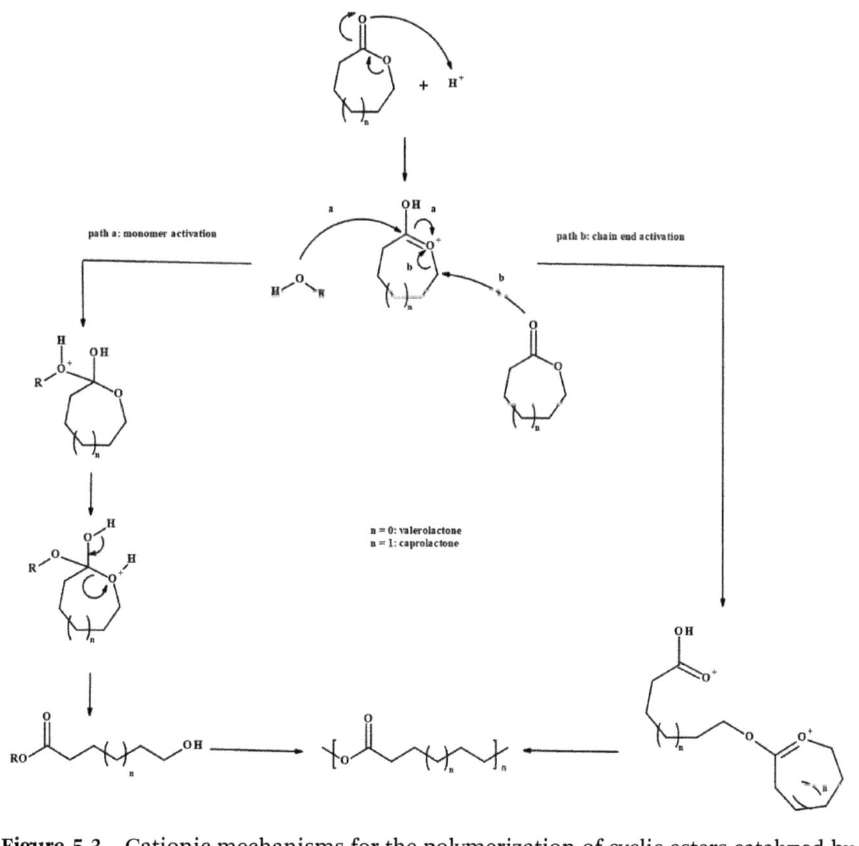

Figure 5.3 Cationic mechanisms for the polymerization of cyclic esters catalyzed by organic acids.

their pK_a should be between 3 and 5 to be efficient catalysts of the polymerization.[17] It must be noted that stronger carboxylic acids could accelerate propagation even more but, at the same time, could also accelerate transesterification reactions at the expense of the selectivity and thus, at the expense of the control of the polymerization. The experience shows that the choice of the acid is important for the control, but many studies do not report a direct correlation between the pK_a and the kinetic constant of propagation. Finally, the cationic mechanism shown in Figure 5.3 assumes the absence of any effect of the anionic counter ion (the carboxylate) beyond its influence on the pK_a.

A particular class of organocatalysts are hydroxyacids such ascorbic acid[18]) and amino acids,[17,19,20] which are able to initiate the polymerization through the nucleophilic group (amine or alcohol) and catalyze the polymerization through the carboxylic acids. With these systems, no additional carboxylic acid must be added in the reactor where the polymerization takes place. It must be noted that hydroxyacids, such as lactic acid,[16,21] tartaric acid[21] and salicylic acid,[22] where the alcohol is secondary, do not belong to this class of organocatalysts because the secondary alcohol does not initiate the polymerization, at least under common experimental conditions.

As far as carboxylic acids are used as catalysts, ROP is often carried out in bulk and at quite high temperature. The control of the polymerization is then limited most of the time, due to the presence of transesterification reactions. In addition, in many papers, when a degree of control is claimed, the reported molar mass remains quite low (often below 10 000 g mol^{-1}) Nevertheless, exceptions are reported and good control was observed for the ring-opening polymerization of δ-valerolactone and ε-caprolactone by benzyl alcohol in the presence of salicylic acid (pK_{a1} = 2.98).[22] No clear explanation is given why this carboxylic should be more efficient than carboxylic acids with similar values of pK_a. A more detailed mechanistic study should be helpful to understand the reasons behind the good control imparted to this polymerization.

Recently, squalene hopene cyclases were reported to protonate alkenes.[23] In the active site of the enzyme, the acidity of aspartic acid was increased by two H-bond donors, histidine and tyrosine. This discovery raises the question whether the concept can be extended to carboxylic acids for the ROP of cyclic esters. Guo achieved the polymerization of ε-caprolactone and δ-valerolactone by an alcohol in the presence of a sulfonyl guanidine and a series of substituted benzoic acids (Figure 5.4).[24] Among them, 4-(trifluoromethyl)benzoic acid turned out to be particularly efficient and nice control of the molar mass and disparities close to 1.2 were observed. The proposed mechanism relies on the increase of the acidity of the carboxylic acid by the establishment of two interactions by hydrogen bonds between the two co-catalysts (Figure 5.4). This work is another nice example showing that concepts used by nature in enzymatic reactions can inspire chemists for the development of efficient organocatalysts.

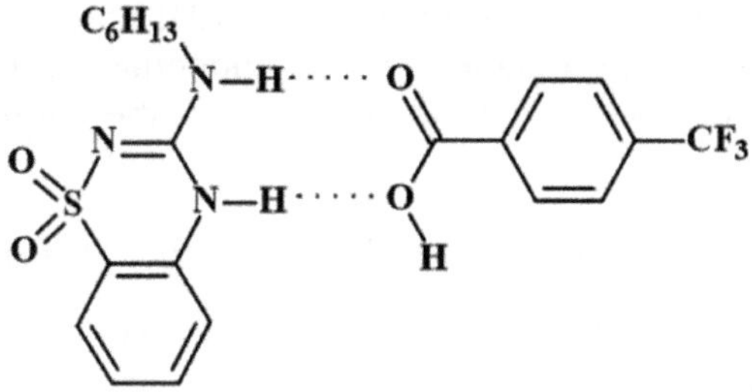

Figure 5.4 Interaction between a sulfonyl guanidine and 4-(trifluoromethyl)benzoic acid.

5.2.2 Polymerization Catalyzed by Sulfonic and Dialkyl Phosphates

Other acids than carboxylic acids were investigated to catalyze the ring-opening polymerization of cyclic monoesters. Inorganic Brønsted and Lewis acids are beyond the scope of this review. Quite recently, this field witnessed a rejuvenation by the discovery that sulfonic[25–28] and phosphonic acids[29,30] are efficient and selective organocatalysts for the ring-opening polymerization of cyclic monoesters. In the 2000s, some control was observed when n-propyl sulfonic acid supported on porous and non-porous silica[25] as well as trifluoromethane sulfonic acid[26] were used as catalysts for the polymerization of ε-caprolactone initiated by alcohols. Bourissou and Martín-Vaca compared the efficiency of methanesulfonic acid and triflic acid for the ring-opening polymerization of ε-caprolactone initiated by primary and secondary alcohols. It was proposed that the polymerization takes place by the cationic activated monomer mechanism.[26] Quite surprisingly, methanesulfonic acid ($pK_a = -1.9$) turned out to be as active as triflic acid ($pK_a = -14.7$), despite its lower acidity, at a temperature as low as 30 °C.[27] This experimental result shows that other parameters than acidity must be considered to understand fully the catalytic activity and that the mechanism needs to be revised. The cationic mechanism implies that the acid is strong enough to protonate the monomer. When this is not the case, another possibility relies on the activation of the monomer by establishing H-bonds. It is worth noting that, if some control is observed, the reported molar masses remain quite low (often lower than 10 000 $g\,mol^{-1}$). As an alternative, the polymerization of ε-caprolactone and δ-valerolactone was initiated by alcohols in the presence of diphenyl phosphate as an organoctalyst.[29,30] Good control was observed at least up to molar masses around 25 000 $g\,mol^{-1}$.

Recently, Guo investigated the polymerization by alcohols in the presence of methanesulfonic acid with the assistance of thiophosphoric triamide as a H-bond donor.[31] The polymerization takes place at room temperature with a good control of the molar mass and dispersities around 1.1.

The interaction between the Brønsted acid (AH) and the H-bond donor (HBD) facilitates the protonation of the monomer. The initiation and propagation should take place through a cationic mechanism.

5.2.3 Polymerization Catalyzed by H-bond Donor

When acids are not strong enough to protonate the cyclic ester and to trigger the cationic polymerization, another possible activation mechanism is based on the establishment of an H-bond between the catalyst and the monomer, enabling an increase in the electrophilicity of the cyclic ester. A typical example is the thiourea (TU) developed by the team of Hedrick (Figure 5.5).[32] TU is used, most of the time, in the presence of a co-catalyst, a Lewis base, prone to activate the alcohol as an initiator. Section 5.2.5 will be devoted to the dual catalysis. Nevertheless, the use of a co-catalyst is not mandatory. For instance, Bourissou catalyzed the ring-opening polymerization of ε-caprolactone by pyridinium salts of camphorsulfonic acid (Figure 5.5).[33] A good control and low dispersity were observed to molar mass up to 17 000 g mol^{-1}. The proposed mechanism is based only on an activation of the monomer by the pyridinium cation through H-bonds. No activation of the alcohol by the counter-ion is reported.

5.2.4 Polymerization Catalyzed by Lewis Bases

Beside the use of Lewis and Brønsted acids as catalysts for the ROP of cyclic esters, Lewis bases were also considered. Roughly, three distinct activation mechanisms can be identified. Firstly, when the bases are strong enough to deprotonate alcohols, alkoxides are generated *in situ* and the polymerization takes then place by an anionic mechanism (Figure 5.6a). Secondly, when the

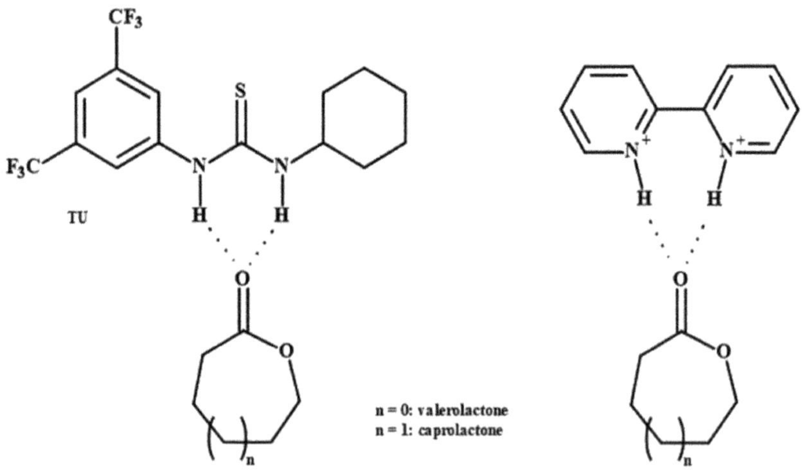

Figure 5.5 H-bond donors.

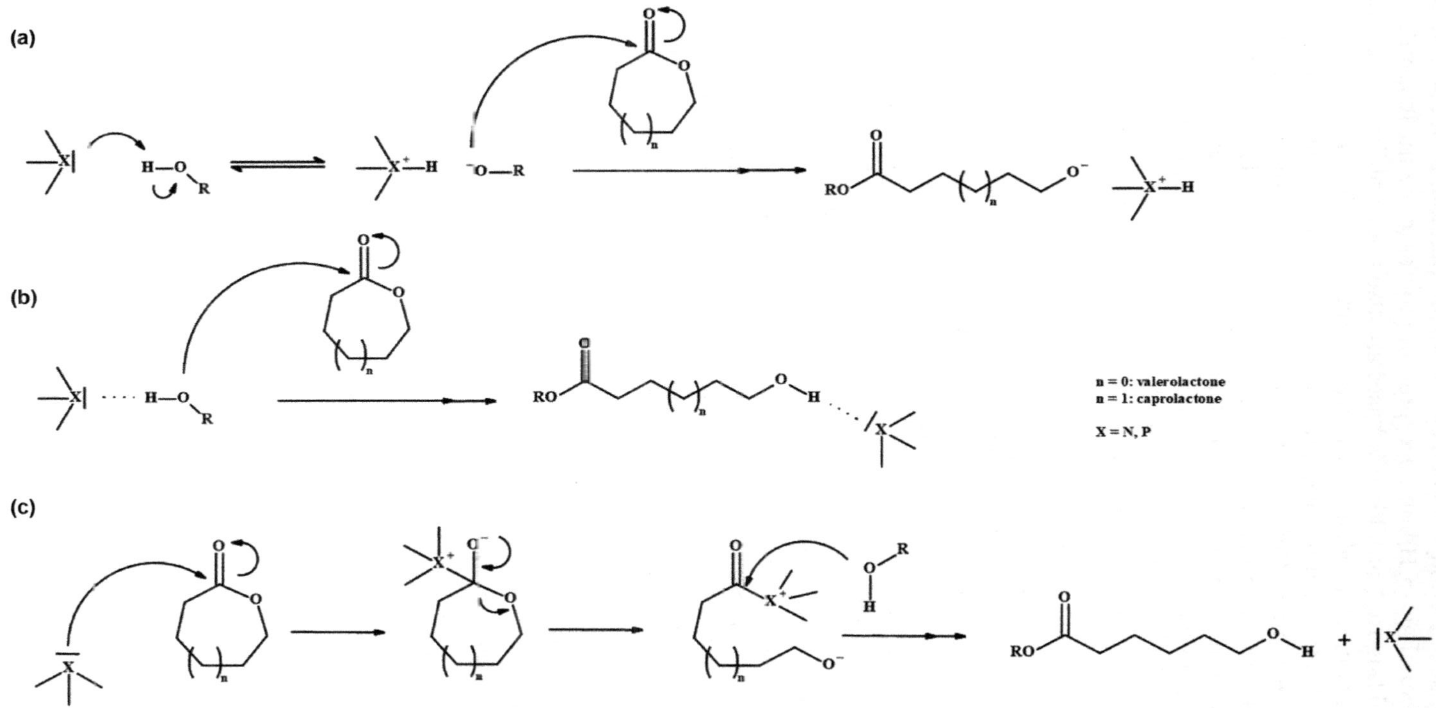

Figure 5.6 Activation of cyclic monoesters: (a) by bases, (b) by H-bond interactions, (c) nucleophilic catalysis (only the main steps are shown).

base is not strong enough to achieve the deprotonation, it can still interact with an alcohol as an H-bond acceptor and increase the nucleophilicity of the alcohol (Figure 5.6b). Finally, bases are also nucleophilic and can react with the monomer to yield a reactive intermediate able to react with the alcohol without any other catalysis (Figure 5.6c). This last mechanism is basically the same as the one involved in lipase-mediated polymerization, serine being the catalyst. The zwitterionic mechanism is observed as far as tertiary amines or other equivalent nucleophilic species are involved as organocatalysts. This mechanism will be discussed in Section 5.2.5. It is obvious that the reactivity of the catalyst and the polymerization mechanism depends on steric factors and on electronic factors, and thus also on pK_a.

The polymerization of ε-caprolactone or δ-valerolactone was initiated by alcohols in the presence of organophosphazenes (Figure 5.7), which are commonly reported as non-nucleophilic superbases because they are more basic than the hydroxide anion. A good control of the polymerization was obtained with the following organophosphazenes: BEMP (^{MeCN}p$K_a = 27.6$),[34] *t*Bu-P$_1$ (^{MeCN}p$K_a = 26.9$),[34] *t*Bu-P$_2$ (^{MeCN}p$K_a = 33.5$),[35] *t*Bu-P$_4$ (^{MeCN}p$K_a = 42.7$),[36] CTPB (^{MeCN}p$K_a = 33.3$).[37] The name "superbase" brings some confusion because the mechanism is not systematically the anionic one shown in Figure 5.6a and the non-ionic mechanism of Figure 5.6b remains possible. Indeed, when the pK_a of the catalyst is higher than the pK_a of the monomer, the alcohol is deprotonated into an alkoxide and the polymerization takes place according to the anionic mechanism (Figure 5.6a). Conversely, when the pK_a of the catalyst is lower than the pK_a of the monomer, the alcohol is no longer deprotonated and the polymerization takes place by the mechanism shown in Figure 5.6b. When the two pK_as are very close, a mixture of alcohol and alkoxides are present and a more complex mixture of mechanisms is the rule. This conclusion remains valid for other classes of catalysts such as amines and N-heterocyclic carbenes. When more nucleophilic catalysts are used, both mechanisms shown in Figure 5.6b and c become possible in the absence of deprotonation. It is worth noting that the use of too strong phosphazenes increases the risk of transesterification reactions. For instance, quite high dispersities are observed when *t*Bu-P$_4$ is used as a catalyst. Most authors have proposed a mechanism for the activation of the alcohol through H-bonds, provided that the base is not basic enough to deprotonate the alcohol.

5.2.5 Dual Catalysts

One main drawback of organophosphazenes is their toxicity, which is a severe limitation for many applications and particularly for biomedical applications. It is mandatory to investigate other families of Lewis bases and among them, amines are good candidates because of their availability. The ring-opening polymerization of ε-caprolactone and δ-valerolactone by alcohols was investigated in the presence of amines such as DBU and

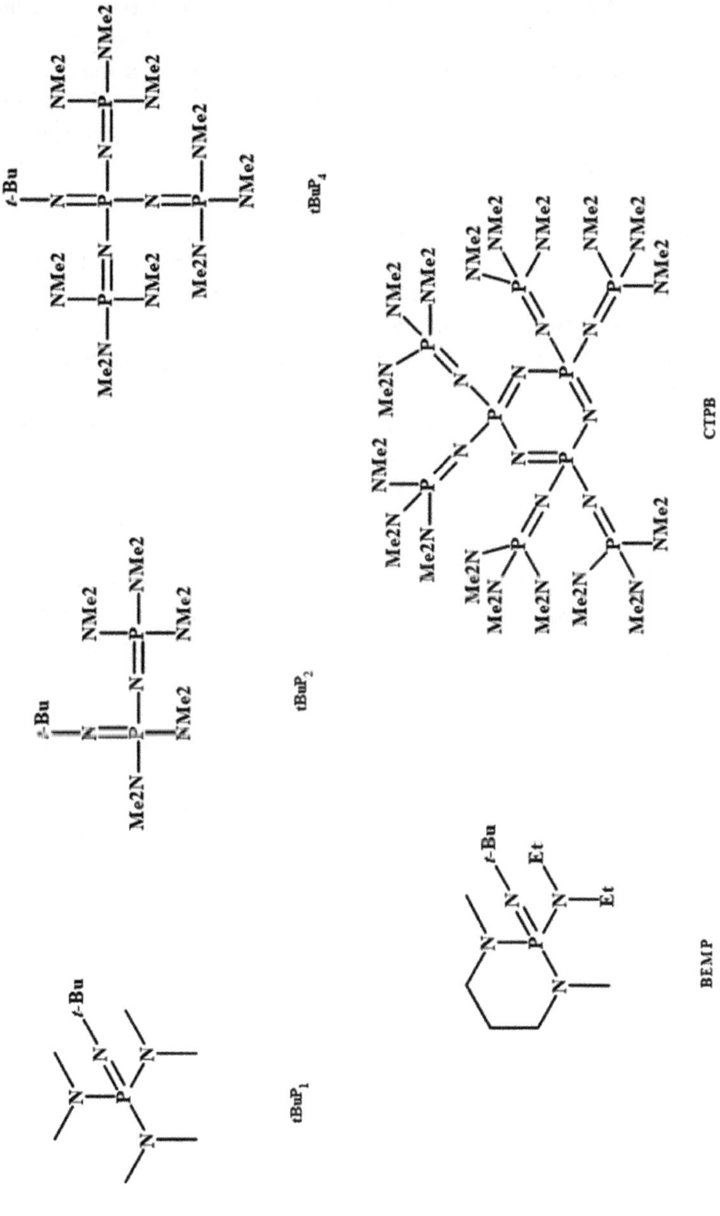

Figure 5.7 Typical organophosphazenes.

MTBD (Figure 5.7). Nevertheless, no polyester was formed.[32] The experiment was repeated in the presence of 1-[3,5-bis(trifluoromethyl)phenyl]-3-cyclohexyl-thiourea (TU), with the hope to activate the monomer. This time polyesters were obtained with a good control of the molar mass and a low dispersity. A mechanism based on the dual activation of the alcohols by the amine and on the activation of the cyclic ester by TU through H-bonds was proposed.[32] A second mechanism is based on the deprotonation of the TU by the amine into a thioimidate as a dual catalyst.[38,39] The influence of the solvent on the polymerization of δ-valerolactone was investigated and it is proposed that non-polar solvent favors the neutral mechanism while polar solvent favors the anionic thioimidate mechanism.[39] A last alternative mechanism relies on the nucleophilic addition of the amine onto the cyclic ester to an intermediate, reactive enough to react with alcohols. As a rule, the presence of the alcohol as an initiator favors the H-bond mechanism rather the nucleophilic mechanism. Since this discovery, chemists modified the nature of the amine and of the (thio)urea to optimize the delicate balance between fast polymerization and good control. As H-bond donors, ureas, bisureas, bisthioureas, trisureas, tristhioureas[40] and trichlocarban (TCC)[41] were investigated (Figure 5.9). As H-bond acceptor, Me_6TREN[41] is a good example (Figure 5.8). Each of them can be combined, which opens the way to a plethora of possible catalysts, which still needs to be explored.

Nowadays, the amine/TU is widely used for synthetic purposes. The concept has been extended to other families of H-bond donors and acceptors. As H-bond acceptor, BEMP[40] is a typical example. As H-bond donor, squaramides are reported (Figure 5.9).[42] The H-bond donor was replaced by a Lewis acid and, for instance, Dove achieved the polymerization of δ-valerolactone and ε-caprolactone by alcohols in the presence of DMAP and MgI_2 or YCl_3.[43]

Rather than using two co-catalysts, a simplification relies on the use of a bifunctional molecule made up of a H-bond donor activating the cyclic ester and a H-bond acceptor activating the alcohol. The typical example of this strategy is the TBD implemented by Hedrick to polymerize δ-valerolactone

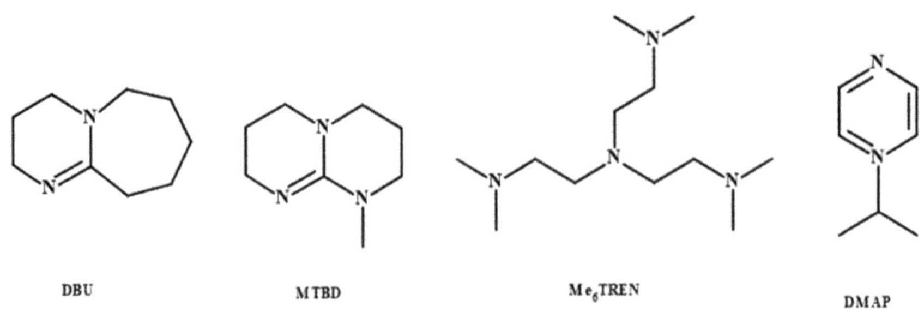

Figure 5.8 Amines used for the polymerization of δ-valerolactone and ε-caprolactone.

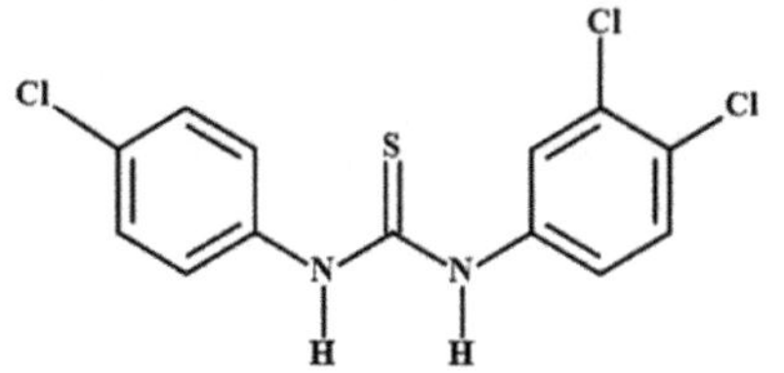

TCC

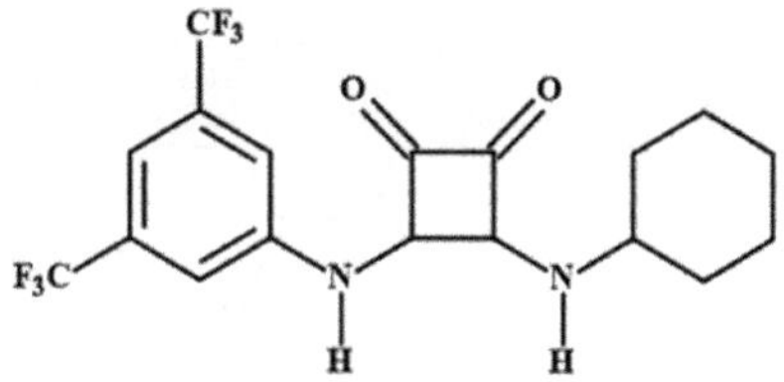

squaramide

Figure 5.9 H-bond donors used for the polymerization of δ-valerolactone and ε-caprolactone.

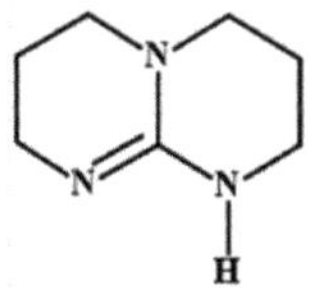

TBD

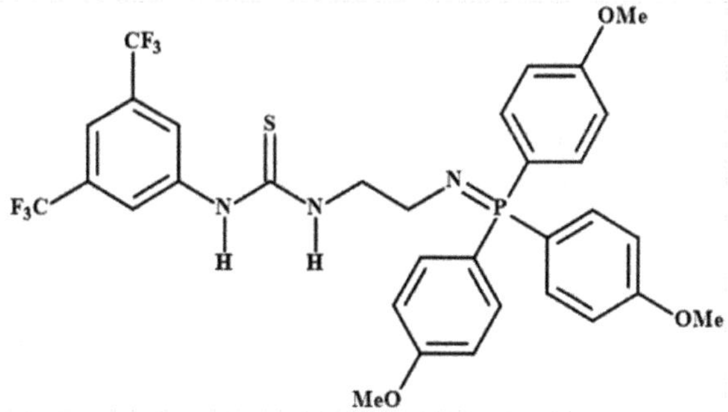

Figure 5.10 Dual organocatalysts.

and ε-caprolactone with no need of thiourea (Figure 5.10).[32,44] The polymerization is very fast but the polymerization is not 100% selective and the polymerization must be stopped before the onset of transesterification reactions. At the moment, TBD remains one of the most popular organocatalyst due to its efficiency and its availability. The "dual catalyst" concept was extended to other families of molecules such as a catalyst made up of a thiourea and a iminophosphorane (Figure 5.10).[45]

Recently, Waymouth achieved the polymerization of ε-caprolactone and δ-valerolactone in the presence of TU and potassium methoxide, rather than MTBD, through the anionic thioimidate route (Figure 5.11).[46,47] The system turned out to achieve very fast polymerization within seconds at room temperature while maintaining an excellent control. The selectivity is improved compared to the use of the TBD. Strictly speaking, this system is not metal-free. Nevertheless, the potassium cation is not a transition metal and is acceptable for many applications and even biomedical applications. The mechanism relies on the deprotonation of the TU by the methoxide to release the alcohol initiating the polymerization and the TU anion as a dual activator by H-bond interactions. It is worth noting that this mechanism must also be considered when organic bases are used instead of potassium methoxide provided they are strong enough to deprotonate the H-bond donor.

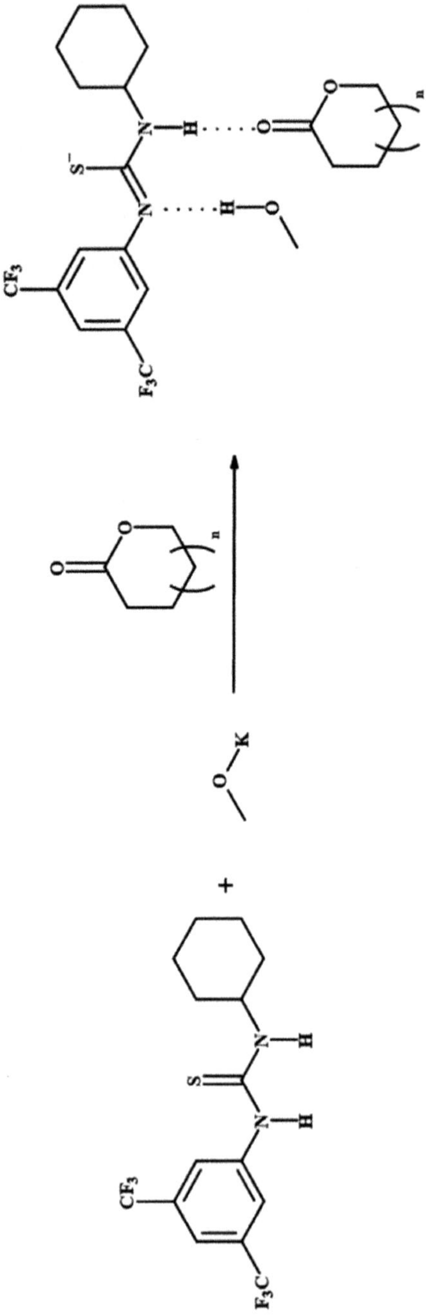

Figure 5.11 Polymerization in the presence of TU and potassium methoxide.

5.2.6 Zwitterionic Polymerization

In all the examples reported up to now, an alcohol was added to initiate the polymerization. What is happening in the absence of any alcohol or equivalent molecule? When tertiary amines are nucleophilic enough or when any other equivalent nucleophilic species is used, the nucleophilic species is capable of directly initiating the polymerization. The mechanism implies the presence of a cation at the α-chain end and of an anionic alkoxide at the ω-chain end. This mechanism is reported as the zwitterionic mechanism and has been known for a very long time.[4] Until recently, less attention was paid to these systems due to the lack of control of ionic processes. Quite recently, more attention was paid to zwitterionic polymerization because several examples showed that this process affords cyclic chains with no need of high dilution. For instance, Hedrick prepared cyclic PCL chains by the ring-opening polymerization initiated by N-heterocyclic carbenes, for instance 1,3,4,5-tetra-methylimidazol-2-ylidene, in the absence of any alcohol (Figure 5.12).[48] The choice of the carbene is important and more particularly, the substituents must be carefully chosen. Indeed, aryl-substituted carbenes are not active for the ROP of caprolactone whereas alkyl-substituted alkenes are well active. A kinetic study was undertaken to better understand the mechanism.[49] The formation of the zwitterion is slow and equilibrated and is followed by a fast propagation.[49,50] A series of transfer reactions takes place. At one side, the shuffling of chain lengths by intermolecular reactions accounts for the increase of the dispersity. At the other side, backbiting reactions provides cyclic chains. High molar mass PCL are obtained by the reaction of the zwitterion on cyclized chains.[49] The mechanism is thus quite complex and results on the combination of chain-growth and step-growth processes. As far as the polymerization of δ-valerolactone is initiated by 1,3-diisopropyl-4,5-dimethylimidazol-2-ylidene (Figure 5.12), the role of the electrostatic interaction taking place during polymerization were indirectly investigated by the addition of LiCl in the polymerization medium.[51] LiCl improves the control of the polymerization as witnessed by high initiation efficiency, good correlation between the theoretical and the experimental molar mass as well as decreased dispersity. Unfortunately, when the amount of LiCl increases, the

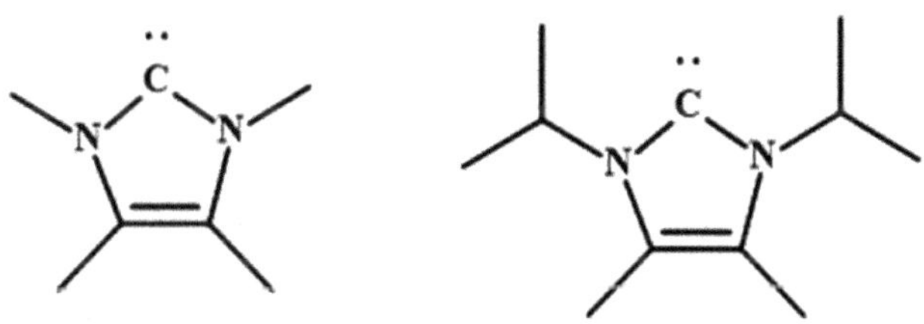

Figure 5.12 N-heterocyclic carbenes used for the polymerization of δ-valerolactone and ε-caprolactone.

cyclization efficiency decreases, and a mixture of cyclic and linear chains are formed. These experiences shed light on the importance of electrostatic interactions in these processes.

5.3 Polymerization of Five-membered Lactones

The ring-opening homopolymerization of five-membered cyclic monoesters, typically tetrahydrofuran-2-one, most commonly reported under the name γ-butyrolactone (γBL) is very difficult to carry out and affords only oligomers. The poor results were accounted for by non-favorable thermodynamics.[52] Indeed, although the direct ring-opening reaction takes place, the backwards ring-closing reaction is faster, resulting in a low ceiling temperature. Several strategies could be considered to tackle thermodynamics. A first strategy is based on copolymerization with other polymerizable cyclic esters such as δ-valerolactone ε-caprolactone.[53,54] This approach is not typical of organocatalysis and was previously implemented by classical coordination polymerization.[55] Another route relies on the achievement of ROP at a temperature low enough, ideally lower than the ceiling temperature. The challenge is to find a catalyst efficient at such a low temperature. The polymerization of γBL initiated by an alcohol is reported at −40 °C in the presence of TBD,[56] tBuP$_4$[57] and CTPB[58] as organocatalysts. Remarkably, yields close to 100% were observed for the best catalysts. DBU was also tested but then no polymer was obtained at −40 °C.[56] When the polymerization is carried out in the absence of alcohol, the deprotonation of γBL by tBuP$_4$ in its corresponding enolate was observed.[57] The enolate is then the initiator of the polymerization.

5.4 Polymerization of Four-membered Small-size Cyclic Monoesters

The polymerization of four-membered cyclic monoesters has been known for a long time, employing ionic processes. Let us mention the zwitterionic polymerization by pyridine[59] and the cationic polymerization by alkylation[60,61] and acylation agents.[61] Recently, the polymerization of 4-methyloxetan-2-one, commonly reported under the name β-butyrolactone (βBL) turned out to be controlled when using the organocatalysts previously implemented for the ROP of ε-caprolactone and δ-valerolactone, and among them, let us cite sulfonic acids,[62] phosphonic acids,[63] BEMP,[64] TBD,[64,65] DBU[64] and NHC.[66–68]

The ring-opening polymerization of four-membered ring cyclic monoesters, typically β-butyrolactone, initiated by alcohols was successful when catalyzed by N-heterocyclic carbenes[68–70] but were not capable to afford the polyester in a controlled fashion when catalyzed by TBD[32,44] even though these catalysts are very efficient for the controlled ring-opening polymerization of cyclic monoesters of larger size such as δ-valerolactone and ε-caprolactone. The formation of an eight-membered ring (Figure 5.13) was

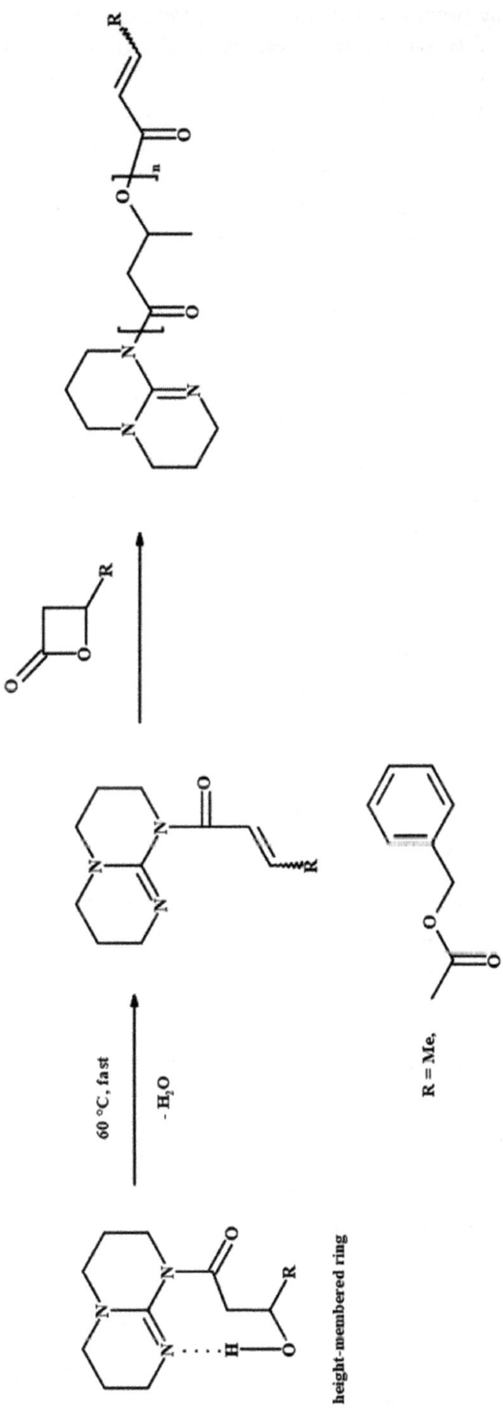

Figure 5.13 Ring opening polymerization of four-membered cyclic esters in the absence of alcohols.

supported by DFT methods[71] and competes for hydrogen-bond formation, which prevents polymerization from taking place efficiently.[32] Later on, the polymerization of β-butyrolactone was carried out efficiently and under control in the absence of any alcohol and in bulk.[64] The polymerization was extended to the polymerization of benzyl 4-oxooxetane-2-carboxylate and to other organocatalysts such as DBU and BEMP.[72] The characterization of the chemical structure of the chain ends enabled detection of the presence of unsaturations at the α chain end. A possible mechanism relies on an elimination of the eight-membered ring into an unsaturated adduct, which then acts as an initiator for ring-opening polymerization (Figure 5.13). Such an elimination reaction is a first specific behavior typical of four-membered cyclic monoesters. This reaction is also reported as a crotonization reaction in the particular case of β-butyrolactone. The elimination reaction is observed due to the high pKa of enolisable protons (pK_a of β-butyrolactone = 4.5 in THF) compared to the pK_a value of TBD (= 26 in THF).

An alternative mechanism was recently proposed and relies on the crotonization reaction taking place directly onto the monomer to afford a crotonate, which then initiates ROP (Figure 5.14).[73] This mechanism relies on a second specific behavior of four-membered cyclic esters. Carboxylate are capable of initiating and propagating the ring-opening polymerization of four-membered cyclic monoesters by the cleavage of the C_{alkyl}–O bond rather than by the C_{acyl}–O bond (Figure 5.14). This reaction is reported with other organocatalysts than TBD such as N-heterocyclic carbenes[68] and phosphazenes.[74]

5.5 Polymerization of Large-size Macrocyclic Monoesters

The polymerization of macrocyclic monoesters remains challenging at the time being due to the low ring strain. Ring-opening polymerization does not benefit anymore from the release of the ring strain as a driving force. Accordingly, the ring-opening polymerization of macrocyclic monoesters is less common than the one of medium and small size cyclic monoesters. A classical approach for their polymerization relies on the resort to enzymatic polymerization. Macrocyclic monoesters being more hydrophobic than medium and low size cyclic monoesters, they are better complexed with enzymes. Although enzymes have some advantages in terms of toxicity and stereoselectivity, the control of the polymerization remains limited. Chemists are thus searching for more efficient catalysts. A first challenge is to increase the activity of the catalyst to reach full conversion. A second challenge is to find a selective catalyst, which is more difficult due to the lower difference of reactivity between cyclic and non-cyclic esters.

Pentadecalactone (PDL) is a typical example of macrocyclic monoester and has been more and more studied in the last few years. The bulk ring-opening polymerization of PDL is initiated by alcohols but requires high temperature

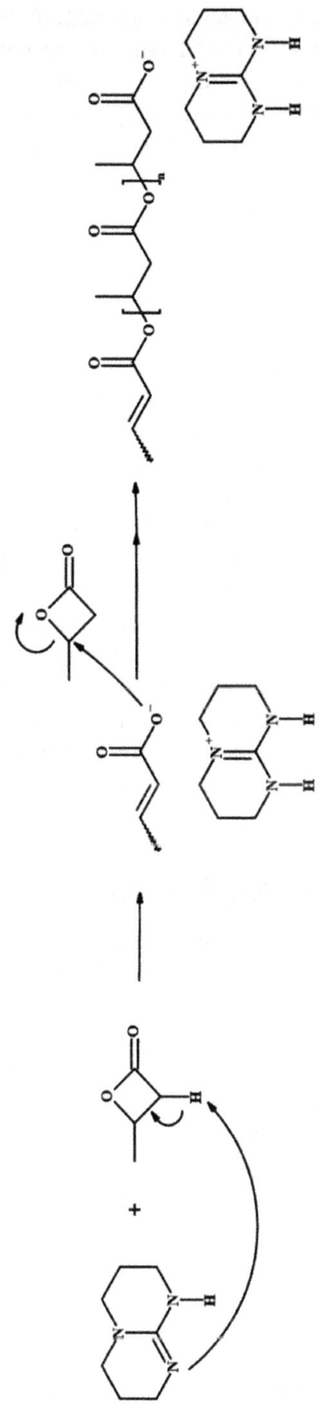

Figure 5.14 Alternative mechanism for the ring-opening polymerization of four-membered cyclic esters in the absence of alcohols.

(80 °C) and long times (24 h) in the presence of Brønsted organocatalysts to reach full conversion.[75] Many classical organocatalysts active for the polymerization of ε-caprolactone and δ-valerolactone turned out to be completely inactive for the polymerization of PDL.[76] Nevertheless, the polymerization of PDL in toluene (0.4 M < [PDL] < 1.5 M) catalyzed by *t*Bu-P_4 ($^{MeCN}pK_a = 42.7$) provides the polyester with high conversion at 80 °C.[77] The dispersity ($1.4 < Đ < 1.7$) is higher compared to the polymerization of medium size cyclic monoesters due to a less selective process, which is expected due to the lower difference of reactivity between cyclic esters and other esters present all along the chains. *t*Bu-P_2 ($^{MeCN}pK_a = 33.5$) was also investigated and turned out to be less active but high conversions are obtained provided that the reaction is performed in bulk at 80 °C. Dove and Naumann achieved the polymerization of PDL in the presence of N-heterocyclic olefins, typically 2-isopropylidene-1,3,4,5-tetramethyl-imidazole, in the absence[78] or in the presence[54] of a Lewis acid as a co-catalyst.

Dual catalysts are especially good candidates as organocatalysts for the polymerization of cyclic esters of low reactivity such as PDL. Among dual catalysts, TBD was found to catalyze the ROP of PDL initiated by alcohols in bulk or in solution at high temperature (100 °C).[76] Again, the polymerization has a moderate selectivity ($1.4 < Đ < 1.9$). Dual activation is important because MTBD and DBU, which activate the alcohol but not the cyclic monoester, are not efficient organocatalysts. Alternatively, the polymerization of PDL initiated by an alcohol was catalyzed by a dual organocatalyst made up of a Lewis acid and nucleophilic compound such as DMAP, DBU or an N-heterocyclic carbene.[43] For instance, the polymerization can be carried out with very simple, cheap and benign catalysts such as DMAP and a magnesium halide.

5.6 Macromolecular Engineering

A steadily increasing number of catalysts are available for the polymerization of cyclic monoesters. Nowadays, all these catalysts are more and more used for the polymerization of a wider range of cyclic monoesters, substituted and functionalized, for the copolymerization with other cyclic esters or other families of monomers. All these processes can be applied not only for the synthesis of linear polymers but also for the synthesis of star-shaped, grafted, hyperbranched and cross-linked polymers. All these techniques are well known whatever the catalyst used and are widely reviewed elsewhere.[8] A full description of this macromolecular engineering is beyond the scope of this review.

Nevertheless, the availability of a wide range of organocatalysts is very useful. Indeed, the polymerization of any new monomer could require an optimization of the structure of the catalyst to reach fast kinetics and maintain a good selectivity. As far as functionalized cyclic esters are polymerized, it is mandatory that the catalyst tolerates the presence of the functional group. For instance, α-halogeno-cyclic esters cannot be polymerized by nucleophilic catalysts and bases prone to trigger either detrimental quaternization or elimination reactions. Conversely, acids are suitable catalysts as shown by the ring-opening polymerization of 3-bromooxepan-2-one by diphenyl phosphate.[79]

Regarding the copolymerization approach, it is well-known that the copolymerization of a mixture of at least two comonomers affords copolymers with a distribution of the repetitive units depending on the reactivity ratio of each comonomers and the presence of transfer reactions such as transesterification reactions in the specific field of polyesters.

As far as a mixture of cyclic esters is copolymerized, organocatalysts are very useful to synthesize copolyesters with a distribution of the repeating units difficult to obtain by other conventional catalysts like organometallic complexes. The copolymerization of γ-butyrolactone with δ-valerolactone or ε-caprolactone is a typical example. Chen initiated the polymerization by an alcohol in the presence of $La[N(SiMe_3)_2]_3$ at 25 °C.[53] It turned out to be very difficult to incorporate more than 35 mol% of γ-butyrolactone in the copolyester. The entropic penalty was overcome by achieving the polymerization at −40 °C, which was possible by using the much more reactive tBuP$_4$ instead of $La[N(SiMe_3)_2]_3$. Higher incorporation of γ-butyrolactone, up to 80 mol%, was observed.

Copolymerization chemistry is very rich because cyclic monoesters can be copolymerized with other families of heterocyclic monomers, typically, cyclic diesters such as lactide and glycolide, cyclic carbonates, cyclic phosphoesters, cyclic anhydrides and cyclic ethers to synthesize a very wide range of copolymers. An example is provided by the alternate copolymerization of a mixture of epoxides and cyclic monoester to synthesize poly(ether-*co*-ester)s. Indeed, the copolymerization of 3,4-dihydrocoumarin, which is a non-homopolymerizable cyclic monoester, with various common epoxides was initiated by benzyl alcohol or by 1,1,1-trihydroxymethylpropane in the presence of tBuP$_4$ to obtain the corresponding linear or star-shaped copolymer.[80]

Regarding the synthesis of block copolyesters by successive polymerization of two different monomers, two different catalysts can be required for the synthesis of each block. An elegant approach is based on a catalyst switch after the synthesis of the first block and before the synthesis of the second block.[81]

In the future, more and more studies will use organocatalysts rather than metallic complexes. The influence of the chemical structure of the organocatalyst on the distribution of the repetitive units in the chains will have to be more systematically investigated with the aim to prepare copolymers difficult to prepare with other catalysts. Monomers are chiral when R or S atoms are present. For instance, cyclic monoesters bearing a substituent on the ring are chiral. In nature, enzymes are also chiral and can achieve stereoselective polymerization. It is worth recalling that the mechanisms involved in enzymatic polymerization are similar to the ones used in organocatalysis, especially when H-bond interactions are concerned. These considerations should prompt chemists to synthesize new organocatalysts to undertake stereoselective ring-opening (co)polymerization and to prepare copolymers with tacticities difficult to reach by coordination polymerization.

5.7 Conclusions

Organocatalysts occupy a key position to catalyze the polymerization of cyclic monoesters. The strong basic or acidic catalysts promote ionic polymerizations. Although these catalysts are very active, they are not very selective and many of them are toxic. Less acidic and basic H-bond donors and acceptors are an interesting alternative. Some of them turned out to be quite selective. In addition, catalytic activity can be increased by using dual catalysts made up of both initiators and monomer activators. This chemistry is still in its infancy because a plethora of combinations between various initiator and monomer activators can be imagined by the chemist and have to be studied. Many studies need to be achieved to better understand how the change of the chemical structure of organocatalysts affects the mechanism of the polymerization. If enzymes are chiral and can bring about stereoselective polymerizations, practically no examples are reported for the stereoselective ring-opening polymerization of cyclic monoesters by organocatalysts. Obviously, this field has a huge potential for new developments in the future.

Acknowledgements

Philippe Lecomte is research associate by the FRS-FNRS. The authors thank the "Belgian Science Policy" in the frame of the "Interuniversity Attraction Poles Program (IAP VII/5) – Functional Supramolecular Systems".

References

1. W. H. Carothers, G. L. Dorough and F. J. Van Natta, *J. Am. Chem. Soc.*, 1932, **54**, 761–772.
2. F. Natta, J. Hill and W. H. Carothers, *J. Am. Chem. Soc.*, 1934, **56**, 455–457.
3. M. Szwarc, *Nature*, 1956, **178**, 1168–1169.
4. A. Löfgren, A.-C. Albertsson, P. Dubois and R. Jérôme, *J. Macromol. Sci., Polym. Rev.*, 1995, **35**, 379–418.
5. K. M. Stridsberg, M. Ryner and A.-C. Albertsson, *Adv. Polym. Sci.*, 2002, 41–64.
6. A.-C. Albertsson and I. K. Varma, *Biomacromolecules*, 2003, **4**, 1466–1486.
7. N. E. Kamber, W. Jeong, R. M. Waymouth, R. C. Pratt, B. G. G. Lohmeijer and J. L. Hedrick, *Chem. Rev.*, 2007, **107**, 5813–5840.
8. C. Jérôme and P. Lecomte, *Adv. Drug Delivery Rev.*, 2008, **60**, 1056–1076.
9. A. Kowalski, A. Duda and S. Penczek, *Macromol. Rapid Commun.*, 1998, **19**, 567–572.
10. H. Uyama and S. Kobayashi, *Chem. Lett.*, 1993, **22**, 1149–1150.

11. D. Knani, A. L. Gutman and D. H. Kohn, *J. Polym. Sci., Part A: Polym. Chem.*, 1993, **31**, 1221–1232.
12. A.-C. Albertsson and R. K. Srivastava, *Adv. Drug Delivery Rev.*, 2008, **60**, 1077–1093.
13. R. A. Gross, A. Kumar and B. Kalra, *Chem. Rev.*, 2001, **101**, 2097–2124.
14. S. Kobayashi and A. Makino, *Chem. Rev.*, 2009, **109**, 5288–5353.
15. B. A. Rozenberg, *Makromol. Chem., Macromol. Symp.*, 1992, **60**, 177–186.
16. P. Valdemar Persson, J. Scroder, K. Whckholm, E. Hedenstrom and T. Iversen, *Macromolecules*, 2004, **37**, 5889–5893.
17. J. Casas, P. V. Persson, T. Iversen and A. Córdova, *Adv. Synth. Catal.*, 2004, **346**, 1087–1089.
18. F. Li, B. Chen, Z. Ping, X. Jia, X. Guo, L. Xiao and L.-J. Liu, *Macromol. Biosci.*, 2014, **14**, 280–288.
19. M. Sobczak, E. Oledzka and W. L. Kolodziejski, *J. Macromol. Sci., Part A: Pure Appl. Chem.*, 2008, **45**, 872–877.
20. J. Liu and L.-J. Liu, *Macromolecules*, 2004, **37**, 2674–2676.
21. P. V. Persson, J. Casas, T. Iversen and A. Córdova, *Macromolecules*, 2006, **39**, 2819–2822.
22. J. Xu, J. Song, S. Pispas and G. Zhang, *J. Polym. Sci., Part A: Polym. Chem.*, 2014, **52**, 1185–1192.
23. S. C. Hammer, A. Marjanovic, J. M. Dominicus, B. M. Nestl and B. Hauer, *Nat. Chem. Biol.*, 2015, **11**, 121–126.
24. H. Sun, S. Xu, Z. Li, J. Xu, J. Liu, X. Wang, H. Wang, H. Dong, Y. Liu and K. Guo, *Polym. Chem.*, 2017, **8**, 5570–5579.
25. B. C. Wilson and C. W. Jones, *Macromolecules*, 2004, **37**, 9709–9714.
26. M. Baśko and P. Kubisa, *J. Polym. Sci., Part A: Polym. Chem.*, 2006, **44**, 7071–7081.
27. S. Gazeau-Bureau, D. Delcroix, B. Martín-Vaca, D. Bourissou, C. Navarro and S. Magnet, *Macromolecules*, 2008, **41**, 3782–3784.
28. N. Stanley, G. Bucataru, Y. Miao, A. Favrelle, M. Bria, F. Stoffelbach, P. Woisel and P. Zinck, *J. Polym. Sci., Part A: Polym. Chem.*, 2014, **52**, 2139–2145.
29. K. Makiguchi, T. Satoh and T. Kakuchi, *Macromolecules*, 2011, **44**, 1999–2005.
30. T. Saito, Y. Aizawa, K. Tajima, T. Isono and T. Satoh, *Polym. Chem.*, 2015, **6**, 4374–4384.
31. X. Li, Q. Zhang, Z. Li, S. Xu, C. Zhao, C. Chen, X. Zhi, H. Wang, N. Zhu and K. Guo, *Polym. Chem.*, 2016, **7**, 1368–1374.
32. B. G. G. Lohmeijer, R. C. Pratt, F. Leibfarth, J. W. Logan, D. A. Long, A. P. Dove, F. Nederberg, J. Choi, C. G. Wade, R. M. Waymouth and J. L. Hedrick, *Macromolecules*, 2006, **39**, 8574–8583.
33. G. Gontard, A. Amgoune and D. Bourissou, *J. Polym. Sci., Part A: Polym. Chem.*, 2016, **54**, 3253–3256.
34. L. Zhang, F. Nederberg, R. C. Pratt, R. M. Waymouth, J. L. Hedrick and C. G. Wade, *Macromolecules*, 2007, **40**, 4154–4158.

35. H. Alamri, J. Zhao, D. Pahovnik and N. Hadjichristidis, *Polym. Chem.*, 2014, **5**, 5471–5478.
36. H. Yang, J. Xu, S. Pispas and G. Zhang, *Macromolecules*, 2012, **45**, 3312–3317.
37. H. Li, N. Zhao, C. Ren, S. Liu and Z. Li, *Polym. Chem.*, 2017, **8**, 7369–7374.
38. J. U. Pothupitiya, N. U. Dharmaratne, T. M. M. Jouaneh, K. V. Fastnacht, D. N. Coderre and M. K. Kiesewetter, *Macromolecules*, 2017, **50**, 8948–8954.
39. J. U. Pothupitiya, R. S. Hewawasam and M. K. Kiesewetter, *Macromolecules*, 2018, **51**, 3203–3211.
40. K. V. Fastnacht, S. S. Spink, N. U. Dharmaratne, J. U. Pothupitiya, P. P. Datta, E. T. Kiesewetter and M. K. Kiesewetter, *ACS Macro Lett.*, 2016, **5**, 982–986.
41. N. U. Dharmaratne, J. U. Pothupitiya, T. J. Bannin, O. I. Kazakov and M. K. Kiesewetter, *ACS Macro Lett.*, 2017, **6**, 421–425.
42. J. Liu, J. Xu, Z. Li, S. Xu, X. Wang, H. Wang, T. Guo, Y. Gao, L. Zhang and K. Guo, *Polym. Chem.*, 2017, 7054–7068.
43. S. Naumann, P. B. V. Scholten, J. A. Wilson and A. P. Dove, *J. Am. Chem. Soc.*, 2015, **137**, 14439–14445.
44. R. C. Pratt, B. G. G. Lohmeijer, D. A. Long, R. M. Waymouth and J. L. Hedrick, *J. Am. Chem. Soc.*, 2006, **128**, 4556–4557.
45. A. M. Goldys and D. J. Dixon, *Macromolecules*, 2014, **47**, 1277–1284.
46. X. Zhang, G. O. Jones, J. L. Hedrick and R. M. Waymouth, *Nat. Chem.*, 2016, **8**, 1047–1053.
47. B. Lin and R. M. Waymouth, *J. Am. Chem. Soc.*, 2017, **139**, 1645–1652.
48. E. J. Shin, W. Jeong, H. A. Brown, B. J. Koo, J. L. Hedrick and R. M. Waymouth, *Macromolecules*, 2009, **44**, 2773–2779.
49. H. A. Brown, S. Xiong, G. Medvedev, Y. A. Chang, M. M. Abu-Omar, J. M. Caruthers and R. M. Waymouth, *Macromolecules*, 2014, **47**, 2955–2963.
50. L. Falivene and L. Cavallo, *Macromolecules*, 2017, **50**, 1394–1401.
51. Y. A. Chang and R. M. Waymouth, *Polym. Chem.*, 2015, **6**, 5212–5218.
52. P. Olsén, K. Odelius and A.-C. Albertsson, *Biomacromolecules*, 2016, **17**, 699–709.
53. M. Hong, X. Tang, B. S. Newell and E. Y.-X. Chen, *Macromolecules*, 2017, **50**, 8469–8479.
54. P. Walther and S. Naumann, *Macromolecules*, 2017, **50**, 8406–8416.
55. A. Duda, S. Penczek, P. Dubois, D. Mecerreyes and R. Jérôme, *Macromol. Chem. Phys.*, 1996, **197**, 1273–1283.
56. M. Hong and E. Y.-X. Chen, *Nat. Chem.*, 2016, **8**, 42–49.
57. M. Hong and E. Y.-X. Chen, *Angew. Chem., Int. Ed.*, 2016, **55**, 4188–4193.
58. N. Zhao, C. Ren, H. Li, Y. Li, S. Liu and Z. Li, *Angew. Chem., Int. Ed.*, 2017, **56**, 12987–12990.
59. H. R. Kricheldorf, M. Garaleh and G. Schwarz, *J. Macromol. Sci., Part A: Pure Appl. Chem.*, 2005, **42**, 139–148.
60. H. R. Kricheldorf, J. M. Jonté and R. Dunsing, *Die Makromol. Chem.*, 1986, **187**, 771–785.

61. A. Hofman, R. Szymański, S. Słomkowski and S. Penczek, *Die Makromol. Chem.*, 1984, **185**, 655–667.
62. F. A. Jaipuri, B. D. Bower and N. L. Pohl, *Tetrahedron: Asymmetry*, 2003, **14**, 3249–3252.
63. K. Makiguchi, T. Saito, T. Satoh and T. Kakuchi, *J. Polym. Sci., Part A: Polym. Chem.*, 2014, **52**, 2032–2039.
64. C. G. Jaffredo, J.-F. Carpentier and S. M. Guillaume, *Macromol. Rapid Commun.*, 2012, **33**, 1938–1944.
65. G. Barouti, K. Jarnouen, S. Cammas-Marion, P. Loyer and S. M. Guillaume, *Polym. Chem.*, 2015, **6**, 5414–5429.
66. E. F. Connor, G. W. Nyce, M. Myers, A. Möck and J. L. Hedrick, *J. Am. Chem. Soc.*, 2002, **124**, 914–915.
67. G. Nyce, T. Glauser, E. F. Connor, A. Möck, R. M. Waymouth and J. L. Hedrick, *J. Am. Chem. Soc.*, 2003, **125**, 3046–3056.
68. O. Coulembier, X. Delva, J. L. Hedrick, R. M. Waymouth and P. Dubois, *Macromolecules*, 2007, **40**, 8560–8567.
69. O. Coulembier, B. G. G. Lohmeijer, A. P. Dove, R. C. Pratt, L. Mespouille, D. A. Culkin, S. J. Benight, P. Dubois, R. M. Waymouth and J. L. Hedrick, *Macromolecules*, 2006, **39**, 5617–5628.
70. O. Coulembier, L. Mespouille, J. L. Hedrick, R. M. Waymouth and P. Dubois, *Macromolecules*, 2006, **39**, 4001–4008.
71. L. Simón and J. M. Goodman, *J. Org. Chem.*, 2017, **72**, 9656–9662.
72. C. G. Jaffredo, J.-F. Carpentier and S. M. Guillaume, *Macromolecules*, 2013, **46**, 6765–6776.
73. S. Moins, C. Henoumont, J. De Winter, A. Khalil, S. Laurent, S. Cammas-Marion and O. Coulembier, *Polym. Chem.*, 2018, **9**, 1840–1847.
74. J. De Winter, O. Coulembier, P. Gerbaux and P. Dubois, *Macromolecules*, 2010, **43**, 10291–10296.
75. A. Pascual, J. R. Leiza and D. Mecerreyes, *Eur. Polym. J.*, 2013, **49**, 1601–1609.
76. M. Bouyahyi, M. P. F. Pepels, A. Heise and R. Duchateau, *Macromolecules*, 2012, **45**, 3356–3366.
77. V. Ladelta, P. Bilalis, Y. Gnanou and N. Hadjichristidis, *Polym. Chem.*, 2017, **8**, 511–515.
78. S. Naumann, A. W. Thomas and A. P. Dove, *ACS Macro Lett.*, 2016, **5**, 134–138.
79. P. Bexis, A. W. Thomas, C. A. Bell and A. P. Dove, *Polym. Chem.*, 2016, **7**, 7126–7134.
80. S. Hu, G. Dai, J. Zhao and G. Zhang, *Macromolecules*, 2016, **49**, 4462–4472.
81. X. Wang, J. Liu, S. Xu, J. Xu, X. Pan, J. Liu, S. Cui, Z. Li and K. Guo, *Polym. Chem.*, 2016, **7**, 6297–6308.

CHAPTER 6

Organic Catalysis for the Polymerization of Lactide and Related Cyclic Diesters

SOPHIE M. GUILLAUME

Univ Rennes, CNRS, Institut des Sciences Chimiques de Rennes – UMR6226, F-35042 Rennes, France
Email: sophie.guillaume@univ-rennes1.fr

6.1 Introduction

Poly(lactic acid)/poly(lactide) (PLA) is an aliphatic polyester that is renewable (derived from biomass: *e.g.*, corn, sugar cane, sugar beet, *etc.*), (bio)degradable, bioassimilable (the polyester backbone is sensitive to hydrolysis releasing lactic acid which is eliminated *via* the Krebs cycle), and biocompatible. Over the past few decades, PLA has received a significant amount of interest both in academic and industrial research. Due to its physical properties (high tensile strength and stiffness, versatile processing through extrusion, thermoforming, injection molding or fiber spinning), and to its industrial implementation at low cost, PLA is nowadays emerging as the most popular and promising bio-sourced, sustainable, biodegradable and ecological/environmentally friendly "plastic" candidate, which can potentially and increasingly substitute traditional petrochemically derived polymeric materials as a viable alternative in many short-time commodity and engineering applications.[1–15]

PLA can be synthesized either by polycondensation of lactic acid (step-growth polymerization affording poly(lactic acid)) or by ring-opening

Polymer Chemistry Series No. 31
Organic Catalysis for Polymerisation
Edited by Andrew Dove, Haritz Sardon and Stefan Naumann

Published by the Royal Society of Chemistry, www.rsc.org

polymerization (ROP; chain-growth polymerization giving poly(lactide)) of lactide (LA), the cyclic dimer of lactic acid (Scheme 6.1). However, the latter ROP approach is more appropriate to prepare well-defined PLA with a greater degree of control over the macromolecular parameters of the resultant polymer (controlled and tunable molar mass values, low dispersities, end-group fidelity, stereocontrolled and tunable microstructure (tacticity)).[1,2,5,8–10,16–18]

Organocatalysis of LA ROP, in which small organic molecules promote the polymerization reaction acting either as true catalysts or as direct initiators, emerged in the early 2000s with the extension of organic catalysis to controlled polymerizations.[19] Following the traditional metal-catalyzed ROP, the

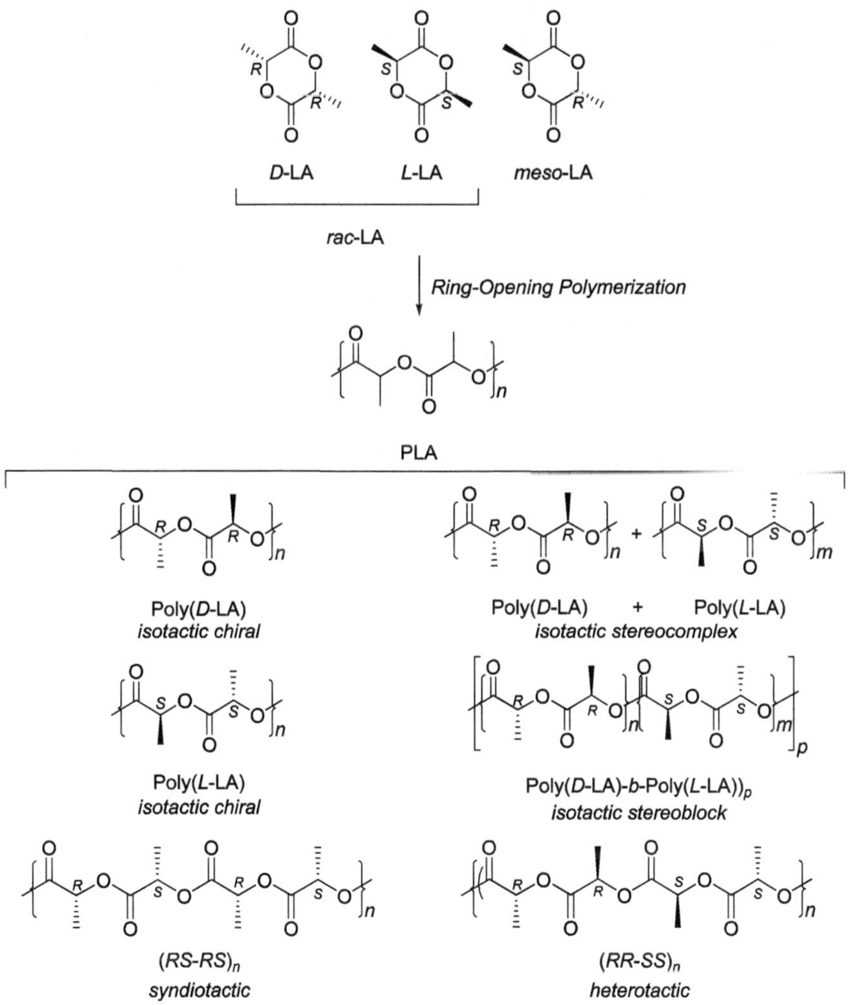

Scheme 6.1 Lactide isomers and their subsequent ROP into PLA with different microstructures.

organocatalyzed ROP (OROP) of LA (similarly to cyclic esters), thus afforded – under mild and highly selective polymerization conditions – PLA with predictable/tunable molar mass values and extremely narrow dispersities (*ca.* 1.1), characteristics that are typical of a living polymerization.[19] Such organocatalysis valuably provides less toxic catalytic residues within the resulting PLA, as compared to the use of the ubiquitous tin(II) 2-ethylhexanoate ($Sn(Oct)_2$), which commonly gives PLA with residual metal traces. This advantage, combined with the cost, stability and availability of organocatalysts, then opened up new applications of such high performance PLA in areas such as microelectronics, biomedicine and industrial food packaging.

Organic activators (*i.e.* catalysts or initiators) used in the ROP of LA (similarly used in the ROP of cyclic esters) basically include Brønsted/Lewis acids or bases, and mono- or bi-component bifunctional catalytic systems most often based on commercially available (natural or not) molecules. These can be classified according to the nature of their functional group (*i.e.* acid or basic compounds, or activators featuring hydrogen-bonds with a donating or an accepting ability). The families of (i) single organic initiators, (ii) Brønsted/Lewis acids, Brønsted/Lewis bases including (iii) alkyl (aryl) amines and pyridine derivatives, (iv) amidines and guanidines, and (v) N-heterocyclic carbenes (NHCs), (vi) phosphines and phosphazenes, and (vii) mono- or multicomponent dual catalytic systems involving hydrogen-bond interactions such as thiourea-amino and phenolic derivatives, are most typically encountered in the OROP of LA.[10,20–35]

Since the initial report in 2001 by Hedrick *et al.* of the first organocatalytic approach to the living ROP of LA using 4-(dimethylamino)pyridine (DMAP) and the related 4-pyrrolidinopyridine (PPY) strongly basic amines as transesterification catalysts,[19] over 270 publications have been published (Figure 6.1). This chapter will focus on recent advances in LA OROP

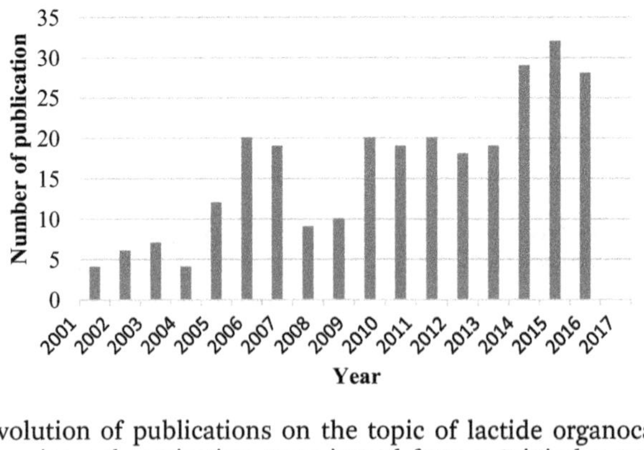

Figure 6.1 Evolution of publications on the topic of lactide organocatalyzed ring-opening polymerization. Data issued from a SciFinder search (2017-12-07) using English documents (including journal articles and reviews, patents and conference abstracts) with the keywords "lactide and ring opening polymerization" refined with "organo* catal* or organic catal*".

developed over the past decade, with a special highlight on polymerization mechanism aspects and monomer activation modes, and also on catalyst development, methodologies, stereocontrol, and functional group incorporation. The following sections address, according to the nature of the activator, the key mechanisms encountered in the OROP of LA, and the most significant advances that have appeared in the literature, although appropriate earlier results are also mentioned so as to provide a relevant overview. This chapter thus focuses on selected studies highlighted with illustrating examples and important milestones. For potential applications of the polymers cited, the reader should refer to the original publication as the emphasis is herein placed on the chemistry.

Whereas this Chapter 6 essentially focuses on PLA, other cyclic diesters such as *O*-carboxy anhydrides (OCAs)), which similar to LA produce poly(α-hydroxy acid)s, have also been investigated in OROP enabling, under mild conditions, to introduce functional groups along the polyester backbone and thereby to modify and finely tune their physicochemical properties.[36] Such relevant work is also mentioned herein.

6.2 Polymerization Mechanisms in the Organocatalyzed ROP of LA

The mechanism involved in the OROP of LA is closely dependent on the acidity/basicity and/or electrophilicity/nucleophilicity of the organic activator. While a single mechanism may be operating during the OROP, different mechanisms may also be taking place concomitantly.[8,17,22,23,29–35]

The activated monomer mechanism (AMM) operates by activation of LA through its oxygen atom of the carbonyl group, following either an electrophilic or a nucleophilic pathway. The greater electrophilicity of the monomer activated by an electrophile E (such as Brønsted acids, alkylating, or acylating agents), **1**, as compared to LA, promotes the nucleophilic addition of the initiator or the propagating chain end to **1**, eventually resulting in the ring opening of the monomer into the mono-adduct ω-alcohol **2** along with the regeneration of the acidic catalyst E (Scheme 6.2a). Direct attack onto LA by a nucleophilic catalyst Nu (such as amines, phosphines, or NHCs) proceeds with LA ring opening into an alkoxide-type zwitterionic activated monomer intermediate **1′**, which is subsequently protonated by the alcohol initiator (ROH) to give **1″**. The Nu^+ α-chain end of **1″** is then displaced by the alkoxide counter-anion resulting in a ring-opened alcohol mono-adduct **2**, concomitant with the release of the organic Nu catalyst. Propagation then proceeds upon reaction of this dormant alcohol mono-adduct **2** with the zwitterionic intermediate **1′** by proton transfer from **2** to **1′** (Scheme 6.2b). Note that in the absence of an alcohol initiator, the Nu moiety remains bound to the PLA α-chain end in the zwitterionic species **1′** and **2**, as well as in the final recovered PLA, resulting in a nucleophilic non-catalytic AMM (Scheme 6.2c).

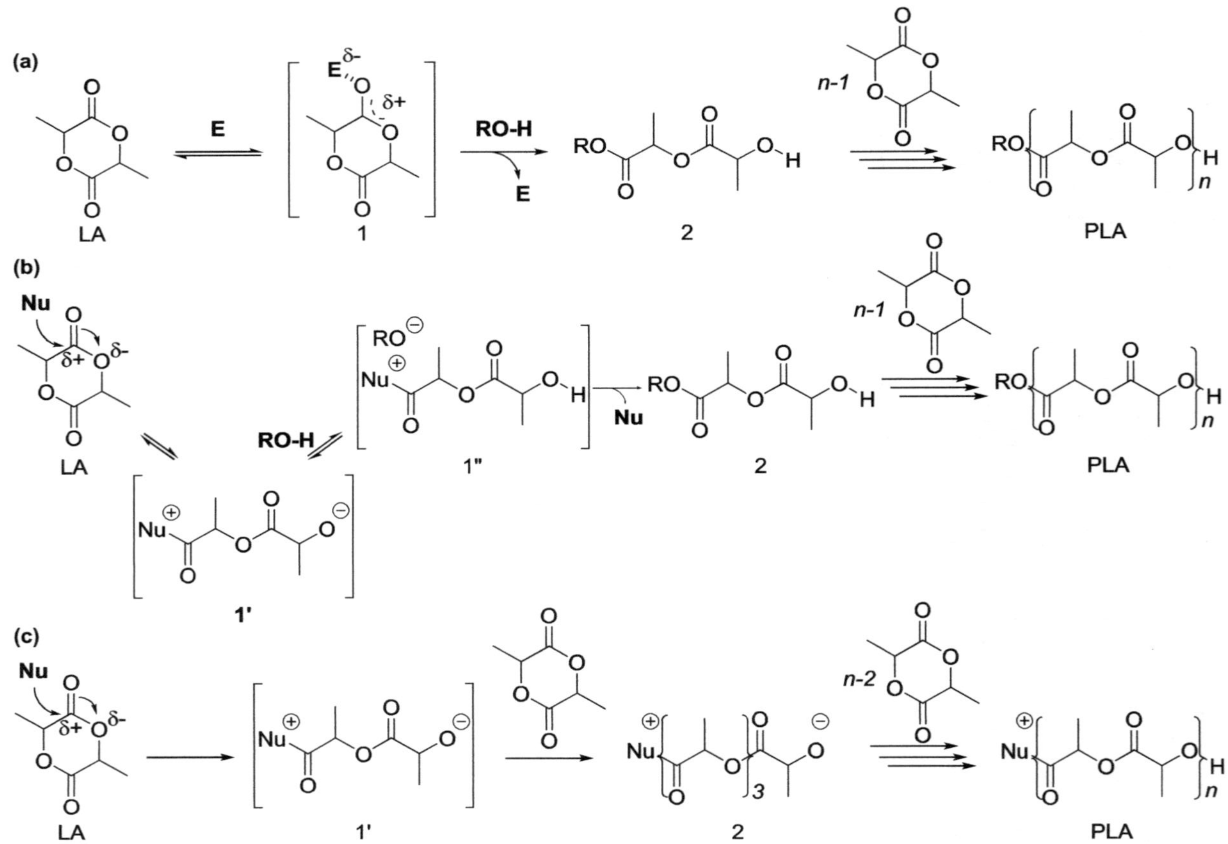

Scheme 6.2 Electrophilic (a), nucleophilic (b), and nucleophilic non-catalytic (c) AMMs involved in the OROP of LA.

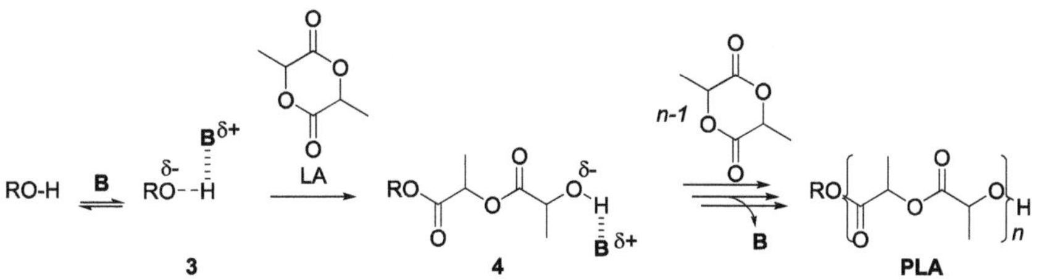

Scheme 6.3 Basic active initiator/chain-end mechanism (ACEM) involved in the OROP of LA.

Polymerization proceeding by activation of the initiator or of the polymer chain end is referred to as the activated chain-end mechanism (ACEM). The basic catalyst B (such as Brønsted basic or H-bonding catalysts) activates the initiating alcohol or polymer chain end through H-bonding giving the activated initiating species **3**, which then displays enhanced nucleophilicity thereby promoting its attack onto LA and the subsequent ring opening of the cyclic monomer. The resulting hydroxyl-terminated mono-adduct **4** then propagates the ROP of further incoming LA units upon its activation by the catalyst B (Scheme 6.3). This ACEM is sensitive to the pK_a difference between the initiating alcohol and the base catalyst.

Finally, the cooperative dual activation of both the monomer and the initiator/polymer chain end with specific mono- or multicomponent hetero-bifunctional organic catalysts (such as Brønsted acids, Brønsted bases, and/or H-bonding catalysts) may also be occurring (Scheme 6.4). Typically, an electrophilic moiety (usually a proton) activates the oxygen atom of LA, while the base interacts with the proton of the alcohol initiator/polymer chain end, to generate the propagating species **4**, ultimately affording PLA.

6.3 Polymerization of LA Directly Induced by Single Organic Initiators

Small organic molecules have been used directly as single nucleophilic initiators (*n.b.*: not as catalysts) in the OROP of LA. These essentially revolve around nitrogen-containing Brønsted bases, including amines, amidines, and NHCs and mostly afforded cyclic PLA.[29–32]

(+)-Sparteine successfully initiated the OROP of optically pure L-lactide (L-LA) under aprotic conditions (CH_2Cl_2, 21 °C) affording poly(L-LA) (PLLA) macrocycles (number average molar mass as determined by size-exclusion chromatography (SEC) ($M_{n,SEC}$) = 6000–20 300 g mol^{-1}; dispersity ($Đ_M$) = 1.13–1.47).[37] The OROP was however much slower than a polymerization initiated by an exogenous alcohol and the control was only achieved at higher L-LA-to-sparteine feed ratio. The cyclic PLLAs formation was rationalized by a backbiting process from the *in situ* generated (upon reaction of sparteine with the first two incoming monomer units) tertiary amine-containing symmetrical binary sparteine/LA zwitterion (Scheme 6.5).

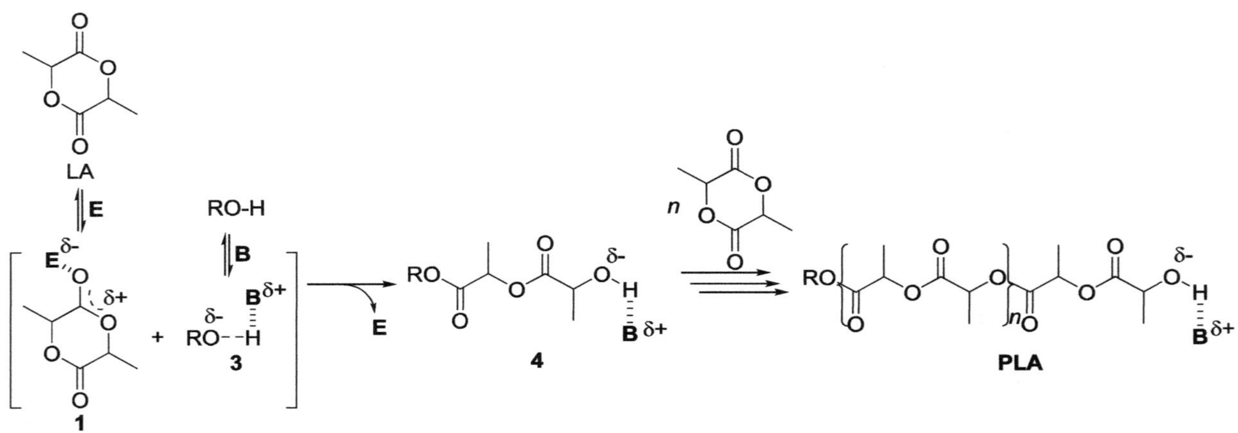

Scheme 6.4 Dual monomer and initiator/polymer chain-end activation mechanism (AM/ACEM) involved in the OROP of LA.

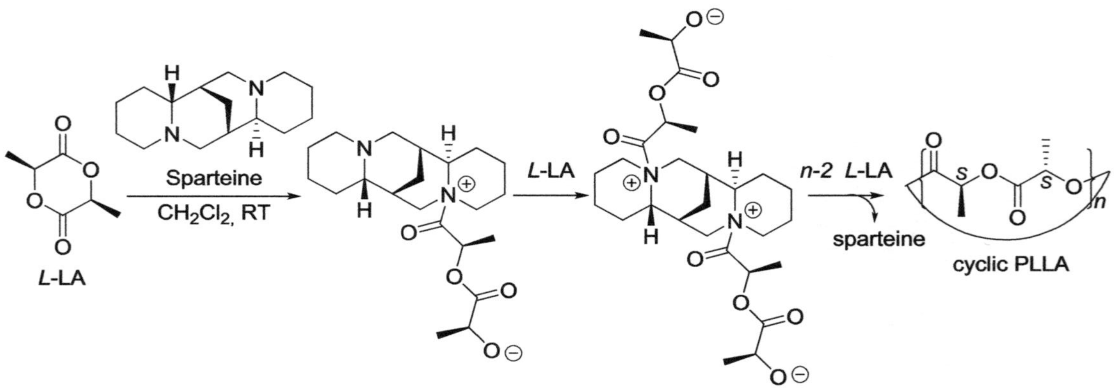

Scheme 6.5 Sparteine mediated OROP of L-LA.[37]

The amidine mediated OROP of *racemic*-LA (D,L-LA, *rac*-LA) with the neutral amine nucleophilic initiators 1,8-diazabicyclo[5.4.0] undec-7-ene (DBU) or 1,5-diazabicyclo[4.3.0]non-5-ene (DBN), in the absence of any initiator, afforded (CH_2Cl_2, or CH_2Cl_2/THF, room temperature (RT)) high molar mass cyclic PLA ($M_{n,SEC} = 32\,000–56\,000$ g mol^{-1}; $Đ_M = 1.25–1.63$). DFT insights supported a mechanism involving a zwitterionic acyl amidinium intermediate.[38] A unified mechanistic scheme for the DBU-catalyzed OROP of *rac*-LA was recently proposed along with a quantitative kinetic analysis that could reliably reproduce the experimentally observed monomer, M_n, and $Đ_M$ time-dependent profiles.[39]

Bicyclic isothioureas also mediated the controlled nucleophilic zwitterionic OROP of L-LA in the absence of protic initiators (CH_2Cl_2, RT) to selectively give high molar mass PLLA macrocycles ($M_{n,SEC} = 38\,000–66\,000$ g mol^{-1}; $Đ_M = 1.39–1.65$).[40] This is in contrast to the DBU initiated alike OROP of LA, isothioureas being more selective towards the formation of cyclic PLA without linear contaminants. Mechanistic studies revealed that the acylated amidines are readily deprotonated to the ketene aminal from which the linear chains obtained in the DBU-mediated OROP of LA originates, whereas the acylated isothioureas are not deprotonated. Also, the rates of polymerization with these isothioureas were significantly slower than those with DBU or NHCs.[38,40]

The zwitterionic controlled OROP of *rac*-LA promoted by 1,3-dimesitylimidazol-2-ylidene (IMes) NHC in the absence of an alcohol initiator (THF, 25 °C), also afforded cyclic PLAs ($M_{n,SEC} = 4200–15\,000$ g mol^{-1}; $Đ_M = 1.15–1.35$).[41] Likewise, the similar OROP of L-LA generated crystalline cyclic PLLA thus highlighting the retention of stereochemistry during the polymerization. Similarly, in the absence of any alcohol initiator, IMes was demonstrated to promote the highly efficient OROP of *rac*-LA (THF, 25 °C). Well-defined cyclic PLAs were thus generated selectively ($M_{n,SEC} = 5300–31\,400$; $Đ_M = 1.14–1.35$) with a high rate of propagation relative to cyclization and chain transfer.[42]

Alcohol adducts of some imidazolin-2-ylidene carbenes were successfully used as highly efficient single-component catalyst/initiators for the OROP of *rac*-LA or L-LA under mild conditions (turnover frequency (TOF) = 933 h^{-1}; THF, 25 °C).[43] The PLAs formed, α-end-capped by the ester functionality of the alkoxy group of the catalyst, displayed controlled molar mass values and narrow dispersities (number average molar mass as determined by NMR ($M_{n,NMR}$) = 12 100–22 500 g mol^{-1}; $Đ_M = 1.16–1.34$). Multifunctional adducts were used to prepare PLAs of more complex architectures such as linear or three-arm star hydroxyl telechelic polymers.

The OROP of *rac*-LA was also successfully mediated by some imidazolium carboxylates (THF or/and toluene, 60 °C) generating PLA ($M_{n,NMR} = 6300–17\,000$ g mol^{-1}; $Đ_M = 1.21–1.83$) (Scheme 6.6).[44] Among the NHC carboxylates evaluated, namely 1,3-bis(2,4,6-trimethylphenyl) imidazolium-2-carboxylate (IMes.CO_2), 1,3-bis(2,6-diisopropylphenyl) imidazolium-2-carboxylate (IPr.CO_2), or 1,3-bis(2,6-bis(diphenylmethyl)-4-methylphenyl) imidazolium-2-carboxylate (IPr*.CO_2), the latter two bulkiest ones were revealed to be less

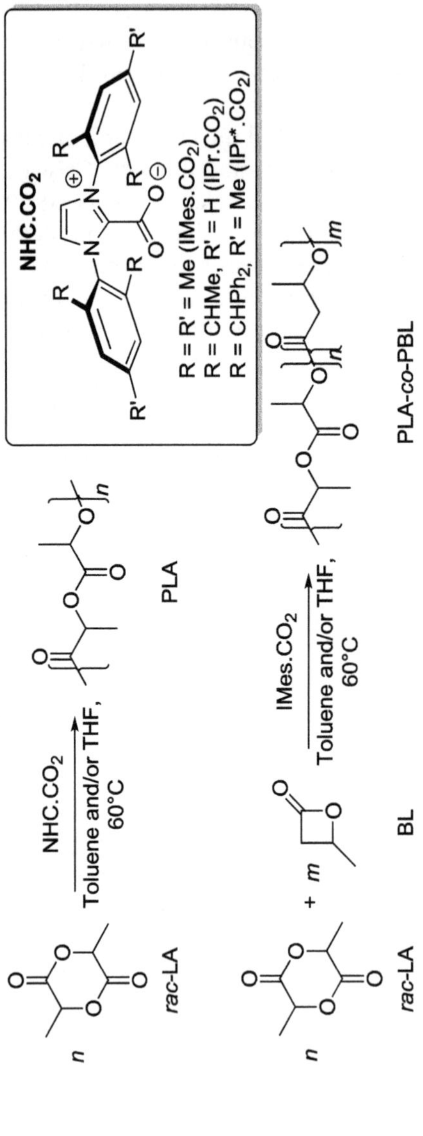

Scheme 6.6 NHC carboxylates mediated OROP of *rac*-LA and its copolymerization with BL.[44]

reactive than the former one, resulting in a poorer control over molar mass values. Similarly, the simultaneous copolymerization of *rac*-LA or L-LA with β-butyrolactone (BL) enabled the preparation of random PBL-*co*-PLA (PBL = poly(BL)) copolymers ($M_{n,NMR} = 5900–9500\ g\,mol^{-1}$; $Đ_M = 1.15–1.38$). Interestingly, these NHC carboxylates are easily synthesized and remarkably robust, thus allowing the use of a bench-top reaction setup.

The zwitterionic OROP of D-lactide (D-LA) or L-LA efficiently mediated by the (+)/(−)1-methyl-3-menthoxymethyl imidazol-2-ylidene enantiomeric NHC catalyst (in the absence of an alcohol initiator; THF, 25 °C) gave cyclic PLAs ($M_{n,SEC} = 12\,800–36\,500\ g\,mol^{-1}$; $Đ_M = 1.23–2.02$).[45] Addition of methanol resulted in the NHC deactivation into an O–H insertion product. Whereas NHC catalysts are generally very sensitive to temperature, oxygen, and moisture, these NHCs were found to be stable and highly active (TOF = 5839 h^{-1}) in the OROP of L-LA up to 55 °C.

6.4 Polymerization of LA Catalyzed by Brønsted and Lewis Acids

Typically, the OROP of LA mediated by acids proceeds *via* an electrophilic activation of LA using protic alcohol initiators.[29–31]

The trifluoromethane sulfonic Brønsted acid CF_3SO_3H (TfOH) effectively catalyzed the controlled living cationic OROP of *rac*-LA and L-LA in the presence of a protic initiator (2-propanol, 1-pentanol, or water) *via* an AMM (CH_2Cl_2, RT). *Atactic* and *isotactic* PLAs were thus obtained *via* an AMM, respectively ($M_{n,NMR} = 1700–16\,400\ g\,mol^{-1}$; $Đ_M = 1.13–1.48$). Subsequently, the copolymerization of L-LA with ε-caprolactone (CL) proceeded with a reverse order of reactivity than that observed in the corresponding homopolymerizations, with L-LA being preferentially incorporated into the poly(CL) (PCL)/PLLA copolymer ($M_{n,NMR}$ *ca.* 3700 $g\,mol^{-1}$; $Đ_M = 1.22$). Dihydroxy end-capped PCL/PLLA random copolymers ($M_{n,SEC} = 7200–23\,000\ g\,mol^{-1}$; $Đ_M$ *ca.* 1.4) were similarly obtained in the presence of 1,4-butanediol or ethylene glycol as initiator (CH_2Cl_2, 35 °C).[46–48]

The one-pot cationic ring-opening copolymerization (ROcP) of *racemic*-BL and L-LA initiated by *iso*-propanol, was shown to be catalyzed by the Brønsted acid TfOH, and to proceed through an AMM (CH_2Cl_2, RT).[49] Interestingly, although both monomers were simultaneously introduced, a block copolymer ($M_{n,SEC} = 1100–3400\ g\,mol^{-1}$; $Đ_M = 1.2–1.7$) was formed as the result of the different reactivity of the two monomers, with PBL being formed during the first stage and next acting as a macroinitiator for the subsequent polymerization of L-LA.

Chiral binaphthol (BINOL)-derived monophosphoric Brønsted acids were efficient organocatalysts for the highly enantiomer-selective OROP of *rac*-LA in the presence of PPA as initiator (toluene, 75 °C) (Scheme 6.7). Depending on the electronic contribution (prevailing over steric hindrance) of the 3,3′-substituents of the binaphtyl backbone, a maximum selectivity factor (k_D/k_L)

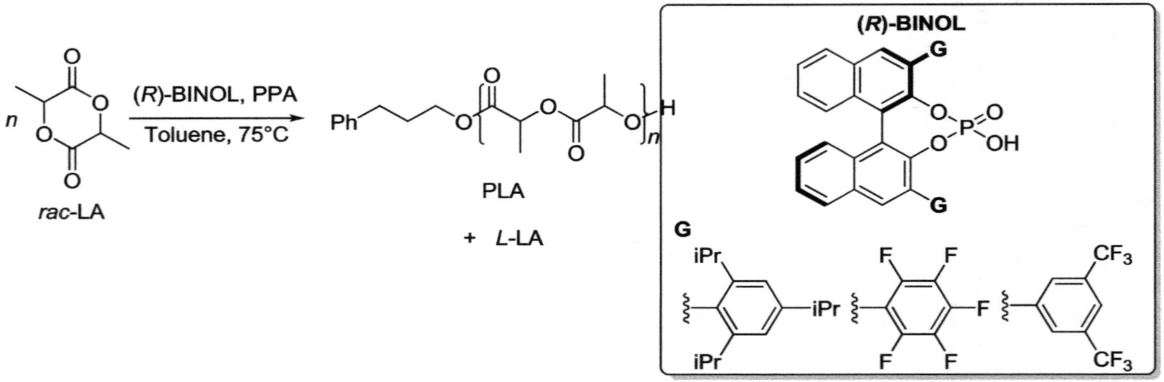

Scheme 6.7 Enantiomer-selective OROP of *rac*-LA catalyzed by chiral BINOL phosphoric acids/PPA.[50]

of 28.3 was reached at 49% of *rac*-LA conversion with an enantiomeric excess *ee* of unreacted LA of 80.6% with the (*R*)-3,3′-bis(pentafluorophenyl)-1,1′-binaphthyl-2,2′-diyl-hydrogenphosphate catalyst ($M_{n,NMR} = 3700$ g mol^{-1}; $Đ_M = 1.13$). The preferential polymerization of D-LA arose from a dual activation of the monomer and the propagating chain-end due to the function of the chiral phosphoric acid catalyst.[50] This strategy for the enantiomer-selective OROP of *rac*-LA using chiral phosphoric acid provides a new method for synthesizing a range of stereocontrolled polymers *via* the selective activation of a monomer and/or propagating chain-end.

The controlled OROP of *rac*-LA was initiated by 3-phenyl-1-propanol or propargyl alcohol (ROH) using the bis(trifluoromethane)sulfonimide (triflimide = Tf_2NH) Lewis acid as catalyst (CH_2Cl_2, RT), to afford α-OR end-capped PLAs ($M_{n,NMR}$ up to 6800 g mol^{-1}; $Đ_M = 1.15–1.19$) (Scheme 6.8).[51]

1,1′-Binaphthyl-2,2′-diyl hydrogen phosphate (BNPH) was recently used to catalyze the ROcP of TMC and L-LA initiated with BnOH (toluene, RT/85 °C) in a one pot approach.[52] The PTMC-*b*-PLLA diblock copolymer was thus prepared ($M_{n,NMR} = 4600$ g mol^{-1}, $Đ_M = 1.20$). The BNPH-catalyzed OROP proceeded with a plausible bifunctional activation mechanism in which BNPH behaved as a Brønsted acid/base pair.

The triarylsulfonium hexafluorophosphate salts effectively catalyzed the ROcP of L-LA and δ-valerolactone (VL) in the presence of 1,4-butanediol and DBU under UV irradiation (365 nm, propylene carbonate, RT), thus affording PVL-*b*-PLLA-*b*-PVL (poly(VL) = PVL) triblock copolymers ($M_n = 47\,400$ g mol^{-1}, $Đ_M = 1.15$) in a one-pot reaction. Indeed, following the formation of first PLA ($M_n = 34\,700$ g mol^{-1}, $Đ_M = 1.05$) in the absence of UV irradiation, the UV-triggered decomposition of the photoactive acidic sulfonium into sulfide species concomitantly produced the *in situ*-generated strong HPF_6 acid (Scheme 6.9). The thus formed H^+ acted both to neutralize DBU and as the actual active species to promote the cationic OROP of VL.[53]

6.5 Polymerization of LA and OCAs Catalyzed by Nitrogen-containing Brønsted/Lewis Bases

6.5.1 Polymerization of LA Catalyzed by Amines and Pyridine Derivatives

Alky and aryl amines as well as pyridine derivatives are Brønsted bases (less basic than amidine and guanidines, and relatively moderate bases *vs.* NHCs), which can behave as strong nucleophiles in the OROP of LA and related OCAs (*vide infra*).[29–31] Amines have been rarely reported as a single component catalyst[37] or in a multicomponent catalyst,[54] whereas they have been often used in bifunctional systems (refer to Section 6.7).

Amide end-capped PLA oligomers were efficiently prepared from amine promoted OROP of *rac*-LA, through a two-step one-pot strategy in which primary or secondary amines first ring-opened one *rac*-LA molecule prior to

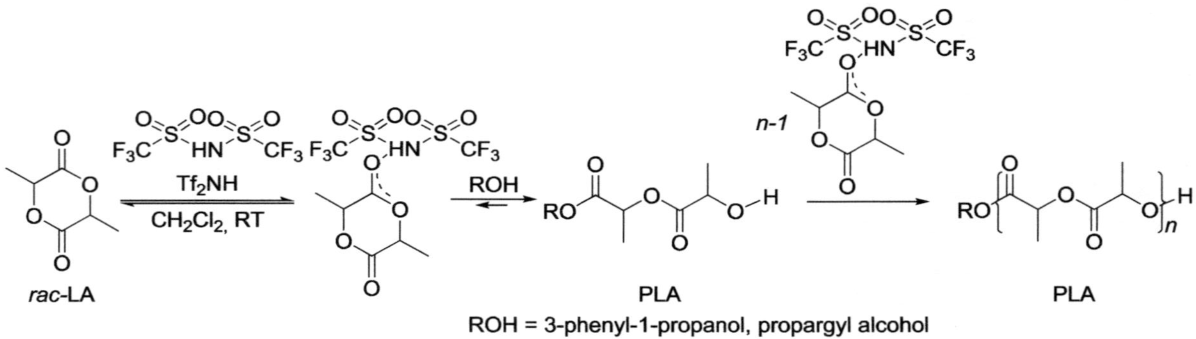

Scheme 6.8 Tf$_2$NH/ROH mediated OROP of *rac*-LA.[51]

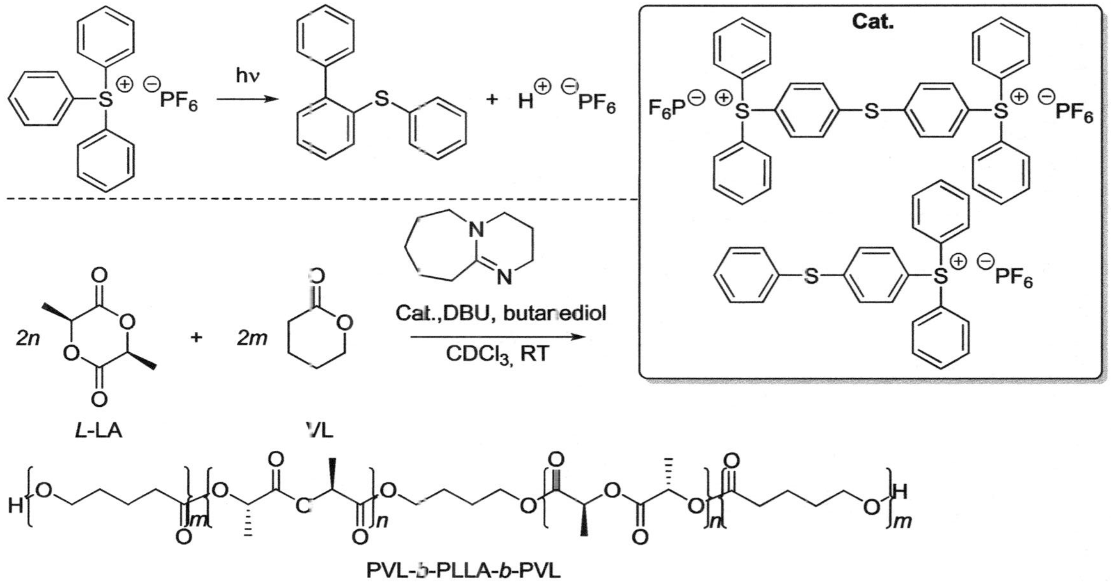

Scheme 6.9 H^+ generation upon UV irradiation of a triarylsulfonium hexafluorophosphate salt (top) and ROcP of L-LA and VL mediated by triarylsulfonium hexafluorophosphate salts in the presence of DBU/1,4-butanediol (bottom).[53]

the addition of the DBU catalyst from which polymerization then proceeded (CH_2Cl_2 or $CDCl_3$, 30–60 °C) (Scheme 6.10).[54] Polyfunctional amine initiators generated amide end-capped PLA of different structures (linear, star-shaped) ($M_{n,NMR}=500$–8000 g mol^{-1}; $Đ_M=1.10$–1.62). Also, primary amines supported on poly(styrene) (PS) resins readily reacted with *rac*-LA in the presence of TfOH/dodecan-1-ol, thus enabling the easy removal of unreacted monomer from PLA samples without affecting the polymer properties.

The adenine nucleobase, which is present as a constituent of important energy resources in the organism, successfully enabled the quantitative OROP of L-LA (bulk, 100–135 °C).[55] Adenine end-capped PLLA oligomers along with PLA macrocycles were thus obtained from a simple one-step procedure involving significant transesterification reactions ($M_{n,SEC}=2000$–4900 g mol^{-1}; $Đ_M=1.36$–2.05). DFT calculations showed that the polymerization occurs *via* hydrogen-bond catalysis.

DMAP and the related PPY bases were the first organocatalysts reported to promote the OROP of *rac*-LA and L-LA in the presence of an alcohol (EtOH, benzyl alcohol (BnOH)) nucleophilic initiator (CH_2Cl_2 or bulk, 35–185 °C). The amines showed a comparable activity affording well-defined α-alkoxy,ω-hydroxy telechelic PLA ($M_{n,NMR}=750$–17 300 g mol^{-1}; $Đ_M=1.08$–1.18). Also, the OROP of the enantiomerically pure L-LA clearly showed the absence of racemization.[19]

Similarly, using BnOH as initiator, DMAP successfully catalyzed the ROcP of trimethylene carbonate (TMC) and L-LA to give PTMC-*b*-PLLA (poly(TMC) = PTMC) block copolymers (toluene, 130 °C).[56] The OROP of L-LA proceeded from the pre-synthesized HO–PTMC–OBn macroinitiator in a two-step sequential procedure, or through the direct one-pot, two-step sequential copolymerization of the two monomers ($M_{n,SEC}=4300$–19 300 g mol^{-1}; $Đ_M=1.13$–1.40).

Associated with various carbohydrate initiators such as glucose or cyclodextrin derivatives, DMAP effectively catalyzed the OROP of *rac*-LA (bulk, 120 °C, or chlorinated solvents, RT).[57] Carbohydrate end-capped PLA, and 2–21-arm star PLA featuring a carbohydrate core were thus synthesized ($M_{n,NMR}=300$–30 300 g mol^{-1}; $Đ_M=1.07$–1.48).

6.5.2 Polymerization of LA Catalyzed by Amidines and Guanidines

Amidines and guanidines are stronger bases than alkyl (aryl) amines, and in some cases stronger bases and in other cases weaker than phosphazenes. They usually act as deprotonating agents of initiators featuring –OH, –SH or C–H bonds. While the amidine DBU only activates the initiating alcohol, the 1,5,7-triazabicyclo[4.4.0]dec-5-ene (TBD) guanidine may also act as dual activator for both the monomer and the alcohol. In contrast to the OROP of lactones, only a few reports describe the OROP of LA catalyzed by guanidines, with only TBD being investigated.[29–31]

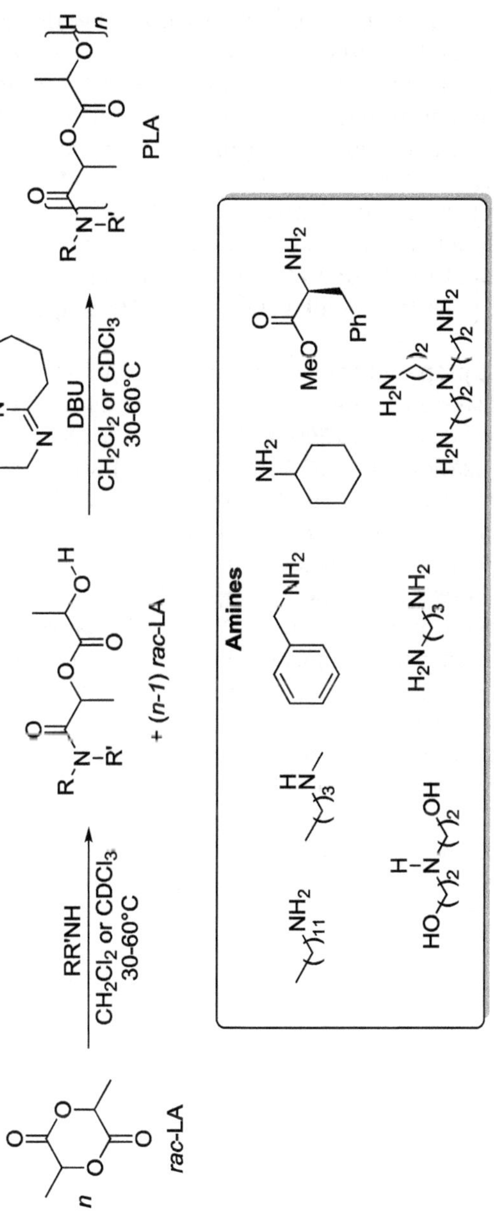

Scheme 6.10 Two-step one-pot synthesis of α-amide,ω-hydroxy end-capped PLA *via* amine ring-opening and DBU-catalyzed OROP of *rac*-LA.[54]

The Brønsted basic DBU amidine has been often reported as an effective catalyst in the OROP of LA in combination with various alcohol initiators. DBU was reported to be effective in the OROP of L-LA in the presence of PyBuOH as the initiator (TOF = 4 min^{-1}), however to a lower extend than TBD (TOF = 475 min^{-1}; see below).[58] PLLA with a low dispersity was thus obtained ($M_{n,NMR}$ = 21 000–85 000 g mol^{-1}; $Đ_M$ = 1.05–1.08) under mild conditions ($CDCl_3$, RT). Similarly, in the presence of BnOH or poly(ethylene glycol) (PEG)–OH (macro)initiator (THF or CH_2Cl_2, RT), the DBU catalyzed OROP of *rac*-LA generated α-alkoxy,ω-hydroxy telechelic PLA or PLA-*b*-PEG (co)polymers ($M_{n,SEC}$ = 6800–17 300 g mol^{-1}; $Đ_M$ = 1.06–1.35).[38,39]

Following the reported DBU/BnOH homopolymerization of the alkyne-functionalized phospholane, namely butynyl phospholane (BYP), the sequential one-pot BYP/L-LA copolymerization afforded the diblock poly(phosphoester)-*b*-PLLA copolymer, poly(BYP)-*b*-PLLA ($M_{n,SEC}$ = 15 000 g mol^{-1}; $Đ_M$ = 1.17) (Scheme 6.11).[59]

The eutectic melt prepared from L-LA and TMC (50 : 50 wt%) was used for the homopolymerization of L-LA catalyzed by DBU using BnOH as an initiator (bulk, 23 °C) ($M_{n,SEC}$ = 15 500–44 000 g mol^{-1}; $Đ_M$ = 1.33–1.65). Subsequent copolymerization of TMC (CH_2Cl_2, 23 °C) afforded P(LLA-*g*-TMC) gradient copolymers ($M_{n,SEC}$ = 22 000 g mol^{-1}; $Đ_M$ = 1.64).[60]

TBD, in the presence of PyBuOH as the initiator, was found to be very active (TOF up to 475 min^{-1}) to catalyze the OROP of L-LA (CH_2Cl_2, RT). Well-defined PLLA was thus formed ($M_{n,SEC}$ up to 62 600 g mol^{-1}; $Đ_M$ = 1.11).[61] The enhanced activity of TBD (TOF = 297 min^{-1}) relative to *N*-methylated TBD (7-methyl-1,5,7-triazabicyclo[4.4.0]dec-5-ene (MTBD); TOF = 184 min^{-1} under similar operating conditions) was attributed to its bifunctionality, enabling the simultaneous activation of both the cyclic ester monomer and the alcohol group of the initiator/propagating species.[58,61] Minimal epimerization of the monomer was observed by ^{1}H NMR during polymerization of L-LA, and no significant difference in the selectivity of the stereochemistry in the OROP of *rac*-LA was observed between TBD, MTBD, and DBU (P_m = the probability of forming a new *isotactic* dyad = 0.58–0.60).[58] Solution copolymerization of *rac*-LA with VL or CL from TBD/PyBuOH afforded the corresponding block copolymers ($M_{n,SEC}$ = 23 300, 26 700 g mol^{-1}; $Đ_M$ = 1.21, 1.17, respectively), whereas PLA block copolymers were similarly prepared form the OROP of *rac*-LA catalyzed by TBD/hydroxyl functional macroinitiator such as PEG–OH, PS–OH or poly(methyl methacrylate) (PMMA)–OH (CH_2Cl_2, RT) ($M_{n,SEC}$ = 11 000–27 700 g mol^{-1}; $Đ_M$ = 1.03–1.17).[58]

Structural modulation by incorporation of electronically active substituents on aromatic acyclic guanidine-based organocatalysts (namely, *N*-(*R*)-*N*′-cyclohexylpyrrolidine guanidines with R = 4-trifluoromethylphenyl, 4-methoxyphenyl, dimethylaminophenyl, phenyl) enabled tuning of the activity of the catalysts for the living OROP of L-LA in the presence of PyBuOH (CH_2Cl_2, RT) (Scheme 6.12).[62] Strongly electron-donating dimethylamino groups, enhanced the catalytic activity by two orders of magnitude as compared to electron-withdrawing trifluoromethyl groups (NMe_2 > OMe > H > CF_3).

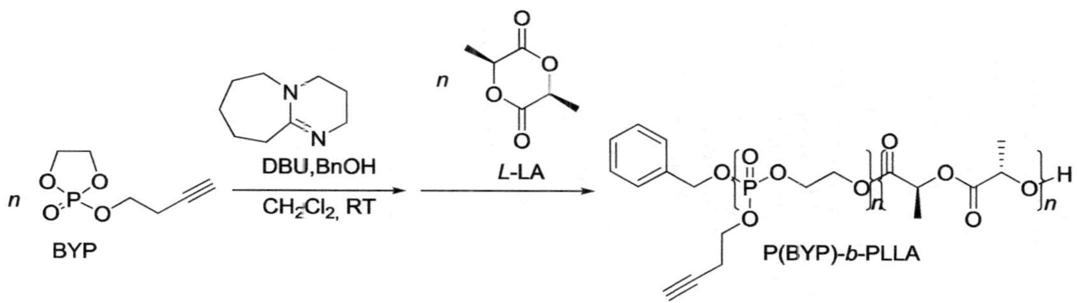

Scheme 6.11 One-pot sequential ROcP synthesis of poly(BYP)-*b*-PLLA mediated by DBU/BnOH.[59]

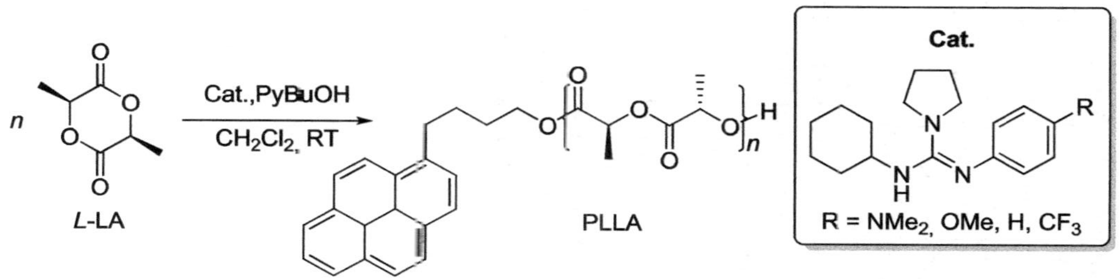

Scheme 6.12 OROP of L-LA mediated by *N*-(R)-*N*′-cyclohexylpyrrolidine guanidines/PyBuOH.[62]

DFT calculations revealed that the formation of an adduct consisting of the catalyst and initiator/chain end was the rate limiting step. The formed PLLA ($M_{n,SEC}$ = 7500–26 100 g mol^{-1}; $Đ_M$ = 1.04–1.18) featured high end-group fidelity.

The TBD guanidine organocatalyzed copolymerization of L-LA and TMC in the presence of BnOH afforded P(LLA-*ran*-TMC) random copolymers (on the basis of detailed $^{13}C\{^1H\}$ NMR analyses) as the result of the similar reactivity of the monomers, which were converted as fast in copolymerization as they were in their homopolymerization (toluene, 100–110 °C). The ROcP proceeded through an AMM without decarboxylation of TMC units, yet with significant transesterification reactions upon prolonged reaction times ($M_{n,SEC}$ = 3500–11 000 g mol^{-1}; $Đ_M$ = 1.4–2.1).[63]

The use of continuous flow microreactor technology was recently reported for the controlled OROP of L-LA catalyzed by TBD in the presence of various alcohols (4-*tert*-butylbenzyl alcohol, PEG–OH, ((1*R*,8*S*,9r)-bicyclo[6.1.0]non-4-yn-9-yl)methanol, and 4-(6-methyl-1,2,4,5-tetrazin-3-yl)phenyl)methanol) as initiators.[64] Well-defined PLA was thus prepared (CH_2Cl_2, −10 to +30 °C; $M_{n,NMR}$ = 22 000–34 600 g mol^{-1}; $Đ_M$ = 1.17–1.22), while using bicyclononyne- and tetrazine-containing initiators gave access to PLLA prone to click chemistry.

Sequential addition of δ-decalactone (DL) followed by *rac*-LA produced PLA-*b*-PDL-*b*-PLA triblock copolymers (poly(DL) = PDL) using TBD and 1,4-benzenedimethanol diol as catalyst and initiator, respectively (CH_2Cl_2, RT). The copolymers ($M_{n,SEC}$ = 134 500 g mol^{-1}; $Đ_M$ = 1.4) displayed a microphase segregation as suggested by the presence of two distinct glass transition temperatures corresponding to the PDL and PLA domains.[65]

In a similar approach, the one-pot two-step TBD, DMAP/trifuoroacetic acid (TFA)-catalyzed OROP of *rac*-LA or L-LA and CL using either 1-pyrenemethanol or PEG–OH as (macro)initiator (bulk, 140 °C), generated PCL-*b*-PLLA, PEG-*b*-P(L)LA diblock or PEG-*b*-PCL-*b*-P(L)LA triblock amphiphilic copolymers, respectively ($M_{n,NMR}$ = 4500–13 400 g mol^{-1}; $Đ_M$ = 1.1–1.5).[66]

A closely related work describes the preparation of well-defined tri- and hepta-PEG side-chain functionalized lactide, PEG-*graft*-PLLA ($M_{n,NMR}$ = 44 000–3 200 000 g mol^{-1}; $Đ_M$ = 1.4–2.1), from the OROP of L-LA-end-capped PEG units using the catalytic system TBD/BnOH (CH_2Cl_2, RT) (Scheme 6.13).[67] Preliminary biological investigations indicated that PEG-*graft*-PLLA reduces cell adhesion when compared to PLA.

6.5.3 Polymerization of LA Catalyzed by N-heterocyclic Carbenes

N-heterocyclic carbenes (NHCs), as neutral bases and nucleophiles, have been successfully used in both chain-growth and step-growth polymerizations following the initial work on the synthesis of polymeric 1,4-diketones.[68] The σ-donor properties of NHCs promote a Brønsted basic or a

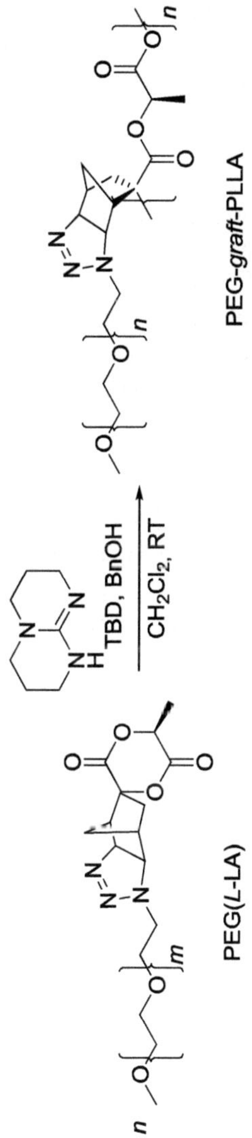

Scheme 6.13 Synthesis of PEG-*graft*-PLLA upon OROP of L-LA mediated by TBD/BnOH.[67]

nucleophilic-type catalysis. NHC-mediated polymerizations typically involve "free" NHCs derived from imidazol(in)ium salts while more recent efforts have been devoted to the development of less air-sensitive "protected" NHCs (also referred to as "masked" NHCs).[28–32] The OROP of LA has been investigated with a variety of NHCs. Indeed, the NHC catalyst platform is extremely versatile (sterically, electronically, *etc.*) with the nature of the substituents having a pronounced effect upon the catalyst stability, activity, ability to influence the rate and the control of the polymerization.

The OROP of L-LA mediated by NHCs such as 1,3-bis-(2,4,6-trimethylphenyl)imidazol-2-ylidene in the presence of BnOH as initiator (THF, 20–25 °C), was found to be considerably faster than that catalyzed by DMAP, showing living and controlled features to generate α-hydroxy, ω-benzyloxy telechelic PLLA ($M_{n,NMR}$ = 1600–28 400 g mol^{-1}; $Đ_M$ = 1.09–1.52).[69,70] The 1-isopropyl-3-(4-methoxyphenyl) imidazol-2-carbene catalyst also revealed highly active (TOF = 399 h^{-1}) for the OROP of L-LA (THF, 15 °C) in the presence of BnOH as initiator (M_n up to 26 800 g mol^{-1}; $Đ_M$ = 1.44), operating through an AMM.[71] Similarly, alcohol adducts of the 1,3,4-triphenyl-4,5-dihydro-1*H*-1,2-triazol-5-ylidene NHC, operated as single-component catalyst/initiators for the OROP of L-LA in the presence of PyBuOH or MeOH at higher temperature (THF, 50 °C, or toluene, 90 °C) to similarly generate ω-alkoxy end-capped PLLA ($M_{n,NMR}$ = 2500–14 700 g mol^{-1}; $Đ_M$ = 1.08–1.29), yet with a lower efficiency (TOF = 23 h^{-1}).[72,73] A variety of macromolecular architectures were also prepared from macroinitiators (PEG–OH, PS–OH, PMMA–OH), multifunctional or dendritic initiators, thus evidencing the versatility of the NHC platform.

Primary amines were found to act as bifunctional initiators for the OROP of *rac*-LA in the presence of the commercially available 1,3,4-triphenyl-4,5-dihydro-1*H*-1,2,4-triazol-5-ylidene to generate imide end groups and two chains per initiating amine.[74] Thus, 4-pyrene methylamine, PEG terminated with 4 or 8 primary amines (PEG bis(3-aminopropyl) (PEG–$(NH_2)_4$) and PEG octa(3-aminopropyl) (PEG–$(NH_2)_8$), respectively) were used as macroinitiators (C_6D_6, 90 °C) for the formation of PLA ($M_{n,SEC}$ = 10 600 g mol^{-1}; $Đ_M$ = 1.1), PLA_2-*b*-PEG-*b*-PLA_2 and PLA_4-*b*-PEG-*b*-PLA_4 copolymers, respectively.

Interestingly, the stereoselective OROP of *rac*-LA mediated by IMes NHC in the presence of BnOH (CH_2Cl_2, −20 to 25 °C) gave (TOF up to 320 h^{-1}) *isotactic* enriched (P_m = 0.75) PLA ($M_{n,SEC}$ = 15 900 g mol^{-1}; $Đ_M$ = 1.26).[75]

Also, some sterically encumbered, unsaturated chiral (*R*,*R*)- and achiral *rac*-NHC catalysts have been reported to stereoselectively polymerize *rac*-LA in the presence of PyBuOH as initiator (CH_2Cl_2), to give highly *isotactic* PLA at low temperature (P_m = 0.83–0.90 at −70 °C), while *meso*-LA afforded *heterotactic* PLA (P_m = 0.58–0.83 at −40 °C) (Scheme 6.14).[76] This stereoselective OROP of LA was shown to follow a chain-end mechanism (CEM) ($M_{n,NMR}$ *ca.* 13 700 g mol^{-1}; $Đ_M$ = 1.19–1.48).

The solvent-free OROP of *rac*-LA catalyzed by 1-*n*-butyl-3-methylimidazolium-2-carboxylate (BMIM-2-CO_2), decarboxylated *in situ* upon heating under

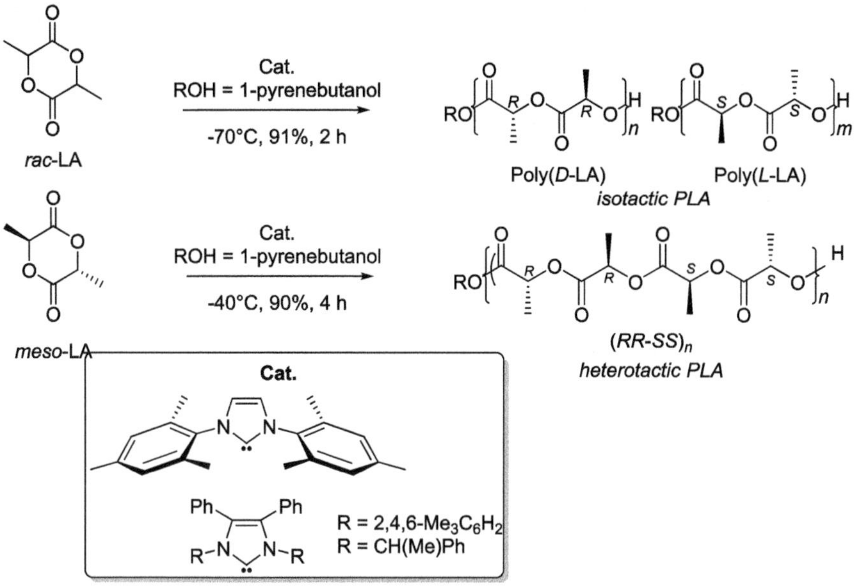

Scheme 6.14 Stereoselective OROP of *rac*-LA and *meso*-LA promoted by unsaturated chiral and achiral NHC catalysts in the presence of PyBuOH.[76]

vacuum or by addition of $NaBPh_4$, afforded, in the presence of BnOH, ethylene glycol, glycerol or pentaerythritol as initiator (75 °C), linear and star-branched PLA oligomers ($M_{n,NMR}$ = 500–1000 g mol^{-1}; $Đ_M$ = 1.16–1.62).[77] The advantage of BMIM-2-CO_2 lies in its easy one-step preparation (66% yield) from commercial *n*-butylimidazole and dimethyl carbonate.

Likewise, related imidazolium-2-carboxylates, (benz)imidazolium hydrogen carbonates, or free NHCs, successfully polymerized *rac*-LA in the presence of BnOH as initiator (THF, RT or 80 °C) to generate PLA ($M_{n,SEC}$ =3200–28 000 g mol^{-1}; $Đ_M$ = 1.07–1.30) (Scheme 6.15). The imidazolium-2-carboxylates were found to be less effective than free NHCs, but approximately three times more efficient than their hydrogen carbonate counterparts, except for 1,3-dicyclohexylbenzimidazolium hydrogen carbonate for which similar catalytic performances were observed. Again, the imidazolium hydrogen carbonates are advantageously easily prepared in a one-step reaction and they are also air-stable pre-catalysts.[78,79] Similar NHC.CO_2 adducts enabled the synthesis of low molar mass propylene oxide/L-LA block or random copolymers by the sequential or simultaneous monomer addition, respectively (diglyme, 100 °C) ($M_{n,SEC}$ = 600–4000 g mol^{-1}; $Đ_M$ = 1.38–1.77).[80]

Finally, the N-heterocyclic olefins (NHOs), a newly emerging class of nitrogen-containing organo-polymerization catalysts, containing a strongly polarized double bond that renders the exocyclic carbon partially anionic in character and reactivity, were found active in the OROP of L-LA in the presence of BnOH (Scheme 6.16).[81] While NHOs bearing a non-substituted methylene moiety did not quantitatively polymerize L-LA, the introduction

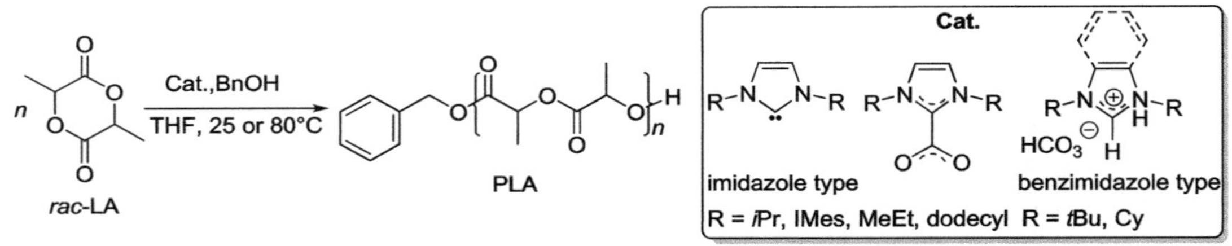

Scheme 6.15 OROP of *rac*-LA mediated by either free NHC, imidazolium-2-carboxylates, or (benz)imidazolium hydrogen carbonates.[78]

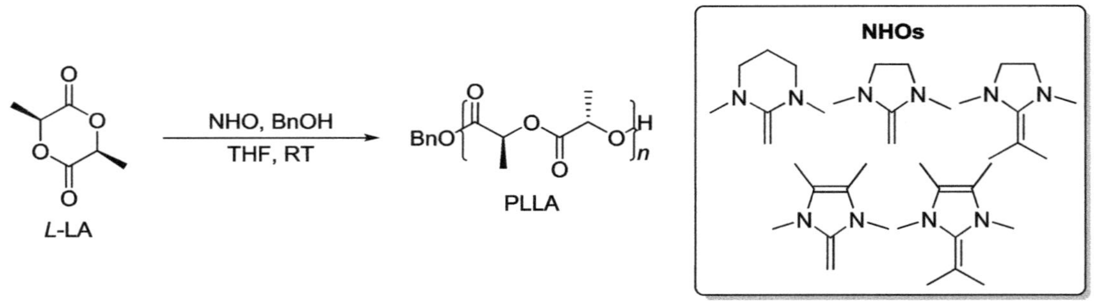

Scheme 6.16 OROP of L-LA mediated by N-heterocyclic olefins (NHOs)/BnOH.[81]

of methyl substituents on the exocyclic carbon enabled improving the activity. However, while 2-isopropylidene-1,3,4,5-tetramethylimidazoline bearing exocyclic methyl groups was effective in the presence of BnOH (THF, RT; non quantitative L-LA conversion; TOF = 95 h^{-1}), its strong basicity induced the abstraction of a proton from L-LA, inducing enolate-based initiation and side reactions, thus precluding the control of the polymerization ($M_{n,SEC}$ = 4300 g mol^{-1}; $Đ_M$ = 1.73). Eventually, an imidazolinium-derived NHO analogue featuring a saturated heterocyclic ring so as to decrease the polarization of the olefinic bond, enabled the highly active (TOF = 594 h^{-1}) controlled OROP of L-LA (THF, RT; $M_{n,SEC}$ = 4400 g mol^{-1}; $Đ_M$ = 1.32), in the presence of BnOH.

6.5.4 Polymerization of OCAs Catalyzed by Pyridine Derivatives and N-heterocyclic Carbenes

DMAP was also reported to successfully promote the controlled and living OROP of lactide equivalents such as the α-lactone 1,3-dioxolane-2,4-dione (the so-called OCA equivalent, lacOCA), in the presence of various protic initiators (both primary and secondary alcohols such as *neo*-pentanol, *iso*-propanol, cholesterol, 2-BrEtOH) with a remarkably higher reactivity (TOF = 240 h^{-1}) as compared with L-LA (TOF = 0.16 h^{-1}). Polymers of controlled molar mass values and narrow dispersities were typically obtained ($M_{n,NMR}$ = 1500–85 300 g mol^{-1}; $Đ_M$ = 1.16–1.22) under mild conditions (CH_2Cl_2, 25 °C).[82] Computational insights into the polymerization mechanism next revealed that the DMAP basic activation of the initiating/propagating alcohol was a more energetically favorable pathway than the nucleophilic activation of the monomer through an acylpyridinium intermediate.[36,83] Furthermore, this favored route involved a weak non-classical H-bond between the ester carbonyl of the OCA monomer and an *ortho*-H of DMAP. Accordingly, DMAP was then considered to behave as a bifunctional catalyst, simultaneously activating both the propagating chain end and the monomer through its basic nitrogen center and an acidic *ortho*-H atom.

Under similar mild operating conditions, the DMAP/*n*- or *neo*-pentanol OROP of the functionalized OCA derived from glutamic acid (L-gluOCA) proceeded with a high level of control (CH_2Cl_2, 25 °C), giving access to homopolymers ($M_{n,NMR}$ = 1400–14 800 g mol^{-1}; $Đ_M$ = 1.14–1.25). Block and random copolymers featuring pendant carboxyl groups with the OCA derived from lactic acid (lacOCA) comonomer were similarly reported ($M_{n,SEC}$ up to 11 600 g mol^{-1}; $Đ_M$ = 1.30) (Scheme 6.17).[84]

Following the same strategy, side-chain aminated poly(α-hydroxy acids) were prepared from the OROP of 5-(4-(prop-2-yn-1-yloxy)benzyl)-1,3-dioxolane-2,4-dione (Tyr-(alkynyl)-OCA), an OCA derived from tyrosine, using DMAP in the presence of 1-pyrenebutanol (PyBuOH) as initiator (CH_2Cl_2, RT).[85] Side-chain modification of the resulting polyesters ($M_{n,SEC}$ = 5800–116 200 g mol^{-1};

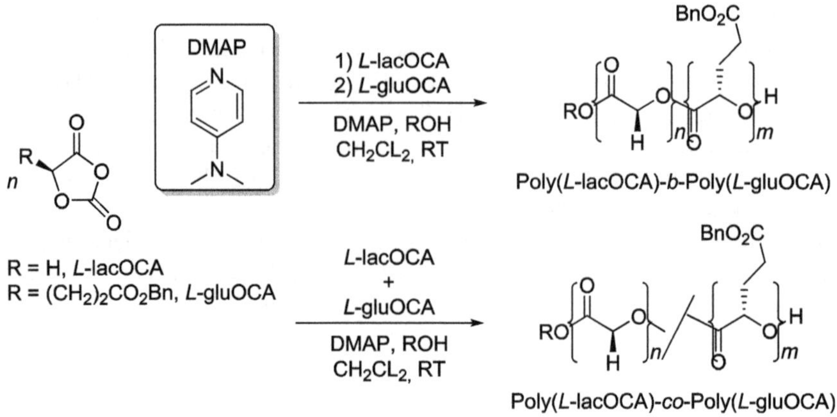

Scheme 6.17 Block and random ROcP of L-gluOCA and L-lacOCA mediated by DMAP/ROH.[84]

$Đ_M = 1.02–1.15$) *via* thiol-yne photochemistry generated polyesters bearing pendant amine groups, which displayed excellent cell penetration and gene delivery properties. Due to their easy availability and high polymerizability, OCAs are promising monomers for the preparation of tailored architectures derived from well-defined poly(hydroxyacid)s.

Investigation of symmetrical imidazol-2-ylidenes and imidazolin-2-ylidenes NHCs in the OROP of the OCA of L-LA (L-lacOCA) or of the OCA of L-mandelic acid (L-manOCA), in the presence of BnOH as initiator (THF, RT), did not reveal any significant difference in the reactivity of imidazolium (unsaturated) *vs.* imidazolinium (saturated) carbenes.[86] The less sterically demanding/less stable NHCs were found more active in the OROP of L-lacOCA than their sterically encumbered analogs. When using polyol initiators such as 1,3-propanediol, trimethylol propane, or pentaerythritol, α,ω-di-, tri-, tetra-hydroxy telechelic three- and four-arm star shaped (co)polymers of L-lacOCA (and L-manOCA) were formed, respectively (Scheme 6.18). Well-defined poly(L-lacOCA) were thus prepared ($M_{n,NMR} = 2000–14\,400\ g\ mol^{-1}$; $Đ_M = 1.10–1.41$).

6.6 Polymerization of LA Catalyzed by Phosphorus-containing Brønsted/Lewis Bases: Phosphines and Phosphazenes

In comparison to nitrogen-containing catalysts, phosphines and phospazenes have been less investigated as basic catalysts in the OROP of LA.[29–32]

6.6.1 Polymerization of LA Catalyzed by Phosphines

Only rare examples of the use of phosphines in the OROP of LA have been reported.

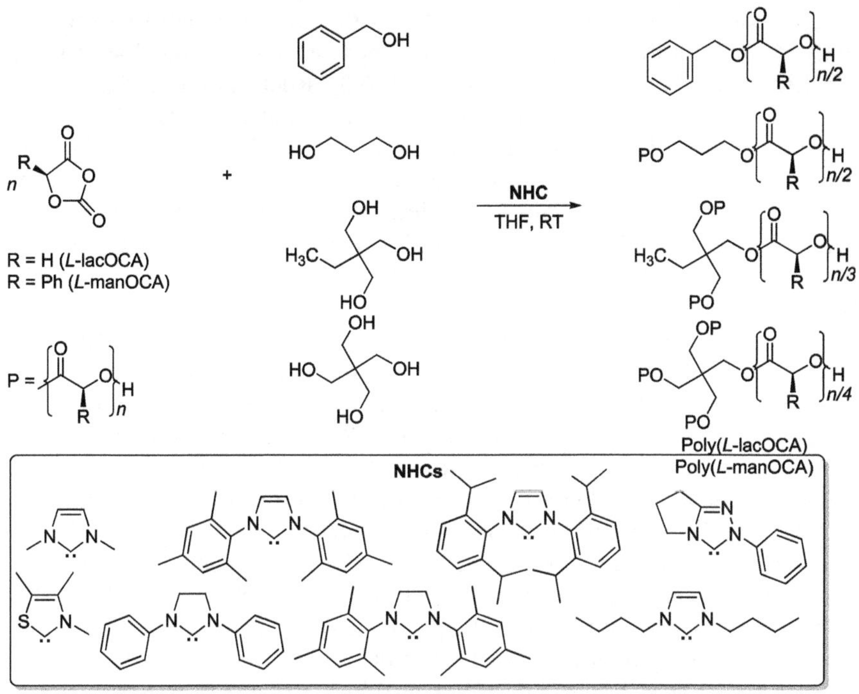

Scheme 6.18 Alcohol-initiated OROP of L-lacOCA and L-manOCA catalyzed by NHCs.[86]

Various alkyl or aryl nucleophilic phosphines such as tributylphopshines ($P(n\text{-}Bu)_3$, $P(tert\text{-}Bu)_3$), dimethylphenylphosphine ($PPhMe_2$), methyldiphenylphosphine (PPh_2Me), triphenylphosphine (PPh_3), and related phosphines, were found to effectively catalyze the OROP of *rac*-LA or L-LA in the presence of a nucleophilic initiator BnOH (bulk, 135 °C or 180 °C), generating well-defined α-OBn,ω-OH PLA ($M_{n,NMR} = 3700–13\,000$ g mol^{-1}; $Đ_M = 1.11–1.40$) through a nucleophilic AMM.[87] The phosphine substituents enabled to control the reactivity, with alkyl-substituted phosphines, as the most basic and most nucleophilic ones, being more active than their aryl-containing counterparts ($P(n\text{-}Bu)_3 > P(tert\text{-}Bu)_3 > PPhMe_2 > PPh_2Me > PPh_3 > P(MeO)_3$-unreactive).

6.6.2 Polymerization of LA Catalyzed by Phosphazenes

As strong neutral Brønsted bases and weak nucleophiles, phosphazenes typically operate in the presence of a protic (usually alcohol) initiator through an ACEM.[29–32]

Phosphazene bases, such as 2-*tert*-butylimino-2-diethylamino-1,3-dimethylperhydro-1,3,2-diazaphosphorine (BEMP) or *N*′-*tert*-butyl-*N*,*N*,*N*′,*N*′,*N*″,*N*″-hexamethylphosphorimidic triamide (P_1-*t*-Bu) revealed active organocatalysts for the living OROP of *rac*-LA or L-LA in the presence of PyBuOH or BnOH as the initiator (toluene, RT).[88] PLLA and PLA prepared through this organocatalytic route showed predictable molar mass values, narrow dispersities, and high end-group fidelity ($M_{n,SEC}$ = 10 700–18 000 g mol^{-1}; $Đ_M$ = 1.06–1.14). Mechanistic insights suggested that the intermolecular H-bonding of the alcohol initiator to the phosphazene base activates the alcohol for the OROP of the cyclic diesters. Polymerization of *rac*-LA gave *isotactic*-enriched PLA (P_m = 0.70). Also, a variety of PLA block copolymers were prepared form the OROP of *rac*-LA catalyzed by BEMP/hydroxyl functional macroinitiator such as PEG–OH, PS–OH or PMMA–OH (CH_2Cl_2, RT) ($M_{n,SEC}$ up to 38 600 g mol^{-1}; $Đ_M$ = 1.17).

The one-pot, two-step sequential polymerization of TMC and L-LA catalyzed by the BEMP phosphazene in the presence of BnOH as initiator (toluene, 100 °C), successfully afforded PTMC-*b*-PLLA diblock copolymers ($M_{n,SEC}$ = 36 2100 g mol^{-1}; $Đ_M$ = 1.32).[56] The ROcP proceeded with first the formation of α-benzyloxy,ω-hydroxy telechelic PTMC which next acted as a macroinitiator in the OROP of L-LA. BEMP was found more effective than DMAP, most likely as the result of its greater basicity.

The dimeric phosphazene base 1-*tert*-butyl-2,2,4,4,4-pentakis(dimethylamino)-2Λ^5,4Λ^5-catenadi(phosphazene) (P_2-*t*-Bu) catalyst also enabled the stereoselective OROP of *rac*-LA in the presence of PyBuOH as the initiator at low temperature (toluene, −75 °C) to form highly *isotactic* PLA stereocomplex (P_m = 0.95; $M_{n,SEC}$ = 27 200 g mol^{-1}; $Đ_M$ = 1.11).[89] Microstructural analysis of PLA using homodecoupled ^{1}H NMR spectroscopy revealed the formation of a stereoblock architecture containing long *isotactic* sequences of poly(*R*) and poly(*S*) segments. A mechanism involving chain-end control with stereoerrors was proposed to rationalize the stereoselectivity of the OROP, mostly attributed to the high basicity, steric hindrance, and high activity (TOF = 33 h^{-1}) of P_2-*t*-Bu at low temperature.

The relatively mild phosphazene P_2-*t*-Bu base combined to PyBuOH as initiator, also highly effectively (TOF = 312 h^{-1}) catalyzed the sequential block copolymerization (toluene/THF, RT) of CL and L-LA to give PCL-*b*-PLLA ($M_{n,NMR}$ up to 17 300 g mol^{-1}; $Đ_M$ = 1.15) copolymers (Scheme 6.19).[90] The incomplete consumption of CL (39–87%) did not proceed further during the OROP of L-LA which rapidly proceeded to full conversion (>99%) regardless of the solvent.

In a related work, in the presence of 3-phenyl-1-propanol (PPA) or even water as monohydroxyl or difunctional initiator, respectively, P_2-*t*-Bu enabled the one-pot synthesis of PEG-*b*-PCL-*b*-PLLA ($M_{n,SEC}$ = 13 300 g mol^{-1}; $Đ_M$ = 1.10) and PLLA-*b*-PCL-*b*-PEG-*b*-PCL-*b*-PLLA ($M_{n,SEC}$ = 16 100 g mol^{-1}; $Đ_M$ = 1.11) triblock and pentablock terpolymers, respectively.[91] The sequential copolymerization of ethylene oxide (THF, 50 °C, 3 d), CL (RT, 17–22 h) and L-LA (RT, 10 min) successfully proceeded, provided this order of monomers addition was followed so as to avoid any inter- or intramacromolecular transesterification side reactions, which may occur during the ROP of ethylene oxide from PCL or PLLA macroinitiators.

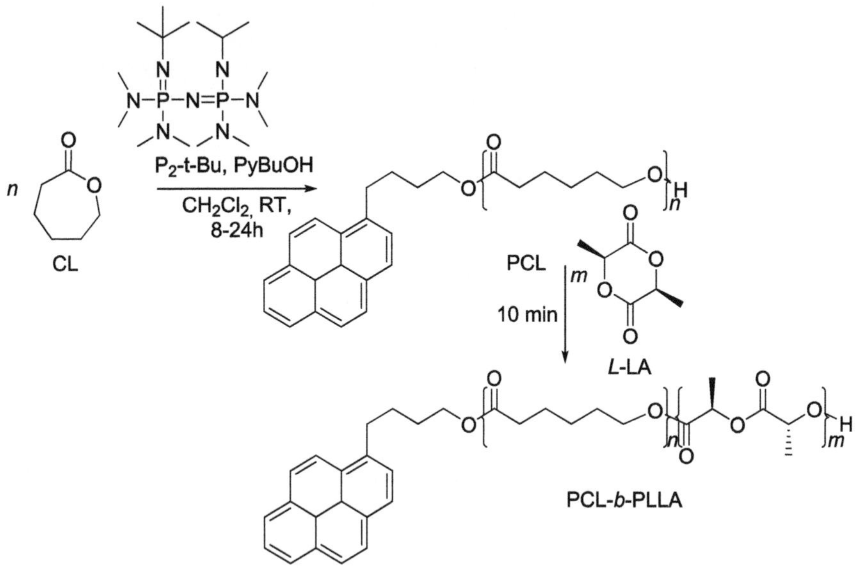

Scheme 6.19 P_2-*t*-Bu/BnOH mediated OROP of CL and the sequential ROP of L-LA.[90]

6.7 Polymerization of LA Catalyzed by Mono- or Multicomponent Dual Catalytic Systems

In the OROP of cyclic esters, the monomer and the initiator/polymer chain end can be concomitantly activated by implementation of an electrophile and a nucleophile within a bimolecular catalytic system composed by the association of a weak Brønsted acid A and a weak Brønsted base B resulting, in the formation of hydrogen-bond (weak electrophilic interactions) A-monomer and B-initiator/polymer chain end complexes, respectively. A unimolecular AB compound with ambiphilic features can also act as a "two in one" dual catalytic system (Scheme 6.20). Amines, amidines and phospohoric acids can be used in bifunctional systems.[29–31]

6.7.1 Polymerization of LA Catalyzed by Monocomponent Dual Catalytic Systems

The living and controlled OROP of LA catalyzed by hydroxyalkylated organic bases featuring the initiator and the catalyst within one molecule, namely 6-(2-hydroxyethyl)-aminopurine or 3-[(4,5-dihydro-1*H*-imidazol-2-yl)amino]-propanol, afforded PLA end-capped with the organic base ($M_{n,SEC} = 4900$–$56\,500$ g mol^{-1}; $Đ_M = 1.07$–1.63).[92]

Cinchona alkaloids such as β-isocupreidine (ICD) consisting of both a chiral nucleophilic amine catalyst site and an electrophilic hydroxy moiety,

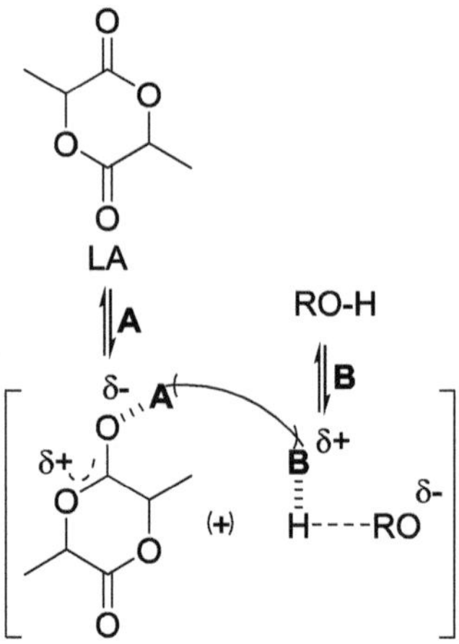

Scheme 6.20 AB mono- and [A + B] bicomponent dual activation systems used in the OROP of LA.

behaved as bifunctional stereoselective H-bonding donor and acceptor organocatalysts for the OROP of L-LA and *rac*-LA (CH_2Cl_2, 25 °C).[93] The OROP of L-LA promoted by ICD proceeded slowly with a poor control of the molar mass due to a sluggish initiation, yet without noticeable epimerization thus affording *isotactic* PLLA. Addition of BnOH as an external protic initiator improved the efficiency (high L-LA conversion, no undesired transesterification reactions) ($M_{n,SEC} = 14\,400–43\,200$ g mol^{-1}; $Đ_M = 1.12–1.43$). More significantly, the OROP of *rac*-LA mediated by ICD/BnOH afforded crystalline *isotactic*-rich stereo-gradient PLA as a result of a partial kinetic resolution polymerization, which preferentially polymerized L-LA and kinetically resolved D-LA (*ee* up to 71.8%; P_m up to 0.75; $M_{n,SEC} = 2400–21\,100$ g mol^{-1}; $Đ_M = 1.11–1.32$). This was the first reported kinetic resolution polymerization of *rac*-LA by an organocatalyst.

6.7.2 Polymerization of LA Catalyzed by Multicomponent Dual Catalytic Systems

An imidazolium trifluoroacetate salt, formed through a simple acid–base reaction between an imidazole base and one equivalent of TFA, enabled the controlled OROP of D-LA or L-LA in the presence of BnOH as initiator (bulk, 140 °C) to afford poly(D-LA) ($M_{n,SEC} = 8900$ g mol^{-1}; $Đ_M = 1.13$) and PLLA ($M_{n,SEC} = 3400–8700$ g mol^{-1}; $Đ_M = 1.14–1.78$), respectively. The polymerization was shown to proceed through a bifunctional mechanism with the imidazolium electrophilically activating the monomer carbonyl through H-bonding, while the trifluorocarboxylate nucleophilically activates hydroxyl initiating/propagating groups (Scheme 6.21).[94]

The ionic nucleophilic character of ammonium betaine catalysts such as *meta*-(trimethylammonio)phenolate betaine, allows the controlled OROP of

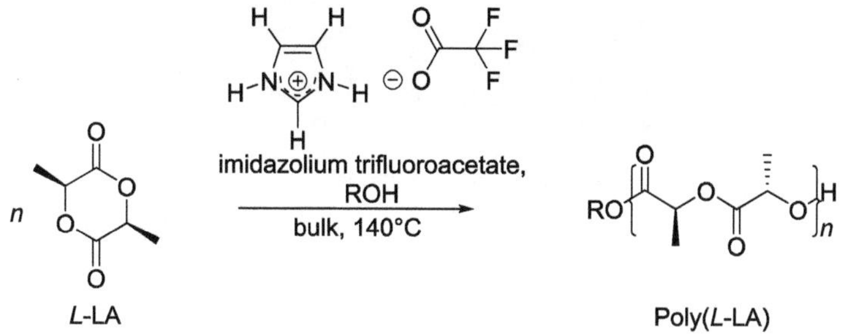

Scheme 6.21 OROP of L-LA catalyzed by an imidazolium-salt in the presence of ROH as initiator.[94]

L-LA ($CHCl_3$, 21 °C) using 1-pyrenemethanol or PEG–OH as a(n) (macro)-initiator to afford PLLA or PEG-*b*-PLLA (co)polymers, respectively ($M_{n,SEC}$ = 6400–97 000 g mol^{-1}; $Đ_M$ = 1.04–1.18).[95] The *meta* isomer was shown to induce a more pronounced ionic nucleophilic activation of the initiating and propagating alcohol site, leading to a and well-controlled polymerization (TOF = 37 h^{-1}).

The Lewis pair 1,4-diazabicyclo [2.2.2]octane (DABCO = triethylenediamine)/$B(C_6F_5)_3$ was reported to catalyze the quantitative epimerization of *meso*-LA into *rac*-LA, which was subsequently polymerized in the presence of BnOH into *isotactic* PLLA ($M_{n,SEC}$ = 6400–11 300 g mol^{-1}; $Đ_M$ = 1.10–1.16) and optically resolved D-LA, with a high stereoselectivity factor of 53 and an *ee* value of 91% at 51% monomer conversion (CH_2Cl_2 or toluene, 25 °C).[96] The epimerization and kinetic resolution (*i.e.* enantioselective) polymerization could also be coupled into a one-pot process, allowing to directly generate *isotactic* PLLA/D-LA from *meso*-LA (P_m = 0.94, at −75 °C).

Ionic H-bonding catalysis, especially in polymerization, has hardly been explored. Guanidinium hexahydro-2*H*-pyrimido[1,2-*a*]pyrimidin-1-ium tetrafluoroborate ($(HppH_2)^+BF_4^-$) and tertiary amine (−)-sparteine, cooperatively played the role of ionic H-bond donor and H-bond acceptor binary catalysts, respectively, to efficiently promote the OROP of *rac*-LA in the presence of an amine (macro)initiator (CH_2Cl_2, RT), towards the formation of PLA ($M_{n,NMR}$ = 2900–17 100 g mol^{-1}; $Đ_M$ = 1.09–1.23) and amphiphilic poly(sarcosine)-*b*-PLA diblock copolymer ($M_{n,NMR}$ = 5700–14 800 g mol^{-1}; $Đ_M$ = 1.16–1.21) (Scheme 6.22).[97] This same binary catalysts in the presence of BnOH as initiator, enabled the OROP of L-LA (CH_2Cl_2, RT) with fair activity (TOF = 109 h^{-1}) affording well-defined PLLA ($M_{n,NMR}$ = 2600–17 900 g mol^{-1}; $Đ_M$ = 1.12–1.29).[98] Among the various commercially available tertiary amines evaluated (*N*,*N*-dimethylcyclohexylamine ($NCyMe_2$), *N*,*N*,-*N*′,*N*″,*N*″-pentamethyl diethylenetriamine (PMDETA), DMAP, DABCO, *N*,*N*-diisopropylethylamine, *N*,*N*,*N*′,*N*′-tetramethylethylenediamine (TMEDA)), (−)-sparteine, revealed as the optimal base to be combined to $(HppH_2)^+BF_4^-$. The controlled/living nature of the OROP next enabled the successful synthesis of PTMC-*b*-PLLA ($M_{n,NMR}$ = 7200 g mol^{-1}; $Đ_M$ = 1.05) copolymers highlighting that such ternary catalysis system is generally applicable.

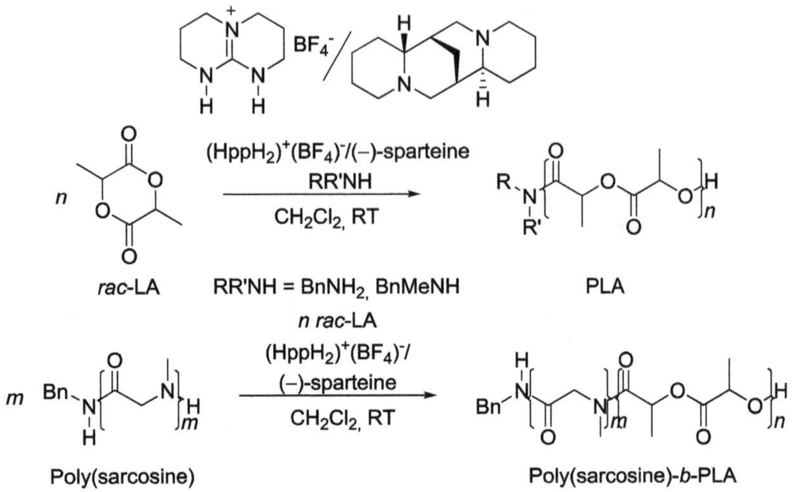

Scheme 6.22 OROP of *rac*-LA catalyzed by $(HppH_2)^+BF_4^-$/(−)-sparteine in the presence of an amine (macro)initiator.[97]

Also, a dibenzyl-functionalized bispidine has been demonstrated to be an excellent catalyst (yet with a lower rate than (−)-sparteine but with a faster rate than tris[2-(dimethylamino)ethyl]amine (Me_6TREN)) in combination with 1-(3,5-bis(trifuoromethyl)phenyl)-3-cyclohexyl TU as a co-catalyst ($CDCl_3$, RT) for the controlled OROP of L-LA in the presence of 1-phenylethanol as initiator (negligible transesterification or epimerization reactions) affording highly crystalline PLLA ($M_{n,SEC} = 3300–61\,600$ g mol^{-1}; $Đ_M = 1.03–1.18$).[99] The OROP of *rac*-LA mediated by this same bispidine/TU/1-phenylethanol catalytic system preferentially formed *isotactic* PLA ($P_m = 0.74$), a modest degree of stereocontrol similar to that obtained from the OROP of *rac*-LA mediated by the alike (−)-sparteine/TU catalyst system ($P_m = 0.74–0.77$).[99,100] Screening of a range of alternative hydrogen bond donor co-catalysts revealed that 1-(3,5-bis(trifuoromethyl)phenyl)-3-cyclohexyl TU in combination with the dibenzyl-functionalized bispidine enabled the highest polymerization rates (TOF = 39 h^{-1}).

Following the identification of (−)-sparteine as an effective organocatalyst for the controlled OROP of L-LA into well-defined PLLA,[37] and given its worldwide shortage, efforts were next devoted to identify possible commercially available cost-effective alternative molecules. Thus, among TMEDA, PMDETA, Me_6TREN, DABCO, triethylamine (TEA), and 1,4,7-trimethyl-1,4,7-triazacyclononane (TACN) amine catalysts, evaluated in the OROP of L-LA in combination with 1-(3,5-bis(trifluoromethyl)phenyl)-3-cyclohexyl TU in the presence of PyBuOH as an initiator (CH_2Cl_2, RT), Me_6TREN displayed the most similar performances to (−)-sparteine with respect to the polymerization rate, low dispersity, and the absence of any observable transesterification reactions ($M_{n,SEC} = 23\,000$ g mol^{-1}; $Đ_M = 1.07$) (Scheme 6.23).[100]

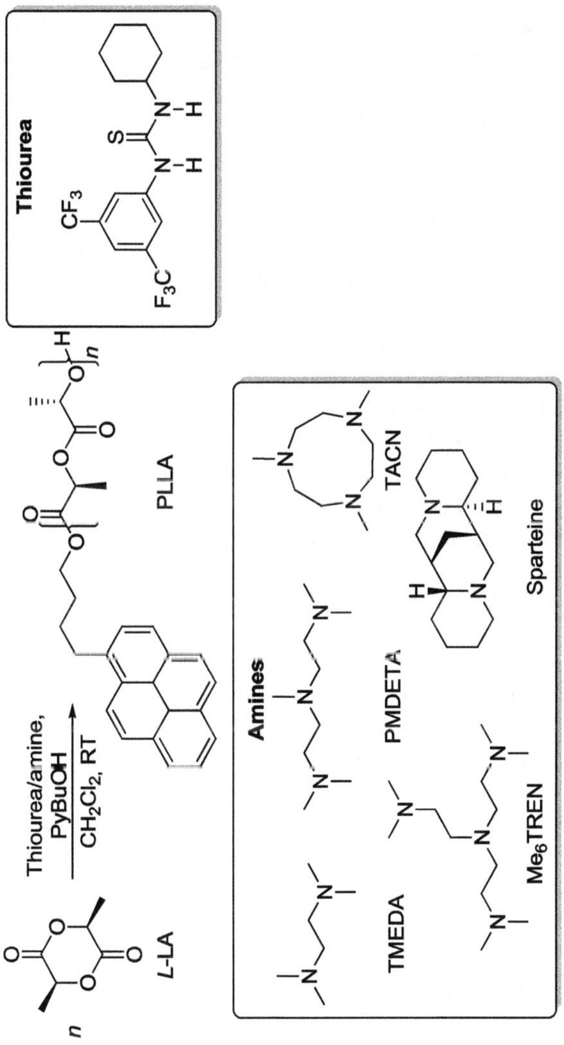

Scheme 6.23 OROP of L-LA catalyzed by TU/amine in the presence of PyBuOH.[100]

Similarly, using a tricomponent organocatalyst composed of 1-(3,5-bis(trifluoromethyl)phenyl)-3-cyclohexyl TU and various tertiary amine cocatalysts ($NCyMe_2$, PMDETA, 1,1,4,7,10,10-hexamethyltriethylenetetramine (HMTETA), Me_6TREN, or DBU), in the presence of BnOH or CO_2-phobic hydroxyl end-capped macroinitiators (PEG–OH, PCL diol, poly(butylene succinate) (PBS) diol, or poly(phosphoester) (PPE) diol), well-defined PLA homopolymers ($M_{n,SEC} = 1800–44\,000$ g mol^{-1}; $Đ_M = 1.06–1.25$) and PLA-based block copolymers were synthesized (PEG-*b*-PLA, PLA-*b*-PCL-*b*-PLA, PLA-*b*-PBS-*b*-PLA, PLA-*b*-PPE-*b*-PLA; $M_{n,SEC} = 4000–21\,500$ g mol^{-1}; $Đ_M = 1.06–1.45$) by a heterogeneous solvent and metal-free green OROP of *rac*-LA or L-LA in supercritical carbon dioxide ($scCO_2$, 300 bar, 50–90 °C).[101] Control over the molar mass and dispersity was achieved although PLA and PLLA are insoluble in the polymerization medium. The reactivity order of the amino catalyst was determined in the order $Me_6TREN > HMTETA > PMDETA > NCyMe_2$.

Block copolymers of 5-methyl-5-allyloxycarbonyl-1,3-dioxan-2-one (MAC) and *rac*-LA were synthesized ($CDCl_3$, RT), either by a one-pot OROP of *rac*-LA from a hydroxy-terminated poly(MAC) (PMAC) prepolymer, or by OROP of MAC from a PLA-OH macroinitiator, similarly catalyzed by the dual 1-(3,5-bis(trifluoromethyl)phenyl)-3-cyclohexyl TU/(−)-sparteine system, and initiated with BnOH ($M_{n,SEC} = 6600–7100$ g mol^{-1}; $Đ_M = 1.17–1.28$).[102]

1-(3,5-bis(trifluoromethyl)phenyl)-3-cyclohexyl TU/amine bifunctional organocatalysts were also demonstrated as selective in the presence of PyBuOH initiator for the controlled solution OROP of *rac*-LA and L-LA into well-defined homopolymers and block copolymers ($CDCl_3$, 25 °C).[103,104] Both the H-bond donating TU and the H-bond accepting amine were shown to be necessary to reach high activity, while the TU and the amine do not have to be incorporated into a single-molecule catalyst; however, an electron withdrawing group next to the TU moiety was required to improve the hydrogen bonding ability to the substrate. Among the various tertiary amines studied (pyridine, *N*,*N*-dimethylaniline, Proton Sponge (1,8-bis(dimethylamino)-naphthalene), DMAP, (−)-sparteine, DABCO, TMEDA, PMDETA, Me_6TREN, or TACN), (−)-sparteine was found as the most effective to activate the alcohol of both initiating and propagating species, while maintaining a good control of macromolecular characteristics and of the stereochemistry of the resulting PLA ($M_{n,SEC} = 16\,700–20\,700$ g mol^{-1}; $Đ_M = 1.05–1.07$) and PLLA ($M_{n,SEC} = 12\,600–26\,900$ g mol^{-1}; $Đ_M = 1.04–1.14$). (−)-sparteine also provided a high stereocontrol of *rac*-LA polymerization (P_m up to 0.77) *via* a CEM. The use of several protic functional groups as initiators including alcohols, thiols, silanols, and macroinitiators (*e.g.* PEG–OH, PS–OH, PMMA–OH) enabled the preparation of a series of α-alkoxy-end capped PLAs as well as block copolymers, respectively.[103] The alkylamine bases were demonstrated to be weakly associated with a TU cocatalyst in solution, and the nature of cocatalyst interactions varied with the nature of the alkylamine.[104]

Similarly, achiral bis(1-(3,5-bis(trifluoromethyl)phenyl)-3)-propylthiourea (bisTU) H-bond donating organocatalyst revealed highly effective (TOF up to 890 h^{-1}), at fractional percent catalyst loadings, in the presence of an alkylamine cocatalyst and BnOH, in the living and controlled OROP of L-LA.[105] The rate acceleration of the bisTU/alkylamine system *vs.* monoTU, regardless of the alkylamine cocatalyst (TMEDA, PMDETA, Me_6TREN, TEA, and TACN), was postulated to arise from the activation at a single monomer ester by both TU moieties. Well-defined PLLAs were thus obtained ($M_{n,SEC} = 9200–32\,200$ g mol^{-1}; $Đ_M = 1.02–1.11$).

Contrary to other lactones, TUs were found to be more superior catalysts in terms of both activity and control than their urea analogues in the solvent-free H-bond mediated OROP of L-LA (100 °C). In particular, the same bisTU plus PMDETA, in the presence of BnOH initiator, was found as the most active (TOF = 544 h^{-1}) and controlled organic co-catalyst system affording PLLA ($M_{n,SEC} = 10\,700$ g mol^{-1}; $Đ_M = 1.06$) with the highest *isotacticity* ($P_m = 0.94$) (Scheme 6.24).[106] In comparison, the solvent-free TBD/BnOH OROP of L-LA under the same conditions was much less effective (TOF = 11 h^{-1}) and gave PLLA ($M_{n,SEC} = 19\,900$ g mol^{-1}; $Đ_M = 1.30$) with a significantly lower *isotacticity* ($P_m = 0.78$). One-pot block copolymerization of L-LA and VL, which was previously shown inaccessible in solution phase OROP, was also achieved from this same ternary catalytic system under these one-pot reaction conditions ($M_{n,SEC} = 27\,600$ g mol^{-1}; $Đ_M = 1.57$).

Remarkably, the first examples of chirality transfer from a catalyst to a polymer was recently reported in the OROP of *rac*-LA from the combination of the diastereomeric *N*-protected densely substituted amino acids (2*S*,3*R*,4*S*,5*S*)-1-methyl-4-nitro-3,5-diphenylpyrrolidine-2-carboxylic acid (*endo* configuration) and (2*S*,3*S*,4*R*,5*S*)-1-methyl-4-nitro-3,5-diphenylpyrrolidine-2-carboxylic acid (*exo* configuration) with DBU as a cocatalyst and BnOH as initiator (CH_2Cl_2, RT) (Scheme 6.25).[107] Both diastereoisomers provided *isotactic* enriched PLA ($M_{n,SEC} = 1900–23\,000$ g mol^{-1}; $Đ_M = 1.1–1.5$) with a probability of *meso* linkage between monomer units P_m up to 0.96, and these prolines were also able to preferentially promote the polymerization of one of the isomers (L or D) with respect to the other. Thus, the *exo* proline preferentially polymerized L-LA as opposed to the *endo* one, which preferred the D-LA substrate from a *racemic* mixture of L and D stereoisomers. In comparison, the Boc-L-proline/DBU complex showed a slight *isotactic* selectivity ($P_m = 0.65$), while DBU alone did not exhibit any significant stereocontrol affording an essentially *atactic* PLA ($P_m = 0.50$). These densely substituted amino acids were thus highly efficient organic catalysts enabling the stereocontrolled OROP of *rac*-LA. Density functional theory calculations enabled to assess the origin of this unique stereocontrol in the OROP of LA, explaining why the chirality of the catalyst was able to define the stereochemistry of the monomer insertion.

The dual catalyst system made of two H-bonding components, a phenol derivative (*o*-CF_3-phenol, 4-*tert*-butyl-catechol, pyrogallol or resorcinol) to activate *rac*-LA, and DBU to enhance the nucleophilicity of the propan-1-ol initiator and the propagating chain end, in the presence of an alcohol, enabled the preparation of block copolyesters PVL-*b*-PLA and PCL-*b*-PLA (CH_2CL_2, 20 °C) ($M_{n,SEC} = 3900–6100$ g mol^{-1}; $Đ_M = 1.22–1.48$).[108]

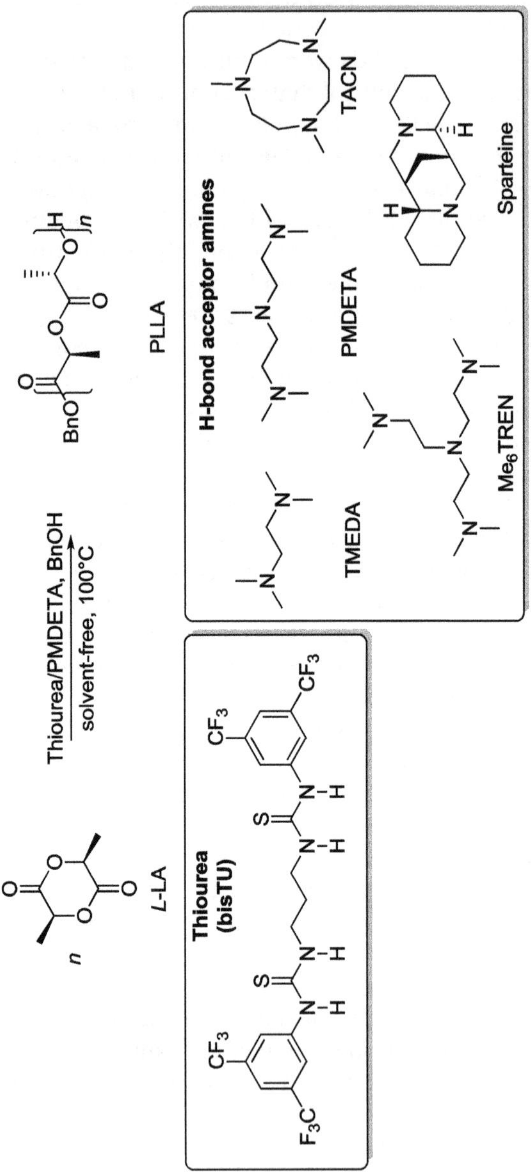

Scheme 6.24 Solvent-free OROP of L-LA catalyzed by TU/amine in the presence of an alcohol initiator.[106]

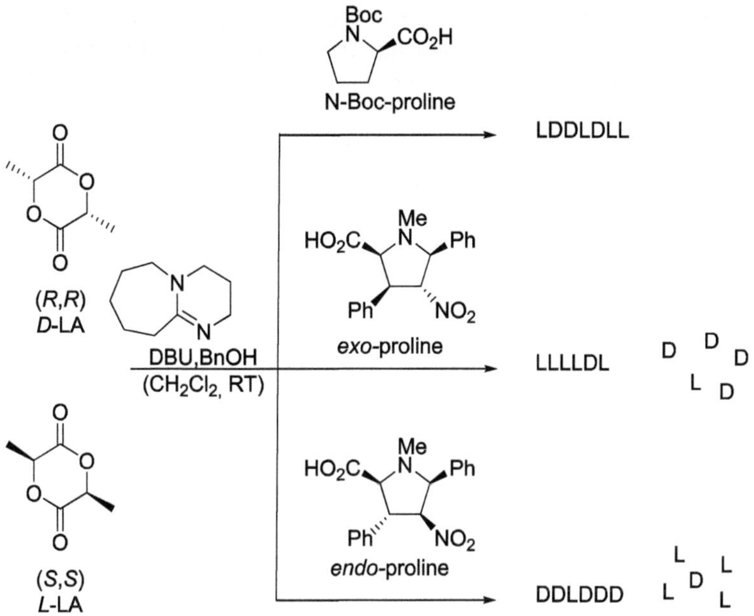

Scheme 6.25 Schematic representation of *rac*-LA OROP in the presence of three different N-protected prolines showing higher preference toward one of the L-LA or D-LA isomers.[107]

L-LA was also successfully ring-open polymerized from DBU associated with benzoic acid using BnOH as initiator (CH_2CL_2, RT) with good control ($M_{n,SEC}$ = 6800–17 700 g mol^{-1}; $Đ_M$ = 1.06–1.50).[109] Corroboration of the experimental results with computational investigations supported a bifunctional catalytic mechanism wherein both the monomer and the propagating nucleophilic hydroxyl group are activated by the amidinium carboxylate catalyst *via* hydrogen bonding.

The OROP of L-LA using the dual DBU/1-(3,5-bis(trifuoromethyl)phenyl)-3-cyclohexyl TU catalyst system in the presence of BnOH as initiator ($CDCl_3$, RT) afforded α-benzyloxy,ω-hydroxy telechelic PLLA oligomers ($M_{n,NMR}$ = 2700 g mol^{-1}; $Đ_M$ = 1.09). The subsequent ROcP of L-LA with 5-methyl-5-propargyloxycarbonyl-1,3-dioxan-2-one (MPC) similarly gave BnO–PLLA-*b*-poly(MPC)–OH copolymers ($M_{n,NMR}$ = 6700 g mol^{-1}; $Đ_M$ = 1.08).[110] This same DBU/TU catalytic system, combined to MeO–PEG–OH as macroinitiator, enabled the random ROcP of the benzyl-protected dihydroxylated bicyclic carbonate, namely, 9-phenyl-2,4,8,10-tetraoxaspiro[5,5]undecan-3-one (PTO) with *rac*-LA (CH_2Cl_2, RT), to afford PEG-*b*-P(PTO-*co*-DL-LA) terpolymers ($M_{n,NMR}$ = 9600 g mol^{-1}; $Đ_M$ = 1.07).[111] The similar OROP of *rac*-LA from PEG-*b*-PPTO (PPTO = poly(PTO)) macroinitiator also using DBU/TU then gave the triblock copolymer PEG-*b*-PPTO-*b*-PLA ($M_{n,NMR}$ = 9000 g mol^{-1}; $Đ_M$ = 1.09).[111] Finally, the catalyst systems MTBD/TU and DBU/TU showed a high selectivity in the copolymerization of CL, VL, and LA, with LA as the fastest propagating monomer ($k_{LA}>>k_{VL}>k_{CL}$) being ring-opened first (>95% conversion) before ring opening of the second monomer.[58]

The acid-base catalytic system based on DMAP and a protic acid, efficient at RT in solution to provide PLLA oligomers (CH_2Cl_2, RT; $M_{n,NMR}=900$–3000 g mol^{-1}; $Đ_M=1.06$–1.17),[112] was also demonstrated to efficiently catalyze the OROP of L-LA in bulk at 100 °C. Thus, DMAP with DMAP.HCl, DMAP.CH_3SO_3H (methanesulfonic acid, MSA), or DMAP.TfOH, in the presence of BnOH, *p*-phenyl benzyl alcohol, trimethylolpropane, or pentaerythritol as an initiator, afforded linear ($M_{n,NMR}=2900$–24 700 g mol^{-1}; $Đ_M=1.09$–1.25) and three- or four-arm star PLLA ($M_{n,NMR}=14\,800$–136 700 g mol^{-1}; $Đ_M=1.10$–1.16) in less than 1 h. PCL-*b*-PLLA ($M_{n,NMR}=$ 12 800 g mol^{-1}; $Đ_M=1.36$) and PVL-*b*-PLLA ($M_{n,NMR}=14\,300$ g mol^{-1}; $Đ_M=1.16$) block copolyesters were also synthesized by the sequential polymerization of the lactone followed by L-LA.[113]

Amine-functionalized squaramides were recently shown to catalyze the controlled OROP of L-LA in the presence of an alcohol initiator (BnOH) (CH_2Cl_2, RT) to give PLLA ($M_{n,SEC}=800$–14 600 g mol^{-1}; $Đ_M=1.02$–1.16) (Scheme 6.26).[114] The squaramides behaved as bifunctional hydrogen-bonding catalysts, activating both the monomer and the initiator. The weaker H-bond donor aminosquaramide without electron-withdrawing NH substituents, was found to be less effective. In comparison, related monofunctional squaramides/BnOH, inactive in the OROP of L-LA, were revealed to be active in the presence of NEt_3 as an external H-bond acceptor. A cooperative dual activation with an AMM/ACEM was then suggested. PLLA oligomers ($M_{n,NMR}$ up to 14 700 g mol^{-1}; $Đ_M=1.04$–1.16) were similarly obtained from alike tertiary aminosquaramides in the presence of PyBuOH (dichloroethane, 50 °C) (Scheme 6.26).[115] Also, multiple combinations of other various squaramides and amines (DBU revealed as the most efficient) co-catalysts, associated with a series of alcohol initiators, enabled the preparation (CH_2Cl_2, 25 °C) of PLLA ($M_{n,NMR}=4100$–16 250 g mol^{-1}; $Đ_M=1.05$–1.09) and PTMC-*b*-PLA ($M_{n,NMR}=7500$ g mol^{-1}; $Đ_M=1.10$) with high end-group fidelity (Scheme 6.26).[116,117] The squaramide-based OROP of L-LA however generally remained not so efficient.[114–117]

The ROcP of L-LA and TMC was also investigated using an acid-base ternary catalytic system made of the MTBD guanidine and the Brønsted acid TFA, with BnOH used as the initiator (CH_2Cl_2, RT). A PTMC-*b*-PLLA diblock copolymer was thus obtained by sequential polymerization of TMC followed by L-LA ($M_{n,SEC}=5100$–13 000 g mol^{-1}; $Đ_M=1.09$–1.12).[118]

Well-defined PLAs ($M_n=2700$–13 800 g mol^{-1}; $Đ_M=1.05$–1.10) were prepared from the OROP of *rac*-LA using DMAP in combination with acid-base salts such as (*R*)-(+)-binaphthyl-diyl hydrogen phosphate (BNPH)/DBU or (1*R*)-(−)-10-camphorsulfonic acid (CSA)/DMAP as catalytic systems, in the presence of BnOH as initiator (CH_2Cl_2, 35 °C) (Scheme 6.27).[119]

The combination of DMAP with diphenyl phosphate (DPP) catalysts in the presence of PPA as initiator promoted the OROP of L-LA (CH_2CL_2, RT)

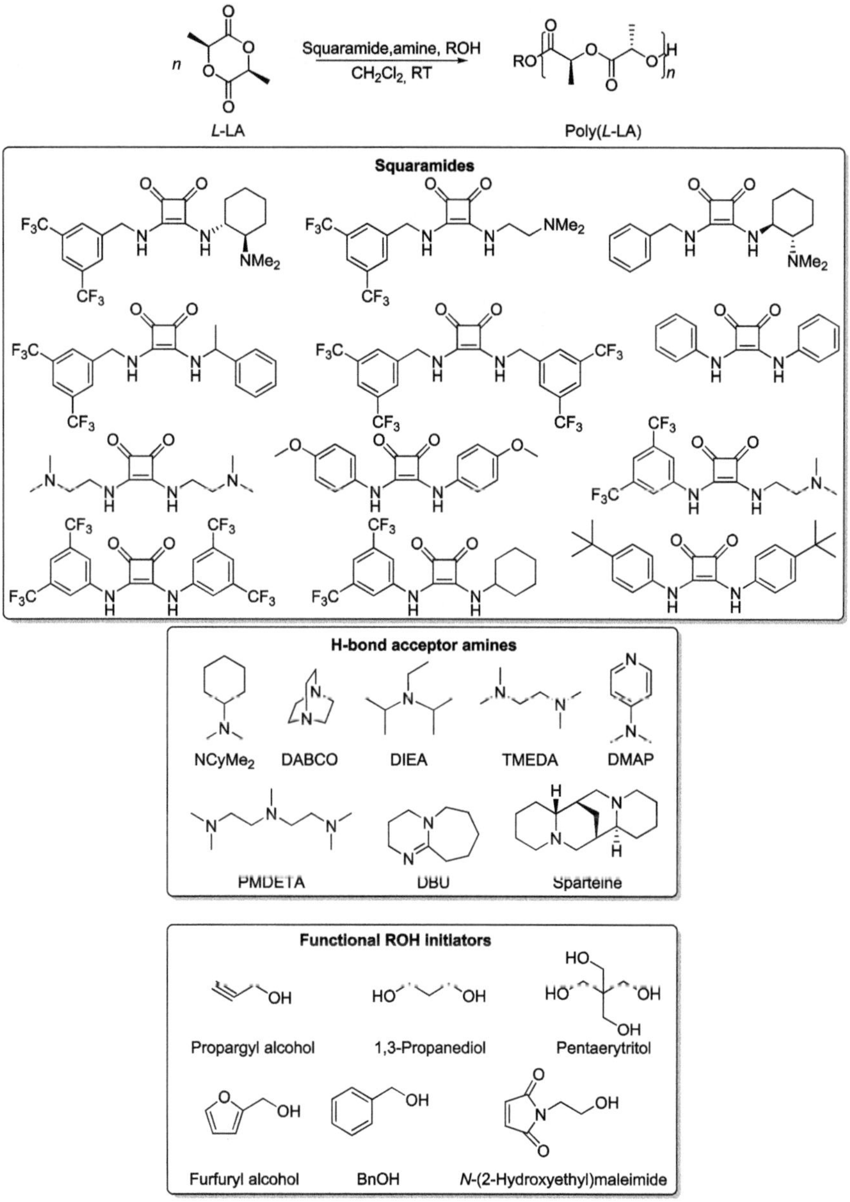

Scheme 6.26 OROP of L-LA mediated by various squaramide/amine/ROH systems.[114–117]

($M_{n,NMR}$ = 5000–19 200 g mol^{-1}; $Đ_M$ = 1.09–1.24), as well as the sequential ROcP with VL, CL, TMC or 1,5-dioxepan-2-one into the corresponding diblock copolyesters (M_n *ca.* 12 000 g mol^{-1}; $Đ_M$ = 1.08–1.16). The bifunctional activation from DPP/DMAP of L-LA and its polymer chain end was evidenced.

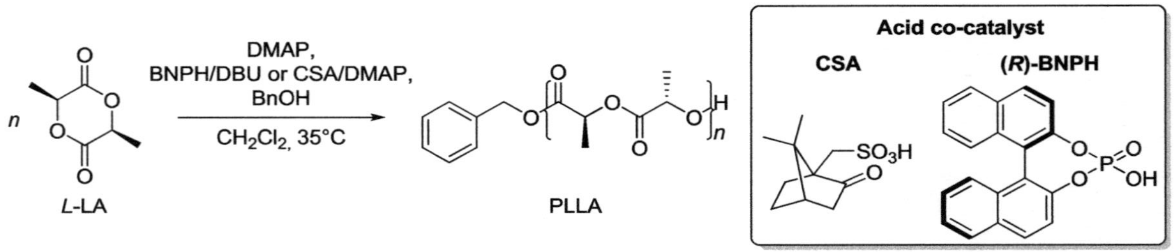

Scheme 6.27 OROP of *rac*-LA using DMAP combined with acid–base salts as catalytic systems in the presence of BnOH as initiator.[119]

Various end-functionalized PLLAs were similarly prepared using functional alcohols (6-azido-1-hexanol, propargyl alcohol, 2,3,4,5,6,-pentafluorobenzyl alcohol, 2-hydroxyethyl methacrylate, or 4-vinylbenzyl alcohol).[120]

A traceless switch organocatalysis swapping from a Brønsted acid cationic mechanism to a Brønsted base/conjugate-acid bifunctional mechanism, simply by addition of a Brønsted base to the Brønsted acid, was proposed for the OROP of L-LA.[121] ROcP of L-LA with VL, CL, and/or TMC, took place through an acid plus a base, *i.e.* "one plus one", approach in one pot (CH_2Cl_2, RT), using a MSA/DBU or a DPP/DBU acid/base pair catalysis (Scheme 6.28). While MSA sluggishly catalyzed the OROP of L-LA, addition of DBU enabled to switch the system into DBU/DBUH$^+$, a base/baseH$^+$ pair catalyst, which polymerized L-LA much faster, and enabled its ROcP with CL and/or TMC from the chain-end of the prior block, giving PCL-*b*-PLLA ($M_{n,NMR} = 13\,100$ g mol^{-1}; $Đ_M = 1.06$) and PCL-*b*-PTMC-*b*-PLLA ($M_{n,NMR} = 15\,700$ g mol^{-1}; $Đ_M = 1.37$) copolymers, respectively. Likewise, "DPP plus DBU" switched DPP into DPP/2DBU catalyzed bifunctional OROP of L-LA, giving PVL-*b*-PLLA ($M_{n,NMR} = 12\,300$ g mol^{-1}; $Đ_M = 1.09$), and PVL-*b*-PTMC-*b*-PLLA ($M_{n,NMR} = 142\,700$ g mol^{-1}; $Đ_M = 1.03$) (Scheme 6.28).

Bifunctional moisture-stable and non-hygroscopic triaryl iminophosphorane catalysts featuring a TU moiety have been found highly active in the OROP of L-LA using PyBuOH as initiator (CH_2Cl_2, RT) (Scheme 6.29).[122] PLLA within short reaction times, with excellent monomer conversion (TOF = 594 h^{-1}), low dispersity and high end-group fidelity were thus prepared ($M_{n,SEC} = 6400$–$32\,900$ g mol^{-1}; $Đ_M$ *ca.* 1.05). PVL-*b*-PLLA ($M_{n,SEC} = 7500$ g mol^{-1}; $Đ_M = 1.13$) and PCL-*b*-PLLA ($M_{n,SEC} = 5700$ g mol^{-1}; $Đ_M = 1.16$) diblock copolymers were also similarly obtained by sequential addition of VL or CL and then L-LA, while the use of MeO–PEG–OH as a macroinitiator led to PEG-*b*-PLLA copolymers ($M_{n,SEC} = 16\,800$ g mol^{-1}; $Đ_M = 1.07$). Notably, mono-functional iminophosphoranes without a TU moiety did not enable the OROP thus supporting the necessity to introduce both the imminophosphorane and the TU catalysts for the monomer activation. Also, significant epimerization of L-LA was observed when using an iminophosphorane featuring an *N*-alkyl substituent rather than a linked H-bond donor (1,1,1-tris(4-methoxyphenyl)-*N*-phenethyl-λ^5-phosphanimine)and the resulting polymer showed a P_m of 0.66 – 0.68.

Some protonated phosphazenium $[(NR_2)_2(NR'_2)P{=}NHR'']^+(X)^-$ salts (ITPP-H$^+$·X$^-$) behaved as H-bond donor organocatalysts in the presence of a tertiary amine basic co-catalyst such as sparteine or DBU, to promote the OROP of *rac*-LA (CH_2Cl_2, 20 °C) in the presence of 4-biphenylmethanol as initiator ($M_{n,SEC} = 1000$–3200 g mol^{-1}; $Đ_M = 1.08$–1.25).[123] Only imino-tri(pyrrolidino)phosphonium derivatives with X being bis(trifluoromethane)-sulfonimide (Tf_2N), or tetrakis[3,5-bis(trifluoromethyl)-phenyl]borate (BArF), were found to be as efficient as the standard TU catalyst, due to the availability of a double H-bonding $(R_2N)_3P{=}NH_2^+$ moiety, without any steric hindrance.

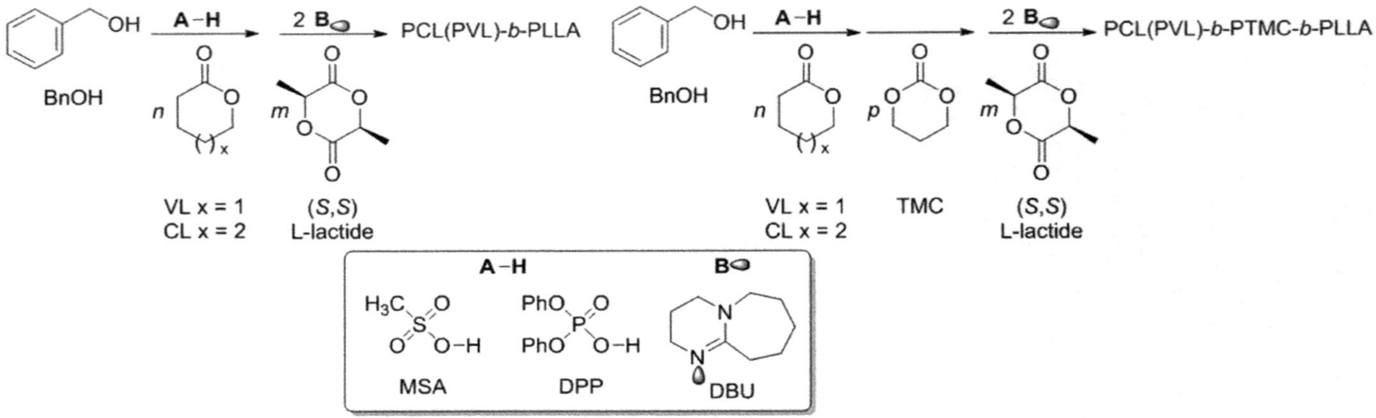

Scheme 6.28 Synthesis of PCL(PVL)-*b*-PLLA copolymers (left) and PCL(PVL)-*b*-PTMC-*b*-PLLA terpolymers (right) through one-pot OROP using the "traceless switch" organocatalysis strategy.[121]

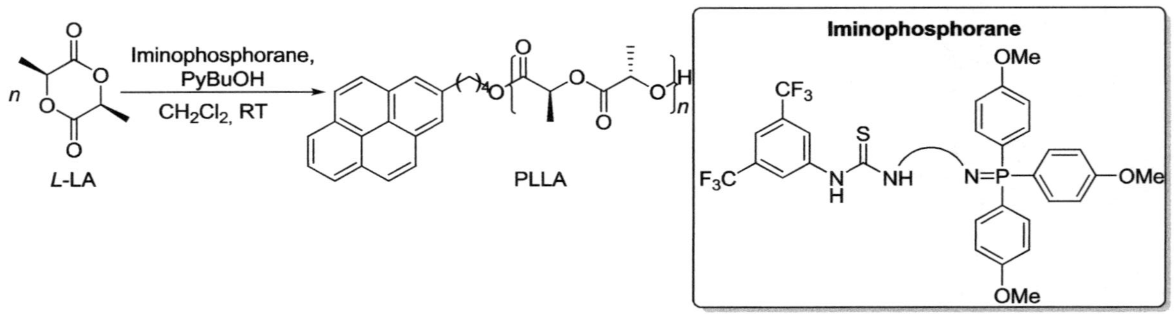

Scheme 6.29 OROP of L-LA catalyzed by iminophosphoranes in the presence of PyBuOH.[122]

6.8 Conclusion

Organocatalysis has now become an important tool in polymer science because of its versatility and specificity. Significant progress has been achieved in the OROP of LA – and, although to a much lesser extent, of the related functionalized OCAs – in terms of catalysts accessibility, robustness and activity as well as in the understanding of their mode of action. A range of organocatalyst acids and bases have thus been successfully established.

The most efficient organocatalysts unveiled in the OROP of LA essentially involve nitrogen-containing bases (which, actually, have been the most investigated as compared to phosphorous containing bases). Among these, the readily commercially available TBD guanidine revealed the most active enabling the OROP of L-LA with only 0.1 mol% of TBD relative to monomer and 1 mol% of PyBuOH initiator (targeted degree of polymerization = 500; conversion of L-LA = 95%) to afford well-defined PLLA ($M_{n,SEC}$ up to 62 600 g mol^{-1}; $Đ_M = 1.11$) in 1 min at RT (CH_2Cl_2). This enhanced activity (TOF up to 475 min^{-1}), most likely arising from its bifunctionality enabling the simultaneous activation of both the cyclic ester monomer and the alcohol group of the initiator/propagating species, competes well with that of the most active metal catalysts.[61] Following TBD and MTBD, the NHC enantiomeric catalyst (+)/(−)1-methyl-3-menthoxymethyl imidazol-2-ylidene also showed a high activity (TOF = 5839 h^{-1}; TBD > MTBD ≫ NHC) on its own (*i.e.* without any alcohol initiator), cnabling to polymcrizc 973 cquiv. of L-LA at 0.1 mol% of NHC relative to monomer in 10 min (THF, 25 °C), into cyclic PLAs ($M_{n,SEC}$ = 17 600 g mol^{-1}; $Đ_M = 1.43$).[45]

Given that PLA features stereogenic centers along its backbone chain, their relative stereochemistry influences the physical properties of the polymer. Indeed, crystalline polymers, which often display mechanical properties superior to those of the corresponding non-stereoregular polymers, necessarily feature some stereoregularity. Whereas the stereoselective polymerization of the chiral LA monomer has been successfully achieved from a variety of metal catalysts, stereoselectivity in the OROP of LA is critically important. To date, a limited number of organic catalysts such as DABCO/B(C_6F_5)$_3$ ($P_m = 0.94$ at −75 °C),[96] IMes NHC ($P_m = 0.75$ at −25 °C),[75] the chiral (*R*,*R*-) and achiral (*rac*-)NHCs ($P_m = 0.90$ at −70 °C),[76] phosphazene (P_2-*t*-Bu: $P_m = 0.95$ at −75 °C),[89] or (*R*)-binaphthol-derived phosphoric acids (*ee* = 80.6% at +75 °C),[50] have recently demonstrated the ability, in the presence of an alcohol, to promote the stereocontrolled polymerization of *rac*-LA, some of them being even stereoselective at RT, such as substituted amino acids ($P_m = 0.96$),[107] BEMP ($P_m = 0.70$),[88] P_2-*t*-Bu ($P_m = 0.72$),[89] dibenzyl bispidine/TU (P_m of 0.74), or[99] TU/(−)-sparteine (P_m up to 0.77).[103] The stereocontrol is most often shown to be imparted by the CEM. The identification of a wide organic catalyst family, easily synthetically available/accessible, able to control the monomer insertion (*i.e.* D-LA or L-LA from a *racemic* mixture) as a function of the chirality of the catalyst, while highly active, still remains a topical challenge. On the other hand, the stereo-controlled/stereoselective OROP of OCAs has not been reported.

Finally, functionality along the polyester backbone has been successfully introduced trough the OROP of OCAs. Basically, only DMAP and symmetrical imidazol-2-ylidenes and imidazolin-2-ylidenes NHCs have been used concomitantly in the presence of an alcohol/polyol initiator.[82,84–86] This domain thus remains relatively unexplored.

Abbreviations

ACEM	activated chain-end mechanism
AMM	activated monomer mechanism
BEMP	2-*tert*-butylimino-2-diethylamino-1,3-dimethylperhydro-1,3,2-diazaphosphorine
BINOL	binaphthol
BL	β-butyrolactone
BMIM-2-CO_2	1-*n*-butyl-3-methylimidazolium-2-carboxylate
BnOH	benzyl alcohol
BNPH	1,1′-binaphthyl-2,2′-diyl hydrogen phosphate
BYP	butynyl phospholane
CEM	chain-end mechanism
CL	ε-caprolactone
CSA	(1*R*)-(−)-10-camphorsulfonic acid
DABCO	1,4-diazabicyclo [2.2.2]octane = triethylenediamine
DBN	1,5-diazabicyclo[4.3.0]non-5-ene
DBU	1,8-diazabicyclo[5.4.0] undec-7-ene
DL	δ-decalactone
D-LA	D-lactide
$Đ_M$	dispersity
DMAP	4-(dimethylamino)pyridine
DPP	diphenyl phosphate
ee	enantiomeric excess
HMTETA	1,1,4,7,10,10-hexamethyltriethylenetetramine
$(HppH_2)^+$	hexahydro-2*H*-pyrimido[1,2-*a*]pyrimidin-1-ium
ICD	β-isocupreidine
IMes	1,3-dimesitylimidazol-2-ylidene
IMes.CO_2	1,3-bis(2,4,6-trimethylphenyl) imidazolium-2-carboxylate
IPr.CO_2	1,3-bis(2,6-diisopropylphenyl) imidazolium-2-carboxylate
IPr*.CO_2	1,3-bis(2,6-bis(diphenylmethyl)-4-methylphenyl) imidazolium-2-carboxylate
LA	lactide
lacOCA	*O*-carboxyanhydride derived from lactic acid
L-gluOCA	*O*-carboxyanhydride derived from glutamic acid
L-LA	L-lactide
L-manOCA	OCA of L-mandelic acid
MAC	5-methyl-5-allyloxycarbonyl-1,3-dioxan-2-one

Me_6TREN	tris[2-(dimethylamino)ethyl]amine
M_n	number average molar mass
$M_{n,NMR}$	number average molar mass as determined by NMR analysis
$M_{n,SEC}$	number average molar mass as determined by SEC analysis
MPC	5-methyl-5-propargyloxycarbonyl-1,3-dioxan-2-one
MSA	methanesulfonic acid, CH_3SO_3H
MTBD	7-methyl-1,5,7-triazabicyclo[4.4.0]dec-5-ene
$NCyMe_2$	*N,N*-dimethylcyclohexylamine
NHC	N-heterocyclic carbene
NMR	nuclear magnetic resonance
OCA	*O*-carboxy anhydride
OROP	organocatalyzed ring-opening polymerization
P_1-*t*-Bu	*N′-tert*-butyl-*N,N,N′,N′,N″,N″*-hexamethylphosphorimidic triamide
P_2-*t*-Bu	1-*tert*-butyl-2,2,4,4,4-pentakis(dimethylamino)-$2\Lambda^5$,$4\Lambda^5$-catenadi(phosphazene)
PBL	poly(β-butyrolactone) = poly(BL)
PBS	poly(butylene succinate)
PBYP	poly(butynyl phospholane) = poly(BYP)
PCL	poly(ε-caprolactone) = poly(CL)
PDL	poly(δ-decalactone) = poly(DL)
PEG	poly(ethylene glycol) = poly(EG)
PLA	poly(lactic acid)/poly(lactide)
PLLA	poly(L-lactide) = poly(L-LA)
P_m	probability of forming a new *isotactic* dyad
PMAC	poly(5-methyl-5-allyloxycarbonyl-1,3-dioxan-2-one) = poly(MAC)
PMDETA	*N,N,-N′,N″,N″*-pentamethyl diethylenetriamine
PMPC	poly(5-methyl-5-propargyloxycarbonyl-1,3-dioxan-2-one) = poly(MPC)
PPA	3-phenyl-1-propanol
PPE	poly(phosphoester)
PPTO	poly(9-phenyl-2,4,8,10-tetraoxaspiro[5,5]undecan-3-one) = poly(PTO)
PTMC	poly(trimethylene carbonate) = poly(TMC)
PTO	9-phenyl-2,4,8,10-tetraoxaspiro[5,5]undecan-3-one
PPY	4-pyrrolidinopyridine
PS	poly(styrene)
PVL	poly(δ-valerolactone) = poly(VL)
PyBuOH	1-pyrenebutanol
rac	*racemic*
ROcP	ring-opening copolymerization
ROP	ring-opening polymerization
RT	room temperature
SEC	size-exclusion chromatography
$Sn(Oct)_2$	tin(II) 2-ethylhexanoate

TACN	1,4,7-trimethyl-1,4,7-triazacyclononane
TBD	1,5,7-triazabicyclo[4.4.0]dec-5-ene
TEA	triethylamine
TFA	trifuoroacetic acid
Tf_2NH	bis(trifluoromethane)sulfonimide (triflimide)
TfOH	trifluoromethane sulfonic Brønsted acid (CF_3SO_3H)
TMC	trimethylene carbonate
TMEDA	*N*,*N*,*N*′,*N*′-tetramethylethylenediamine
TOF	turnover frequency
TU	thiourea
Tyr-(alkynyl)-OCA	5-(4-(prop-2-yn-1-yloxy)benzyl)-1,3-dioxolane-2,4-dione
VL	δ-valerolactone

Acknowledgements

The CNRS and my (former) group members who contributed to the projects as coauthors are acknowledged.

References

1. K. Masutani and Y. Kimura, *Polym. Int.*, 2017, **66**, 260–276.
2. H. Tsuji, *Adv. Drug Delivery Rev.*, 2016, **107**, 97–135.
3. M. Murariu and P. Dubois, *Adv. Drug Delivery Rev.*, 2016, **107**, 17–46.
4. E. Castro-Aguirre, F. Iniguez-Franco, H. Samsudin, X. Fang and R. Auras, *Adv. Drug Delivery Rev.*, 2016, **107**, 333–366.
5. K. Masutani and Y. Kimura, PLA Synthesis. From the Monomer to the Polymer, in *Poly(lactic acid) Science and Technology: Processing, Properties, Additives and Applications*, ed. A. Jiménez, M. Peltzer and R. Ruseckaite, *RSC Polymer Chemistry Series*, 2015, pp. 1–36.
6. I. Armentano, N. Bitinis, E. Fortunati, S. Mattioli, N. Rescignano, R. Verdejo, M. A. Lopez-Manchado and J. M. Kenny, *Prog. Polym. Sci.*, 2013, **38**, 1720–1747.
7. G. Kfoury, J.-M. Raquez, F. Hassouna, J. Odent, V. Toniazzo, D. Ruch and P. Dubois, *Front. Chem.*, 2013, **1**, 32.
8. J. M. Becker and A. P. Dove, Poly(lactide) as robust renewable materials, in *Green Polymerization: Methods Renewable Starting Materials, Catalysis and Waste Reduction*, ed. R. T. Mathers and M. A. R. Meier, WILEY-VCH Verlag GmbH & Co. KGaA, Weinheim, 2011, pp. 201–220.
9. P. J. Dijkstra, H. Du and J. Feijen, *Polym. Chem.*, 2011, **2**, 520–527.
10. M. J. Stanford and A. P. Dove, *Chem. Soc. Rev.*, 2010, **39**, 486–494.
11. A. J. R. Lasprilla, G. A. R. Martinez, B. H. Lunelli, A. L. Jardini and F. R. Maciel, *Biotechnol. Adv.*, 2012, **30**, 321–328.
12. S. Inkinen, M. Hakkarainen, A.-C. Albertsson and A. Sodergard, *Biomacromolecules*, 2011, **12**, 523–532.

13. J. M. Becker, R. J. Pounder and A. P. Dove, *Macromol. Rapid Commun.*, 2010, **31**, 1923–1937.
14. H. Tsuji, Hydrolytic degradation, in *Poly (Lactic Acid)*, ed. R. Auras, 2010, pp. 345–381.
15. Total Corbion PLA, Producer of biobased and biodegradable PLA polymers, https://www.total-corbion.com/.
16. S. Slomkowski, S. Penczek and A. Duda, *Polym. Adv. Technol.*, 2014, **25**, 436–447.
17. O. Dechy-Cabaret, B. Martin-Vaca and D. Bourissou, Polyesters from dilactones, in *Handbook of Ring-Opening Polymerization*, ed. P. Dubois, O. Coulembier and J.-M. Raquez, Wiley-VCH, 2009, pp. 255–286.
18. S. M. Guillaume, E. Kirillov, Y. Sarazin and J.-F. Carpentier, *Chem. – Eur. J.*, 2015, **21**, 7988–8003.
19. F. Nederberg, E. F. Connor, M. Moeller, T. Glauser and J. L. Hedrick, *Angew. Chem., Int. Ed.*, 2001, **40**, 2712–2715.
20. L. Mezzasalma, A. P. Dove and O. Coulembier, *Eur. Polym. J.*, 2017, **95**, 628–634.
21. S. Hu, Z. Junpeng, G. Zhang and H. Schlaad, *Prog. Polym. Sci.*, 2017, **74**, 34–77.
22. W. N. Ottou, H. Sardon, D. Mecerreyes, J. Vignolle and D. Taton, *Prog. Polym. Sci.*, 2016, **56**, 64–115.
23. J. Pretula, S. Slomkowski and S. Penczek, *Adv. Drug Delivery Rev.*, 2016, **107**, 3–16.
24. S. P. Dubey, H. A. Abhyankar, V. Marchante, J. L. Brighton and K. Blackburn, *Int. Res. J. Pure App. Chem.*, 2016, **12**, 1–20.
25. S. Naumann and A. P. Dove, *Polym. Chem.*, 2015, **6**, 3185–3200.
26. D. J. Liu and E. Y.-X. Chen, *Green Chem.*, 2014, **16**, 964–981.
27. H. A. Brown and R. M. Waymouth, *Acc. Chem. Res.*, 2013, **46**, 2585–2596.
28. M. Fèvre, J. Pinaud, Y. Gnanou, J. Vignolle and D. Taton, *Chem. Soc. Rev.*, 2013, **42**, 2142–2172.
29. M. Fèvre, J. Vignolle, Y. Gnanou and D. Taton, *Polym. Sci.*, 2012, **42**, 67–115.
30. A. P. Dove, *ACS Macro. Lett.*, 2012, **1**, 1409–1412.
31. M. K. Kiesewetter, E. J. Shin, J. L. Hedrick and R. M. Waymouth, *Macromolecules*, 2010, **43**, 2093–2107.
32. S. Naumann and M. R. Buchmeiser, *Catal. Sci. Technol.*, 2014, **4**, 2466–2479.
33. S. Boileau and N. Illy, *Prog. Polym. Sci.*, 2011, **36**, 1132–1151.
34. N. E. Kamber, W. Jeong, R. M. Waymouth, R. C. Pratt, B. G. G. Lohmeijer and J. L. Hedrick, *Chem. Rev.*, 2007, **107**, 5813–5840.
35. D. Bourissou, S. Moebs-Sanchez and B. Martin-Vaca, *C. R. Chim.*, 2007, **10**, 775–794.
36. B. Martin Vaca and D. Bourissou, *ACS Macro Lett.*, 2015, **4**, 792–798.
37. O. Coulembier, J. De Winter, T. Josse, L. Mespouille, P. Gerbaux and P. Dubois, *Polym. Chem.*, 2013, **5**, 2103–2108.

38. H. A. Brown, A. G. De Crisci, J. L. Hedrick and R. M. Waymouth, *ACS Macro Lett.*, 2012, **1**, 1113–1115.
39. N. J. Sherck, H. Chang Kim and Y.-Y. Won, *Macromolecules*, 2016, **49**, 4699–4713.
40. X. Zhang and R. M. Waymouth, *ACS Macro Lett.*, 2014, **3**, 1024–1028.
41. D. A. Culkin, W. Jeong, S. Csihony, E. D. Gomez, N. P. Balsara, J. L. Hedrick and R. M. Waymouth, *Angew. Chem., Int. Ed.*, 2007, **46**, 2627–2630.
42. W. Jeong, E. J. Shin, D. A. Culkin, J. L. Hedrick and R. M. Waymouth, *J. Am. Chem. Soc.*, 2009, **131**, 4884–4891.
43. S. Csihony, D. A. Culkin, A. C. Sentman, A. P. Dove, R. M. Waymouth and J. L. Hedrick, *J. Am. Chem. Soc.*, 2005, **127**, 9079–9084.
44. E. Brule, V. Guerineau, P. Vermaut, F. Prima, J. Balogh, L. Maron, A. M. Z. Slawin, S. P. Nolan and C. M. Thomas, *Polym. Chem.*, 2013, **4**, 2414–2423.
45. A. V. Prasad, L. P. Stubbs, Z. Ma and Z. Yinghuai, *J. Appl. Polym. Sci.*, 2012, **123**, 1568–1575.
46. M. Basko and P. J. Kubisa, *J. Polym. Sci., Part A: Polym. Chem.*, 2006, **44**, 7071–7081.
47. D. Bourissou, B. Martin-Vaca, A. Dumitrescu, M. Graullier and F. Lacombe, *Macromolecules*, 2005, **38**, 9993–9998.
48. M. Basko and P. Kubisa, *J. Polym. Sci., Part A: Polym. Chem.*, 2007, **45**, 3090–3097.
49. M. Basko, A. Duda, S. Kazmierski and P. Kubisa, *J. Polym. Sci., Part A: Polym. Chem.*, 2013, **51**, 4873–4884.
50. K. Makiguchi, T. Yamanaka, T. Kakuchi, M. Terada and T. Satoh, *Chem. Commun.*, 2014, **50**, 2883–2885.
51. K. Makiguchi, S. Kikuchi, T. Satoh and T. Kakuchi, *J. Polym. Sci., Part A: Polym. Chem.*, 2013, **51**, 2455–2463.
52. J. Liu, S. Cui, Z. Li, S. Xu, J. Xu, X. Pan, Y. Liu, H. Dong, H. Sun and K. Guo, *Polym. Chem.*, 2016, **7**, 5526–5535.
53. I. A. Barker and A. P. Dove, *Chem. Commun.*, 2013, **49**, 1205–1207.
54. A. Alba, O. Thillaye du Boullay, B. Martin-Vaca and D. Bourissou, *Polym. Chem.*, 2015, **6**, 989–997.
55. G. Nogueira, A. Favrelle, M. Bria, J. P. Prates Ramalho, P. J. Mendes, A. Valente and P. Zinck, *React. Chem. Eng.*, 2016, **1**, 508–520.
56. W. Guerin, M. Helou, J. F. Carpentier, M. Slawinski, J. M. Brusson and S. M. Guillaume, *Polym. Chem.*, 2013, **4**, 1095–1106.
57. Y. Miao, C. Rousseau, A. Mortreux, P. Martin and P. Zinck, *Polymer*, 2011, **52**, 5018–5026.
58. B. G. G. Lohmeijer, R. C. Pratt, F. Leibfarth, J. W. Logan, D. A. Long, A. P. Dove, F. Nederberg, J. Choi, C. Wade, R. M. Waymouth and J. L. Hedrick, *Macromolecules*, 2006, **39**, 8574–8583.
59. Y. H. Lim, G. S. Heo, S. Cho and K. L. Wooley, *ACS Macro Lett.*, 2013, **2**, 785–789.

60. O. Coulembier, V. Lemaur, T. Josse, A. Minoia, J. Cornil and P. Dubois, *Chem. Sci.*, 2012, **3**, 723–726.
61. R. C. Pratt, B. G. Lohmeijer, D. A. Long, R. M. Waymouth and J. L. Hedrick, *J. Am. Chem. Soc.*, 2006, **128**, 4556–4557.
62. F. Eisenreich, P. Viehmann, F. Mueller and S. Hecht, *Macromolecules*, 2015, **48**, 8729–8732.
63. W. Guerin, M. Helou, M. Slawinski, J.-M. Brusson, S. M. Guillaume and J.-F. Carpentier, *Polym. Chem.*, 2013, **4**, 3686–3693.
64. S. A. van den Berg, H. Zuilhof and T. Wennekes, *Macromolecules*, 2016, **49**, 2054–2062.
65. M. T. Martello, A. Burns and M. Hillmyer, *ACS Macro Lett.*, 2012, **1**, 131–135.
66. B. Guillerm, V. Lemaur, B. Ernould, J. Cornil, R. Lazzaroni, J.-F. Gohy, P. Dubois and O. Coulembier, *RSC Adv.*, 2014, **4**, 10028–10038.
67. J. A. Castillo, D. E. Borchmann, A. Y. Cheng, Y. Wang, C. Hu, A. Garcia and M. Weck, *Macromolecules*, 2012, **45**, 62–69.
68. R. A. Jones, M. Karatza, T. N. Voro, U. P. Cilcir, A. Franck, O. Ozturk, J. P. Seaman, A. P. Whitmore and D. J. Williamson, *Tetrahedron*, 1996, **52**, 8707–8724.
69. E. F. Connor, G. W. Nyce, M. Myers, A. Mock and J. L. Hedrick, *J. Am. Chem. Soc.*, 2002, **124**, 914–915.
70. A. P. Dove, R. C. Pratt, B. G. G. Lohmeijer, D. A. Culkin, E. C. Hagberg, G. W. Nyce, R. M. Waymouth and J. L. Hedrick, *Polymer*, 2006, **47**, 4018–4025.
71. Y. Wang, L. Zhang, X. Guo, R. Zhang and J. Li, *J. Polym. Res.*, 2013, **20**, 87, (1–6).
72. O. Coulembier, B. G. G. Lohmeijer, A. P. Dove, R. C. Pratt, L. Mespouille, D. A. Culkin, S. J. Benight, P. Dubois, R. M. Waymouth and J. L. Hedrick, *Macromolecules*, 2006, **39**, 5617–5628.
73. O. Coulembier, A. P. Dove, R. C. Pratt, A. C. Sentman, D. A. Culkin, L. Mespouille, P. Dubois, R. M. Waymouth and J. L. Hedrick, *Angew. Chem., Int. Ed.*, 2005, **44**, 4964–4968.
74. O. Coulembier, M. K. Kiesewetter, A. Mason, P. Dubois, J. L. Hedrick and R. M. Waymouth, *Angew. Chem., Int. Ed.*, 2007, **46**, 4719–4721.
75. T. R. Jensen, L. E. Breyfogle, M. A. Hillmyer and W. B. Tolman, *Chem. Commun.*, 2004, 2504–2505.
76. A. P. Dove, H. Li, R. C. Pratt, B. G. G. Lohmeijer, D. A. Culkin, R. M. Waymouth and J. L. Hedrick, *Chem. Commun.*, 2006, **27**, 2881–2883.
77. A. Hoppe, F. Sadaka, C.-H. Brachais, G. Boni, J.-P. Couvercelle and L. Plasseraud, *Beilstein J. Org. Chem.*, 2013, **9**, 647–654.
78. M. Fèvre, J. Vignolle and D. Taton, *Polym. Chem.*, 2013, **4**, 1995–2003.
79. M. Fèvre, J. Pinaud, A. Leteneur, Y. Gnanou, J. Vignolle, D. Taton, K. Miqueu and J.-M. Sotiropoulos, *J. Am. Chem. Soc.*, 2012, **134**, 6776–6784.

80. R. Lindner, M. L. Lejkowski, S. Lavy, P. Deglmann, K. T. Wiss, S. Zarbakhsh, L. Meyer and M. Limbach, *ChemCatChem*, 2014, **6**, 618–625.
81. S. Naumann, A. W. Thomas and A. P. Dove, *ACS Macro Lett.*, 2016, **5**, 134–138.
82. O. Thillaye Du Boullay, E. Marchal, B. Martin-Vaca, F. P. Cossio and D. Bourissou, *J. Am. Chem. Soc.*, 2006, **128**, 16442–16443.
83. C. Bonduelle, B. Martin-Vaca, F. P. Cossio and D. Bourissou, *Chem. – Eur. J.*, 2008, **14**, 5304–5312.
84. O. Thillaye du Boullay, C. Bonduelle, B. Martin-Vaca and D. Bourissou, *Chem. Commun.*, 2008, 1786–1788.
85. Z. Zhang, L. Yin, Y. Xu, R. Tong, Y. Lu, J. Ren and J. Cheng, *Biomacromolecules*, 2012, **13**, 3456–3462.
86. H. Xia, S. Kan, Z. Li, J. Chen, S. Cui, W. Wu, P. Ouyang and K. Guo, *J. Polym. Sci., Part A: Polym. Chem.*, 2014, **52**, 2306–2315.
87. M. Myers, E. F. Connor, T. Glauser, A. Mock, G. Nyce and J. L. Hedrick, *J. Polym. Sci., Part A: Polym. Chem.*, 2002, **40**, 844–851.
88. L. Zhang, F. Nederberg, R. C. Pratt, R. M. Waymouth, J. L. Hedrick and C. G. Wade, *Macromolecules*, 2007, **40**, 4154–4158.
89. L. Zhang, F. Nederberg, J. M. Messman, R. C. Pratt, J. L. Hedrick and C. G. Wade, *J. Am. Chem. Soc.*, 2007, **129**, 12610–12611.
90. H. Alamri, J. Zhao, D. Pahovnik and N. Hadjichristidis, *Polym. Chem.*, 2014, **5**, 5471–5478.
91. N. Hadjichristidis, J. Zhao, D. Pahovnik and Y. Gnanou, *Polym. Chem.*, 2014, **5**, 3750–3753.
92. P. Lewinski, S. Sosnowski and S. Penczek, *Polymer*, 2017, **108**, 265–271.
93. G. M. Miyake and E. Y.-X. Chen, *Macromolecules*, 2011, **44**, 4116–4124.
94. O. Coulembier, T. Josse, B. Guillerm, P. Gerbaux and P. Dubois, *Chem. Commun.*, 2012, **48**, 11695–11697.
95. B. Guillerm, V. Lemaur, J. Cornil, R. Lazzaroni, P. Dubois and O. Coulembier, *Chem. Commun.*, 2014, **50**, 10098–10101.
96. J. B. Zhu and E. Y. X. Chen, *J. Am. Chem. Soc.*, 2015, **137**, 12506–12509.
97. S. Chen, Y. Liu, Z. Li, X. Wang, H. Dong, H. Sun, K. Yang, H. Gebru and K. Guo, *Eur. Polym. J.*, 2017, **97**, 389–396.
98. X. Zhi, J. Liu, Z. Li, H. Wang, X. Wang, S. Cui, C. Chen, C. Zhao, X. Li and K. Guo, *Polym. Chem.*, 2016, **7**, 339–349.
99. R. Todd, G. Rubio, D. J. Hall, S. Tempelaar and A. P. Dove, *Chem. Sci.*, 2013, **4**, 1092–1097.
100. D. J. Coady, A. C. Engler, H. W. Horn, K. M. Bajjuri, K. Fukushima, G. O. Jones, A. Nelson, J. E. Rice and J. L. Hedrick, *ACS Macro Lett.*, 2012, **1**, 19–22.
101. B. Grignard, J. De Winter, P. Gerbaux, B. Gilbert, C. Jerome and C. Detrembleur, *Eur. Polym. J.*, 2017, **95**, 635–649.
102. S. Tempelaar, L. Mespouille, P. Dubois and A. P. Dove, *Macromolecules*, 2011, **44**, 2084–2091.

103. R. C. Pratt, B. G. G. Lohmeijer, D. A. Long, P. N. P. Lundberg, A. P. Dove, H. B. Li, C. G. Wade, R. M. Waymouth and J. L. Hedrick, *Macromolecules*, 2006, **39**, 7863–7871.
104. O. I. Kazakov and M. K. Kiesewetter, *Macromolecules*, 2015, **48**, 6121–6126.
105. S. S. Spink, O. I. Kazakov, E. T. Kiesewetter and M. K. Kiesewetter, *Macromolecules*, 2015, **48**, 6127–6131.
106. J. U. Pothupitiya, N. U. Dharmaratne, T. M. M. Jouaneh, K. V. Fastnacht, D. N. Coderre and M. K. Kiesewetter, *Macromolecules*, 2017, **50**, 8948–8954.
107. A. Sanchez-Sanchez, I. RiviL-LA, M. Agirre, A. Basterretxea, A. Etxeberria, A. Veloso, H. Sardon, D. Mecerreyes and F. P. Cossío, *J. Am. Chem. Soc.*, 2017, **139**, 4805–4814.
108. C. Thomas, F. Peruch and B. Bibal, *RSC Adv.*, 2012, **2**, 12851–12856.
109. D. J. Coady, K. Fukushima, H. W. Horn, J. E. Rice and J. L. Hedrick, *Chem. Commun.*, 2011, **47**, 3105–3107.
110. S. Tempelaar, I. A. Barker, V. X. Truong, D. J. Hall, L. Mespouille, P. Dubois and A. P. Dove, *Polym. Chem.*, 2013, **4**, 174–183.
111. Y. E. Aguirre-Chagala, J. L. Santos, R. Herrera-Najera and M. Herrera-Alonso, *Macromolecules*, 2013, **46**, 5871–5881.
112. J. Kadota, D. Pavlovic, J.-P. Desvergne, B. Bibal, F. Peruch and A. Deffieux, *Macromolecules*, 2010, **43**, 8874–8879.
113. J. Kadota, D. Pavlovic, H. Hirano, A. Okada, Y. Agari, B. Bibal, A. Deffieux and F. Peruch, *RSC Adv.*, 2014, **4**, 14725–14732.
114. D. Specklin, F. Hild, L. Chen, L. Thévenin, M. Munch, F. Dumas, F. Le Bideau and S. Dagorne, *ChemCatChem*, 2017, **9**, 3041–3046.
115. A. Rostami, E. Sadeh and S. Ahmadi, *J. Polym. Sci., Part A: Polym. Chem.*, 2017, **55**, 2483–2493.
116. J. Liu, J. Xu, Z. Li, S. Xu, X. Wang, H. Wang, T. Guo, Y. Gao, L. Zhang and K. Guo, *Polym. Chem.*, 2017, **8**, 7054–7068.
117. J. Liu, C. Chen, Z. Li, W. Wu, X. Zhi, Q. Zhang, H. Wu, X. Wang, S. Cui and K. Guo, *Polym. Chem.*, 2015, **6**, 3754–3757.
118. X. Wang, S. Cui, Z. Li, S. Kan, Q. Zhang, C. Zhao, H. Wu, J. Liu, W. Wu and K. Guo, *Polym. Chem.*, 2014, **5**, 6051–6059.
119. Y. Miao, N. Stanley, A. Favrelle, T. Bousquet, M. Bria, A. Mortreux and P. Zinck, *J. Polym. Sci., Part A: Polym. Chem.*, 2015, **53**, 659–664.
120. K. Makiguchi, S. Kikuchi, K. Yanai, Y. Ogasawara, S.-I. Sato, T. Satoh and T. Kakuchi, *J. Polym. Sci., Part A: Polym. Chem.*, 2014, **52**, 1047–1054.
121. X. Wang, J. Liu, S. Xu, J. Xu, X. Pan, J. Liu, S. Cui, Z. Li and K. Guo, *Polym. Chem.*, 2016, **7**, 6297–6308.
122. A. M. Goldys and D. J. Dixon, *Macromolecules*, 2014, **47**, 1277–1284.
123. D. Jardel, C. Davies, F. Peruch, S. Massip and B. Bibal, *Adv. Synth. Catal.*, 2016, **358**, 1110–1118.

CHAPTER 7

ROP of Cyclic Carbonates

KAZUKI FUKUSHIMA

Yamagata University, Graduate School of Organic Materials Science, 4-3-16 Jonan, Yonezawa, Yamagata 992-8510, Japan
Email: fukushima@yz.yamagata-u.ac.jp

7.1 Introduction

As represented by poly(bisphenol-A carbonate), typical examples of engineering polycarbonates have been produced by the condensation approach, and they are employed in optical and impact-resistant materials.[1] Most polymers obtained through the ring-opening polymerization (ROP) of cyclic carbonates are aliphatic and thus demonstrate biodegradability. Ethylene carbonate (EC), a simple five-membered cyclic carbonate (5CC), is the first cyclic carbonate to be applied in ROP. However, it is still challenging to obtain a high molecular weight polymer with no structural defect by this methodology, due to high thermodynamic stability of the five-membered ring monomer.[2] The most studied cyclic carbonate is a six-membered cyclic carbonate (6CC), namely trimethylene carbonate (TMC).[3,4] Poly(TMC) has been widely applied in practical biomedical devices such as suture and bone repair materials.[5,6] In addition, as described below, 6CCs that have functional groups as a substituent have been recently reported with an exponential rate for applications in high-valued biomaterials including nanomedicine and regenerative medicine.[7] This is attributed to reasonable handling property under ambient condition and facile and diverse synthetic methodologies of these monomers, which are quite advantageous over other cyclic monomer platforms such as cyclic esters, anhydrides, and amides. Controlled and regulated formation of microstructures directly influence

Polymer Chemistry Series No. 31
Organic Catalysis for Polymerisation
Edited by Andrew Dove, Haritz Sardon and Stefan Naumann

Published by the Royal Society of Chemistry, www.rsc.org

functions and performance for these biomaterials. For instance, molecular weight and dispersity of a polymer determine the size of a drug carrier such as polymeric micelles of amphiphilic block copolymers.[8] Topology of block copolymers may affect the shape of self-assemblies. Moreover, available catalyst/initiator systems could vary depending on target applications and monomers. Thus, reviewing and understanding ROP of cyclic carbonates should be supportive for those addressing the developing polycarbonate-based smart materials.

The ROP of cyclic carbonates is typically performed in bulk or in solution. Most of cyclic carbonates show a melting point below 100 °C, therefore severe heating conditions are scarcely required for their bulk polymerization.[9] Even though bulk polymerization is a greener approach, solution polymerization is more frequently adopted due to its better control, such as narrow distribution of molecular weight ($Đ_M$) and fewer structural defects in the resultant polymer by decarboxylation. Although not very popular, some examples of microwave-assisted ROP of cyclic carbonates are reported, proving its efficacy compared with conventional heating.[10,11] As the reaction media for solution polymerization, aprotic solvents are preferred, such as toluene, benzene, methylene chloride (CH_2Cl_2), THF, acetonitrile, and DMF, because protic solvents such as methanol can be involved in the initiation step. Solvation sometimes affects the reactivity of monomers and catalysts/initiators. Thus, low or non-polar solvents are favoured. Other than conventional organic solvents, supercritical carbon dioxide and ionic liquids (ILs) are recognized as alternative media for the ROP of cyclic carbonates with a green approach.[11,12]

This chapter describes topics particularly focusing on recent developments in the mechanism of ROP of cyclic carbonates including intrinsic decarboxylation, catalysts, initiators, and regioselective ROP, copolymerization, and polymerizable cyclic carbonates.

7.2 Classical Mechanism

As with lactones and cyclic esters, ROP of cyclic carbonates proceeds based on transesterification *via* anionic, cationic or coordination–insertion mechanisms.[13] The main driving force for the ring-opening is provided by ring-strain, and thus the mechanistic aspects are analogous to the ROP of lactones and lactides (LAs), except for 5CCs without a fused structure,[14] which occurs only when followed by decarboxylation.[15] Mostly studied cyclic carbonates are six- and seven-membered rings and the polymerization mechanism of the most popular TMC is described below as the typical example. General thermodynamics and kinetics of ROP are omitted here and can be found in the other chapters.

7.2.1 Anionic Pathway

Anionic ROP begins with a nucleophilic attack by anionic initiators typically represented by potassium *tert*-butoxide (*t*BuOK), sec-butyl lithium (*sec*-BuLi), sodium naphthalene and tertiary amines including 1,8-diazabicyclo-[5.4.0]undec-7-ene (DBU).[16–19] There are possibly two sites subjected to attack in the cyclic carbonate: (a) the carbonyl carbon or (b) the methylene carbon adjacent to oxygen (Figure 7.1). Initiation by attacking the carbonyl carbon gives the corresponding alkoxide anion (path a) as the active species for propagation, which ideally attacks the carbonyl carbon consecutively to result a polycarbonate sequence without structural defect. Initiation occurring at the other site generates a carbonate anion (path b) that often decarboxylates to form an alkoxide anion as the propagating species. If this pathway takes place for propagation, the resultant polymer will contain ether linkages in the backbone. However, unlike alkyl-carbonyl cleavage of lactone, the methylene group adjacent to carbonate has low electrophilicity, thereby the latter type of initiation would not frequently take place. A major side reaction for the anionic route is backbiting in which the active alkoxide chain end attacks to carbonyl group of a linear carbonate bond in the propagating chain, resulting in the formation of the cyclic monomer and oligomers (Figure 7.2).[19]

7.2.2 Cationic Pathway

Cationic ROP starts from the activation of the monomer by protonation or alkylation of the carbonyl oxygen by cationic initiators such as Lewis acids and then proceeds *via* two different possible pathways (Figure 7.3).[20] The activated monomer (AM) mechanism continues by an attack on the carbonyl carbon by a nucleophile such as alcohols that proceeds through the tetrahedral intermediate to generate an ω-hydroxy acyclic carbonate (Figure 7.3a). Alternatively, the activated monomer can be subjected to a nucleophilic attack at the methylene carbon adjacent to oxygen by a nucleophilic carbonyl oxygen of another monomer. This forms an acyclic carbonate attaching the activated chain end (ACE) in a form of a carbenium type cation derived from a cyclic carbonate (Figure 7.3b).

A common problem in cationic ROP of cyclic carbonates is CO_2 elimination from the carbonate sequences, generating ether bonds that are often recognized as structural defect in the resultant polycarbonate. This indicates there are further aspects to be considered in the reaction pathway in the cationic ROP of cyclic carbonates. Possible patterns of CO_2 elimination during the ROP are shown in Figure 7.4.[21,22] When the nucleophile (R′-OH) is water in the AM pathway, an ω-hydroxy carbonic acid is generated, and the subsequent decarboxylation results in an α,ω-diol that serves as a nucleophile for the ROP to form α,ω-dihydroxyopolycarbonates. In a similar manner, when the acidic initiator (R–X) is a protic acid (H–X) at the ACE pathway, the carbonic acid end undergoes CO_2 elimination to generate a hydroxyl group, which then proceeds *via* the AM pathway. Alternatively, a nucleophilic attack on the carbenium end by the oxygen adjacent to a methylene group of a cyclic carbonate in the ACE pathway eventually affords polyether sequences.[23]

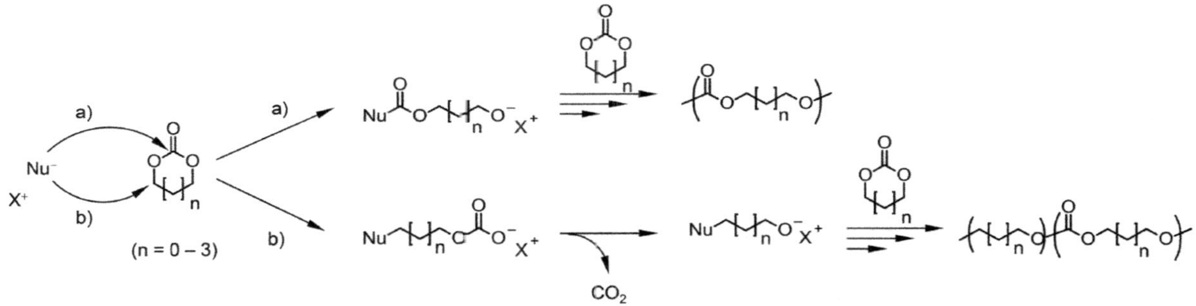

Figure 7.1 Anionic pathways for ROP of cyclic carbonates: (a) acyl-oxygen cleavage forming alkoxide chain end; (b) alkyl-oxygen cleavage forming carbonate chain end.

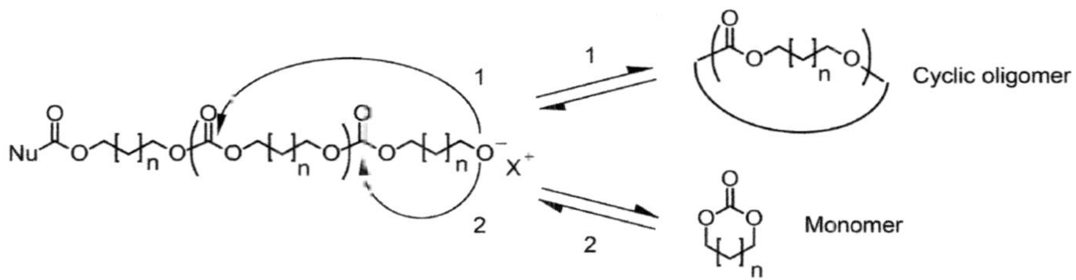

Figure 7.2 Possible backbiting reaction in the anionic ROP of cyclic carbonates forming cyclic oligocarbonate (1) and back to monomer (2).

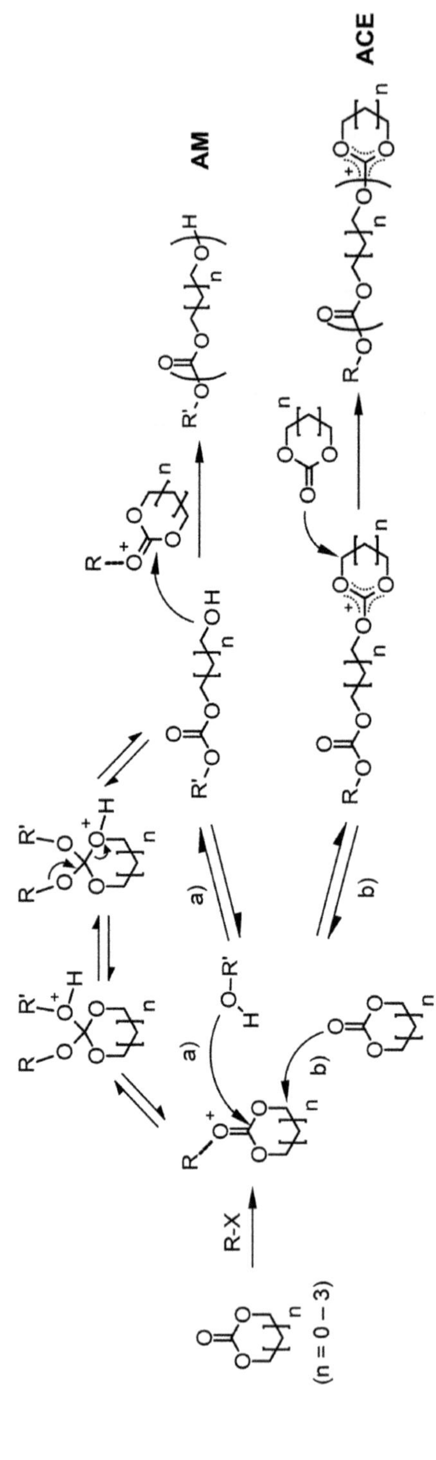

Figure 7.3 Cationic pathways for ROP of cyclic carbonates. (a) Activated monomer (AM) mechanism and (b) active chain end (ACE) mechanism.

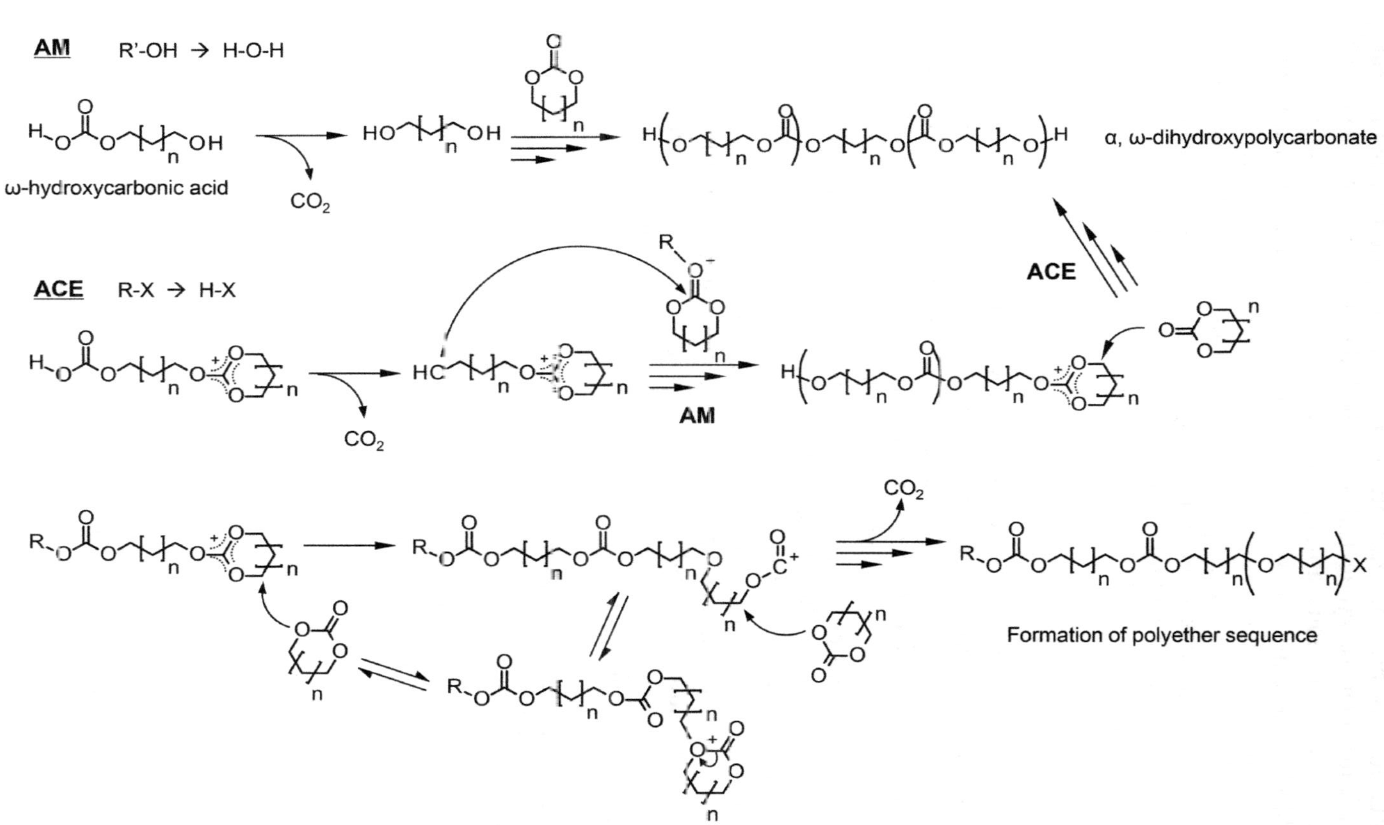

Figure 7.4 Complex pathways of cationic ROP of cyclic carbonates for possible involvement of an ether unit. Adapted from ref. 21 with permission from American Chemical Society, Copyright 2013.

Although a recent DFT study of zirconocene catalysis for ROP of TMC supports the ACE mechanism,[24] most of the experimental results indicate that cationic ROP of cyclic carbonate includes both the AM and ACE mechanisms. In addition, carbonyl groups in the propagating chain can also be subjected to activation by protonation or alkylation. The carbenium ion is converted to an oxionium ion by rearrangement of the alkylene group, leading to the formation of an ether linkage by decarboxylation (Figure 7.5).[25] Thus, it is appropriate to discuss the ratio of carbonate bonds to ether bonds in the resultant polymer chain when cationic ROP is applied for cyclic carbonate.

7.2.3 Coordination–Insertion Pathway

The coordination–insertion mechanism for ROP of cyclic carbonates progresses similarly to that of cyclic esters. In the typical case using tin(II) 2-ethylhexanoate, so-called stannous octanoate ($Sn(Oct)_2$), which is officially approved by the Food and Drug Administration (FDA) in the United States, the carbonyl oxygen is coordinated to the tin atom, and either 2-ethylhexanoyl ligand or alkoxide ligand derived from the alcohol initiator attacks the carbonyl carbon to let the ring open. The former yields a carbonylcarbonate that easily decomposes to convert to 2-ethylhexanoyl ester as an initiating end group *via* decarboxylation (Figure 7.6). The latter is provided by two possible ways based on proposed mechanisms for ROP of cyclic esters by Kricheldolf *et al.* and Penczek *et al.* (Figure 7.7): coordination of the alcohol initiator to the tin atom (Figure 7.7a),[26,27] and ligand exchange of 2-ethylhexanoyl ligand to alcohol (Figure 7.7b).[28,29] In this case, the ring-opened structure contains an alkyl carbonate as an initiating end. Kricheldorf and Stricker concluded that a reasonable heating condition operating usually at 100–120 °C requires alcohol initiators, and the attack by a 2-ethylhexanoate ligand often occurs at higher temperature conditions.[30]

7.3 Recent Trends in Catalysts and Initiators

7.3.1 Organometallics

Organometallic catalysts/initiators, which are applied to the ROP of lactones and LAs, usually can also be used for ROP of cyclic carbonates. Typically, $Sn(Oct)_2$,[30–32] dibutyltinbisoctanoate ($Bu_2Sn(Oct)_2$),[33] and aluminium alkoxides (Al-OR)[34] are used.[35] There are few catalyst/initiator systems to be specialized for the ROP of cyclic carbonates because the reactivity is comparable to that of lactones and often less than that of LAs. In addition, as compared to polylactides (PLAs), for aliphatic polycarbonates there is an overwhelming gap in terms of the extent for application, social recognition, and scientific curiosity. Recent trends in organometallics for the ROP of cyclic carbonates lean to targeting higher efficiency, greener processes, and more precise control. To this end, a wide variety of transition metals

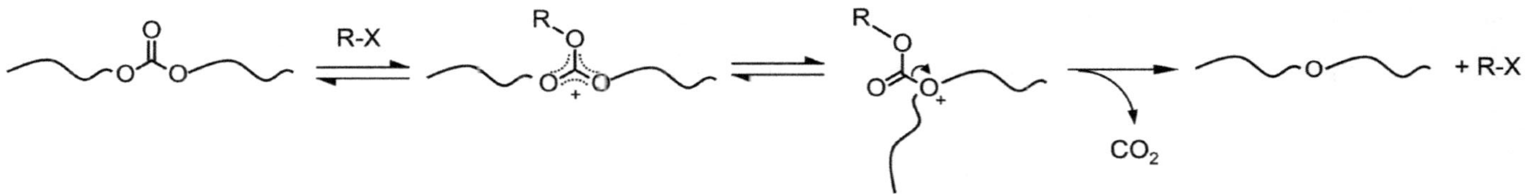

Figure 7.5 Formation of ether bonds in the polymer chain by activation of acyclic carbonate and the subsequent decarboxylation.

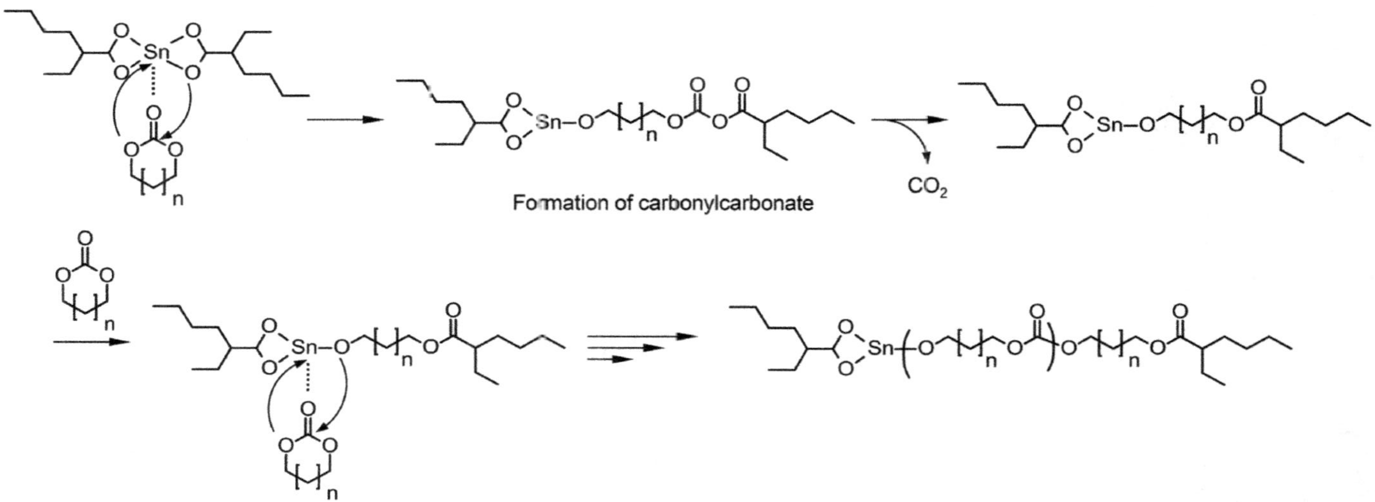

Figure 7.6 Coordination–insertion mechanism for ROP of cyclic carbonates using $Sn(Oct)_2$: initiation by 2-ethylhexanoate ligand. Adapted from ref. 30 with permission from Jon Wiley and Sons, © 2000 WILEY-VCH Verlag GmbH, Weinheim, Fed. Rep. of Germany.

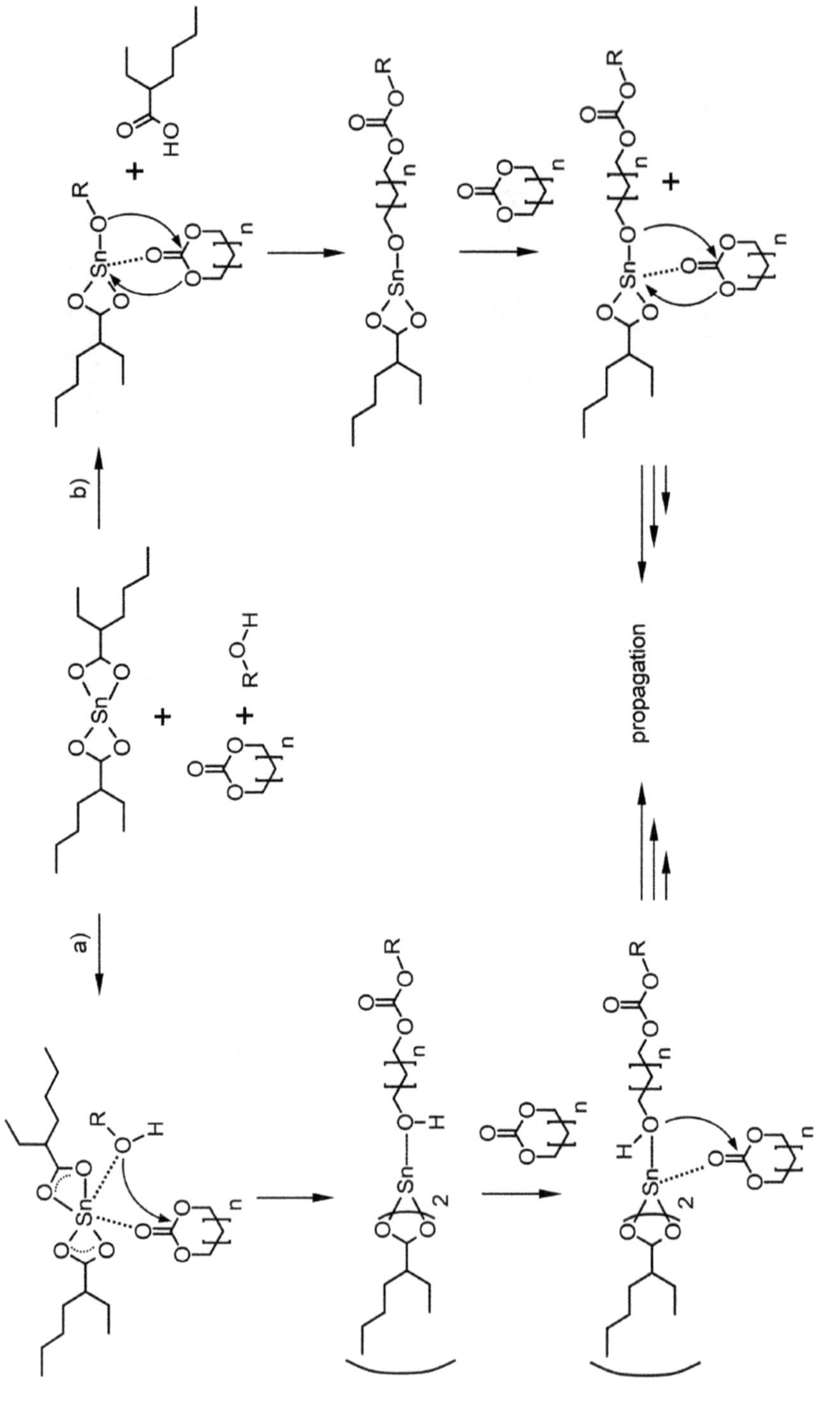

Figure 7.7 Coordination–insertion mechanism for ROP of cyclic carbonates using $Sn(Oct)_2$: initiation by alcohol initiators. (a) Coordinated alcohol and (b) alkoxide as a result of ligand exchange.
Adapted from ref. 30 with permission from John Wiley and Sons, © 2000 WILEY-VCH Verlag GmbH, Weinheim, Fed. Rep. of Germany.

including rare-earth metals and ancillary ligands has been investigated for their potential capability of promoting bulk and solution polymerization of cyclic carbonates.

In particular, over the last decade, revisited immortal polymerization, redox-control polymerization, and formation of α-hydroxy-ω-formate functional polymers have often been reported as a trendy study for the organometallic ROP of cyclic carbonates,[36–38] in addition to sporadic reports regarding highly active catalysts.[39,40] Shen *et al.* reported mixed-metal or multinuclear alkoxide clusters including lanthanide and alkali metals as highly active catalysts for the ROP of TMC, *e.g.* 6000–8000 equiv. of TMC is converted in 1 min in toluene solution at room temperature.[39,41] These are usually obtained by simply mixing of inexpensive reagents and the catalytic activity is just tuned by the ratio of alkali metal to lanthanide. Thus, these catalyst systems are suitable for industrial applications, although the mechanistic insight is not well understood due to the complicated structure.

One of the advantages of cyclic carbonates, represented by TMC, is their low melting point, which enables bulk polymerization at a low temperature. Bergman *et al.* developed an aluminium complex supported by a phenylenediamine ligand that is inexpensive and readily available, demonstrating a quite high catalytic activity for ROP of TMC.[40] The turn over frequency (TOF) is often adopted as a parameter for catalytic activity. The highest TOF of 36 900 h^{-1} is confirmed in the bulk polymerization performed at 70 °C for 2 min, forming poly(TMC) with a number-average molecular weight (M_n) up to 68 000 $g\,mol^{-1}$.

Biocompatibility and low toxicity of catalysts must be concerned, especially when the resultant polymers are expected to be applied in biomedical applications. Although $Sn(Oct)_2$ is a FDA-approved ROP catalyst, it also highly promotes transesterification, resulting in poor control. Al-ORs are highly active catalysts and also appropriate for controlled/living polymerization.[42] However, toxicity and health concerns remain. From this point of view, controlled polymerization using organometallics based on biometals such as calcium, magnesium, zinc, and iron is promising for applications in biomaterials.[43,44] Darensbourg *et al.* revealed that calcium Salen-complexes **1** are more active than magnesium and zinc complexes for the ROP of TMC (Figure 7.8).[45] In addition, several Schiff-base derivatives **2** based on calcium were explored for effective ROP of cyclic carbonates.[46]

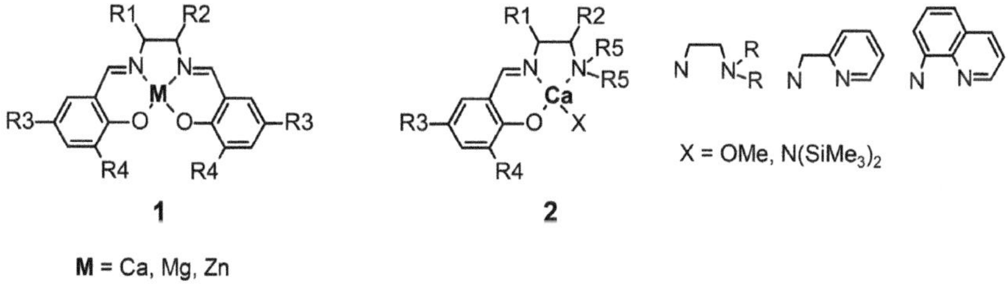

Figure 7.8 Biometal Schiff-base complexes for ROP of TMC.

7.3.1.1 Immortal ROP

Immortal ROP of heterocycles including lactone, epoxides, and cyclic carbonates was first proposed by Inoue *et al.* in 1990, using metalloporphyrins as a highly active catalyst.[47,48] As illustrated in Figure 7.9, ROP using metal alkoxide as an initiator is regarded as "classical" ROP where the metal-alkoxide complex is always active and resides in each chain end. In immortal ROP the molar ratio of a protic initiator (typically alcohol) to monomer determines the targeted molecular weight in the same manner as classical ROP, but not all chain ends are active because catalyst loading is much lower than that of the initiator, which also serves as a chain transfer agent (CTA), according to Guillaume and Carpentier.[49] Only the alcohol and hydroxy chain end activated by or formed with the catalyst play a role of active chain end in the polymerization. To maintain the immortal nature and minimize undesired termination, the catalyst needs to be robust and highly active, and the chain transfer reaction should be reversible between the active (OH–catalyst complex) and dormant (free OH) species (Figure 7.10). Such catalysts enable minimal amount of loading, which is desirable for practical production and could be a solution to concerns about residual metals including toxicity and deterioration of the polymers.[49]

The low reactivity and inherent issues with decarboxylation might have impeded the study of immortal ROP of cyclic carbonates. The revisitation by Carpentier and Guillaume in 2008 began with zinc-β-diiminate amido complexes (BDI)Zn{N($SiMe_3$)$_2$} (M = Zn, **3** in Figure 7.11),[36] previously having confirmed the living nature and regio- and stereo-selectivity in the ROP of LAs and lactones.[50,51] The zinc complex affords efficient ROP of TMC in bulk under mild heating condition at 60–110 °C with benzyl alcohol as a CTA by which loading molecular weight is well-controlled as targeted. For instance, poly(TMC) with M_n of around 40 000 g mol^{-1} is formed within 10 min at 60 °C. The highest TOF for this zinc complex was around 30 000 h^{-1} in the ROP carried out at 110 °C for 40 min. This value is reasonably high as compared to the aluminium complex reported by Bergman *et al.* (TOF 36 900 h^{-1}),[40] except for Guillaume and Carpentier's neutral magnesium complex **6** described later (TOF 240 000 h^{-1}).[52]

Since zinc is an essential mineral for the human body, zinc-based catalysts are considered to be less toxic.[53] Then, metal-β-diiminate complexes **3** (BDI)M{N($SiMe_3$)$_2$} with iron and zinc and amido-chelated biometals such as iron, calcium, and magnesium, namely M{N($SiMe_3$)$_2$}$_x$ (**4**, **5**), were evaluated as a catalyst for the immortal ROP of TMC (Figure 7.11). As with Schiff base complexes **1** containing those metals reported by Darensberg *et al.* in which calcium complexes showed higher activity than others,[46,54] the amido-chelated complexes M{N($SiMe_3$)$_2$}$_x$ (**4**, **5**) exhibited a similar trend for catalytic activity (Ca > Zn > Mg).[53] In contrast, any iron complexes are less active than others irrespective of ligands. The zinc-β-diiminate complex (BDI)Zn{N($SiMe_3$)$_2$} demonstrated a comparable catalytic performance with Ca{N($SiMe_3$)$_2$}$_2$(THF)$_2$, but with a somewhat better control.[53] However, these

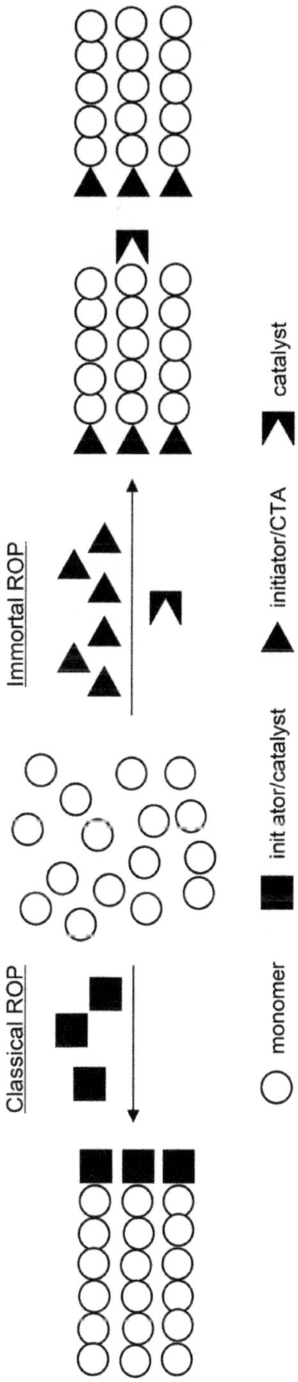

Figure 7.9 Conceptual representation of immortal ROP.
Adapted from ref. 49 with permission from the Royal Society of Chemistry.

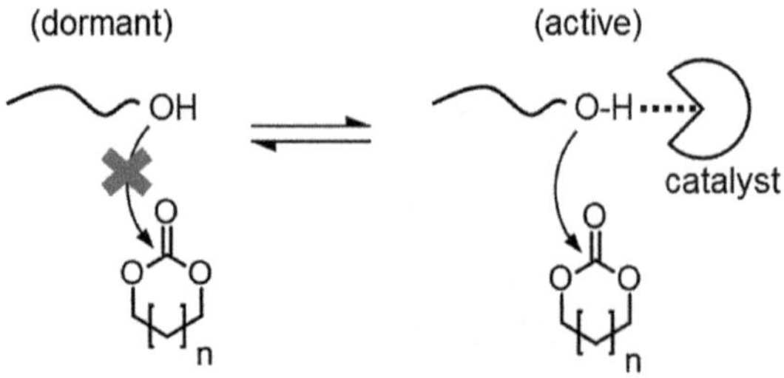

Figure 7.10 Dormant and active species in the immortal ROP.

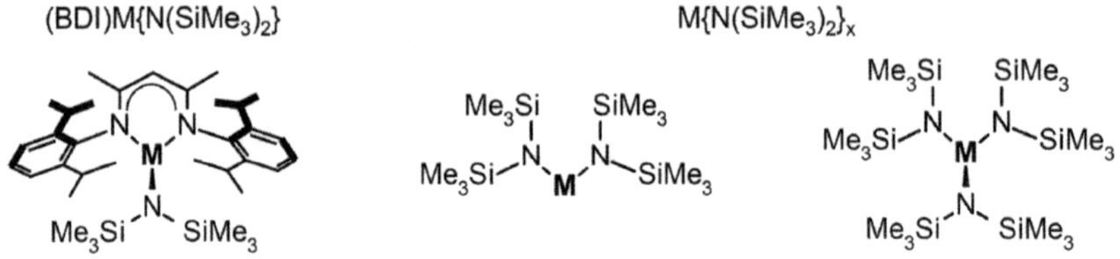

3 (M = Zn, Fe)

4 (M = Zn, Fe, Mg, Ca)

5 (M = Fe(III), Y)

Figure 7.11 Amido-chelated metal complexes for immortal ROP of cyclic carbonates.

catalysts are sensitive to impurities, air, and moisture, resulting in deterioration and deactivation of the catalysts. Then, Carpentier and Guillaume *et al.* explored the use of metal triflate, $M(OTf)_x$, in particular using calcium, scandium, zinc, aluminium, and bismuth.[55] Rare-earth metal triflates such as $Sc(OTf)_3$ and $Bi(OTf)_3$ are known to be tolerant to hydrolysis.[56] These metal triflates could catalyse ROP of TMC in bulk at 110–150 °C with maintained activity even using unpurified monomer. $Al(OTf)_3$ also displays a similar behaviour and a comparable activity (TOF 27 600 h^{-1}) with $Bi(OTf)_3$ showing the highest TOF of 28 200 h^{-1}. For these Lewis acid-based catalysts, an AM mechanism is suggested, in contrast with the metal-β-diiminate complexes **3** (BDI)M{N(SiMe$_3$)$_2$} in which the alkoxide formed by complexation between metal and the hydroxyl group of an alcohol initiator or chain ends is the active species, supporting a coordination–insertion mechanism.[55]

Guillaume and Carpentier have reported zinc and magnesium complexes with a *N,N,O*-phenolate ligand (**6**–**8**) as ROP catalysts of TMC (Figure 7.12).[52] The catalytic activity of the complexes are previously confirmed for the ROP of LAs, especially along with tertiary amines, suggesting a double activation mechanism through ionic and hydrogen bonding where a metal Lewis acid activates a carbonyl group of LA and a tertiary amine activates a hydroxy group of an initiator.[57] In a form of neutral metal complex **6**, both zinc and magnesium catalysts promote immortal ROP of TMC in bulk and 1 M toluene solution, exhibiting high activities with TOF of 29 600 h^{-1} for the zinc complex (110 °C, 1.5 min) and 223 200 h^{-1} for the magnesium complex

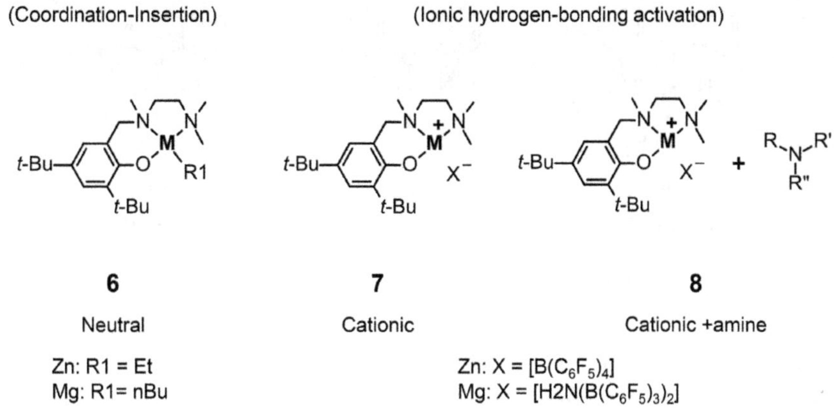

Figure 7.12 Mg and Zn complexes with a *N,N,O*-ligand effective for immortal ROP of TMC.

(60 °C, 0.25 min).[52] In sharp contrast to the neutral metal complex **6** that is believed to act through a coordination–insertion mechanism, the catalytic activities of the cationic metal complexes **7** are drastically reduced by almost 100 times. Then, adding tertiary amines as a co-catalyst (**8**) accelerates the ROP, indicating that the activation of TMC by cationic metal complexes is not sufficient to undergo a nucleophilic attack by an alcohol initiator or the hydroxyl chain end. However, this so-called double activation system is only applicable to the zinc complex. The cationic magnesium complex is more acidic than the cationic zinc complex, and thus the amines mitigated by the stronger Lewis acid could not sufficiently activate the initiator.

For industrial applications, the catalytic activity and efficiency are generally more relevant than a well-defined control such as end group fidelity and narrow distribution of molecular weight. However, studies focusing on the latter are also indispensable to construct complex macromolecular architecture, typically for drug delivery vehicles in nanomedicine field. Dagorne *et al.* developed a *N,O,N*-supported tetracoordinate aluminium(III) complex **9** (M = Al) with increased Lewis acidity, exhibiting a high catalytic activity and good control of chain length under a mild condition. For instance, poly(TMC) with M_n up to 20 000 g mol^{-1} ($Đ_M$ <1.20) was formed by the ROP conducted in a 1 M solution of CH_2Cl_2 at room temperature for 30 min (Figure 7.13).[58]

Since bulk ROP of TMC is usually carried out under heating conditions, vigorous chain transfer based on transesterification often occurs and thus results in moderate control and polymers typically with $Đ_M$ over 1.5. Although solution ROP operated at ambient condition is desired for controlled polymerization, the catalytic activity tends to be attenuated in such conditions. Thus, catalysts with balanced activity and controllability are required. Aluminium complexes are susceptible to moisture and become less active compared with other Lewis acidic or oxophilic metal complexes

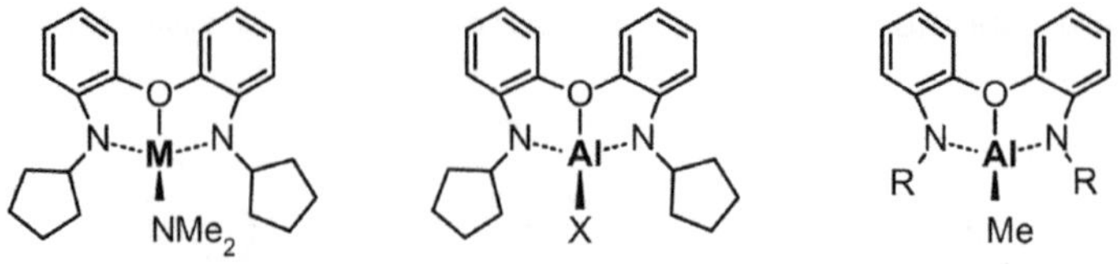

9 M = Al, Ga

10 X = NMe_2, Me

11 R = C_6F_5, C_5H_9

Figure 7.13 *N,O,N*-supported tetracoordinate metal complexes.

including Sn(II), Zn(II), Mg(II), Ca(II), and Ln(III).[59] Moreover, potential concerns about toxicity of Al(III) is still in controversy, which could limit the industrial applications of aluminium-based catalysts, despite the advantage in cost and availability.[60] Lower group 13 metals such as gallium and indium are regarded as being more biocompatible and stable in protic media including water. The gallium analogue supported by the same *N,O,N*-constrained tetracoordinate ligand (**9**; M = Ga) was tested for the ROP of TMC (Figure 7.13).[59] However, the gallium complex demonstrated no pragmatic catalytic activity compared to the aluminium counterpart (**9**; M = Al), unlike that for ROP of *rac*-lactide (*rac*-LA) showing controlled, immortal, and stereoselective nature. The study also proved that the amido ligand is more efficient to activate the aluminium complex than methyl ligand (**10**), and pentaflluorophenyl ligands are more active than cyclopentyl ligands (**11**) for the ROP of TMC (Figure 7.13).

Carpentier and Sarazin *et al.* developed several tin(II) complexes supported by amino-ether phenolate ligands (**12–14**, Figure 7.14) to examine as catalysts for the immortal ROP of LA and TMC and reported that a coordination–insertion mechanism is operative.[61] This is a contrast to Dagorne's aluminium catalysts **9** that a nucleophilic amido ligand is not easily converted to alcoholate derived from an alcohol initiator, supporting the AM mechanism as a more plausible pathway.[58] Although these tin complexes **12–14** demonstrate a good chain length control for the ROP of LA conducted in toluene at 60–100 °C ($Đ_M \sim 1.10$), their catalytic activity is low (TOF < 500 h^{-1}) and becomes lower again for ROP of TMC. In addition, these tin catalysts are not active for TMC in the presence of LA, and thereby formation of the copolymer only occurs when TMC is polymerized first.[61] They have then evaluated the catalytic activity of ion pairs based on several cationic rare-earth metal (Yb^{II} and Eu^{II}) complexes with the same amino-ether phenolate ancillary ligand.[62] The cationic metal complexes embracing ytterbium(II) **15, 16** (Figure 7.14) are inactive for ROP of LAs because the metal centre is oxidized by lactate, whereas the complexes efficiently facilitate ROP of TMC, revealing a high activity (TOF 5000–7000 h^{-1}) and a high level of control ($Đ_M < 1.10$).[62] This is an unusual system to only catalyse ROP of TMC despite low activity against LAs.

Immortal and controlled ROP of TMC using other zinc-based catalysts are reported by Dagorne and Aviles *et al.* The binary system BnOH/Zn(C_6F_5)$_2$

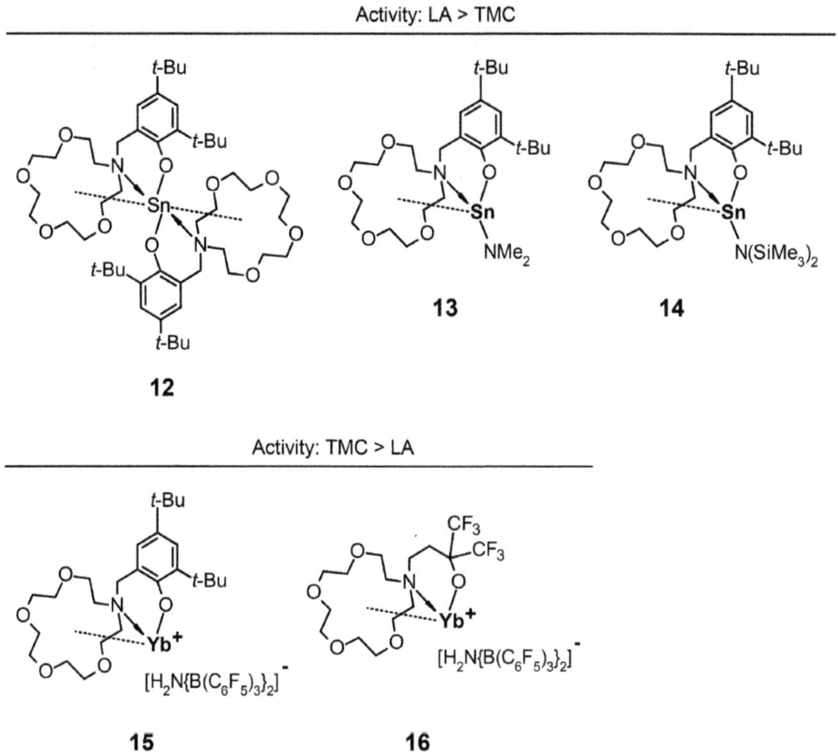

Figure 7.14 Amino-phenolate supported metal complexes.

mediates the ROP of TMC conducted in 1 M CH_2Cl_2 solution at room temperature, providing polymers with a narrow $Đ_M$ (~1.10) in a few hours.[63] Additional experiments and analysis proved that Zn-OR species were hardly formed, which corroborates an activated monomer mechanism (Figure 7.15). This is also supported by the following: the Lewis acidity of zinc is increased by the highly electron-withdrawing C_6F_5 ligands which strongly activates the monomer. The adducts of $Zn(C_6F_5)_2$ with N-heterocyclic carbenes (NHCs) are expected to act as a bicomponent system that dissociates *in situ* into a NHC nucleophile and a Lewis acid $Zn(C_6F_5)_2$. The NHC-Zn adducts are effective for ROP of TMC but with a moderate control ($Đ_M > 1.4$). This is explained by the fact that the initiation step by a NHC nucleophile is slower than the propagation, where the hydroxy chain end is mainly involved in an AM mechanism as the nucleophile (Figure 7.15). Then, a dimeric NHC zinc alkoxide **17**, which is designed to offer more Lewis acidity to the zinc centre, is found to be more effective and provide narrowly dispersed poly(TMC) with $Đ_M \sim 1.10$ through a coordination–insertion mechanism (Figure 7.16).[64] A zinc complex supported by *P,O*-phosphinophenolate ligand (**18**) is designed to present high catalytic activity by stabilizing the metal with O-based hard donor and form a labile coordination bond between zinc and soft phosphine donor, based on

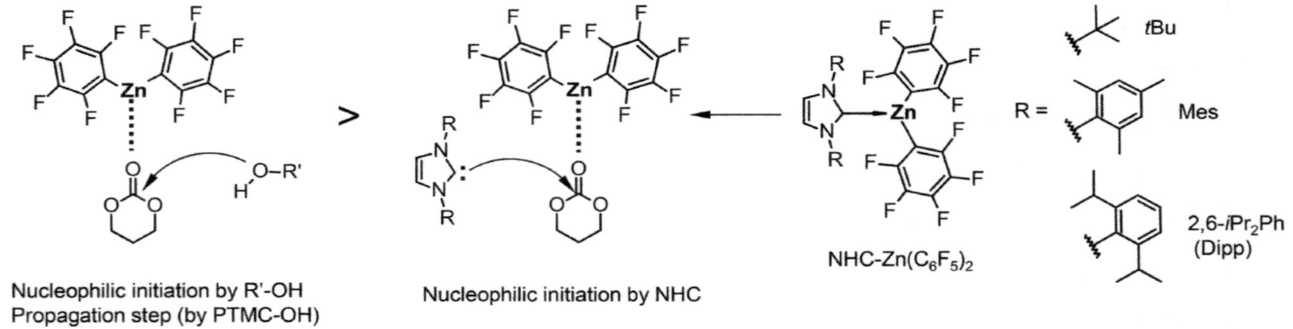

Figure 7.15 Initiation step for ROP of TMC in the presence of $Zn(C_6F_5)_2$ through the AM mechanism.

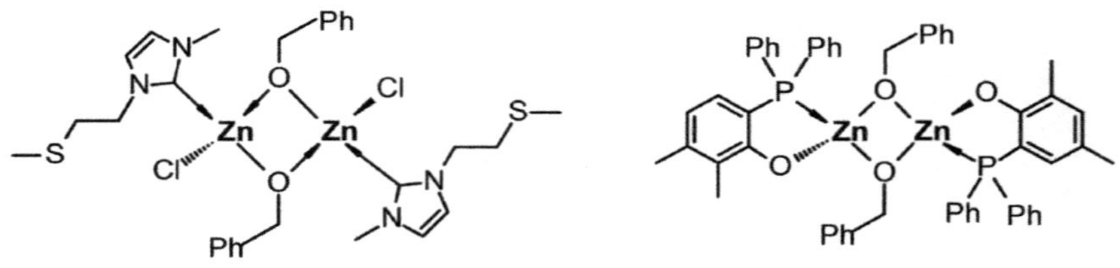

17 **18**

Figure 7.16 Dimeric NHC-zinc alkoxide complexes.
Adapted from ref. 64 and 66 with permission from John Wiley & Sons, Copyright © 2014 John Wiley and Sons, Ltd, and the Royal Society of Chemistry.

previous study for the ROP of lactones and epoxides.[65] However, in this case Zn-OBn plays a role in initiation through a coordination–insertion mechanism and *P,O*-ligands are not involved in the polymerization mechanism.[66] In addition, the catalyst activity is rather lower for the ROP of TMC compared to those for LAs and lactones.

7.3.1.2 Borohydride Ligand: Formation of α-hydroxy,ω-formate telechelic Polycarbonates

Borohydride (BH_4) ligand has previously been found to yield α,ω-dihydroxy-telechelic polymers in the ROP of ε-caprolactone (CL).[67,68] Guillaume *et al.* reported the formation of α-hydroxy,ω-formate telechelic poly(TMC) in 2007, using a samarium borohydride complex $Sm(BH_4)_3(THF)_3$ that presented a reasonable activity, converting *ca.* 200 equiv. of TMC in several hours with a moderate control of molecular weight distribution with $1.2 < Đ_M < 1.4$. The ROP proceeded according to a coordination–insertion mechanism where acyl-oxygen cleavage predominantly occurs and an opened species inserts into a Sm-HBH_3 bond to form Sm-O and formate/BH_3 (Figure 7.17).[38]

Some other heteroleptic lanthanide borohydride complexes supported by a popular ancillary ligand β-diketimine (**19, 20**) have also shown a reasonable catalytic activity for the ROP of TMC, affording α-hydroxy,ω-formate telechelic poly(TMC).[69] Considering that the reaction media is different (THF *vs.* toluene), the samarium complex **20** (Ln = Sm) converts 250 equiv. of TMC in 30 min, which is much faster than $Sm(BH_4)_3(THF)_3$. Nonetheless, the reaction is much slower than ROP of CL using the same catalysts. In the case of ytterbium complexes, monoborohydride **20** (Ln = Yb) presented higher activity than bis-borohydride **19**. In contrast, the concomitant formation of α,ω-dihydroxyt-elechelic poly(TMC) is not ruled out when bis(phosphinimino)methanide bisborohydride complexes of rare-earth metals such as lanthanum, yttrium, and lutetium are used as the ROP initiator for TMC (**21, 22** in Figure 7.18).[70] These complexes exhibit higher catalytic activity than the samarium complex, reaching quantitative conversion in 2 min for 100 equiv. of TMC in toluene at 20 °C (TOF = 3000–7800 h^{-1}). The phosphoiminomethanide ancillary ligand is expected to trap the BH_3 group during the reaction, which might affect the concomitant formation of dihydroxy ends resulting from reduction of formate and hemiacetal chain ends. More recently, (iminophosphoranyl)(thiophosphoranyl)methane ytterbium and samarium borohydride complexes **23, 24**

Figure 7.17 Ring-opening of TMC by $Sm(BH_4)_3(THF)_3$.
Adapted from ref. 38 with permission from John Wiley & Sons Inc.

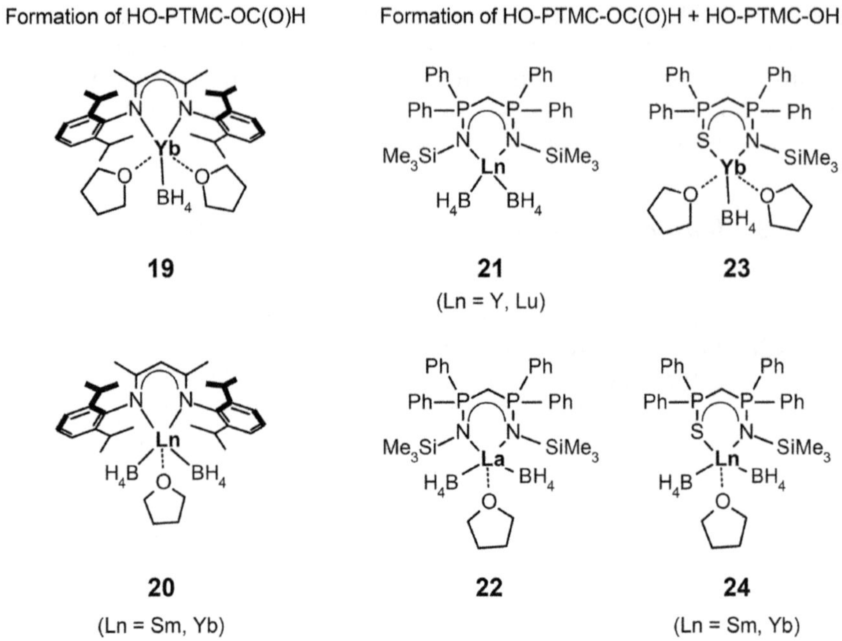

Figure 7.18 Rare-earth metal complexes with BH_4 ligands forming α-hydroxy,ω-formate telechelic or/and α,ω-dihydroxytelechelic polycarbonates.

have been revealed to form both α,ω-dihydroxytelechelic and α-hydroxy,ω-formate telechelic poly(TMC)s with a fair molecular weight distribution ($1.34 < Đ_M < 1.82$) as a result of a moderate catalysis.[71] In the same way, the trapping effect of BH_3 group by the phosphoimino ligand might be attributed to formation of different types of chain end.

7.3.1.3 *Redox-switchable Catalyst*

Selective ROP of different monomers by redox control of the metal centre or metal-containing ligand in organometallic catalysts has been widely investigated,[72–74] since the first report in 2011 by Diaconescu *et al.*[37] In the case of redox control by a ligand, the catalytic activity is believed to depend on the changed electron-donating or electron-withdrawing abilities of the ligand. Although TMC was tested to confirm the availability of the redox control catalysts, a limited number of studies have been described so far. In Diaconescu's first report, the yttrium alkoxide complex **25** (M = Y) supported by a ferrocene-based ligand was shown to polymerize TMC in both oxidized and reduced states at room temperature, whereas for L-lactide (LLA), it proceeded only in its reduced state (Figure 7.19). However, at −78 °C, the reduced state demonstrated higher activity than the oxidized state for the ROP of TMC (monomer conv. 29% *vs.* 3.5%). In contrast, the indium complex **25** (M = In) exhibited an opposite behaviour, with the oxidized state being more active than the reduced one.[37]

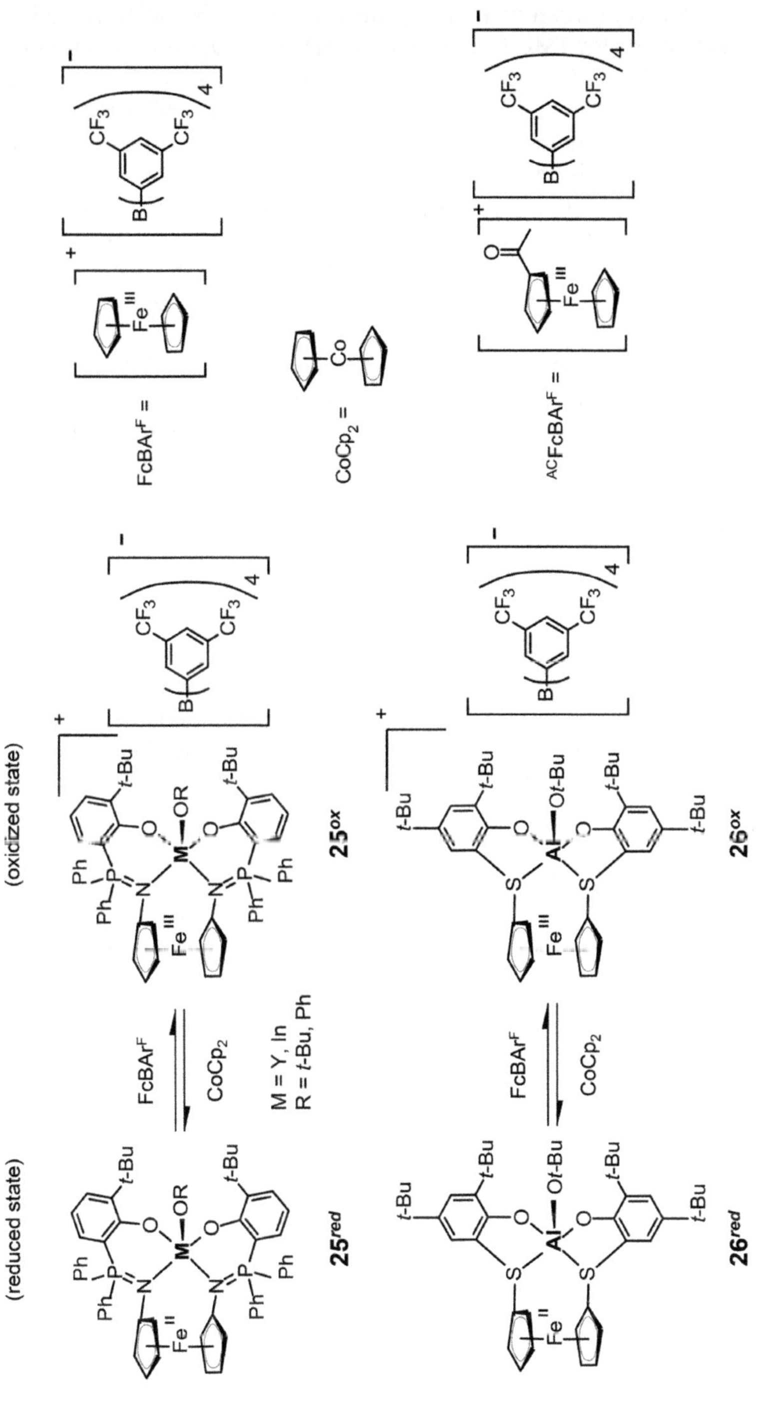

Figure 7.19 Redox-switchable catalysts effective for ROP of TMC. Adapted from ref. 37 and 75 with permission from American Chemical Society, Copyright 2017.

A similar redox-switchable system containing aluminium at the centre (**26**) exhibits a significant difference in polymerization rate between the reduced and oxidized states for TMC (room temperature in C_6D_6), and this tendency is identical to LLA and CL (reduced state > oxidized state).[75] In contrast to the ROP of LLA and lactones, the polymerization of TMC affords less chain control, forming polymers with $Đ_M$ over 1.4. The ROP using these redox switchable catalysts is believed to proceed through the coordination–insertion mechanism initiated from metal alkoxides (M-OR). A DFT calculation result conducted in parallel with the experiments revealed that the extent of chelation of the first inserted monomer at the initiation step determines the insertion of a second monomer for propagation.[75] The metal centre at the oxidized state is more electron deficient, which strongly interacts with a carbonyl group of the ring-opened species. This disfavours the insertion of a second monomer, resulting in a retarded propagation and slower polymerization when compared to the reduced state catalyst. The inserted LLA favours the formation of a stable five-membered ring with aluminium, while inserted TMC provides weaker chelation, which allows a fast propagation and leads to a low level of chain control.

7.3.2 Organocatalysts

A metal-free approach is desirable for the production of polymers that are expected to be used in the biomedical field due to health concerns caused by potentially toxic metal catalyst residues trapped in the polymer. Ionic ROP using nitrogenated bases and organic protonic acids as initiators has been studied for a long time. After extensive efforts by Hedrick, Waymouth, and coworkers to explore a broad range of organic bases and NHCs,[76,77] organocatalysts have now been recognized as useful catalytic systems with respect to efficiency and controllability. In addition, organic acids and acid–base complexes have been currently reviewed.[78,79] Nevertheless, as with the metal complexes, studies focused on the ROP of TMC are not as many as those for LA.

7.3.2.1 Organic Bases

The first representative study for the organocatalytic ROP of TMC was reported by Nederberg in 2007, where amidine and guanidine superbases, NHCs, and thiourea-amine system (TU/A) were evaluated (Figure 7.20).[80] In most cases, polymerization is operated under mild conditions in solution (typically at room temperature) to evaluate chain length control and end group fidelity with emphasis. Among them, a TU-2/A system, especially using DBU, is often used as a preferred catalyst for controlled ROP of TMC and analogues,[81] with respect to accessibility, bench stability, handling, and balance of reactivity and control. The mechanism for these catalyst systems is believed to go through a double hydrogen bonding activation, in which thiourea acts as a Lewis acid to activate the carbonyl group of TMC and DBU as a Lewis base activates the hydroxy group of the alcohol initiator or propagating chain (Figure 7.21).[80,82]

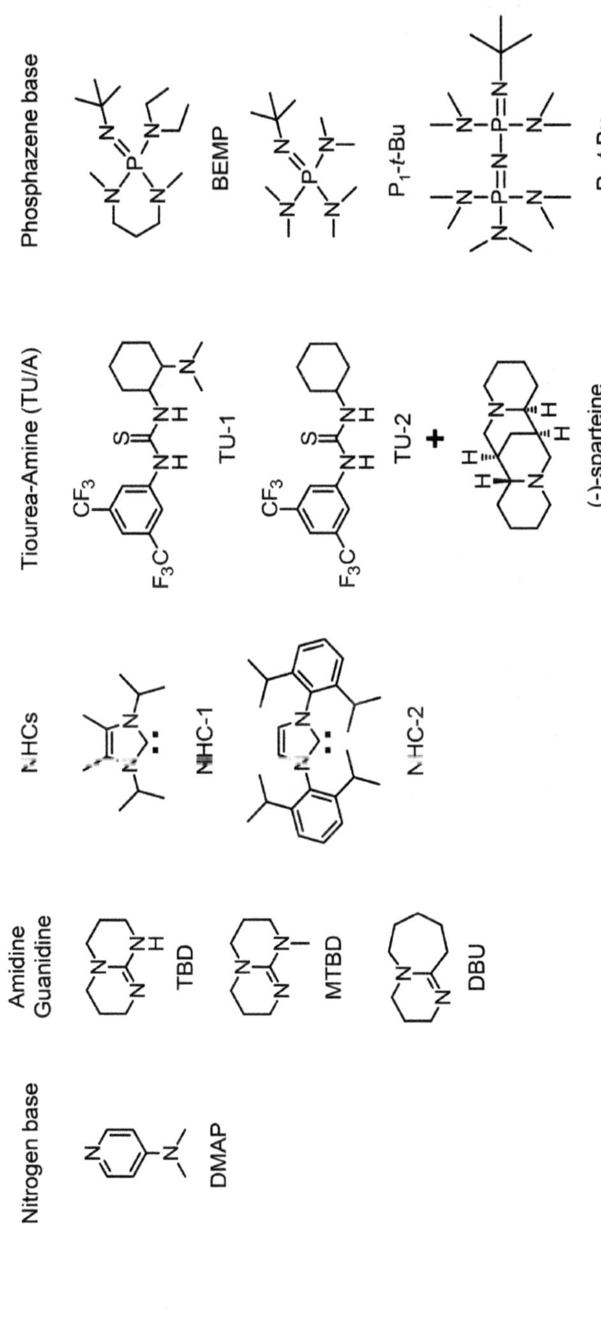

Figure 7.20 Basic organocatalysts applied to ROP of cyclic carbonates.

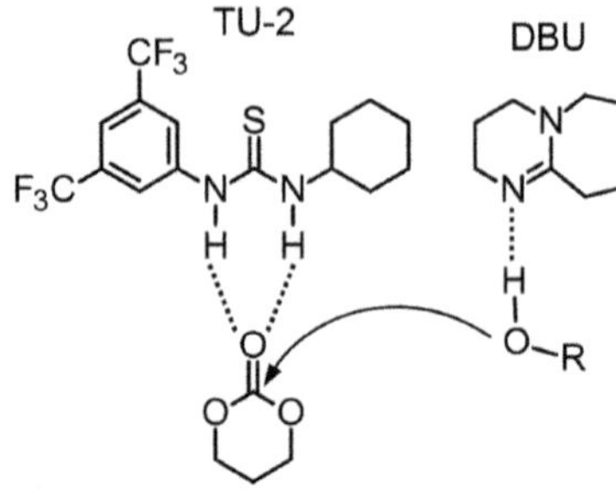

Figure 7.21 Double hydrogen bonding activation mechanism by TU-2 and DBU for ROP of TMC.
Adapted from ref. 80 and 82 with permission from American Chemical Society, Copyright 2007.

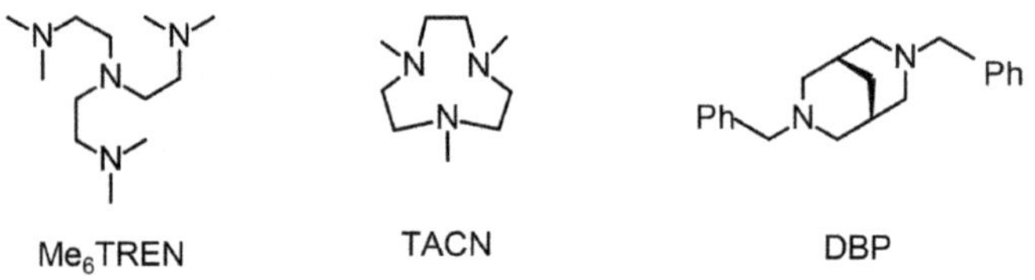

Figure 7.22 Tertiary amines considered as alternatives to (−)-sparteine.
Adapted from ref. 84 with permission from the Royal Society of Chemistry.

This system is accomplished in moderate reaction rates in exchange for controlled chain length whilst suppressing transesterification. The catalytic activity of the organic bases is mostly in accordance with the basicity, namely pK_a.[82] (−)-Sparteine is an alkaloid that can also be applied as a Lewis base in the TU-2/A system. Due to a decreased availability of (−)-sparteine in the early 2010s, several alternatives were explored and developed, which also revealed that two nitrogen atoms of (−)-sparteine activate an alcohol in a bidentate form, proving a higher catalytic activity than that estimated from pK_a (Figure 7.22).[83] *N*,*N*′-dibenzyl bispidine (DBP) seems comparable with (−)-sparteine in terms of activity and controllability among them.[84] However, the actual utility for ROP of cyclic carbonates could be limited for substituted 6CCs because the polymerization rate of TMC using this binary system (TU-2/DBP) is not reasonably high compared with the (−)-sparteine/TU-2 system, taking a few days to convert 10 equiv. of TMC in 2 M $CDCl_3$. Coulembier *et al.* found that *m*-(trimethylammonio)-phenolate betaine (*m*-BE) is effective for ROP of TMC and substituted 6CCs.[85] The ROP is believed to be mediated by a bifunctional activation of an initiator alcohol through ionic nucleophilic interaction (Figure 7.23).[86]

TBD and phosphazene bases are considerably potent and thereby tend to be avoided as this generally translates to a poor control over the polymerization.[87,88] Guillaume *et al.* evaluated the catalytic activity of some organic bases for bulk ROP of TMC under the immortal conditions.[89] TBD and DMAP

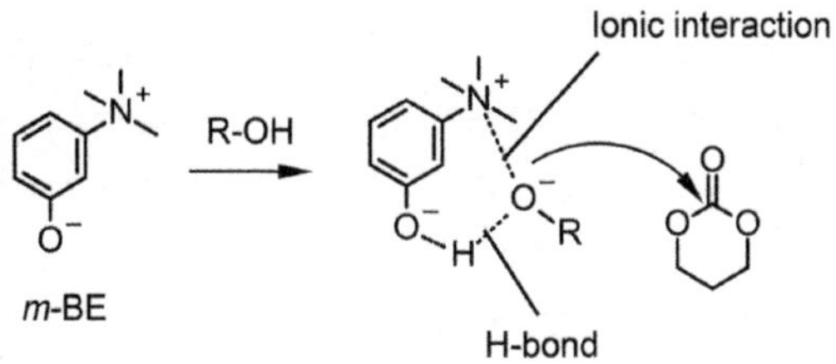

Figure 7.23 Bifunctional activation of an alcohol initiator by m-BE. Adapted from ref. 85 with permission from the Royal Society of Chemistry.

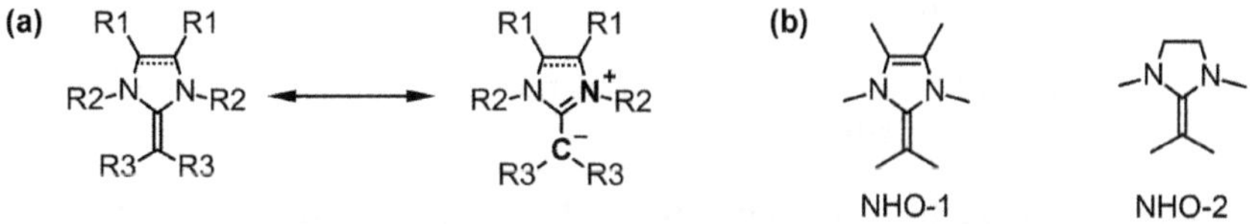

Figure 7.24 (a) Resonance structure of NHO. (b) NHOs effective for ROP of TMC. Adapted from ref. 96, http://d7.doi.org/10.1021/acsmacrolett.5b00873, under the terms of the CC BY 4.0 license, https://creativecommons.org/licenses/by/4.0/.

demonstrated relatively high activity with TOF over 50 000 h^{-1}. However, in comparison with a zinc-β-diiminate complex (BDI)Zn{N(SiMe$_3$)$_2$} and aluminium triflate Al(OTf)$_3$ in similar conditions, the catalytic activity of these organocatalysts still appears modest. Nonetheless, this result proves that organic bases can achieve great performance even in the bulk ROP of TMC operated at temperatures as a high as 150 °C.[89]

NHCs are recognized to be highly active ROP organocatalysts that provides cyclic polymers from LA,[90] lactones,[91,92] carbosiloxanes,[93] and cyclic phosphates[94] by a zwitterionic pathway. However, when applying for the ROP of TMC only linear polymers have been typically obtained.[80] On the other hand, Waymouth and coworkers recently report the formation of a cyclic polycarbonate, although a *N*-substituted eight-membered cyclic carbonate (*N*-8CC) was utilized in the presence of NHC-1 as a catalyst (Figure 7.20).[95] In parallel, N-heterocyclic olefins (NHOs) have drawn attention as a new class of highly active organocatalysts for ROP of heterocycles.[96] NHOs are reported to be stronger electron-donating ligands than NHCs,[97] which translates to an expected higher nucleophilicity and basicity. Dove and coworkers have explored several NHOs (Figure 7.24) for ROP of different cyclic monomers, revealing that the TMC polymerization proceeds differently from that of cyclic esters.[98] Due to the strong basicity of NHOs, the acidic α-carbonyl protons of the cyclic esters are more susceptible to side reactions such as transesterification, enolate formation, and deactivation. As TMC does not possess α-protons, the monomer is consumed in seconds in a well-controlled polymerization with $Đ_M$ as low as 1.12.

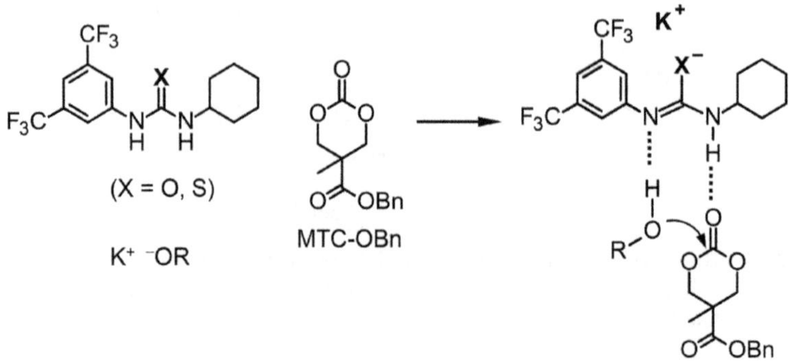

Figure 7.25 Bifunctional activation through dual hydrogen bonding of (thio)urea anions.
Adapted from ref. 100 with permission from American Chemical Society, Copyright 2017.

More recently, inspired by the high levels of control provided by the TU/A system described before, Waymouth *et al.* proved that the use of an alkoxide as initiator in the presence of TU-2 generates a thiourea anion that serves as a bifunctional catalyst for fast and selective ROP of cyclic esters and carbonates (Figure 7.25).[99] A substituted 6CC (MTC-OBn) was converted into the corresponding polymer in 30 min by using 1% of $KOCH_3$ and 5% of TU-2 relative to the monomer in CH_2Cl_2 solution ([M] = 1.0 M), which provided a M_n of 23 000 g mol^{-1} and $Đ_M$ of 1.11. A similar strategy utilizing urea compounds was later reported, which showed to be highly active and selective for the ROP of a series of cyclic monomers.[100] In particular, the urea anion derived from a urea analogue of TU-2 enabled an ultrafast yet controlled ROP of MTC-OBn in THF solution, yielding a polymer with M_n 17 700 g mol^{-1} and $Đ_M$ of 1.14 in only 5 s. Although the catalytic activity was tuned by changing the substituent on the aryl moiety, the ROP of cyclic carbonates was not investigated in depth at this point. Interestingly, these (thio)urea anions presented high catalytic activity in THF, which is an unusual solvent for ROP: other neutral nitrogen bases have the performance decreased due to solvation that mitigates the hydrogen bonding activation. Poor solubility of these anions in CH_2Cl_2 and benzene could also be another reason why THF was adopted.

7.3.2.2 *Organic Acids*

Acid-mediated ROP of cyclic carbonates was first reported by Endo *et al.*, simply using HCl-Et_2O in the presence of H_2O, BCl_3, and alcohols as initiator to form polymers with narrow molecular weight distribution ($Đ_M < 1.20$),[101–103] although trifluoroacetic acid (TFA) was already evaluated for ROP of CL before then.[104] However, the polymerization and the molecular weights of the resultant polymers were shown to be below 10 000 g mol^{-1} when carried out for 24 h at room temperature in CH_2Cl_2. After these reports, increased attention to acidic organocatalysts has been demonstrated after the evolutionary growth of basic organocatalysts by Hedrick and Waymouth (Figure 7.26). In particular for TMC, Martin-Vaca and Bourissou *et al.* revealed that methanesulfonic acid (MSA) afforded a better control than trifluoromethanesulfonic acid, namely triflic acid (HOTf), as a consequence of the lower acidity of the former acid.[23] In addition, the MSA-catalysed ROP demonstrated the absence of ether linkages in poly(TMC), which have often been problematic in cationic ROP of cyclic carbonates. The ROP proceeds through both AM and ACE mechanisms, thus affording a bimodal molecular weight distribution in the resultant polymer. Ribeiro and Peruch *et al.* later reported formation of narrowly dispersed poly(TMC) with monomodal distribution ($Đ_M \sim 1.10$) by simply using diol initiators in the MSA-catalysed ROP of TMC to inhibit the ACE pathway.[22,105] Bourissou and Maron *et al.* also manifested that these sulfonic acids activate both monomer and initiator through a dual hydrogen bonding where a S=O oxygen and an O–H hydrogen act as a hydrogen bonding acceptor and donor, respectively, based on a DFT calculation.[106]

Decarboxylation to form undesired ether linkages and bimodal molecular weight distributions are usually attributed to strong acidity of those catalysts. Kakuchi *et al.* evaluated a weaker organoacidic catalyst diphenyl phosphate (DPP) for the ROP of TMC,[107] a catalyst that has previously demonstrated a controlled/living polymerization of CL.[108] The ROP successfully proceeded only through the AM pathway with mono-alcohol initiators in a toluene solution, providing poly(TMC) with M_n up to 10 000 g mol^{-1} and $Đ_M \sim 1.13$. Li *et al.* also demonstrated the use of phosphoramidic acid (PPA), which is slightly more active than DPP but kept the same level of control ($Đ_M \sim 1.14$).[109] Bourissou and Li revealed that these phosphoric and phosphoramidic acids perform as a bifunctional hydrogen bonding catalyst, by experimental and theoretical studies, respectively.[109,110]

Similarly, phosphorimides (imidodiphosphoric acid (IDPA)) and sulfonimides (o-benzenedisulfonimide (OBS)) are examined in the ROP of TMC, exhibiting comparable catalytic activities and controllability to those of PPA.[111,112] Nevertheless, the catalytic activity of these acidic catalysts is not competing with even moderate basic organocatalysts such as a TU/A system with respect to the polymerization rate. Considering that the acid-catalysed ROP is usually performed in non-polar toluene, the relative activities would be further reduced when operated in slightly more polar solvents such as CH_2Cl_2, which is commonly utilized with basic catalysts. Indeed, the basic organocatalysts based on hydrogen bonding activation are more active for the ROP of lactones and cyclic carbonates in benzene and toluene rather than CH_2Cl_2 and $CHCl_3$.[76,82]

According to a recent report by Satoh *et al.*,[113] bulk ROP using DPP provides multiple benefits as compared to the solution condition using toluene:[107] reduction of catalyst loading (1 *vs.* 0.05 equiv. relative to an initiator) and polymerization time for quantitative conversion (36 h *vs.* 17 h for 50 equiv. of TMC relative to an initiator), yet preserving control over the molecular weight ($Đ_M$ <1.10), which indicates immortal nature. In particular, this high level of control has been scarcely achieved for organocatalytic

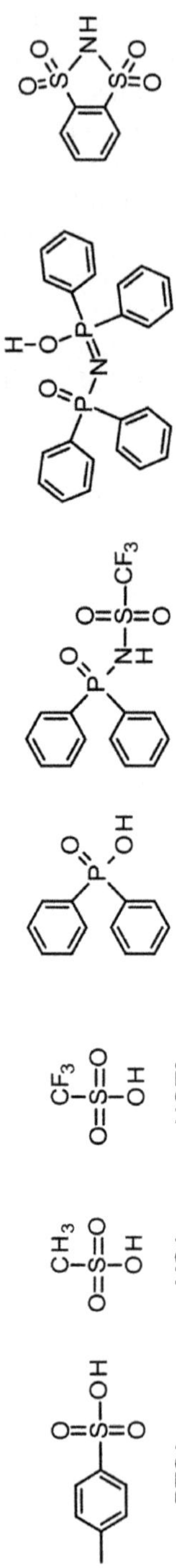

PTSA MSA HOTf DPP PPA IDPA OBS

Figure 7.26 Acidic organocatalysts applied to ROP of cyclic carbonates.

bulk ROP of cyclic esters and carbonates since the DBU-catalysed one,[80] and thus is a notable result. Coady *et al.* found another benefit in the use of these acidic catalysts for ROP of functional cyclic carbonates containing a base sensitive moiety such as an acidic proton of an amide linkage as well as utility of *p*-toluenesulfonic acid (PTSA) for ROP of TMC.[21] This also has opened an efficient post-polymerization functionalization strategy of polycarbonates, using a pentafluorophenyl-functionalized 6CC (MTC-PF, see Figure 7.38) as described later.[114,115]

7.3.2.3 Conjugate Acid–Base Pairs

Salts of strong bases with acids have been used as latent catalysts that sometimes are designed to mitigate the activity. For instance, DBU salts are used in aza-Michael addition and epoxy curing.[116,117] Coady *et al.* first reported a conjugate acid–base pair of DBU and benzoic acid (BA) for controlled ROP.[79] Such acid–base complexes have been recognized as a deliberately mitigated catalyst or a latent catalyst activated by heat. Therefore, the controlled characteristics observed can be attributed to the restrained activity. However, a DFT calculation reveals that the actual catalytic species are the conjugate base and acid; DBU-H^+ activates the carbonyl in the monomer, whereas the benzoate anion (BA^-) interacts with the hydroxy group in the initiator and a propagating chain end. Thus, this conjugate system serves as a dual functional catalyst. Although this DBU-BA system is not sufficiently active for ROP of TMC, Coady's work has opened new class of organic catalysts for controlled ROP as demonstrated by further development of other organic acid/base complexes.[118,119]

In the case of a pair of MTBD and trifluoroacetic acid (TFA) developed by Guo *et al.*, the optimal pairing ratio for the efficient and controlled ROP of TMC is 2 : 1 (Figure 7.27).[120] This indicates that the conjugate acid of MTBD (MTB-H^+) acts as a Lewis acid for activation of TMC and another equiv. of MTBD, which is not sufficiently active to promote ROP of TMC alone, activates the alcohol initiator through double hydrogen bonding.[120] With a similar concept, Guo *et al.* recently reported the use of TBD-BF_4 salt as an alternative Lewis acid to TU-2.[121] Although a role of these organic acid-base complexes in the ROP is not identical to DBU-BA and DMAP-DPP systems,[119] it is significant to have commercially available Lewis acid activators of cyclic carbonyl monomers comparable to TU-2 that has to be synthesized.

Other representative example of acid-base complexes as a ROP catalyst is the NHC-CO_2 adducts that generate free carbene by decarboxylation upon heating.[122,123] The CO_2-adducts are stable under air and moisture conditions, which allows handling in an open atmosphere.[124] It is known that the catalytic reactivity changes with the electronic and steric structure of the carbene. For instance, sterically unencumbered NHCs are highly active.[125,126] As the other base-CO_2 adducts are used as a latent catalyst, the NHC-CO_2 adducts are used in the ROP of TMC with heating at usually above 60 °C in both bulk or solution (Figure 7.28). Since the actual catalytic species

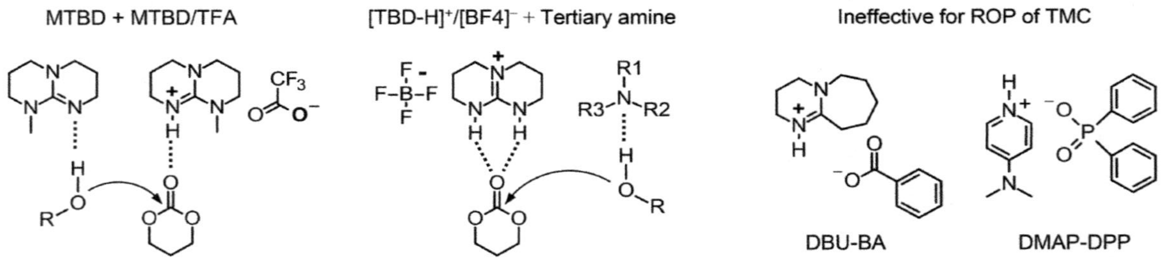

Figure 7.27 Conjugate acid-base pairs used for ROP of TMC and LAs.
Adapted from ref. 120 and 121 with permission from the Royal Society of Chemistry.

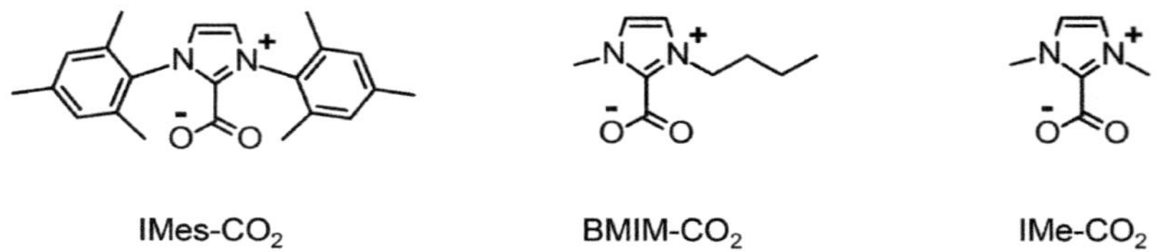

Figure 7.28 NHC-CO_2 adducts efficiently promoting ROP of TMC.
Adapted from ref. 128 with permission from John Wiley and Sons, © 2016 Wiley Periodicals, Inc.

is a carbene and the heating condition promotes transesterification, the extent of chain length control is moderate, yielding the poly(TMC) with $1.2 < Đ_M < 3.0$.[127] The NHC-CO_2 mediated ROP of TMC is also affected by the reaction media, in which CH_2Cl_2 is among the best solvent exhibiting a rate constant comparable with that for bulk polymerization.[128]

7.3.3 Enzymes

Enzymes such as lipase could be an option as a greener ROP catalyst for cyclic carbonates as with lactones and LAs with respect to use of biological resources and low impact to environment. There are only a few reports regarding enzymatic ROP of cyclic carbonates so far. However, Matsumura's extensive and dedicated studies have proved the feasibility of enzymatic ROP of TMC that affords polymers with a weight-average molecular weight (M_w) of over 150 000 g mol^{-1} under reasonable conditions (100 °C, 24 h, in bulk).[129] For ROP of TMC, porcine pancreatic lipase (PPL) and PPL immobilized on Celite (PPL-IM) are more effective than any other enzymes, whereas Novozym-435, which is often applied to ROP of lactones, is not active at the same condition. With an increase of the enzyme concentration, ROP of TMC proceeds in use of Novozym-435, but with low control.[130] Immobilized lipase from *Candida antarctica* (lipase CA) is found to efficiently promote ROP of tetramethylene- or hexamethylene dicarbonates to form high molecular weight polycarbonates over 350 000 g mol^{-1} in M_w.[131] In this study, Matsumura *et al.* have also demonstrated that the lipase CA is also capable of forming macrocyclic dicarbonates from 1,4-butanediol and 1,6-hexanediol in high yields, which had been difficult using triphosgene or diphenyl carbonate in the presence of K_2CO_3.[132–134] A higher number of cyclic compounds is less strained, which is adverse for ROP. As the $Đ_M$ of the resultant polymers is usually broad, enzymatic ROP is not suitable for controlled polymerization. Nonetheless, enzymatic ROP could be an option to produce high molecular weight polycarbonates and an effective alternative route to transform macrocyclic dicarbonates.

7.4 Regioselective ROP of Cyclic Carbonates

Stereoselective polymerization has not been successful in ROP of cyclic carbonates despite many efforts to develop substituted cyclic carbonates for functionalization of polycarbonates as described later. This might be attributed to the lack of commercial availability of substituted cyclic carbonates. Carpentier and Guillaume *et al.* have explored several metal catalysts for ROP of methyl substituted 6 and 7CC to evaluate immortal nature as well as stereo- and regioselectivities arising from dissymmetry of the monomers (Figure 7.29).[135,136] The methyl substituted seven-membered cyclic carbonates (Me7CC) are theoretically derived from renewable resources such as levulinic acid and itaconic acid through 1- or 2-methyl-1,4-butanediols, affording α- and β-substituted Me7CC. No stereoselectivity was observed for

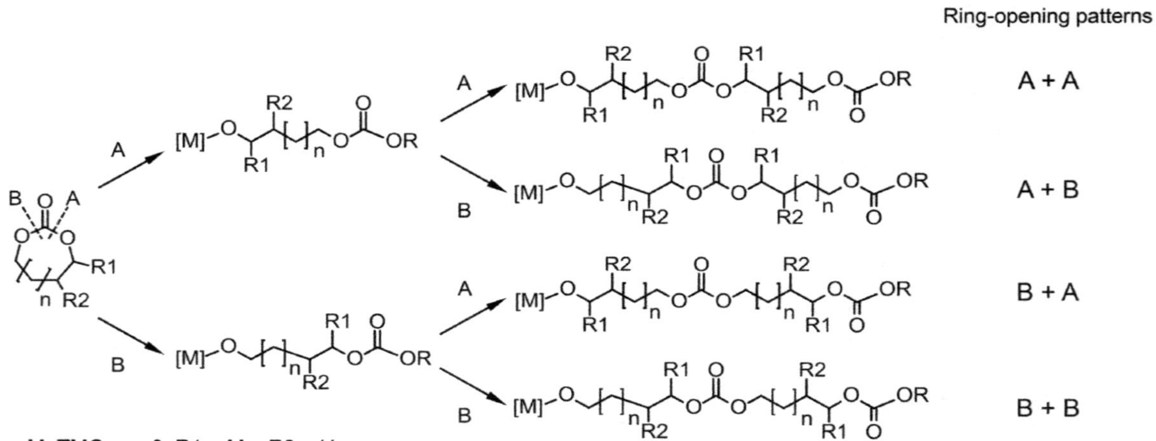

α-MeTMC: n = 0, R1 = Me, R2 = H
α-Me7CC: n = 1, R1 = Me, R2 = H
β-Me7CC: n = 1, R1 = H, R2 = Me

Figure 7.29 Possible patterns of ring-opening of methyl substituted 6- and 7-CCs. Adapted from ref. 135 with permission from American Chemical Society, Copyright 2010.

any of the monomers tested, whereas α-Me7CC exhibited regioselectivity that was dependent on the catalyst employed, although ROP of β-Me7CC proceeded statistically irrespective of catalyst.[135] The lack of the regioselectivity for β-Me7CC was explained by the remote position of the methyl substituent that inhibits the discrimination between the two O–C(O)O bonds by the catalyst (A and B in Figure 7.29). The catalyst-depending order of regioselectivity (X_{reg}) observed for α-Me7CC followed the same trend for stereoselective ROP of *rac*-LA forming heterotactic PLA;[50,137,138] (BDI)Zn{N(SiMe$_3$)$_2$} ($X_{reg}=0.71$) > Y(ONOO^{tBu}){N(SiHMe$_2$)$_2$}(THF) (**27** in Figure 7.30; $X_{reg}=0.58$) > Al(OTf)$_3$ ($X_{reg}=0.27$). These experimental results were supported by a DFT study, which revealed that the activation barriers involved in the ROP of both enantiomers (*R* and *S*) are very close, and a cleavage of oxygen–acyl bond close to methyl substituent is preferred (pattern A + A in Figure 7.29).[139]

α-Methyl substituted six-membered cyclic carbonate, namely α-methyltrimethylene carbonate (α-MeTMC), is more frequently investigated regarding the regioselectivity of the ROP catalysts. In bulk conditions, (BDI)Zn{N(SiMe$_3$)$_2$} reported by Guillaume *et al.* is among the best ($X_{reg}=0.98$), along with a high level of activity (23 °C, 90 min for quantitative conversion) and control ($Đ_M \sim 1.12$).[136] In addition, a series of magnesium and zinc complexes supported by *N,N,O* tridentate binaphthyl- or biphenyl-based iminophenolate ligands (**28–32**), developed by Ma *et al.*, also presented a high regioselectivity ($X_{reg}>0.80$), especially when operated in toluene solution (Figure 7.30).[140–142] In contrast, several basic organocatalysts such as BEMP, TBD, and DMAP (Figure 7.20) examined at the same time indicate neither stereo- nor regioselectivity in the ROP of α-MeTMC.

7.5 Copolymerization

7.5.1 Copolymerization of TMC and LLA

Copolymerization with other cyclic monomers provides a variety of sequences depending on the comonomers and catalysts utilized, which can be suitable for a series of high value applications. A random copolymer of TMC and LLA, for example, has been used in suture and commercialized as Resormer LT® and Inion Optima®, in which TMC is incorporated to decrease the brittleness intrinsic in poly(LLA).[7] On the other hand, block copolymers of TMC and LLA are recognized as biodegradable thermoplastic elastomers and are promising to be applied in biomedical devices such as flexible implants, drug delivery matrices, substrates for cell culture and scaffolds for tissue engineering.[143,144]

As Dobrzynski *et al.* manifested, there is a large gap in the reactivity ratio between TMC and LLA; $r_{TMC}=0.53$, $r_{LLA}=13.0$ in use of Zr(acac)$_4$.[145] Similarly, with SmI$_2$/Sm and lanthanide chlorides, $r_{TMC}/r_{LLA}=0.25/7.24$ and 0.19/15.5 were determined, respectively, by Fineman–Ross or/and Kelen–Tudos methods.[146,147] Curiously, despite these large differences in reactivity,

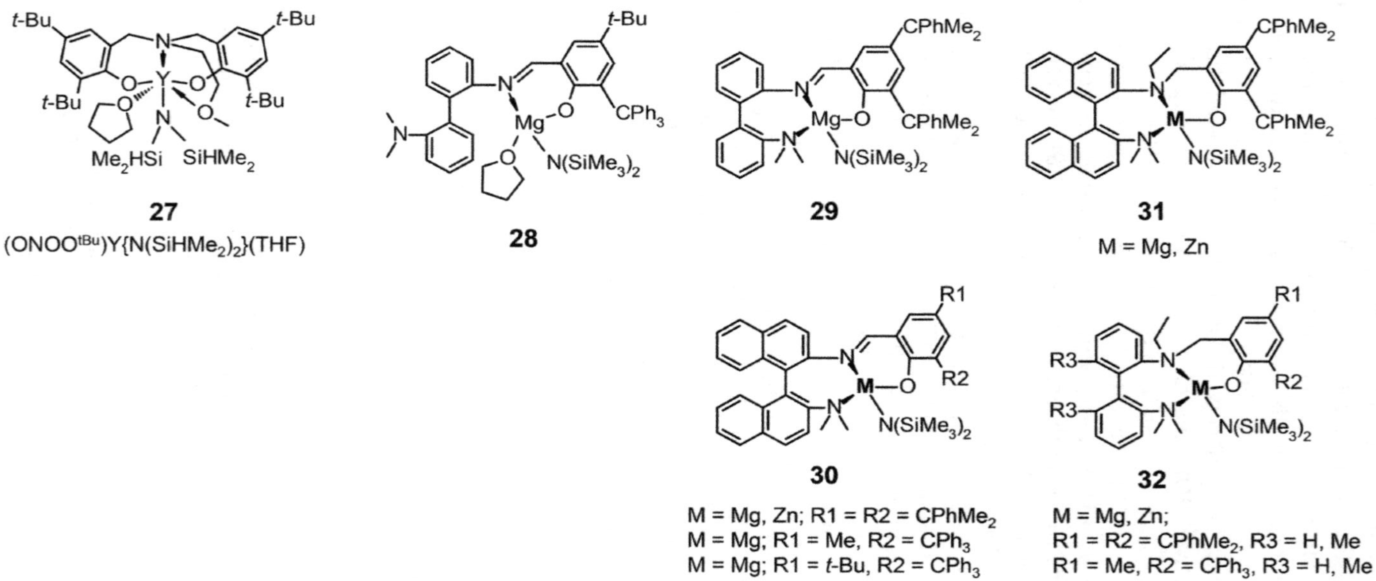

Figure 7.30 Organometallic catalysts enabling regioselective ROP of α-methyl substituted cyclic carbonates.

formation of blocky (gradient) copolymers with a long LLA sequence by simultaneous copolymerization has only been successful using the catalysts mentioned above. Lipik *et al.* reported that α-methylstyrene inhibits transesterification-based chain scrambling in synthesizing block copolymers of LLA and TMC by sequential ROP using $Sn(Oct)_2$ that often yields randomized sequence.[148] Coulembier *et al.*, recently found a unique route to obtain blocky gradient copolymers composed of LLA and TMC exploiting selective organocatalysis of an eutectic melt complex of LLA and TMC that presents a melting point at 21.3 °C.[149]

As described above, since using common catalysts random copolymers are usually obtained, and sequential ROP is often applied to obtain block copolymers of TMC and LLA. In particular, TMC needs to be polymerized first prior to LLA in most of organometallic ROP due to strong chelation of chain end of poly(LLA) to the metal centre that inhibits insertion of TMC for the second block formation.[150] The pentagonal stable lactate chelate formation with tin(II)-complexes and the mechanistic insight for its inhibition of ROP of TMC have recently been elucidated by a DFT study (Figure 7.31). This could also extend to interpret the large gap of the reactivity between TMC and LLA, as mentioned in the section of redox-switchable catalysts, when other organometallic catalysts are used.[151]

Thus, one-pot formation of block copolymers is a challenging theme, but industrially attractive. Although the average block length and reactivity ratios are not mentioned, Spassky *et al.* reported the formation of a nearly perfect block structure of LLA and TMC using an yttrium alkoxide complex $Y(OCH_2CH_2OR)_3$ (R = Me, *i*Pr) as an initiator.[152] In the study by Dobrzynski *et al.* mentioned above, using $Zr(acac)_4$ as a catalyst for the copolymerization, the resultant copolymers were found to show an average block length of 3 to 5 for each segment and was considered to be a multiblock structure.[143] However, the polymers indicated a clear melting endotherm attributed to PLLA segment on DSC, which is associated with a much longer block structure and possibly gradient sequence. Guillaume and Carpentier *et al.* recently attempted simultaneous copolymerization of TMC and LLA using $(BDI)Zn\{N(SiMe_3)_2\}$, $Al(OTf)_3$, and TBD, affording the copolymers composed of almost equimolar amounts of monomers with an average block length of

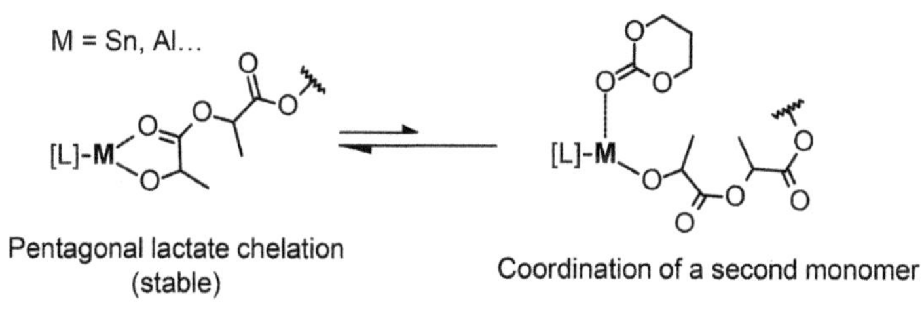

Figure 7.31 Proposed pentagonal chelation of lactate chain end inhibiting coordination of TMC as a second monomer.
Adapted from ref. 151 with permission from John Wiley and Sons, Copyright © 2013 WILEY-VCH Verlag GmbH & Co. KGaA, Weinheim.

6, 11, and 3, respectively.[153] Although these block lengths are still far from being sufficient to exhibit thermoplastic elastomeric properties, $Al(OTf)_3$ develops the potential of one-pot formation of the block copolymer. In addition, this catalyst presents a unique polymerization behaviour that reacts TMC faster than LLA, which is the opposite to most catalysts.

7.5.2 Copolymerization of TMC and CL

Formation of gradient copolymers of TMC and CL has been accomplished by simultaneous ROP of both monomers catalysed by MSA, according to Bourissou.[154] In each homopolymerization with MSA, TMC polymerizes much slower than CL. For the simultaneous ROP, the reactivity ratios of each monomer are determined as $r_{TMC} = 1.2$ and $r_{CL} = 3.4$, which is consistent with those expected from each polymerization rate.[23,78] The reactivity ratios have not often been determined and they vary depending on the catalytic system used; *e.g.* $r_{TMC}/r_{CL} = 0.2/2.4$, 1.6/0.7 and 7.3/0.4 with $Sn(Oct)_2$,[155] $NdCl_3$-5PO,[156] and ZnEtOEt,[157] respectively. Nevertheless, due to a large gap in the reactivity ratio, the microstructure of the resultant copolymers are mentioned as being blocky or gradient.

7.5.3 Copolymerization of TMC and Other Six-membered Cyclic Carbonates

Copolymerization of TMC and substituted cyclic carbonates, in particular with functional groups linked by an ester group, as described later in Section 7.6.2 as MTCs (see Figure 7.36), usually proceeds in a gradient manner where MTC polymerizes first followed by TMC when a moderate catalyst such as a DBU/TU-2 system is used.[81] Although the reactivity ratios are not determined, Hedrick *et al.* have revealed the difference in the polymerization rate between TMC and MTCs and exploited it to form blocky amphiphilic copolymers by simultaneous polymerization.[158] In contrast, it is also possible to construct branched or cross-linked structure based on TMC and MTCs by using catalysts that promotes non-specific transesterification actively, such as lanthanum(III) acetylacetonate.[159]

7.6 Cyclic Carbonates as Polymerizable Monomers

Preparation of cyclic carbonates is categorized in three ways: ring closing, ring expansion, and backbiting (Figure 7.32). The most common way is the ring closing of α, ω-diols with carbonylation reagents such as phosgene and its derivatives.[160,161] However, the toxicity concerns are still controversial even in the solid derivative, triphosgene. Therefore, there are many alternatives developed as carbonylative ring-closing reagents such as carbodiimides and bis-aryl carbonates.[162,163] The ring-expansion of cyclic ethers including epoxides and oxetanes with CO_2 has now drawn attention as a use of renewable feedstock.[164,165] However, in most cases, the formation of

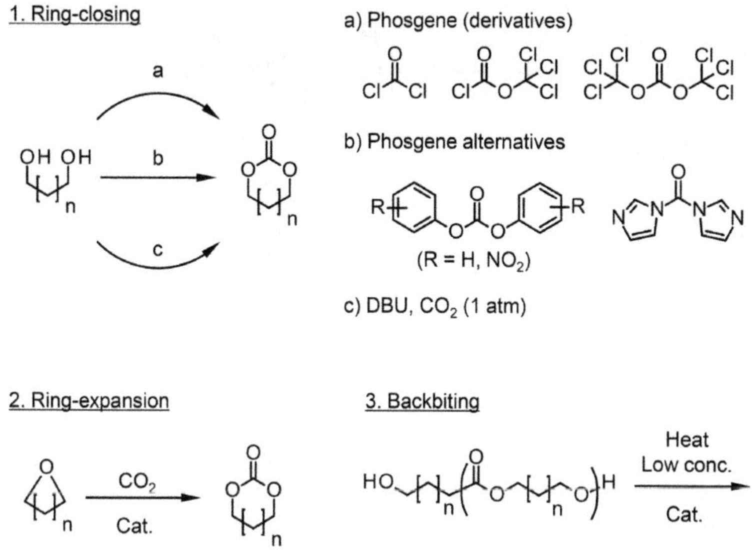

Figure 7.32 Typical routes to prepare cyclic carbonates.

carbonate bonds from cyclic ether and CO_2 is applied to obtain linear polycarbonates by alternative copolymerization using organometallic catalysts.[166,167] Recently, Buchard *et al.* reported organocatalytic approaches to form 6CCs directly from various 1,3-diols and CO_2.[168] Backbiting is not very common but sometimes useful as another option to produce cyclic carbonate as known as ring-closing depolymerization usually performed by adjusting the ring-chain equilibrium.[169]

7.6.1 Five-membered Cyclic Carbonates

As a small numbered ring, 5, 6, 7, and 8 membered aliphatic cyclic carbonates are known to polymerize. However, this does not mean that the all polymerized products are polycarbonates. Typical examples are ethylene carbonate (EC) and propylene carbonate (PC) that are currently not only used as solvent and electrolyte but are also promising as ecological and economical building blocks. Since the enthalpy (ΔH) and entropy (ΔS) for ring opening of most heterocyclic monomers are both negative, they can polymerize below the ceiling temperature (T_c). For EC and PC, the T_c is below 25 °C, but they can polymerize above 100 °C. Although Kadokawa *et al.* demonstrated the ROP of EC below 100 °C using ILs, it is hard to attain molecular weights over 2000 g mol^{-1}.[170] Indeed, ΔH of EC and PC are positive ($\Delta H_p{}^\circ$ 125.6 kJ mol^{-1} for EC), and high temperatures are necessary so the free energy becomes negative. In addition, the ROP proceeds along with decarboxylation, which allows ΔS to be positive and thus drives the polymerization forward. This is why the resultant polymers contain both ether and carbonate linkages in the main chain, as shown in Figure 7.33, and usually are low molecular weight.[171–173]

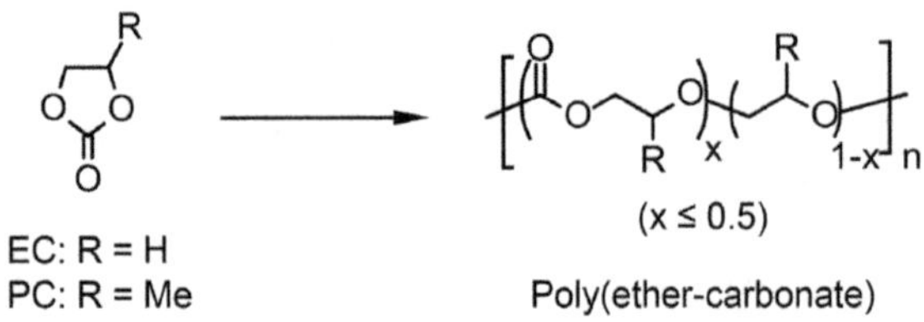

Figure 7.33 ROP of EC and PC forming poly(ether-carbonate)s.
Adapted from ref. 172 with permission from American Chemical Society, Copyright 1992.

The ratio of carbonate to ether bonds in the main chain of the resultant polymers varies depending on polymerization condition such as a catalyst and temperature. Organo-tin or -zinc and acidic catalysts afford poly(ether-carbonate) (PEC) with 40–50 mol% of carbonate unit,[172,174,175] while alkaline or basic catalysts such as KOH, $KHCO_3$, TBD, and phosphazenes provide PEC with 10–20 mol% of carbonate unit as a result of anionic ROP of EC and PC.[15,176] There have been few attempts to obtain high molecular weight polymers by the ROP of EC and PC. Recently, Zhang *et al.* demonstrated the organocatalytic ROP of EC conducted at 180 °C for 2 h using a phosphazene base P_4-*t*-Bu, affording poly(ethylene carbonate-*co*-ethylene oxide) (PECEO) with M_n of 14 000 g mol^{-1} containing 21 mol% of EC unit.[177]

PEC involves two characteristics: biodegradability of aliphatic carbonates and biocompatibility of alkylene ethers. From the viewpoint of biodegradability, there have been several efforts to control decarboxylation, especially by increasing carbonate linkages. However, the formation of poly(EC) and poly(PC) without ether units has never been accomplished by ROP of EC and PC, except for alternating copolymerization of ethylene oxide and propylene oxide with CO_2.[178] No CO_2 abstraction during the ROP of EC and PC has been confirmed in copolymerization with lactones, in particular using specific catalysts (($C_6Me_5)_2SmMe(THF)$ and **6**) at a certain feed ratios.[179,180] Copolymerization of EC/PC with cyclic urea such as tetramethylene urea (TMU) also avoids decarboxylation during the ROP that yields polyurethanes through alternating insertion of each monomer in the presence of dibutyl magnesium Bu_2Mg (Figure 7.34).[181,182] In contrast, the PECEO with a low carbonate composition is soluble in water and demonstrates poly(ethylene glycol) (PEG)-like properties, being used as a hydrophilic block of amphiphilic drug carriers.[183] Thus, ether-rich PEC is expected to be a PEG alternative, and preparation of PEG is also being investigated by anionic ROP of EC with actively promoted decarboxylation.[184] Nonetheless, the molecular weight control and end group fidelity should be improved for these PEC to be feasibly considered as a PEG alternative.

EC and PC, namely aliphatic 5CCs, do not favour the formation of a consecutive sequence of carbonate units by ROP as described above. This is due to the thermodynamic stability of the 5CC structure and the eventual ring-chain equilibrium unfavourable to ring opening. However, the ring-opening

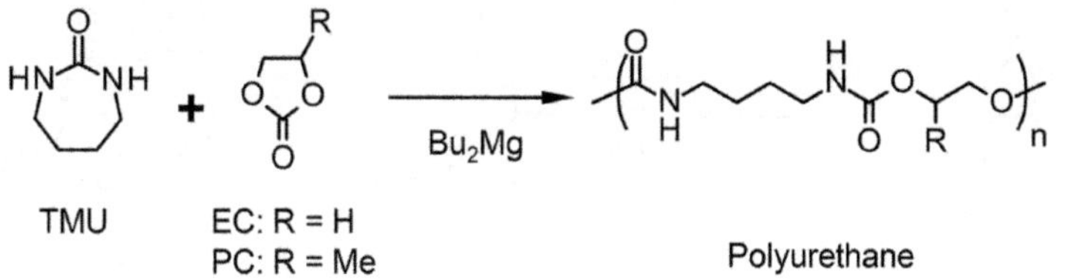

Figure 7.34 Formation of polyurethane by alternating ROP of TMU and EC/PC. Adapted from ref. 181 with permission from John Wiley and Sons, Copyright © 2004 WILEY-VCH Verlag GmbH & Co. KGaA, Weinheim.

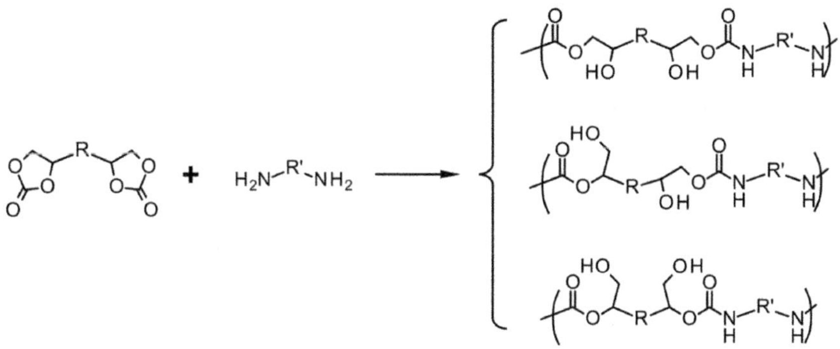

Figure 7.35 Ring-opening aminolysis polycondensation of bis-5CC compound with diamine forming poly(hydroxy urethane).
Adapted from ref. 187 with permission from Springer Nature, Copyright 2012.

reaction by nucleophiles other than alcohol is thermodynamically possible as verified by polyurethane formation with cyclic urea.[181] Thus, 5CC-bearing compounds such as (meth)acrylates and vinyl monomers are often used as a reactive building block, mostly against amine reagents.[185] 5CC-derived polyurethanes are prepared by step-growth reaction of bis-5CC compounds and diamines, which is regarded as ring-opening aminolysis polycondensation.[186] This reaction has drawn attention as an isocyanate-free approach for polyurethane synthesis (Figure 7.35).[187,188] The resultant polyurethanes have hydroxy side chains that contain both primary and secondary hydroxy groups.

Other 5CCs include mostly strained ring by fusing with cycloalkanes/alkenes and sugars, which significantly favours the ROP to form polycarbonates without ether linkages. Haba and Endo *et al.* first reported a successful ROP of 5CC without CO_2 elimination using a glucopyranoside fused ring 5CC, namely methyl 4,6-*O*-benzylidene-2,3-*O*-carbonyl-α-D-glucopyranoside (MBCG) (Figure 7.36).[18] The further model reaction using 5CCs fused with cyclohexane (CHC) revealed that only a 5CC attached to a cyclohexyl or pyranoside ring in a *trans* fashion is capable of ring-opening.[189,190] Guillaume *et al.* then demonstrated stereoselective ROP of a racemic mixture of *trans*-CHCs, yielding

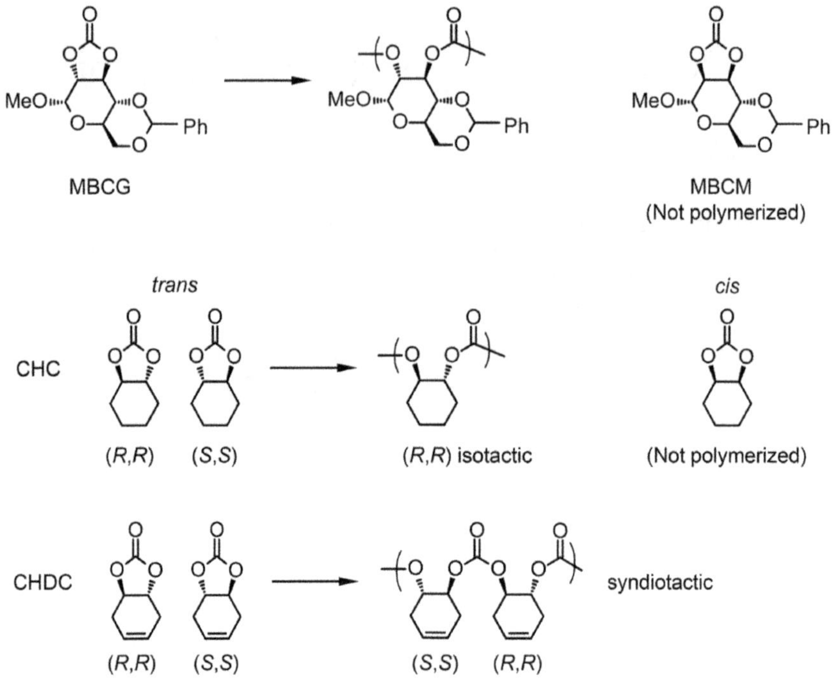

Figure 7.36 Polymerizable 5CCs with fused ring.

highly isotactic polymers majorly containing *trans*-(*R*,*R*)-CHC.[191] This enantioselectivity was attributed to the steric character of the monomer rather than catalyst, as demonstrated by the successful performance of various catalysts. A similar 5CC fused with a cyclohexene ring (CHDC) also afforded successful ROP, although the reactivity was lower than that of CHC. The resultant polymers derived from a racemic mixture of *trans*-CHDC exhibited a highly syndiotactic microstructure that was impractical by alternating ring-opening copolymerization of cyclohexadiene oxide and CO_2.[192] These fused-ring polycarbonates showed high thermal properties, and so are a promising candidate for engineering applications utilizing renewable feedstock.

7.6.2 Six-membered Cyclic Carbonates

The most popular 6CC is TMC, and poly(TMC) is the most widely studied polycarbonate as biomaterials such as suture, scaffold and bone fixation materials.[7] As TMC is derived from 1,3-propanediol and synthesized by a carbonylative ring-closing reaction using phosgene derivatives and equivalents, a variety of TMC analogues have been developed using similar strategy from substituted 1,3-diols, many of which are commercially available (Figure 7.37). These TMC analogues have been shown to be ring-opening polymerized to provide functional aliphatic polycarbonates.[7] Among them, 6CCs derived from 2,2-bis(hydroxymethyl)propionic acid (bis-MPA) are now

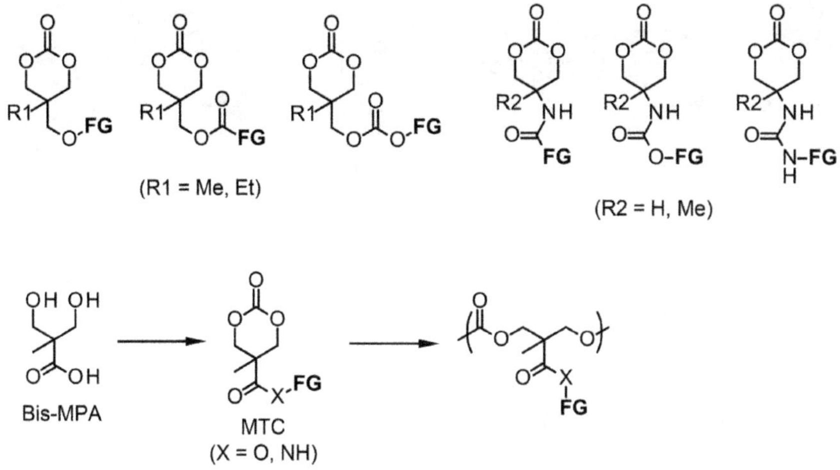

Figure 7.37 TMC analogues with a functional group (FG) as a substituent.

widely investigated by many researchers since a comprehensive synthetic strategy was generalized by Hedrick *et al.*[81,193] Unlike substituted LAs/glycolides and α-substituted lactones, the installed substituent (FG: functional group) apart from the centre of ring-opening scarcely affects the polymerizability of the bis-MPA-derivatives, namely MTCs, which allows statistical sequences for copolymerization of differently substituted MTCs.[194] In contrast, probably due to the inductive effect of the ester linkage, the relative reactivity of MTCs to TMC is high, manifesting formation of blocky structure when TMC and a MTC are copolymerized simultaneously.[158]

Since most MTCs possess an ester group at the side chain, the polymerization requires optimization to avoid transesterification that possibly occurs between main chain carbonate linkages and side chain ester groups. Organocatalysed ROP conducted in solution is thus often adopted, which allows controlled polymerization as well as no side reaction such as cross-linking as a result of the transesterification. In fact, Dobrzynski *et al.* reported formation of branched polycarbonates by copolymerization of methyl-functionalized MTC (X = O, FG = Me in Figure 7.37) and TMC performed in bulk at 120 °C in the presence of lanthanum acetylacetonate $La(acac)_3$.[159] In contrast, Harth *et al.* demonstrated that $Sn(OTf)_2$ enables bulk ROP of MTCs (X = O, FG = ethyl, allyl) at 70 °C with good control, maintaining a high level of end group fidelity and narrow $Đ_M$ of less than 1.20.[195]

Typical organometallic compounds such as $Sn(Oct)_2$ and $Al(i\text{PrO})_3$ are often used for ROP of other functionalized 6CCs without ester linkage at the side chain both in bulk and solution, resulting in successful polymerization with no cross-linking.[196–198] Amide group in the side chain of 6CCs including MTCs (X = NH) is somewhat problematic in ROP using basic organocatalysts, due to its acidic amide proton or hydrogen bonding of amide group that potentially hampers the catalysis based on hydrogen bonding

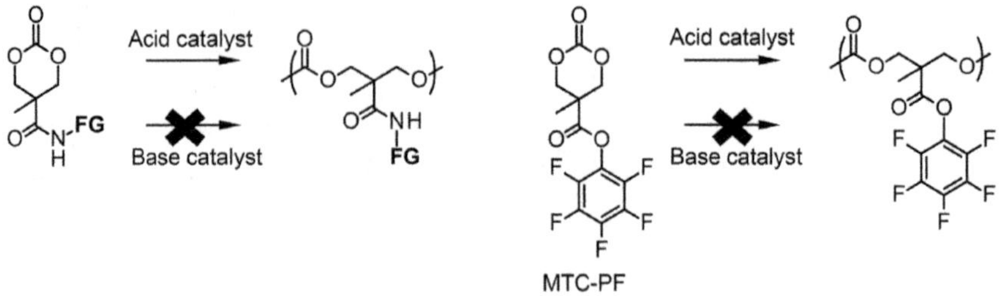

Figure 7.38 Amide-containing 6CC and MTC-PF favouring acidic organocatalysis for ROP.

activation (Figure 7.38). Hedrick *et al.* reported that acidic organocatalysts are able to promote the ROP of MTCs functionalized with amide or pentafluorophenyl group (MTC-PF).[21,114]

From the viewpoint of bio-based production, glycerol-based 6CCs are an alternative including those derived from dihydroxyacetone (DHA), as shown in Figure 7.39.[199] In addition, there are sugar-fused 6CCs that are derived from carbohydrates such as glucose, mannose, xylose, deoxyribose, and their derivatives.[200–202] Recent advance in the cyclization technologies and CO_2 chemistry allows these sugar-based polycarbonates to fully comprise natural feedstock (carbohydrates and CO_2).[168] These sugar-based polycarbonates exhibit high glass transition temperature (T_g) ranging from over 58 to 152 °C, depending on substituents on the fused sugar ring that contribute to restricting chain mobility.[200,201,203] Thus, these sugar-based polycarbonates are expected to be applied in a new class of engineering polycarbonates as with poly(CHC) and poly(CHDC) (Figure 7.36).

7.6.3 Seven-membered Cyclic Carbonates

Generally, 7-CCs are more easily polymerized by ROP than 6CCs due to more strained ring structure that facilitates ring-opening towards formation of linear chains.[204] Endo and co-workers' extensive works have revealed that the simplest 7CC, 1,3-dioxepan-2-one is polymerized 100 times faster than TMC under a cationic ROP condition using 1 mol% of MeOTf as an initiator in CH_2Cl_2 at 20 °C. In addition, this 7CC is not subjected to decarboxylation at the cationic ROP in which elimination of CO_2 is somewhat observed in the case of TMC.[132] Nevertheless, not many functional 7CC analogues are studied in contrast to 6CCs, probably due to poor commercial availability of functionalized 1,4-diols that are considered as starting materials for the substituted 7CCs. Instead, some aromatic 7CCs have been reported, and they can be polymerized by anionic initiators such as *t*BuOK.[205,206] The 7CCs with biphenyl structure are highly reactive due to the ring strain caused by the biphenyl ring that also provides chirality of the resultant polycarbonates (Figure 7.40).[207] These aromatic

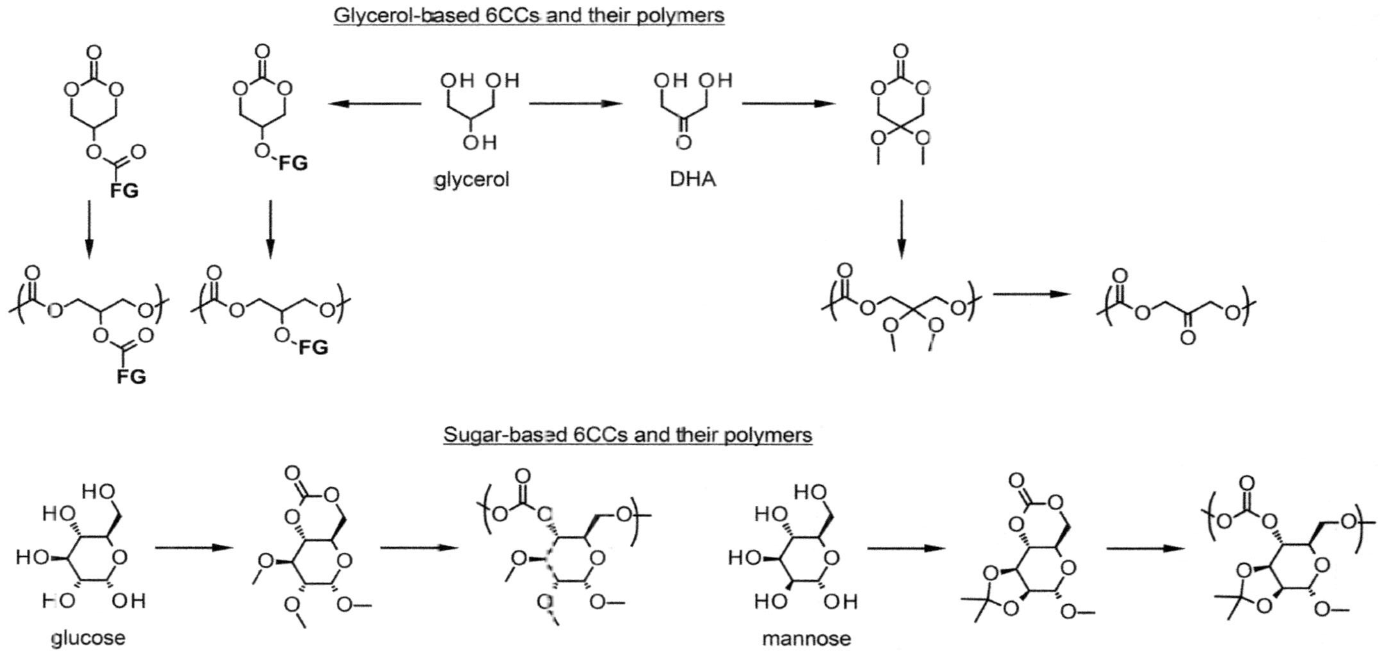

Figure 7.39 Bio-based 6CCs derived from glycerol and sugars.
Adapted from ref. 199–201 with permission from American Chemical Society, Copyright 2016.

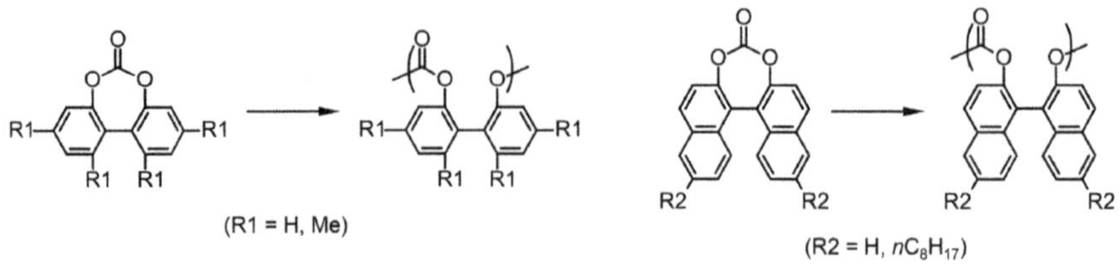

Figure 7.40 7CCs with biphenyl structure and their polymers.
Adapted from ref. 207 with permission from Springer Nature, Copyright 1999.

polycarbonates have been expected to be alternatives to bisphenol-A (BPA) polycarbonate.[208] However, the thermal properties of these polymers have not been reported. Considering the need of severe heating over 200 °C for bulk ROP, the polymers derived from these aromatic 7CCs are estimated to have high thermal stability.

7.6.4 Eight-membered Cyclic Carbonates

Cyclic carbonates with more than seven-membered rings are relatively uncommon.[31,32,133] Recently, Hedrick and Yang *et al.* developed *N*-substituted eight-membered aliphatic cyclic carbonates (N-8CCs) and successfully obtained the corresponding polymers with narrow molecular weight distribution ($Đ_M$ < 1.3) using basic organocatalysts such as TBD and DBU.[209] In addition, Waymouth *et al.* demonstrated successful formation of cyclic polycarbonates from N-8CCs in the presence of NHC-1 (Figure 7.20) with no initiator.[95] *N*-alkyl groups in the main chain could be subsequently quaternized with alkyl iodides (R-I, Figure 7.41) to yield cationic polycarbonates that can be employed for antimicrobials and gene carriers.[210] Interestingly, more recently Sardon *et al.* reported that direct ROP of quaternized monomer has been successfully processed. This is a rare report for ROP of heterocyclic monomers in ionic form.[211]

7.6.5 Cyclic Oligo-/Polycarbonates

At the reaction to form cyclic carbonates from diols through a ring-closing reaction, cyclic dicarbonates and larger rings could be obtained along with desired cyclic monocarbonates, especially when the target ring size is over seven.[31,32,133] As with polyesters, a transesterification-driven polycondensation method to prepare polycarbonates often involves ring-closing depolymerization reaction at the late stage of the reaction.[212] It is known that ROP of macrocyclic oligomers affords higher molecular weight of BPA polycarbonate compared to that prepared by the conventional condensation approach.[212,213] Ring strain does not much reside in these macrocyclic oligomers, changes in enthalpy upon the ring-opening are minimal, and thus polymerization

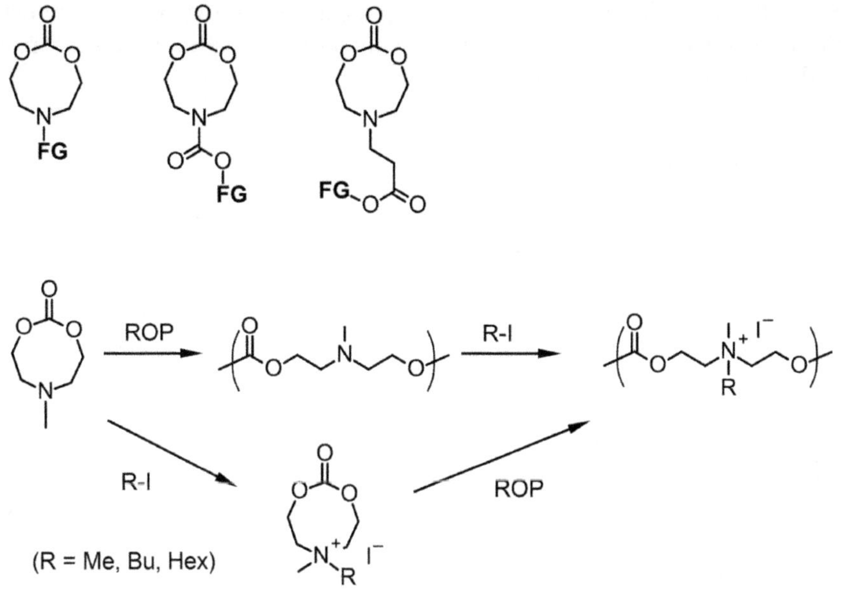

Figure 7.41 N-8CCs and their transformation to cationic polycarbonates. Adapted from ref. 209–211 with permission from American Chemical Society, Copyright 2017, and John Wiley and Sons, © 2017 WILEY-VCH Verlag GmbH & Co. KGaA, Weinheim.

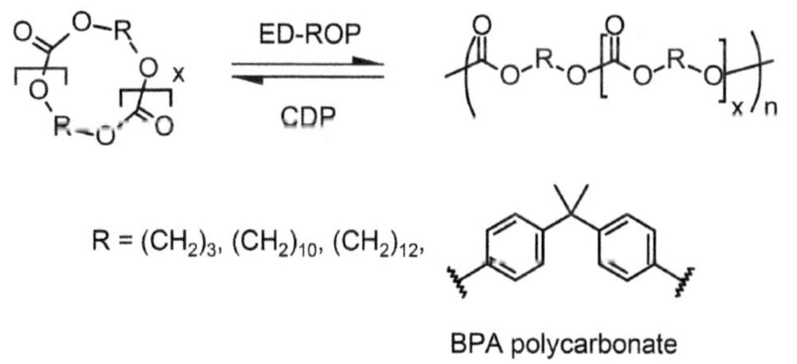

Figure 7.42 ED-ROP and CDP of macrocyclic oligocarbonates.

becomes entropy-driven through an increase of conformational freedom. Entropically driven ROP (ED-ROP) is thereby carried out for these macrocyclic oligocarbonates with heating in the presence of catalysts (Figure 7.42). ED-ROP is based on classical ring-chain equilibria, which are sensitive to concentration: leaning to macrocycles at diluted condition (preferably <2–3 w/v%), and linear chain (polymer) at high concentration (preferably bulk).[214] BPA polycarbonate and hexa/deca/dodeca-methylene polycarbonates have been successfully subjected to cyclo-depolymerization (CDP), which is an opposite reaction of ED-ROP. Eventually, the ED-ROP is employed to yield high molecular weight polycarbonates from the macrocycles resulting from CDP of linear polycarbonates with lower molecular weights. Since the macrocycles are usually a mixture of several sizes of rings, the molar-mass distribution is as broad as that of condensation products.[215]

7.7 Conclusion

ROP of cyclic carbonates is summarized, in particular recent progress in catalysts and monomer synthesis in the last decade in addition to fundamental information regarding mechanistic insights. Aliphatic polycarbonates, which are mostly obtained by ROP of corresponding cyclic carbonates, are a relatively new class of biodegradable polymers but have been drawing increasing attention due to the progress in catalysts and functionalization approaches including monomer designs. Cyclic carbonates with fused-ring including sugars enable formation of alternative engineering polycarbonates with high thermal stability. In addition, direct use of CO_2 for cyclization of α,ω-diols is now being established, which provides a greener alternative to access ROP-derived polycarbonates. Thus, learning and advancing the ROP of cyclic carbonates would be a key to further expand the potentials and applications of the polycarbonates as both biomaterials and green engineering materials. For instance, catalyst developments are still desired in the future for production of polycarbonate-based materials with high economic impact.

References

1. D. C. Clagett and S. J. Shafer, in *Comprehensive Polymer Science and Supplements*, ed. G. Allen and J. C. Bevington, Pergamon, Amsterdam, 1989, pp. 345–356.
2. L. Vogdanis, B. Martens, H. Uchtmann, F. Hensel and W. Heitz, *Makromol. Chem.*, 1990, **191**, 465–472.
3. K. J. Zhu, R. W. Hendren, K. Jensen and C. G. Pitt, *Macromolecules*, 1991, **24**, 1736–1740.
4. A. C. Albertsson and M. Sjoeling, *J. Macromol. Sci., Pure Appl. Chem.*, 1992, **A29**, 43–54.
5. A. C. van Leeuwen, H. Yuan, G. Passanisi, J. W. van der Meer, J. D. de Bruijn, T. G. van Kooten, D. W. Grijpma and R. R. Bos, *Eur. Cells Mater.*, 2014, **27**, 81–97.
6. A. C. van Leeuwen, R. R. M. Bos and D. W. Grijpma, *J. Biomed. Mater. Res., Part B*, 2012, **100B**, 1610–1620.
7. K. Fukushima, *Biomater. Sci.*, 2016, **4**, 9–24.
8. S. Venkataraman, J. L. Hedrick, Z. Y. Ong, C. Yang, P. L. R. Ee, P. T. Hammond and Y. Y. Yang, *Adv. Drug Delivery Rev.*, 2011, **63**, 1228–1246.
9. M. Sepulchre, M.-O. Sepulchre, M.-A. Dourges and M. Neblai, *Macromol. Chem. Phys.*, 2000, **201**, 1405–1414.

10. F. Liu, L.-L. Mei, Z.-L. Tan, G.-P. Yan, J.-F. Guo, Q. Zhang, H. Liu and J. Yang, *Chin. J. Polym. Sci.*, 2016, **34**, 1330–1338.
11. L. Liao, C. Zhang and S. Gong, *J. Polym. Sci., Part A: Polym. Chem.*, 2007, **45**, 5857–5863.
12. X. Zhang, M. Chen, C. Liu, A. Zhang and R. Sun, *Molecules*, 2015, **20**, 6033.
13. T. Endo, *Handbook of Ring-Opening Polymerization*, 2009.
14. A. Duda and A. Kowalski, in *Handbook of Ring-opening Polymerization*, ed. P. Dubois, O. Coulembier and J. Marie Raquez, 2009, pp. 1–51.
15. J.-C. Lee and M. H. Litt, *Macromolecules*, 2000, **33**, 1618–1627.
16. M. Murayama, F. Sanda and T. Endo, *Macromolecules*, 1998, **31**, 919–923.
17. S. Kühling, H. Keul and H. Höcker, *Makromol. Chem.*, 1989, **15**, 9–13.
18. O. Haba, H. Tomizuka and T. Endo, *Macromolecules*, 2005, **38**, 3562–3563.
19. H. Keul, R. Baecher and H. Hoecker, *Makromol. Chem.*, 1986, **187**, 2579–2589.
20. T. Endo, Y. Shibasaki and F. Sanda, *J. Polym. Sci., Part A: Polym. Chem.*, 2002, **40**, 2190–2198.
21. D. J. Coady, H. W. Horn, G. O. Jones, H. Sardon, A. C. Engler, R. M. Waymouth, J. E. Rice, Y. Y. Yang and J. L. Hedrick, *ACS Macro Lett.*, 2013, **2**, 306–312.
22. J. M. Campos, M. R. Ribeiro, M. F. Ribeiro, A. Deffieux and F. Peruch, *Macromol. Chem. Phys.*, 2013, **214**, 85–93.
23. D. Delcroix, B. Martín-Vaca, D. Bourissou and C. Navarro, *Macromolecules*, 2010, **43**, 8828–8835.
24. J. Jitonnom and W. Meelua, *J. Organomet. Chem.*, 2017, **841**, 48–56.
25. H. R. Kricheldorf and J. Jenssen, *J. Macromol. Sci., Chem.*, 1989, **26**, 631–644.
26. H. R. Kricheldorf, I. Kreiser-Saunders and C. Boettcher, *Polymer*, 1995, **36**, 1253–1259.
27. H. R. Kricheldorf, I. Kreiser-Saunders and A. Stricker, *Macromolecules*, 2000, **33**, 702–709.
28. A. Kowalski, A. Duda and S. Penczek, *Macromol. Rapid Commun.*, 1998, **19**, 567–572.
29. K. Majerska, A. Duda and S. Penczek, *Macromol. Rapid Commun.*, 2000, **21**, 1327–1332.
30. H. R. Kricheldorf and A. Stricker, *Macromol. Chem. Phys.*, 2000, **201**, 2557–2565.
31. H. R. Kricheldorf, M. Lossin and A. Mahler, *Macromol. Chem. Phys.*, 1997, **198**, 3559–3570.
32. H. R. Kricheldorf, A. Mahler and S.-R. Lee, *J. Macromol. Sci., Pure Appl. Chem.*, 1997, **A34**, 417–428.
33. H. R. Kricheldorf and A. Stricker, *Polymer*, 2000, **41**, 7311–7320.
34. B. Wurm, H. Keul and H. Hocker, *Macromol. Chem. Phys.*, 1994, **195**, 3489–3498.
35. H. Hoecker, H. Keul, S. Kuehling, W. Hovestadt, A. Mueller and B. Wurm, *Makromol. Chem., Macromol. Symp.*, 1993, **73**, 1–5.

36. M. Helou, O. Miserque, J.-M. Brusson, J.-F. Carpentier and S. M. Guillaume, *Chem. - Eur. J.*, 2008, **14**, 8772–8775.
37. E. M. Broderick, N. Guo, C. S. Vogel, C. Xu, J. Sutter, J. T. Miller, K. Meyer, P. Mehrkhodavandi and P. L. Diaconescu, *J. Am. Chem. Soc.*, 2011, **133**, 9278–9281.
38. I. Palard, M. Schappacher, B. Belloncle, A. Soum and S. M. Guillaume, *Chem. - Eur. J.*, 2007, **13**, 1511–1521.
39. H. Sheng, F. Xu, Y. Yao, Y. Zhang and Q. Shen, *Inorg. Chem.*, 2007, **46**, 7722–7724.
40. J. Koller and R. G. Bergman, *Organometallics*, 2011, **30**, 3217–3224.
41. H. Sheng, J. Shi, Y. Feng, H. Wang, Y. Jiao, H. Sheng, Y. Zhang and Q. Shen, *Dalton Trans.*, 2012, **41**, 9232–9240.
42. A. Duda, *Macromolecules*, 1996, **29**, 1399–1406.
43. B. Wurm, H. Keul, H. Hoecker, G. Sylvester, E. Leitz and K. H. Ott, *Makromol. Chem., Rapid Commun.*, 1992, **13**, 9–14.
44. P. Dobrzynski, M. Pastusiak and M. Bero, *J. Polym. Sci., Part A: Polym. Chem.*, 2005, **43**, 1913–1922.
45. D. J. Darensbourg, W. Choi, P. Ganguly and C. P. Richers, *Macromolecules*, 2006, **39**, 4374–4379.
46. D. J. Darensbourg, W. Choi, O. Karroonnirun and N. Bhuvanesh, *Macromolecules*, 2008, **41**, 3493–3502.
47. S. Inoue, *J. Polym. Sci., Part A: Polym. Chem.*, 2000, **38**, 2861–2871.
48. H. Sugimoto, T. Aida and S. Inoue, *Macromolecules*, 1990, **23**, 2869–2875.
49. S. M. Guillaume and J.-F. Carpentier, *Catal. Sci. Technol.*, 2012, **2**, 898–906.
50. B. M. Chamberlain, M. Cheng, D. R. Moore, T. M. Ovitt, E. B. Lobkovsky and G. W. Coates, *J. Am. Chem. Soc.*, 2001, **123**, 3229–3238.
51. L. R. Rieth, D. R. Moore, E. B. Lobkovsky and G. W. Coates, *J. Am. Chem. Soc.*, 2002, **124**, 15239–15248.
52. P. Brignou, S. M. Guillaume, T. Roisnel, D. Bourissou and J.-F. Carpentier, *Chem. - Eur. J.*, 2012, **18**, 9360–9370.
53. M. Helou, O. Miserque, J.-M. Brusson, J.-F. Carpentier and S. M. Guillaume, *Adv. Synth. Catal.*, 2009, **351**, 1312–1324.
54. D. J. Darensbourg, W. Choi and C. P. Richers, *Macromolecules*, 2007, **40**, 3521–3523.
55. M. Helou, O. Miserque, J.-M. Brusson, J.-F. Carpentier and S. M. Guillaume, *ChemCatChem*, 2010, **2**, 306–313.
56. S. Kobayashi and C. Ogawa, *Chem. - Eur. J.*, 2006, **12**, 5954–5960.
57. E. Piedra-Arroni, P. Brignou, A. Amgoune, S. M. Guillaume, J.-F. Carpentier and D. Bourissou, *Chem. Commun.*, 2011, **47**, 9828–9830.
58. F. Hild, L. Brelot and S. Dagorne, *Organometallics*, 2011, **30**, 5457–5462.
59. F. Hild, N. Neehaul, F. Bier, M. Wirsum, C. Gourlaouen and S. Dagorne, *Organometallics*, 2013, **32**, 587–598.
60. C. C. Willhite, N. A. Karyakina, R. A. Yokel, N. Yenugadhati, T. M. Wisniewski, I. M. F. Arnold, F. Momoli and D. Krewski, *Crit. Rev. Toxicol.*, 2014, **44**, 1–80.

61. V. Poirier, T. Roisnel, S. Sinbandhit, M. Bochmann, J.-F. Carpentier and Y. Sarazin, *Chem. - Eur. J.*, 2012, **18**, 2998–3013.
62. B. Liu, T. Roisnel, L. Maron, J.-F. Carpentier and Y. Sarazin, *Chem. - Eur. J.*, 2013, **19**, 3986–3994.
63. G. Schnee, C. Fliedel, T. Aviles and S. Dagorne, *Eur. J. Inorg. Chem.*, 2013, **2013**, 3699–3709.
64. C. Fliedel, S. Mameri, S. Dagorne and T. Aviles, *Appl. Organomet. Chem.*, 2014, **28**, 504–511.
65. M. Haddad, M. Laghzaoui, R. Welter and S. Dagorne, *Organometallics*, 2009, **28**, 4584–4592.
66. C. Fliedel, V. Rosa, F. M. Alves, A. M. Martins, T. Aviles and S. Dagorne, *Dalton Trans.*, 2015, **44**, 12376–12387.
67. S. M. Guillaume, M. Schappacher and A. Soum, *Macromolecules*, 2003, **36**, 54–60.
68. I. Palard, A. Soum and S. M. Guillaume, *Macromolecules*, 2005, **38**, 6888–6894.
69. M. Schmid, S. M. Guillaume and P. W. Roesky, *Organometallics*, 2014, **33**, 5392–5401.
70. S. M. Guillaume, P. Brignou, N. Susperregui, L. Maron, M. Kuzdrowska, J. Kratsch and P. W. Roesky, *Polym. Chem.*, 2012, **3**, 429–435.
71. M. Schmid, P. Ona-Burgos, S. M. Guillaume and P. W. Roesky, *Dalton Trans.*, 2015, **44**, 12338–12348.
72. E. M. Broderick, N. Guo, T. Wu, C. S. Vogel, C. Xu, J. Sutter, J. T. Miller, K. Meyer, T. Cantat and P. L. Diaconescu, *Chem. Commun.*, 2011, **47**, 9897–9899.
73. C. K. A. Gregson, V. C. Gibson, N. J. Long, E. L. Marshall, P. J. Oxford and A. J. P. White, *J. Am. Chem. Soc.*, 2006, **128**, 7410–7411.
74. L. A. Brown, J. L. Rhinehart and B. K. Long, *ACS Catal.*, 2015, **5**, 6057–6060.
75. J. Wei, M. N. Riffel and P. L. Diaconescu, *Macromolecules*, 2017, **50**, 1847–1861.
76. N. E. Kamber, W. Jeong, R. M. Waymouth, R. C. Pratt, B. G. G. Lohmeijer and J. L. Hedrick, *Chem. Rev.*, 2007, **107**, 5813–5840.
77. M. K. Kiesewetter, E. J. Shin, J. L. Hedrick and R. M. Waymouth, *Macromolecules*, 2010, **43**, 2093–2107.
78. S. Gazeau-Bureau, D. Delcroix, B. Martín-Vaca, D. Bourissou, C. Navarro and S. Magnet, *Macromolecules*, 2008, **41**, 3782–3784.
79. D. J. Coady, K. Fukushima, H. W. Horn, J. E. Rice and J. L. Hedrick, *Chem. Commun.*, 2011, **47**, 3105–3107.
80. F. Nederberg, B. G. G. Lohmeijer, F. Leibfarth, R. C. Pratt, J. Choi, A. P. Dove, R. M. Waymouth and J. L. Hedrick, *Biomacromolecules*, 2007, **8**, 153–160.
81. R. C. Pratt, F. Nederberg, R. M. Waymouth and J. L. Hedrick, *Chem. Commun.*, 2008, 114–116.
82. B. G. G. Lohmeijer, R. C. Pratt, F. Leibfarth, J. W. Logan, D. A. Long, A. P. Dove, F. Nederberg, J. Choi, C. Wade, R. M. Waymouth and J. L. Hedrick, *Macromolecules*, 2006, **39**, 8574–8583.

83. D. J. Coady, A. C. Engler, H. W. Horn, K. M. Bajjuri, K. Fukushima, G. O. Jones, A. Nelson, J. E. Rice and J. L. Hedrick, *ACS Macro Lett.*, 2012, **1**, 19–22.
84. R. Todd, G. Rubio, D. J. Hall, S. Tempelaar and A. P. Dove, *Chem. Sci.*, 2013, **4**, 1092–1097.
85. B. Guillerm, V. Lemaur, J. Cornil, R. Lazzaroni, P. Dubois and O. Coulembier, *Chem. Commun.*, 2014, **50**, 10098–10101.
86. Y. Tsutsumi, K. Yamakawa, M. Yoshida, T. Ema and T. Sakai, *Org. Lett.*, 2010, **12**, 5728–5731.
87. R. C. Pratt, B. G. G. Lohmeijer, D. A. Long, R. M. Waymouth and J. L. Hedrick, *J. Am. Chem. Soc.*, 2006, **128**, 4556–4557.
88. L. Zhang, F. Nederberg, J. M. Messman, R. C. Pratt, J. L. Hedrick and C. G. Wade, *J. Am. Chem. Soc.*, 2007, **129**, 12610–12611.
89. M. Helou, O. Miserque, J.-M. Brusson, J.-F. Carpentier and S. M. Guillaume, *Chem. - Eur. J.*, 2010, **16**, 13805–13813.
90. D. A. Culkin, W. Jeong, S. Csihony, E. D. Gomez, N. P. Balsara, J. L. Hedrick and R. M. Waymouth, *Angew. Chem., Int. Ed.*, 2007, **46**, 2627–2630.
91. Y. A. Chang and R. M. Waymouth, *Polym. Chem.*, 2015, **6**, 5212–5218.
92. H. A. Brown, S. Xiong, G. A. Medvedev, Y. A. Chang, M. M. Abu-Omar, J. M. Caruthers and R. M. Waymouth, *Macromolecules*, 2014, **47**, 2955–2963.
93. H. A. Brown, Y. A. Chang and R. M. Waymouth, *J. Am. Chem. Soc.*, 2013, **135**, 18738–18741.
94. T. S. Stukenbroeker, D. Solis-Ibarra and R. M. Waymouth, *Macromolecules*, 2014, **47**, 8224–8230.
95. Y. A. Chang, A. E. Rudenko and R. M. Waymouth, *ACS Macro Lett.*, 2016, **5**, 1162–1166.
96. S. Naumann, A. W. Thomas and A. P. Dove, *Angew. Chem., Int. Ed.*, 2015, **54**, 9550–9554.
97. A. Fürstner, M. Alcarazo, R. Goddard and C. W. Lehmann, *Angew. Chem. Int. Ed.*, 2008, **47**, 3210–3214.
98. S. Naumann, A. W. Thomas and A. P. Dove, *ACS Macro Lett.*, 2016, **5**, 134–138.
99. X. Zhang, G. O. Jones, J. L. Hedrick and R. M. Waymouth, *Nat. Chem.*, 2016, **8**, 1047.
100. B. Lin and R. M. Waymouth, *J. Am. Chem. Soc.*, 2017, **139**, 1645–1652.
101. Y. Shibasaki, H. Sanada, M. Yokoi, F. Sanda and T. Endo, *Macromolecules*, 2000, **33**, 4316–4320.
102. Y. Shibasaki, F. Sanda and T. Endo, *Macromol. Rapid Commun.*, 2000, **21**, 489–492.
103. Y. Shibasaki, F. Sanda and T. Endo, *Macromolecules*, 2000, **33**, 3590–3593.
104. J. Matsuo, S. Nakano, F. Sanda and T. Endo, *J. Polym. Sci., Part A: Polym. Chem.*, 1998, **36**, 2463–2471.

105. J. M. Campos, M. R. Ribeiro, M. F. Ribeiro, A. Deffieux and F. Peruch, *Eur. Polym. J.*, 2013, **49**, 4025–4034.
106. N. Susperregui, D. Delcroix, B. Martin-Vaca, D. Bourissou and L. Maron, *J. Org. Chem.*, 2010, **75**, 6581–6587.
107. K. Makiguchi, Y. Ogasawara, S. Kikuchi, T. Satoh and T. Kakuchi, *Macromolecules*, 2013, **46**, 1772–1782.
108. K. Makiguchi, T. Satoh and T. Kakuchi, *Macromolecules*, 2011, **44**, 1999–2005.
109. J. Chen, S. Kan, H. Xia, F. Zhou, X. Chen, X. Jiang, K. Guo and Z. Li, *Polymer*, 2013, **54**, 4177–4182.
110. D. Delcroix, A. Couffin, N. Susperregui, C. Navarro, L. Maron, B. Martin-Vaca and D. Bourissou, *Polym. Chem.*, 2011, **2**, 2249–2256.
111. X. He, Y. Ji, Y. Jin, S. Kan, H. Xia, J. Chen, B. Liang, H. Wu, K. Guo and Z. Li, *J. Polym. Sci., Part A: Polym. Chem.*, 2014, **52**, 1009–1019.
112. H. Wu, Y. Ji, Z. Li, X. Wang, Q. Zhang, S. Cui, W. Wu, J. Liu and K. Guo, *J. Polym. Sci., Part A: Polym. Chem.*, 2015, **53**, 729–736.
113. T. Saito, Y. Aizawa, K. Tajima, T. Isono and T. Satoh, *Polym. Chem.*, 2015, **6**, 4374–4384.
114. A. C. Engler, J. M. W. Chan, D. J. Coady, J. M. O'Brien, H. Sardon, A. Nelson, D. P. Sanders, Y. Y. Yang and J. L. Hedrick, *Macromolecules*, 2013, **46**, 1283–1290.
115. A. C. Engler, X. Ke, S. Gao, J. M. W. Chan, D. J. Coady, R. J. Ono, R. Lubbers, A. Nelson, Y. Y. Yang and J. L. Hedrick, *Macromolecules*, 2015, **48**, 1673–1678.
116. A.-G. Ying, L. Liu, G.-F. Wu, G. Chen, X.-Z. Chen and W.-D. Ye, *Tetrahedron Lett.*, 2009, **50**, 1653–1657.
117. G. Kim Whan, G. Yoon Ho and Y. Lee Jun, *J. Appl. Polym. Sci.*, 2001, **81**, 2711–2720.
118. J. Kadota, D. Pavlović, J.-P. Desvergne, B. Bibal, F. Peruch and A. Deffieux, *Macromolecules*, 2010, **43**, 8874–8879.
119. K. Makiguchi, S. Kikuchi, K. Yanai, Y. Ogasawara, S.-I. Sato, T. Satoh and T. Kakuchi, *J. Polym. Sci., Part A: Polym. Chem.*, 2014, **52**, 1047–1054.
120. X. Wang, S. Cui, Z. Li, S. Kan, Q. Zhang, C. Zhao, H. Wu, J. Liu, W. Wu and K. Guo, *Polym. Chem.*, 2014, **5**, 6051–6059.
121. X. Zhi, J. Liu, Z. Li, H. Wang, X. Wang, S. Cui, C. Chen, C. Zhao, X. Li and K. Guo, *Polym. Chem.*, 2016, **7**, 339–349.
122. J. D. Holbrey, W. M. Reichert, I. Tkatchenko, E. Bouajila, O. Walter, I. Tommasi and R. D. Rogers, *Chem. Commun.*, 2003, 28–29.
123. M. Smiglak, C. C. Hines and R. D. Rogers, *Green Chem.*, 2010, **12**, 491–501.
124. M. Fevre, J. Pinaud, A. Leteneur, Y. Gnanou, J. Vignolle, D. Taton, K. Miqueu and J.-M. Sotiropoulos, *J. Am. Chem. Soc.*, 2012, **134**, 6776–6784.
125. N. E. Kamber, W. Jeong, S. Gonzalez, J. L. Hedrick and R. M. Waymouth, *Macromolecules*, 2009, **42**, 1634–1639.

126. A. P. Dove, R. C. Pratt, B. G. G. Lohmeijer, D. A. Culkin, E. C. Hagberg, G. W. Nyce, R. M. Waymouth and J. L. Hedrick, *Polymer*, 2006, **47**, 4018–4025.
127. A. Reitz, R. Wilhelm and D. Kuckling, *Macromol. Symp.*, 2013, **334**, 92–97.
128. A. K. Reitz, Q. Sun, R. Wilhelm and D. Kuckling, *J. Polym. Sci., Part A: Polym. Chem.*, 2017, **55**, 820–829.
129. S. Matsumura, K. Tsukada and K. Toshima, *Macromolecules*, 1997, **30**, 3122–3124.
130. F. Deng and R. A. Gross, *Int. J. Biol. Macromol.*, 1999, **25**, 153–159.
131. Y. Yamamoto, S. Kaihara, K. Toshima and S. Matsumura, *Macromol. Biosci.*, 2009, **9**, 968–978.
132. J. Matsuo, F. Sanda and T. Endo, *Macromol. Chem. Phys.*, 1998, **199**, 97–102.
133. H. R. Kricheldorf and A. Mahler, *Polymer*, 1996, **37**, 4383–4388.
134. R. Kricheldorf Hans and A. Mahler, *J. Polym. Sci., Part A: Polym. Chem.*, 1996, **34**, 2399–2406.
135. P. Brignou, M. Priebe Gil, O. Casagrande, J.-F. Carpentier and S. M. Guillaume, *Macromolecules*, 2010, **43**, 8007–8017.
136. P. Brignou, J.-F. Carpentier and S. M. Guillaume, *Macromolecules*, 2011, **44**, 5127–5135.
137. C.-X. Cai, A. Amgoune, C. W. Lehmann and J.-F. Carpentier, *Chem. Commun.*, 2004, 330–331.
138. A. Amgoune, C. M. Thomas, T. Roisnel and J.-F. Carpentier, *Chem. – Eur. J.*, 2006, **12**, 169–179.
139. I. del Rosal, P. Brignou, S. M. Guillaume, J.-F. Carpentier and L. Maron, *Polym. Chem.*, 2015, **6**, 3336–3352.
140. M. Huang, C. Pan and H. Ma, *Dalton Trans.*, 2015, **44**, 12420–12431.
141. W. Yi and H. Ma, *Dalton Trans.*, 2014, **43**, 5200–5210.
142. M. Huang and H. Ma, *Eur. J. Inorg. Chem.*, 2016, **2016**, 3791–3803.
143. Z. Zhang, D. W. Grijpma and J. Feijen, *Macromol. Chem. Phys.*, 2004, **205**, 867–875.
144. W. Guerin, M. Helou, J.-F. Carpentier, M. Slawinski, J.-M. Brusson and S. M. Guillaume, *Polym. Chem.*, 2013, **4**, 1095–1106.
145. P. Dobrzynski and J. Kasperczyk, *J. Polym. Sci., Part A: Polym. Chem.*, 2006, **44**, 3184–3201.
146. S. Agarwal, M. Puchner, A. Greiner and J. H. Wendorff, *Polym. Int.*, 2005, **54**, 1422–1428.
147. Q. Huang, Z. Shen, Y. Zhang, Y. Shen, L. Shen and H. Yuan, *Polym. J.*, 1998, **30**, 168–170.
148. V. T. Lipik and M. J. M. Abadie, *Int. J. Biomater.*, 2012, **390947**, 390911.
149. O. Coulembier, V. Lemaur, T. Josse, A. Minoia, J. Cornil and P. Dubois, *Chem. Sci.*, 2012, **3**, 723–726.
150. H. R. Kricheldorf and A. Stricker, *Macromol. Chem. Phys.*, 1999, **200**, 1726–1733.

151. L. Wang, C. E. Kefalidis, S. Sinbandhit, V. Dorcet, J.-F. Carpentier, L. Maron and Y. Sarazin, *Chem. – Eur. J.*, 2013, **19**, 13463–13478.
152. V. Simic, S. Pensec and N. Spassky, *Macromol. Symp.*, 2000, **153**, 109–121.
153. W. Guerin, M. Helou, M. Slawinski, J.-M. Brusson, S. M. Guillaume and J.-F. Carpentier, *Polym. Chem.*, 2013, **4**, 3686–3693.
154. A. Couffin, D. Delcroix, B. Martin-Vaca, D. Bourissou and C. Navarro, *Macromolecules*, 2013, **46**, 4354–4360.
155. A.-C. Albertsson and M. Eklund, *J. Polym. Sci., Part A: Polym. Chem.*, 1994, **32**, 265–279.
156. Y. Shen, Z. Shen, Y. Zhang, Q. Huang, L. Shen and H. Yuan, *J. Appl. Polym. Sci.*, 1997, **64**, 2131–2139.
157. M. Pastusiak, P. Dobrzynski, J. Kasperczyk, M. Sobota, B. Kaczmarczyk and H. Janeczek, *Polym. Int.*, 2017, **66**, 1259–1268.
158. F. Nederberg, Y. Zhang, J. P. K. Tan, K. Xu, H. Wang, C. Yang, S. Gao, X. D. Guo, K. Fukushima, L. Li, J. L. Hedrick and Y.-Y. Yang, *Nat. Chem.*, 2011, **3**, 409–414.
159. M. Pastusiak, J. Jaworska, M. Kawalec, J. Kasperczyk and P. Dobrzynski, *J. Polym. Sci., Part A: Polym. Chem.*, 2017, **55**, 808–819.
160. A. Hamdach, E. M. El Hadrami, S. Gil, R. J. Zaragozá, E. Zaballos-García and J. Sepúlveda-Arques, *Tetrahedron*, 2006, **62**, 6392–6397.
161. L. Cotarca, T. Geller and J. Répási, *Org. Process Res. Dev.*, 2017, **21**, 1439–1446.
162. H. Matsukizono and T. Endo, *J. Appl. Polym. Sci.*, 2015, **132**, 41956/41951–41956/41910.
163. J. V. Olsson, D. Hult, Y. Cai, S. Garcia-Gallego and M. Malkoch, *Polym. Chem.*, 2014, **5**, 6651–6655.
164. M. North, R. Pasquale and C. Young, *Green Chem.*, 2010, **12**, 1514–1539.
165. D. J. Darensbourg, A. I. Moncada and S.-H. Wei, *Macromolecules*, 2011, **44**, 2568–2576.
166. K. Nakano, S. Hashimoto, M. Nakamura, T. Kamada and K. Nozaki, *Angew. Chem., Int. Ed.*, 2011, **50**, 4868–4871.
167. M. Winkler, C. Romain, M. A. R. Meier and C. K. Williams, *Green Chem.*, 2015, **17**, 300–306.
168. G. L. Gregory, M. Ulmann and A. Buchard, *RSC Adv.*, 2015, **5**, 39404–39408.
169. P. Olsen, J. Undin, K. Odelius, H. Keul and A.-C. Albertsson, *Biomacromolecules*, 2016, **17**, 3995–4002.
170. J.-I. Kadokawa, Y. Iwasaki and H. Tagaya, *Macromol. Rapid Commun.*, 2002, **23**, 757–760.
171. R. F. Storey and D. C. Hoffman, *Polym. Bull.*, 1991, **27**, 267–274.
172. R. F. Storey and D. C. Hoffman, *Macromolecules*, 1992, **25**, 5369–5382.
173. A. G. Patil, U. R. Kapadi and D. G. Hundiwale, *J. Sci. Ind. Res.*, 2005, **64**, 364–366.
174. K. Soga, Y. Tazuke, S. Hosoda and S. Ikeda, *J. Polym. Sci., Polym. Chem. Ed.*, 1977, **15**, 219–229.

175. L. Vogdanis and W. Heitz, *Makromol. Chem., Rapid Commun.*, 1986, **7**, 543–547.
176. S. Keki, J. Toeroek, G. Deak and M. Zsuga, *Macromolecules*, 2001, **34**, 6850–6857.
177. H. Yang, M. Yan, S. Pispas and G. Zhang, *Macromol. Chem. Phys.*, 2011, **212**, 2589–2593.
178. M. Acemoglu, F. Nimmerfall, S. Bantle and G. H. Stoll, *J. Controlled Release*, 1997, **49**, 263–276.
179. H. Shirahama, A. Kanetani and H. Yasuda, *Polym. J.*, 2000, **32**, 280.
180. W. Guerin, M. Helou, M. Slawinski, J.-M. Brusson, J.-F. Carpentier and S. M. Guillaume, *Polym. Chem.*, 2015, **6**, 1972–1985.
181. L. Ubaghs, C. Novi, H. Keul and H. Hoecker, *Macromol. Chem. Phys.*, 2004, **205**, 888–896.
182. F. Schmitz, H. Keul and H. Hocker, *Macromol. Rapid Commun.*, 1997, **18**, 699–706.
183. S. Panja, S. Nayak, S. K. Ghosh, M. Selvakumar and S. Chattopadhyay, *RSC Adv.*, 2014, **4**, 51766–51775.
184. G. Kapiti, H. Keul and M. Moeller, *Polym. Chem.*, 2016, **7**, 5050–5059.
185. D. C. Webster, *Prog. Org. Coat.*, 2003, **47**, 77–86.
186. F. Camara, S. Benyahya, V. Besse, G. Boutevin, R. Auvergne, B. Boutevin and S. Caillol, *Eur. Polym. J.*, 2014, **55**, 17–26.
187. J.-Z. Hwang, S.-C. Wang, P.-C. Chen, C.-Y. Huang, J.-T. Yeh and K.-N. Chen, *J. Polym. Res.*, 2012, **19**, 1–10.
188. H. Sardon, A. Pascual, D. Mecerreyes, D. Taton, H. Cramail and J. L. Hedrick, *Macromolecules*, 2015, **48**, 3153–3165.
189. K. Tezuka, K. Komatsu and O. Haba, *Polym. J.*, 2013, **45**, 1183–1187.
190. K. Tezuka, K. Koda, H. Katagiri and O. Haba, *Polym. Bull.*, 2015, **72**, 615–626.
191. W. Guerin, A. K. Diallo, E. Kirilov, M. Helou, M. Slawinski, J.-M. Brusson, J.-F. Carpentier and S. M. Guillaume, *Macromolecules*, 2014, **47**, 4230–4235.
192. A. K. Diallo, E. Kirillov, M. Slawinski, J.-M. Brusson, S. M. Guillaume and J.-F. Carpentier, *Polym. Chem.*, 2015, **6**, 1961–1971.
193. D. P. Sanders, D. J. Coady, M. Yasumoto, M. Fujiwara, H. Sardon and J. L. Hedrick, *Polym. Chem.*, 2014, **5**, 327–329.
194. K. Fukushima, R. C. Pratt, F. Nederberg, J. P. K. Tan, Y. Y. Yang, R. M. Waymouth and J. L. Hedrick, *Biomacromolecules*, 2008, **9**, 3051–3056.
195. D. M. Stevens, H. A. Watson, M.-A. LeBlanc, R. Y. Wang, J. Chou, W. S. Bauer and E. Harth, *Polym. Chem.*, 2013, **4**, 2470–2474.
196. M. Helou, J.-M. Brusson, J.-F. Carpentier and S. M. Guillaume, *Polym. Chem.*, 2011, **2**, 2789–2795.
197. L.-S. Wang, S.-X. Cheng and R.-X. Zhuo, *Polym. Sci., Ser. B*, 2013, **55**, 604–610.
198. J. Qiu, M. Lu, Y. Yao, Y. Zhang, Y. Wang and Q. Shen, *Dalton Trans.*, 2013, **42**, 10179–10189.

199. J. Simon, J. V. Olsson, H. Kim, I. F. Tenney and R. M. Waymouth, *Macromolecules*, 2012, **45**, 9275–9281.
200. K. Mikami, A. T. Lonnecker, T. P. Gustafson, N. F. Zinnel, P.-J. Pai, D. H. Russell and K. L. Wooley, *J. Am. Chem. Soc.*, 2013, **135**, 6826–6829.
201. G. L. Gregory, L. M. Jenisch, B. Charles, G. Kociok-Kohn and A. Buchard, *Macromolecules*, 2016, **49**, 7165–7169.
202. Y. Shen, X. Chen and R. A. Gross, *Macromolecules*, 1999, **32**, 2799–2802.
203. G. L. Gregory, G. Kociok-Kohn and A. Buchard, *Polym. Chem.*, 2017, **8**, 2093–2104.
204. T. Endo and F. Sanda, *Macromol. Symp.*, 2000, **159**, 1–7.
205. T. Takata, H. Matsuoka, T. Hirasa, J. Matsuo, T. Endo and Y. Furusho, *Kobunshi Ronbunshu*, 1997, **54**, 974–981.
206. H. R. Kricheldorf and J. Jenssen, *Eur. Polym. J.*, 1989, **25**, 1273–1279.
207. T. Takata, K.-I. Murakawa and Y. Furusho, *Polym. J.*, 1999, **31**, 1051–1056.
208. D. G. LeGrand, in *Mechanical Properties of Polycarbonates*, ed. J. T. Bendler, CRC Press, 1999, pp. 107–130.
209. S. Venkataraman, V. W. L. Ng, D. J. Coady, H. W. Horn, G. O. Jones, T. S. Fung, H. Sardon, R. M. Waymouth, J. L. Hedrick and Y. Y. Yang, *J. Am. Chem. Soc.*, 2015, **137**, 13851–13860.
210. J. P. K. Tan, D. J. Coady, H. Sardon, A. Yuen, S. Gao, S. W. Lim, Z. C. Liang, E. W. Tan, S. Venkataraman, A. C. Engler, M. Fevre, R. Ono, Y. Y. Yang and J. L. Hedrick, *Adv. Healthcare Mater.*, 2017, **6**, 1601420.
211. A. Y. Yuen, E. Lopez-Martinez, E. Gomez-Bengoa, A. L. Cortajarena, R. H. Aguirresarobe, A. Bossion, D. Mecerreyes, J. L. Hedrick, Y. Y. Yang and H. Sardon, *ACS Biomater. Sci. Eng.*, 2017, **3**, 1567–1575.
212. D. J. Brunelle, E. P. Boden and T. G. Shannon, *J. Am. Chem. Soc.*, 1990, **112**, 2399–2402.
213. R. Nagahata, J.-I. Sugiyama, M. Goyal, M. Asai, M. Ueda and K. Takeuchi, *Polym. J.*, 2000, **32**, 854–858.
214. S. Strandman, J. E. Gautrot and X. X. Zhu, *Polym. Chem.*, 2011, **2**, 791–799.
215. P. Hodge, S. D. Kamau, A. Ben-Haida and R. T. Williams, *React. Funct. Polym.*, 2012, **72**, 868–877.

CHAPTER 8

Metal-free Polyether Synthesis by Organocatalyzed Ring-opening Polymerization

DANIEL TATON

CNRS-Univ. Bordeaux, Laboratoire de Chimie des Polymères Organiques (LCPO), UMR 5629, 16 Av. Pey-Berland, F-33607 Pessac Cedex, France
Email: taton@enscbp.fr

8.1 Introduction

Epoxides are three-membered heterocyclic compounds that are particularly useful as monomeric building blocks to produce technically important polymeric materials. The highly strained ring structure of the epoxide group enables a ring-opening reaction in the presence of various types of reagents and/or catalysts following different modes of activation. Notably, the lone pair of electrons in the oxygen atom of the epoxide is prone to activation by either a Lewis acid or a Brønsted acid, while nucleophilic species can readily attack the carbon atom of the epoxide, owing to the polarization of the carbon–oxygen bond (Scheme 8.1). Epoxides are also amenable to ring-opening copolymerization (ROcP) reactions involving comonomers, including CO_2, anhydrides, vinyl ethers, cyclic esters, or non-epoxide cyclic ethers, such as oxetanes, or other epoxides as well. A diversity of copolymers, mostly of copolyether and copolyester type, are thus accessible, broadening the potential of epoxides as monomer substrates.[1–6]

Polymer Chemistry Series No. 31
Organic Catalysis for Polymerisation
Edited by Andrew Dove, Haritz Sardon and Stefan Naumann

Published by the Royal Society of Chemistry, www.rsc.org

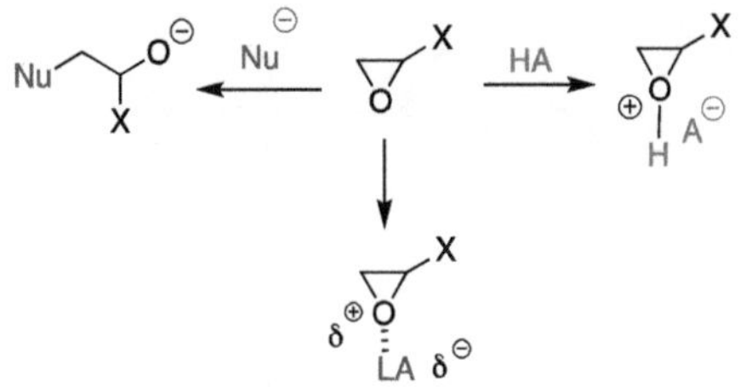

Scheme 8.1 Possible activation modes of the epoxy group (Nu = nucleophile, *e.g.* an alkoxide; HA = Brønsted acid; LA = Lewis acid).

Organometallic catalysts/initiators—typically based on aluminum or on zinc or on rare earths—are the most employed to produce aliphatic (co)polyethers by ring-opening polymerization (ROP) and ROcP of epoxides, following either an ionic, or a coordinated-ionic or a coordination-insertion mechanism.[1–6] The same is not true at all regarding the catalytic homopolymerization of epoxy resins, nor for their step-growth polymerization with hardeners (see Section 8.3). Indeed, these reactions commonly utilize nonmetallic catalysts, including Lewis bases, *e.g.* a tertiary aliphatic amine or a *N*-alkylimidazole, or Lewis acids, *e.g.* boron trifluoride complexes or strong organic acids, to achieve the cured network. In other words, metal-free polyepoxide synthesis was an established method long before the emergence of organocatalyzed polymerization in the early 2000s and its adoption by many groups in polymer synthesis.[7–15]

Organocatalyzed ring-opening polymerization (OROP) of epoxides, which is the topic of this book chapter, finds its roots in works by Kubisa, Penczek *et al.* in the late 1980s, with the use of non-metallic acids, such as BF_3 or HPF_6, to catalyze the ROP of ethylene oxide and other epoxides, following a cationic-type mechanism.[1,16–19]

As highlighted in this book and in recent reviews,[7–15] the last two decades have witnessed the advent of organocatalysts in polymer synthesis to access polymeric materials free of any metallic residues, with a potential use in high-value applications, including nanomedicine, cosmetics, microelectronics or food packaging. Despite significant advances in this area, with a range of monomers amenable to polymerization by organocatalysis that has been largely expanded beyond cyclic esters, there are still challenging monomers to be polymerized by an organocatalyzed pathway. Non-polar alkenes or dienes, such as styrene, ethylene or butadiene are typical examples of monomers that cannot be polymerized from the benchmark organocatalysts. To some extent, epoxide monomers can be distinguished from other heterocyclics, such as lactide, lactones or cyclic carbonates, as just a few organocatalysts allow for their efficient ROP, as described hereafter. The main epoxide monomer substrates and main organocatalysts used in OROP, with their abbreviations that will be subsequently used, are displayed in Figure 8.1 and in Figure 8.2, respectively.

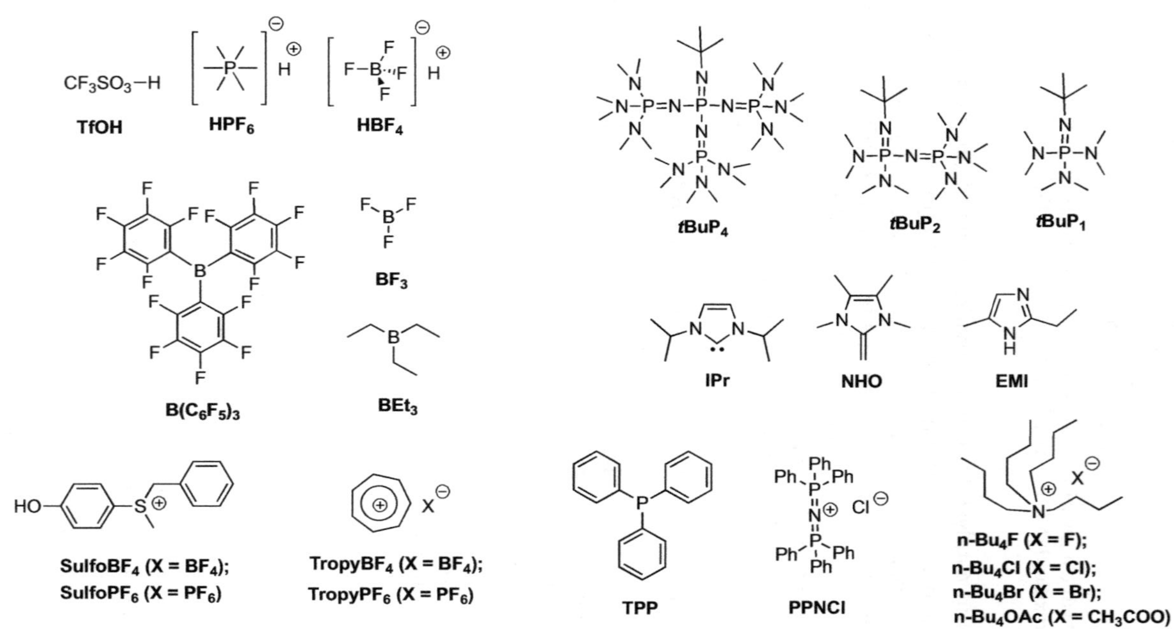

Figure 8.1 Main catalytic systems used for metal-free ROP of epoxides; compounds on the left side induce a cationic ROP process, while catalysts on the right side trigger ROP *via* an anionic-type pathway.

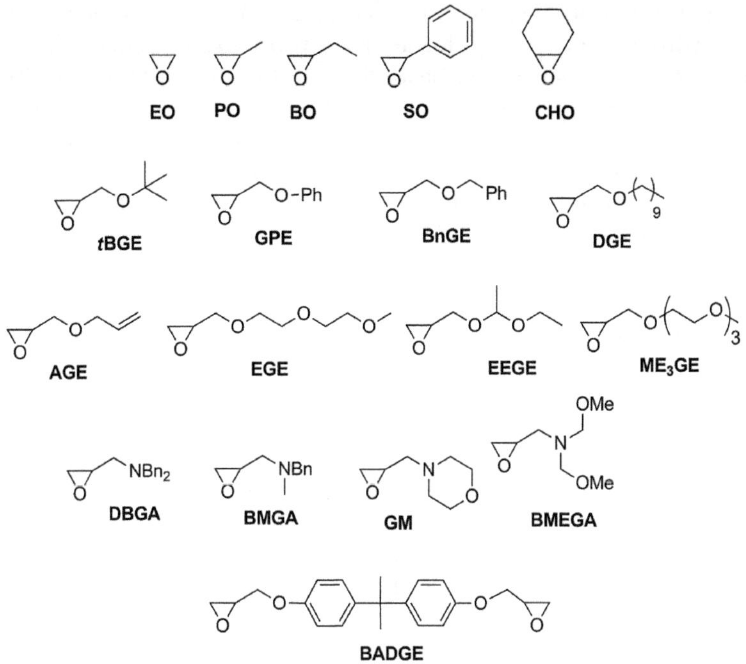

Figure 8.2 The main epoxides studied by ORO(c)P.

8.2 Metal-free Synthesis of Aliphatic Polyethers by ROP of Epoxides

8.2.1 Industrial Importance

An important family of polymers resulting from the ROP of epoxides, which involves a chain-growth reaction pathway, is that of aliphatic polyethers, including poly(ethylene oxide) (PEO), poly(propylene oxide) (PPO), and, to a lesser extent, poly(butylene oxide) (PBO). Production of polyether-based materials from these monomers indeed reaches 33 million tons per year.[1–4] Aliphatic polyethers are characterized by good chemical stability and high flexibility, with a glass transition temperature often below −60 °C. ROP of epoxides can follow different types of mechanisms, including initiation by a base, an acid, or by coordination. Synthesis of EO and PO, which represent the most important epoxide monomer substrates, can be readily achieved by direct oxidation of ethylene and propylene. In contrast, butylene oxide (BO), which is also industrially available, is synthesized by oxidation of butadiene into vinyloxirane, followed by a hydrogenation step. Mono-substituted epoxides, such as glycidyl ethers, epichlorohydrine, longer alkylene epoxides, and glycidyl amines (Figure 8.2), are also amenable to an ROP process, and have gained increasing interest in research for the synthesis of functional (co)polyethers for specific applications.

The first report on the ROP of EO dates back to 1863 with the use of alkali metal hydroxide or zinc chloride as an initiating system.[20] In 1929, Staudinger and Schweitzer synthesized poly(ethylene oxide)s (PEOs) from EO and undertook a separation as a function of the molecular weight of the different populations.[21] PEO made from the ROP of EO using ethylene glycol under basic conditions was then commercialized in the 1930s, whereas industrial synthesis of liquid poly(propylene oxide) (PPO) under similar conditions, for a use as hydraulic fluids and as lubricants was established in the 1940s. In 1940, Flory established the mechanism taking place during the base-initiated ROP of EO.[22]

PEO, often referred to as PEG for poly(ethylene glycol), has many potential applications in biomedical and pharmaceutical areas owing to its water solubility, non-toxicity, ion-transporting ability, presence of functional group(s) for the attachment of biologically active molecules and non-recognition by the immune system, known as the "stealth effect" of PEG.[1,23–25] Thus, the "PEGylation reaction" of biologically active molecules increases their water solubility and their stability against enzymatic degradation and facilitates their pharmacological administration. High molecular weight PEOs are semi-crystalline thermoplastics with a melting point of 65 °C. Blending PEOs of different molecular weights allows tuning the melting point of PEO, which is implemented for specific applications, *e.g.* for creams, ointments, and suppositories. It is important to note, however, that special care should be taken when handling EO as this monomer is a highly carcinogenic gas requiring specific pressure-resistant equipment.

As for PPO, it is a hydrophobic and amorphous flexible polyether with a glass transition temperature around −70 °C.[1,2] However, low molecular weight PPOs exhibit a lower critical solution temperature, *i.e.* a cloud point, at around 15 °C in aqueous solution. Applications of PPO include antifoaming agents, softeners, and rheology modifiers. In this context, PPO often exhibits a star-like architecture, referred to as polyether-polyols, when synthesized by ROP utilizing multifunctional initiators, such as glycerol, pentaerythritol, or sorbitol, under basic conditions. These star-like PPOs are particularly useful for the manufacture of polyurethane flexible foam.[26] However, it is most of all in its copolymeric form with EO segments that PPO finds many applications, for instance, as non-ionic surfactants with PPO-*b*-PEO block copolymers, or as soft segments for polyurethane foams, or as lubricants.[1,4]

A major complication during anionic ROP of PO, and that of higher epoxides, such as BO and glycidyl ethers as well, is the occurrence of a chain transfer reaction to the monomer, owing to the high nucleophilicity of the propagating alkoxides (Scheme 8.2). This side reaction obviously limits the molecular weights and sometimes makes the synthesis of block copolymers challenging.[1,4]

Poly(butylene oxide) (PBO) exhibits properties similar to those of PPO discussed above, although PBO is more hydrophobic. In this regard, BO often serves as comonomer to tailor the properties of (co)polyethers, *i.e.* make them more apolar.

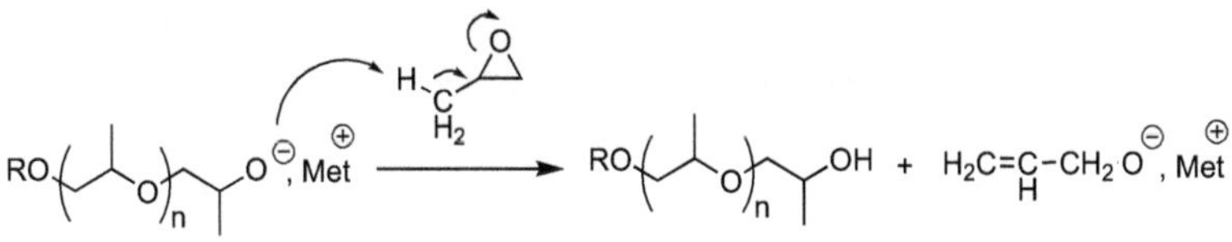

Scheme 8.2 Chain transfer reaction to the monomer of a propagating alkoxide, shown here for the ROP of PO, forming an allyloxy group that can initiate new PPO chains.

8.2.2 Brønsted and Lewis acids

The cationic ring-opening polymerization (CROP) of cyclic ethers, including epoxides, generally proceeds by an active chain-end mechanism (ACEM), when initiated by a Lewis or a Brønsted acid. CROP processes are usually much faster than AROP ones, with propagation rate constants in the range 10^4–10^6 *vs.* 1–10^1 L mol^{-1} s^{-1}, respectively.[1,19,27] Therefore, controlling the molecular features of polyethers can be complicated by the exothermy of the reaction. Moreover, CROP of epoxides, such as EO or PO, is often accompanied by a side reaction, namely, intramolecular chain transfer of the growing oxonium to the polyether chain, also referred to as the "backbiting" reaction, forming cyclic ether byproducts (Scheme 8.3). Intermolecular chain transfer is also possible. In the case of the CROP of EO, substantial amounts of 1,4-dioxane generated by backbiting has been evidenced, contaminating the resulting low molecular weight PEO.[1,19,28] Similarly, the occurrence of intra- and/or intermolecular transfer reactions does not allow controlling the CROP of PO and other substituted epoxides.

The Penczek group in Lodz partly solved this issue, *i.e.* to minimize the content in cyclic oligomers in CROP of epoxides, by resorting to simple alcohols (ROH) and non-metallic Brønsted or Lewis acid catalysts, such as trifluoromethanesulfonic acid (TfOH), tetrafluoroboric acid (HBF_4) or trifluoroboron (BF_3) or hexafluorophosphoric acid (HPF_6).[1,16–19,27] In this case, the so-called activated monomer mechanism (AMM), becomes predominant over the ACEM discussed above, the alcohol playing the role of a reversible chain transfer agent/initiator, which controls the polymer chain length.

AMM involves oxonium-type active centers that are located on the monomer, whereas the polymer chain exists in a (neutral) dormant form with a terminal OH-functionality (Scheme 8.3). Under such conditions, backbiting is minimized, allowing for the synthesis of well-defined polyethers, including telechelics and macromonomers derived from ethylene and propylene oxides, epichlorohydrin and glycidol, and cyclic acetals.[16–19,29] However, a relatively limited molecular weight can be achieved in this way. To favor the AMM over the ACEM, a slow monomer addition process is beneficial, as the instantaneous concentration of monomer is kept low. In the ideal case where the ACEM is negligible, the molecular weight of the final polymer is controlled by the initial [ROH]/[monomer] ratio and a linear relationship is observed between M_n and this ratio. These early works

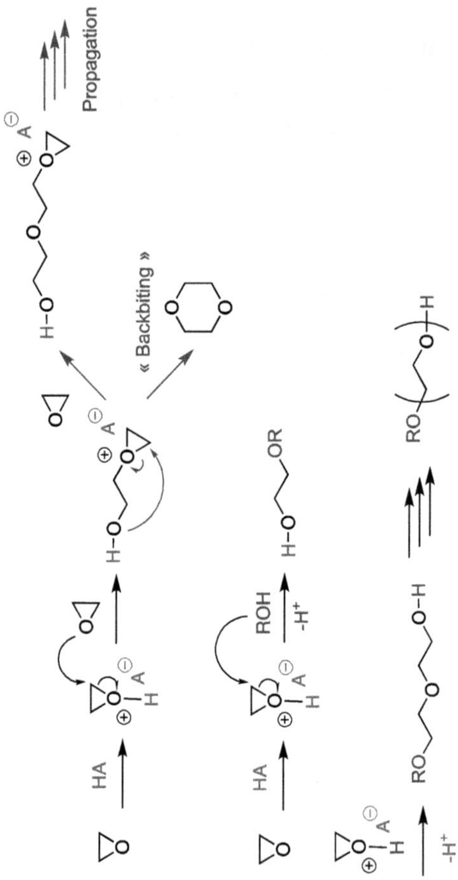

Scheme 8.3 ACEM *vs.* AMM during the CROP of an epoxide (here EO, forming 1,4-dioxane by backbiting).[1,16–19,27]

eventually paved the way of a metal-free approach for precision polyether synthesis.

This strategy was later applied to macromolecular engineering, for instance, for the synthesis of amphiphilic block copolymers based on PEO and poly(glycidyl methacrylate).[30] The cationic ring-opening copolymerization (CROcP) of EO and THF carried out in the presence of diols, forming copolyethers, was also described.[31] Statistical copolymers were also synthesized by CROcP of *n*-butyl glycidyl ether and 3-isochromanone (ICM) at 100 °C, using TfOH as direct initiator, *i.e.* in the absence of any alcohol, as reported by Endo *et al.*[32] The authors showed that ICM did not follow the same mechanism to that of its homopolymerization. Active propagating cations indeed reacted with the carbonyl group of ICM during the CROcP process, whereas they reacted with the aromatic ring of ICM by homopolymerization.

Chen *et al.* resorted to relatively strong Lewis acid-type organoborane catalysts, *e.g.* tris(perfluorophenyl) borane, $B(C_6F5)_3$, combined with hydroxy-containing controlling agent (ROH or H_2O) for the ROP of PO.[33] In the absence of any alcohol or water, propionaldehyde was mainly formed in hexane or in toluene, which was rationalized by the isomerization of PO. In contrast, use of hydroxy compounds led to PPOs of rather low molecular weights (1.1–2.3 $kg\,mol^{-1}$) and a dispersity ($Đ$) < 1.25. The presence of a small proportion of cyclic oligomers was evidenced by MALDI-ToF MS. However, the authors did not qualify this borane-catalyzed ROP of PO as a CROP process, strictly speaking.

Glycidyl phenyl ether (GPE) has received special attention as an epoxide monomer to be directly polymerized by CROP, *i.e.* in the absence of an alcohol. In particular, the Endo group investigated various cationically charged salts as thermally or photochemically latent initiators, including benzyl *p*-hydroxyphenyl methylsulfonium (sulfoBF_4 and sulfoPF_6),[34] pyridinium,[35] phosphonium,[36] and tropylium salts[37] for the CROP of epoxides (Scheme 8.4A). For instance, tropylium tetrafluoroborate (tropyBF_4) and

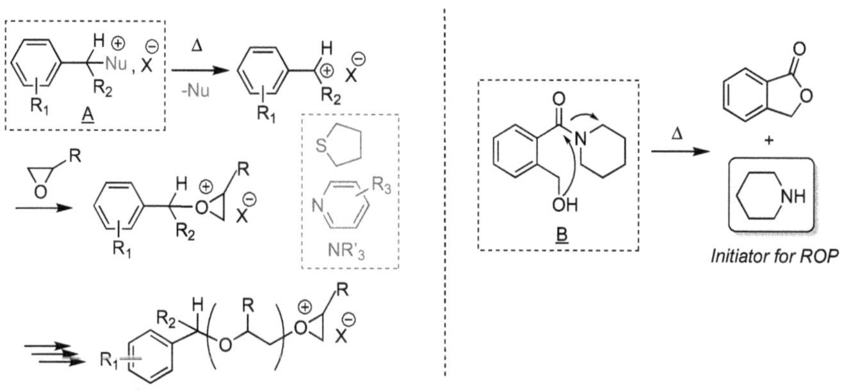

Scheme 8.4 Latent initiators for the metal-free CROP (left) and AROP (right) of epoxides.[34–39]

tropylium hexafluorophosphate (tropyPF_6), used for the CROP of glycidyl phenyl ether (GPE) gave oligomers with molecular weights <3.3 kg mol^{-1}.[37]

The same group employed hydroxylamides as neutral thermally latent organic initiators to polymerize GPE (or bisphenol-A diglycidyl ether).[38,39] These hydroxylamides were devised in a way that their hydroxyl group could react with the amide moiety in an intramolecular manner upon heating, forming the corresponding lactone and releasing pyrrolidine that could serve as initiator, in this case, of the AROP of epoxides (Scheme 8.4B). Initiation eventually took place by reaction between pyrrolidine and two epoxide monomers, forming the propagating quaternary ammonium alkoxide. For instance, the temperature at which the ROP of bisphenol-A diglycidyl ether occurred was found to strongly depend on the structure of the latent initiator.

More recently, Barroso-Bujans *et al.* described the one-step CROP of GPE using $B(C_6F_5)_3$ as Lewis acid-type initiator.[40] When performed in bulk or dichloromethane at room temperature, polymers with molecular weights up to 12 kg mol^{-1} and dispersities lower than 2.0 were achieved. Either cyclic or a mixture of cyclics and linear polyethers were finally produced, in the absence of or in the presence of water, respectively. Chain growth occurred in this case following a zwitterionic ROP (ZROP) mechanism, generating oxonium borate intermediate species and involving initiation by the direct attack of GPE onto $B(C_6F_5)_3$ and further propagation. Depending on the conditions (anhydrous or not), the zwitterionic polyether could be subjected to cyclization at the completion of the reaction, *via* a nucleophilic attack of the oxygen adjacent to the borane onto the electron-deficient α-carbon atom of the oxonium moiety, thus regenerating the $B(C_6F_5)_3$ initiator.

The same group reported the $B(C_6F_5)_3$-mediated zwitterionic ROcP (ZROcP) of GPE and THF at room temperature, to derive cyclic copolyethers of high molecular weights (in the range 15–330 kg mol^{-1}).[41] Under these conditions, THF was not able to homopolymerize but could eventually copolymerize with GPE in a broad range of compositions. Analysis of the copolyethers by MALDI-ToF mass spectrometry revealed that instead of cyclics, linear copolymers were generated at high THF content. In contrast, copolymers obtained from a feed containing 20 mol% in THF were not only constituted of the cyclic copolyethers, but also of some cyclic homoPGPE. To explain these differences, the authors invoked the formation of zwitterionic intermediate species of different reactivity toward propagation and termination. Kinetic investigations by quantitative ^{13}C NMR showed that, from a copolymerization reaction involving equal molar amounts of the co-monomers, THF was incorporated more rapidly than GPE. The glass transition temperature of the copolyethers varied in a broad range of temperatures, from −84 to −4 °C, as a function of the composition and molecular weight. These copolyethers were amorphous at high GPE content, but proved semi-crystalline as the amount of THF increased. Finally, rheological measurements showed that copolyethers with a rubber-like behavior could be obtained below room temperature by manipulating the degree of crystallization. In a subsequent contribution, Grayson, Barroso-Bujans *et al.* could quantify that the major impurities in cyclic poly(GPE) derived by this $B(C_6F_5)_3$-mediated ZROP method were tadpoles.[42] They also succeeded in removing these impurities by developing a "click-scavenging" technique, taking advantage of the presence of a functional handle in the tadpole impurities.

Another borane as non-metallic Lewis acid catalyst, namely, triethyl borane (TEB, Figure 8.1), was employed in conjunction with organic cations to promote the alternating copolymerization between CO_2 and PO or cyclohene oxide (CHO).[43] For instance, the TEB/n-Bu_4NCl catalytic system proved efficient to achieve copolymers from PO and CO_2 with a high carbonate content and molecular weights up to 50 $kg\,mol^{-1}$. Instead of CO_2, carbon sulfide, COS, was copolymerized with PO in THF at 25 °C, in the presence of TEB and dodecyltrimethylammonium bromide, producing also in this case perfectly alternating colorless copolymers.[44]

Of particular interest, Aoshima *et al.* exploited the use of $B(C_6F_5)_3$ to catalyze the concurrent cationic copolymerization between epoxides and vinyl ethers (VE), involving a ring-opening and a vinyl-addition mechanism, respectively.[45] Such a strategy is very specific to the cationic growth pathway, as propagating oxonium species of the CROP of epoxides shows a partially carbocationic character (see Scheme 8.5). In these specific copolymerizations, initiation was thought to occur by adventitious water or by coordination of $B(C_6F_5)_3$ with one or the other monomer.[45–49] As an example, isopropyl vinyl ether (IPVE) was copolymerized with isobutylene oxide (IBO), chain ends being switched from a carbocation derived from IPVE to an oxonium after ring-opening of IBO, then evolving to a tertiary carbocation that in turn add onto IPVE, thus regenerating the secondary carbocation arising from IPVE.[46] Multiple cross-over events thus yielded multiblock-like copolymers. However, copolymer chains of various structures depending on the catalyst and monomer combination, including alternating-rich and multiblock-like copolymers could be designed (Scheme 8.5).[47,48] Thus, epoxides giving rise to stabilized, allyl-type carbocations (*e.g.* isoprene monoxide and butadiene monoxide) could be efficiently copolymerized with IPVE, leading to copolymers constituted of short sequences of each type of monomer units. Concurrent copolymerization involving monomers forming tertiary carbocations (*e.g.* isobutylene oxide) led to less frequent crossover events, resulting in a multiblock-like copolymer. Finally, cross over proved limited for epoxide comonomers possibly forming secondary carbocations after ring opening (*e.g.* 3,3-dimethyl-1,2-butylene oxide and 1,2-butylene oxide). Solvent polarity as well as use of weak Lewis bases such as ethyl acetate or 1,4-dioxane, might also strongly affect the crossover reaction from VE-derived carbocation to the epoxide.[48]

Aoshima *et al.* extended this concept to the concurrent cationic terpolymerization of VE, epoxide, and ketone proceeding *via* vinyl-addition, ring-opening, and carbonyl-addition mechanisms for three types of monomers, respectively.[49] Terpolymers constituted of (IPVE-2-CHO-2-methyl ethyl

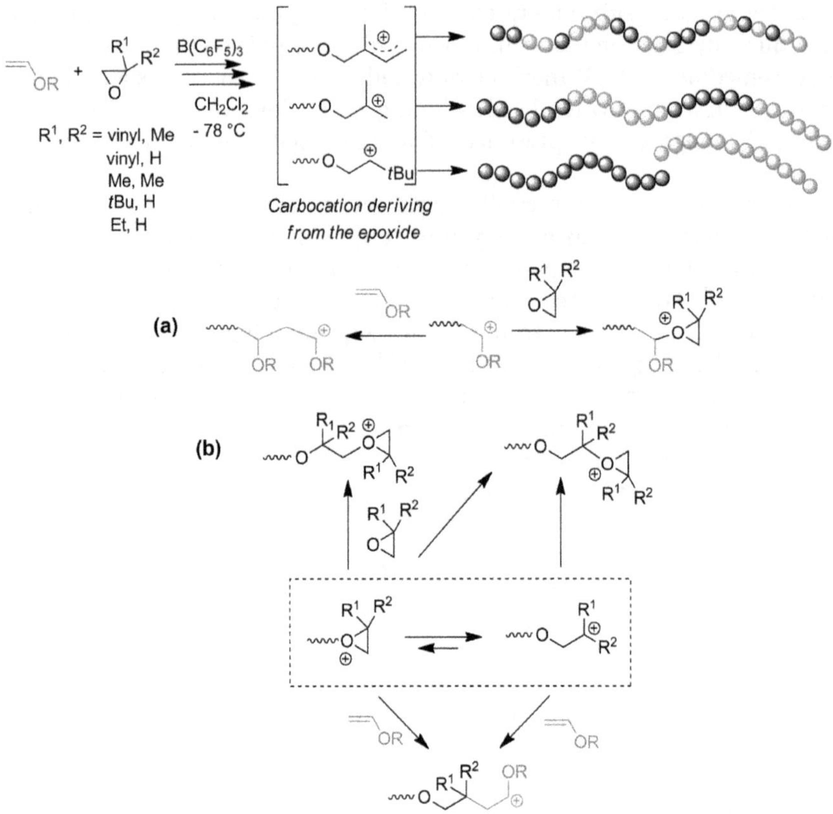

Scheme 8.5 Proposed mechanisms for the concurrent copolymerization of a vinyl ether (VE) and an epoxide; propagation from (a) VE-derived and from (b) epoxide-derived species.[45–49]

ketone) sequences, *i.e.* sequence-regulated terpolymers, could thus be achieved following a one-way copolymerization cycle, only as follows: IPVE → CHO, CHO → ketone, and ketone → IPVE.

8.2.3 Phosphazenes, Phosphazenium Salts, Phosphines and Phosphonium Salts

Phosphazenes, also known as Schwesinger bases, are "super strong" neutral and organic Brønsted bases ($pK_a = 26–47$ in MeCN),[50,51] and are also poor nucleophiles. In the presence of protic initiators (*e.g.* an alcohol, a thiol, an amide), phosphazenes operate *via* a basic mechanism, also referred to as the active chain-end mechanism (ACEM), consistent with their high Brønsted basicity. The deprotonation of a protic alcohol initiator results in the formation of a strongly nucleophilic alkoxide, as it is associated with a soft and bulky phosphazenium cation. This allows the metal-free polymerization to

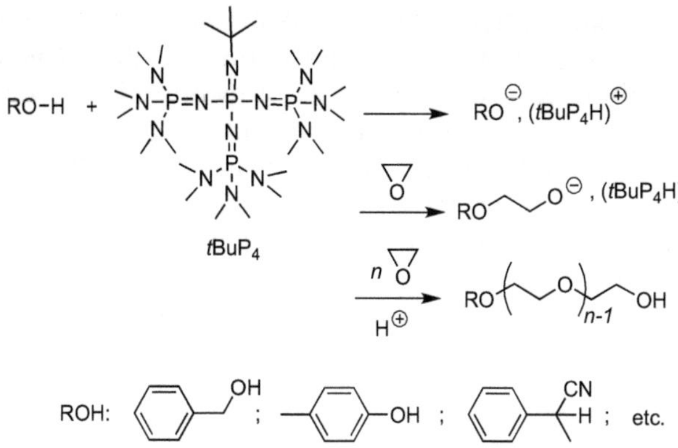

Scheme 8.6 OROP of EO using ROH as initiator and tBuP$_4$ as organic base.[52,56]

be controlled (see Scheme 8.6).[51] The underlying mechanism thus shows features of an anionic-type ROP of epoxides employing alkali metal-based initiators.

Möller *et al.* reported a metal-free anionic ROP of EO using $[t\text{BuP}_4\text{H}]^+$ as counter-cation.[52] Reactions were performed at 80 °C in a glass autoclave, or at room temperature in solution in THF or in toluene. Methanol or octan-1-ol served as initiators, enabling access to well-defined PEOs with M_n values ranging from 4.4 to 6.6 kg mol^{-1} and a dispersity <1.13. In addition, pentaerythritol and poly[ethylene-*co*-(vinyl alcohol)] were used as tetrafunctional and multifunctional polymeric initiator, respectively, to achieve tetraarm star PEOs and poly{[ethylene-*co*-(vinyl alcohol)]-*graft*-PEO} graft copolymers. In comparison to metal-based alkali alkoxides, the presence of the soft and hindered $[t\text{BuP}_4\text{H}]^+$ improved the solubility of the cation of the multiple propagating alkoxides. Thus, although initiation of ROP of EO from pentaerythritol took place under heterogeneous conditions in THF, the reaction mixture became homogeneous after 30% of monomer conversion. The as-obtained star-like PEOs thus exhibited a relatively well-defined structure. Similarly, graft copolymer synthesis from the poly[ethylene-*co*-(vinyl alcohol)] precursor initially occurred under heterogeneous conditions in xylene at room temperature, but complete solubilization was noted after 25% conversion (2–3 h), leading to a homogeneous solution. The least basic phosphazene base, tBuP$_1$, did not enable polymerizing EO—or PO—in THF at 60 °C, while SO could be homopolymerized in bulk at 80 °C, in the presence of this phosphazene base.[53,54] Use of tBuP$_2$ required higher loadings and longer reaction times, compared to tBuP$_4$.[55]

In 2001, Schlaad *et al.* reported the synthesis of heterobifunctional PEOs using either (R)-(α-methylbenzyl) cyanide or *p*-cresol as initiators, in presence of equimolar amounts of tBuP$_4$ (Scheme 8.6).[56] OROP of EO was carried out in THF at 45 °C for 20 h, achieving PEOs ($M_n < 3$ kg mol^{-1}; $Đ < 1.1$). A post-functionalization step of α-PEO chain-ends yielded α-amino-ω-hydroxy and α-bromo-ω-hydroxy PEOs.

Mulhaupt *et al.* later described the OROP of PO at 100 °C, in the presence of tetrakis[cyclohexyl(methyl)amino]-phosphonium (P_1^+, Scheme 8.7).[57,58] This sterically hindered cationic counter-ion, where the positive charge is delocalized over five atoms, was compared to other cations, such as K^+, P_2^+, for which the positive charge is not delocalized. The deprotonated forms of dipropylene glycol (DPG) served as initiators, *i.e.* with different counter-cations with an extent of deprotonation of 5 mol% only, as the exchange between propagating alkoxides of PPO and dormant OH-containing chains is much faster than chain propagation. Such conditions, *i.e.* $[\text{base}]_0/[\text{hydroxyl}]_0 < 1$, enabled all polyether chains to grow at the same rate, and thus guaranteed the control of the OROP process. The propagation rate was found to increase as follows: $Bu_4P^+ < K^+ < P_1^+ < P_2^+ < tBuP_4H^+$. Despite low dispersities being obtained by SEC (*Đ* in the range 1.03–1.09), unsaturated chain-ends were detected by NMR, due to the occurrence of transfer reactions to monomer, which was the more pronounced as ROP rates increased.

Yang *et al.* investigated various phosphonium salts, namely, tetrakis-(pyrrolidino) phosphonium ($Py_4P_1^+$), tetrakis-(piperidino) phosphonium ($Pi_4P_1^+$), tetrakis-(morpholino) phosphonium ($Mo_4P_1^+$), tetrakis [cyclohexyl (methyl) amino] phosphonium ($Cy_4P_1^+$) and tetrakis [tris (dimethylamino) phosphonoamino] phosphazene (P_5^+) for the same purpose. Better defined PPOs could be obtained from $Cy_4P_1^+$, $Pi_4P_1^+$, and P_5^+ as counter ions, *i.e.* with a very low content of unsaturations and with molecular weights up to 9.0 kg mol^{-1}.[59] In the latter case, PO was polymerized using glycerol as initiator in between 70 and 90 °C.

In a series of papers, Satoh, Kakuchi *et al.* further exploited phosphazene base catalysts for the OROP of various monosubstituted epoxides.[60–69] Use of an alcohol, *e.g.* 3-phenyl-1-propanol (PPA), deprotonated by $tBuP_4$ enabled control of the OROP of 1,2-butylene oxide (BO),[60–62] styrene oxide (SO),[63,64] allyl glycidyl ether (AGE), benzyl glycidyl ether (BnGE), *tert*-butyl glycidyl

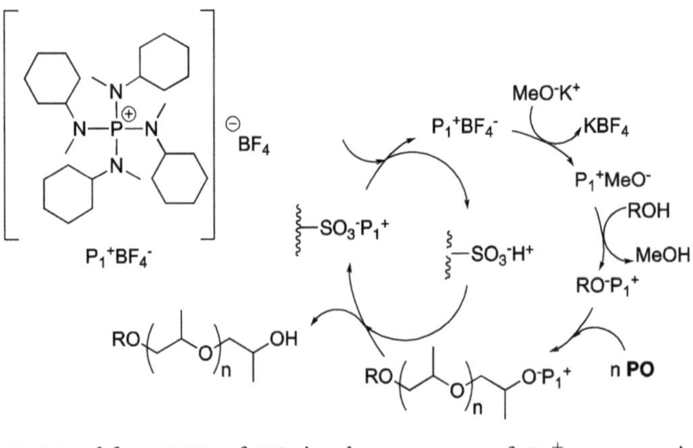

Scheme 8.7 Metal-free ROP of PO in the presence of P_1^+ as organic counter-cation.[57,58]

ether (*t*BuGE) and ethoxyethyl glycidyl ether (EEGE)[65] 2-(2-(2-methoxyethoxy)-ethoxy) ethyl glycidyl ether (EGE) and decyl glycidyl ether (DGE),[61,66–68] and different glycidylamines (see Figure 8.2).[69] For instance, the PPA-initiated *t*BuP$_4$-OROP of SO carried out at room temperature showed the features of "controlled/living" process, as assessed by kinetic investigations, chain-end analysis by MALDI-ToF mass spectrometry, and chain-extension experiments.[63] Poly(styrene oxide)s (PSOs) with a molecular weight ranging from 5.2 to 21.8 kg mol^{-1} and a dispersity <1.14 were obtained in this way. Importantly, no signal due to allylic protons was detected, meaning an absence of chain transfer to the monomer and possibility for controlling the PSO molecular weight from the initial monomer to the initiator ratio. Similar results were reported for the *t*BuP$_4$-OROP of BO, BnGE and *t*BuGE, performed in toluene at a concentration of 5 mol L^{-1}, at 25 °C, using initial [*t*BuP$_4$]$_0$/[hydroxyl]$_0$ ratios from 0.3 to 1. This led, for instance, to PBOs with molecular weights in the range 3.8–12.0 kg mol^{-1} and a *Đ* value lower than 1.09.[60]

In contrast, chain transfer to monomer was evidenced by NMR when polymerizing PO and EEGE under identical conditions,[60,70,71] but the differing behavior between these monomers and the others mentioned above was not explained. In addition, α-functional,ω-hydroxy initiators, such as 4-vinylbenzyl alcohol, 5-hexen-1-ol, 6-azide-1-hexanol, and 3-hydroxymethyl-3-methyloxetane, allowed the authors to access well-defined, end-functionalized and metal-free PSOs (Scheme 8.8).[63] Further analysis by ^{13}C NMR spectroscopy and implementation of model reaction confirmed that ring opening of the epoxy group mainly proceeded by a regioselective β-scission, though α-scissions also occurred to a minor extent.

The *t*BuP$_4$-OROP of BO was also successfully achieved in toluene at room temperature, using secondary amides as initiators.[72] However, side reactions to initiation were evidenced with some initiating systems, for instance those featuring a *tert*-butyl group in the α-position of the carbonyl or the phenylethyl group onto the nitrogen atom. Other amide-type initiators enabled to achieve PBOs with controlled molecular weights (4.7 to 9.4 kg mol^{-1}; *Đ* < 1.2).

The polymerization of EO initiated by carboxylic compounds, *i.e.* 1-pyrenebutyric acid (PyBA) and palmitic acid (PmA), in the presence of *t*BuP$_4$ as catalyst (10–50 mol% relative to the initiator) was also investigated in THF

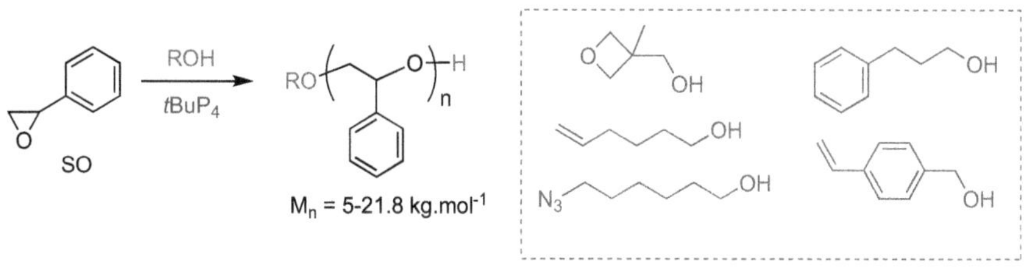

Scheme 8.8 End-functionalized PSOs by *t*BuP$_4$-OROP of SO in presence of an alcohol.[61]

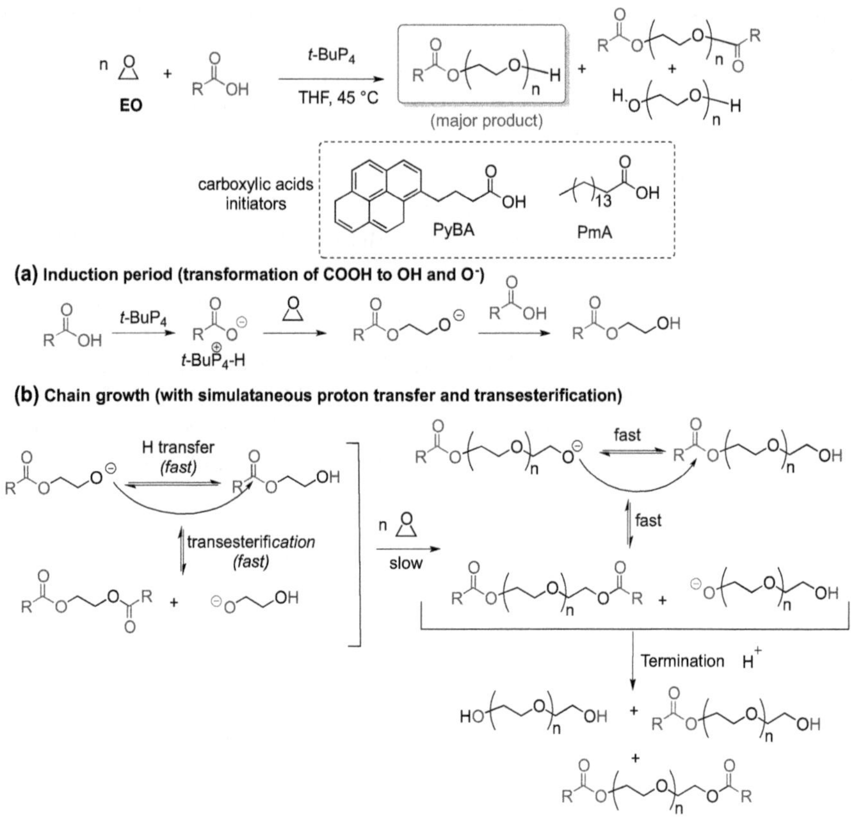

Scheme 8.9 tBuP$_4$ promoted-ROP of ethylene oxide (EO) initiated by carboxylic acids and proposed mechanisms.[73]

at 45 °C (Scheme 8.9).[73] PEOs thus obtained exhibited molecular weights $M_n = 1700–2300\ g\,mol^{-1}$ and dispersity values in the range 1.04–1.13. Kinetic studies revealed an induction period likely resulting from the rather slow ring opening of EO by the phosphazenium carboxylate. The resulting propagating alkoxide allowed the polymer chain to slowly grow with simultaneous fast end-group transesterification and proton transfer (Scheme 8.9).

The same group investigated the influence of solvents over such tBuP$_2$-catalyzed ROP of ε-CL and L-LA at r.t. using primary alcohols as initiators.[74] While a higher polymerization rate was noted in more polar solvents (dichloromethane *vs.* toluene), slower polymerization reactions were noted in cyclic ether solvents (THF, 1,4-dioxane *vs.* toluene). Due to their slight basicity, these solvents might indeed interact with initiating species, thus slowing down the ROP.

Interestingly, use of tBuP$_4$ proved to be well suited for the ROcP between an epoxide and heterocyclic, *e.g.* ε-caprolactone (ε-CL), thus enabling the synthesis of the corresponding copolymers. For instance, ROcP of ε-CL and *t*BuGE organocatalyzed by *t*BuP$_4$ led to statistical copolymers with molecular weights ranging from 12.0 to 49.2 kg mol^{-1}.[75] A random copolymer of CL and *t*BuBGE was also synthesized using *t*BuP$_4$ in the presence of an alcohol initiator, and subsequent hydrolysis afforded the corresponding amphiphilic poly(ε-caprolactone-*co*-glycidol) statistical copolymers.

Inspired by these findings, Zhang *et al.* developed a "non-copolymerization strategy" to derive statistical copolymers based on ether and ester monomer units.[76] PCL macroinitiators were thus employed for the *t*BuP$_4$-OROcP of PO and BO. It was found that both inter- and intramolecularly transesterification reactions involving hydroxyl end groups and the ester moieties of the copolymer backbone occurred much more rapidly than ring-opening reactions of the epoxides. Thus, nearly all ester groups could be cleaved generating a hydroxylated species and a CL-epoxide linkage, favoring the formation of either statistical poly(ether-*co*-ester)s or multiblock-like polyethers constituted of ester linkages, as a function of the initial ratio between the epoxide and the PCL precursor.

Application of phosphazene bases to polyether engineering, in particular, for block copolyether synthesis by one-pot sequential OROP, was described in various examples by Satoh, Kakuchi *et al.* For instance, AGE and EEGE, were sequentially copolymerized at room temperature, using PPA and *t*BuP$_4$ as initiator and organocatalyst, respectively.[65] Although occurrence of chain transfer to monomer was evidenced during the homopolymerization of EEGE, this side reaction was not observed when adding AGE and EEGE in this order, enabling the synthesis of well-defined PAGE-*b*-PEEGE diblock copolymers at room temperature. Subsequent deprotection of EEGE monomer units and/or post-chemical modification methods yielded miscellaneous functionalized metal-free aliphatic diblock copolyethers, the self-assembly of which led to various nanostructures.

This strategy employing *t*BuP$_4$ as organic base and an alcohol, namely, butanol, as initiator was implemented to the sequential ROP of glycidylamine derivatives with various types of *N*-substituents (see Figure 8.2), *e.g.* *N*,*N*′-dibenzylglycidylamine (DBGA), *N*-benzyl-*N*-methylglycidylamine (BMGA), *N*-glycidylmorpholine (GM), and *N*,*N*′-bis(2-methoxyethyl)glycidylamine (EGE).[69] Not only block copolymers, but also random and gradient-type structures were thus synthesized in toluene at room temperature as a function of the N-substituents determining the reactivity ratios of glycidylamine co-monomers. Deprotection of poly(DBGA) and poly(BMGA) by debenzylation using hydrogen in the presence of Pd/C led to polyethers with pendant primary and secondary amino groups, *i.e.*, poly(glycidylamine) and poly(glycidylmethylamine), respectively. A copolyether featuring both hydroxyl and primary amino side groups was also synthesized by copolymerizing DBGA and BGE, and subsequent deprotection.

Similarly, various thermoresponsive homopolymers as well as statistical and block copolymers were synthesized from the *t*BuP$_4$-catalyzed BnOH-initiated RO(c)P of 2-methoxyethyl glycidyl ether and 2-ethoxyethyl glycidyl ether.[68]

Synthesis of well-defined PEO-*b*-PCL diblock and PCL-*b*-PEO-*b*-PCL triblock copolymers of $M_n = 10\,900$ ($Đ = 1.09$) and 13.3 kg mol^{-1} ($Đ = 1.11$), could be achieved by sequential tBuP$_2$-OROP of EO and ε-CL, using PPA or even water as initiators.[55] Pentablock copolymers, PLLA-*b*-PCL-*b*-PEO-*b*-PCL-*b*-PLLA ($M_n = 16.1$ kg mol^{-1}; $Đ = 1.11$), were also synthesized by the introduction of LLA (nearby complete conversion of LLA within 10 min). It was however essential to follow the EO/ε-CL/L-LA addition in order to avoid any inter- or intramacromolecular side reactions, which could occur during the ROP of EO with PCL or PLLA macroinitiators.

Block copolymer synthesis involving different types of monomers by sequential tBuP$_4$ base-OROP met with more limited success. For instance, attempts to sequentially polymerize EO and ε-CL in the presence of an alcohol and tBuP$_4$ yielded ill-defined copolymers based on PEO and PCL.[77] However, the less basic tBuP$_2$ phosphazene base proved more appropriate, and well-defined PEO-*b*-PCL-*b*-PLLA triblock terpolymers and PLLA-*b*-PCL-*b*-PEO-*b*-PCL-*b*-PLLA pentablock terpolymers could be synthesized using a monofunctional alcohol and water, respectively, by sequential OROP of the corresponding monomers.[62] A more general catalyst switch strategy was later developed by Zhao, Hadjichristidis *et al.* to derive miscellaneous block copolymers emanating from epoxides and monomers of differing reactivity by sequential polymerization (see below).[77,78]

Non-linear metal-free (co)polyethers were also designed by organocatalysis utilizing phosphazene bases, including star homopolymers, star-block and miktoarm star copolymers, cyclic, figure-eight-shaped, tadpole-shaped block copolymers, graft copolymers, as well as trefoil, tetrafoil and cage-shaped polyethers derived thereof.[61,62,79,80] For instance, multihydroxy-containing initiators were deprotonated by a phosphazene base to achieve star-shaped polyethers. Thus, the ROP of EO utilizing tBuP$_4$ and pentaerythritol in THF at room temperature led to narrowly distributed four-arm star PEOs, each arm having a degree of polymerization around 50, following the "core-first" (divergent) method (Scheme 8.9).[62,79] Similarly, use of sucrose as octafunctional initiator in conjunction with either tBuP$_4$ or tBuP$_2$ as a base, in THF as solvent, led to well-defined eight-arm star PEOs, with a number of arms ranging from 10 to 110 on average and a dispersity <1.12.[80] Although the deprotonated form of sucrose was hardly soluble in the reaction mixture, a homogeneous solution was obtained upon heating after a few propagation steps, ensuring a good control over molecular weights and dispersities. Beyond the precise synthesis of PEO stars, this study revealed that residual phosphazene bases showed some toxicity. Residual phosphazenium salts found in the compounds indeed proved cytotoxic, tBuP$_4$ showing a higher cytotoxicity than tBuP$_2$, witnessing the need for carefully removing traces of phosphazene organocatalysts as far as sensitive applications (*e.g.* cosmetics or drug delivery) would be concerned.

A previous report by Kakuchi, Satoh *et al.* had demonstrated the possibility to access amphiphilic star-block copolyethers (A_nB_n and B_nA_n; $n = 3$ and 4;

total molecular weight = 22 200 g mol^{-1}) *via* the *t*BuP$_4$-OROP using tri- and tetrafunctional alcohols as initiators.[67] Hydrophilic (A) monomer units, namely, 2-(2-(2-methoxyethoxy)-ethoxy)ethyl glycidyl ether were associated to hydrophobic (B) ones, namely, decyl glycidyl ether. The well-defined character of the star-like polyether structures was verified by analysis of the arms after their selective cleavage from the core. Mikto-arm star copolyethers (A$_2$B$_2$, AB$_2$, and A$_2$B) were also synthesized by combining the *t*BuP$_4$-OROP method and azido-alkyne click chemistry and using both functional initiators and a chain terminator. Further investigations into the self-assembly in water of these amphiphilic branched copolymers showed that the size of the micelle-like structures varied with the block arrangement and the type of architecture.

Synthesis of sets of cyclic, figure-eight-shaped and tadpole-shaped amphiphilic block copolyethers based on poly(decyl glycidyl ether) and poly[2-(2-(2-methoxyethoxy)ethoxy)ethyl glycidyl ether], again by combining the *t*BuP$_4$-OROP method and azido-alkyne click chemistry, was reported by the same group (Schemes 8.10 and 8.11).[62,66] Cyclization was achieved under very diluted conditions, the resulting branched copolymers being well defined. The authors then evidenced that the cyclic topology had a dramatic effect on the cloud point of the amphiphilic copolyether solution, but little effect was observed on the critical micelle concentration and the morphology of aggregates formed by self-assembly.

The synthesis of graft copolymers constituted of a vinylic-type polymer backbone, and grafts grown by metal-free ROP, was also reported.[79–84]

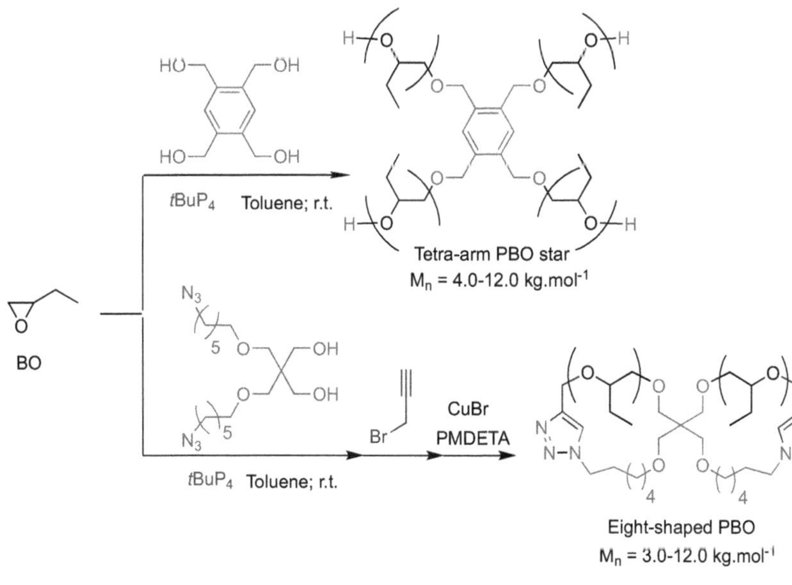

Scheme 8.10 Synthetic strategy to star- and eight-shaped PBOs by *t*BuP$_4$-OROP in the presence of multifunctional alcohols.[62]

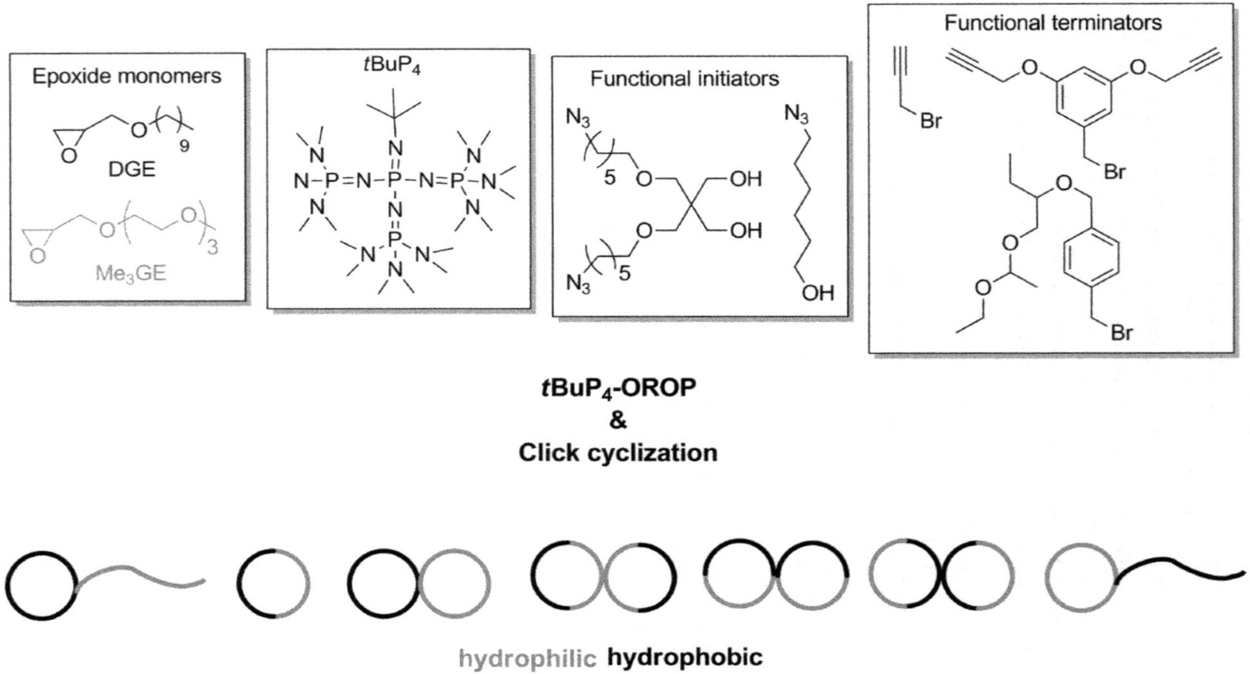

Scheme 8.11 Synthetic strategy to cyclic, eight- and tadpole-shaped amphiphilic block copolyethers *via* *t*BuP4-OROP of glycidyl ethers.[66]

In early works, Möller *et al.* employed an EVA copolymer, *i.e.* a poly[ethylene-*co*-(vinyl alcohol)], as a multifunctional initiator for the *t*BuP$_4$-OROP of EO, yielding graft copolymers made of a polyvinylic backbone with PEO side chains.[79] In this case, proton exchange between active alkoxides and dormant hydroxyls was not effective, so that only hydroxyls that were originally deprotonated by *t*BuP$_4$, namely, one-third, contributed to the growth of PEO grafts.

Later on, Zhao *et al.* described various examples of graft copolymers, *e.g.* polystyrene-*b*-poly(p-hydroxystyrene-*g*-ethyleneoxide), PS-*b*-(PHOS-*g*-PEO),[82] PHOS-*g*-(PEO-*b*-PPO), PHOS-*g*-(PPO-*b*-PEO)[70] or PHOS-*g*-(PEO-*stat*-PPO).[71] The overall synthetic strategy combined (i) conventional lithium-based "living" anionic polymerization of protected hydroxystyrene, (ii) post-chemical modification hydrolysis to introduce the multiple OH side groups, and (iii) metal-free *t*BuP$_4$-OROP following the "grafting from" method. Generally, low *t*BuP4/phenol ratios yielded copolymers of lower grafting density and longer PEO grafts. Solution properties in water of all these thermoresponsive graft copolymers were investigated in detail.

Alternatively, amide groups of a well-defined poly(*N*-isopropylacrylamide) (PNIPAM) were also used to grow PEO side chains by the *t*BuP$_4$-OROP method.[71] The grafting density was adjusted by the initial ratio between *t*BuP$_4$ and the amide groups, which had a strong impact onto the thermo-responsive water solution properties of the graft copolymers.

By using a poly(*N*,*N*-dimethylacrylamide-*co*-acrylamide) as particular macroinitiator (Scheme 8.12),[84] Hadjichristidis *et al.* employed *t*BuP$_4$ to trigger the anionic ring-opening graft polymerization (AROGP) of various epoxides, including EO, PO, BO and *t*BGE. While EO and PO, BO enabled achieving targeted graft copolymers with various side chains in toluene, at 45 °C, including single- or double-grafted homopolymers, block and statistical copolymers (M_n in the range 600–1700 kg mol^{-1}; $Đ = 1.17–1.31$), *t*BGE did not, likely due to steric hindrance. In these syntheses, initiation was possible thanks to the deprotonation of pendant primary amides by *t*BuP$_4$.

As mentioned, to overcome the difficulty, in some cases, of achieving block copolymers from an epoxide and another category of monomer, that is, showing a different reactivity/selectivity (*e.g.* a cyclic ester or a cyclic carbonate), Zhao, Hadjichristidis *et al.* set up a catalyst switch strategy.[77,78,85,86] For instance, a "base-to-acid" organocatalyst change allowed the authors to synthesize well-defined polyether-*b*-polyester/polycarbonate block copolymers (Scheme 8.13).[77,78] Following this strategy, the epoxide (EO or BO) was polymerized first by the ROH-initiated *t*BuP$_4$-OROP method, before crossing over to the ROP of ε-CL or δ-valerolactone or trimethylene carbonate. This was achieved by adding—a generally large—excess of diphenyl phosphate (DPP), *i.e.* an organic acid serving both to quench the phosphazenium alkoxide of the first block and catalyze chain growth of the polyester from the OH-terminus of the polyether block. A switch from the butanediol-initiated *t*BuP$_4$-OROP of EO, forming a PEO diol, to the DPP-organocatalyzed synthesis of polyurethane by step-growth polymerization,

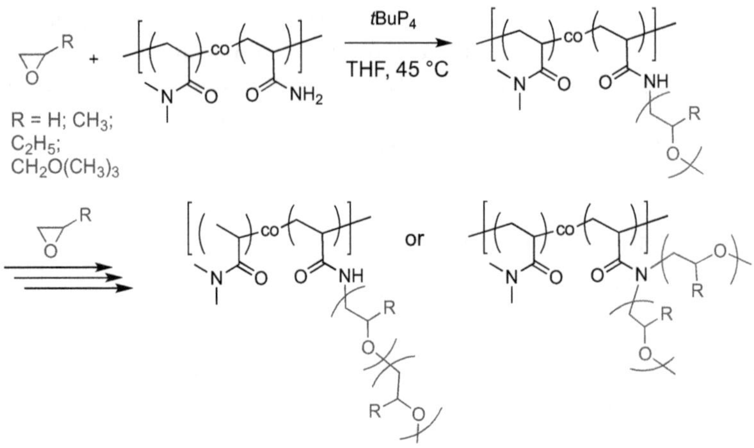

Scheme 8.12 tBuP$_4$-OROP of epoxides for the synthesis of graft copolymers.[84]

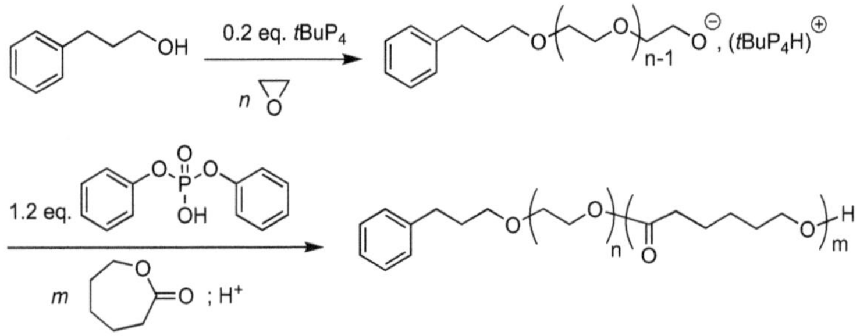

Scheme 8.13 Block copolymer synthesis by sequential OROP involving a base-to-acid organocatalyst switch.[77,78]

was also applied to access block copolymers based on polyether and polyurethane.

Although broadening the scope of polymer synthesis by an organocatalysis pathway, a switch from tBuP$_4$ to DPP did not enable polymerization of lactide (LA) as the second monomer. Moreover, the phosphazenium phosphate salt formed by addition of the excess of DPP was found to cause some retardation during DPP-OROP of the cyclic ester.[77] A strategy involving switch from tBuP$_4$ to a thiourea-based organocatalyst was therefore developed.[87] In the latter case, the phosphazenium alkoxide was shown to deprotonate the thiourea, generating a thiourea anion combining both a basic character and a hydrogen-bonding donor capability of the remaining NH group. This subtle effect enabled to efficiently polymerize either CL or LLA by OROP under controlled conditions. The catalytic activity of the thiourea anion could be tuned by the structural variation of the *N*-substituents of the thiourea precursor. Although lacking of mechanistic investigations, this strategy was successfully applied to the tandem OROP and step-growth polymerization.

In a recent contribution, Guo *et al.* reported a similar strategy by switching from the *n*-BuNF$_4$-OROP of GPE to the deprotonated thiourea-catalyzed OROP of LA, thus yielding a well-defined PGPE-*b*-PLA diblock copolymer one-pot.[88]

Several examples of alternating copolymers derived by an organocatalyzed pathway were also described. This was achieved by using non-homopolymerizable comonomers, such as 3,4-dihydrocoumarin (DHC) and its derivatives,[53,54,89–92] cyclic anhydrides[81,93,94] or carbon dioxide (CO_2),[43,44,95] which were found to readily copolymerize with epoxides in an alternating manner by organocatalysis. The ROH-initiated *t*BuP$_4$-OROP method proved relevant for that purpose, as reported by Zhao, Zhang *et al.*[53,54,89]

Pioneering works by Endo *et al.*, however, utilized 2-ethyl-4 methylimidazole (EMI, see Figure 8.1) as initiator for the OROcP of GPE and DHC.[89,90] In contrast to DHC, GPE could be directly homopolymerized by EMI, and mixtures of the two comonomers gave much higher reaction rates. An equimolar $[GPE]_0/[DHC]_0$ feed ratio yielded a copolymer showing 90% alternating sequence, and alternation was found to increase as this feed ratio decreased. The copolymerization was thought to involve initiation by nucleophilic attack of EMI onto two eq. of GPE, forming a EMI-(GPE)2 zwitterionic intermediate, and further propagation by a zwitterionic mechanism. Each of the active chain ends deriving from each monomer would selectively react with the other monomer, *i.e.* the alkoxide deriving from GPE would add onto DHC, while the phenoxide formed from DHC would preferentially react with GPE.

This strategy was applied to obtain alternating copolymers from various epoxides and DHC derivatives.[89–92] OROcP reactions involving DHC and epoxides most often led to a mixture of linear and cyclic alternating copolymers (Scheme 8.14),[53,54,89–92] witnessing the occurrence of intramolecular transesterification during synthesis. However, the less basic *t*BuP$_2$ and *t*BuP$_1$ allowed minimizing this side reaction, favoring the formation of linear cyclic alternating copolymers. This was attributed to a proton exchange between the phosphazenium and the alkoxide, forming dormant hydroxylated species thus reducing the probability of transesterification from the alkoxides. Hydroxy-terminated alternating copolymers could then serve as macroinitiators for the growth of either a poly(δ-valerolactone) or a polyurethane block.[53] To illustrate the versatility of their approach, the authors also derived star-like polymers and graft copolymers using multi-hydroxylated initiators.[89]

Organocatalysis by phosphazene bases was also efficient for alternating copolymer synthesis by ROcP of EO with phthalic anhydride (PA).[94,95] Use of *t*BuP$_4$ in the presence of a hydroxy initiator caused extensive transesterification reactions and a loss of control of the ROcP process. Although not suitable to promote the homopolymerization of any of the two monomers separately, the least basic *t*BuP$_1$ enabled preparation of a much better defined poly(PA-alt-EO)s, with molecular weights up to 50 $kg\,mol^{-1}$ and dispersities lower than 1.1 (Scheme 8.15). A similar proton exchange, as discussed previously for the ROcP of DHC and epoxides, was put forward to

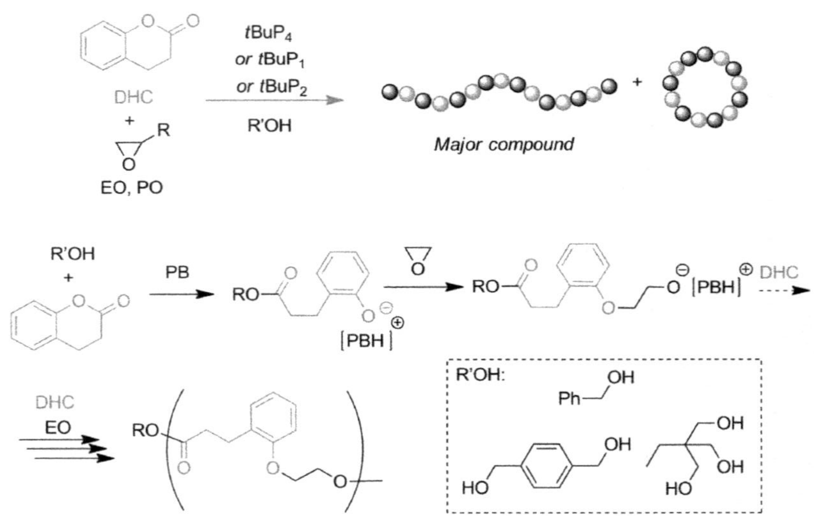

Scheme 8.14 Alternating ROcP of DHC and epoxides using a phosphazene base (PB) in the presence of an alcohol initiator (R′OH).[53,54,89–92]

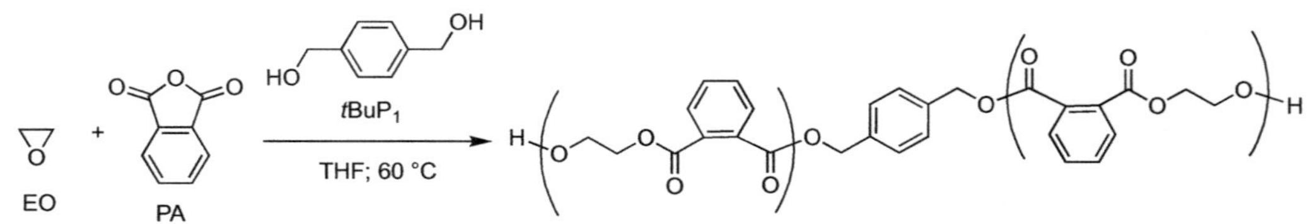

Scheme 8.15 Synthesis of poly(EO-*alt*-PA) by tBuP$_1$-OROcP using a diol-type initiator.[94,95]

explain these results. Very recently, various mono-, di-, and trisubstituted epoxides were copolymerized with PA in the presence of both $t\text{BuP}_1$ and either an alcohol or a carboxylic acid-containing initiator, followed by chain extension experiments to obtain not only block copolymers consisting of an alternating block, but also statistical-alternating copolymers by copolymerizing PA and two mixed epoxides.[96]

Note that phosphines also proved efficient to promote the zwitterionic ROcP of bicyclic bis(γ-butyrolactone) (BBL) and epoxides, forming linear alternating copolymers.[97] Reactive alternating polyesters featuring α,β-unsaturated ketone or sulfide functions were even achieved using BBL bearing an isopropenyl group.[98,99]

8.2.4 Dual Activation from a Phosphazene Base and a Metallic Lewis Acid

Several studies accounted for the possibility to combine two distinct catalysts for polyether synthesis. Although not related to organocatalyzed synthesis, strictly speaking, since they are based on the association of a metallic complex with an organic catalyst, these examples are here mentioned as they have enabled enhancement of the catalytic activity and/or the selectivity of the ROP process. Many examples of such a dual catalysis strategy can be found in molecular chemistry to achieve complex molecules,[100–103] and it has also emerged in polymer synthesis as well. Both catalysts operate by activating the reaction partners in a dual/cooperative manner.[104–108] In the context of the ROP of epoxides, phosphazene bases were thus used in conjunction with metal-based initiators/catalysts.[109,110] For instance, it was shown that dissociation of lithium alkoxide ion pairs could be favored, enhancing its nucleophilic character, upon mixing $t\text{BuP}_4$ with *n*-butyllithium (n-BuLi) for the ROP of EO. In the latter case, good control over the molecular weights could be achieved. Different results were obtained using sec-butyllithium (sec-BuLi), which is more basic than n-BuLi: when combined with $t\text{BuP}_4$: (i) an induction period was observed, (ii) and formation of aldehyde end-groups followed by reaction with living chains were evidenced by monitoring the ROP process by Fourier-transform near-infrared (FT-NIR) fiber-optic spectroscopy online. This afforded PEOs with a higher molecular weight than expected.[110] Control of the ROP process was eventually found to depend on various parameters, including the temperature, the concentration in $t\text{BuP}_4$, and the order of addition of the components.

The ability of $t\text{BuP}_4$ to complex with lithium cations and favor polyether chain growth from lithium alkoxides was exploited for sequential block copolymer synthesis. When polymerizing styrene first using s-BuLi as initiator, and by then adding EO in excess, a polystyrene (PS) block terminated by lithium alkoxide moiety was obtained. As such an ion pair is unable to further propagate PEO chains, addition of $t\text{BuP}_4$ allowed dissociating the ion pair, enhancing its reactivity to give access to well-defined PS-*b*-PEO block copolymers.[109,110]

In other words, crossover from one block to the other was here achieved by switching the active end group from a highly active carbanion to a less reactive alkoxide. The same strategy could be applied for the synthesis of other block copolymers involving the prior anionic polymerization of a vinylic monomer, including styrene, butadiene or isoprene. In this way, PB-*b*-PEO, PI-*b*-PEO, PB-*b*-PI-*b*-PEO, PS-*b*-PEO-*b*-PEEGE, (PB and PI standing for polybutadiene and polyisoprene, respectively) were successfully synthesized, either in THF or in benzene.[111–115]

Use of triisobutylaluminum (TIBA) as a strong metallic Lewis acid in combination with $t\text{BuP}_4$ proved efficient for the ROP of PO in the presence of alcohols as initiators (Scheme 8.16).[116] Good control of the reaction and minimization of chain transfer reactions to the monomer were explained both by the formation of a weakly nucleophilic "ate complex" formed between TIBA, the alkoxide and the $t\text{BuP}_4\text{H}$ phosphazenium cation, and by a very high activation of PO by TIBA leading to high ROP rates. PPOs with molecular weights up to 80 kg mol^{-1} and low dispersities could be obtained after 5 h at 20 °C in toluene as apolar solvent.

TIBA combined with $t\text{BuP}_4$ or $t\text{BuP}_2$ also proved efficient for the ROP of BO in the presence of specific carbamates as initiators.[117] The cooperative effect of the two activators was required to cap the polyether chains with carbamate end groups. When using $t\text{BuP}_4$ alone, carbamate anions proved indeed unstable and more prone to degradation than in the case of a dual activation with TIBA. In the latter case, stabilization was provided by interaction of TIBA with carbonyl end-groups, enabling achievement of PBOs fitted with a carbamate chain ends. However, control of the ROP reactions was possible only with the carbamate/$t\text{BuP}_2$/TIBA system (1/1/1 molar, in 2-methyltetrahydrofuran at 25 °C), enabling a quantitative initiation by the carbamate anion; in contrast, chain transfer reactions were observed using carbamate/$t\text{BuP}_4$/TIBA system.

8.2.5 N-heterocyclic carbenes (NHCs) and N-heterocyclic olefins (NHOs)

Owing to their broad modularity, N-heterocyclic carbenes (NHCs) have gained increasing popularity in the last three decades, not only in molecular chemistry, but also in polymer synthesis.[118–121] These neutral divalent species of carbon possess only six-electron valency (four electrons are involved in σ-bonds and two remain at the central carbon) and are strongly stabilized by interaction of the two nitrogen lone pairs with the empty p orbital of the carbene center. The structural diversity of NHCs allows fine-tuning their catalytic activity and selectivity. The range of monomers amenable to polymerization by NHC organocatalysis has been significantly expanded, beyond cyclic esters (D,L-lactide and lactones), in recent years, and thus achieve a variety of metal-free (co)polymers. NHCs can serve as organocatalysts or as direct nucleophilic initiators for various polymerization reactions.[120,121]

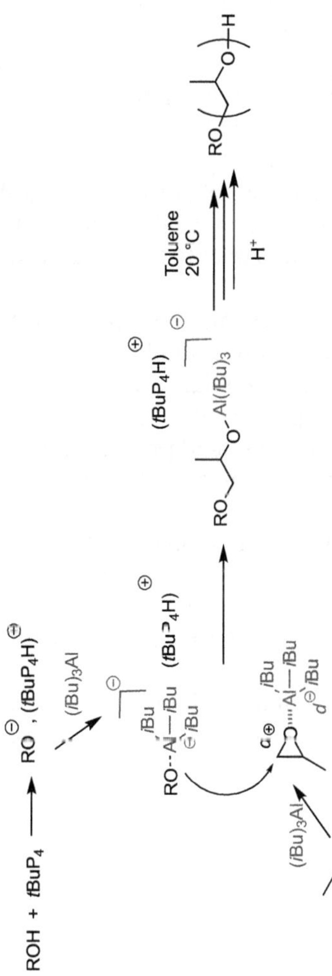

Scheme 8.16 Combination of TIBA as a metallic Lewis acid and *t*BuP4 as an organic base for the rapid and controlled synthesis of PPO.[116]

Gnanou, Taton *et al.* reported that NHCs were able to directly open ethylene oxide (EO) and served as direct initiators in a ZROP. Their use as true organocatalysts was also reported when used in conjunction with NuE-type initiators, with Nu and E = the nucleophilic and the electrophilic part, respectively; *e.g.* Nu = $PhCH_2O$, HC≡CCH_2O, N_3 and E = H, $SiMe_3$.[122] Thus, 1,3-diisopropylimidazol-2-ylidene (**Ipr**) directly initiated the metal-free ROP of EO at 50 °C in DMSO following a ZROP process. In that case, the PEO molecular weight was controlled by the EO to the NHC molar ratio ([EO]/[NHC] = 100/1). Chain growth occurred by a 1,3-diisopropylimidazol-2-ylidinium alkoxide, and quenching of the ZROP process by the NuE functionalizing terminator afforded well-defined linear α-Nu,ω-OH (or α-Nu,ω-$OSiMe_3$)-difunctionalized PEOs (Scheme 8.17). For instance, difunctionalized α,ω-dihydroxy-telechelic, α-benzyl,ω-hydroxy and α-azido,ω-hydroxy-PEOs were synthesized by NHC-initiated ZROP, using H_2O, $PhCH_2OH$ and N_3SiMe_3 as terminating agents, respectively. A well-defined PEO-*b*-PCL diblock co-polymer was also synthesized by sequential ZROP in DMSO, using the same NHC as organic initiator.

Instead of using NHCs as direct initiators, NHCs were employed as true organocatalysts in conjunction with alcohols or with trimethylsilylated initiators.[123] Using typically 10 mol% relative to NuE, α-Nu,ω-OH PEOs with molecular weights in the range 1.8–10.5 kg mol^{-1} and dispersities <1.15 could be obtained. Control over the ROP process could be explained by the occurrence of either a basic or a nucleophilic mechanism, or the combination of both.

Both strategies were implemented in the ROP of PO, *i.e.* use of IPr either as direct initiator in ZROP or as organocatalyst in the presence of NuE-type reagents.[124] However, yields were limited to 30–40% and required long reaction times (3 d) at 50 °C. In both cases, α,ω-difunctionalized PPOs could be derived under metal-free and solvent-free conditions. When NHC was used as a direct organic initiator, the polymerization was quenched by H_2O, leading to a dihydroxytelechelic PPO with molecular weights up to

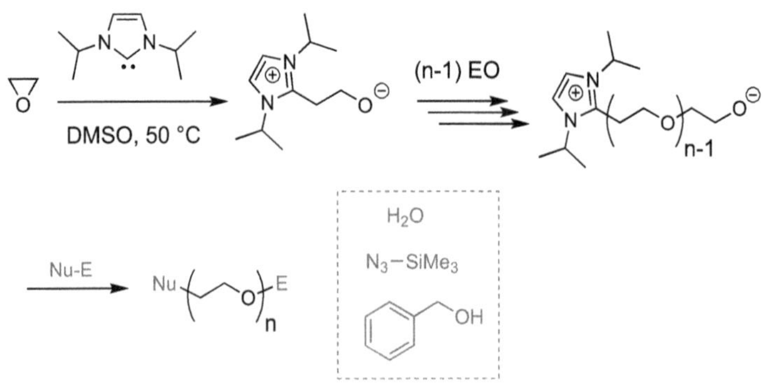

Scheme 8.17 ZROP of EO and synthesis of PEO telechelics using functional terminators.[122]

4.5 kg mol^{-1}. However targeting PPOs of higher molecular weights revealed the presence of a small proportion of allyl-ended PPO chains, as detected by ^{1}H NMR, owing to the occurrence of the chain transfer to monomer of active PPO chains. With the NHC used as organocatalyst in presence of NuE initiators, at 50 °C, PPOs of molecular weights up to 8.0 kg mol^{-1} and a dispersity lower than 1.18 could be obtained.

BASF researchers reported the bulk NHC-OROP of PO at 120 °C, from masked-NHCs, namely, imidazolium and benzimidazolium carboxylates (=NHC–CO_2 adducts) serving as latent organocatalysts.[125] These adducts were decarboxylated upon heating to release the catalytically active free NHC. In this case, diethylene glycol (DEG) was employed as initiator, enabling to access diol-functionalized PPO oligomers (M_n = 0.3–1.2 kg mol^{-1} and $Đ$ = 1.08–1.23), after 4 h, with 40–74% monomer conversion. Consistent with previous findings by Taton *et al.*, imidazolium pre-catalysts prove the most active, with the least bulky substituents affording the higher catalytic activity. DFT calculations and *in situ* IR spectroscopy suggested that decarboxylation was slow, affording a continuous delivery of the free NHC and a better control over the NHC-OROP process. NHC-CO_2 adducts would thus play the role of a reservoir of free NHCs. Interestingly, this procedure allowed synthesizing PPO-*b*-polyester copolymers (M_n = 0.75–3.7 kg mol^{-1} and $Đ$= 1.38–1.77), by sequential ROP of PO and ε-CL (or L-LA). The authors finally proposed a free anionic-type mechanism, as highlighted in Scheme 8.18,

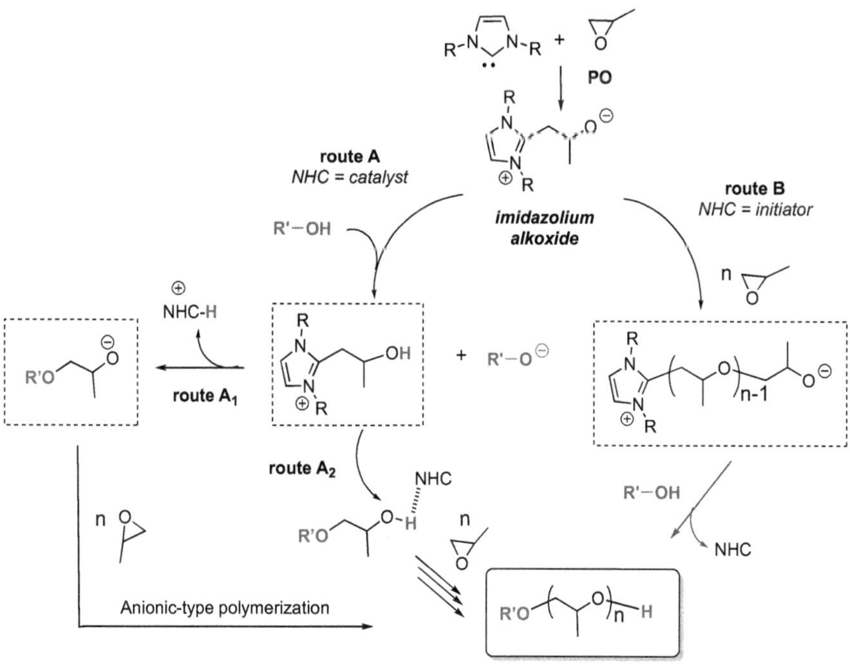

Scheme 8.18 Proposed mechanisms for the NHC-OROP of PO using an alcohol.[124,125]

based on their DFT calculations. In the anionic mechanism, the zwitterionic imidazolium alkoxide would evolve into an imidazolium alkoxide ion-pair (route A_1).

Instead of NHCs, Dove *et al.* resorted to N-heterocyclic olefins (NHOs) as organocatalysts to overcome the shortcomings pertaining to the metal-free and solvent-free synthesis of PPO, and thus reported the first example of an NHO-catalyzed organopolymerization.[126] Compared to NHCs, NHOs feature a reactive C=C double bond between the C_2-position of the imidazole ring and a exocyclic carbon atom, making them highly nucleophilic carbanionic-type species. By properly manipulating the NHO structure, their catalytic activity towards PO polymerization could be fine tuned. Thus, a NHO with a five-membered unsaturated ring (= imidazole-type) and possessing two methyl groups on the exocyclic carbon showed a high performance when used in presence of an alcohol initiator, enabling the suppression of the side chain transfer reaction to the monomer, thus producing well-defined PPOs in high yields with high turnover (TON > 2000), with molecular weights up to 12 kg mol^{-1} and a dispersity <1.09.

This NHO-organocatalysis method was recently applied by Naumann *et al.* to the synthesis of structurally well-defined PPO-*b*-PEO-*b*-PPO and PBO-*b*-PEO-*b*-PBO triblock copolymers.[127] The former compounds could serve as structure-directing precursors of ordered mesoporous carbon-based materials.

Cavallo *et al.* further performed DFT calculations,[128] which supported experimental findings by Dove *et al.* and the possible occurrence of two competitive mechanisms, namely, the anionic and zwitterionic pathways involved in the NHO-OROP of PO initiated by BnOH. In the anionic-type mechanism, the acid–base reaction between BnOH and NHO would favor ring opening of PO by the benzyloxy group, while the zwitterionic mechanism would involve the nucleophilic attack of the exocyclic carbon atom of the NHO to PO, and formation of a zwitterionic intermediate species after ring opening. In the anionic pathway, the energetic barrier of the initiation step was found to depend on the steric and electronic effects of the NHO, which was not the case for the zwitterionic pathway. These theoretical calculations thus seemed to indicate that a particular mechanism might be favored by fine tuning the NHO structure.

8.2.6 Other Organic Salts

The metal-free ROP of glycidyl phenyl ether (GPE) was investigated using, tetra-*n*-butylammonium fluoride (Bu_4NF) and the less hygroscopic tetra-*n*-butylammonium acetate (Bu_4NOAc), as a means to synthesize telechelics.[129–132] Well-defined oligomers with a molecular weight (M_n) of 2.5 and 2.9 kg mol^{-1}, $Đ = 1.28$ and 1.12, were achieved from Bu_4NF[129] and Bu_4NOAc,[133] respectively. In the former case, a strong C–F bond was created in α-position, but it could not be converted into any other

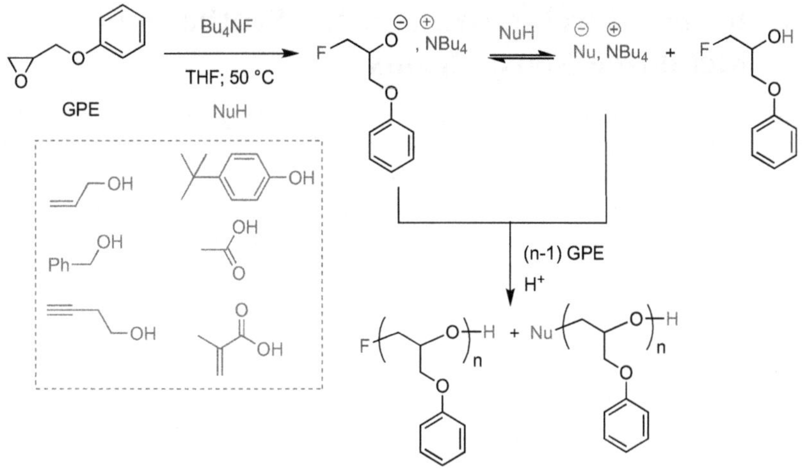

Scheme 8.19 ROP of GPE using Bu_4NF and functional CTAs.[129–134]

functional group by post-modification. In the latter case, the presence of the CH_3COO-end group was verified, and subsequent hydrolysis allowed Endo *et al.* to achieve α-hydroxy-ended oligo (PGE). When used in the presence of water or ethanol as chain transfer agents (CTAs), Bu_4NF was found to play both the role of an organic catalyst and of an initiator.[130] Analysis of the resulting oligomers by NMR spectroscopy indeed showed the presence of both $F–CH_2$– and C_2H_5O– moieties as chain ends. Similar investigations by Kawakami *et al.* utilizing Bu_4NF combined with protic functional CTAs enabled the synthesis of metal-free oligo(PGE)s featuring various end groups, including an alkene, a benzyl ether, an alkyne, an ester and a methacrylate group (Scheme 8.19).

Endo *et al.* applied a similar strategy by combining Bu_4NF and various amounts of a poly(ethylene glycol) (PEG) monomethyl ether for the metal-free ROP of GPE,[132] which led to amphiphilic PEG-*b*-oligo(GPE) diblock copolymers. Contamination with a small proportion of F-CH_2-ended oligo(GPE) was evidenced, requiring a purification step to obtain chemically pure diblock copolymers. These compounds were found to self-assemble into micelle-like structures with a size of 58 and 140 nm, for a composition of 62 : 38 and 53 : 47, respectively. Block copolymers, namely, poly(GPE)-block-poly(TC), exhibiting a well-defined structure could also be prepared by sequential Bu_4NF-mediated ROP of GPE and TC.[134]

Norbornene and phthalic anhydrides, NA[93] and PA,[94] respectively, were found to produce alternating structures with molecular weights in the range 3.0–9.3 kg mol^{-1}, when copolymerized with CHO in toluene at 60–110 °C, in the presence of bis(triphenylphosphine)iminium chloride (PPNCl), or 4-dimethylaminopyridine (DMAP), tetra-*n*-butylammonium chloride (n-Bu_4Cl), tetra-*n*-butylammonium bromide (n-Bu_4Br), and triphenylphosphine (TPP), as peculiar organocatalytic systems.

8.3 Recent Developments in the Synthesis of Metal-free Epoxy Resins

Besides aliphatic polyethers, polyepoxides emanating from so-called epoxy resins represent another category of industrially important polymeric materials, involving the polymerization of epoxides. Epoxy resins are reactive monomers or oligomers possessing at least two epoxide functional groups. A typical example is bisphenol A diglycidyl ether (BADGE). Epoxy resins can thus be crosslinked either through a catalyzed chain-growth homopolymerization, or following a step-growth polymerization pathway when reacted with co-monomer partners, often named hardeners, such as polyfunctional aliphatic, cycloaliphatic and aromatic amines, acids, acid anhydrides, alcohols or thiols.[135–137] The crosslinking reaction, also referred to as the curing reaction, leads to a three-dimensional network called thermoset with a glass transition temperature higher than room temperature. Owing to the broad structural diversity of both the epoxy resin and the hardener, a wide range of polyepoxides can be achieved, the properties of which can also be fine tuned by curing conditions and the stoichiometric ratio between the co-monomers. Polyepoxides are usually characterized by a high thermal and chemical resistance and good mechanical properties and are used in various applications, such as paints and structural adhesives, in electrical and electronic components, or as fiber-reinforced plastic materials and as polymer matrices of composite materials for aeronautics.

Buchmeiser *et al.* described the NHC-mediated thermal curing of anhydride hardened epoxy resins, *e.g.* based on bisphenol A diglycidyl ether (BADGE) and on phthalic acid anhydride (PhA) or hexahydrophthalic anhydride (HHPA).[138,139] The authors resorted to thermally latent NHC-CO_2 adducts as well as to metal-protected NHCs in a process that was shown to be compatible with vacuum-assisted resin infusion. Mixtures of the epoxy compound, the anhydride and protected NHC formed homogeneous and stable formulations, such mono-component mixtures could be handled and stored under air without any particular precautions. Upon heating to 140–160 °C, free NHCs (0.1–1 mol%) were released and proved more efficient in comparison with commonly used accelerators (*e.g.* tertiary amines or *N*-alkylimidazoles) for warm-curing epoxy resins. Thereby, use of NHCs allowed reducing catalyst loadings, while affording fast polymerization kinetics and milder curing temperatures. Interestingly, the polymerization rate could be modulated by the chemical structure of the NHC. A NHC-triggered ring-opening of cyclic anhydrides, rather than the generally admitted ring opening of oxirane moieties by conventional imidazole-based accelerators, has been postulated (Scheme 8.20).[138,139]

By carefully tuning the different components, the authors were able to achieve a complete curing process within minutes at lower curing temperatures, *i.e.* at $T = 100–120$ °C.[140] It was found that aliphatic epoxides combined with low-melting or liquid anhydrides provided the highest

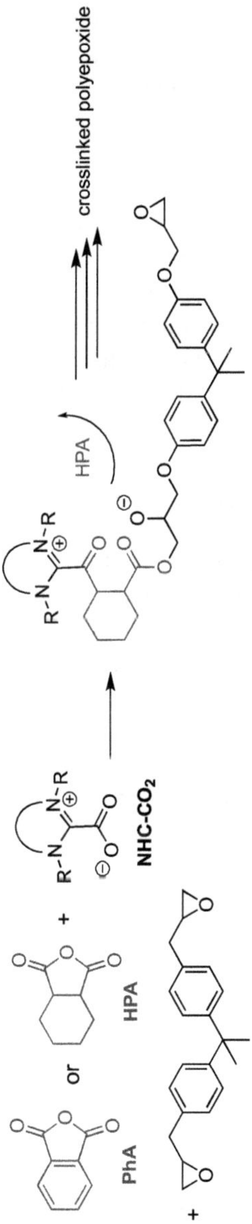

Scheme 8.20 Curing of epoxide/anhydride resins using NHCs.[138–140]

polymerization rate, while the initial viscosity remained very low (<100 mPa s at room temperature), enabling the manipulation of a sprayable mixture.

Some urea derivatives[141,142] were also found to be effective as thermally latent initiators for the ROP of DGEBA. The 4-aminopyridine generated from the dissociation of the urea could initiate the ROP of DGEBA and the isocyanates could react with epoxide through cycloaddition. It was also considered that the urea-derivatives could participate in the polymerization by activating the epoxide through hydrogen-bonding.[142]

8.4 Conclusion

The diverse reactivity of commercially available or easily accessible organocatalysts has been successfully applied to precision polyether synthesis through the design of a broad range of well-defined structures. These include not only end-functionalized linear polyethers, but also block, alternated, and statistical copolymers, as well as many branched architectures, such as macrocyclics, graft copolymers, star-like homopolyethers, star-block and miktoarm star copolymers, and figure-eight-shaped and tadpole-shaped block copolymers. In this context, phosphazene bases have certainly been the most investigated, but other organocatalytic systems, including N-hetrocyclic carbenes and N-hetrocyclic olefins, or "super-strong" Lewis or Brønsted acids, or even simple organic salts, also provide excellent control over molecular weights, dispersities with high-chain-end fidelity of the resulting (co)polyethers. Epoxide substrates thus appear today as less challenging monomers to be polymerized in a controlled fashion by organocatalysis. In particular, side reactions originally observed, either in cationic ROP (*e.g.* backbiting forming cyclic oligomeric by-products), or in anionic ROP (*e.g.* chain transfer to the monomer in the case of mono-substituted epoxides, such as propylene and butylene oxides, or with glycidyl ethers), can be almost totally suppressed through the use of appropriate organocatalysts. Some of these exhibit an adapted selectivity for polyether synthesis, sometimes exhibiting performances that compare with some common organometallic catalysts. This allows accessing industrially important polyethers free of any metallic residues under relatively mild conditions, including, PEO, PPO, PBO various poly(glycidyl ethers) and polymers deriving from epoxy resins.

However, among the remaining challenges to tackle in the utilization of organocatalysis for polyether synthesis, there is probably the need for further investigations on the toxicity of some organocatalytic systems. For instance, preliminary studies have shown that residual phosphazenium salts in polyethers derived from phosphazene bases exhibit some cytotoxicity. Therefore, efforts are probably required towards the organocatalyst recycling in general, and to drastically reduce the catalyst loading. Moreover, the organocatalytic platform still lacks of versatility as far as block copolymer synthesis from epoxides and other heterocyclics of different reactivity is concerned. The concept of a "dual/cooperative catalysis", that is, combining the activation of an organocatalyst with that of a metal-based catalyst in a synergistic manner, also appears very promising, as it allows dramatically increasing both the polymerization activity and selectivity.

References

1. S. Penczek, M. Cypryk, A. Duda, P. Kubisa and S. Slomkowski, *Prog. Polym. Sci.*, 2007, **32**, 247–282.
2. A.-L. Brocas, C. Mantzaridis, D. Tunc and S. Carlotti, *Prog. Polym. Sci.*, 2013, **38**, 845–873.
3. M. I. Childers, J. M. Longo, N. J. Van Zee, A. M. LaPointe and G. W. Coates, *Chem. Rev.*, 2014, **114**, 8129–8152.
4. R. Klein and F. R. Wurm, *Macromol. Rapid Commun.*, 2015, **36**, 1147–1165.
5. Y. Sarazin and J.-F. Carpentier, *Chem. Rev.*, 2015, **115**, 3564–3614.
6. J. Herzberger, K. Niederer, H. Pohlit, J. Seiwert, M. Worm, F. R. Wurm and H. Frey, *Chem. Rev.*, 2016, **116**, 2170–2243.
7. N. E. Kamber, W. Jeong, R. M. Waymouth, R. C. Pratt, B. G. G. Lohmeijer and J. L. Hedrick, *Chem. Rev.*, 2007, **107**, 5813–5840.
8. M. K. Kiesewetter, E. J. Shin, J. L. Hedrick and R. M. Waymouth, *Macromolecules*, 2010, **43**, 2093–2107.
9. M. Fèvre, J. Vignolle, Y. Gnanou and D. Taton, *Polymer Science: A Comprehensive Reference*, 2012, **4**, 67–115.
10. A. P. Dove, *ACS Macro Lett.*, 2012, **1**, 1409–1412.
11. K. Fuchise, Y. G. Chen, T. Satoh and T. Kakuchi, *Polym. Chem.*, 2013, **4**, 4278–4291.
12. L. Mespouille, O. Coulembier, M. Kawalec, A. P. Dove and P. Dubois, *Prog. Polym. Sci.*, 2014, **39**, 1144–1164.
13. C. Thomas and B. Bibal, *Green Chem.*, 2014, **16**, 1687–1699.
14. W. Nzahou Ottou, H. Sardon, D. Mecerreyes, J. Vignolle and D. Taton, *Prog. Polym. Sci.*, 2016, **56**, 64–115.
15. S. Hu, J. Zhao, G. Zhang and H. Schlaad, *Prog. Polym. Sci.*, 2017, **74**, 34–77.
16. K. Brzezinska, R. Szymanski, P. Kubisa and S. Penczek, *Makromol. Chem., Rapid Commun.*, 1986, **7**, 1–4.
17. P. Kubisa, *Makromol. Chem., Macromol. Symp.*, 1988, **13–14**, 203–210.
18. T. Biedron, P. Kubisa and S. Penczek, *J. Polym. Sci., Part A: Polym. Chem.*, 1991, **29**, 619–628.
19. P. Kubisa and S. Penczek, *Prog. Polym. Sci.*, 1999, **24**, 1409–1437.
20. C. A. Wurtz, *Ann. Chim. Phys.*, 1863, **69**, 317–354.
21. H. Staudinger and O. Schweitzer, *Ber. Dtsch. Chem. Ges. B*, 1929, **62**, 2395–2405.
22. P. J. Flory, *J. Am. Chem. Soc.*, 1940, **62**, 1561–1565.
23. S. Zalipsky, *Adv. Drug Delivery Rev.*, 1995, **16**, 157–182.
24. M. J. Roberts, M. D. Bentley and J. M. Harris, *Adv. Drug Delivery Rev.*, 2002, **54**, 459–476.

25. M. S. Thompson, T. P. Vadala, M. L. Vadala, Y. Lin and J. S. Riffle, *Polymer*, 2008, **49**, 345–373.
26. D. Chattopadhyay and K. Raju, *Prog. Polym. Sci.*, 2007, **32**, 352–418.
27. S. Penczek, *J. Polym. Sci., Part A: Polym. Chem.*, 2000, **38**, 1919–1933.
28. S. Kobayashi, K. Morikawa and T. Saegusa, *Polym. J.*, 1979, **11**, 405–412.
29. T. Biedron, K. Brzezinska, P. Kubisa and S. Penczek, *Polym. Int.*, 1995, **36**, 73–80.
30. W. Huang, Y. Zhou and D. Yan, *J. Polym. Sci., Part A: Polym. Chem.*, 2005, **43**, 2038–2047.
31. M. Bednarek, P. Kubisa and S. Penczek, *Macromolecules*, 1999, **32**, 5257–5263.
32. A. Sudo, A. Suzuki and T. Endo, *J. Polym. Sci., Part A: Polym. Chem.*, 2013, **51**, 4213–4220.
33. D. Chakraborty, A. Rodriguez and E. Y. X. Chen, *Macromolecules*, 2003, **36**, 5470–5481.
34. F. Hamazu, S. Akashi, T. Koizumi, T. Takata and T. Endo, *J. Polym. Sci., Part A: Polym. Chem.*, 1991, **29**, 1675–1680.
35. B. L. Sang, T. Takata and T. Endo, *Macromolecules*, 1991, **24**, 2689–2693.
36. T. Endo and F. Sanda, *Macromol. Symp.*, 1996, **107**, 237–242.
37. T. Hino and T. Endo, *J. Polym. Sci., Part A: Polym. Chem.*, 2004, **42**, 2166–2170.
38. Y. Wang, M. Kimura, A. Sudo and T. Endo, *J. Polym. Sci., Part A: Polym. Chem.*, 2016, **54**, 2611–2617.
39. F. Sanda, T. Kaizuka, A. Sudo and T. Endo, *Macromolecules*, 2003, **36**, 967–968.
40. I. Asenjo-Sanz, A. Veloso, J. I. Miranda, J. A. Pomposo and F. Barroso-Bujans, *Polym. Chem.*, 2014, **5**, 6905–6908.
41. I. Asenjo-Sanz, A. Veloso, J. I. Miranda, A. Alegria, J. A. Pomposo and F. Barroso-Bujans, *Macromolecules*, 2015, **48**, 1664–1672.
42. F. M. Haque, A. Alegria, S. M. Grayson and F. Barroso-Bujans, *Macromolecules*, 2017, **50**, 1870–1881.
43. D. Zhang, S. K. Boopathi, N. Hadjichristidis, Y. Gnanou and X. Feng, *J. Am. Chem. Soc.*, 2016, **138**, 11117–11120.
44. J.-L. Yang, H.-L. Wu, Y. Li, X.-H. Zhang and D. J. Darensbourg, *Angew. Chem., Int. Ed.*, 2017, **56**, 5774–5779.
45. A. Kanazawa and S. Aoshima, *Polym. J.*, 2016, **48**, 679–687.
46. A. Kanazawa, S. Kanaoka and S. Aoshima, *J. Am. Chem. Soc.*, 2013, **135**, 9330–9333.
47. A. Kanazawa, S. Kanaoka and S. Aoshima, *Macromolecules*, 2014, **47**, 6635–6644.
48. A. Kanazawa and S. Aoshima, *Polym. Chem.*, 2015, **6**, 5675–5682.
49. A. Kanazawa and S. Aoshima, *ACS Macro Lett.*, 2015, **4**, 783–787.
50. R. Schwesinger and H. Schlemper, *Angew. Chem., Int. Ed.*, 1987, **26**, 1167–1169.
51. S. Boileau and N. Illy, *Prog. Polym. Sci.*, 2011, **36**, 1132–1151.

52. B. Esswein, N. M. Steidl and M. Möller, *Macromol. Rapid Commun.*, 1996, **17**, 143–148.
53. H. Zhang, S. Hu, J. Zhao and G. Zhang, *Macromolecules*, 2017, **50**, 4198–4205.
54. H. Zhang, S. Hu, J. Zhao and G. Zhang, *Eur. Polym. J.*, 2017, **95**, 693–701.
55. J. Zhao, D. Pahovnik, Y. Gnanou and N. Hadjichristidis, *Polym. Chem.*, 2014, **5**, 3750–3753.
56. A. Molenberg and M. Möller, *Macromol. Rapid Commun.*, 1995, **16**, 449–453.
57. O. Rexin and R. Mülhaupt, *J. Polym. Sci., Part A: Polym. Chem.*, 2002, **40**, 864–873.
58. O. Rexin and R. Mülhaupt, *Macromol. Chem. Phys.*, 2003, **204**, 1102–1109.
59. J. Zhang, Q. Liu, H. Ren, N. Zhang, P. Li and K. Yang, *J. Mol. Struct.*, 2017, **1148**, 421–428.
60. H. Misaka, E. Tamura, K. Makiguchi, K. Kamoshida, R. Sakai, T. Satoh and T. Kakuchi, *J. Polym. Sci., Part A: Polym. Chem.*, 2012, **50**, 1941–1952.
61. Y. Satoh, H. Matsuno, T. Yamamoto, K. Tajima, T. Isono and T. Satoh, *Macromolecules*, 2017, **50**, 97–106.
62. T. Isono, K. Kamoshida, Y. Satoh, T. Takaoka, S. Sato, T. Satoh and T. Kakuchi, *Macromolecules*, 2013, **46**, 3841–3849.
63. H. Misaka, R. Sakai, T. Satoh and T. Kakuchi, *Macromolecules*, 2011, **44**, 9099–9107.
64. Y. Chen, N. Xiao, T. Satoh and T. Kakuchi, *Polym. Chem.*, 2014, **5**, 4993–5001.
65. W. Kwon, Y. Rho, K. Kamoshida, K. H. Kwon, Y. C. Jeong, J. Kim, H. Misaka, T. J. Shin, J. Kim, K.-W. Kim, K. S. Jin, T. Chang, H. Kim, T. Satoh, T. Kakuchi and M. Ree, *Adv. Funct. Mater.*, 2012, **22**, 5194–5208.
66. T. Isono, Y. Satoh, K. Miyachi, Y. Chen, S. Sato, K. Tajima, T. Satoh and T. Kakuchi, *Macromolecules*, 2014, **47**, 2853–2863.
67. Y. Satoh, K. Miyachi, H. Matsuno, T. Isono, K. Tajima, T. Kakuchi and T. Satoh, *Macromolecules*, 2016, **49**, 499–509.
68. T. Isono, K. Miyachi, Y. Satoh, S.-I. Sato, T. Kakuchi and T. Satoh, *Polym. Chem.*, 2017, **8**, 5698–5707.
69. T. Isono, S. Asai, Y. Satoh, T. Takaoka, K. Tajima, T. Kakuchi and T. Satoh, *Macromolecules*, 2015, **48**, 3217–3229.
70. J. Zhao, G. Mountrichas, G. Zhang and S. Pispas, *Macromolecules*, 2010, **43**, 1771–1777.
71. J. Zhao, G. Zhang and S. Pispas, *J. Polym. Sci., Part A: Polym. Chem.*, 2010, **48**, 2320–2328.
72. L. Dentzer, C. Bray, S. Noinville, N. Illy and P. Guegan, *Macromolecules*, 2015, **48**, 7755–7764.

73. J. Zhao, D. Pahovnik, Y. Gnanou and N. Hadjichristidis, *Macromolecules*, 2014, **47**, 1693–1698.
74. H. Alamri, J. Zhao, D. Pahovnik and N. Hadjichristidis, *Polym. Chem.*, 2014, **5**, 5471–5478.
75. J. Xu, J. Yang, X. Ye, C. Ma, G. Zhang and S. Pispas, *J. Polym. Sci., Part A: Polym. Chem.*, 2015, **53**, 846–853.
76. S. Hu, J. Zhao and G. Zhang, *ACS Macro Lett.*, 2016, **5**, 40–44.
77. J. Zhao, D. Pahovnik, Y. Gnanou and N. Hadjichristidis, *Macromolecules*, 2014, **47**, 3814–3822.
78. J. Zhao and N. Hadjichristidis, *Polym. Chem.*, 2015, **6**, 2659–2668.
79. B. Esswein, N. M. Steidl and M. Möller, *Macromol. Rapid Commun.*, 1996, **17**, 143–148.
80. Y. Xia, J. Shen, H. Alamri, N. Hadjichristidis, J. Zhao, Y. Wang and G. Zhang, *Biomacromolecules*, 2017, **18**, 3233–3237.
81. S.-D. Lee, F. Sanda and T. Endo, *J. Polym. Sci., Part A: Polym. Chem.*, 1999, **37**, 1041–1048.
82. J. Zhao, G. Mountrichas, G. Zhang and S. Pispas, *Macromolecules*, 2009, **42**, 8661–8668.
83. J. Zhao and H. Schlaad, *Macromolecules*, 2011, **44**, 5861–5864.
84. J. Zhao, H. Alamri and N. Hadjichristidis, *Chem. Commun.*, 2013, **49**, 7079–7081.
85. H. Alamri and N. Hadjichristidis, *Polym. Chem.*, 2016, **7**, 3225–3228.
86. J. Zhao, D. Pahovnik, Y. Gnanou and N. Hadjichristidis, *J. Polym. Sci., Part A: Polym. Chem.*, 2015, **53**, 304–312.
87. Y. Xia, Y. Chen, Q. Song, S. Hu, J. Zhao and G. Zhang, *Macromolecules*, 2016, **49**, 6817–6825.
88. Y. Liu, X. Wang, Z. Li, F. Wei, H. Zhu, H. Dong, S. Chen, H. Sun, K. Yang and K. Guo, *Polym. Chem.*, 2018, **9**, 154–159.
89. S. Hu, G. Dai, J. Zhao and G. Zhang, *Macromolecules*, 2016, **49**, 4462–4472.
90. A. Sudo, K. Uenishi and T. Endo, *J. Polym. Sci., Par A: Polym. Chem.*, 2007, **45**, 3798–3802.
91. K. Uenishi, A. Sudo and T. Endo, *Macromolecules*, 2007, **40**, 6535–6539.
92. K. Uenishi, A. Sudo and T. Endo, *J. Polym. Sci., Part A: Polym. Chem.*, 2008, **46**, 4092–4102.
93. B. Han, L. Zhang, B. Liu, X. Dong, I. Kim, Z. Duan and P. Theato, *Macromolecules*, 2015, **48**, 3431–3437.
94. Z. Hostalek, O. Trhlíkova, Z. Walterova, T. Martinez, F. Peruch, H. Cramail and J. Merna, *Eur. Polym. J.*, 2017, **88**, 433–447.
95. H. Li, J. Zhao and G. Zhang, *ACS Macro Lett.*, 2017, **6**, 1094–1098.
96. H. Li, H. Luo, J. Zhao and G. Zhang, *Macromolecules*, 2018, **51**, 2247–2257.
97. S. Ohsawa, K. Morino, A. Sudo and T. Endo, *Macromolecules*, 2010, **43**, 3585–3588.
98. S. Ohsawa, K. Morino, A. Sudo and T. Endo, *Macromolecules*, 2011, **44**, 1814–1820.

99. S. Ohsawa, K. Morino, A. Sudo and T. Endo, *J. Polym. Sci., Part A: Polym. Chem.*, 2012, **50**, 4666–4673.
100. A. E. Allen and D. W. C. MacMillan, *Chem. Sci.*, 2012, **3**, 633–658.
101. Z. Du and Z. Shao, *Chem. Soc. Rev.*, 2013, **42**, 1337–1378.
102. Y. J. Park, J.-W. Park and C.-H. Jun, *Acc. Chem. Res.*, 2008, **41**, 222–234.
103. Z. Shao and H. Zhang, *Chem. Soc. Rev.*, 2009, **38**, 2745–2755.
104. P. Degee, P. Dubois and R. Jérôme, *Macromol. Chem. Phys.*, 1997, **198**, 1973–1984.
105. Y. Zhang, G. M. Miyake and E. Y. X. Chen, *Angew. Chem., Int. Ed.*, 2010, **49**, 10158–10162.
106. E. Piedra-Arroni, C. Ladavière, A. Amgoune and D. Bourissou, *J. Am. Chem. Soc.*, 2013, **135**, 13306–13309.
107. S. Naumann, P. B. V. Scholten, J. A. Wilson and A. P. Dove, *J. Am. Chem. Soc.*, 2015, **137**, 14439–14445.
108. W. Nzahou Ottou, E. Condé, D. Bourichon, J. Vignolle, A.-L. Wirotius, F. Robert, Y. Landais, J.-M. Sotiropoulos, K. Miqueu and D. Taton, *Macromolecules*, 2017, **50**, 762–774.
109. B. Esswein and M. Möller, *Angew. Chem., Int. Ed.*, 1996, **35**, 623–625.
110. H. Schmalz, M. G. Lanzendörfer, V. Abetz and A. H. E. Müller, *Macromol. Chem. Phys.*, 2003, **204**, 1056–1071.
111. S. Pispas and N. Hadjichristidis, *Langmuir*, 2003, **19**, 48–54.
112. S. Forster and E. Kramer, *Macromolecules*, 1999, **32**, 2783–2785.
113. S. Pispas, *J. Polym. Sci., Part A: Polym. Chem.*, 2006, **44**, 606–613.
114. H. Schmalz, A. Knoll, A. J. Müller and V. Abetz, *Macromolecules*, 2002, **35**, 10004–10013.
115. A. A. Toy, S. Reinicke, A. H. E. Müller and H. Schmalz, *Macromolecules*, 2007, **40**, 5241–5244.
116. A.-L. Brocas, A. Deffieux, N. Le Malicot and S. Carlotti, *Polym. Chem.*, 2012, **3**, 1189–1195.
117. L. Hassouna, N. Illy and P. Guegan, *Polym. Chem.*, 2017, **8**, 4005–4013.
118. D. Bourissou, O. Guerret, F. P. Gabbai and G. Bertrand, *Chem. Rev.*, 2000, **100**, 39–92.
119. M. N. Hopkinson, C. Richter, M. Schedler and F. Glorius, *Nature*, 2014, **510**, 485–496.
120. M. Fèvre, J. Pinaud, Y. Gnanou, J. Vignolle and D. Taton, *Chem. Soc. Rev.*, 2013, **42**, 2142–2172.
121. S. Naumann and A. P. Dove, *Polym. Chem.*, 2015, **6**, 3185–3200.
122. J. Raynaud, C. Absalon, Y. Gnanou and D. Taton, *J. Am. Chem. Soc.*, 2009, **131**, 3201–3209.
123. J. Raynaud, C. Absalon, Y. Gnanou and D. Taton, *Macromolecules*, 2010, **43**, 2814–2823.
124. J. Raynaud, W. Nzahou Ottou, Y. Gnanou and D. Taton, *Chem. Commun.*, 2010, **46**, 3203–3205.
125. R. Lindner, M. L. Lejkowski, S. Lavy, P. Deglmann, K. T. Wiss, S. Zarbakhsh, L. Meyer and M. Limbach, *Chem. Cat. Chem.*, 2014, **6**, 618–625.

126. S. Naumann, A. W. Thomas and A. P. Dove, *Angew. Chem., Int. Ed.*, 2015, **54**, 9550–9554.
127. A. Balint, M. Papendick, M. Clauss, C. Muller, F. Giesselmann and S. Naumann, *Chem. Commun.*, 2018, **54**, 2220–2223.
128. M. Al Ghamdi, L. Cavallo and L. Falivene, *J. Phys. Chem. C*, 2017, **121**, 2730–2737.
129. H. Morinaga, B. Ochiai and T. Endo, *Macromolecules*, 2007, **40**, 6014–6016.
130. H. Morinaga, Y. Ujihara, N. Yuto, D. Nagai and T. Endo, *J. Polym. Sci., Part A: Polym. Chem.*, 2011, **49**, 5210–5216.
131. H. Morinaga, F. Tsuneishi, S. Taniguchi and G. Kawakami, *Tetrahedron Lett.*, 2014, **55**, 3768–3770.
132. H. Morinaga, Y. Ujihara and T. Endo, *J. Polym. Sci., Part A: Polym. Chem.*, 2013, **51**, 4451–4458.
133. H. Morinaga, Y. Ujihara, T. Yamanaka, D. Nagai and T. Endo, *J. Polym. Sci., Part A: Polym. Chem.*, 2011, **49**, 4092–4097.
134. H. Morinaga, Y. Ujihara and T. Endo, *J. Polym. Sci., Part A: Polym. Chem.*, 2012, **50**, 3461–3465.
135. I. Hamerton, Recent developments in epoxy resins; Rapra Technology Ltd.: Shawbury, Shrewsbury, Shropshire, U.K., 1996.
136. R. Auvergne, S. Caillol, G. David, B. Boutevin and J.-P. Pascault, *Chem. Rev.*, 2014, **114**, 1082–1115.
137. T. Vidil, F. Tournilhac, S. Musso, A. Robinson and L. Leibler, *Prog. Polym. Sci.*, 2016, **62**, 126–179.
138. S. Naumann, M. Speiser, R. Schowner, E. M. Giebel and M. R. Buchmeiser, *Macromolecules*, 2014, **47**, 4548–4556.
139. M. R. Buchmeiser, J. A. Kammerer, S. Naumann, J. Unold, R. Ghomeshi, S. K. Selvarayan, P. Weichand and R. Gadow, *Macromol. Mater. Eng.*, 2015, **300**, 937–943.
140. H. J. Altmann, S. Naumann and M. R. Buchmeiser, *Eur. Polym. J.*, 2017, **95**, 766–774.
141. N. Makiuchi, A. Sudo and T. Endo, *J. Polym. Sci., Part A: Polym. Chem.*, 2014, **52**, 2518–2522.
142. N. Makiuchi, A. Sudo and T. Endo, *J. Polym. Sci., Part A: Polym. Chem.*, 2015, **53**, 2569–2574.

CHAPTER 9

Ring-opening Polymerization of N-*carboxyanhydrides Using Organic Initiators or Catalysts*

DAVID SIEFKER AND DONGHUI ZHANG*

Department of Chemistry and Macromolecular Studies Group,
Louisiana State University, Baton Rouge, LA 70803, USA
*Email: dhzhang@lsu.edu

9.1 Introduction

Synthesis of amino acid-derived *N*-carboxyanhydrides (NCAs) was first reported by Hermann Leuchs in the early 1900s.[1–3] The first polymerization of NCAs was accidental due to the thermal instability of the monomers. As the concept of macromolecules became widely accepted around 1920–1930, there was a renewed interest in developing synthetic strategies to access polypeptides by polymerization and characterizing their physical properties.[4] Significant advances were made prior to 1970s in understanding the NCA polymerization mechanism and kinetics as described in the reviews by Bamford[5] and Szwarc.[6] Over the course of about 100 years, a variety of amino acid-derived NCAs have been synthesized and investigated as substrates for polymerization.[7–9] In the last 20 years or so, a number of initiator/catalyst systems have been shown to mediate the ring-opening polymerizations (ROPs) of amino acid-derived NCAs in a controlled manner, yielding polypeptides with predictable molecular weight and narrow molecular weight distribution (Scheme 9.1).[8,10] The scope of NCAs has been expanded to include monomers bearing different functional sidechains, which can either

Polymer Chemistry Series No. 31
Organic Catalysis for Polymerisation
Edited by Andrew Dove, Haritz Sardon and Stefan Naumann

Published by the Royal Society of Chemistry, www.rsc.org

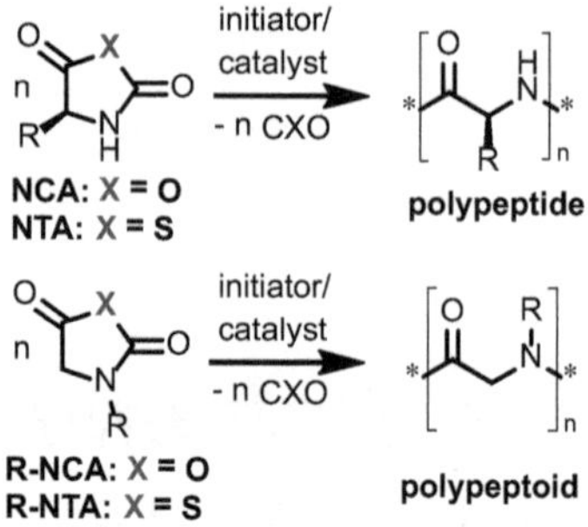

Scheme 9.1 A generic scheme depicting polymerizations of amino acid-derived NCAs/NTAs and *N*-substituted glycine-derived R-NCAs/R-NTAs to produce polypeptide and polypeptoids respectively.

be further derivatized readily by "click" chemistry to confer unique physiochemical (*e.g.*, stimuli-responsiveness)[11] or biological properties (*e.g.*, membrane activity, antigenicity).[9,10] A variety of *N*-substituted glycine-derived *N*-carboxyanhydrides (R-NCAs) have also emerged in the last ten years or so and have been studied as substrates for polymerization to access *N*-substituted polyglycines, *a.k.a.* polypeptoids (Scheme 9.1).[12,13] In addition to NCAs or R-NCAs, controlled ROPs of hydrolytically more stable monomers such as amino acid-derived *N*-thiocarboxyanhydrides (NTAs) or *N*-substituted glycine-derived *N*-thiocarboxyanhydrides (R-NTAs) (Scheme 9.1) have recently been developed to yield well-defined polypeptides[14] or polypeptoids.[15–19]

This chapter is not intended to provide a comprehensive review of recent studies on the synthesis of various polypeptides and polypeptoids by ROPs of (R)-NCAs, which can be found in several recent reviews.[9,10,12,13,19] Instead, we will focus our discussion on the ROPs of (R)-NCAs that are initiated or mediated by organic molecules. The order of the discussion is organized based on the unique mechanisms by which the ROPs of (R)-NCAs are operated when different organic initiators/catalysts are used. Examples of organo-initiated ROPs of (R)-NTAs, mercapto analogs of (R)-NCAs, will also be briefly discussed. We will provide our viewpoints as to the challenge and perspective of organo-mediated ROPs of (R)-NCAs towards the end of the chapter.

9.2 Synthesis of NCAs, R-NCA, NTA and R-NTA Monomers

Amino acid-derived NCAs are typically synthesized by the Fuchs–Farthing method that involves acylation and cyclization of amino acid precursors with phosgene or its derivatives (Scheme 9.2). Alternatively, NCAs can also be accessed by installation of an *N*-carboxyalkyl group at the nitrogen followed by cyclization with acylating agents such as PCl_3, PBr_3, $SOCl_2$, which is often referred to as the Leuchs method (Scheme 9.2). For the synthesis of R-NCAs, the Leuchs method is the preferred route in the literature as it tends to

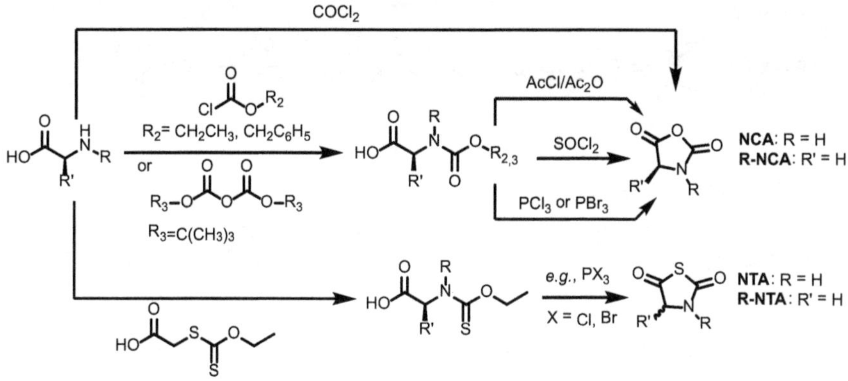

Scheme 9.2 Synthesis of NCA, R-NCA, NTA and R-NTA monomers.

produce the monomers in higher yields and purity. NTAs and R-NTAs are similarly prepared *via* the Leuchs route by first *N*-protection of the amino acid or *N*-substituted glycine precursors with the thiocarboxyalkyl group followed by cyclization promoted by acylating agents (*e.g.*, Ac_2O, $SOCl_2$, PX_3, PX_5, X = Br or Cl) (Scheme 9.2).[20–22] The optical purity of the resulting NTAs varies and is dependent on the reaction condition and the structure of the amino acid precursors.[21] Due to their sensitivity towards moisture, NCAs and R-NCAs are typically purified by recrystallization, sublimation or flash chromatography under rigorously anhydrous conditions (*e.g.*, using glovebox or Schlenk line). For NTAs and R-NTAs that are much more hydrolytically stable, purification can be conducted on benchtop without the need of rigorously anhydrous conditions.

9.3 Polymerization of NCAs, NTAs, R-NCAs or R-NTAs by the Normal Amine Mechanism (NAM) and/or Activated Monomer Mechanism (AMM)

9.3.1 ROPs of NCAs by the Normal Amine Mechanism Using Protic Nucleophilic Initiators

Due to the electrophilic nature of NCAs, organic molecules with sufficient nucleophilicity (*e.g.*, water, alcohol, primary amine, secondary amine, aniline, imidazoline) can initiate the ring-opening polymerization of NCAs by a mechanism commonly referred to as the normal amine mechanism (NAM). In this mechanism, the nucleophilic initiators initiate the polymerization by ring-opening addition to the C_5 carbonyl of the NCA monomer followed by decarboxylation to generate the initiating species bearing a primary amino terminus, from which enchainment ensues by the same sequence of events (Scheme 9.3). The most commonly used nucleophilic initiators are sterically unhindered primary amines. Water,[1,23–27] alcohols,[17,24,28] aniline,[29–31] secondary amines,[32,33] and thiols[34] can also initiate the ROPs of NCAs by

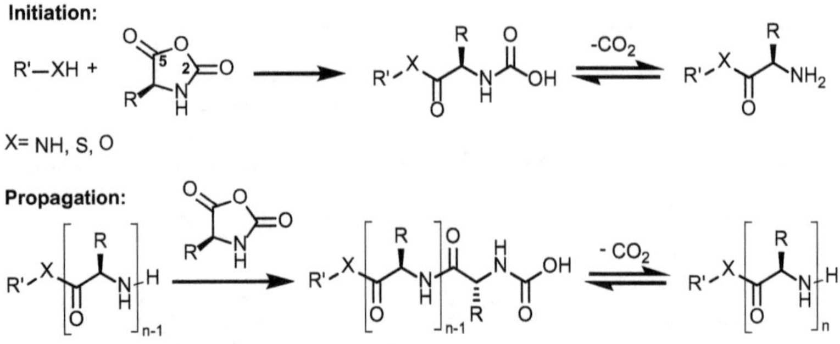

Scheme 9.3 The normal amine mechanism for the ROPs of NCAs using organic nucleophiles.

the same mechanism except that they are less efficient initiators relative to unhindered primary amines due to their reduced nucleophilicity, resulting in either no reaction or long reaction time, and poor molecular weight control due to slow initiation relative to propagation.

Imidazoline can initiate the ROPs of NCAs similarly, as a typical secondary amine to form a growing polypeptide having an imidazoline-derived amide at the α-terminus and a primary amine at the ω-terminus.[35] The imidazoline-derived amide terminus is significantly more electrophilic than other amide bonds along the chain. This allows the polypeptide chain ends to react with one another *via* a kinetically controlled polycondensation pathway, yielding cyclic polypeptides.

9.3.2 Side Reactions in the ROPs of NCAs Bearing the *N*–H Proton

Due to the presence of electrophilic (C_2, C_5 carbonyl) and latent nucleophilic sites (*N*–H and α-C–H) in the NCA monomers, primary amine-initiated ROPs of NCAs bearing the *N*–H proton are susceptible to various side reactions that can lead to chain transfer and termination, resulting in diminished control over molecular weight and molecular weight distribution. For example, the labile *N*–H proton in NCAs is susceptible to deprotonation by amines to yield an isocyanate species that is slightly more electrophilic than NCAs and thus can terminate the chain growth *via* the formation of ureido acid chain ends (Scheme 9.4A). Addition at the C_2 carbonyl position of the NCAs can also lead to inactive ureido acid chain ends, although the extent of C_2 addition is much limited (0.15 mol%) (Scheme 9.4B).[33] Termination *via* the formation of hydantoin chain ends has also been reported in the ROPs of DL-alanine NCA and glycine NCA in water (Scheme 9.4C).[36] Chain transfer *via* intramolecular transamidation can also occur, resulting in the formation of a mixture of linear and cyclic polypeptides (Scheme 9.4D).[35,37–39] For specific NCA monomers, secondary reactions with the sidechain functionality can occur, resulting in chain termination or structural alteration of polypeptide chains. For example, a substantial presence of polypeptides with pyroglutamate

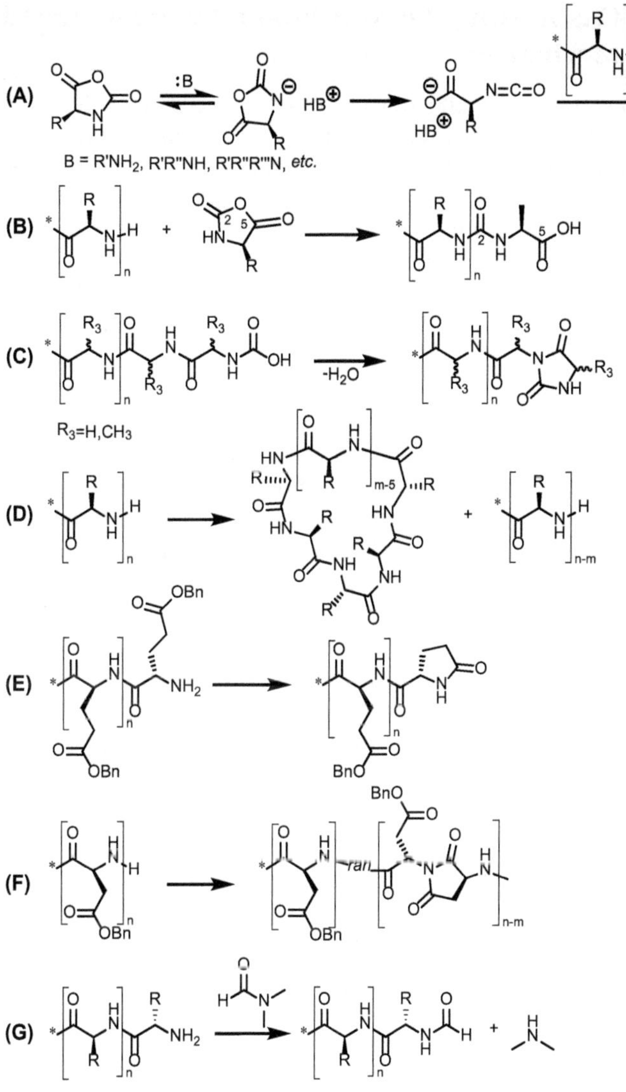

Scheme 9.4 Various side reactions that can lead to chain transfer or termination in the primary amine-initiated ROPs of NCAs bearing the *N*–H proton.

chain terminus has been observed as a major mode of termination in the primary amine-initiated ROPs of γ-benzyl-L-glutamate derived NCAs at room temperature (Scheme 9.4E).[39,40] The formation of succinimide repeating units has also been observed for the primary amine-initiated ROPs of β-benzyl-L-aspartate derived NCAs (Scheme 9.4F).[39,40] In addition, side reactions with solvents can lead to chain termination. For example, polymerizations of NCAs that are conducted in DMF often produce *N*-formyl-terminated polypeptide chains in the final product (Scheme 9.4G).[39–41]

9.3.3 ROPs of NCAs Bearing the *N*–H Proton by the Activated Monomer Mechanism

In addition to polymerization by the normal amine mechanism, ROPs of NCAs using nucleophilic initiators (*e.g.*, primary or secondary amines) can also proceed by a different mechanism commonly referred to as the ''activated monomer'' mechanism (AMM) (Scheme 9.5). In this pathway, the primary amines serve as an organic base and deprotonate the NCAs to generate NCA anions **1**, *a.k.a.*, the activated monomers, which react with another NCA at the C_5 carbonyl position by nucleophilic ring-opening addition followed by decarboxylation to form the acylated NCA initiating species **3** (Scheme 9.5).[33,42] The chain propagation involves further addition of NCA anions to the acylated NCA species **4**. As the acylated NCAs **4** are more electrophilic than the NCAs and the decarboxylation from the carboxyamide intermediate **5** is more facile than the carbamate species **2**, this leads to faster propagation relative to initiation. As a result, the polymerizations by AMM often yield polypeptides having much higher molecular weight than theoretical prediction based on a living polymerization. In general, amines with enhanced basicity and reduced nucleophilicity (*e.g.*, hindered secondary or tertiary amines) tend to favor the AMM for the polymerization of NCAs bearing the *N*–H proton.

It should also be mentioned here that Blout *et al.* has proposed an alternative polymerization mechainsm when studying the polymerization of NCAs using sodium alkoxide.[43] This mechanism, commonly referred to as the carbamate mechanism, involves the nuclephilic ring-opening addition of the NCAs by a propagating carbamate species (Scheme 9.6). This reaction pathway is later considered to be highly unlikely for NCAs bearing *N*–H using organic amine initiators.[33,42] However, it maybe operative for the polymerization of R-NCAs using nuclephilic initiators.

9.3.4 Towards Controlled ROPs of NCAs Bearing the *N*–H Proton by Optimization of Reaction Conditions

Due to the presence of complex side reactions and the competing polymerization mechanism in the primary amine-initiated ROPs of NCAs bearing the *N*–H proton, reaction conditions have to be optimized in order to produce well-defined polypeptides with controlled molecular weights. Lowering the reaction temperature,[39,40,44,45] and conducting the reaction in high vacuum conditions[41] or under a constant nitrogen flow[46] have been demonstrated to be effective in suppressing side reactions and enhancing the polymerization control by the NAM.

It has been shown that removal of CO_2 under reduced pressure (1×10^{-5} bar)[40] or by blowing nitrogen through the reaction mixture can accelerate the rate of polymerization of γ-benzyl-L-glutamate, ε-*N*-benzyloxycarbonyl-L-lysine, L-alanine derived NCAs using primary amine initiators.[46] This has been attributed to shifting the equilibrium from the carbamic acid/carbamate propagating intermediates towards the active amino

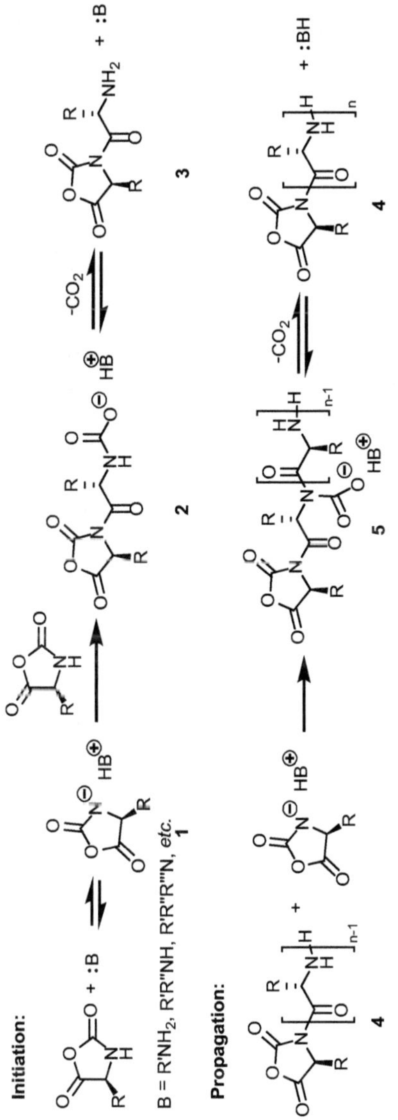

Scheme 9.5 Activated monomer mechanism (AMM) for the ROPs of NCAs bearing the *N*–H proton in the presence of an organic base (*e.g.*, primary, secondary and tertiary amines).

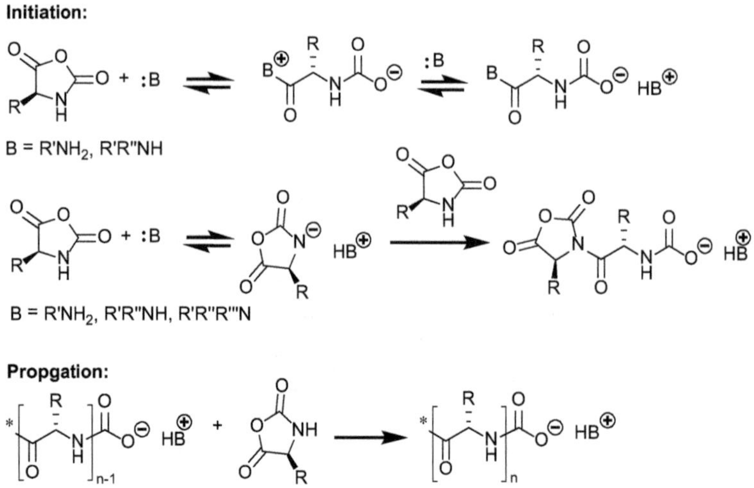

Scheme 9.6 The proposed carbamate mechanism for the ROPs of NCAs using an organic nucleophile (*e.g.*, primary or secondary amines).

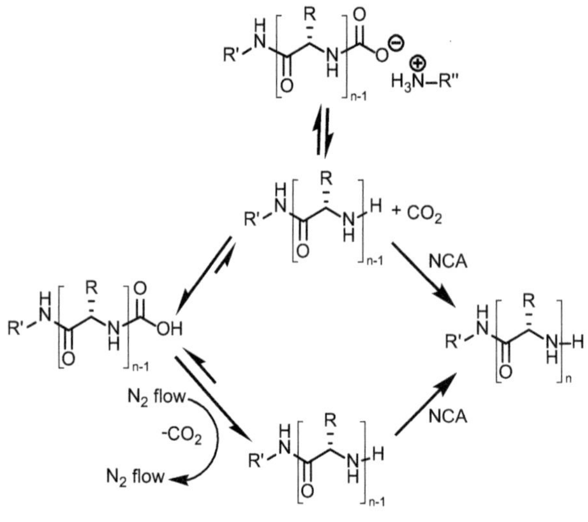

Scheme 9.7 Purging CO_2 using nitrogen flow shift the equilibrium of carbamic acid/carbamate intermediates towards the active amine propagating species.

propagating species, thus resulting in an enhanced polymerization rate (Scheme 9.7). However, there are exceptions where varying CO_2 pressure has no notable effect on the polymerization rate.[40,47] For example, Heise and coworker studied the ROPs of β-benzyl-L-aspartate, *O*-benzyl-L-serine, and *O*-benzyl-L-threonine derived NCAs using benzyl amine initiators in DMF and found the polymerization rate to be comparable at the ambient pressure or under vacuum (1×10^{-5} bar).[40] Bamford and coworker studied the polymerization of DL-leucine and DL-phenylalanine derived NCAs using pre-synthesized polypeptide initiators in nitrobenzene under different CO_2 pressure and found the polymerization rate to be independent of the CO_2 concentration in the solvent.[47]

Early studies revealed that the CO_2 partial pressure can influence the polymerization rate of selected NCA monomers *via* the formation of carbamic acid species *via* equilibrium with the propagating amine species (eqn (2), Scheme 9.8).[5,47,48] A kinetic model, which involves an acid-catalyzed ring-opening decarboxylation as the rate-determining step (RDS) following an amine addition to NCAs pre-equilibrium, has been proposed to account for the effect (eqn (1) and (3), Scheme 9.8). While the CO_2 effect is autocatalytic by this scheme, a great excess of CO_2 can also slow down polymerization due to the decrease of the effective concentration of active amine propagating species *via* the formation of carbamate salt (eqn (4), Scheme 9.8).[47] For the ROPs of some NCAs where the CO_2 has no effect on the polymerization rate, the amine addition to the NCAs becomes the RDS. A recent DFT calculation has shown that the RDS for the reaction between L-alanine or sarcosine derived NCAs with a primary amine (*i.e.*, ethylamine) or a secondary amine (*i.e.*, *N*,*N*-dimethylamine) is in fact the nucleophilic addition of the respective amine to the NCAs ($\Delta G^{\neq} = 39.96–41.19$ kcal mol^{-1}), not the ring-opening ($\Delta G^{\neq} = 16.38–20.33$ kcal mol^{-1}) nor decarboxylation ($\Delta G^{\neq} = 29.38–31.95$ kcal mol^{-1}).[49,50]

9.3.5 Towards Controlled ROPs of NCAs Bearing the *N*–H Proton by Modulating the Reactivity of Propagating Species

Apart from optimizing the reaction conditions, an alternative strategy to suppress the undesired "activated monomer" polymerization pathway that competes with the NAM is to modulate the reactivity of the propagating amino

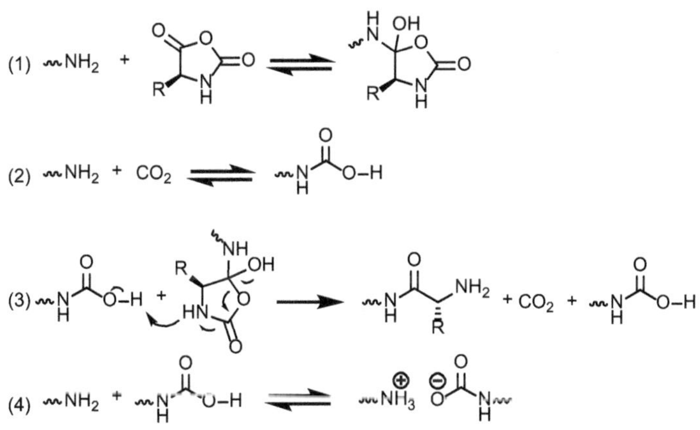

Scheme 9.8 A proposed reaction model that accounts for the CO_2 effect on the polymerization rate.

chain ends *via* the formation of equilibrium with a dormant propagating ammonium species[51,52] or frustrated Lewis acid–base pairs.[30]

It has been shown that primary amine hydrochloride salt can mediate the ROPs of NCAs in a controlled manner, producing polypeptides with improved molecular weight control (DP ≈ 30, $M_n \approx 18$ kg mol^{-1}) and narrower molecular weight distribution (PDI = 1.03) relative to those obtained by using the corresponding primary amine (Scheme 9.9).[52] This has been attributed to the formation of equilibrium between the active amine propagating species and the dormant ammonium propagating species, which significantly limits the undesired side reactions and non-NAM polymerization pathways (*i.e.*, the activated monomer pathway) that are promoted by the presence of base. The enhanced polymerization control (*i.e.*, controlled M_n and low PDI) is at the cost of reduced polymerization rate due to the decreased effective concentration of the active amino propagating species. The equilibrium between the active amino and dormant ammonium propagating species is dependent on the pH, solvent and monomer structure. In some cases, the equilibrium can be overly biased towards the ammonium species, rendering the method ineffective towards polypeptide synthesis. To enhance control of the equilibrium, Schlaad and coworkers investigated the primary ammonium in conjunction with a tertiary amine (*i.e.*, triethylamine (TEA)) as a base catalyst for the ROPs of NCAs.[52,53] It was found that the polymerization rate can be controlled by adjusting the ratio of the primary ammonium and TEA *via* the equilibrium between the active and dormant propagating species. Polymerization can also be paused and restarted *via* addition of HCl and TEA, respectively. Despite the fact that the polymerization was found to proceed by a combination of NAM and AM mechanism, polypeptides with defined end-group structure and controlled molecular weight (DP ≈ 150, $M_n \approx 25$ kg mol^{-1}) and low dispersity (PDI = 1.07–1.12) can be obtained when the molar ratio of tertiary amine relative to the primary ammonium is not too high (*i.e.*, 0.2 : 1–1.1 1).[52]

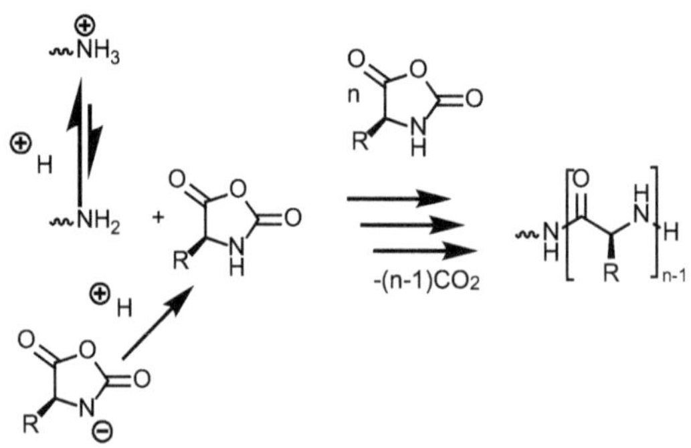

Scheme 9.9 The equilibrium between the active amine propagating species and dormant ammonium species that serves to diminish or suppress base-promoted side reactions and polymerization *via* the activated monomer mechanism.

It has been shown that anilines when complexed with a sterically hindered borane (*e.g.*, $Fmes_2BF$, Fmes = 1,3,5-trifluoromethylphenyl) can initiate and mediate the controlled ROPs of NCAs (Scheme 9.10), yielding polypeptides with controlled molecular weight (DP ≈ 50, $M_n \approx 10$ kg mol^{-1}) and moderate molecular weight distribution (PDI = 1.22–1.36) in solvents (*e.g.*, DCM) that do not compete for binding to borane.[30] By contrast, anilines alone are poor initiators for the ROPs of NCAs due to the reduced nucleophilicity relative to primary amines, yielding polypeptide with broad molecular weight distribution (PDI = 1.63–1.81). The formation of a frustrated Lewis acid–base pair between the aniline initiator or the propagating amino species and bulky borane is critical and has been attributed to the enhanced polymerization control (Scheme 9.10). A strongly interacting Lewis acid–base pair between aniline and $B(C_6F_5)_3$ resulted in no polymerization activity, and the use of aniline with $Fmes_2BH$ afforded low conversions and polypeptides with broad molecular weight distribution (PDI = 1.53). While the complex formation reduces the polymerization rate, the rate of propagation is more notably

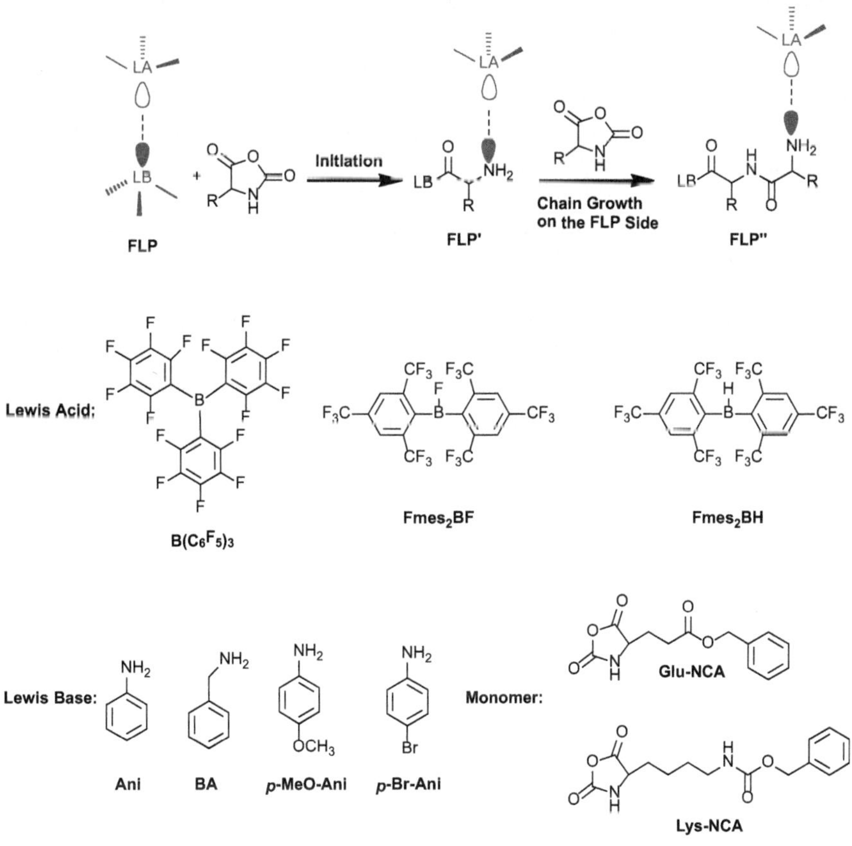

Scheme 9.10 Controlled ROPs of NCAs using aniline initiators *via* the formation of frustrated Lewis acid base pairs.

decreased relative to that of initiation due to the varying strength of complexation in the frustrated Lewis acid–base pair of the aniline and borane *vs.* that of the propagating amino chain end and borane. The polymerization rate was found to be first-order dependent on the monomer concentration. The observed polymerization rate constant was shown to increase ten-fold as the borane concentration was decreased by a factor of 3 in the range 0.005–0.015 M (*i.e.*, 0.5–1.5 equivalents relative to $[\text{aniline}]_0$). The detailed rate law for the polymerization is not fully determined. In addition, the formation of frustrated Lewis acid–base pair also suppressed the "activated monomer" polymerization pathway by rendering the propagating chain ends significantly less basic.

9.3.6 Towards Controlled ROPs of NCAs Bearing the *N*–H Proton by Modulating the Reactivity of Propagating Species and Activation Of Monomers

It was recently reported that trimethylaminetriamine (TREN), which has one tertiary amino and three primary amino groups in the same molecule, can initiate and mediate the controlled ROPs of γ-benzyl-L-glutamate derived NCAs and ε-*N*-carboxybenzyl-L-lysine derived NCAs at room temperature in DMF to reach quantitative conversions in 3 h, producing well-defined tri-arm star-shaped polypeptides having tunable molecular weight (DP = 50–200, M_n = 10–45 kg mol^{-1}) and narrow molecular weight distribution (PDI = 1.13–1.19) (Scheme 9.11).[54] The molecular weight control is comparable to that obtained when using ethylenediamine (EDA) to yield linear polypeptides. By contrast, ROPs of γ-benzyl-L-glutamate derived NCAs using tetramethylethylendiamine (TMEDA) and *N*,*N*-dimethylethylenediamine (DMEDA) (Scheme 9.11) yielded polypeptides with broad bimodal distributions and molecular weights (M_n) that are significantly higher than the theoretical values based on a living polymerization from the amine groups, which is consistent with polymerization by the activated monomer mechanism. The polymerization using TREN is slower than that of TMEDA and

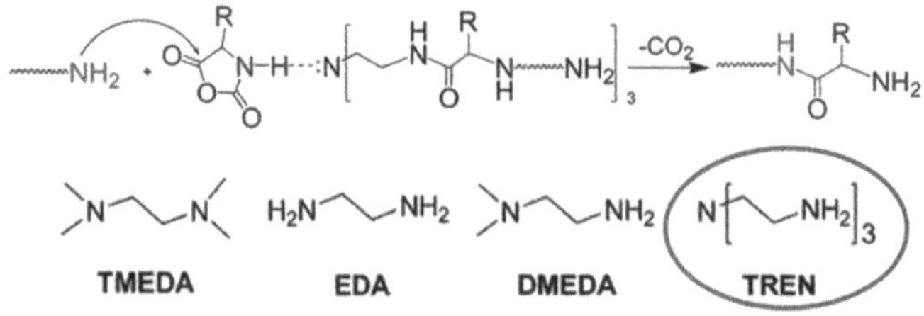

Scheme 9.11 The proposed accelerated amine mechanism through monomer activation (AAMMA) in which the tertiary amine in TREN activates the NCA monomer by hydrogen bonding.
Adapted from ref. 28 with permission from the Royal Society of Chemistry.

DMEDA but slightly faster than that of EDA. The polymerization exhibited a first-order dependence on the monomer concentration. As the TREN concentration increases, the polymerization is accelerated. The dependence of the polymerization rate on the TREN concentration is not unambiguously determined. A ^{15}N-NMR study has revealed that the tertiary amine in TREN is more electron-deficient relative to that in TMEDA. This may have contributed to the suppression of any side reactions that require a base and the competing "activated monomer" polymerization pathways. The authors proposed an accelerated amine mechanism through monomer activation (AAMMA) to explain the enhanced polymerization control (Scheme 9.11). This mechanism involves the hydrogen bonding interaction between the tertiary amine of TREN and the *N*–H proton of NCAs. If this mechanism were correct, the polymerization rate should be second-order dependent on the TREN concentration, which requires additional kinetic studies to confirm.

A subsequent study has shown that a series of oligomeric amines (*e.g.*, triethylenetetramine (TETA)) (Scheme 9.12) containing both secondary amino and primary amino groups can also initiate and mediate ROPs of γ-benzyl-L-glutamate derived NCAs in a controlled fashion in room temperature DMF, reaching quantitative conversions in 3 h and producing the corresponding well-defined polypeptides (DP ≈ 10–200, $M_n = 2.4$–$46\ \text{kg mol}^{-1}$, PDI = 1.04–1.29).[55] Polymerizations using the oligomeric amines are faster than those with primary amines (*e.g.*, hexamethylenediamine (HMDA) and 2-phenylethylamine (PEA)). All polymerizations using oligomeric amine initiators exhibited a first-order dependence of the polymerization rate on the monomer concentration. The polymerization rate also increased with increasing the TETA concentration with 0.8 order dependence, suggesting that nearly all active propagating species are monomeric without a significant extent of chain aggregation. The author proposed a similar AAMMA mechanism that involves the activation of NCAs *via* hydrogen bonding with the secondary amino functionality in the oligomeric amine initiators or α-polypeptide chain ends. The proposed mechanism would give rise to a polymerization kinetic that is second-order dependent on the oligomeric amine concentration, which is not consistent with the kinetic results. Further investigation is required to fully establish the reaction mechanism.

9.3.7 Towards Controlled ROPs of NCAs Bearing the *N*–H Proton by Activating the Alcohol Initiators and Monomers, and Modulating the Reactivity of Propagating Species

Alcohols are poor initiators due to their reduced nucleophilicity relative to the primary amines. As a result, they cannot efficiently initiate the ROPs of NCAs to yield well-defined polypeptides. It has recently been demonstrated that tertiary aminoalcohol (*e.g.*, DMEA, TEA, MDEA, THEED) (Scheme 9.13) in the presence of *N*,*N*′-bis[3,5-bis(trifluoromethyl)phenyl]thiourea (TU-S)

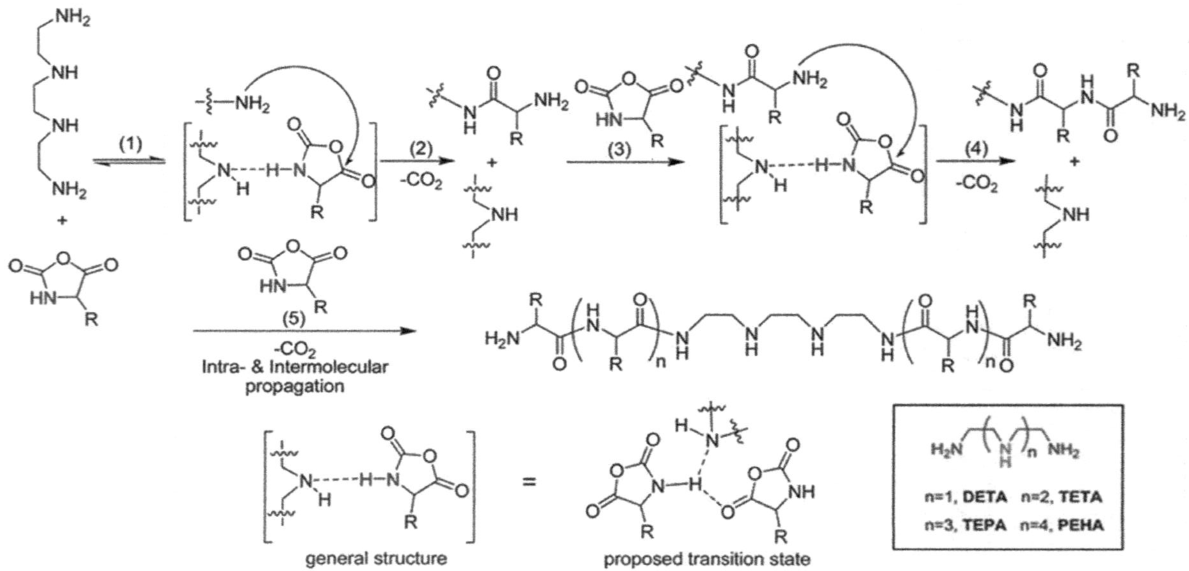

Scheme 9.12 The proposed accelerated amine mechanism through monomer activation (AAMMA) mechanism in which the secondary amino groups in the oligomeric amines (*e.g.*, TETA) activate the NCA monomer by hydrogen bonding. Adapted from ref. 55 with permission from the American Chemical Society, 2015.

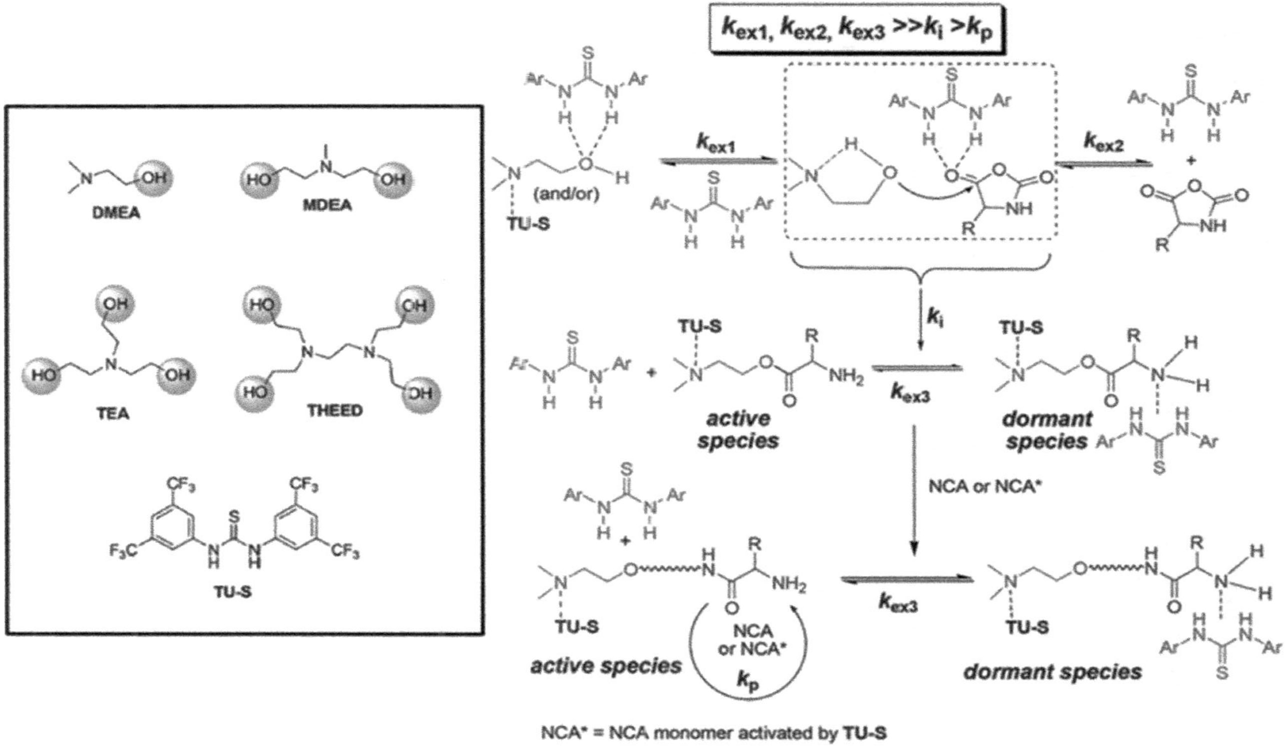

Scheme 9.13 The proposed thiourea-promoted controlled ROPs of NCAs using tertiary aminoalcohol initiators in which the activation of alcohol groups in the initiators and NCA monomers, and modulation of the propagating chain end reactivity are achieved *via* the hydrogen bonding with the thiourea.
Adapted from ref. 28 with permission from the Royal Society of Chemistry.

catalyst (0.8 mol% relative to initial monomer concentration) can promote the rapid and controlled polymerization of γ-benzyl-L-glutamate derived NCAs under mild conditions (*i.e.*, room temperature, in DCM).[28] The reactions reach quantitative conversion in 10–300 min, producing high molecular weight polypeptides with controlled molecular weights ($M_n = 30–180$ kg mol^{-1}) and narrow molecular weight distribution (PDI = 1.02–1.05). The molecular weights of the resulting polypeptides (M_n) agree well with the theoretical values based on a living polymerization. ^{1}H NMR analysis confirmed that the polymerization was initiated by the alcohol functionality in the initiators. It was found that the polymerization is first-order dependent on the monomer concentration with a short induction period at the beginning stage; the polymerization rate decreased 150-fold as the TU-S loading was increased from 0.8 mol% to 8.0 mol%. The alcohol is activated through intermolecular hydrogen bonding interaction with the tertiary amine. Replacement of the tertiary amine with a secondary amine or amide such as in the case of MEA or Boc-EA initiators (Scheme 9.13) failed to fully activate the alcohol functionality, resulting in either no polymerization or reduced polymerization control due to slow initiation relative to propagation. It has been shown that TU-S influences the polymerization *via* hydrogen bonding interactions with the initiator, monomer and propagating amino chain ends. Specifically, TU-S activates the NCA monomer towards nucleophilic ring-opening addition, and deactivates the growing amino chain ends and the tertiary amine functionality in the initiator by reversible binding, thereby suppressing the undesired activated monomer polymerization pathway and various base-promoted side reactions. By using oligomeric aminoalcohols in conjunction with TU-S, this method has been successfully extended towards the synthesis of well-defined multi-arm star-shaped polypeptides and linear block copolypeptides.[56]

9.3.8 Towards the Controlled ROPs of NTAs Bearing the *N*–H Proton by NAM

Amino acid-derived NTAs are structural analogs of NCAs where the endocyclic oxygen is replaced by sulfur. These monomers were first developed and investigated as substrates for aqueous phase synthesis of oligomeric peptides.[20–22,57–59] NTAs are thermally and hydrolytically more stable than the NCA analogs,[14] rendering their purification, storage and polymerization to be conducted on the benchtop without the need for rigorously anhydrous conditions. However, polymerization of amino acid-derived NTAs using primary amine initiators prove to be challenging and often found to be terminated at low conversions.[60,61] A recent study has shown that ROPs of γ-benzyl-glutamate, ε-*N*-carboxybenzyl-lysine, and L-leucine derived NTAs can occur in a controlled manner when the reaction is conducted in a heated non-solvent such as hexanes or heptane using a soluble primary amine initiator (*e.g.*, hexylamine).[14] All reactions reached quantitative conversions

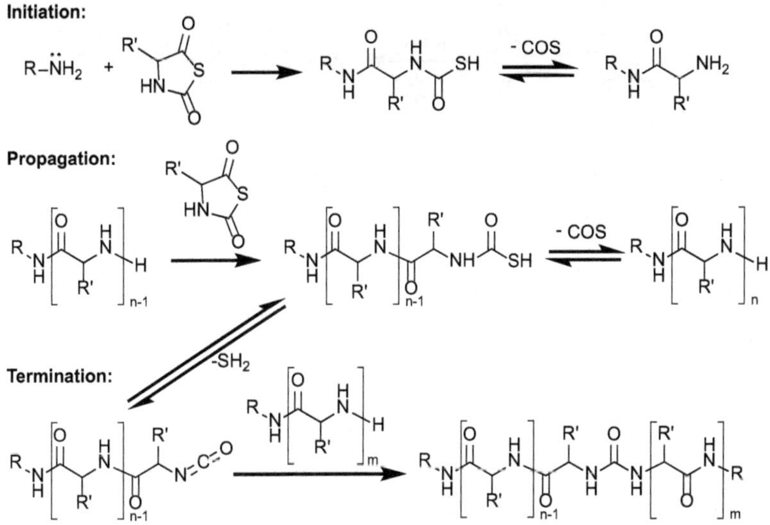

Scheme 9.14 Primary amine-initiated ROPs of NTAs by NAM and side reactions that can cause the termination of polymerization.

in two days, producing polypeptide with M_n in the 2.4–51 kg mol^{-1} range and moderate molecular weight distribution (PDI = 1.21–1.31). ^{1}H NMR and MS analysis of the resulting polypeptides indicates the polymerization proceeds by the normal amine mechanism (Scheme 9.14). In the NAM, the thiocarbamic acid propagating intermediate loses COS to generate the amine propagating species from which the enchainment ensues. It was found that the thiocarbamic acid propagating intermediate is susceptible to H_2S elimination to form isocyanate species, which can terminate the chain growth (Scheme 9.14). The extent of this side reaction is strongly dependent on the solvent. In conventional solvents (*e.g.*, DMF, dioxane) that are typically used for the ROPs of NCAs, the side reaction is competitive relative to the chain propagation by the NAM. By contrast, in hexanes or heptane, the side reaction is suppressed, resulting in enhanced control of the polymerization and quantitative conversions. The reaction was found to occur at the interface of the partially solubilized monomers and the insoluble polypeptide chains whose chain ends remain accessible to the soluble monomer in solution. As a result, the polymer chains can be successfully extended by the sequential addition of new monomers, making it possible to access well-defined polypeptide block copolymers.

9.3.9 Towards the Controlled ROPs of R-NCAs or R-NTAs by NAM

ROPs of R-NCAs can be similarly initiated as those for NCAs by the NAM using primary amines (Scheme 9.15). Primary amine initiates the polymerization by ring-opening addition to the C_5 carbonyl of R-NCAs in a

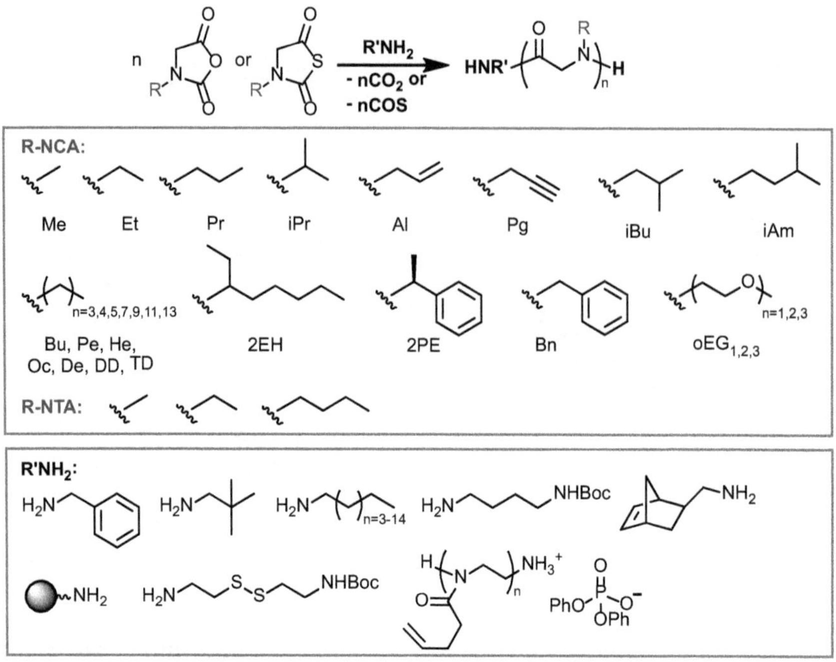

Scheme 9.15 ROPs of various R-NCAs or R-NTAs using different primary amine initiators to yield the linear polypeptoids. (The solid sphere below indicates a solid support bearing primary amine groups).

regioselective fashion followed by decarboxylation to generate the initiating species bearing a secondary amino terminus from which the nucleophilic enchainment ensues. Secondary amines tend to be less nucleophilic than the primary amines due to the steric effect. This ensures faster initiation relative to propagation, an important criterion for a living polymerization. Furthermore, side reactions that plague the primary amine-initiated ROPs of NCAs bearing the *N*–H proton (Scheme 9.4) are notably diminished in the ROPs of R-NCAs. For example, formation of isocyanate or hydantoin species, which terminates the chain growth in the ROPs of NCAs bearing the *N*–H, is suppressed for the ROPs of R-NCAs due to the *N*-substitution in the latter. In addition, the regioselectivity for the nucleophilic addition of amine to C_5 over C_2 carbonyl is enhanced for R-NCAs bearing electron-donating *N*-substituents relative to NCAs bearing the *N*–H, further reducing the probability of termination by this pathway relative to propagation.

It has been demonstrated that ten sequential ROPs of sarcosine derived NCAs (*i.e.*, Me-NCA) using benzylamine initiators can successfully lead to chain extension, yielding polysarcosine polymers with molecular weights ($M_n = 0.78–4.9\ \text{kg mol}^{-1}$) that are in excellent agreement with theoretical values based on living polymerization and narrow molecular weight distributions (PDI = 1.01–1.07).[62] In the last ten years or so, a variety of different primary amines have been investigated as initiators for the ROPs of R-NCAs having different *N*-substituents (Scheme 9.15).[12,13,19] In most cases, polymerizations produce polypeptoids with predictable molecular weight and narrow dispersity with a few exceptions. For example, in the ROPs of *N*-ally NCAs and *N*-2-phenylethyl NCAs using benzylamine initiators (Scheme 9.15), intramolecular transamidation *via* the "backbiting" mechanism was found to compete with chain propagation, limiting the chain length (DP $<$ 100) that can be accessed by this method.[63,64] These studies have highlighted the applicability of primary amine-initiated ROPs of R-NCAs to access well-defined linear polypeptoid polymers having diverse structures.

Kinetic studies of several R-NCAs have revealed first order-dependence of polymerization rate on both the monomer and primary amine initiators, consistent with the NAM mechanism.[65,66] As the *N*-substituents become more sterically bulky, the polymerization rate decreases. This is consistent with nucleophilic addition of the secondary amino propagating species to the R-NCA being the rate-limiting step of the polymerization.

Several *N*-substituted glycine-derived *N*-thiocarboxyanhydrides (*i.e.*, Me-NTA, Ee-NTA and Bu-NTA), mercapto analogs of R-NCAs, have also been synthesized and investigated as substrates for polymerization using primary amine initiators (Scheme 9.15).[15–18] In all cases, polymerization proceeds in a controlled fashion, yielding the corresponding polypeptoids with predictable molecular weight and narrow-to-moderate dispersity. Polymerization of R-NTAs tends to be slower than that of R-NCAs having identical *N*-substituents. This can be attributed to the reduced electrophilicity of the C_5 carbonyl in R-NTA relative to R-NCA and more stable thiocarbamic acid propagating intermediates, resulting in slower nucleophilic ring-opening addition of amines to the R-NTAs and decarboxylation relative to those of R-NCAs.

9.3.10 Towards the Controlled ROPs of R-NCAs or R-NTAs by Activation of Alcohol Initiators

Alcohols do not efficiently initiate the ROPs of NCAs by the NAM due to the reduced nucleophilicity. Activation of alcohols to enhance the nucleophilicity can be accomplished by a number of methods, *e.g.*, deprotonation to generate alkoxide or the formation of hydrogen bonding complexes. Hydrogen bonding has been demonstrated as a mild method of activating alcohols in the studies of organo-mediated ROPs of cyclic esters using alcohol initiators.[67–74]

It was recently reported that benzyl alcohol in conjunction with a catalytic amount of 1,1,3,3-tetramethylguanidine (TMG) can mediate the controlled polymerization of Bu-NCAs in low dielectric solvent (THF or toluene) (Scheme 9.16), yielding the corresponding polypeptoids with M_n in the 3–21 kg mol^{-1} range and low molecular weight distribution (PDI $=$ 1.03–1.08). By contrast, benzyl alcohol alone cannot initiate the polymerization of Bu-NCA under the same conditions.[75] MS and NMR analysis of the resulting polymer revealed that benzyl alcohol is the main initiator for the polymerization, although a small fraction of the polymer product was found to bear a

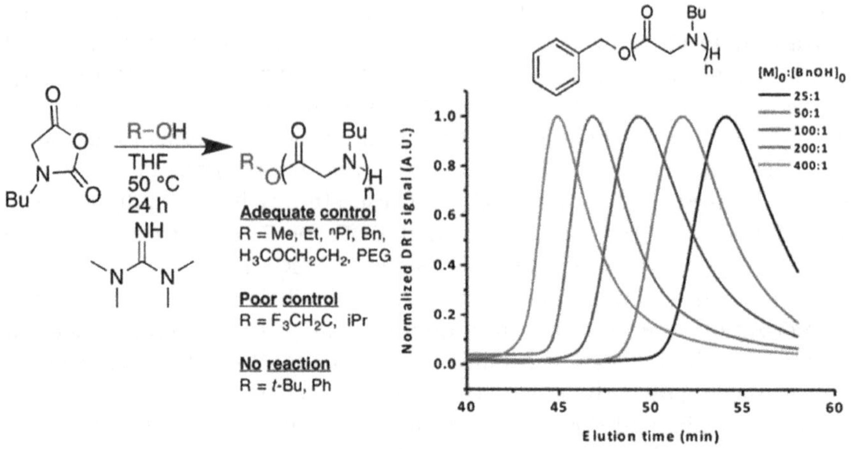

Scheme 9.16 1,1,3,3-tetramethylguanidine (TMG)-promoted ROPs of Bu-NCAs using different alcohol initiators to afford the linear PNBG polypeptoid. Adapted from ref. 75 with permission from the American Chemical Society, Copyright 2016.

TMG end-group, indicating that TMG can also serve to initiate the ROPs of Bu-NCA. The polymerization control appears to be dependent on the solvent. In more polar solvents such as DCM and DMF, only low molecular weight polypeptoids were formed regardless of the initial monomer to benzyl alcohol feed ratios, indicating a lack of controlled polymerization behavior. Kinetic studies have revealed that the polymerization is first-order dependent on the monomer and benzyl alcohol concentration and zero order dependent on the TMG concentration respectively. Further ^{1}H NMR analysis revealed the formation of a hydrogen bonding complex between TMG with the alcohol (Scheme 9.17). These combined results suggest that TMG promotes the polymerization by enhancing the nucleophilicity of alcohols *via* hydrogen bonding, facilitating the nucleophilic ring-opening addition of the alcohols to the Bu-NCA monomer during the initiation of the polymerization.

The extent of alcohol activation by TMG towards initiation is strongly dependent on the steric and electronic properties of the alcohols. For alcohols that are less sterically hindered (*e.g.*, methanol, ethanol, propanol, and benzyl alcohol), the activation of alcohol is sufficient to ensure a fast initiation relative to propagation, giving rise to controlled polymerization behaviors where the polymer molecular weights agree reasonably well with theoretical prediction in the DP < 100 range. As the alcohols become more sterically hindered such as isopropanol or *tert*-butanol, the resulting alcohol–TMG complexes cannot initiate the polymerization efficiently relative to propagation, resulting in diminished control of polymer molecular weight. For example, it was found that the resulting polypeptoids had significantly higher M_n than theoretical values based on the initial monomer to alcohol feed ratios when isopropanol was used as the initiator. No polymerization was observed under the same reaction conditions when *tert*-butanol was used

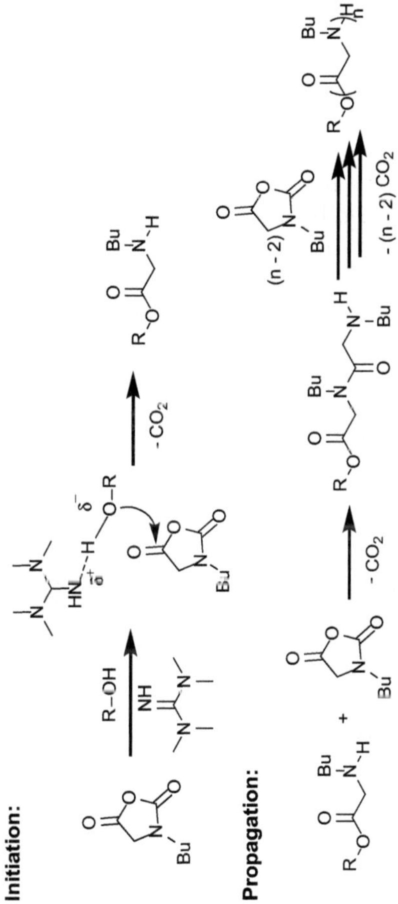

Scheme 9.17 The proposed mechanism of the TMG-promoted ROPs of Bu-NCAs using alcohol initiators.

as initiator. The electronic characteristics of the alcohols are also important in the initiation efficiency. Increasing deviation from controlled polymerization characteristics was observed with alcohols that are increasingly electron deficient (*i.e.*, ethanol < 2-methoxyethanol < 2,2,2-trifluoroethanol). In the case of 2,2,2-trifluoroethanol, M_n is significantly higher than the theoretical prediction based on a controlled polymerization. Clearly, the nature of the alcohols is important to the polymerization characteristics. To use this method towards the synthesis of polypeptoid-containing hetero-block copolymers, it is important that the macro-initiators can be sufficiently activated by TMG to enable efficient initiation for the polymerization of Bu-NCAs. It has been shown that a poly(ethylene glycol) (PEG) bearing a hydroxyl terminus in conjunction with a catalytic TMG can mediate the controlled ROPs of Bu-NCA to produce a well-defined heteroblock copolymer whose molecular weight and composition agree well with the theoretical prediction for a controlled polymerization.

Activation of alcohols *via* hydrogen bonding to enable nucleophilic ROPs of R-NCAs has recently been demonstrated in the synthesis of telechelic water-soluble polypeptoids. A series of aminoalcohols have been investigated as initiators for the ROPs of Me-NTAs or Et-NTAs in acetonitrile or THF at 60 °C (Scheme 9.18).[17] It was found that amino alcohols (**AE**, **AP** and **AMB**, Scheme 9.16), in which the alcohol groups are activated through intermolecular or intramolecular hydrogen bonding, can initiate the polymerizations of R-NTAs at both the amino and alcohol ends, producing a mixture of α-diamino-terminated polypeptoids and α-hydroxyl-ω-amino-terminated polypeptoids (Scheme 9.19). By contrast, for aminoalcohols (**AH** and **AD**, Schemes 9.18 and 9.19) where hydrogen bonding is absent, only the amino end of the initiators initiates the polymerization, producing exclusively α-hydroxyl-ω-amino-terminated polypeptoids with controlled molecular weight (up to DP ≈ 100) and moderate molecular weight distribution (PDI = 1.1–1.3) in good yields.

9.4 Polymerization of NCAs or R-NCAs by the Silyl Group Transfer Mechanism

A new organo-mediated ROPs of NCAs method that involves the use of *N*-trimethylsilyl amines as initiators and silyl carbamate or silyl amine propagating species has been recently developed (Scheme 9.20).[76–78] It was originally proposed that the *N*-trimethylsilyl amines (*e.g.*, hexamethyldisilazane (HMDS)) initiates the ROPs of NCAs by the transfer of a trimethylsilyl (TMS) group to the C_2 carbonyl oxygen of NCAs to form the 2-oxazoline species and a free primary amine (Scheme 9.20a).[76,77] The free primary amine then undergoes ring-opening addition at the C_5 carbonyl position of NCAs to form the *O*-TMS carbamate or *N*-TMS amino initiating species. The chain propagation involves the ring-opening addition of the NCAs by the *O*-TMS carbamate species *via* the silyl group transfer to the NCA monomer *via* a six-membered transition state (Scheme 9.20c). It was subsequently

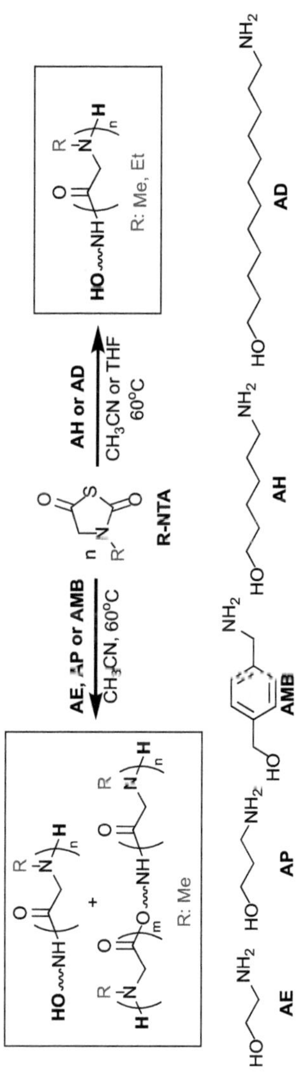

Scheme 9.18 Aminoalcohol-initiated ROPs of Me-NTAs or Et-NTAs to produce the corresponding telechelic polypeptoids.

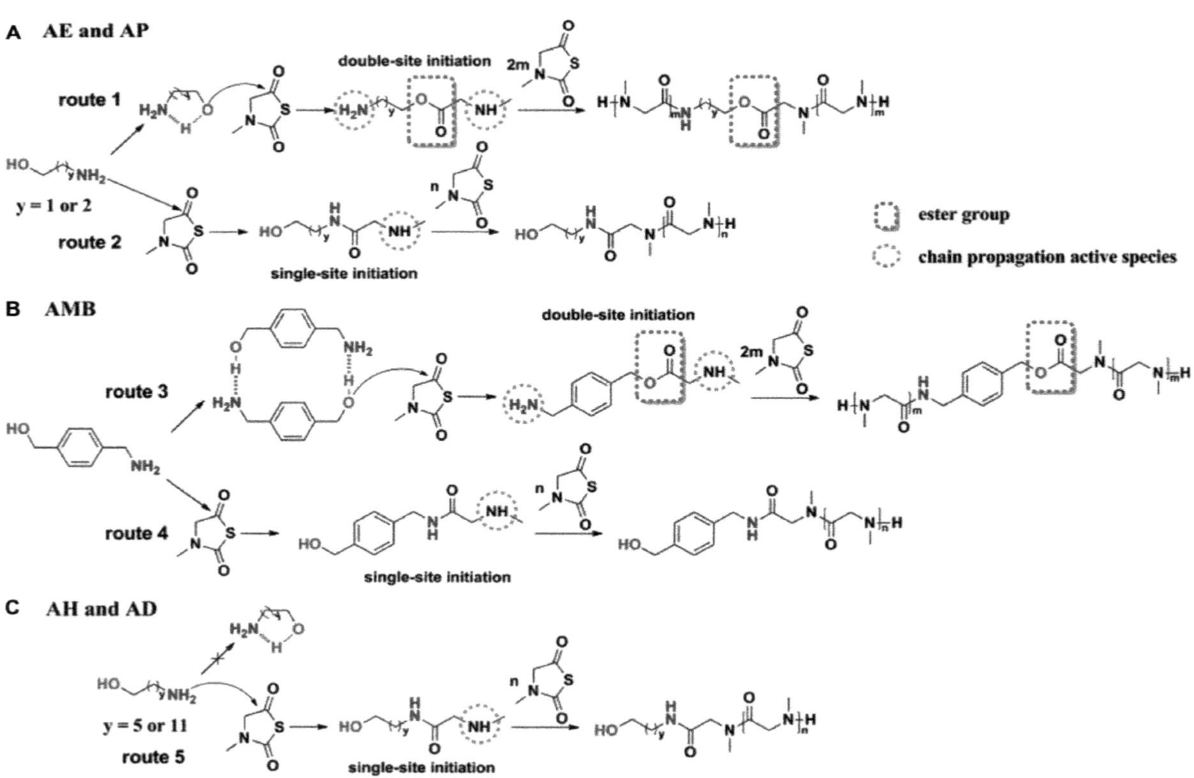

Scheme 9.19 The proposed mechanisms for the ROPs of Me-NTAs using different aminoalcohols. Adapted from ref. 17 with permission from the American Chemical Society, Copyright 2017.

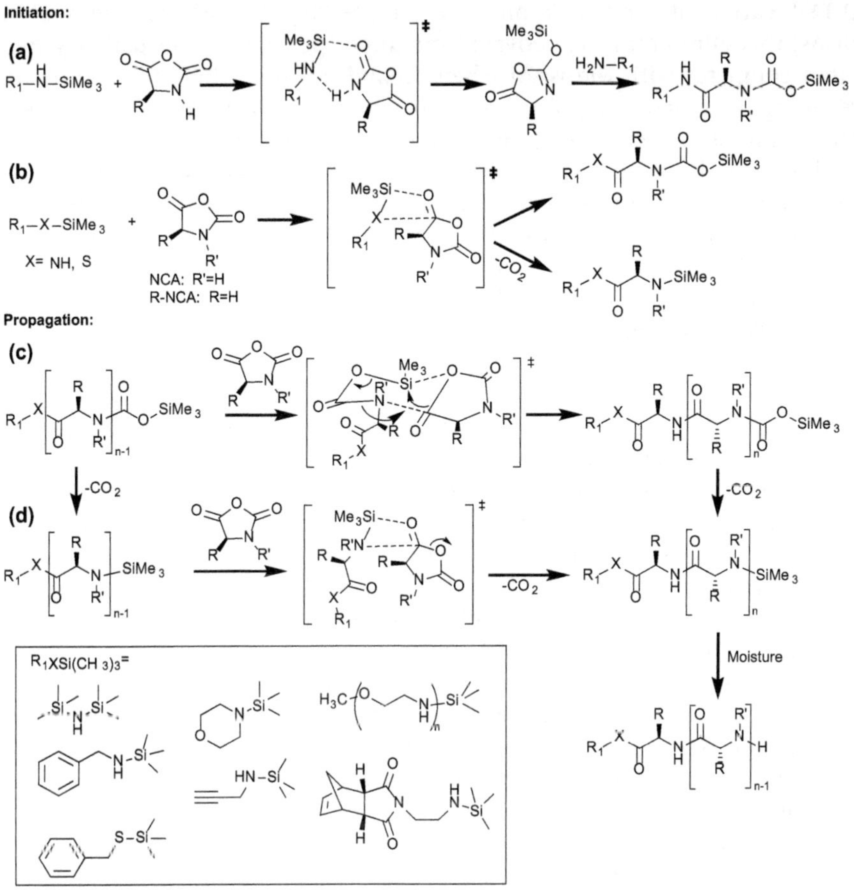

Scheme 9.20 A scheme showing the silyl group transfer mechanism for the ROPs of NCAs using different TMS amine or TMS sulfide initiators.

found that the HMDS can also initiate the ROPs of sarcosine-derived NCAs that do not have the *N*–H proton, which leads to the proposal of a modified mechanism. The modified mechanism involves the direct transfer of TMS group from the HMDS to C_5 carbonyl oxygen of NCAs and simultaneous ring-opening addition of *N*-TMS amine at C_5 carbonyl of NCAs (Scheme 9.20b). A recent computational study has revealed a concerted *syn* addition of C_5 carbonyl of NCAs by the N–Si bond of *N*-TMS methylamine has a lower energy barrier ($\Delta G^{\neq} = 31.76$ kcal mol^{-1}) than that by the N–H bond of the same amine ($\Delta G^{\neq} = 47.79$ kcal mol^{-1}), a key step in the NAM mechanism, which is consistent with the modified mechanism.[79] A variety of *N*-TMS secondary amines has been synthesized and used to initiate the polymerization of γ-benzyl-L-glutamate derived NCAs (Scheme 9.20), producing the corresponding polypeptides with controlled molecular weight and low polydispersity.[77] The enhanced polymerization control has been attributed to the unique polymerization mechanism that involves silyl group transfer. As the

O-TMS carbamate or *N*-TMS amine do not readily deprotonate the *N*–H of NCAs, this eliminates the activated monomer polymerization pathway that often competes with the NAM when secondary amine initiators are used, resulting in enhanced polymerization control as observed when *N*-TMS amine initiators were used. One recent study has shown that when polynorbornenes bearing *N*-TMS amino groups on the sidechains are used as macroinitiators for the ROPs of γ-benzyl-L-glutamate derived NCAs to yield the bottlebrush polypeptides, the polymerization rate is significantly enhanced relative to that using a simple *N*-TMS amine initiator to produce linear polypeptides. For example, the polymerization of γ-ethyl-L-glutamate NCA (ELG) using the macroinitiator reached quantitative conversion in 1 h, whereas using the small molecular initiator only reached full conversion after 10 h under identical conditions ($[M]_0 : [RNH(TMS)]_0 = 50 : 1$, $[M]_0 = 0.5$ M, room temperature). The rate enhancement has been attributed to the cooperative interactions of macrodipoles of the helical polypeptides during the formation of bottlebrush architecture.[78]

Inspired by the early success with *N*-TMS amine initiators, Lu and coworkers recently investigate phenyl trimethylsilyl sulfide as a potential group transfer initiator for the ROPs of NCAs (Scheme 9.20).[80] It was hypothesized that the TMS sulfide would enable more efficient initiation for the polymerization of NCAs than *N*-TMS amine due to the consideration that the sulfide is more nucleophilic than amine, and the S–Si bond is weaker than the N–Si bond. It has been shown that phenyl trimethylsilyl sulfide can initiate and mediate the controlled ROPs of a variety of NCAs bearing the *N*–H proton and Bu-NCA to yield the corresponding polypeptides or polypeptoids. The initiation was found to proceed more efficiently using phenyl trimethylsilyl sulfide than HMDS, consistent with the initiation reaction occurring *via* direct addition of a S–Si bond across the C_5 carbonyl of NCAs. A variety of NCAs bearing different sidechain structures can undergo ROPs using phenyl trimethylsilyl sulfide initiators, producing well-defined polypeptides with controlled molecular weight ($DP \approx 50–100$) and narrow molecular weight distribution ($PDI = 1.03–1.09$). An additional advantage of this method is that the resulting polypeptides or polypeptoids have thioester as the α-chain-end structure, which can be readily derivatized by native chemical ligation to introduce additional functionalities.

9.5 Polymerization of NCAs or R-NCAs by the Zwitterionic Ring-opening Polymerization Mechanism

NCAs can also undergo zwitterionic ring-opening polymerizations (ZROPs) using non-protic nucleophilic initiators (*e.g.*, pyridine and its derivative (picoline, lutidine and DMAP),[81,82] N-heterocyclic carbene,[83–88] cyclic amidine).[89] In this mechanism, the non-protic nucleophiles initiate the ROPs of NCAs by ring-opening addition of NCAs at the C_5 carbonyl position to form a

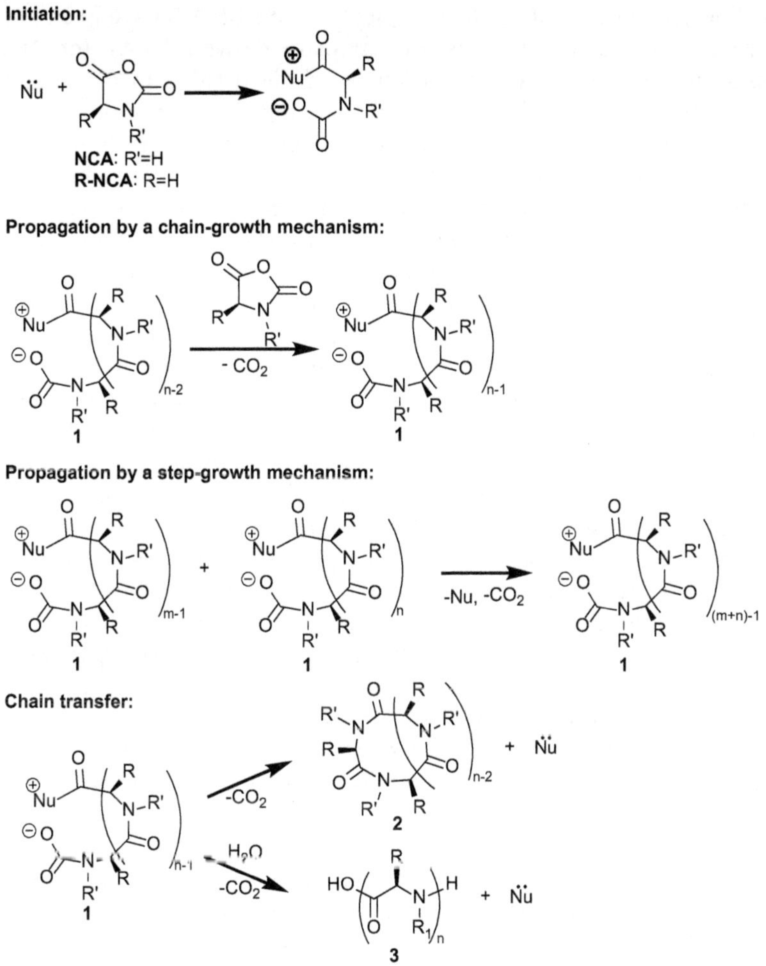

Scheme 9.21 Zwitterionic ring-opening polymerizations of NCAs or R-NCAs using organic initiators by a chain-growth or step-growth mechanism. The chain transfer may occur by macrocyclization or reaction with residual water in the system.

zwitterionic initiating species (Scheme 9.21). The propagation can occur either by a chain-growth or step-growth mechanism. The former involves nucleophilic ring-opening addition of the zwitterionic propagating species to the C_5 carbonyl of NCAs accompanied by the CO_2 elimination (Scheme 9.21). Alternatively, the zwitterionic species can react with each other by a step-growth polymerization (Scheme 9.21). While ZROPs tend to yield a substantial cyclic polypeptides or polypeptoids by macrocyclization of the zwitterionic propagating species, the relative content of cyclic *vs.* linear chains are strongly dependent on the solvent polarity, polymer conformation and solubility in the reaction medium.[81,82]

Pyridine and its derivatives (*e.g.*, α-picoline and DMAP) have been shown to initiate the polymerizations of sarcosine-derived NCAs (or NTAs), L-alanine, DL-phenylalanine, and DL-leucine derived NCAs in dioxane, pyridine and *N*-methyl pyrrolidone (NMP), yielding the corresponding cyclic polypeptides/polypeptoids together with water-initiated/terminated linear polymers.[82] In the case of sarcosine and DL-phenyalanine derived NCAs, the cyclic polymers are the major products. The fraction of cyclic polymers is particularly high when NMP is the solvent. Zwitterionic species are proposed to form during the initiation of all polymerizations. The polymerization of sarcosine derived NCAs is most likely to proceed by the ZROP mechanism as shown in Scheme 9.21. For other NCAs, in addition to the ZROP mechanism, the zwitterionic propagating species (**1**, Scheme 9.22) can also react with NCAs to generate polypeptide chains with an *N*-acylated NCA α-terminus and a primary amino ω-terminus (**2**, Scheme 9.22), which is identical to those obtained by the activated monomer mechanism when tertiary amines are used as initiators. This species **2** can proceed to grow either in a chain growth or step growth fashion. All polypeptides exhibit broad molecular weight distribution, suggesting that all polymerizations are most likely to occur by a step-growth mechanism. In addition, ZROPs of sarcosine, L-alanine, DL-phenylalanine, DL-leucine and DL-valine derived NCAs can also be initiated with appropriate nucleophilic solvents (*e.g.*, DMF, DMSO and NMP)[38] or thermally,[37] producing mainly oligomeric cyclic and linear polypeptides or polypeptoids in a step growth mechanism.

Controlled ZROPs of NCAs to access cyclic polypeptides with predictable molecular weight and narrow dispersity is challenging due to the rapid chain transfer, termination by various base-promoted side reactions and competing non-ZROP pathways. Several recent studies have shown some encouraging development in the controlled ZROPs of R-NCAs with different *N*-substituents, enabling access to well-defined cyclic polypeptoids.

ZROPs of Bu-NCAs using N-heterocyclic carbenes (NHCs) as nucleophilic initiators have been shown to produce the corresponding cyclic polypeptoids (*i.e.*, poly(*N*-butyl glycine)) with narrow dispersity (PDI = 1.04–1.12) and tunable ring sizes (DP ≈ 25–250) (Scheme 9.23).[83,86] The polymer molecular weight and the ring size can be adjusted by controlling the initial monomer to NHC feed ratio. The polymerization is initiated by two mechanisms: (1) nucleophilic ring-opening addition of R-NCAs by a NHC molecule to form a zwitterionic

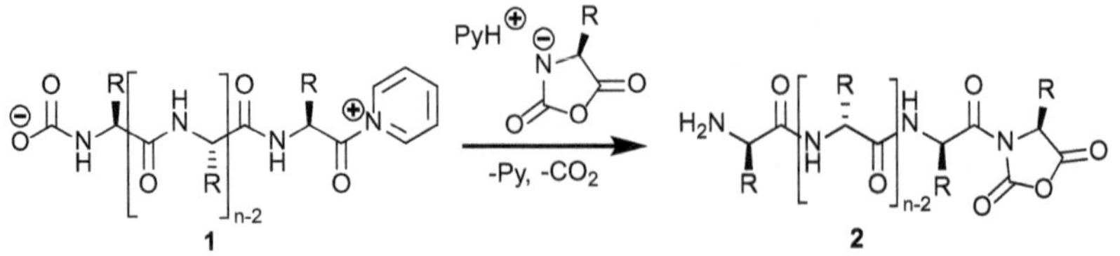

Scheme 9.22 Side reactions for the zwitterionic propagating species in the pyridine-initiated ZROPs of NCAs.

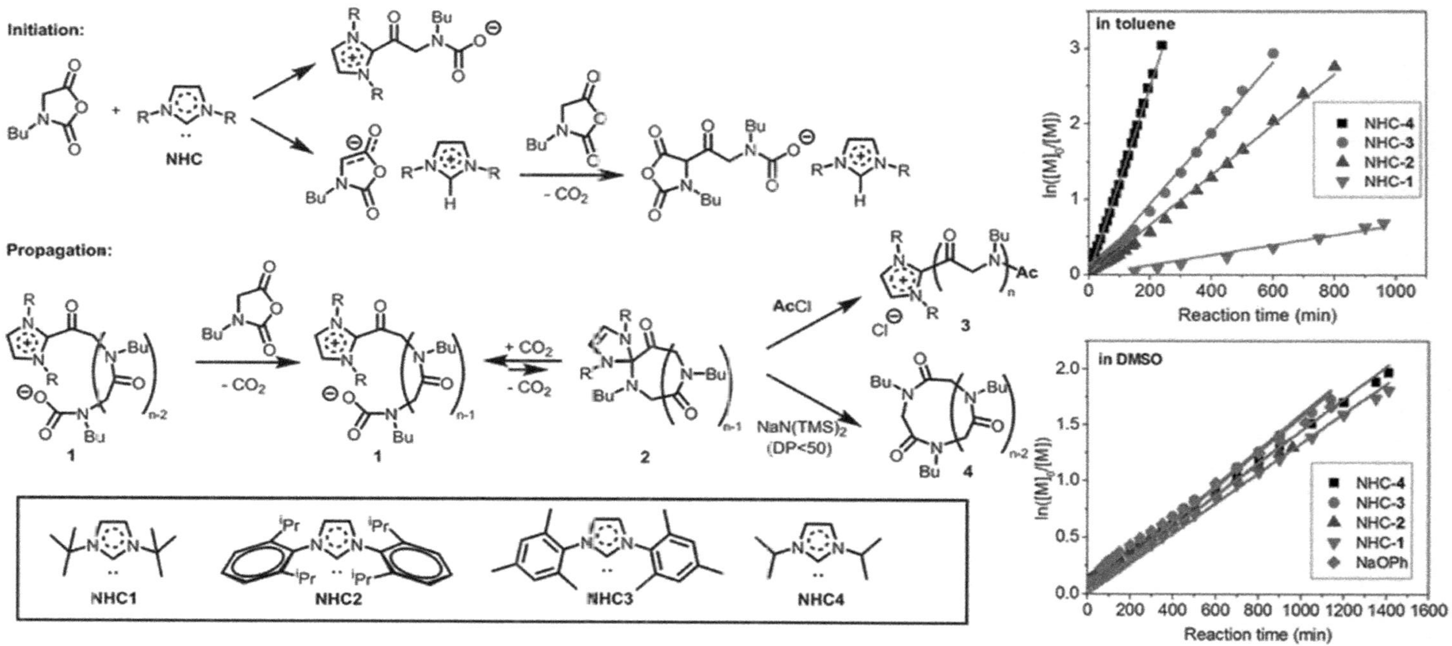

Scheme 9.23 NHC-mediated controlled ZROPs of R-NCAs to afford cyclic PNBGs and post-polymerization conversion of the cyclic PNBGs into the linear and NHC-free cyclic polymeric analogs and the polymerization kinetic analysis in toluene *vs.* DMSO using NHCs with varying steric hindrance using a pseudo first-order kinetic plot.
Adapted from ref. 83 and 86 with permission from the American Chemical Society, Copyright 2012.

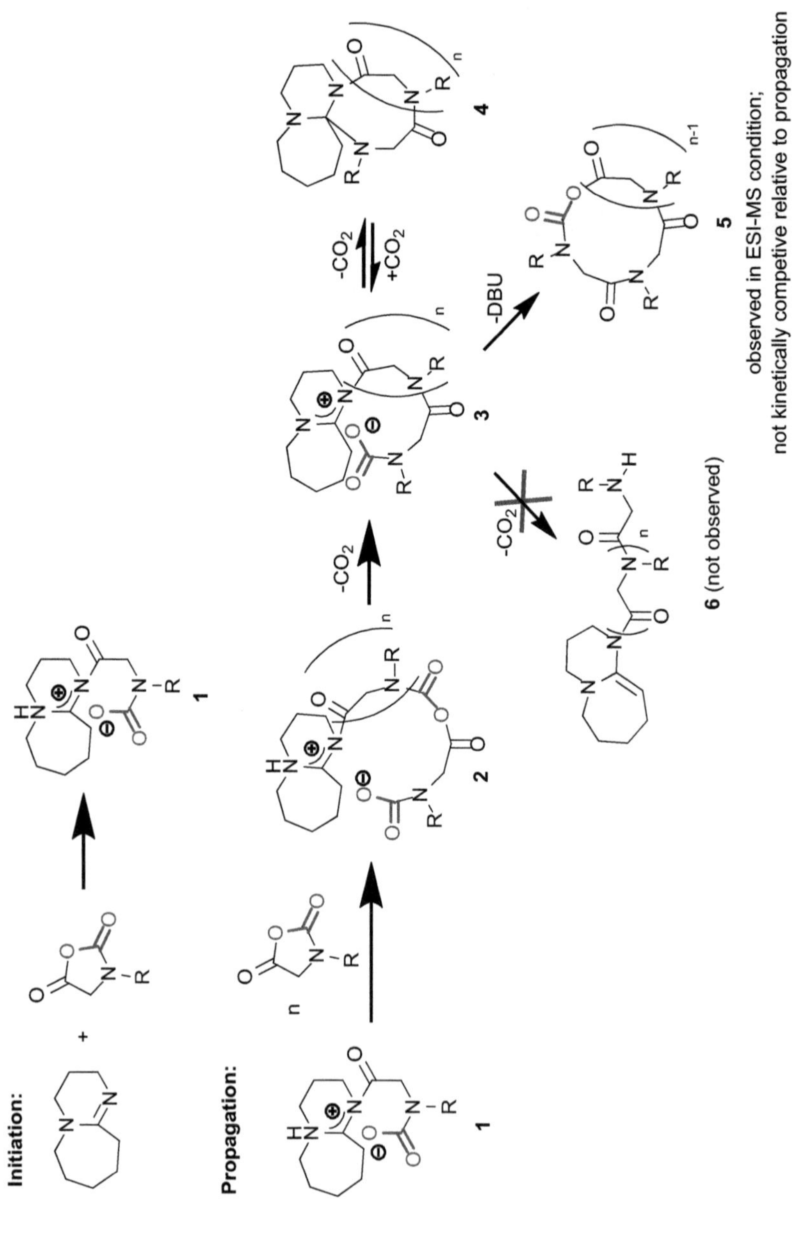

Scheme 9.24 DBU-mediated controlled ZROPs of Bu-NCAs to afford cyclic PNBGs.

initiating species (*i.e.*, zwitterionic initiation); (2) deprotonation of α-C–H of R-NCAs by the NHC to form R-NCA anions which then react with another NCA to form an anionic initiating species (*i.e.*, anionic initiation). A recent DFT calculation has shown that the zwitterionic initiation has a much lower energy barrier ($\Delta G^{\neq} = 7.0$ kcal mol^{-1}) than that of the anionic initiation ($\Delta G^{\neq} = 16.6$ kcal mol^{-1}), thus favoring the former as the major mode of initiation.[90] The polymerization mainly proceeds through a zwitterionic propagating intermediate where the two oppositely charged chain ends are held in proximity through electrostatic interaction (Scheme 9.23). The zwitterionic species is in equilibrium with a dormant spirocyclic propagating species. The reaction exhibited a pseudo-first order polymerization kinetic and the polymerization rate is dependent on the steric characteristic of NHCs due to the intramolecular counter-ion effect (Scheme 9.23). The reactions proceed in a controlled manner in low dielectric solvents such as THF and toluene. In more polar solvents (*e.g.*, DMF, DMSO, and nitrobenzene), only low molecular weight mixtures of cyclic and linear polypeptoids were obtained, due to the competitive intramolecular transamidation relative to chain propagation. The zwitterionic species can be isolated and treated with $NaNTMS_2$ to give the corresponding NHC-free cyclic polypeptoids. This transformation is limited to polypeptoids with low MW (DP < 50); the efficiency of $NaNTMS_2$-induced macrocyclization varies for zwitterionic polypeptoid precursors with DP > 50. In addition, treatment of the zwitterionic species with electrophiles (*e.g.*, AcCl) afforded the corresponding linear polypeptoids. This method has been used to polymerize a variety of R-NCA monomers (*e.g.*, Me-NCA, Et-NCA, Pg-NCA, Bu-NCA, Hex-NCA, Oct-NCA, De-NCA, 2EH-NCA, 2PE-NCA, *etc.*) (Scheme 9.15) to produce well-defined cyclic polypeptoids with different *N*-substituents including those adopting polyproline I (PPI) helical conformations[64,91] and random and block copolymers having cyclic architectures.[83–85,87,88] The cyclic architecture of these polypeptoids has been verified by a combination of microscopic, scattering and intrinsic viscosity measurements and comparison to their linear counterparts.[64,83,84,89,92]

It was recently reported that 1,8-diazabicyclo[5.4.0]undec-7-ene (DBU), a bicyclic amidine, can also mediate the controlled ZROPs of Bu-NCAs similarly to NHCs (Scheme 9.24).[89] The reaction occurs in a controlled manner in low dielectric solvents (*e.g.*, THF and toluene), allowing access of polypeptoids having low to medium molecular weights (DP ≈ 25–300) and narrow molecular weight distribution (PDI = 1.02–1.12), similar to what has been reported for NHCs. By contrast, in DMF, only low molecular weight polymers (DP < 40) were formed regardless of the initial monomer to DBU ratios. In toluene, the initiation rate was found to be comparable to that of the propagating rate. The active propagating species has a zwitterionic structure with a positively charged DBU moiety and a negatively charged carbamate group at each chain end, and the electrostatic interaction keeps the propagating chain in a cyclic form. It is in equilibrium with a dormant spirocyclic species 4. The cyclic polypeptoid architecture was verified by a combination of end-group analysis with NMR, MS, AFM and SANS analysis.[89,92] While the ZROPs of Bu-NCAs using DBU initiators is comparable to that with NHC initiators in the polymerization rate, DBU has the advantage of being air and moisture stable, in contrast to NHCs. This makes DBU a more robust initiator relative to NHCs towards the ZROPs of R-NCAs.

9.6 Concluding Remarks

Recent advances in the development of controlled ROPs of amino acid-derived NCAs using organic initiators/catalysts have highlighted the potential to control complex polymerization kinetic pathways of the NCAs bearing *N*–H *via* weak or reversible interactions (*e.g.*, hydrogen bonding interactions[28,56] or the formation of frustrated Lewis acid–base pairs).[30] These interactions modulate the reactivity of monomers, initiators and the propagating chain ends, enabling polymerization to occur in controlled manner to produce polypeptides with predictable molecular weight, narrow molecular weight distribution and various molecular architectures. The general strategy of using weak and reversible interactions to modulate the polymerization activity remains to be fully investigated for the ROPs of NCAs bearing different functional sidechains, R-NCAs and their mercapto counterparts (*i.e.*, NTA or R-NTAs). The development of organo-promoted controlled ROPs of NTAs bearing *N*–H will be particularly attractive in view of the high hydrolytic and thermal stability of NTAs relative to the analogous NCAs but the significantly reduced reactivity towards polymerization.[14]

The emergence and development of silyl group transfer polymerization[76,77,80] and organo-mediated zwitterionic ring-expansion polymerization,[83,86,89] which have enabled the controlled ROPs of NCAs and R-NCAs, have demonstrated the importance of continuing efforts in exploring the reactivity of new organic initiators/catalysts and elucidation of the polymerization mechanism that provide guidance in the development of more efficient and robust initiating systems. As polypeptides and polypeptoids are increasingly investigated as biomaterials for different applications, the simultaneous development of efficient catalysis and new monomers that are easy to handle will be of utmost importance for the wide adoption of the synthetic methods by a broader community.

Acknowledgements

This work is supported by the National Science Foundation (CHE 1609447) and Louisiana State University.

References

1. H. Leuchs, Ueber die Glycin-carbonsäure, *Ber. Dtsch. Chem. Ges.*, 1906, **39**(1), 857–861.
2. H. Leuchs and W. Manasse, Über die Isomerie der Carbäthoxyl-glycyl glycinester, *Ber. Dtsch. Chem. Ges.*, 1907, **40**(3), 3235–3249.

3. H. Leuchs and W. Geiger, Über die Anhydride von α-Amino-N-carbonsäuren und die von α-Aminosäuren, *Ber. Dtsch. Chem. Ges.*, 1908, **41**(2), 1721–1726.
4. H. Block, *Poly(γ-benzyl-L-Glutamate) and other Glutamic Acid Containing Polymers*, Gordon and Breach, New York, 1983.
5. C. H. Bamford and H. Block, The Polymerization of N-Carboxy-α-Amino Acid Anhydrides, in *Comprehensive Chemical Kinetics*, ed. C. H. Bamford and C. F. H. Tipper, Elsevier, 1976, ch. 8, vol. 15, pp. 583–637.
6. M. Szwarc, The Kinetics and Mechanism of N-carboxy-a-amino-acid Anhydride (NCA) Polymerisation to Poly-amino Acids, *Adv. Polym. Sci.*, 1965, **4**, 1–65.
7. H. R. Kricheldorf, Polypeptides and 100 years of chemistry of alpha-amino acid N-carboxyanhydrides, *Angew. Chem., Int. Ed.*, 2006, **45**(35), 5752–5784.
8. N. Hadjichristidis, H. Iatrou, M. Pitsikalis and G. Sakellariou, Synthesis of Well-Defined Polypeptide-Based Materials via the Ring-Opening Polymerization of α-Amino Acid N-Carboxyanhydrides, *Chem. Rev.*, 2009, **109**(11), 5528–5578.
9. C. Deng, J. Wu, R. Cheng, F. Meng, H.-A. Klok and Z. Zhong, Functional polypeptide and hybrid materials: Precision synthesis via α-amino acid N-carboxyanhydride polymerization and emerging biomedical applications, *Prog. Polym. Sci.*, 2014, **39**(2), 330–364.
10. H. Lu, J. Wang, Z. Song, L. Yin, Y. Zhang, H. Tang, C. Tu, Y. Lin and J. Cheng, Recent advances in amino acid N-carboxyanhydrides and synthetic polypeptides: chemistry, self-assembly and biological applications, *Chem. Commun.*, 2014, **50**(2), 139–155.
11. C. He, X. Zhuang, Z. Tang, H. Tian and X. Chen, Stimuli-Sensitive Synthetic Polypeptide-Based Materials for Drug and Gene Delivery, *Adv. Healthcare Mater.*, 2012, **1**(1), 48–78.
12. D. Zhang, S. H. Lahasky, L. Guo, C.-U. Lee and M. Lavan, Polypeptoid Materials: Current Status and Future Perspectives, *Macromolecules*, 2012, **45**(15), 5833–5841.
13. N. Gangloff, J. Ulbricht, T. Lorson, H. Schlaad and R. Luxenhofer, Peptoids and Polypeptoids at the Frontier of Supra- and Macromolecular Engineering, *Chem. Rev.*, 2016, **116**(4), 1753–1802.
14. J. Cao, D. Siefker, B. A. Chan, T. Yu, L. Lu, M. A. Saputra, F. R. Fronczek, W. Xie and D. Zhang, Interfacial Ring-Opening Polymerization of Amino-Acid-Derived N-Thiocarboxyanhydrides Toward Well-Defined Polypeptides, *ACS Macro Lett.*, 2017, **6**(8), 836–840.
15. X. Tao, C. Deng and J. Ling, PEG-amine-initiated polymerization of sarcosine N-thiocarboxyanhydrides toward novel double-hydrophilic PEG-b-polysarcosine diblock copolymers, *Macromol. Rapid Commun.*, 2014, **35**(9), 875–881.
16. Y. Deng, T. Zou, X. Tao, V. Semetey, S. Trepout, S. Marco, J. Ling and M. H. Li, Poly(epsilon-caprolactone)-block-polysarcosine by Ring-Opening Polymerization of Sarcosine N-Thiocarboxyanhydride: Synthesis and Thermoresponsive Self-Assembly, *Biomacromolecules*, 2015, **16**(10), 3265–3274.

17. X. Tao, B. Zheng, T. Bai, B. Zhu and J. Ling, Hydroxyl Group Tolerated Polymerization of N-Substituted Glycine N-Thiocarboxyanhydride Mediated by Aminoalcohols: A Simple Way to α-Hydroxyl-ω-aminotelechelic Polypeptoids, *Macromolecules*, 2017, **50**(8), 3066–3077.
18. X. Tao, B. Zheng, H. R. Kricheldorf and J. Ling, AreN-substituted glycineN-thiocarboxyanhydride monomers really hard to polymerize?, *J. Polym. Sci., Part A: Polym. Chem.*, 2017, **55**(3), 404–410.
19. B. A. Chan, S. Xuan, A. Li, J. M. Simpson, G. L. Sternhagen, T. Yu, O. A. Darvish, N. Jiang and D. Zhang, Polypeptoid Polymers: Synthesis, Characterization and Properties, *Biopolymers*, 2018, **109**, e23070.
20. P. Aubert, R. A. Jeffreys and E. B. Knott, 477. Thiazolid-2: 5-dione, *J. Chem. Soc.*, 1951, 2195–2197.
21. R. Hirschmann, R. S. Dewey, E. F. Schoenewaldt, H. Joshua, W. J. Paleveda, H. Schwam, H. Barkemeyer, B. H. Arison and D. F. Veber, Synthesis of peptides in aqueous medium. VII. Preparation and use of 2,5-thiazolidinediones in peptide synthesis, *J. Org. Chem.*, 1971, **36**(1), 49–59.
22. J. L. Bailey, 679. The synthesis of simple peptides from anhydro-N-carboxyamino-acids, *J. Chem. Soc.*, 1950, 3461–3466.
23. E. Katchalski, *Advances in Protein Chemistry*, Academic Press, Inc, New York, NY, 1951, vol VI, pp. 123–185.
24. H. R. Kricheldorf, *α-Amino acid-N-carboxy-anhydrides and Related Heterocycles: Syntheses, Properties, Peptide Synthesis, Polymerization*, Springer-Verlag, Berlin, 1987, ch. 2, pp. 60–64.
25. R. B. Woodward and C. H. Schramm, Synthesis of Protein Analogs, *J. Am. Chem. Soc.*, 1947, **69**(6), 1551–1552.
26. V. Go and H. Tani, Über die hochkondensierten Produkte aus einigen Aminosäuren, *Bull. Chem. Soc. Jpn.*, 1939, **14**(11), 510–516.
27. F. Wessely, Untersuchungen über α-Amino-N-Carbonsäureanhydride. I, *Z. Physiol. Chem.*, 1925, **146**, 72.
28. W. Zhao, Y. Gnanou and N. Hadjichristidis, Organocatalysis by hydrogen-bonding: a new approach to controlled/living polymerization of α-amino acid N-carboxyanhydrides, *Polym. Chem.*, 2015, **6**(34), 6193–6201.
29. F. Sigmund and F. Wessely, α-Amino-N-carboxylic acid anhydrides. II, *Z. Physiol. Chem.*, 1926, **157**, 91–105.
30. H. Zhang, Y. Nie, X. Zhi, H. Du and J. Yang, Controlled ring-opening polymerization of alpha-amino acid N-carboxy-anhydride by frustrated amine/borane Lewis pairs, *Chem. Commun.*, 2017, **53**(37), 5155–5158.
31. F. Sigmund and F. Wessely, Untersuchungen über α-Amino-N-Carbonsäureanhydride. II, *Z. Physiol. Chem.*, 1926, **157**, 91.
32. C. H. Bamford and H. Block, 982. The initiation step in the polymerization of N-carboxy-[small alpha]-amino-acid anhydrides. Part II. Effects related to the structure of amine initiators, *J. Chem. Soc.*, 1961, 4992–4995.
33. M. Goodman and J. Hutchison, The Mechanisms of Polymerization of N-Unsubstituted N-Carboxyanhydrides1, *J. Am. Chem. Soc.*, 1966, **88**(15), 3627–3630.

34. X. Zhang, M. Oddon, O. Giani, S. Monge and J.-J. Robin, Novel Strategy for ROP of NCAs Using Thiols As Initiators: Synthesis of Diblock Copolymers Based on Polypeptides, *Macromolecules*, 2010, **43**(6), 2654–2656.
35. H. R. Kricheldorf, C. Von Lossow and G. Schwarz, Imidazole-initiated polymerizations of α-amino acid N-carboxyanhydrides - simultaneous chain-growth and step-growth polymerization, *J. Polym. Sci., Part A: Polym. Chem.*, 2005, **43**(22), 5690–5698.
36. L. A. A. Sluyterman and B. Labruyere, Side reactions in the polymerization of α-amino acid N-carbonic anhydrides. Titration data of polyglycine and polyalanine, *Recl. Trav. Chim. Pays-Bas*, 1954, **73**(4), 347–354.
37. H. R. Kricheldorf, C. V. Lossow, N. Lomadze and G. Schwarz, Cyclic polypeptides by thermal polymerization of α-amino acid N-carboxyanhydrides, *J. Polym. Sci., Part A: Polym. Chem.*, 2008, **46**(12), 4012–4020.
38. H. R. Kricheldorf, C. von Lossow and G. Schwarz, Cyclic Polypeptides by Solvent-Induced Polymerizations of α-Amino Acid N-Carboxyanhydrides, *Macromolecules*, 2005, **38**(13), 5513–5518.
39. G. J. M. Habraken, M. Peeters, C. H. J. T. Dietz, C. E. Koning and A. Heise, How controlled and versatile is N-carboxy anhydride (NCA) polymerization at 0 °C? Effect of temperature on homo-, block- and graft (co)polymerization, *Polym. Chem.*, 2010, **1**(4), 514–524.
40. G. J. M. Habraken, K. H. R. M. Wilsens, C. E. Koning and A. Heise, Optimization of N-carboxyanhydride (NCA) polymerization by variation of reaction temperature and pressure, *Polym. Chem.*, 2011, **2**(6), 1322–1330.
41. D. L. Pickel, N. Politakos, A. Avgeropoulos, J. M. Messman and A. Mechanistic, Study of α-(Amino acid)-N-carboxyanhydride Polymerization: Comparing Initiation and Termination Events in High-Vacuum and Traditional Polymerization Techniques, *Macromolecules*, 2009, **42**(20), 7781–7788.
42. M. Goodman and J. Hutchison, Propagation Mechanism in Strong Base Initiated Polymerization of α-Amino Acid N-Carboxyanhydrides, *J. Am. Chem. Soc.*, 1965, **87**(15), 3524–3525.
43. M. Idelson and E. R. Blout, Polypeptides. XVIII.1 A Kinetic Study of the Polymerization of Amino Acid N-Carboxyanhydrides Initiated by Strong Bases, *J. Am. Chem. Soc.*, 1958, **80**(10), 2387–2393.
44. W. Vayaboury, O. Giani, H. Cottet, S. Bonaric and F. Schué, Mechanistic Study of α-Amino Acid N-Carboxyanhydride (NCA) Polymerization by Capillary Electrophoresis, *Macromol. Chem. Phys.*, 2008, **209**(15), 1628–1637.
45. W. Vayaboury, O. Giani, H. Cottet, A. Deratani and F. Schué, Living Polymerization of α-Amino Acid N-Carboxyanhydrides(NCA) upon Decreasing the Reaction Temperature, *Macromol. Rapid Commun.*, 2004, **25**(13), 1221–1224.
46. J. Zou, J. Fan, X. He, S. Zhang, H. Wang and K. L. Wooley, A facile glovebox-free strategy to significantly accelerate the syntheses of

well-defined polypeptides by N-carboxyanhydride (NCA) ring opening polymerizations, *Macromolecules*, 2013, **46**(10), 4223–4226.

47. D. G. H. Ballard and C. H. Bamford, Studies in polymerization - VII. The polymerization of N-carboxy-α-amino acid anhydrides, *Proc. R. Soc. London, Ser. A*, 1954, **223**(1155), 495–520.
48. D. Thunig, J. Semen and H.-G. Elias, Carbon Dioxide Influence on NCA Polymerizations, *Makromol. Chem.*, 1977, **178**(2), 603–607.
49. J. Ling and Y. Huang, Understanding the Ring-Opening Reaction of α-Amino Acid N-Carboxyanhydride in an Amine-Mediated Living Polymerization: A DFT Study, *Macromol. Chem. Phys.*, 2010, **211**(15), 1708–1711.
50. J. Liu and J. Ling, DFT Study on Amine-Mediated Ring-Opening Mechanism of alpha-Amino Acid N-Carboxyanhydride and N-Substituted Glycine N-Carboxyanhydride: Secondary Amine versus Primary Amine, *J. Phys. Chem. A*, 2015, **119**(27), 7070–7074.
51. I. Dimitrov and H. Schlaad, Synthesis of nearly monodisperse polystyrene–polypeptide block copolymers via polymerisation of N-carboxyanhydrides, *Chem. Commun.*, 2003, **23**, 2944–2945.
52. C. D. Vacogne and H. Schlaad, Primary ammonium/tertiary amine-mediated controlled ring opening polymerisation of amino acid N-carboxyanhydrides, *Chem. Commun.*, 2015, **51**(86), 15645–15648.
53. C. D. Vacogne and H. Schlaad, Controlled ring-opening polymerization of α-amino acid N -carboxyanhydrides in the presence of tertiary amines, *Polymer*, 2017, **124**, 203–209.
54. W. Zhao, Y. Gnanou and N. Hadjichristidis, From competition to cooperation: a highly efficient strategy towards well-defined (co)polypeptides, *Chem. Commun.*, 2015, **51**(17), 3663–3666.
55. W. Zhao, Y. Gnanou and N. Hadjichristidis, Fast and living ring-opening polymerization of alpha-amino acid N-carboxyanhydrides triggered by an "alliance" of primary and secondary amines at room temperature, *Biomacromolecules*, 2015, **16**(4), 1352–1357.
56. W. Zhao, Y. Gnanou and N. Hadjichristidis, Well-defined (co)polypeptides bearing pendant alkyne groups, *Polym. Chem.*, 2016, **7**(21), 3487–3491.
57. H. Kato, T. Higashimura and S. Okamura, Condensation polymerization of N-dithiocarbonyl alkoxyearbonyl amino acids. Part V. Studies on reaction mechanism, *Makromolekul. Chem.*, 1967, **109**(1), 9–21.
58. F. J. Vinick and S. Jung, Concerning the preparation of optically pure N-(thiocarboxy)-L-aspartic anhydride, *J. Org. Chem.*, 1982, **47**(11), 2199–2201.
59. R. S. Dewey, E. F. Schoenewaldt, H. Joshua, W. J. Paleveda, H. Schwam, H. Barkemeyer, B. H. Arison, D. F. Veber, R. G. Denkewalter and R. Hirschmann, Synthesis of peptides in aqueous medium. V. Preparation and use of 2,5-thiazolidinediones (NTA's). Use of the 13C-H nuclear magnetic resonance signal as internal standard for quantitative studies, *J. Am. Chem. Soc.*, 1968, **90**(12), 3254–3255.
60. H. R. Kricheldorf, M. Sell and G. Schwarz, Primary Amine-Initiated Polymerizations of α–Amino Acid N-Thiocarbonic Acid Anhydrosulfide, *J. Macromol. Sci., Part A: Pure Appl.Chem.*, 2008, **45**(6), 425–430.

61. H. R. Kricheldorf, ÜBer die Polymerisation von α-Aminosäure-N-carboxyanhydriden (1,3-Oxazolidin-2,5-dionen) und α-Aminosäure-N-thiocarboxyanhydriden (1,3-Thiazolidin-2,5-dionen), *Makromol. Chem.*, 1974, **175**(12), 3325–3342.
62. C. Fetsch and R. Luxenhofer, Highly Defined Multiblock Copolypeptoids: Pushing the Limits of Living Nucleophilic Ring-Opening Polymerization, *Macromol. Rapid Commun.*, 2012, **33**(19), 1708–1713.
63. J. W. Robinson and H. Schlaad, A versatile polypeptoid platform based on N-allyl glycine, *Chem. Commun.*, 2012, **48**(63), 7835–7837.
64. L. Guo, J. Li, Z. Brown, K. Ghale and D. Zhang, Synthesis and characterization of cyclic and linear helical poly(α-peptoid)s by N-heterocyclic carbene-mediated ring-opening polymerizations of N-substituted N-carboxyanhydrides, *Pept. Sci.*, 2011, **96**(5), 596–603.
65. C. Fetsch, A. Grossmann, L. Holz, J. F. Nawroth and R. Luxenhofer, Polypeptoids from N-Substituted Glycine N-Carboxyanhydrides: Hydrophilic, Hydrophobic, and Amphiphilic Polymers with Poisson Distribution, *Macromolecules*, 2011, **44**(17), 6746–6758.
66. S. Xuan, S. Gupta, X. Li, M. Bleuel, G. J. Schneider and D. Zhang, Synthesis and Characterization of Well-Defined PEGylated Polypeptoids as Protein-Resistant Polymers, *Biomacromolecules*, 2017, **18**(3), 951–964.
67. R. C. Pratt, B. G. G. Lohmeijer, D. A. Long, R. M. Waymouth and J. L. Hedrick, Triazabicyclodecene: A Simple Bifunctional Organocatalyst for Acyl Transfer and Ring-Opening Polymerization of Cyclic Esters, *J. Am. Chem. Soc.*, 2006, **128**(14), 4556–4557.
68. A. Chuma, H. W. Horn, W. C. Swope, R. C. Pratt, L. Zhang, B. G. G. Lohmeijer, C. G. Wade, R. M. Waymouth, J. L. Hedrick and J. E. Rice, The Reaction Mechanism for the Organocatalytic Ring-Opening Polymerization of L-Lactide Using a Guanidine-Based Catalyst: Hydrogen-Bonded or Covalently Bound?, *J. Am. Chem. Soc.*, 2008, **130**(21), 6749–6754.
69. L. Zhang, R. C. Pratt, F. Nederberg, H. W. Horn, J. E. Rice, R. M. Waymouth, C. G. Wade and J. L. Hedrick, Acyclic Guanidines as Organic Catalysts for Living Polymerization of Lactide, *Macromolecules*, 2010, **43**(3), 1660–1664.
70. O. I. Kazakov, P. P. Datta, M. Isajani, E. T. Kiesewetter and M. K. Kiesewetter, Cooperative Hydrogen-Bond Pairing in Organocatalytic Ring-Opening Polymerization, *Macromolecules*, 2014, **47**(21), 7463–7468.
71. O. I. Kazakov and M. K. Kiesewetter, Cocatalyst Binding Effects in Organocatalytic Ring-Opening Polymerization of L-Lactide, *Macromolecules*, 2015, **48**(17), 6121–6126.
72. S. S. Spink, O. I. Kazakov, E. T. Kiesewetter and M. K. Kiesewetter, Rate Accelerated Organocatalytic Ring-Opening Polymerization of L-Lactide via the Application of a Bis(thiourea) H-bond Donating Cocatalyst, *Macromolecules*, 2015, **48**(17), 6127–6131.
73. K. V. Fastnacht, S. S. Spink, N. U. Dharmaratne, J. U. Pothupitiya, P. P. Datta, E. T. Kiesewetter and M. K. Kiesewetter, Bis- and Tris-Urea

H-Bond Donors for Ring-Opening Polymerization: Unprecedented Activity and Control from an Organocatalyst, *ACS Macro Lett.*, 2016, **5**(8), 982–986.

74. J. U. Pothupitiya, N. U. Dharmaratne, T. M. M. Jouaneh, K. V. Fastnacht, D. N. Coderre and M. K. Kiesewetter, H-Bonding Organocatalysts for the Living, Solvent-Free Ring-Opening Polymerization of Lactones: Toward an All-Lactones, All-Conditions Approach, *Macromolecules*, 2017, **50**(22), 8948–8954.
75. B. A. Chan, S. Xuan, M. Horton and D. Zhang, 1,1,3,3-Tetramethylguanidine-Promoted Ring-Opening Polymerization of N-Butyl N-Carboxyanhydride Using Alcohol Initiators, *Macromolecules*, 2016, **49**(6), 2002–2012.
76. H. Lu and J. Cheng, Hexamethyldisilazane-mediated controlled polymerization of alpha-amino acid N-carboxyanhydrides, *J. Am. Chem. Soc.*, 2007, **129**(46), 14114–14115.
77. H. Lu and J. Cheng, N-Trimethylsilyl amines for controlled ring-opening polymerization of amino acid N-carboxyanhydrides and facile end group functionalization of polypeptides, *J. Am. Chem. Soc.*, 2008, **130**(38), 12562–12563.
78. R. Baumgartner, H. Fu, Z. Song, Y. Lin and J. Cheng, Cooperative polymerization of α-helices induced by macromolecular architecture, *Nat. Chem.*, 2017, **9**(7), 614.
79. T. Bai and J. Ling, NAM-TMS Mechanism of alpha-Amino Acid N-Carboxyanhydride Polymerization: A DFT Study, *J. Phys. Chem. A*, 2017, **121**(23), 4588–4593.
80. J. Yuan, Y. Sun, J. Wang and H. Lu, Phenyl Trimethylsilyl Sulfide-Mediated Controlled Ring-Opening Polymerization of alpha-Amino Acid N-Carboxyanhydrides, *Biomacromolecules*, 2016, **17**(3), 891–896.
81. H. R. Kricheldorf and K. Bösinger, Mechanismus der NCA-Polymerisation, 3. Über die Amin katalysierte Polymerisation von Sarkosin-NCA und -NTA, *Makromolekul. Chem.*, 1976, **177**(5), 1243–1258.
82. H. R. Kricheldorf, C. Von Lossow and G. Schwarz, Tertiary amine catalyzed polymerizations of α-amino acidN-carboxyanhydrides: The role of cyclization, *J. Polym. Sci., Part A: Polym. Chem.*, 2006, **44**(15), 4680–4695.
83. L. Guo and D. Zhang, Cyclic Poly(α-peptoid)s and Their Block Copolymers from N-Heterocyclic Carbene-Mediated Ring-Opening Polymerizations of N-Substituted N-Carboxylanhydrides, *J. Am. Chem. Soc.*, 2009, **131**(50), 18072–18074.
84. S. H. Lahasky, W. K. Serem, L. Guo, J. C. Garno and D. Zhang, Synthesis and Characterization of Cyclic Brush-Like Polymers by N-Heterocyclic Carbene-Mediated Zwitterionic Polymerization of N-Propargyl N-Carboxyanhydride and the Grafting-to Approach, *Macromolecules*, 2011, **44**(23), 9063–9074.
85. C.-U. Lee, A. Li, K. Ghale and D. Zhang, Crystallization and Melting Behaviors of Cyclic and Linear Polypeptoids with Alkyl Side Chains, *Macromolecules*, 2013, **46**(20), 8213–8223.

86. L. Guo, S. H. Lahasky, K. Ghale and D. Zhang, N-Heterocyclic Carbene-Mediated Zwitterionic Polymerization of N-Substituted N-Carboxyanhydrides toward Poly(α-peptoid)s: Kinetic, Mechanism, and Architectural Control, *J. Am. Chem. Soc.*, 2012, **134**(22), 9163–9171.
87. S. H. Lahasky, X. Hu and D. Zhang, Thermoresponsive Poly(α-peptoid)s: Tuning the Cloud Point Temperatures by Composition and Architecture, *ACS Macro Lett.*, 2012, **1**(5), 580–584.
88. C.-U. Lee, T. P. Smart, L. Guo, T. H. Epps and D. Zhang, Synthesis and Characterization of Amphiphilic Cyclic Diblock Copolypeptoids from N-Heterocyclic Carbene-Mediated Zwitterionic Polymerization of N-Substituted N-Carboxyanhydride, *Macromolecules*, 2011, **44**(24), 9574–9585.
89. A. Li, L. Lu, X. Li, L. He, C. Do, J. C. Garno and D. Zhang, Amidine-Mediated Zwitterionic Ring-Opening Polymerization of N-Alkyl N-Carboxyanhydride: Mechanism, Kinetics, and Architecture Elucidation, *Macromolecules*, 2016, **49**(4), 1163–1171.
90. L. Falivene, M. Al Ghamdi and L. Cavallo, Mechanistic Insights into the Organopolymerization of N-Methyl N-Carboxyanhydrides Mediated by N-Heterocyclic Carbenes, *Macromolecules*, 2016, **49**(20), 7777–7784.
91. L. Guo and D. Zhang, Synthesis and Characterization of Helix-Coil Block Copoly(α-peptoid)s. *Non-Conventional Functional Block Copolymers*, ACS Symposium Series, 2011, vol. 1066, pp. 71–79.
92. P. Du, A. Li, X. Li, Y. Zhang, C. Do, L. He, S. W. Rick, V. T. John, R. Kumar and D. Zhang, Aggregation of cyclic polypeptoids bearing zwitterionic end-groups with attractive dipole-dipole and solvophobic interactions: a study by small-angle neutron scattering and molecular dynamics simulation, *Phys. Chem. Chem. Phys.*, 2017, **19**(22), 14388–14400.

CHAPTER 10

Organocatalytic Ring-opening Polymerization Towards Poly(cyclopropane)s, Poly(lactame)s, Poly(aziridine)s, Poly(siloxane)s, Poly(carbosiloxane)s, Poly(phosphate)s, Poly(phosphonate)s, Poly(thiolactone)s, Poly(thionolactone)s and Poly(thiirane)s

THOMAS WOLF AND FREDERIK R. WURM*

Max Planck-Institut für Polymerforschung, Ackermannweg 10, 55128 Mainz, Germany
*Email: wurm@mpip-mainz.mpg.de

Polymer Chemistry Series No. 31
Organic Catalysis for Polymerisation
Edited by Andrew Dove, Haritz Sardon and Stefan Naumann

Published by the Royal Society of Chemistry, www.rsc.org

10.1 C–C Bond Forming Monomer Units *via* Metal-free Ring-opening Polymerization Poly(cyclopropane)s

The C–C bond is by far the most common backbone forming bond in polymer chemistry. It is the basis of multimillion ton scale materials like polyethylene, poly(methyl methacrylate), or poly(styrene), and many more. Typically, these polymers are synthesized from vinyl monomers or unsaturated olefins.

The use of ring-opening polymerization (ROP) to produce C–C bonds is much less common. Apart from prominent transition metal catalyzed systems like ROMP, the ring-opening of cyclopropanes provides an elegant way to produce C–C bonds. However, unlike the metathesis systems, this reaction is only being actively studied by a few groups worldwide.

The first two reports of the polymerization of vinyl cyclopropanes were published in 1949 by Volkenburgh *et al.* who, after successful synthesis of the compound, observed a polymerization in the presence of radical initiators (see Scheme 10.1a).[1] In the following years, many attempts have been

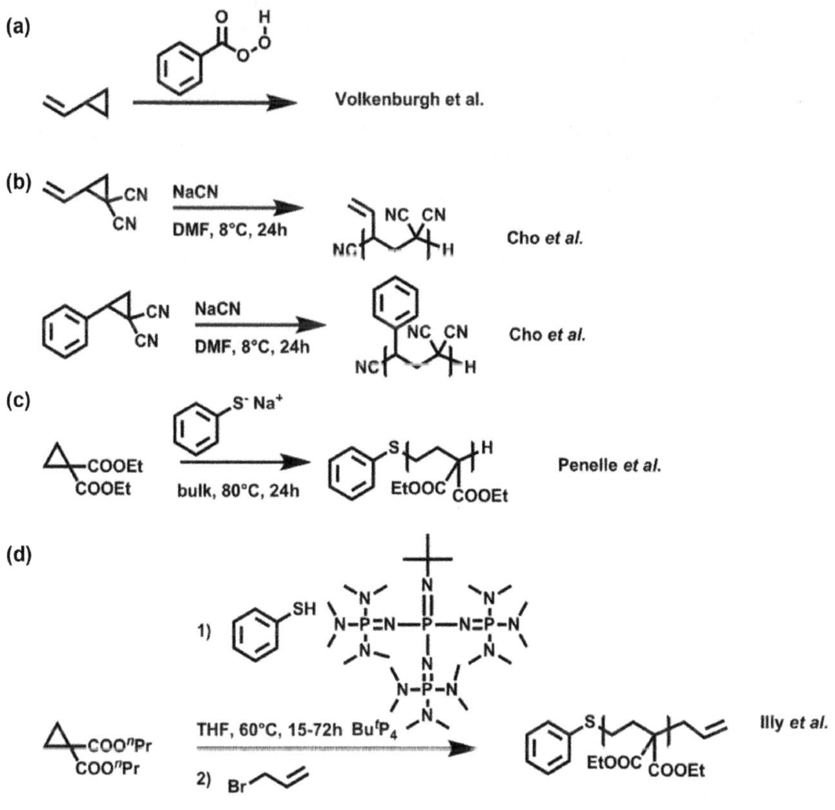

Scheme 10.1 Synthetic scheme of the ROP of some selected cyclopropanes.

made to polymerize vinyl cyclopropanes either *via* a radical, catalytic or cationic pathway. However, all of these studies presented low conversions and several side-reactions due to competition between the 1,2- and the 1,5-polymerization pathway, resulting in low control over molecular weight or polymer microstructure.[2,3]

These problems were avoided *via* a change to the anionic pathway: in 1979, Cho *et al.* presented the sodium cyanide initiated anionic ring-opening polymerization (AROP) of 1,1-disubstituted 2-vinylcycloporpanes as well as 2-substituted cyclopropane-1-1-dicarbonitirles under preservation of the double bond (Scheme 10.1b). They investigated in detail the effect of the substituents on the polymerizability and concluded that a push–pull system was needed for the polymerization. Furthermore, only soft nucleophiles like cyanide anion and thiols were able to initiate the polymerization.[4–6] While not considered as being "organocatalytic" these studies represent the first attempt to produce poly(cyclopropane)s *via* AROP.

In 1998, Penelle *et al.* were able to polymerize electron deficient 1,1-dicarboxylated cyclopropanes by initiation with thiolates (see Scheme 10.1c). They managed to optimize the process in a follow-up study in the year 2000. The reaction proceeded with excellent control ($1.06 < Đ < 1.13$) and up to full conversion in 16 h in DMSO at 140 °C and exhibited linear first-order kinetics, indicating a livingness of the polymerization.[7,8]

However, due to the use of alkali metal thiolates as initiators, these reactions cannot be considered purely organocatalytic. This changed, in the year 2009, when Illy *et al.* presented the first truly organocatalytic anionic ring-opening polymerization of di-*n*-propyl cyclopropane-1,1-dicarboxylate (Scheme 10.1d). Again, soft and highly nucleophilic thiols were used as initiating species; however, Illy *et al.* used the phosphazene base Bu^tP_4 to generate their initiating thiolate *in situ*, thus avoiding the use of metals.[9] They achieved near quantitative conversion for an initiator to monomer ratio of up to 1 : 50, within 40 h at 60 °C in THF. All polymers had narrow molecular weight distributions ($Đ < 1.15$) and the polymerization still proceeded in a living fashion. In addition to the respective kinetic studies, livingness was further demonstrated by end-capping experiments with allyl bromide, which worked nearly quantitatively. In the following years, Illy *et al.* further broadened the scope of their initiators towards phenols, carbazoles, and even malonates, and furthermore introduced allyl-pendant groups for future side-chain modifications.[10,11]

10.2 Nitrogen-containing Monomers

10.2.1 Polylactams

Lactams, cyclic amides, are important and well-studied precursors towards poly(amide)s.[12,13] The most relevant polylactam is poly(ε-caprolactam) (perlon), which was developed by Farben in 1938 and is produced on a multimillion ton scale per year today (Scheme 10.2).[14] Lactams typically

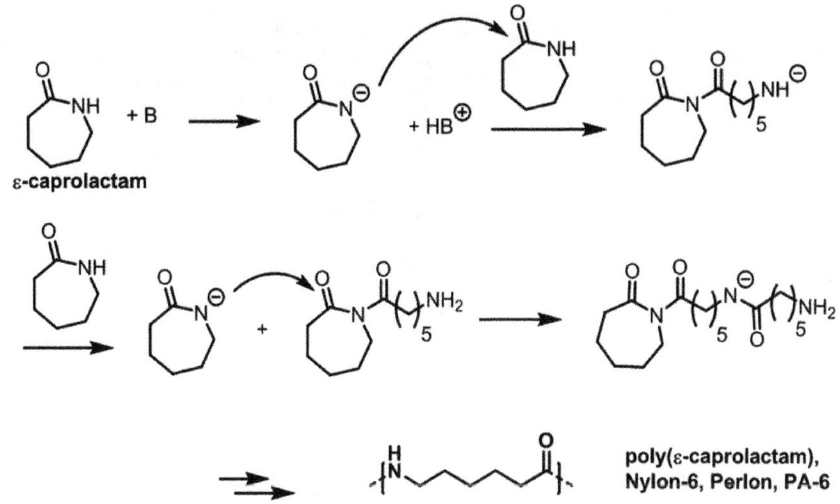

Scheme 10.2 Chemical structure of ε-caprolactam and poly(ε-caprolactam) (Perlon), the most industrially relevant poly(lactam).

polymerize *via* the activated monomer mechanism by the addition of a strong base. Usually, *N*-acylated lactam derivatives serve as an initiator for the polymerization and metals or metal alcoholates as the catalytic base.[15]

The first metal-free organocatalytic ROP of lactams was presented in the 1970s by Sekiguchi *et al.* They presented the synthesis of shelf-life stable quaternary ammonium salts of lactams and used them to catalyze the polymerization of α-pyrrolidone and α-piperidone. α-pyrrolidone was polymerized at 30 °C over a period of 24 h and produced, depending on the catalyst loading (ranging from 1 to 10%), polymers with reduced viscosity up to 5.00 and intrinsic viscosities up to 2.35. The polymerization of α-piperidone needed reaction times up to 120 h and slightly elevated temperatures (45 °C). Still, at high catalyst loadings (of 8%), conversions did not exceed 60%. However, with sufficiently low initiator concentrations, high molecular weight materials were synthesized (reduced viscosity up to 4.58).[16]

The first organic base to catalyze the polymerization of lactams was introduced in 1978 by Fiala *et al.* They presented the use of guanidine derivatives to polymerize α-pyrrolidone and ε-caprolactam. However, compared to the established alkali metal lactam salts initiated systems the polymerization kinetics were significantly slower (25% conversion after 24 h at 40 °C, 80% conversion after 24 h at 175 °C) and molecular weights did not exceed 2000 g mol^{-1} (determined by viscometry, Figure 10.1).[17]

Following the idea of using a storage stable neutral organic base for the polymerization, Memeger *et al.* were the first to use several phosphazene bases to catalyze the polymerization of lactams. After optimization of the reaction conditions, conversions up to 85% were achieved and polymers with molecular weights up to 50 000 g mol^{-1} were obtained at 270 °C

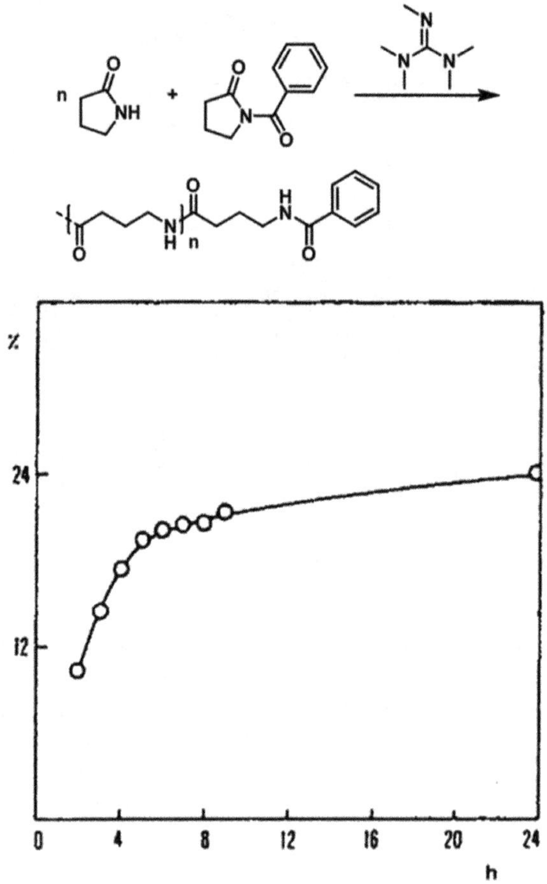

Figure 10.1 Top: schematic polymerization of α-pyrrolidone initiated with *N*-benzoyl pyrrolidone and catalyzed with penta methyl guanidine. Bottom: time-conversion plot of the polymerization of α-pyrrolidone presented by Fiala *et al.*
Reproduced from ref. 17 with permission from John Wiley and Sons, Copyright Wiley-VCH Verlag GmbH & Co. KGaA.

with P_4-^{t}Bu. Due to crystallization, polyamides have to be synthesized at high temperatures, *i.e.* above their melting point. However, with this approach molecular weight distributions between $1.4 < Đ < 3.2$ indicated significant side-reactions, most likely transamidation reactions, during the polymerization.[18]

In 2011, Yang *et al.* optimized the system presented by Memeger *et al.* towards the polymerization of β-lactams in the presence of phosphazene bases to produce nylon-3. The addition of LiCl to the polymerization solution (DMAc) was necessary to increase the solubility of the otherwise insoluble polymers. Under these conditions, at 50 °C, polymers with molecular weights up to 100 000 g mol^{-1} were obtained within 3 h (M_w, determined by static light scattering). The polymerization followed linear first-order kinetics until 50% conversion. Afterwards, transamidation reactions dominate the reaction and produce ill-defined, branched polymers, as indicated by a loss of first-order kinetics. Unfortunately, no SEC data was presented.[19]

In order to broaden the range of suitable catalysts and further provide thermally labile catalyst system, Naumann *et al.* introduced latent NHCs for the polymerization of ε-caprolactam in 2013 (Scheme 10.3).

At room temperature, the monomer and the latent NHCs coexist, without initiation of the polymerization. Increasing the reaction temperature to 180 °C initiated the polymerization by thermal decomposition of the latent NHC into the active NHC. Polymers with molecular weights up to $M_n = 420\,000$ g mol^{-1} were obtained within 45 min in the bulk. NHC to monomer ratios up to 1 : 300 were investigated and the polymer's molecular weight increased accordingly. The polymerization was delayed by a short induction period followed by rapid increase in melt viscosity. This delay was shortened by pre-incubation of the catalyst at 90 °C for several minutes. Finally, screening of several NHC catalysts enabled Naumann *et al.* to present a potential polymerization mechanism. They proved that the NHC functions as a strong base to deprotonate the lactam and induce the polymerization *via* the known active monomer mechanism.[20,21]

As a follow-up, Naumann *et al.* presented the NHC-catalyzed copolymerization of laurolactam with ε-caprolactam. For the homopolymerization of laurolactam, quantitative yield and molecular weights up to 14 000 g mol^{-1} were obtained in bulk at 180 °C if the utilized NHC had a p$K_a > 24$. The copolymerization with ε-caprolactam produced gradient polymers with ε-caprolactam being incorporated faster than laurolactam (Figure 10.2).[22]

In 2017, Chen *et al.* presented polymerization of substituted ε-caprolactam derivatives by the use of phosphazenes. They synthesized four caprolactam derivatives derived from *N,N′*-alkylated lysine and polymerized them in the presence of ^{t}BuP$_4$ and the *N*-acylated derivative of the respective monomer as an initiator (Figure 10.2). 98% monomer conversion was achieved in anisole at 140 °C within 6 h and polymers with molecular weights up to 12 000 g mol^{-1} were obtained. Molecular weight distribution in the range $1.44 < Đ < 3.31$ indicated transamidation reactions during the polymerization.[23]

Also organic acids have been used for the polymerization of lactams. Sanchez-Sanchez *et al.* studied the ROP of ε-CLa in bulk at 180 °C.[24] Among the evaluated organic acids, sulfonic acids were found to be the most effective for the polymerization of ε-CLa, being the Brønsted acid ionic

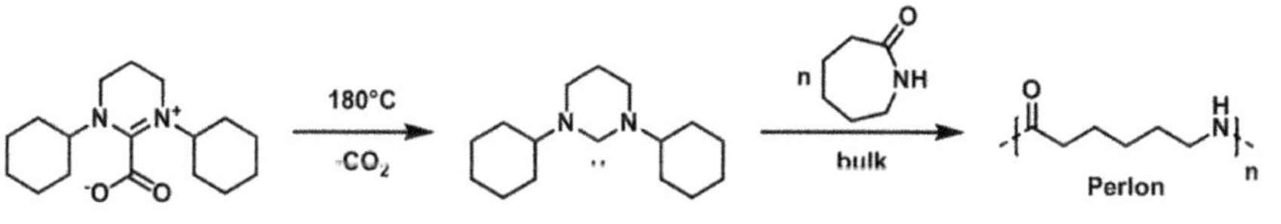

Scheme 10.3 Schematic polymerization of ε-caprolactam catalyzed by latent NHCs presented by Naumann *et al.*

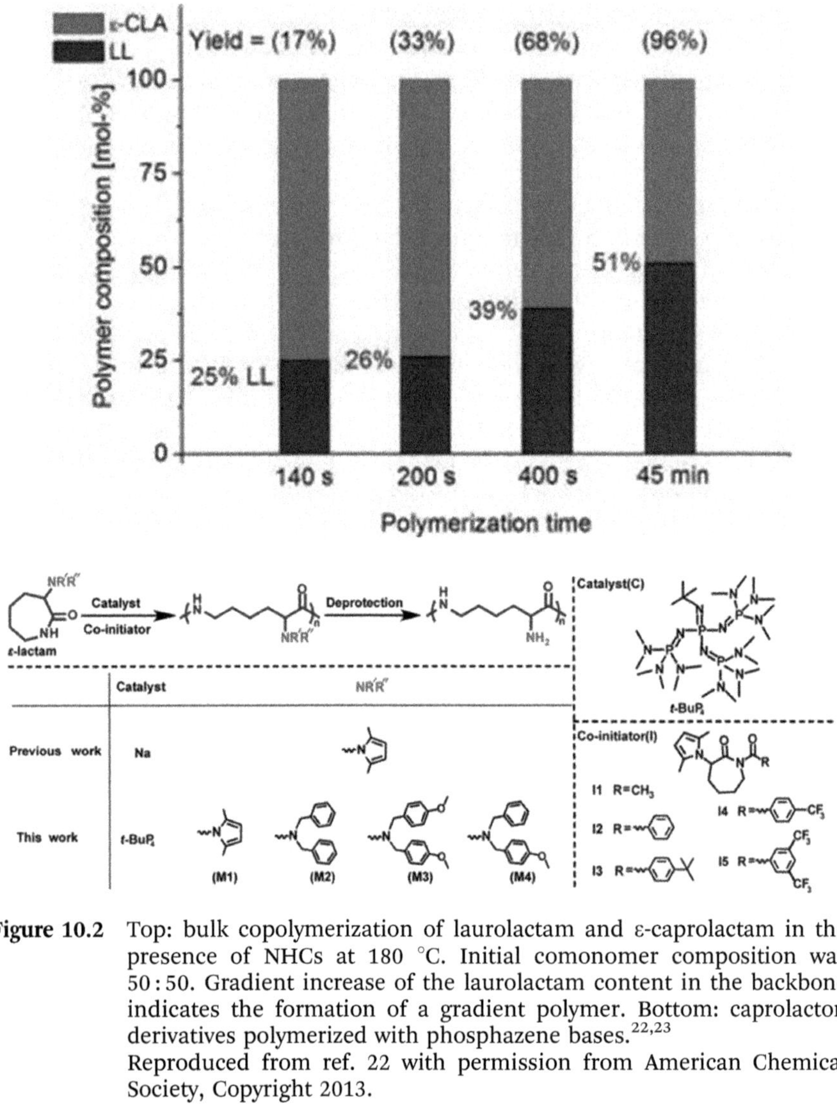

Figure 10.2 Top: bulk copolymerization of laurolactam and ε-caprolactam in the presence of NHCs at 180 °C. Initial comonomer composition was 50 : 50. Gradient increase of the laurolactam content in the backbone indicates the formation of a gradient polymer. Bottom: caprolacton derivatives polymerized with phosphazene bases.[22,23]
Reproduced from ref. 22 with permission from American Chemical Society, Copyright 2013.

liquid: 1-(4-sulfobutyl) – 3-methylimidazolium hydrogen sulfate was the most suitable due to its higher thermal stability. The authors suggested that the catalytic activity of sulfonic acids was due to their high acidity and the nucleophilic character of conjugate base as well. This strategy also allowed the copolymerization of lactams and lactones by organocatalysis to produce polyester-*co*-polyamides.[24,25]

To conclude, the organocatalytic ROP of lactams can be performed at high temperatures in the presence of strong bases like NHCs or phosphazene bases. As the polymers crystallize conduction and control of polymerizations to polyamides is difficult, under optimized conditions, polymers with molecular weights up to 400 000 g mol^{-1} can be obtained in under 1 h. In all presented cases to date, the molecular weight distributions of the produced polymers broadens up at high conversions due to transamidation reactions.

10.2.2 Poly(aziridine)s

Aziridines are the nitrogen analogs of epoxides and as such possess a high ring-strain (*ca.* 112 kJ mol^{-1} for ethylene imine), making them ideal candidates for ring-opening polymerizations.[26] The cationic polymerization of aziridine and 2-methyl aziridine has been well-known since the 1940s and produces ill-defined, hyperbranched poly(ethylene imine) (Scheme 10.4). The highly charged polymer is water-soluble and, despite its high toxicity, often used as a DNA complexation agent for transfection applications in gene therapy.[27]

The cationic polymerization of aziridines has been thoroughly investigated in the presence of various organic or inorganic acids, including carbon dioxide and boron trifluoride. Generally, the reaction proceeds vigorously at room temperature and somewhat controlled at −78 °C.[28] The characterization of the resulting hyperbranched polycation is generally challenging and much work has been done to unravel the composition of the final polymers.[29] CROP of aziridines leading to linear polyamines has not been reported to date.

Due to the acidic hydrogen of aziridines, an anionic polymerization of aziridines is not possible. The first controlled polymerization of aziridines leading to well-defined, narrowly distributed polymers was presented in 2005 by Stewart *et al.*. They substituted the acidic proton with a sulfonamide resulting in monomers suitable for a nucleophilic attack, following a living

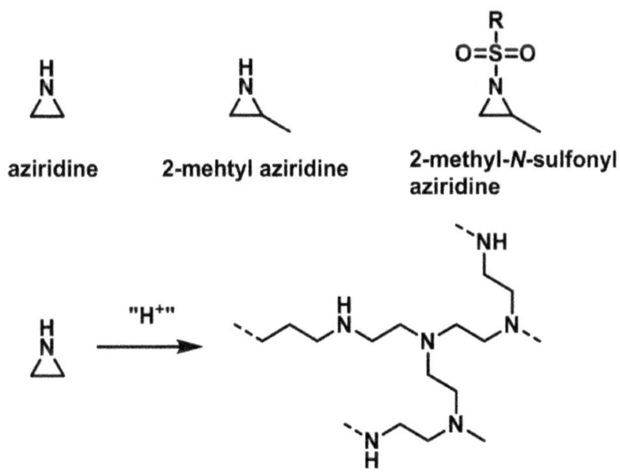

Scheme 10.4 Top: chemical structure of aziridine, 2-methyl aziridine, and *N*-sulfonyl 2-methyl aziridines. Bottom: schematic cationic ring-opening polymerization of aziridine into branched poly(ethylene imine).

anionic polymerization (Scheme 10.4). Introduction of an alkyl chain at the aziridine ring was necessary to prevent precipitation of the polymer during the polymerization, a problem observed when polymerizing *N*-sulfonyl aziridines. Using a *N*-alkyl sulfonamide as initiators and a strong base, potassium bis(trimethylsilyl)amide (KHMDS) at 45 °C in DMF, they produced narrowly distributed polymers ($Đ < 1.10$) with molecular weights up to 20 000 g mol^{-1}.[30] Following this initial work, Wurm and coworkers expanded the range of suitable monomers concerning the 2-alkyl side-chain and the *N*-sulfonyl activation group. Sophisticated ^{1}H NMR online kinetics analysis was performed to get a detailed insight into the polymerization mechanism and copolymerization properties of different 2-alkyl-*N*-sulfonyl aziridines (Figure 10.3). Furthermore, the copolymerization behavior was precisely altered from random to gradient copolymerization by performing the polymerization in a DMSO in cyclohexane emulsion, with the polymerization occurring in the discontinuous phase while the monomers are partitioned between the two phases with different partition coefficients.[31–36]

Reismann *et al.* investigated the polymerization behavior of non-alkylated *N*-sulfonyl aziridines with different activating groups under the same polymerization conditions as presented by Stewart *et al.* Generally, their reports matched that of Rieger *et al.* concerning the production of narrowly distributed polymers ($Đ < 1.10$) with molecular weights up to 20 000 g mol^{-1}. However, the precipitation of low P_n oligomers (up to 10) of poly(*N*-sulfonyl aziridine) was observed. Increasing the steric demands of the activating group from methylsulfonyl- to *sec*-butylsulfonyl only increased the maximum P_n up to 25. A 1 : 1 mixture of methyl- and *sec*-butylsulfonyl activated *N*-sulfonyl aziridine, however, stayed soluble during the course of the polymerization and P_n up to 200 were obtained. Furthermore, chain extension of the polymerization with 2-methyl-*N*-methylsulfonyl aziridine was proven to be possible.[37]

The first organocatalytic ROP of 2-alkyl-*N*-sulfonyl aziridines was presented by Bakkali-Hassani *et al.* in 2016. A sterically hindered N-heterocyclic carbene was used to catalyze the polymerization of 2-alkyl-*N*-sulfonamide-aziridines, initiated from an *N*-alkyl sulfonamide. Polymers with a narrow molecular weight distribution ($1.04 < Đ < 1.15$) and molecular weights up to 21 000 g mol^{-1} were obtained.

Depending on the ring-substituent of the aziridine monomer, reaction times between 1 and 5 days were necessary to reach full conversions at 50 °C in THF.[38] The mechanism of the polymerization, with the NHC potentially functioning either as nucleophile or base, was investigated. The attachment of the initiator sulfonamide on the polymer was proven by MALDI-ToF MS analysis, indicating a basic mechanism of the NHC catalyzed reaction (Figure 10.4). Finally, chain extension of preformed polymers was possible, indicating the livingness of the polymerization.[38]

In a follow-up work in 2017, Bakkali-Hassani *et al.* expanded the scope of NHC catalyzed ROP of aziridines. Utilizing the same sterically hindered NHC, they presented the polymerization of aziridines initiated with a

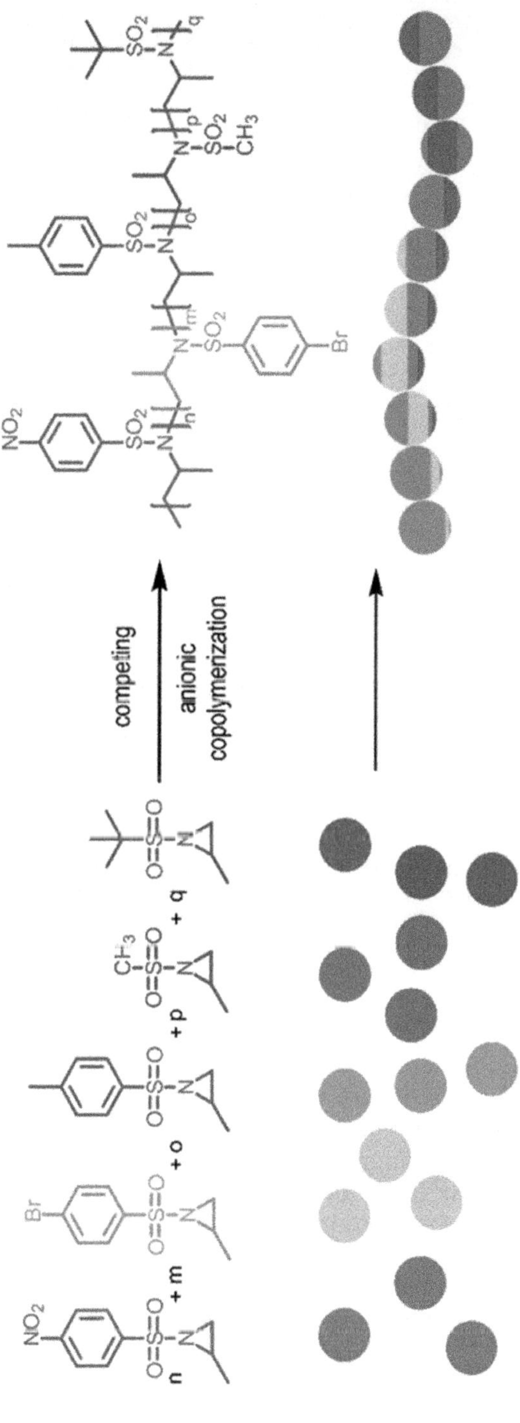

Figure 10.3 Competing anionic copolymerization of aziridines resulting in multi-gradient copolymers as presented by Rieger *et al.* Adapted from ref. 33 with permission from John Wiley and Sons, © 2016 WILEY-VCH Verlag GmbH & Co. KGaA, Weinheim.

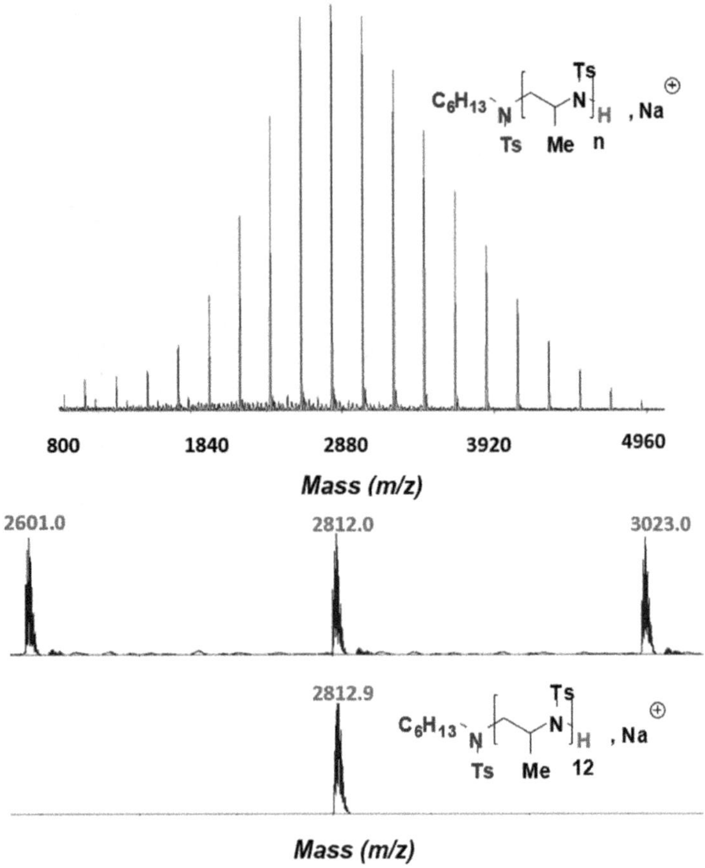

Figure 10.4 MALDI-ToF mass spectrum of poly(2-methyl- *N*-toluenesulfonyl aziridine) polymerized by the NHC catalyzed organocatalytic ROP presented by Bakkali-Hassani *et al.*
Reproduced from ref. 38 with permission from the Royal Society of Chemistry.

functional *N*-alkyl sulfonamide, a non-activated secondary amine and initiation form an azide-anion (Figure 10.5, left). In all cases, conversions up to 100%, molecular weights up to 22 000 g mol^{-1} and molecular weight distributions below 1.2 were obtained within 1–5 d at 50 °C in THF (Figure 10.5, right). Initiation of the polymerization form the respective initiator and not the NHC was again shown by MALDI-ToF MS analysis, as was one exemplary post-polymerization modification per functional group.[39]

Following these first organocatalytic ROP of substituted aziridines, Wang *et al.* presented the first use of non-NHC organocatalysts for the ROP of aziridines in 2017.[40] They screened two phosphazene bases and the organobases DBU, MTBD, and TMG for their potential to catalyze the sulfonamide-initiated polymerization of 2-substituted-*N*-sulfonyl-aziridines

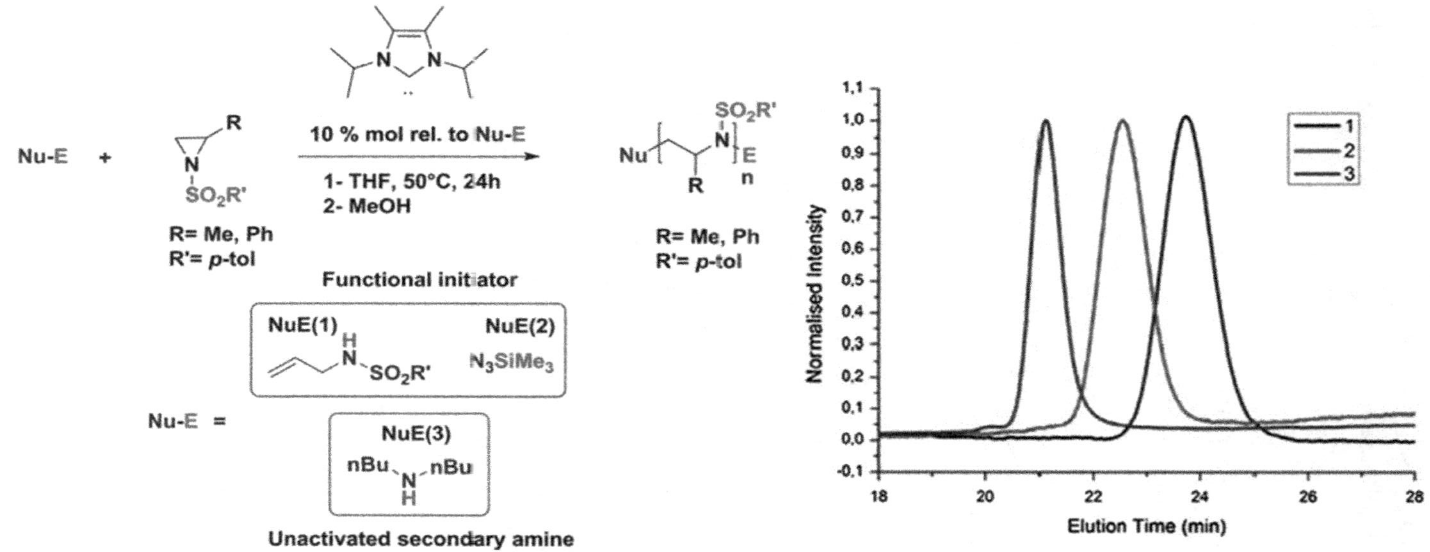

Figure 10.5 Left: schematic presentation of the organocatalytic ROP of 2-substituted-*N*-sulfonyl aziridines catalyzed by NHCs and initiated with varying initiators. Right: SEC traces of the resulting poly(aziridine)s presented by Bakkali-Hassani *et al.* Reproduced from ref. 39 with permission from Elsevier, Copyright 2017.

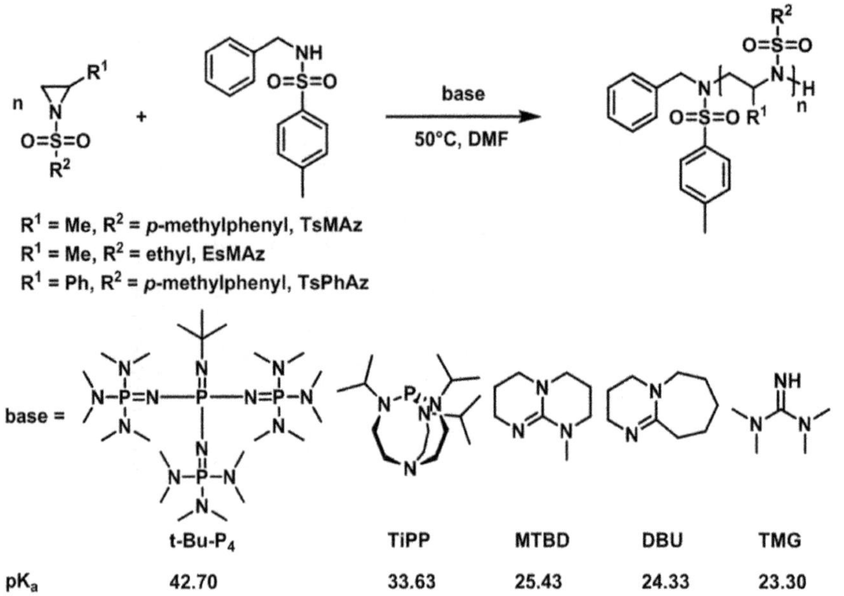

Figure 10.6 Anionic ring-opening polymerization of 2-alkyl-*N*-sulfonyl aziridines catalyzed with various organic bases as presented by Wang *et al.*

(Figure 10.6). All catalysts investigated were able to initiate the polymerization and molecular weight distributions below 1.4 were obtained. They correlated the catalytic activity of their bases with the respective pK_a values. Accordingly, the best results (regarding reaction time, conversion, and dispersity) were obtained with the strongest phosphazene bases (*e.g.* 20 min, 99% conversion, $Đ=1.05$), while the "weaker" bases DBU and TMG needed slightly longer reaction times and produced polymers with broader molecular weight distributions (DBU, 25 min, 96% conversion, $Đ=1.26$). Furthermore, the phosphazene bases were also active in catalytic concentrations. More detailed kinetics analysis of the phosphazene catalyzed reaction proved the livingness of the polymerization for all three tested monomers and at different base concentrations.

To conclude, the controlled polymerization of aziridines is possible by substitution of the acidic NH proton with an electron withdrawing sulfonamide group. This prevents interference of the proton during the polymerization and furthermore activates the aziridine ring to enable efficient polymerization. While initial investigations focused on the use of strong, metal-containing bases, novel studies focus on the use of superbases like NHCs, phosphazene bases or MTBD. Under either condition, however, the polymerization of activated aziridines proceeds in a living manner with high control over molecular weight, molecular weight distributions below 1.2 and high end-group fidelity.

10.2.3 Polyurethanes

Polyurethanes (PUs) are important polymers in many applications and are typically prepared by polyaddition of isocyanates and polyols. Several isocyanate-free routes to PUs have been reported.[41] Höcker and coworkers reported the cationic ring-opening polymerization of trimethylene urethane (tetrahydro-2H-1,3-oxazin-2-one) in the melt at 100 °C with methyl trifluoromethanesulfonate (TfOMe), trifluoromethanesulfonic acid (TfOH), and $BF_3{\cdot}OEt_2$ as initiators to produce poly(trimethylene urethane).

10.3 Silicon-containing Monomers

10.3.1 Poly(cyclosiloxane)s

Poly(siloxane)s are without a doubt the most well-investigated and industrially relevant class of inorganic/organic hybrid polymers with an annual production of more than 400 000 t. Their commercial success is based on their low thermal conductivity, low chemical reactivity, low toxicity, thermal stability, low glass transition, and electrical insulation, giving poly(siloxane)s an exceptionally broad range of applications. Many review articles on the properties and synthesis of poly(siloxane)s have been published in the last few decades.[42–44] Industrial synthesis of poly(siloxanes) usually proceeds *via* the condensation of dialkyl dichlorosilanes (Scheme 10.5), however, the ring-opening polymerization of cyclic siloxanes (D_3 or D_4 in most cases) is also a prominent industrial and academic way to produce poly(siloxane)s (Scheme 10.5).

D_3 and D_4 (nomenclature indicating the number of ($-Si(CH_3)_2O-$) repetition units in the cycle) can be polymerized *via* the ring-opening polymerization process. Due to the higher ring-strain, D_3 usually polymerizes in a more controlled manner with D_4 resulting in more side-reactions, *e.g.* re-cyclization, depolymerization, and condensation reactions.

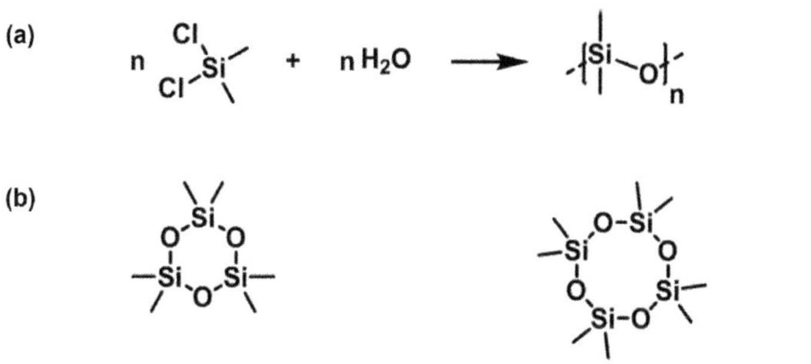

Hexamethylcyclotrisiloxane D3 Octamethylcyclotetrasiloxane D4

Scheme 10.5 (a) Industrially relevant polycondensation of dimethyl dichlorosilane with water to produce poly(dimethyl siloxane) (PDMS) and (b) structures of the cyclic monomers for the ROP towards PDMS, D3 and D4.

The ROP of cyclic siloxanes has been known since the 1950s and has focused on the monomers D_3 and D_4. Early kinetic investigations by Grubb *et al.* have already showed that cyclic and polymeric (linear) siloxanes exist in a polymerization–depolymerization equilibrium.[45] The polymerization, typically in bulk, produces a mixture of high molecular weight polymer and low molecular weight cyclic siloxanes due to the formation of said equilibrium. Furthermore, in bulk, the reaction environment becomes more and more hydrophobic as the polymerization proceeds, leading to the aggregation of the polar chain ends and drastic reduction of the reaction rate.

The ROP can be catalyzed by acids or bases. Chojnowski and coworkers studied organic acid catalyzed ROP in the 1970s and 1980s.[46,47] The reaction of D_3 catalyzed by sulfonic acids occurs mostly by addition of monomer to the ends of growing molecular chains. However, an important role is played by cleavage of a siloxane bond in the monomer by acid or water as well as by reverse processes of silanol and/or +ylester end group condensation to siloxane linkages. The latter account for the formation of large amounts of cyclic products and for the coupling of linear chain fragments.

Originally, the polymerization of D_4 was performed in the presence of low amounts of KOH (typically 0.01%) at high temperatures (>140 °C) within hours. This produced ill-defined polymer mixtures consisting of (partially cyclic) low molecular weight volatiles and polymers with M_n as high as 1×10^6 g mol^{-1} within several hours. Furthermore, the catalyst needs to be removed completely. Otherwise, decomposition reactions will be catalyzed at elevated temperatures (~250 °C) resulting in rapid degradation of the polymer. Completely catalyst-free polysiloxanes, on the other hand, are thermally stable up to ~500 °C.

In order to overcome the difficult removal of KOH from the final polymer gum, Gilbert *et al.* applied quaternary ammonium and phosphonium hydroxides for the ROP of D_4. The biggest benefits of these catalysts compared to KOH are a higher reaction rate even at a lower temperature at the same catalyst loading (0.01%, 110 °C, 5 min), a lower possible catalyst loading (as low as 0.001%) and efficient thermal decomposition of the catalyst at elevated temperatures.[48] They showed the decomposition of their catalysts at 130 °C. The final, catalyst-free polymer remained essentially stable at 250 °C until the end of the experiment after 24 h. The KOH catalyzed polymer showed 90% weight loss after 14 h under the same conditions (Figure 10.7).[49]

Molenberg *et al.* were the first to use phosphazene bases (Scheme 10.6) for the oAROP of D_4 in 1995 and accordingly presented the first truly organocatalytic ROP of cyclic siloxanes. Addition of a methanol phosphazene mixture in toluene to the monomer (bulk) leads to an immediate increase in viscosity. Equilibration of the polymerization was typically reached within 1 min at room temperature. Molecular weights as high as 440 000 g mol^{-1} were obtained with molecular weight dispersities between 1.7 and 1.9. However, the formation of small (D_3 and D_4) and large (up to D12) cycles could not be prevented, resulting in an overall ill-defined mixture.[50]

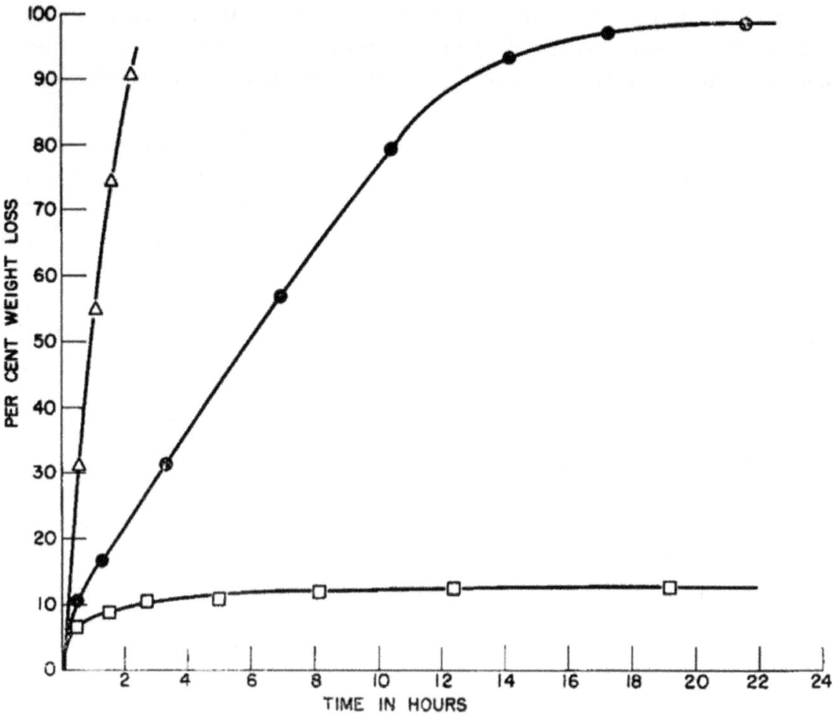

Figure 10.7 Weight loss of PDMS samples at 250 °C prepared with different catalyst systems as presented in the work of Gilbert *et al.*
Copyright © 1959 Wiley Online Library. Adapted from ref. 49 with permission from John Wiley and Sons, Copyright © 1959 Interscience Publishers, Inc., New York.

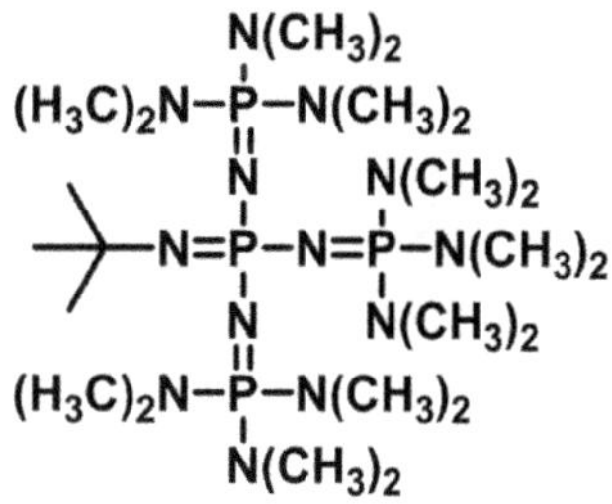

Scheme 10.6 Structure of the phosphazene base used by Molenberg *et al.* for the polymerization of D_4.

In a follow-up work, Molenberg *et al.* used the same phosphazene base as a promoter for the polymerization of D_3. They first produced non-propagating lithium silanolates by addition of *sec*-BuLi to the monomer and initiated the polymerization by adding the phosphazene base at room temperature. Kinetic investigations showed that the polymerization D_3 followed linear first-order kinetic (with narrow molecular weight distributions,

Đ<1.1) until about 70% conversion (Figure 10.8). Afterwards, a deviation from linearity and broadening of the molecular weight distribution is observed due to transetherification reactions. Molecular weights up to 500 000 g mol^{-1} were obtained.[51]

In 1999, Taylor *et al.* performed a large-scale synthesis of PDMS starting from D_4 in bulk with a phosphazene base. They were able to polymerize 1 kg of D_4 at 100 °C within 1 min of reaction after addition of the base. Due to the high reaction speed, the kinetics of the polymerization was highly dependent on process parameters like stirring or moisture residue in the monomer. The degree of polymerization was dependent on the reaction time and the amount of residual water present in the system. Polymers with high molecular weights (up to 4×10^6 g mol^{-1}) and molecular weight distributions around 1.9 were obtained. They concluded that the ROP produces silanol terminated PDMS that slowly condenses over time, resulting in the formation of high molecular weight polymer with broad molecular weight distributions.[52]

Based on the phosphonium hydroxide catalyzed ROP presented by Gilbert *et al.*, Bessmertnykh *et al.* investigated the phosphorus ylide-catalyzed polymerization of D_4 initiated with an alcohol. They substituted the high-cost phosphazene bases with their strongly basic, non-nucleophilic phosphorous ylides.

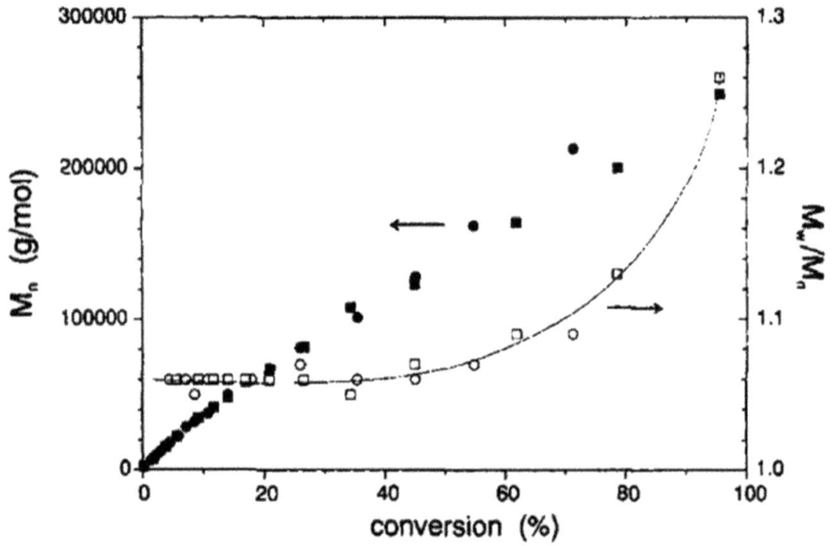

Figure 10.8 Number-average molecular weight (black squares and circles) and molecular weight distribution (hollow squares and circles) *vs.* monomer conversion for the polymerization of D_3 at room temperature with different concentration of base (2.14×10^{-3} mol L^{-1}, squares; 0.57×10^{-3} mol L^{-1}, circles).

The polymerization was considerably slower than under phosphonium hydroxide catalysis, giving only 36% yield and a polymer of around 10 000 g mol^{-1} after 10 h of polymerization initiated with methanol (D_4, room temperature). Initiation with *tert*-butanol, however, proceeded much faster, with 87% conversion within one hour, however, no explanation for this finding is given. The molecular weight distributions of the *tert*-butanol initiated polymers remained rather narrow in all cases ($Đ < 1.5$), indicating a low tendency towards *trans*etherification reactions and condensation of the final polymer.[53]

In 2006, Lohmeijer *et al.* deviated from the use of phosphazene bases and presented the polymerization of D_3 in the presence of either the guanidine base TBD or an N-heterocyclic carbene. In the presence of an alcohol as the initiator, TBD effectively polymerized D_3 within 5 min and produced well-defined polymers with molecular weight distributions of $Đ < 1.2$. The polymerization with the NHC catalysts proceeded with less control and polymers with molecular weight distributions above 1.4 were obtained; however, no detailed information on either polymer was given.[54]

Following these first NHC catalyzed D_3 polymerizations, Rodriguez *et al.* investigated the polymerizability of D_4 with different NHCs in 2007 (Scheme 10.7). They report high control over the molecular weight and a linear dependency with the amount of primary alcohol used as initiators. Molecular weights up to 200 000 g mol^{-1} were obtained and molecular weight distributions were between 1.5 and 1.7. The monomer conversion reached a plateau at around 85–90% conversion, even for reaction times of more than 24 h.[55]

In analogy to early experiments from Gilbert *et al.*, they used heat to inactivate their catalyst after the polymerization. Heating the reaction mixture to 150 °C for 5h completely deactivated their NHC. The rate of the polymerization was found to be mainly dependent on the amount of NHC in the system and not on the initiator concentration. They ascribed this effect on

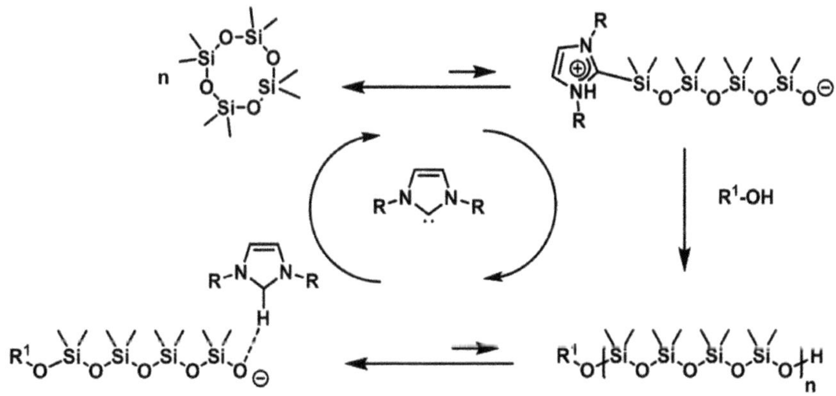

Scheme 10.7 Zwitterionic polymerization of D_4 with NHC catalysis.

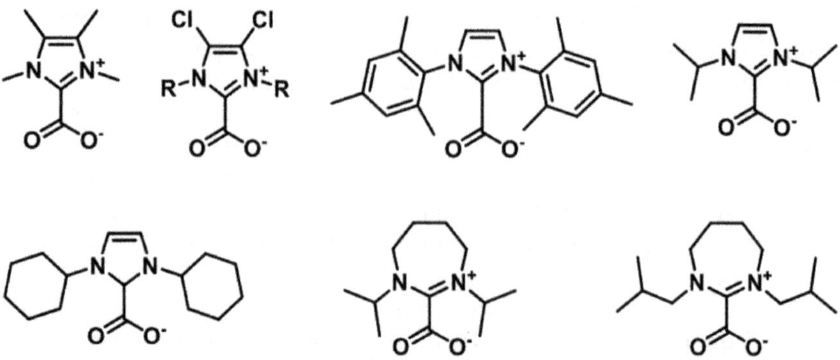

Scheme 10.8 Chemical structure of the latent NHC catalysts screened for the polymerization of D_4 by Naumann *et al.*

the proposed zwitterionic mechanism of the polymerization where the rate-determining step is the initial nucleophilic attack of the NHC on D_4.[55]

Consequently, in 2015, Naumann *et al.* expanded the scope of usable NHCs and latent NHCs for the polymerization of D_4 (Scheme 10.8). They screened several NHCs and latent NHCs for their catalytic activity towards the ROP of D_4. The fast and nearly quantitative conversion was found for sterically less hindered NHCs whereas the more hindered NHCs did not induce polymerization. Molecular weight distributions were between 1.7 and 2.2 in all cases. They also presented the initiator-free polymerization of D_4 only in the presence of an NHC. Taking these two facts into account, they concluded that the polymerization of D_4 *via* NHCs is dependent on the initial ring opening of D_4 *via* nucleophilic attack of the NHC catalyst. Furthermore, they showed the thermal decomposition of latent NHCs to induce a delayed polymerization only after surpassing a certain temperature.[56]

To conclude, the anionic ROP of cyclic siloxanes still presents a challenging reaction due to the formation of a propagation–depropagation equilibrium. Relatively well-defined polymers with high molecular weights (up to 200 000 $g\,mol^{-1}$) are obtained only by the most reactive systems like phosphazene bases, TBD and NHCs.

10.3.2 Poly(cyclocarbosiloxane)s

Like poly(siloxane)s, poly(carbosiloxane)s are a class of silicon-containing inorganic/organic hybrid polymers. They differ from poly(siloxane)s by their additional Si–C and C–C bonds in every repetition unit and are typically synthesized by ROP of the five-membered 1,2,5-oxadisilolane derivatives (cyclocarbosiloxanes, Scheme 10.9). While not relevant in the industry, as opposed to siloxanes, carbosiloxanes have been known in the literature since the 1960s.

One of the first publications for the ROP of cyclic carbosiloxanes mainly concerns the synthesis of the oxadisilolane monomers in 1960

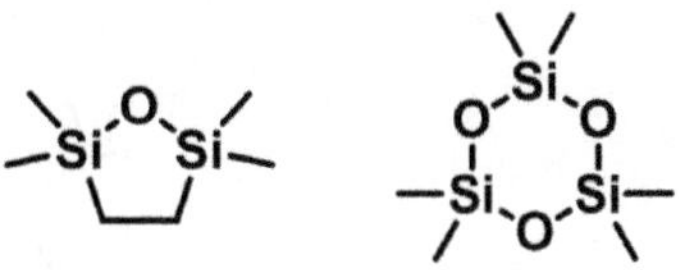

Scheme 10.9 Structure of 2,2,5,5-tetramethyl-1,2,5-oxadisilolane (left) a typical monomer for the ROP towards poly(carbosiloxane)s and a typical cyclic siloxane monomer (hexamethyl cyclotrisiloxan D3, right) for the ROP towards poly(dialkyl siloxane)s.

by Picolli *et al.* The polymerization is only briefly stated and not discussed in detail. However, Piccoli *et al.* had already compared the polymerizability of different ring sizes and mentioned that they obtained high M_n material (up to 870 000 g mol^{-1}, determined from light scattering data) for bulk polymerization at 120–130 °C of the five-membered ring (2,2,5,5-tetramethyl-1,2,5-oxadisilolane) within 20 min (Scheme 10.9). The six- and eight-membered rings polymerized slower than the five-membered ring, and took about 19 h for polymerization, but no information apart from the polymer precipitating during the reaction was given at that time.[57]

Their work was continued by Suryanarayanan *et al.* in 1997, when they investigated the polymerization of 2,2,5,5-tetramethyl-1,2,5-oxadisilolane in the presence of lithium silanolates in THF. They present a very detailed investigation of the polymerization behavior in regard to variation of the initiator to monomer ratio, THF concentration, water concentration and temperature. They report a linear dependency of the monomer:initiator ratio on the molecular weight and a strong effect of the THF concentration on the reaction rate. Monomodal molecular weight distributions were obtained for low initiator-to-water ratios. When higher amounts of water were present, bimodal distributions were reported, indicating water-initiation during the polymerization. Molecular weights up to 36 000 g mol^{-1} (SEC, THF) were obtained in the best cases.[58,59]

In 1999, Loy *et al.* presented the first organocatalytic ROP of this monomer class. They used tetrabutylammonium hydroxide to initiate the gelation of a bifunctional cyclocarbosiloxanes (Scheme 10.10). Due to the insolubility of these gels, however, no decent polymer characterization apart from the thermogravimetric analysis was presented. Their work was motivated by the low shrinkage during the polymerization.[60]

In 2005, this process was continued by Rahimian *et al.* They presented the bulk copolymerization of mono-and bifunctional cyclocarbosiloxanes catalyzed with either TBAH, formic acid or a photoacid. The solvent-free process proceeded within 30 min to gelation in all cases. Due to the non-soluble nature of the cross-linked polymers, again no detailed analytical data are presented for the polymerization.[61]

Lohmeijer *et al.* broadened the scope of usable organocatalysts and performed detailed analytical studies on the ROP of cyclocarbosiloxanes in 2006. They investigated the effect of MTBD, TBD and two NHCs on the

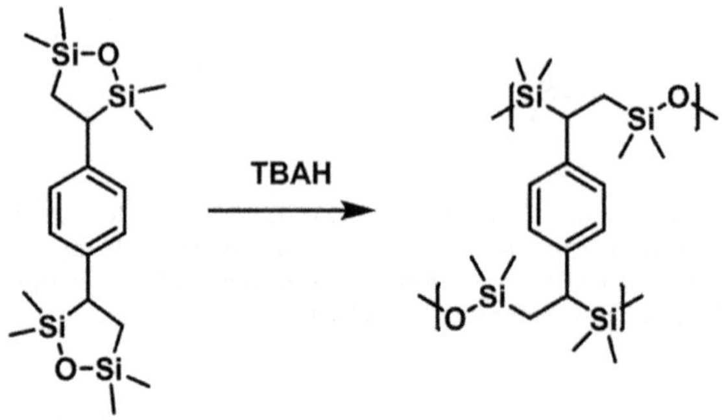

Scheme 10.10 Ring-opening polymerization of bifunctional cyclocarbosiloxanes for gel-formation.

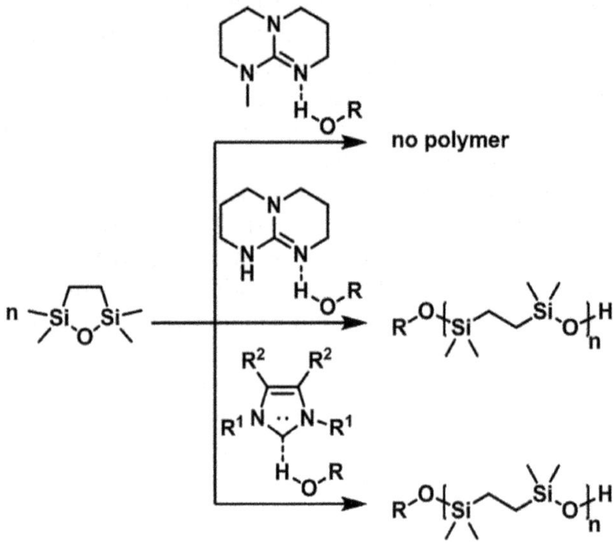

Scheme 10.11 Schematic presentation of the polymerization of 2,2,5,5-tetramethyl-1,2,5-oxadisilolane by either MTBD, TBD, or various NHCs.

polymerization of cyclocarbosiloxanes in the presence of an initiator (Scheme 10.11). MTBD was not able to catalyze the polymerization, whereas TBD catalysis produced narrowly distributed polymers ($Đ < 1.1$). For TBD, the reaction rate and final conversion were found to be dependent on the initial monomer to initiator ratio.[54]

For an initial ratio of 50, *e.g.*, 94% conversion was reached after 3 h, whereas for a ratio of 400, only 76% conversion was reached after 26 h. Nonetheless, molecular weight distributions stayed below 1.1. A linear first-order kinetics was observed, indicating the livingness of the polymerization. End group fidelity was demonstrated by the complete incorporation of a UV active pyrene-containing initiator. Different initiating groups were analyzed;

however, only hydroxyl groups initiated the polymerization. The use of amino or thiol initiators produced no polymeric materials at all. The investigated NHCs were also able to polymerize cyclocarbosiloxanes. The polymerization rate was drastically higher than for TBD but the polymerization control suffered, as evidenced by a slightly broader molecular weight distribution ($1.14 < Đ < 1.19$). The authors attributed this effect to the higher nucleophilicity of the NHCs compared to TBD, which in turn promotes transetherification reactions.[54]

Brown *et al.* further investigated the NHC-catalyzed polymerization of cyclocarbosiloxanes in 2013. They screened a variety of different NHCs in the absence of additional initiator for their ability to form cyclic polymers. Sterically non-hindered NHCs were able to initiate the polymerization. Until around 50% conversion, the polymerization followed a chain growth mechanism, after which the molecular weight distribution broadened up as a result of NHC catalyzed transetherification reactions.

Sterically hindered NHCs did not produce any polymer under these conditions due to the inability for a nucleophilic attack on the monomer. Reaction rates depended on the nucleophilicity of the NHC, with the most reactive systems reaching 95% conversion and molecular weights above 500 000 g mol^{-1} within 5 min at room temperature in THF.[62]

To conclude, cyclic carbosiloxanes readily polymerize under alcohol initiation in the presence of strong guanidine bases like TBD. Narrow molecular weight distributions and linear first-order kinetics are observed, enabling good control over molecular weight and end-groups. The NHC catalyzed reaction either provides linear (in the presence of alcohol) or cyclic structures (no initiator present) but tends towards transetherification reactions. The TBAH catalyzed reactions provides an efficient and fast way towards poly(carbosiloxane)s, however, not much analytical data concerning the polymerization mechanisms are present today and leave a lot of room for future investigations.

10.4 Phosphorus-containing Monomers

10.4.1 Poly(phosphoric acid ester)s, Polyphosphates

Phosphorus compounds are omnipresent in nature, either bound in inorganic salts or organic compounds. Without phosphoric acid esters, in particular, life as we know it would not be possible.[63] The most fundamental requisites for life are all governed by or with the help of phosphoric acid derivatives:

The universal energy storage system preserved in all living organisms, ATP, is based on the potential energy stored in phosphoric acid anhydride bonds, *i.e.* pyrophosphates (Scheme 10.12a). Cleavage of this bond provides energy for active biochemical synthesis, active transport through cells, nerve transport, growth mechanisms, and movement.[63,64]

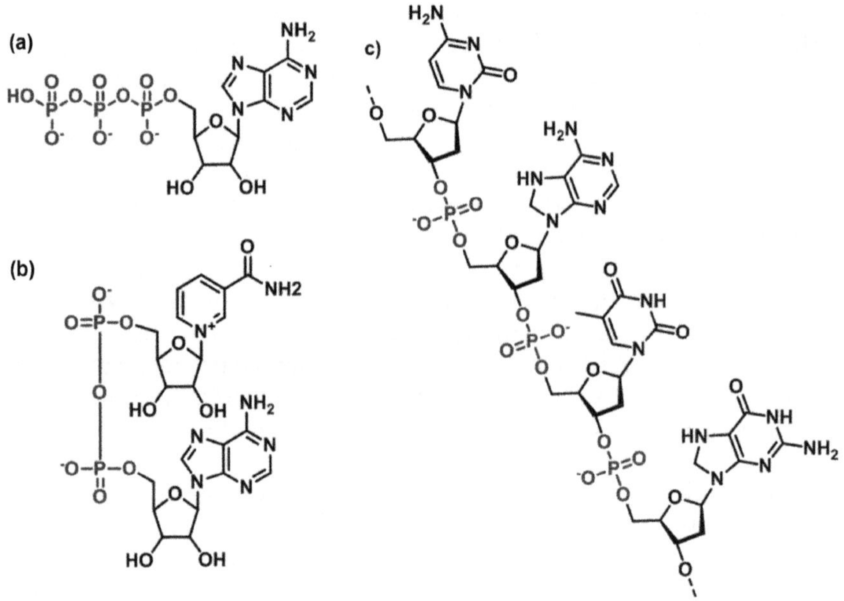

Scheme 10.12 Structures of the most important naturally occurring phosphoesters: (a) universal energy "currency" ATP; (b) redox-responsive system NAD^+; (c) DNA, the only naturally occurring PPE.

Apart from energy storage, the activity of many enzymes is tightly regulated by active phosphorylation or dephosphorylation of key amino acids. In many, but not all, of these cases, ATP serves as a direct phosphorylation agent and simultaneously provides the necessary energy *via* anhydride cleavage.[63,64]

Furthermore, a large part of biological redox reactions depends on the redox potential of the NAD^+/NADH coenzyme (Scheme 10.12b, NAD^+ form). While not being involved in the redox reaction, phosphoric acid esters serve to enhance solubility and link two essential functional parts together in one molecule.[65]

Finally, probably the most important phosphoric acid esters are DNA (Scheme 10.12c) and RNA, respectively. The carriers of the genetic information are sequence defined polycondensates of phosphoric acid and substituted (deoxy-)ribose derivatives. Here, the polyphosphoesters (PPE) perform several tasks: they serve as the backbone to retain the sequence definition and provide the necessary stability towards hydrolysis. Furthermore, the backbone can be cleaved on demand under enzymatic catalysis to repair *e.g.* defects.[66] The negative charge distributed on the backbone prevents passive and undesired diffusion through membranes, enables ionic complexation with components of the transcription apparatus necessary to "read" the genetic information, some enzymes of the DNA replication tools, and histones, positively charged proteins necessary for DNA storage and protein expression.[63]

Due to this omnipresence and significance in nature, chemists started to work on the development of synthetic analogs of these polymers. Their potential biodegradability, biocompatibility, and high structural diversity make them interesting especially in the field of biomedical applications. In comparison to poly(carboxylic acid ester)s, PPEs contain the pentavalent phosphorus atom, increasing the structural diversity by the addition of an additional side-chain. Different binding patterns can be incorporated into the backbone or side-chain (P–C, P–N, *etc.*), further broadening the property bandwidth of these materials (Scheme 10.13).

Today, several reviews concerning the different synthetic routes towards and applications of PPEs are available.[42,67–71] This section will give a short historical background regarding the synthesis of PPEs, but the main focus will lie on the modern organocatalytic ring-opening polymerization procedure.

Phosphates are the most abundant phosphorus derivatives in nature. Also in synthetic polymers, poly(phosphate)s are by far the best-studied sub-class of all PPEs. While the first PPEs were synthesized *via* polycondensation reactions, the pioneering work on the ring-opening polymerization towards poly(phosphate)s was conducted by Penczek *et al.* in the 1970s.[72] They focused their initial investigations on the cationic polymerization of 2-alkoxy-2-oxo-1,3,2-dioxaphosphorinanes, six-membered cyclic phosphates, in order to synthesize synthetic oligonucleotides *via* a controlled route (Scheme 10.14).

ROP was carried out in the presence of triphenyl methyl salts and was prone to transesterification processes, thus resulting in formation oligomers.[72] This work was complemented in 1977 by the detailed kinetic investigation of differently substituted monomers. The same anionic initiators as in the previous studies were used with a strong tendency towards chain-transfer reactions.[73] The anionic polymerization of these monomers initiated with sodium and potassium alkoxides, published in the same year, produced only low molecular weight compounds of around 1000 g mol^{-1}.

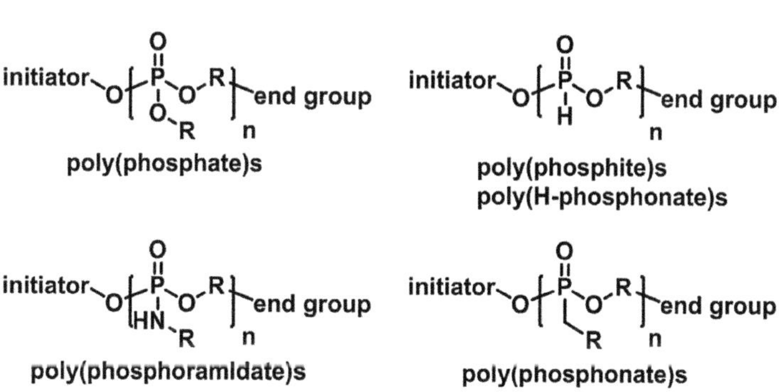

Scheme 10.13 Structures of the most commonly used PPEs with varying P–X bonds in the side-chain.

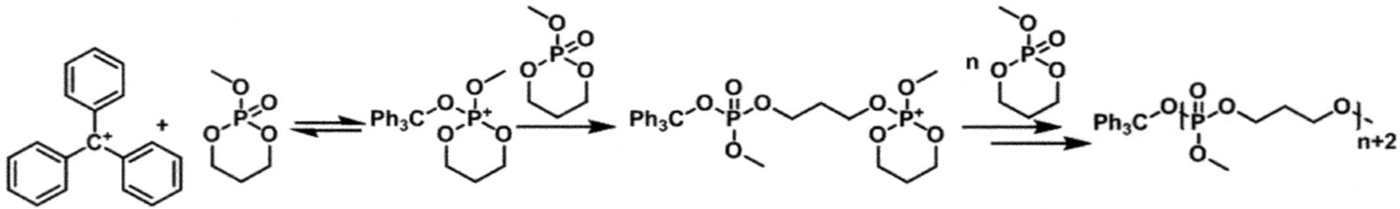

Scheme 10.14 Proposed cationic polymerization of 2-methoxy-2-oxo-1,3,2-dioxaphosphorinane initiated with the triphenyl methyl cation (anions omitted).[72]

High degrees of transesterifications reactions limited the propagation, a result further confirmed by parallel work of Vogt *et al.*[74,75]

Parallel to Penczek's work on six-membered monomers, Vogt *et al.* studied the polymerization of five-membered 2-alkoxy-2-oxo-1,3,2-dioxaphospholanes. Following the initial works of Munoz *et al.* they presented the alkoxide initiated ROP of these monomers. In contrast to the six-membered homologs, rapid monomer conversion was found at room temperature. However, kinetic studies pointed towards a tendency for transesterification reactions.[76]

In 1978, a shift towards the use of the five-membered dioxaphospholane monomers occurred and the six-membered rings vanished from the following literature, probably due to their lower ring strain, making a controlled ROP without chain transfer difficult. In the following years, several five-membered cyclic monomers and detailed mechanistic studies were reported (see Scheme 10.15).

Polymerizations prior to 1998 were performed mainly with organo-aluminum or organo-magnesia reagents resulted in higher molecular weight polymers ($M_n > 10\,000$ g mol^{-1}).[77] As a consequence the use of easy to synthesize organo-aluminum compounds dominated the field for many years.[78]

The first non-metallic catalysis for PPE preparation was reported by Wen *et al.* in 1998 when they used lipase to enzymatically polymerize 2-isopropoxy-2-oxo-1,3,2-dioxaphsopholane (Scheme 10.15, structure 3). Unlike the alkoxide or aluminum-catalyzed reactions, the polymerization needed high temperature and long reaction time (up to 170 h) to proceed to high conversions (94%). Even then, materials with molecular weights below 2000 g mol^{-1} were obtained.[79]

Higher degrees of polymerization (molecular weights up to 5000 g mol^{-1}) were achieved by Wang *et al.* in 2006, when they utilized $Sn(Oct)_2$ as a

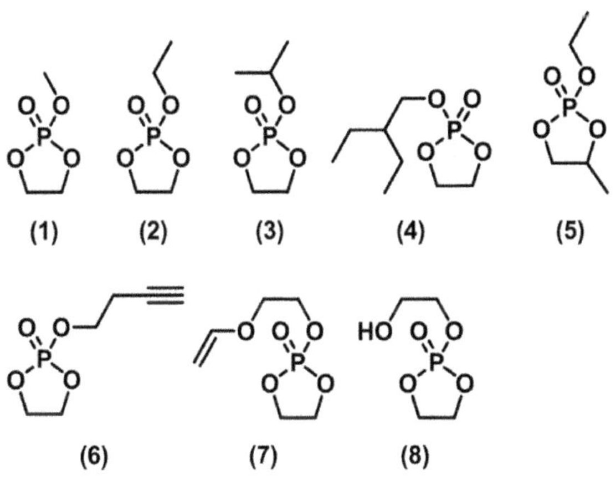

Scheme 10.15 Representative 2-alkoxy-2-oxo-1,3,2-dioxaphosphorinanes used in ring-opening polymerization.

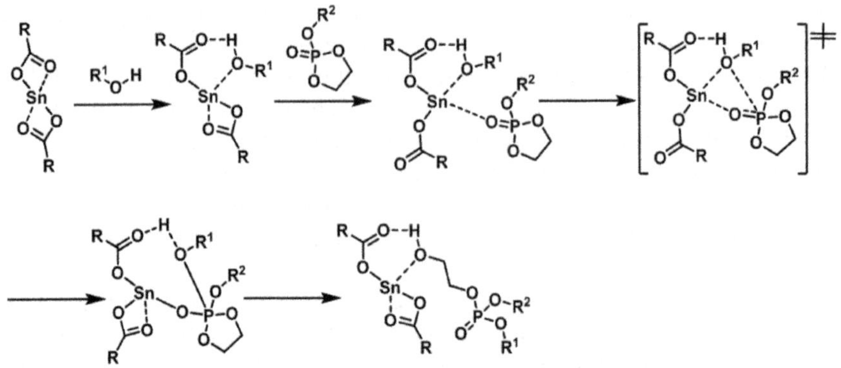

Scheme 10.16 Proposed mechanism of initiation and propagation of EEP polymerization with $Sn(Oct)_2$.
Reproduced from ref. 80 with permission from American Chemical Society, Copyright 2006.

catalyst for the insertion polymerization of 2-ethoxy-2-oxo-1,3,2-dioxaphsopholane (Scheme 10.15, structure 2; Scheme 10.16, mechanism).[80]

The polymerization proceeded to high conversions (90%) within 30 min in THF at 40 °C. Only low conversion of less than 40% was achieved at 0 °C. However, a more detailed analysis of the polymerization and the resulting polymer showed that the reaction was prone to transesterification and depolymerization as side-reactions. The molecular weight distribution broadened significantly over time (1.31 to 1.69) and eventually after 4 h of reaction became bimodal (Figure 10.9). Furthermore, the apparent molecular weight deviated from the initial monomer to initiator feed ratio, indicating side-chain transesterifications and the formation of branched structures.[80]

Around that time, the first attempts to use well-defined PPEs as materials in biomedical applications were undertaken by Leong *et al.* They synthesized copolymers composed of lactide and 2-ethoxy-2-oxo-1,3,2-dioxaphsopholane with yields of 70% and molecular weights of 10 000 g mol^{-1} in order to produce degradable, non-toxic drug carriers. However, no data concerning the microstructure of the polymer was shown. The copolymers showed little to no toxicity towards Hela-cells, indicating the biocompatibility of PPEs. Furthermore, they presented the diffusion-controlled release of a model compound, BSA, from porous microspheres made of their copolymer (Figure 10.10).[81]

The first organocatalytic anionic ring-opening polymerization (oAROP) of cyclic phosphate esters was presented in 2010 by Iwasaki *et al.* They were the first to use the well-known organic bases DBU and TBD to polymerize 2-isopropoxy-2-oxo-1,3,2-dioxaphsopholane in the presence of a primary alcohol as an initiator (Scheme 10.17).

Their investigation showed that this cyclic monomer could be rapidly polymerized with TBD to high conversions (90%) within 20 min at 0 °C. Exceptionally high control over molecular weight and narrow molecular

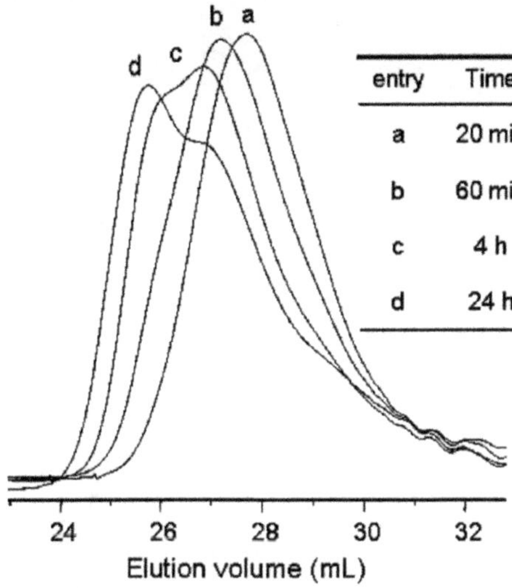

entry	Time	Mn	PDI
a	20 min	3140	1.31
b	60 min	3690	1.41
c	4 h	4300	1.53
d	24 h	4680	1.69

Figure 10.9 SEC elugrams of EEP polymerization with $Sn(Oct)_2$ in THF at 40 °C. Reproduced from ref. 80 with permission from American Chemical Society, Copyright 2006.

weight distributions ($Đ<1.10$) were reported. A linear dependence of the molecular weight evolution of the polymers to the monomer conversion indicated the absence of side-reactions like transesterifications and de-polymerization, a further side-reaction not discussed in the synthesis of PPEs so far.[82] The polymerization with DBU only gave moderate conversions (up to 60%) and needed reaction times up to 6 h, especially for the high monomer to initiator ratios. Nonetheless, for 2-isopropoxy-2-oxo-1,3,2-dioxaphospholane, excellent control over molecular weight and narrow molecular weight distributions were obtained without the occurrence of side-reactions (Figure 10.11).[82]

After this pioneering work on organocatalysis of phosphoesters, bases like DBU and TBD became the state of the art for the polymerization of cyclic phosphate monomers. In the same year, Liu *et al.* presented the synthesis of PPEs with varying topology, ranging from linear diblock copolymers over star diblock copolymers up to hyperbranched PPEs by use of the inimer method (Scheme 10.15, structure 8, inimer).[83] While they synthesized their respective PPE macroinitiator conventional with $Sn(Oct)_2$, the second block was grown *via* modern DBU catalysis. This combination resulted in moderate molecular weight distributions ($Đ<1.4$) and yields (<78%). Furthermore, they investigated the cell-toxicity of such polymers against NIH 3T3 fibroblasts. This cell line is much less resistant than Hela cells and still no significant toxicity was observed for all copolymers.[83]

In 2012, Zhang *et al.* presented their DBU-catalyzed oAROP of an alkynyl functional phosphoester (Scheme 10.15, structure 6), offering the first side-chain modifiable PPE at that time (Figure 10.12). Polymerization was

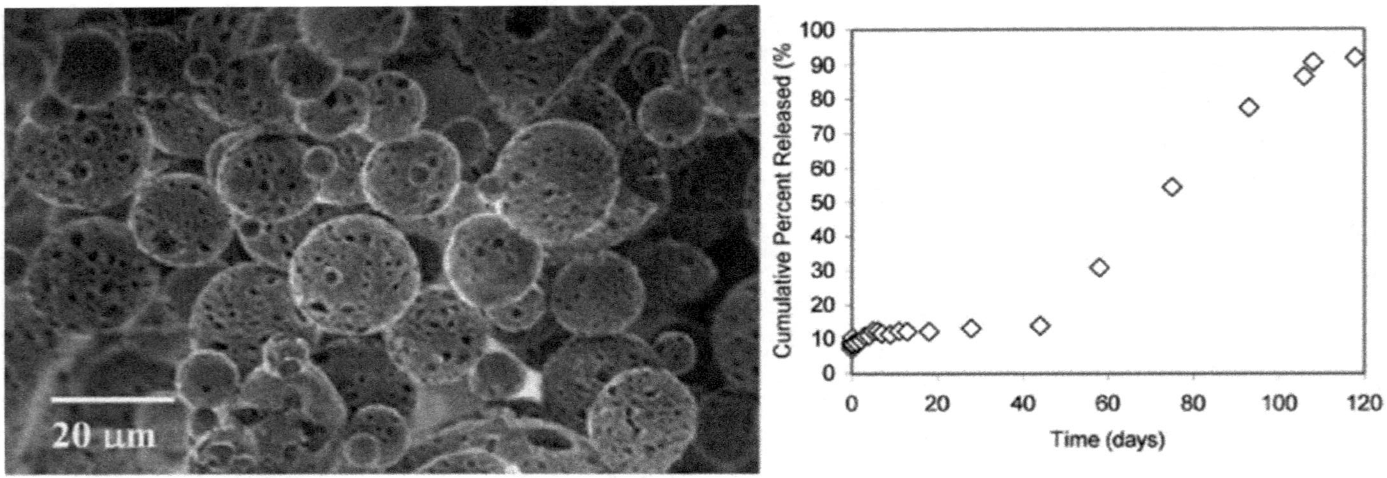

Figure 10.10 SEM image (left) and cumulative release profile of BSA from PLA-EEP microspheres in PBS (pH 7.4) at 37 °C (right). Reproduced from ref. 81 with permission from Elsevier, Copyright 2003 Elsevier.

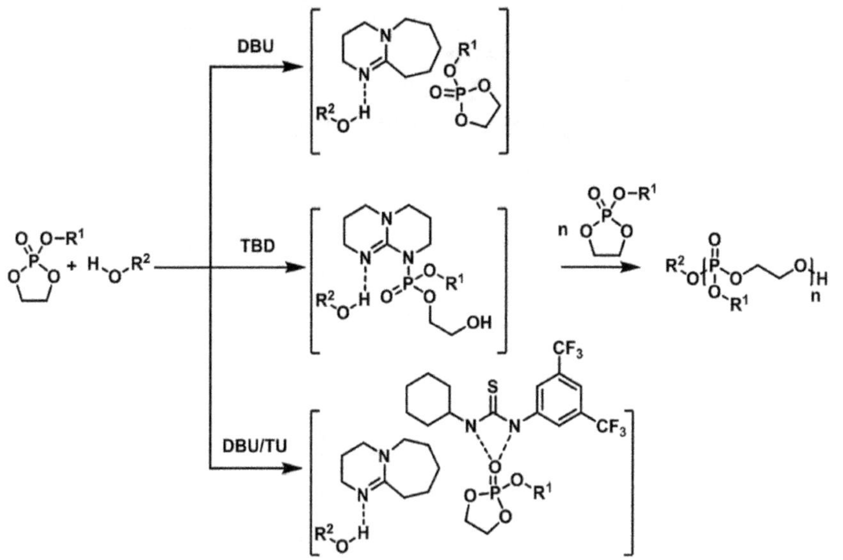

Scheme 10.17 Catalysts for the oAROP of cyclic phosphate monomers and their proposed activation mechanism. Top: activation of initiator/active chain end ROH *via* DBU. Middle: dual-activation of initiator/active chain end and monomer *via* TBD. Bottom: activation of initiator/active chain end *via* DBU and activation of monomer *via* thiourea as a binary catalyst mixture.

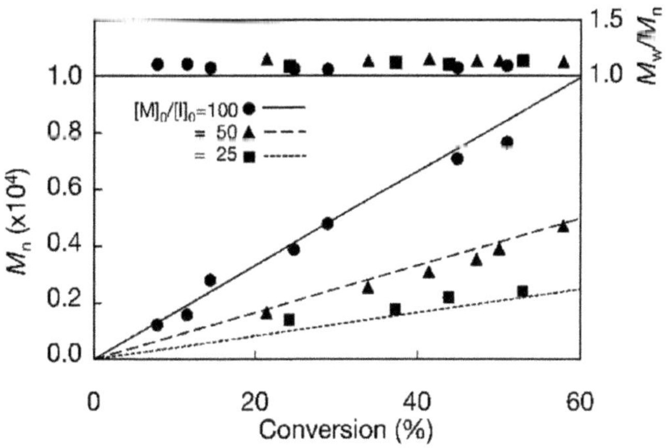

Figure 10.11 Plot of M_w/M_n and M_n *vs.* monomer conversion for the polymerization of 2-isopropoxy-2-oxo-1,3,2-dioxaphospholane catalyzed with DBU for different $[M_0]/[I_0]$ values.
Reproduced from ref. 82 with permission from American Chemical Society, Copyright 2010.

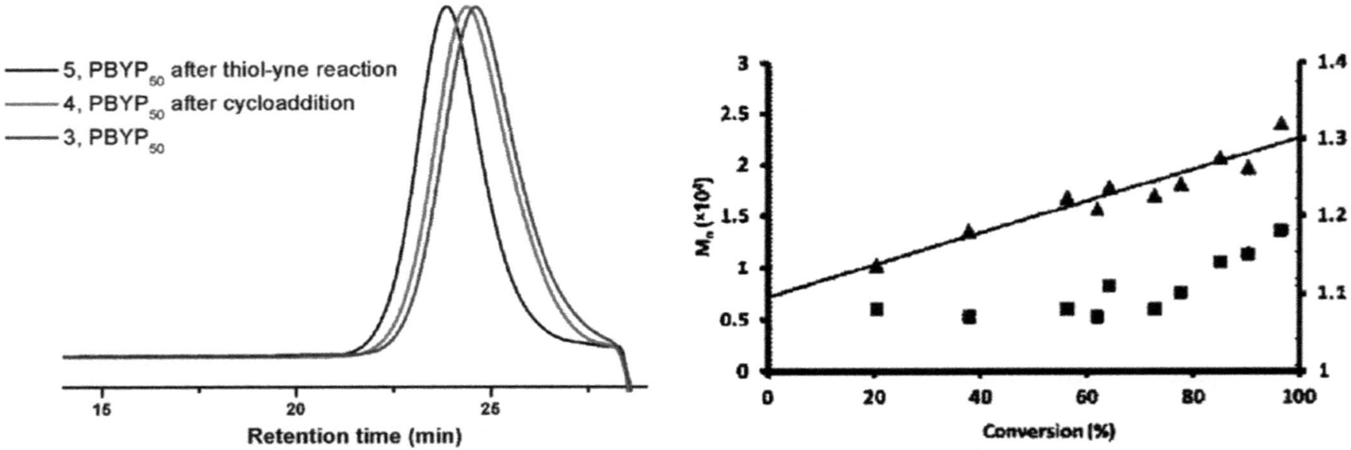

Figure 10.12 Left: SEC traces of alkynyl functional poly(phosphate) before (blue) and after modification *via* thio-yne (black) and Huisgen cycloaddition (blue). Right: plot of M_n and M_w/M_n *vs.* monomer conversion for the polymerization of alkynyl functional monomer catalyzed with DBU for $[M_0]/[I_0] = 100$.
Reproduced from ref. 84 with permission from American Chemical Society, Copyright 2012.

performed in dichloromethane and proceeded up to near quantitative conversion within 6 min following a first-order kinetic, despite the presence of the acidic alkynyl protons. However, using DBU as a base, molecular weight distributions broadened at conversions above 60%.[84]

An essential breakthrough in synthetic procedures was presented in the same year. Clement *et al.* published a detailed kinetic analysis of the oAROP of 2-ethoxy-, 2-isobutoxy-, and 2-butenoxy-2-oxo-1,3,2-dioxaphospholanes.[85] As catalysts, they used DBU, TBD, and a binary DBU/thiourea mixture, which has become popular for the ROP of lactones.[86] Here, the thiourea (Scheme 10.17, right) activates the monomer *via* H-bonding interactions. They impressively showed that these catalytic systems were able to polymerize a variety of dioxaphospholanes following first-order kinetics. However, only the DBU/thiourea mixture effectively suppressed transesterification reactions (Scheme 10.18) at high conversions leading to narrowly distributed ($Đ < 1.10$), high molecular weight (up to 70 000 g mol^{-1}) polymers with conversions above 90% (Figure 10.13).

While being equally fast as the DBU/TU catalyst mixture, TBD-catalyzed reactions tended towards transesterifications even after reaching nearly quantitative conversions if left enough time for all monomers (Figure 10.14).

DBU catalyzed reactions, again, needed to be terminated at low conversions (around 50 to 60%) to avoid side-reactions and broadening of the molecular weight distribution for all monomers (Figure 10.15).[85] Since then, several research groups have focused their work on the synthesis and biomedical applications of PPEs and have either used DBU, TBD or the binary

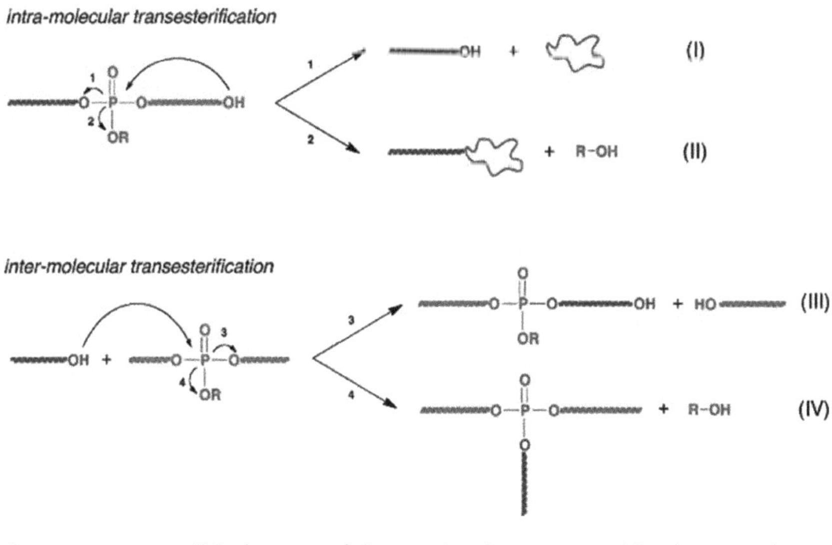

Scheme 10.18 Possible intra- and intermolecular transesterification reactions of poly(phosphoester)s during ROP.
Reproduced from ref. 85 with permission from American Chemical Society, Copyright 2012.

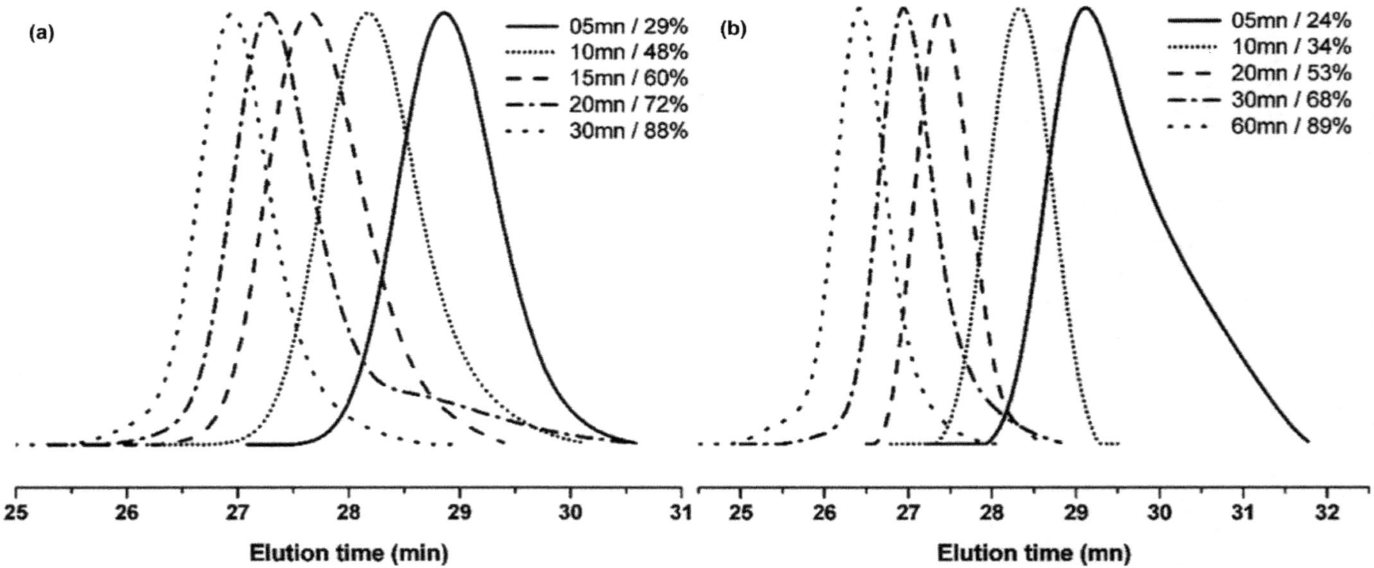

Figure 10.13 SEC traces of the polymerization 2-isobutyl-2-oxo-1,3,2-dioxaphospholane catalyzed with the binary DBU/TU mixture at 0 °C in toluene with $[M_0]/[I_0] = 100$ (left) and $[M_0]/[I_0] = 400$ (right).
Reproduced from ref. 85 with permission from American Chemical Society, Copyright 2012.

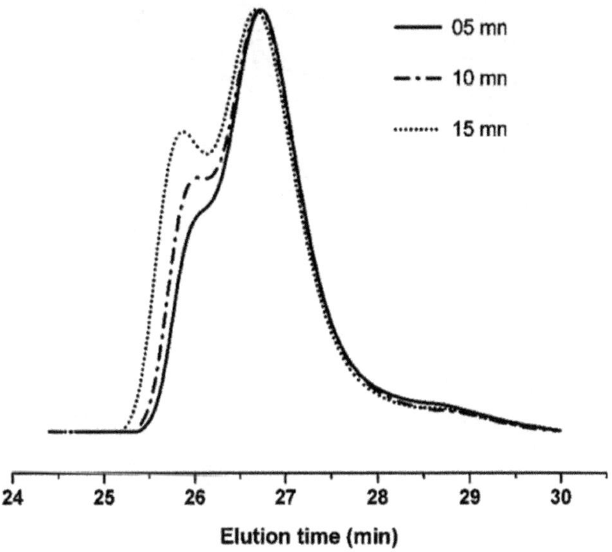

Figure 10.14 SEC traces of the polymerization 2-isobutyl-2-oxo-1,3,2-dioxaphospholane catalyzed with TBD at 0 °C in toluene with $[M_0]/[I_0] = 200$. Reproduced from ref. 85 with permission from American Chemical Society, Copyright 2012.

DBU/thiourea (TU) mixture as the respective catalysts. In all cases, however, the binary catalyst system or TBD have proven to be superior to the (much less expensive) DBU catalysis.

The following paragraph features and briefly discusses a few selected publications showing the potential of oAROP for PPE synthesis and respective applications. In 2013, Steinbach *et al.* investigated the copolymerization behavior of dioxaphospholanes with 4-exo-substituted dioxaphospholanes *via* TBD catalysis to gain insight into the microstructure of such copolymers, proving a random incorporation of 2-ethoxy-2-oxo-1,3,2-dioxaphospholane and the 4-exo-substituted dioxaphospholane.[87] Lim *et al.* expanded the scope of side-chain functional PPEs by the DBU-catalyzed polymerization of a vinyl ether functional dioxaphospholane. They utilized different side-chain modifications, ranging from acetalization over thioacetalization up to thiol-ene reactions. The polymers aggregation behavior into polymeric nanoparticles, their degradation in water and negligible cell-toxicity against RAW 264.7 macrophage cells as well as OVCAR-3 human ovarian adenocarcinoma cells was reported.[88] An interesting work was presented in 2015, when Gao *et al.* used the DBU/TU mixture to catalyze the polymerization of 2-ethoxy-2-oxo-1,3,2-dioxaphospholane initiated with polyethylene with a P_n of 680, showing the effectiveness of this catalytic mixture even for difficult initiator systems. In all cases molecular weight distributions below 1.15 and good control over the molecular weight (polymers synthesized with M_n up to 44 000 g mol^{-1} was achieved. Despite the

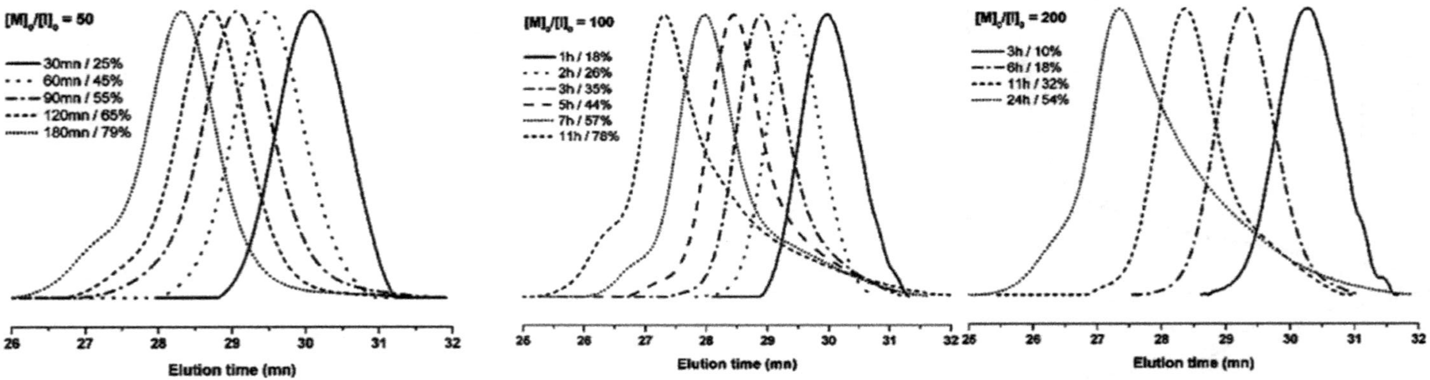

Figure 10.15 SEC traces of the polymerization 2-isobutyl-2-oxo-1,3,2-dioxaphospholane catalyzed with DBU at 0 °C in toluene with $[M_0]/[I_0] = 50$ (left), $[M_0]/[I_0] = 100$ (middle) and $[M_0]/[I_0] = 200$ (right).
Reproduced from ref. 85 with permission from American Chemical Society, Copyright 2012.

amphiphilic structure, the obtained polymers showed no cell-toxicity against Hela-cells.[89] In the same year, Lim *et al.* synthesized a block copolymer composed of an alkynyl substituted PPE and poly(lactic acid) by sequential polymerization.

Their goal was the synthesis of an amphiphilic, degradable polymer bearing anionic charges on the side chain for complexation of antimicrobial active silver complexes (Figure 10.16).[90]

In parallel, they presented the sequential synthesis of a PPE-*b*-PPE block copolymer with hydrophobic and alkynyl side-chains for paclitaxel incorporation and potential for click-side-chain modification. The high polymerization rate of the TBD catalyzed reaction made these polymers available within 10 min of total synthesis time.[91] In 2016, Baeten *et al.* deviated from the traditional batch chemistry in glass reactors. They polymerized dioxaphospholanes under continuous flow conditions either with the DBU/TU mixture or TBD and with subsequent thiol-ene reaction under continuous flow conditions (Figure 10.17).[92]

Wu *et al.* developed a tandem strategy to synthesize ABA triblock copolymers starting with the B block being synthesized by metal-catalyzed CO_2/epoxide polymerization followed by DBU catalyzed oAROP of a dioxaphospholane.[93] Schöttler *et al.* proved in 2016 that poly(2-ethoxy-2-oxo-1,3,2-dioxaphospholane)s can be covalently attached to and stabilize polystyrene nanoparticles. Furthermore, they demonstrated that these polymers exhibit the same stealth behavior as poly(ethylene glycol), both with regard to their cell-uptake and the protein adsorption in human blood plasma.[94] In a following work, Müller *et al.* prepared PPE surfactants and reduced the protein adsorption in a similar manner to the covalently attached PPEs.[95] In 2017, Becker *et al.* introduced furfuryl containing dioxaphospholanes and used this platform for the first reversible side-chain modifications *via* the Diels–Alder reaction. The hydrophilicity of the resulting polymers was analyzed *via* UV-VIS turbidity measurements with regard to the attached side-chain.

Furthermore, they presented the first detailed copolymerization online-NMR-kinetics of dioxaphospholanes, showing a preferred incorporation of the furfuryl monomer (Figure 10.18). The formation of such a gradient microstructure needs to be considered during copolymerization as the microstructure greatly alter the polymer properties, such as solubility or distribution of reactive groups in the backbone.[96]

A potential alternative to the abovementioned catalysts DBU, TBD and DBU/TU was presented in 2014 by Stukenbroeker *et al.* They reported the zwitterionic ring-opening polymerization of 2-isopropoxy-2-oxo-1,3,2-dioxaphospholane initiated with an N-heterocyclic carbene in the absence of alcohol initiators. A rapid monomer conversion within minutes is reported. Molecular weights up to 200 000 $g\,mol^{-1}$ with moderate molecular weight dispersity ($Đ < 1.3$) and mainly cyclic topology were obtained (Scheme 10.19).[97]

In conclusion, poly(phosphate)s are now a class of polymers that can be synthesized with excellent control over molecular weight, narrow molecular weight distributions, and a multitude of functional side-chains with varying

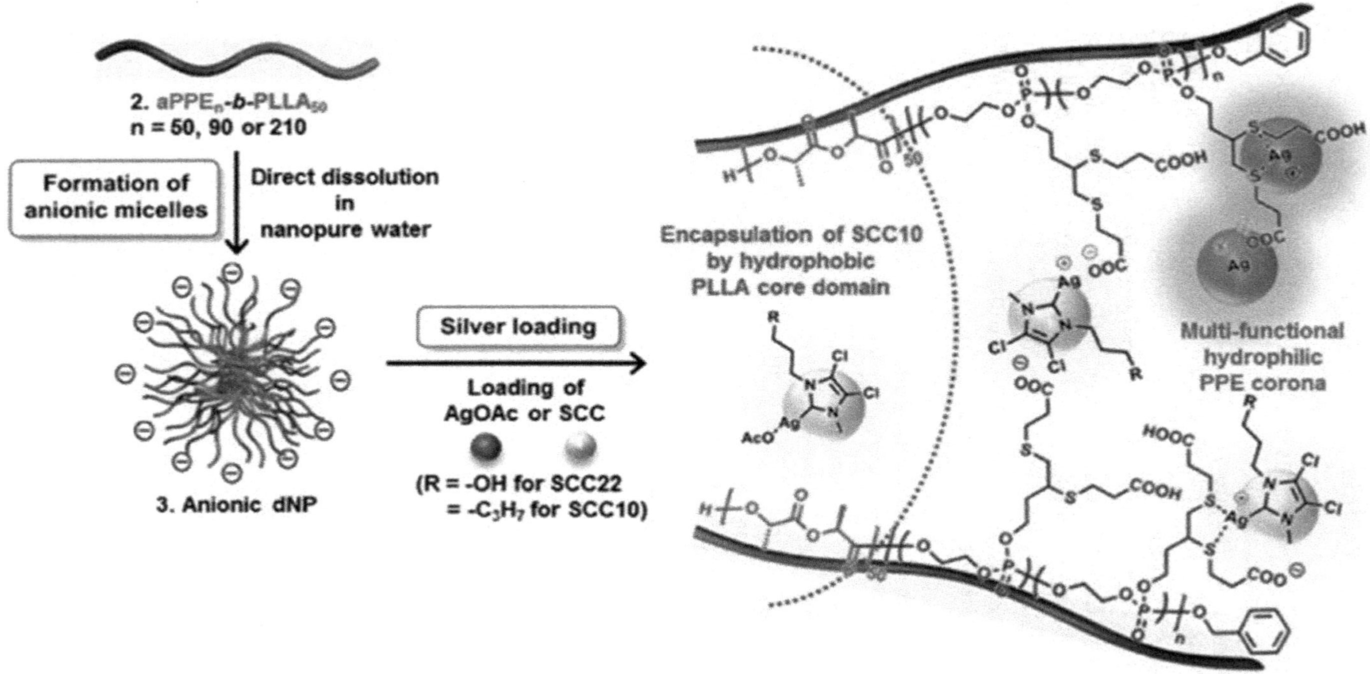

Figure 10.16 Schematic illustration of the work presented by Lim *et al.*: formation of PPE-b-PLLA block copolymer, the formation of micelles in solution and complexation of antimicrobial active silver complexes.
Reproduced from ref. 90 with permission from American Chemical Society, Copyright 2015.

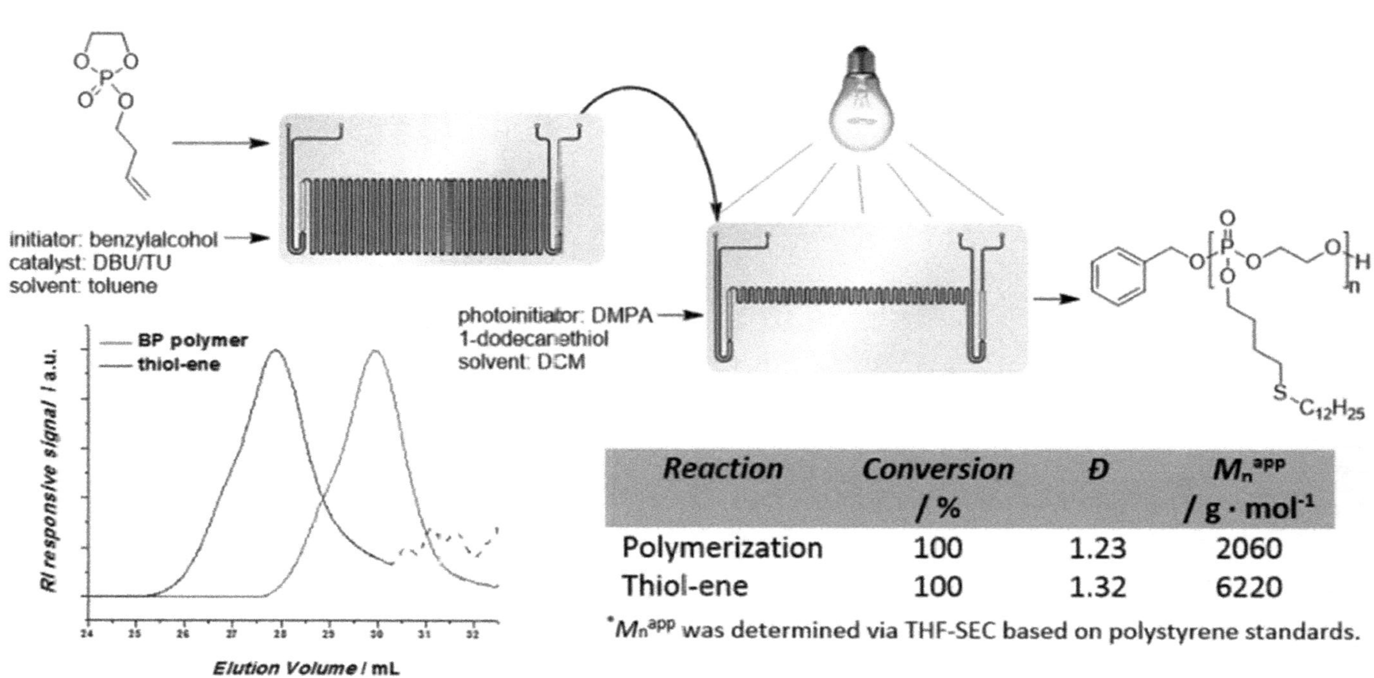

Reaction	Conversion / %	Đ	M_n^{app} / g · mol^{-1}
Polymerization	100	1.23	2060
Thiol-ene	100	1.32	6220

*M_n^{app} was determined via THF-SEC based on polystyrene standards.

Figure 10.17 Schematic representation of the microfluidic cascade consisting of a two-stage microreactor. First, polymerization of the alkene-functional cyclic phosphate with DBU/TU in toluene and subsequent photochemical thiol-ene reaction with 1-dodecane thiol.
Reproduced from ref. 92 with permission from American Chemical Society, Copyright 2015.

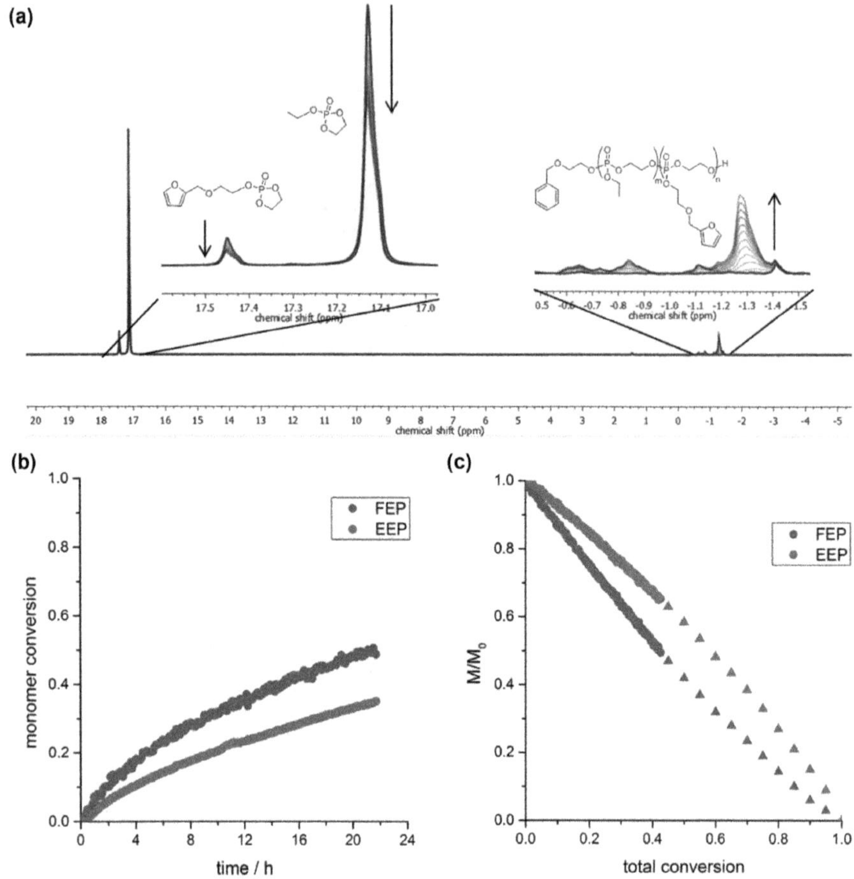

Figure 10.18 Real-time $^{31}P\{H\}$ NMR kinetics of the copolymerization of 2-ethoxy-2-oxo-1,3,2-dioxaphospholane (EEP, red) and the furfuryl functional cyclic phosphate (FEP, green): (a) overlay and zoom-in into $^{31}P\{H\}$ NMR measurements, (b) monomer conversion *vs.* time, and (c) normalized monomer concentrations in the reaction *vs.* total conversion. Reproduced from ref. 96 with permission from American Chemical Society, Copyright 2015.

topologies and microstructures. The polymerization catalyzed with TBD or the binary DBU/TU system proceeds within minutes to high conversions at low temperatures. Only the latter binary system seems to be able to fully suppress side-reactions occurring at high conversions (as in the case of DBU) or longer reaction times (in the case of TBD). The use of an NHC catalyst provides access to the synthesis of high M_n cyclic polymers. A large quantity of different homo-, random-, or block copolymers has been synthesized for varying applications in recent years. The overall low toxicity, degradability, functionality, and relative ease of synthesis makes poly(phosphate)s a seminal class of polymers, especially in the field of biomedical applications.

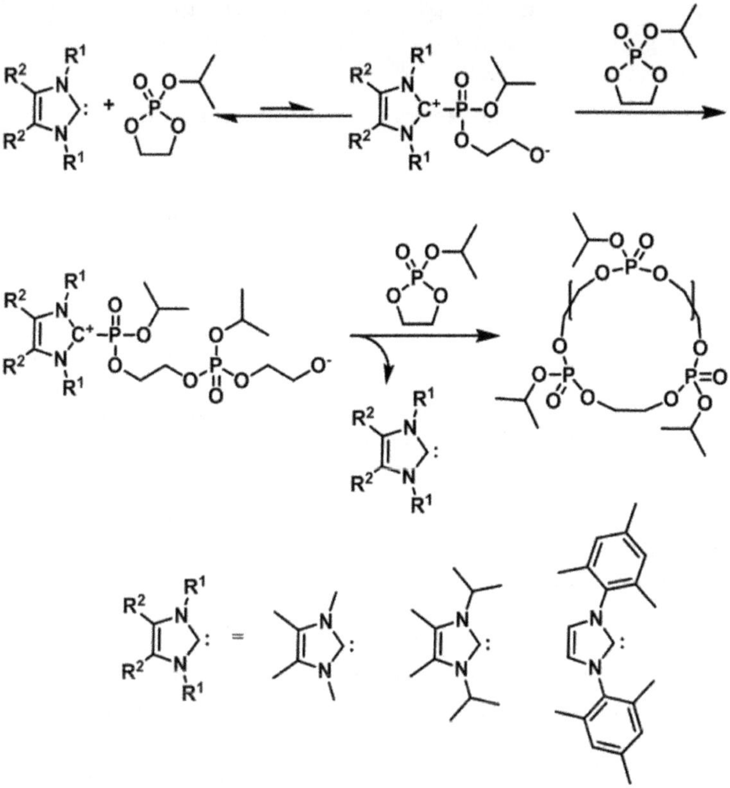

Scheme 10.19 Proposed mechanism for the zwitterionic ring-opening polymerization of 2-isopropoxy 2 oxo-1,3,2-dioxaphopspholane catalyzed with various N-heterocyclic carbenes.
Reproduced from ref. 97 with permission from American Chemical Society, Copyright 2015.

10.4.2 Poly(phosphonic acid ester)s, Poly(phosphonate)s

Poly(phosphonic acid ester)s are a sub-class of poly(phosphoester)s, bearing a single P–C bond, instead of a third P–O bond (Scheme 10.20).This alters their reactivity significantly: the electron withdrawing one R–O–R group is inverted into an electron donor (R–CH_2–R) by hyper-conjugation of the C–H σ-bonds with orbitals of the phosphorus. This increases the electron density at the phosphorous atom, potentially making nucleophilic attacks more difficult. Furthermore, unlike the P–O–R bond, the P–CH_2–R bond is not susceptible towards hydrolysis or transesterification reactions and only cleavable under harsh thermal, radical conditions. Phosphonic acid esters are C–H acidic and are widely used in the stereospecific synthesis of E-olefins (Horner–Wadsworth–Emmons reaction).[98,99] However, despite being a potential issue in anionic polymerizations, the pK_a value of the α-proton of non-activated phosphonic acid esters is in the range of 30.[100,101]

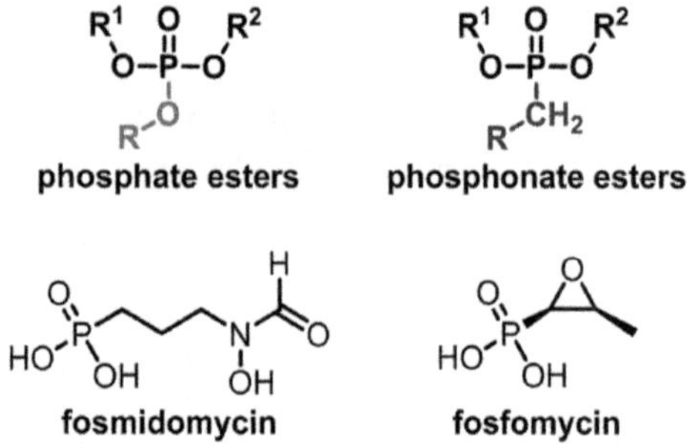

Scheme 10.20 General structure of phosphoric acid esters (third P–O bond colored red) and phosphonic acid esters (essential P–C bond colored blue). Explicit structure of the naturally occurring and clinically used phosphonic acid derivatives fosmidomycin (anti-malaria drug) and fosfomycin (broad-band antibiotic).

Significant deprotonations in the presence of typical organic bases for anionic polymerization ($21 < pK_a < 25$) is unlikely. Additionally, these polymerizations are usually performed in the presence of an alcohol as an initiator, which possesses by far the most acidic proton in the mixture.

Phosphonic acids are not as wildly spread in nature as phosphates but still present a major phosphorus reservoir in nature. Recent ^{31}P NMR studies showed that 20–30% of the maritime phosphorus is bound in the form of phosphonic acid derivatives.[102] Furthermore, naturally occurring phosphonic acid esters have been and are constantly identified. This is due to the high biological activity of several phosphonic acids, *e.g.* the anti-malaria drug fosmidomycin or the broad-band antibiotic fosfomycin (Scheme 10.20).[103]

As a consequence, while not as deeply investigated as phosphates, phosphonic acid esters have since long found interest both in academic and industrial macromolecular research.

The first synthetic poly(phosphonate)s were synthesized by polycondensation reactions of phosphonic acid esters/chlorides with the respective diols. Pioneering work has been reported by Millich *et al.* in a series of publications describing the "interfacial synthesis of polyphosphonate and polyphosphate esters" starting in 1969.[104–107] This led to the development of mainly aromatic poly(phosphonate)s with excellent properties as flame retardant materials, a field in which they are being used as of today (*e.g.* FRX Polymers).[108] Research in more controlled synthetic pathways towards poly(phosphoester)s, however, was nearly exclusively being performed on poly(phosphate)s. A more comprehensive review concerning the history, synthesis, degradation, and applications of poly(phosphoester)s in general and poly(phosphonate)s, in particular, has recently been published by Bauer *et al.*[67]

The first synthesis of poly(phosphonate)s *via* ring-opening polymerization, while not finding much attention at the time, was reported in 1973 by Sharov *et al.* They were the first to describe the thermal ring-opening polymerization

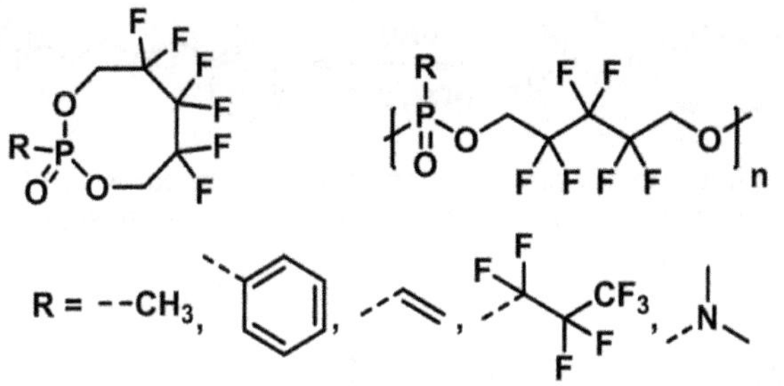

Scheme 10.21 Structure of cyclic phosphonate and proposed structure of poly(phosphonate)s after thermal ring-opening polymerization by Sharov *et al.*[109]

of fluorinated cyclic alkylene alkyl(aryl) phosphonates with varying exocyclic side-chains (Scheme 10.21).[109] However, no detailed mechanistic investigations of the polymerization was presented.

Their goal was the investigation of the polymerizability of the monomers and compare the polymer properties with the respective poly(phosphate) derivatives. They report a reduced tendency towards ring-opening polymerization and a higher susceptibility to hydrolysis compared to the phosphate derivates. While not being investigated extensively, the use of sodium heptafluoro butylate was found to increase the propagation rates. Again, no mechanistic insights were presented as to whether the butylate functioned as initiator or catalysts. Therefore, this publication also presents the first anionic ring-opening polymerization of cyclic phosphonate monomers.[109]

In 1977, during their ongoing work on the polymerization of six-membered cyclic phosphoesters, Lapienis *et al.* published the polymerization of 2-hydro-2-oxo-1,3,2-dioxaphopsphorinane. Unlike six-membered phosphate monomers, this six-membered H-phosphonate polymerized readily in the presence of traditional anionic and cationic initiators. Moderate yields (up to 70%) and rather high molecular weights (up to 10 000 g mol^{-1}) were reported. The authors highlighted the polymers exceptionally high susceptibility to hydrolysis as well as the potential of the P–H bond being transformed into P–C, P–O and P–N derivatives.[110]

Apart from these pioneering works in the 1970s, however, the ring-opening polymerization of cyclic phosphonic acid esters did not receive any attention in academic or industrial research. This gives the unusual opportunity to present an exhaustive discussion of all publications concerning the oAROP of cyclic phosphonates since then.

In 2014, Steinbach *et al.* revived the academic interest in poly(phosphonate)s by presenting the synthesis and polymerization of a five-membered cyclic methyl phosphonic acid ester: 2-methyl-2-oxo-1,3,2-dioxaphospholane (Scheme 10.22).[111] The idea was to eliminate the potential side-chain transesterification reactions occurring during the DBU catalyzed polymerization of 2-alkoxy-2-oxo-1,3,2-dioxaphospholane to produce

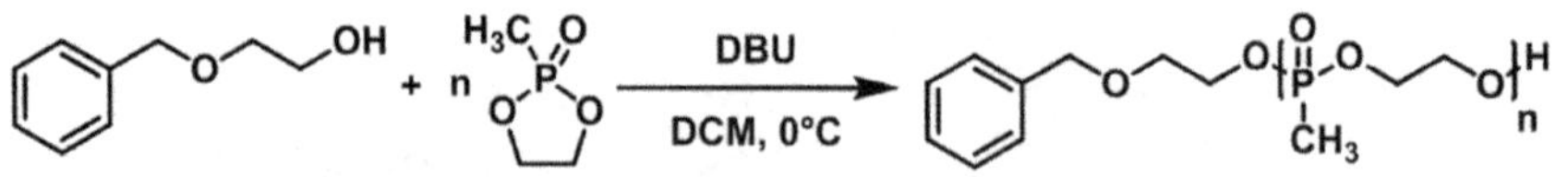

Scheme 10.22 First organocatalytic anionic ring-opening polymerization of five-membered 2-methyl-2-oxo-1,3,2-dioxophospholane with DBU catalysis.[111]

poly(phosphate)s. To this end, the hydrolytically labile P–O–R side-chain was substituted by a hydrolytically stable P–CH_2–R bond.

The polymerization proceeded under DBU catalysis by initiation with a primary alcohol in DCM at 0 °C. However, propagation was considerably slower than the respective cyclic phosphate monomer. Degrees of polymerization of up to 200, resulting in polymers with molecular weights of 24 000 g mol^{-1}, were obtained. Narrow molecular weight distributions ($Đ < 1.10$, Figure 10.19) were achieved even at high conversions up to 90%. Under the same reaction conditions, the respective phosphate polymerization needs to be terminated at 50–60% conversion to prevent transesterification reactions.[85] The polymerization followed linear first-order kinetics, without broadening of molecular weight distributions or loss of control at high conversions, indicating a living polymerization (Figure 10.19).

Furthermore, poly(2-methyl-2-oxo-1,3,2-dioxaphospholane) (PMeEP) was water soluble, non-cell toxic against Hela-cells, and degraded hydrolytically. The degradation rate by hydrolysis was dependent on pH and faster under basic conditions.[111]

PMeEP was also the first poly(phosphonate) from AROP used in the field of biomedical science. Steinbach *et al.* presented the first protein–PMeEP conjugate in 2016.[112] The DBU catalyzed polymerization of 2-methyl-2-oxo-1,3,2-dioxoaphsopholane was terminated with *N,N′*-disuccinimidyl carbamate (DSC) to produce an active-ester functional poly(phosphonate) in one step with excellent yields, a high degree of ω-functionality and narrow molecular weight distribution. The polymers were used to functionalize two enzymes, BSA as a model compound and uricase as a medically relevant drug against gout, with PMeEP and PEG. In a first study, the conjugation efficiency and residual enzyme activity were reported to be comparable to the PEG control model. A significant advantage over PEG, however, was the possibility to degrade the phosphonate backbone (Figure 10.20).[112]

In a second study, the relaxation dynamics of the PMeEP–protein conjugates were analyzed by neutron scattering. Therefore, fully deuterated MeEP was synthesized and used for the polymerization and subsequent conjugation.[113]

Following these results, Wolf *et al.* presented a systematic investigation of the synthesis and polymerization of higher homologs of 2-alkyl-2-oxo-1,3,2-dioxaphospholanes.[114] In general, *n*-alkylated cyclic phosphonates readily polymerize under DBU catalysis (see Scheme 10.23, top and middle row).

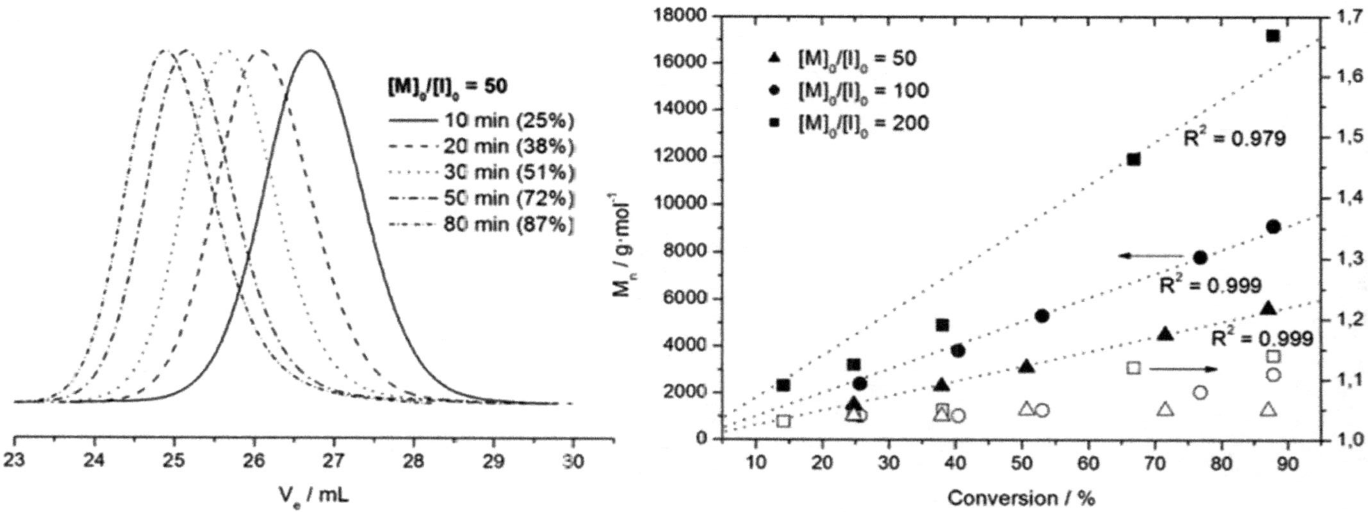

Figure 10.19 Left: kinetic studies, plot of M_n and $Đ$ *vs.* the conversion of MeEP during the oAROP catalyzed with DBU in DCM at 0 °C. $[M_0]/[I_0]$ varied from 50 to 200. Right: SEC analyses of the polymerization MeEP, catalyzed with the DBU at 0 °C in DCM with $[M_0]/[I_0] = 50$ at different times of the polymerization.
Reproduced from ref. 111, http://pubs.acs.org/doi/abs/10.1021%2Fma501764c, with permission from American Chemical Society, Copyright 2014.

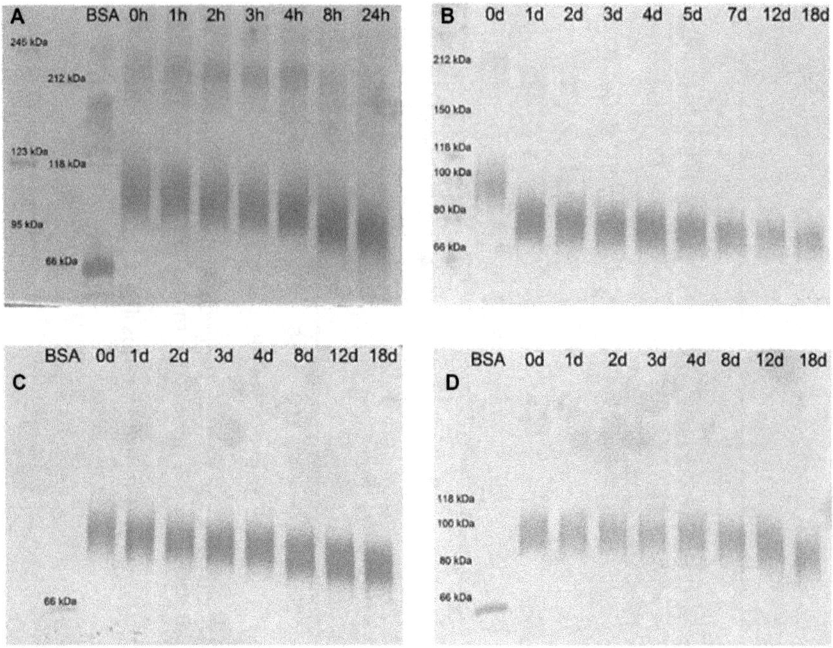

Figure 10.20 SDS-PAGE visualizing the degradation of the PPE conjugated to BSA incubated in aqueous buffer at pH 9.0 (A, B), pH 7.4 (C), and pH 5.0 (D).
Reproduced from ref. 112, http://pubs.acs.org/doi/abs/10.1021/acs.biomac.6b01107, with permission from American Chemical Society, Copyright 2016.

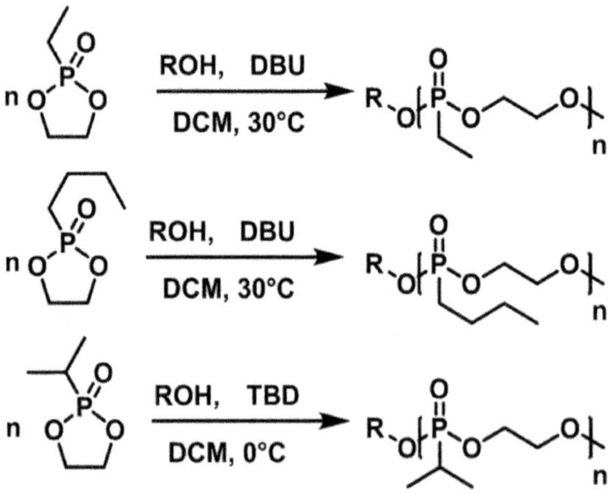

Scheme 10.23 DBU catalyzed anionic ring-opening polymerization of 2-alkyl-2-oxo-1,3,2-dioxaphospholanes.[114]

Due to the stronger inductive effect compared to the methyl derivative, both reaction time and temperature needed to be increased. Under optimized conditions (30 °C, 17 h), however, well-defined polymers with excellent control over molecular weight (up to 6000 g mol^{-1}) and rather narrow molecular weight distributions ($Đ < 1.25$) were available. Again, no transesterification reactions were observed even at conversions above 90%. For the polymerization of 2-isopropyl-2-oxo-1,3,2-dioxaphospholane, the more reactive catalyst TBD was necessary (see Scheme 10.23, bottom row). The need for a catalyst change could be attributed to steric hindrance as well as a more pronounced inductive effect of the secondary alkyl chain, increasing the electron density at the phosphorous thus impeding nucleophilic attacks. Nonetheless, conversions above 90% were obtained and the polymerization proceeded with excellent control over molecular weight and molecular weight distribution. The ethyl and isopropyl substituted polymers were found to be water-soluble and non-cell toxic, whereas the *n*-butyl derivative was somewhat less water-soluble, showed LCST behavior in aqueous solution and a slight toxicity against Hela-cells at high concentrations. Variation of the side-chain further altered the hydrolysis rate of the polymers significantly. While the methyl derivative hydrolyzed readily within hours at pH 9, the ethyl and isopropyl polymers remained stable for days to weeks under the same conditions, respectively.[114]

McDonald *et al.* published a comprehensive study of catalysts for the polymerization of cyclic phosphonate monomers.[115] They used the previously presented 2-methyl-2-oxo-1,3,2-dioxaphospholane as a model monomer. Concerning organic bases, under otherwise identical conditions, they observed a linear relationship between the pK_a of the base and the monomer conversion (Scheme 10.24).

Only TBD, with its known dual-activation of chain end and monomer, did not fit this linear trend as conversion was higher than expected from its pK_a value. Furthermore, a minimum pK_a of 14 was found to be necessary to initiate the polymerization, whereas high molar mass materials were only observed for catalysts with a pK_a of at least 19. Interestingly, apart from slow initiation observed in the presence of DMAP, all able catalysts produced narrowly distributed polymers. This was attributed to the inherent incapability of the monomer for transesterification reactions.[115]

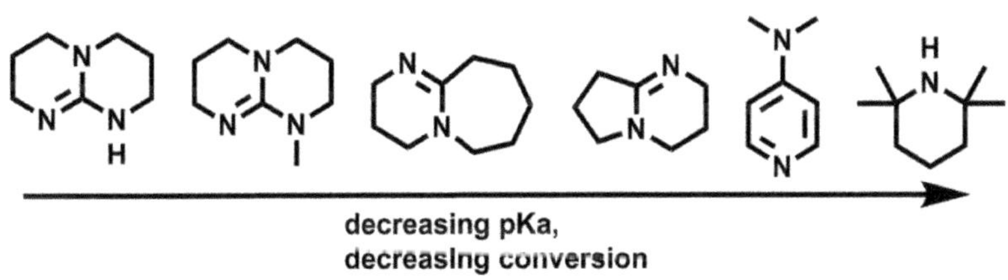

Scheme 10.24 Organocatalysts capable of polymerizing 2-methyl-2-oxo-1,3,2-dioxaphospholane in order of decreasing pK_a value.[115]

Under the same reaction conditions (4 h, room temperature, DCM), polymerization with $Sn(Oct)_2$ was not successful, however, the more reactive aluminum salen and salan complexes produced well-defined polymers in excellent yields and no loss of control. Finally, they tried to expand the monomer scope by synthesizing bicyclic and aromatic dioxaphospholane derivatives. However, these monomers did not polymerize under all investigated conditions.[115]

In order to increase the glass transition temperature of poly(ethylene alkyl phosphonate)s, inherently low T_g materials (around −40 °C) a cyclohexyl substituted monomer was designed by Wolf *et al.* (Scheme 10.25).[116]

Homopolymerization as well as copolymerization with 2-isopropyl-2-oxo-1,3,2-dioxaphospholane was conducted with TBD as a catalyst and produced rather narrowly distributed polymers ($Đ < 1.5$). The respective homopolymer showed a glass transition temperature 50 °C higher compared to *n*-alkylated poly(phosphonate)s. Statistical copolymerization was utilized to adjust the poly(phosphonate)s T_g. The final copolymer composition matched the monomer feed ratio in all cases and a linear trend between copolymer composition and T_g was observed, making for a simple adjustment of the polymer properties (Figure 10.21).[116]

In 2017, copolymerization of cyclic phosphonates was further extended to terpolymerization of 2-ethyl, 2-butyl, and 2-allyl-2-oxo-1,3,2-dioxaphospholanes by Wolf *et al.* While again utilizing DBU as a single catalyst, the reaction temperature needed to be lowered to 0 °C to prevent isomerization reactions of the allyl side-chain. This effect is already known from the AROP of *e.g.* allyl glycidyl ether.[117] Kinetic investigations showed that all monomers incorporated statistically in the polymer backbone, resulting in the formation of a random polymer. This work also presented the first side-chain modification by thiol-ene modification to alter the water-solubility and thermal phase separation behavior of poly(ethylene alkyl phosphonate)s (Scheme 10.26).[118]

The synthesis of pendant carboxylic acid bearing poly(phosphonate)s showing UCST behavior in water was presented in the same year by Wolf *et al.* Copolymerization of 2-ethyl and 2-allyl-2-oxo-1,3,2-dioxaphospholanes was conducted *via* DBU catalysis using either methoxy ethylene glycol or poly(ethylene glycol) as a macroinitiator ($M_n = 5000$ g mol^{-1}) for the polymerization.[119] Narrowly distributed polymers with excellent control over

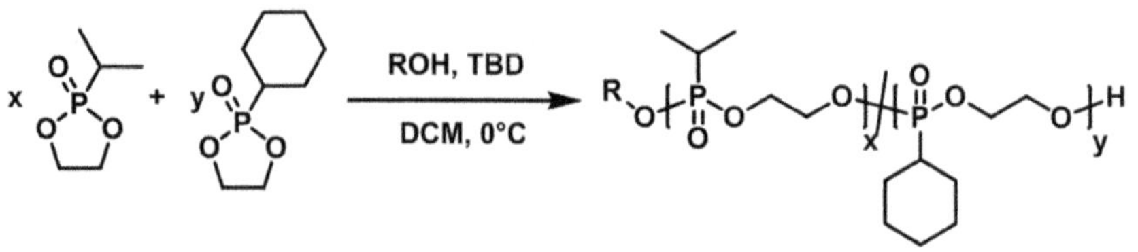

Scheme 10.25 TBD catalyzed, organocatalytic anionic ring-opening copolymerization of 2-isopropyl- and 2-cyclohexyl-2-oxo-1,3,2-dioxaphospholane.

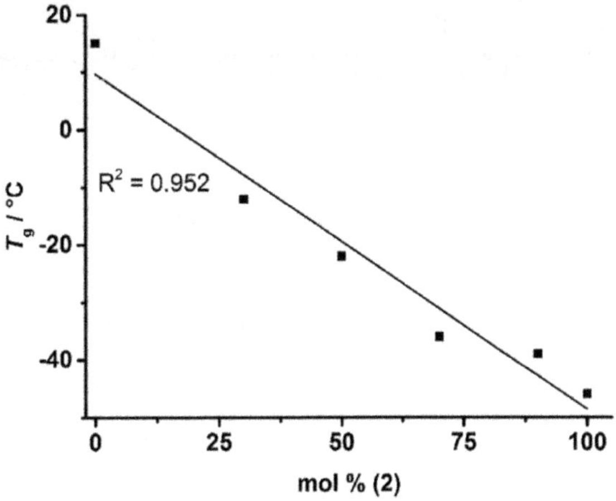

Figure 10.21 Correlation between T_g (from DSC) and the mol fraction of 2-isopropyl-2-oxo-1,3,2-dioxaphospholane (2) in copolymers with 2-cyclohexyl-2-oxo-1,3,2-dioxaphospholanes.
Reproduced from ref. 116 with permission from the Royal Society of Chemistry.

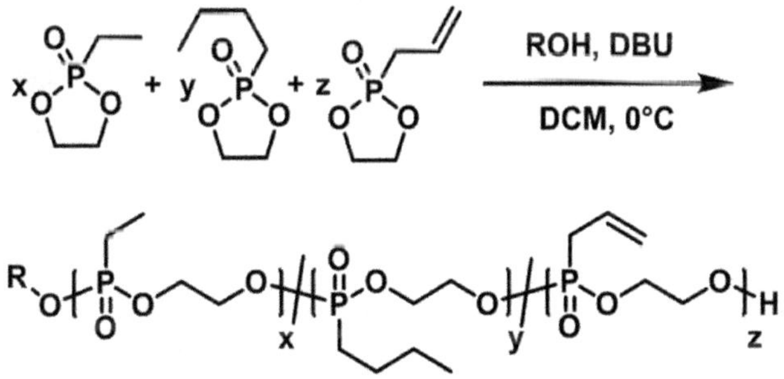

Scheme 10.26 DBU catalyzed terpolymerization towards side-chain reactive, random poly(phosphonate) terpolymers.

molecular weight and molecular weight distributions were obtained. The resulting (block-) copolymers were side-chain modified with 3-mercapto propionic acid *via* thiol-ene reaction to produce UCST thermo-responsive polymers (Scheme 10.27).

The amphiphilic PEG-*b*-poly(phosphonate) block copolymers assembled into vesicular structures of around 200 nm in diameter in water at room temperature. Increasing the temperature above the UCST type phase separation temperature resulted in a disassembling/swelling of the polymersomes, potentially releasing encapsulated cargo (Figure 10.22).

In the same year, Lin *et al.* published the first multi-gram scale synthesis of poly(ethylene alkyl phosphonate)s as well as their use as moderately active kinetic hydrate inhibitors in natural gas pipelines. Furthermore, they

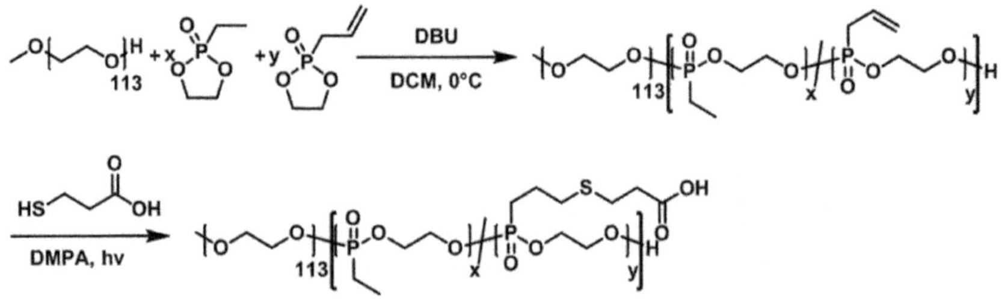

Scheme 10.27 Synthesis of UCST responsive amphiphilic block copolymers.

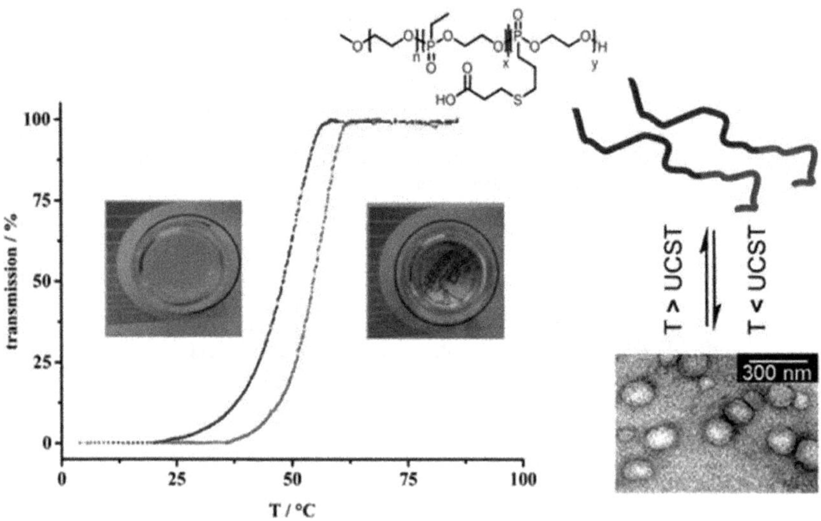

Figure 10.22 UV-vis turbidity measurements of UCST thermo-responsive poly(phosphonate) copolymers (left) and self-assembling of UCST responsive block copolymers into polymersomes in water.
Reproduced from ref. 119 with permission from American Chemical Society, Copyright 2017.

presented the first evaluation of poly(phosphonate) biodegradation in seawater by bacteria. They showed the degradation of up to 31% of polymeric material within 28 d under standardized conditions (OECD 306 protocol), proving poly(phosphonate)s microbial degradability in sea water.[120]

Lastly, Nifant'ev *et al.* presented their work on the [Mg] catalyzed polymerization of dioxaphospholanes.[26] They compared the polymerizability of the ethyl- and a novel *t*-butyl substituted dioxaphospholane with TBD and their magnesia catalysts. Using this magnesia catalyst, the ethyl dioxaphospholane reached 89% (equaling 12 000 $g\,mol^{-1}$) conversion at -20 °C in 1 h. However, molecular weight distributions of 1.46 indicated transesterification reactions. In agreement with previous studies from Wolf *et al.* concerning the lower reactivity of *i*-propyl derivatives compared to the ethyl derivative, the t-butyl dioxaphospholane did not polymerize with TBD at all, even at 100 °C. Using the magnesia catalyst in bulk at 100 °C, however, the polymerization proceeded to 75% within 1 h. Molecular weight distributions of 1.16 indicated the absence of transesterification reactions.

In conclusion, while being forgotten for nearly 40 years since its initial discovery, the ROP of cyclic phosphonate esters has experienced a comeback since 2014 and especially *via* organocatalysis so far. Catalysis *via* DBU is efficient for the polymerization of all to date reported *n*-alkylated five-membered cyclic phosphonate derivatives in the presence of primary alcohols. Polymerization is generally slower than the respective phosphate polymerization due to the inductive effect of the alkyl side-chains and molecular weight distribution broadening. However, high conversions (>95%) can be reached without side-reactions. Molecular weights up to 25 000 g mol^{-1} can be achieved and controlled by variation of the initiator to monomer ratio. Copolymerization generates random copolymers with good control copolymer composition. Due to the growing monomer variety, different chemical and physical polymer properties (*e.g.* T_g, solubility, thermal response) can be addressed and fine-tuned.

10.5 Sulfur-containing Monomers

10.5.1 Poly(thiolactone)s and Poly(thionolactone)s

Poly(thioester)s, thio-derivatives of traditional poly(ester)s are a long known, but a scarcely investigated class of polymers. One reason might be the difficult synthesis, mainly the prevention of undesired thio-thionoester isomerization under basic and acidic conditions (Scheme 10.28). They are, however, highly interesting materials as the sulfur atom in the main chain changes the polymers properties and chemical reactivity, *e.g.*, increase of the melting temperature, compared to the poly(ester) analogs.[121]

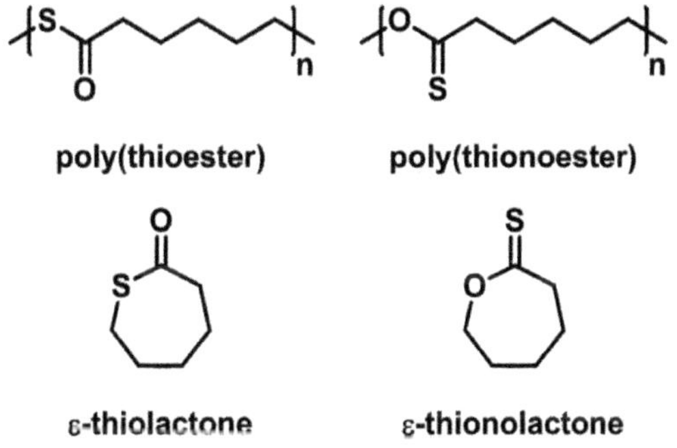

Scheme 10.28 Structure of top: poly(thioester)s and poly(thionoester)s, and bottom: ε-thiolactone, a typical monomer for the ROP towards poly(thioester)s and its isomer, ε-thionolactone.

These polymers can be synthesized from thiolactones and thionolactones (Scheme 10.28) either *via* the anionic or cationic ring-opening process. Early anionic ROP was investigated by Overberger *et al.* in 1967. The reaction of thiolactones with three different ring-sizes with classical anionic initiators, mainly *tert*-BuOK, ^{n}BuLi, and $Al(Et)_3$ was investigated. Their investigations showed that the seven-membered ring, ε-thiocaprolactone, polymerized readily, whereas the smaller rings either did not polymerize at all (γ-thiobutyrolactone) or only produced polymers of low viscosity (δ-thiovalerolactone). Still, high temperatures (>150 °C) and long reaction times (>17 h) were needed to reach conversions of ε-thiocaprolactone between 77 and 100%. Under these conditions, both *tert*-BuOK and ^{n}BuLi produced polymeric material with high intrinsic viscosity, whereas $Al(Et)_3$ produced a cross-linked material.[122] In a follow-up work, the same group presented the polymerization of either racemic or optically pure γ-methylated-ε-thiocaprolactone. ^{n}BuLi was again used for the polymerization.[123] Using ^{n}BuLi as well for the synthesis, Seefried *et al.* presented a comparison of the viscoelastic properties of poly(ester)s and poly(thioester)s in 1974, highlighting the beneficial properties of poly(thioester)s, such as an increased melting point.[124]

In 1999, Sanda *et al.* presented the first publication on the ROP of thionolactones. They highlighted the tendency of the used monomers (ε-thionocaprolactone) towards thio-thiono isomerization and analyzed the degree of isomerization for different initiator systems (Scheme 10.29). Generally, the use of strong Li-containing bases in toluene at 100 °C, like *n*/*sec*/*tert*-BuLi, PhLi, and *tert*-BuOLi, retains the thiono-isomer and produces neat poly(thionocaprolactone) in a nearly quantitative conversion within 20 h. Polymers with molecular weights up to 20 000 g mol^{-1} and molecular weight distributions between 1.54 and 2.67 were obtained. Interestingly, changing the counterion towards potassium (*tert*-BuOK) resulted in a drop of conversion by 70% and 60% isomerization to the thio-isomer.

Furthermore, Sanda *et al.* analyzed the temperature, solvent, and reaction time effects of the polymerization. They observed an increase in reactivity but broadening of molecular weight distributions at increased temperature and longer reaction times due to transesterification/backbiting reactions. Change from toluene to THF resulted in a drastic increase in reaction rate but the formation of large quantities (>30%) of cyclic dimers. Finally, it needs to be mentioned that Sanda *et al.* were the first to report the use of DBU for the polymerization and hence provided the first organocatalytic

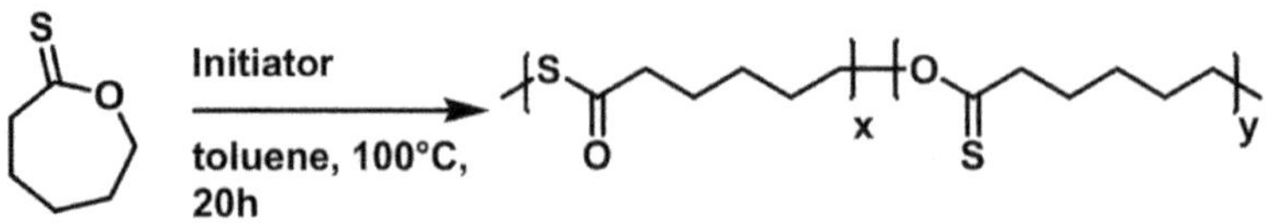

Scheme 10.29 Anionic ROP of ε-thionocaprolactone with different catalysts to produce poly[(ε-thionocaprolactone)-*co*-(ε-thiocaprolactone)] copolymers.[125]

ROP of thionolactones. However, apart from a single experiment, indicating a conversion of only 25%, but relatively narrow molecular weight distribution ($Đ=1.54$), and 90% isomerization to the thio-compound, no further details on the polymerization were given.[125]

In a follow-up work, Sanda *et al.* investigated the cationic ROP of ε-thionocaprolactone in the presence of $BF_3{\cdot}OEt_2$, several Brønsted acids, and methylation agents. Under all conditions, complete isomerization into the poly(thioester) occurred for all tested initiators. Polymers with molecular weights up to 60 000 g mol^{-1} and molecular weight distributions between 1.58 and 2.85 were obtained.[126]

A "green" ROP towards poly(thioester)s was presented by Shimokawa *et al.* in 2011. They employed lipase to first ring-close 6-mercapto hexanoic acid. Afterwards, they used the same lipase for the ROP of the formed ring (Scheme 10.30). Furthermore, they copolymerized ε-thiocaprolactone with ε-caprolactone to produce poly(thioester)-*co*-poly(ester) copolymers, though no information on the microstructure of the final copolymer was given. Nonetheless, copolymers with varying amounts of ε-caprolactone were presented with molecular weights up to 24 000 g mol^{-1} and dispersities in the range $1.4<Đ<2.5$. Thermal analysis showed an increasing melting point and thermal stability with increasing amounts of thioester in the copolymer.[127]

In 2015, Bannin *et al.* presented an essential work regarding the organocatalytic ROP of thiolactones. They investigated the polymerization of ε-thiocaprolactone, initiated with hexadecane thiol in the presence of organic bases (at room temperature in benzene). Generally, the weaker bases DMAP or Me_6TREN were not able to initiate the polymerization under these conditions. Stronger bases like DBU, TBD, and MTBD, however, produced polymers up to 32 000 g mol^{-1} in less than 4 h reaction time. Interestingly, the use of the strong, sterically hindered base BEMP did not initiate the polymerization, which led to the hypothesis that the polymerization is initiated by a nucleophilic attack on the monomer. The polymerization rate was dependent on the basicity of the catalyst, TBD reaching a P_n of 50 (97% conversion) in 30 min, while the same reaction with MTBD and DBU took 80 and 240 min, respectively. Degrees of polymerization up to 200 were reported, but, in the case of MTBD, reaction times of up to 24 h were necessary.

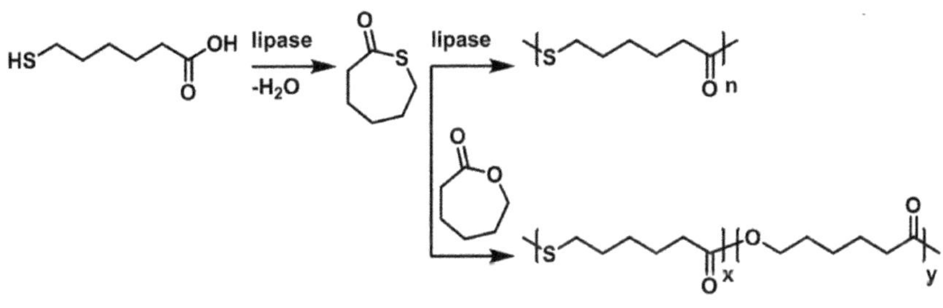

Scheme 10.30 Synthetic scheme for the production of poly(ester)-*co*-poly(thioester)s starting from 6-mercapto hexanoic acid, catalyzed by lipase.

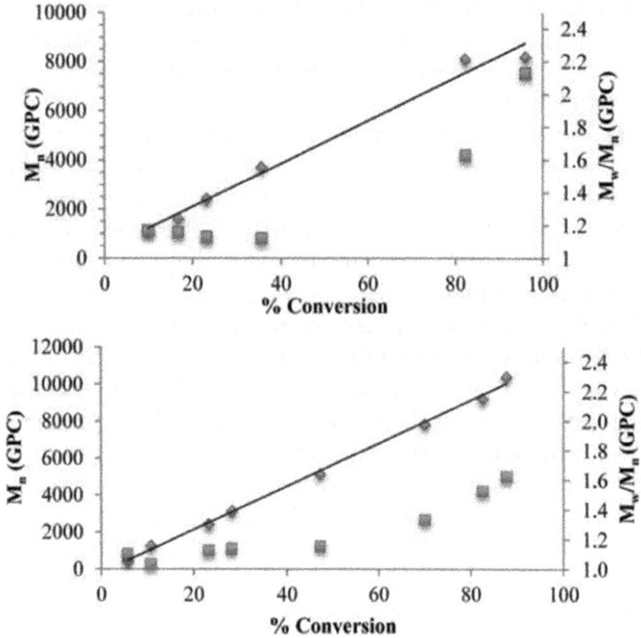

Figure 10.23 Kinetic investigations of the hexadecane thiol initiated polymerization of ε-thiocaprolactone catalyzed with top: 0.05 M MTBD, and bottom: 0.05 M MTBD and 0.05 M thiourea co-catalyst at room temperature. Reproduced from ref. 128 with permission from American Chemical Society, Copyright 2015.

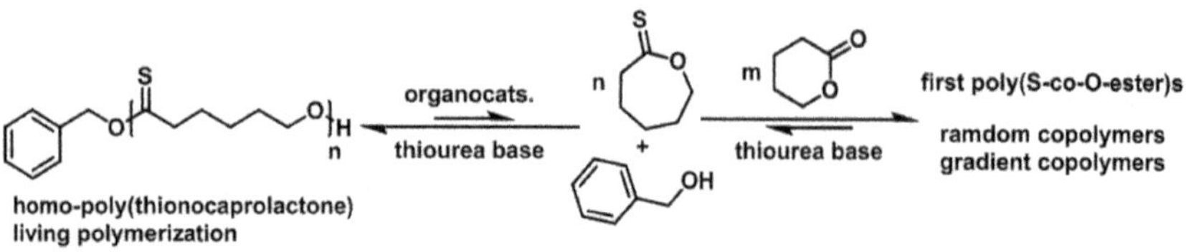

Scheme 10.31 Schematic (co-)polymerization of ε-thionocaprolactone with γ-valerolactone catalyzed with organic bases and a thiourea co-catalyst.

Furthermore, kinetic studies proved a strong tendency towards transesterifications and hence broad molecular weight distributions ($1.40 < Đ < 1.70$) at conversions above 70% (Figure 10.23).

The use of a thiourea co-catalyst, a well-established method to increase the control during ROP of cyclic esters,[86] increased the polymerization rate by a factor of two (and enabled the polymerization with BEMP) but did not prevent transesterification reactions at high conversions.[128]

Following this report, Datta *et al.* investigated the organocatalytic, thiol-initiated ROP of ε-thionocaprolactone and the respective copolymerization with γ-valerolactone (Scheme 10.31). Their findings concerning the use of different catalysts in the ROP of ε-thionocaprolactone were in agreement with previous results on the ε-thiocaprolactone polymerization by Bannin *et al.* At room temperature, in benzene, DBU, TBD, and MTBD were able to polymerize the monomer (without the co-catalyst), while BEMP did not produce any polymeric material.

TBD was the fastest organocatalyst (>99% conversion after 30 min compared to, *e.g.*, 89% conversion after 25 h for DBU). The nearly quantitative conversion was reached in all cases with molecular weights up to 20 000 g mol^{-1}. Molecular weight distributions, however, broadened up after surpassing 50% conversion ($Đ < 1.1$ at 40%, $Đ \approx 1.30$ at 89% conversion, DBU) as a result of transesterification reactions. Unlike the previous findings of Sanda *et al.*, however, no thiono-/thio isomerization was observed and pure poly(thionoester) was obtained. The authors attributed this to the different reaction temperatures, with Sanda *et al.* working at 100 °C, and them performing the reaction at room temperature. When using the thiourea co-catalyst, again, the general reaction time was reduced and BEMP was able to catalyze the polymerization. Under these conditions, well-defined polymers with molecular weights up to 22 000 g mol^{-1} and molecular weight distributions below 1.3 were obtained with either DBU, MTBD, or BEMP. Furthermore, the reaction was no longer limited to initiation with thiols and, *e.g.*, benzyl alcohol was also successfully utilized for the polymerization.

Finally, they used their most efficient system, benzyl alcohol initiation and BEMP/thiourea co-catalyst, to copolymerize ε-thionocaprolactone with γ-valerolactone. Regardless of the monomer feed ratio, well-defined polymers ($Đ < 1.25$) with molecular weights up to 30 000 g mol^{-1} were obtained within 4–7 h. Concerning the microstructure, the authors present a nearly linear decrease of the melting point with increasing lactone content, indicating a random copolymerization.[129]

To conclude, poly(thio/thionoester)s have generally higher melting points than the analog poly(ester)s. The polymerization of the respective monomers, especially the seven-membered rings, have been studied, but other ring-sizes were only scarcely investigated. The use of modern organic bases like DBU and TBD in combination with thiourea co-catalysts provides an elegant way to produce polymers of rather high molecular weight narrow molecular weight distributions. Especially in the case of the thiono-derivatives, thiono-thio scrambling needs to be considered, however, this can be suppressed by operating at room temperature.

10.5.2 Poly(thiirane)s

Thiiranes are the sulfur analogs of epoxides. The high ring-strain (37 kJ mol^{-1}) of the three-membered ring makes them interesting candidates for ROP.[130] Furthermore, the oxidation sensitivity of sulfur and the accompanied change in hydrophilicity when oxidizing a thioether to a sulfoxide and a sulfone make poly(thiirane)s interesting materials for many applications, *e.g.*, in medical applications.

Initial work on the ROP of thiiranes and substituted derivatives was performed in the 1960s by Boileau *et al.* Their work focused on the initiation of propylene sulfide with naphthyl sodium and produced polymers with narrow molecular weight distributions (1.22 < Đ < 1.06) in a living manner with molecular weights up to 300 000 g mol^{-1}.[131]

The first metal-free polymerization was achieved by a direct nucleophilic attack of soft nucleophiles without the need for catalysts. Bonnas *et al.*, *e.g.*, presented the copolymerization of 2-methyl thiirane and 2-(hydroxyl methyl)thiirane initiated with dithio carboxylic acid salts (Scheme 10.32). The polymerization proceeded in DMF to conversions up to 75%. Polymers with molecular weights up to 30 000 g mol^{-1} and molecular weight distributions of Đ = 1.9 were obtained. In a follow-up work, the copolymerization was extended to a carboxylic acid ester-bearing thiirane to obtain carboxylic acid bearing poly(thiirane)s after saponification.[132,133]

In 1999 Nicol *et al.* presented the first organocatalytic polymerization of thiiranes. They utilized DBU as a base to deprotonate the otherwise non-reactive thiols as initiators for the ROP of 2-methyl thiirane. The polymerization was carried out in DMF and quantitative conversion was reached in less than 5 min even at −20 °C. To perform kinetics studies, THF (temperature range from −15 °C to 20 °C) was used as the reaction rate was slower. A linear increase of molecular weight with the monomer conversion was observed and polymers with narrow molecular weight distributions (1.07 < Đ < 1.27) were observed. Molecular weights up to 12 000 g mol^{-1} were obtained. The authors had already mentioned a potential side-reaction in the form of coupling of thiolate end-groups into disulfides, resulting in bimodal distributions. The oxidation was effectively suppressed by increasing the amount of terminating agent after the polymerization and working under O_2 exclusion. Furthermore, the polymerization needed to be performed below 40 °C to prevent polymer degradation (chain-thioether alkylation).[134]

In 2002, Rehor *et al.* presented the robustness of the process by performing this organocatalytic anionic ROP in an oil-in-water emulsion.[135] 1,3-dithiopropane and 2-methyl thiirane were used as the oil phase. After emulsification in the presence of a pluronic-F68, or F127, the polymerization was initiated by the addition of DBU to the water phase (Scheme 10.33). The polymerization stopped after 2 h without reaching full conversion and

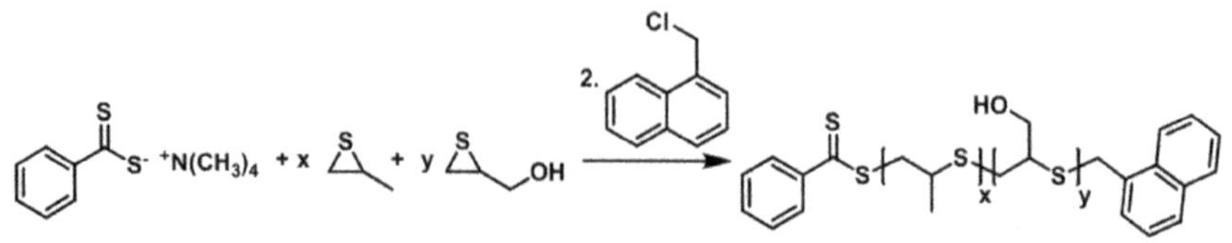

Scheme 10.32 Dithio carboxylic acid initiated copolymerization of 2-methyl thiirane and 2-(hydroxyl methyl)thiirane. The termination was performed with chloro methylnaphthalene.

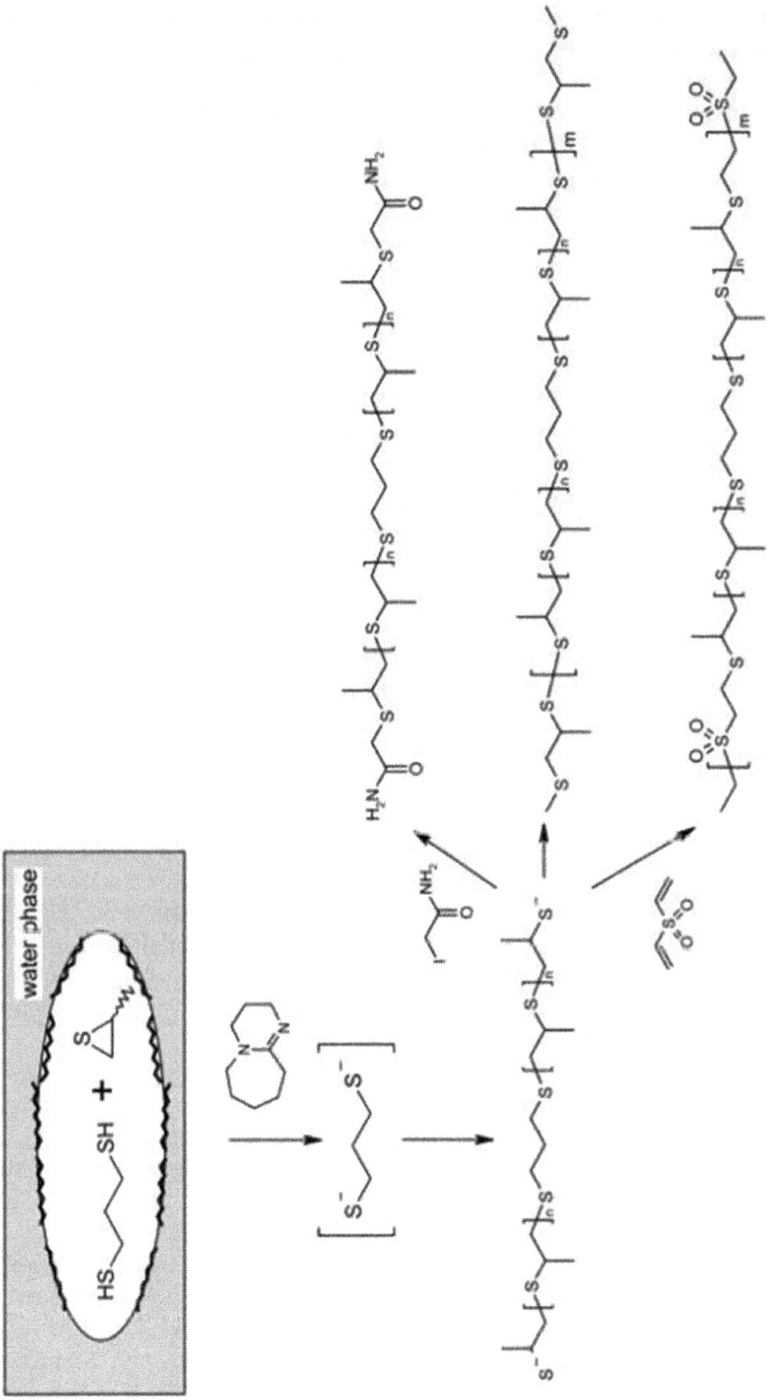

Scheme 10.33 Schematic presentation of the emulsion polymerization of 1,3-dithiopropane and 2-methyl thiirane initiated by DBU. Reproduced from ref. 135 with permission from American Chemical Society, Copyright 2002.

mono-, or bifunctional terminating reagents were again utilized to prevent uncontrolled thiol-dimerization. The polymerization stop was explained by an increased viscosity in the droplets and thus a strongly reduced propagation rate. The molecular weights of the polymers were dependent on the amount of terminating reagent and went as high as 16 000 $g\,mol^{-1}$. The molecular weight distributions of the thio-end-capped polymers were in the range $1.11 < Đ < 1.15$, showing exceptional control during the polymerization even in a heterogeneous system.

In 2001, Napoli *et al.* introduced a process to prevent the usually observed dimerization of thiol initiators. They synthesized thioacetate functionalized PEG and used it as a macroinitiator for the ROP of 2-methyl thiirane. Saponification of the thioacetate generated the desired thiolate for the polymerization *in situ*. Polymers with molecular weights up to 5000 $g\,mol^{-1}$ and molecular weight distributions between 1.3 and 1.2 were obtained.[136]

This system was later used to produce oxidation responsive polymeric vesicles for drug delivery applications.[137–139]

To conclude, poly(thiirane)s are a very interesting class of polymers due to their potential property change upon oxidation. The polymers have narrow molecular weight distributions and molecular weights up to 16 000 $g\,mol^{-1}$. The use of thiolates as initiators, produced by deprotonation of thiols with DBU, presents the most elegant way for ROP of thiiranes. Care must be taken, however, to prevent oxidative dimerization of the reactive chain ends. Usually, a terminating agent is used to prevent this side-reaction.

References

1. R. V. Volkenburgh, K. W. Greenlee, J. M. Derfer and C. E. Boord, A Synthesis of Vinylcyclopropane, *J. Am. Chem. Soc.*, 1949, **71**(11), 3595–3597.
2. T. Takahashi and I. Yamashita, 1,5-Polymerization of vinylcyclopropane, *J. Polym. Sci., Part B: Polym. Lett.*, 1965, **3**(4), 251–255.
3. I. Cho and K.-D. Ahn, Polymerizations of substituted cyclopropanes. I. Radical polymerization of 1,1-disubstituted 2-vinylcyclopropanes, *J. Polym. Sci., Polym. Chem. Ed.*, 1979, **17**(10), 3169–3182.
4. I. Cho and K.-D. Ahn, Polymerizations of substituted cyclopropanes. II. Anionic polymerization of 1,1-disubstituted 2-vinylcyclopropanes, *J. Polym. Sci., Polym. Chem. Ed.*, 1979, **17**(10), 3183–3191.
5. I. Cho and J.-B. Kim, Polymerization of substituted cyclopropanes. III. Anionic polymerizations of 2-substituted cyclopropane-1,1-dicarbonitriles, *J. Polym. Sci., Polym. Chem. Ed.*, 1980, **18**(10), 3053–3057.
6. J. B. Kim and I. H. Cho, Synthesis and electronic effect of the substituents on anionic ring-opening polymerization of para-substituted phenyl cyclopropanes, *Tetrahedron*, 1997, **53**(45), 15157–15166.
7. J. Penelle, H. Hérion, T. Xie and P. Gorissen, Synthesis and characterization of a model carbon-chain polymer substituted by two esters on every third atom via anionic ring-opening polymerization of a

cyclopropane-1,1-dicarboxylate, *Macromol. Chem. Phys.*, 1998, **199**(7), 1329–1336.

8. J. Penelle and T. Xie, Ring-Opening Polymerization of Diisopropyl Cyclopropane-1,1-dicarboxylate under Living Anionic Conditions: A Kinetic and Mechanistic Study, *Macromolecules*, 2000, **33**(13), 4667–4672.
9. N. Illy, S. Boileau, J. Penelle and V. Barbier, Metal-free activation in the anionic ring-opening polymerization of cyclopropane derivatives, *Macromol. Rapid Commun.*, 2009, **30**(20), 1731–1735.
10. N. Illy, S. Boileau, W. Buchmann, J. Penelle and V. Barbier, Control of End Groups in Anionic Polymerizations Using Phosphazene Bases and Protic Precursors As Initiating System (XH-ButP4Approach): Application to the Ring-Opening Polymerization of Cyclopropane-1,1-Dicarboxylates, *Macromolecules*, 2010, **43**(21), 8782–8789.
11. N. Illy, S. Boileau, M. A. Winnik, J. Penelle and V. Barbier, Thiol-ene "clickable" carbon-chain polymers based on diallyl cyclopropane-1,1-dicarboxylate, *Polymer*, 2012, **53**(4), 903–912.
12. S. Russo and E. Casazza, *Ring-Opening Polymerization of Cyclic Amides (Lactams)*, 2012, pp. 331–396.
13. K. Marchildon, Polyamides - Still Strong After Seventy Years, *Macromol. React. Eng.*, 2011, **5**(1), 22–54.
14. S. K. R. Manas Chanda, Industrial Polymers, Specialty Polymers, and Their Applications.
15. K. Hashimoto, Ring-opening polymerization of lactams. Living anionic polymerization and its applications, *Prog. Polym. Sci.*, 2000, **25**(10), 1411–1462.
16. P. K. T. Hikaru Sekiguchi and A. B. Coutin, Anionic Polymerization of A-pyrrolidone and A-piperidone Using their Quaternary Ammonium Salts as Catalysts, *J. Appl. Polym. sci.: Appl. Polym. Symp.*, 1973, **42**, 51–61.
17. F. Fiala and J. Králíček, Die Polymerisation der Lactame. XV. Die Polymerisation der Lactame in Gegenwart von N-Benzoyllactamen und Pentamethylguanidin, *Angew. Makromol. Chem.*, 1978, **71**(1), 29–41.
18. W. Memeger, G. C. Campbell and F. Davidson, Poly(aminophosphazene)s and Protophosphatranes Mimic Classical Strong Anionic Base Catalysts in the Anionic Ring-Opening Polymerization of Lactams, *Macromolecules*, 1996, **29**(20), 6475–6480.
19. H. Yang, J. Zhao, M. Yan, S. Pispas and G. Zhang, Nylon 3 synthesized by ring opening polymerization with a metal-free catalyst, *Polym. Chem.*, 2011, **2**(12), 2888.
20. S. Naumann, S. Epple, C. Bonten and M. R. Buchmeiser, Polymerization of ε-Caprolactam by Latent Precatalysts Based on Protected N-Heterocyclic Carbenes, *ACS Macro Lett.*, 2013, **2**(7), 609–612.
21. K. Jovic, J. Unold, S. Naumann, M. Ullrich, F. G. Schmidt and M. R. Buchmeiser, In Situ Copolymerization of Lactams for Melt Spinning, *Macromol. Mater. Eng.*, 2016, **301**(4), 423–428.

22. S. Naumann, F. G. Schmidt, M. Speiser, M. Böhl, S. Epple, C. Bonten and M. R. Buchmeiser, Anionic Ring-Opening Homo- and Copolymerization of Lactams by Latent, Protected N-Heterocyclic Carbenes for the Preparation of PA 12 and PA 6/12, *Macromolecules*, 2013, **46**(21), 8426–8433.
23. J. Chen, M. Li, W. He, Y. Tao and X. Wang, Facile Organocatalyzed Synthesis of Poly(ε-lysine) under Mild Conditions, *Macromolecules*, 2017, **50**(23), 9128–9134.
24. S. S. Ana, B. Andere, M. Daniele, E. Agustin, E. Cristina, d. l. C. Amaia, G. A. Sonia, S. Haritz and M. David, , Organic-acid mediated bulk polymerization of ε-caprolactam and its copolymerization with ε-caprolactone, *J. Polym. Sci., Part A: Polym. Chem.*, 2016, **54**(15), 2394–2402.
25. A. Basterretxea, E. Gabirondo, A. Sanchez-Sanchez, A. Etxeberria, O. Coulembier, D. Mecerreyes and H. Sardon, Synthesis and characterization of poly (ε-caprolactam-co-lactide) polyesteramides using Brønsted acid or Brønsted base organocatalyst, *Eur. Polym. J.*, 2017, **95**, 650–659.
26. J. B. Sweeney, Aziridines: epoxides' ugly cousins?, *Chem. Soc. Rev.*, 2002, **31**(5), 247–258.
27. S. Taranejoo, J. Liu, P. Verma and K. Hourigan, A review of the developments of characteristics of PEI derivatives for gene delivery applications, *J. Appl. Polym. Sci.*, 2015, **132**(25), 42096.
28. G. D. Jones, A. Langsjoen, S. M. M. C. Neumann and J. Zomlefer, The Polymerization of Ethylenimine, *J. Org. Chem.*, 1944, **09**(2), 125–147.
29. C. R. Dick and G. E. Ham, Characterization of Polyethylenimine, *J. Macromol. Sci., Part A: Pure Appl. Chem.*, 1970, **4**(6), 1301–1314.
30. I. C. Stewart, C. C. Lee, R. G. Bergman and F. D. Toste, Living ring-opening polymerization of N-sulfonylaziridines: synthesis of high molecular weight linear polyamines, *J. Am. Chem. Soc.*, 2005, **127**(50), 17616–17617.
31. E. Rieger, J. Blankenburg, E. Grune, M. Wagner, K. Landfester and F. R. Wurm, Controlling the polymer microstructure in anionic polymerization by compartmentalization, *Angew. Chem., Int. Ed.*, 2018, **57**, 2483–2487.
32. L. Thomi and F. R. Wurm, Living Anionic Polymerization of Functional Aziridines, *Macromol. Symp.*, 2015, **349**(1), 51–56.
33. E. Rieger, A. Alkan, A. Manhart, M. Wagner and F. R. Wurm, Sequence-Controlled Polymers via Simultaneous Living Anionic Copolymerization of Competing Monomers, *Macromol. Rapid Commun.*, 2016, **37**(10), 833–839.
34. E. Rieger, A. Manhart and F. R. Wurm, Multihydroxy Polyamines by Living Anionic Polymerization of Aziridines, *ACS Macro Lett.*, 2016, **5**(2), 195–198.
35. T. Gleede, E. Rieger, T. Homann-Müller and F. R. Wurm, 4-Styrenesulfonyl-(2-methyl)aziridine: The First Bivalent Aziridine-Monomer for Anionic and Radical Polymerization, *Macromol. Chem. Phys.*, 2017, 1700145.

36. E. Rieger, T. Gleede, K. Weber, A. Manhart, M. Wagner and F. R. Wurm, The living anionic polymerization of activated aziridines: a systematic study of reaction conditions and kinetics, *Polym. Chem.*, 2017, **8**(18), 2824–2832.
37. L. Reisman, C. P. Mbarushimana, S. J. Cassidy and P. A. Rupar, Living Anionic Copolymerization of 1-(Alkylsulfonyl)aziridines to Form Poly(sulfonylaziridine) and Linear Poly(ethylenimine), *ACS Macro Lett.*, 2016, **5**(10), 1137–1140.
38. C. Bakkali-Hassani, E. Rieger, J. Vignolle, F. R. Wurm, S. Carlotti and D. Taton, The organocatalytic ring-opening polymerization of N-tosyl aziridines by an N-heterocyclic carbene, *Chem. Commun.*, 2016, **52**(62), 9719–9722.
39. C. Bakkali-Hassani, E. Rieger, J. Vignolle, F. R. Wurm, S. Carlotti and D. Taton, Expanding the scope of N-heterocyclic carbene-organocatalyzed ring-opening polymerization of N-tosyl aziridines using functional and non-activated amine initiators, *Eur. Polym. J.*, 2017, **95**, 746–755.
40. X. Wang, Y. Liu, Z. Li, H. Wang, H. Gebru, S. Chen, H. Zhu, F. Wei and K. Guo, Organocatalyzed Anionic Ring-Opening Polymerizations of N-Sulfonyl Aziridines with Organic Superbases, *ACS Macro Lett.*, 2017, 1331–1336.
41. L. Maisonneuve, O. Lamarzelle, E. Rix, E. Grau and H. Cramail, Isocyanate-Free Routes to Polyurethanes and Poly(hydroxy Urethane)s, *Chem. Rev.*, 2015, **115**(22), 12407–12439.
42. S. Penczek, M. Cypryk, A. Duda, P. Kubisa and S. Slomkowski, Living ring-opening polymerizations of heterocyclic monomers, *Prog. Polym. Sci.*, 2007, **32**(2), 247–282.
43. M. Cypryk, *Polymerization of Cyclic Siloxanes, Silanes, and Related Monomers*. 2012, pp. 451–476.
44. F. Müller and S. Silber, *Polysiloxanes*, 2012, 443–451.
45. W. T. Grubb and R. C. Osthoff, Kinetics of the Polymerization of a Cyclic Dimethylsiloxane1, *J. Am. Chem. Soc.*, 1955, **77**(6), 1405–1411.
46. C. Julian and W. Lech, Mechanism of the polymerization of hexamethylcyclotrisiloxane (D3) in the presence of a strong protonic acid, *Die Makromol. Chem.*, 1979, **180**(1), 117–130.
47. L. Wilczek and J. Chojnowski, Acidolytic ring opening of cyclic siloxane and acetal monomers. Role of hydrogen bonding in cationic polymerization initiated with protonic acids, *Macromolecules*, 1981, **14**(1), 9–17.
48. J. Tanaka, J. E. Dunning and J. C. Carter, Thermal Decomposition of Quaternary Ammonium Hydroxides. Synthesis and Properties of Tri(2H3-methyl)amine, *J. Org. Chem.*, 1966, **31**(10), 3431–3433.
49. A. R. Gilbert and S. W. Kantor, Transient catalysts for the polymerization of organosiloxanes, *J. Polym. Sci.*, 1959, **40**(136), 35–58.
50. A. Molenberg and M. Möller, A fast catalyst system for the ring-opening polymerization of cyclosiloxanes, *Macromol. Rapid Commun.*, 1995, **16**(6), 449–453.

51. A. Molenberg and M. Möller, Polymerization of cyclotrisiloxanes by organolithium compounds and P2-Et base, *Macromol. Chem. Phys.*, 1997, **198**(3), 717–726.
52. P. C. Hupfield and R. G. Taylor, Ring-Opening Polymerization of Siloxanes Using Phosphazene Base Catalysts, *J. Inorg. Organomet. Polym.*, 1999, **9**(1), 17–34.
53. A. Bessmertnykh, F. Ben, A. Baceiredo and G. Mignani, Anionic ring-opening polymerization of cyclic organosiloxanes using phosphorus ylides as strong non-ionic bases, *J. Organomet. Chem.*, 2003, **686**(1–2), 281–285.
54. B. G. Lohmeijer, G. Dubois, F. Leibfarth, R. C. Pratt, F. Nederberg, A. Nelson, R. M. Waymouth, C. Wade and J. L. Hedrick, Organocatalytic living ring-opening polymerization of cyclic carbosiloxanes, *Org. Lett.*, 2006, **8**(21), 4683–4686.
55. M. Rodriguez, S. Marrot, T. Kato, S. Stérin, E. Fleury and A. Baceiredo, Catalytic activity of N-heterocyclic carbenes in ring opening polymerization of cyclic siloxanes, *J. Organomet. Chem.*, 2007, **692**(4), 705–708.
56. S. Naumann, J. Klein, D. Wang and M. R. Buchmeiser, Convenient preparation of high molecular weight poly(dimethylsiloxane) using thermally latent NHC-catalysis: a structure-activity correlation, *Beilstein J. Org. Chem.*, 2015, **11**, 2261–2266.
57. W. A. Piccoli, G. G. Haberland and R. L. Merker, Highly Strained Cyclic Paraffin-Siloxanes, *J. Am. Chem. Soc.*, 1960, **82**(8), 1883–1885.
58. B. Suryanarayanan, B. W. Peace and K. G. Mayhan, Anionic polymerization of 2,2,5,5-tetramethyl-1-oxa-2,5-disilacyclopentane, *J. Polym. Sci., Polym. Chem. Ed.*, 1974, **12**(5), 1089–1107.
59. B. Suryanarayanan, B. W. Peace and K. G. Mayhan, Anionic polymerization of a series of five-membered cyclocarbosiloxanes, *J. Polym. Sci.: Poly. Chem. Ed.*, 1974, **12**(5), 1109–1123.
60. D. A. Loy, K. Rahimian and M. Samara, Phenylene-Bridged Cyclic Siloxanes as Precursors to Nonshrinking Sol-Gel Systems and Their Use as Encapsulants, *Angew. Chem., Int. Ed.*, 1999, **38**(4), 555–557.
61. K. Rahimian, D. A. Loy and P. P. Chen, Nonshrinking, Photopolymerizable Polycarbosiloxanes through Ring-Opening Polymerization of Disilaoxacyclopentane Monomers, *Chem. Mater.*, 2005, **17**(6), 1529–1534.
62. H. A. Brown, Y. A. Chang and R. M. Waymouth, Zwitterionic polymerization to generate high molecular weight cyclic poly(carbosiloxane)s, *J. Am. Chem. Soc.*, 2013, **135**(50), 18738–18741.
63. F. H. Westheimer, Why nature chose phosphates, *Science*, 1987, **235**(4793), 1173–1178.
64. F. Vella, Essential molecular biology review, *Biochem. Educ.*, 1997, **25**(1), 49.
65. C. Canto, K. J. Menzies and J. Auwerx, NAD(+) Metabolism and the Control of Energy Homeostasis: A Balancing Act between Mitochondria and the Nucleus, *Cell Metab.*, 2015, **22**(1), 31–53.

66. Y. Lee, P. C. Klauser, B. M. Brandsen, C. Zhou, X. Li and S. K. Silverman, DNA-Catalyzed DNA Cleavage by a Radical Pathway with Well-Defined Products, *J. Am. Chem. Soc.*, 2017, **139**(1), 255–261.
67. K. N. Bauer, H. T. Tee, M. M. Velencoso and F. R. Wurm, Main-chain poly(phosphoester)s: History, syntheses, degradation, bio-and flame-retardant applications, *Prog. Polym. Sci.*, 2017, **73**, 61–122.
68. T. Steinbach and F. R. Wurm, Poly(phosphoester)s: A New Platform for Degradable Polymers, *Angew. Chem., Int. Ed.*, 2015, **54**(21), 6098–6108.
69. G. Lapienis, *Ring-Opening Polymerization of Cyclic Phosphorus Monomers*, 2012, pp. 477–505.
70. Z. E. Yilmaz and C. Jerome, Polyphosphoesters: New Trends in Synthesis and Drug Delivery Applications, *Macromol. Biosci.*, 2016, **16**(12), 1745–1761.
71. H. Henke, O. Bruggemann and I. Teasdale, Branched Macromolecular Architectures for Degradable, Multifunctional Phosphorus-Based Polymers, *Macromol. Rapid Commun.*, 2017, **38**(4).
72. G. Lapienis and S. Penczek, Kinetics and Thermodynamics of the Polymerization of the Cyclic Phosphate Esters. II. Cationic Polymerization of 2-Methoxy-2-oxo-1,3,2-dioxaphosphorinane (1,3-Propylene Methyl Phosphate), *Macromolecules*, 1974, **7**(2), 166–174.
73. G. Lapienis and S. Penczek, Cationic Polymerization of 2-Alkoxy-2-oxo-1,3,2-dioxaphosphorinanes (1,3-Propylene Alkyl Phosphates), *Macromolecules*, 1977, **10**(6), 1301–1306.
74. G. Łapienis and S. Penczek, Kinetics and thermodynamics of anionic polymerization of 2-methoxy-2-oxo-1,3,2-dioxaphosphorinane, *J. Polym. Sci., Polym. Chem. Ed.*, 1977, **15**(2), 371–382.
75. W. Vogt und R. Siegfried, Polymere ester von säuren des phosphors, 4. Polymerisation des 2-äthoxy-2-oxo-4,5-dihydro-1,3,2-dioxaphosphorins, *Die Makromol. Chem.*, 1976, **177**(6), 1779–1789.
76. W. Vogt and R. Pflüger, Polymere ester von säuren des Phosphors, 3. Polymerisation des 2-Äthoxy-2-oxo-1,3,2-dioxaphospholans, *Die Makromol. Chem.*, 1975, **1**(S19751), 97–110.
77. J. Libiszowski, K. Kałużynski and S. Penczek, Polymerization of cyclic esters of phosphoric acid. VI. Poly(alkyl ethylene phosphates). Polymerization of 2-alkoxy-2-oxo-1,3,2-dioxaphospholans and structure of polymers, *J. Polym. Sci., Polym. Chem. Ed.*, 1978, **16**(6), 1275–1283.
78. S. Sosnowski, J. Libiszowski, S. Słomkowski and S. Penczek, Thermodynamics of the polymerization of ethylene methyl phosphate, *Die Makromol. Chem., Rapid Commun.*, 1984, **5**(5), 239–244.
79. J. Wen and R.-X. Zhuo, Enzyme-catalyzed ring-opening polymerization of ethylene isopropyl phosphate, *Macromol. Rapid Commun.*, 1998, **19**(12), 641–642.
80. C.-S. Xiao, Y.-C. Wang, J.-Z. Du, X.-S. Chen and J. Wang, Kinetics and Mechanism of 2-Ethoxy-2-oxo-1,3,2-dioxaphospholane Polymerization Initiated by Stannous Octoate, *Macromolecules*, 2006, **39**(20), 6825–6831.

81. J. Wen, G. J. A. Kim and K. W. Leong, Poly(d,llactide–co-ethyl ethylene phosphate)s as new drug carriers, *J. Controlled Release*, 2003, **92**(1–2), 39–48.
82. Y. Iwasaki and E. Yamaguchi, Synthesis of Well-Defined Thermoresponsive Polyphosphoester Macroinitiators Using Organocatalysts, *Macromolecules*, 2010, **43**(6), 2664–2666.
83. J. Liu, Y. Pang, W. Huang, X. Zhai, X. Zhu, Y. Zhou and D. Yan, Controlled Topological Structure of Copolyphosphates by Adjusting Pendant Groups of Cyclic Phosphate Monomers, *Macromolecules*, 2010, **43**(20), 8416–8423.
84. S. Zhang, A. Li, J. Zou, L. Y. Lin and K. L. Wooley, Facile Synthesis of Clickable, Water-soluble and Degradable Polyphosphoesters, *ACS Macro Lett.*, 2012, **1**(2), 328–333.
85. B. Clément, B. Grignard, L. Koole, C. Jérôme and P. Lecomte, Metal-Free Strategies for the Synthesis of Functional and Well-Defined Polyphosphoesters, *Macromolecules*, 2012, **45**(11), 4476–4486.
86. B. G. G. Lohmeijer, R. C. Pratt, F. Leibfarth, J. W. Logan, D. A. Long, A. P. Dove, F. Nederberg, J. Choi, C. Wade, R. M. Waymouth and J. L. Hedrick, Guanidine and Amidine Organocatalysts for Ring-Opening Polymerization of Cyclic Esters, *Macromolecules*, 2006, **39**(25), 8574–8583.
87. T. Steinbach, R. Schröder, S. Ritz and F. R. Wurm, Microstructure analysis of biocompatible phosphoester copolymers, *Polym. Chem.*, 2013, **4**(16), 4469.
88. Y. H. Lim, G. S. Heo, Y. H. Rezenom, S. Pollack, J. E. Raymond, M. Elsabahy and K. L. Wooley, Development of a Vinyl Ether-Functionalized Polyphosphoester as a Template for Multiple Post-polymerization Conjugation Chemistries and Study of Core Degradable Polymeric Nanoparticles, *Macromolecules*, 2014, **47**(14), 4634–4644.
89. H. Gao, Y. Tan, Q. Guan, T. Cai, G. Liang and Q. Wu, Synthesis, characterization and micellization of amphiphilic polyethylene-b-polyphosphoester block copolymers, *RSC Adv.*, 2015, **5**(61), 49376–49384.
90. Y. H. Lim, K. M. Tiemann, G. S. Heo, P. O. Wagers, Y. H. Rezenom, S. Zhang, F. Zhang, W. J. Youngs, D. A. Hunstad and K. L. Wooley, Preparation and in vitro antimicrobial activity of silver-bearing degradable polymeric nanoparticles of polyphosphoester-block-poly(L-lactide), *ACS Nano*, 2015, **9**(2), 1995–2008.
91. F. Zhang, S. Zhang, S. F. Pollack, R. Li, A. M. Gonzalez, J. Fan, J. Zou, S. E. Leininger, A. Pavia-Sanders, R. Johnson, L. D. Nelson, J. E. Raymond, M. Elsabahy, D. M. Hughes, M. W. Lenox, T. P. Gustafson and K. L. Wooley, Improving paclitaxel delivery: in vitro and in vivo characterization of PEGylated polyphosphoester-based nanocarriers, *J. Am. Chem. Soc.*, 2015, **137**(5), 2056–2066.
92. E. Baeten, S. Vanslambrouck, C. Jérôme, P. Lecomte and T. Junkers, Anionic flow polymerizations toward functional polyphosphoesters in

microreactors: Polymerization and UV-modification, *Eur. Polym. J.*, 2016, **80**, 208–218.

93. G.-P. Wu and D. J. Darensbourg, Mechanistic Insights into Water-Mediated Tandem Catalysis of Metal-Coordination CO2/Epoxide Copolymerization and Organocatalytic Ring-Opening Polymerization: One-Pot, Two Steps, and Three Catalysis Cycles for Triblock Copolymers Synthesis, *Macromolecules*, 2016, **49**(3), 807–814.
94. S. Schottler, G. Becker, S. Winzen, T. Steinbach, K. Mohr, K. Landfester, V. Mailander and F. R. Wurm, Protein adsorption is required for stealth effect of poly(ethylene glycol)- and poly(phosphoester)-coated nanocarriers, *Nat. Nanotechnol.*, 2016, **11**(4), 372–377.
95. J. Muller, K. N. Bauer, D. Prozeller, J. Simon, V. Mailander, F. R. Wurm, S. Winzen and K. Landfester, Coating nanoparticles with tunable surfactants facilitates control over the protein corona, *Biomaterials*, 2017, **115**, 1–8.
96. G. Becker, T. A. Marquetant, M. Wagner and F. R. Wurm, Multifunctional Poly(phosphoester)s for Reversible Diels–Alder Postmodification To Tune the LCST in Water, *Macromolecules*, 2017.
97. T. S. Stukenbroeker, D. Solis-Ibarra and R. M. Waymouth, Synthesis and Topological Trapping of Cyclic Poly(alkylene phosphates), *Macromolecules*, 2014, **47**(23), 8224–8230.
98. B. E. Maryanoff and A. B. Reitz, The Wittig olefination reaction and modifications involving phosphoryl-stabilized carbanions. Stereochemistry, mechanism, and selected synthetic aspects, *Chem. Rev.*, 1989, **89**(4), 863–927.
99. T. Steinbach; C. Wahlen; F. R. Wurm, Poly(phosphonate)-mediated Horner–Wadsworth–Emmons reactions. *Polym. Chem.* 2015, 6(7), 1192–1202.
100. F. G. Bordwell, Equilibrium acidities in dimethyl sulfoxide solution, *Acc. Chem. Res.*, 2002, **21**(12), 456–463.
101. C. Qi, Y. Du, Y. Lu, X. Sun and X. M. Zhang, Equilibrium acidities and homolytic bond dissociation enthalpies of the acidic C–H bonds in phosphonates and related phosphorus containing compounds, *J. Org. Chem.*, 2009, **74**(21), 8078–8085.
102. L. L. Clark, E. D. Ingall and R. Benner, Marine organic phosphorous cycling; novel insights from nuclear magnetic resonance, *Am. J. Sci.*, 1999, **299**(1), 724–737.
103. P. Kafarski and B. Lejczak, Aminophosphonic acids of potential medical importance, *Curr. Med. Chem. Anticancer Agents*, 2001, **1**(3), 301–312.
104. F. Millich and C. E. Carraher, Interfacial Syntheses of Polyphosphonate and Polyphosphate Esters .I. Effects of Alkaline Medium, *J. Polym. Sci., Part A-1: Polym. Chem.*, 1969, 7(9pa1), 2669.
105. F. Millich and C. E. Carraher, Interfacial syntheses of polyphosphonate and polyphosphate esters. II. Dependence of yield and molecular weight on solvent volumes and concentrations of comers in basic

polymerization of hydroquinone and phenylphosphonic dichloride, *J. Polym. Sci., Part A-1: Polym. Chem.*, 1970, **8**(1), 163–169.

106. F. Millich and C. E. Carraher, Interfacial Syntheses of Polyphosphonate and Polyphosphate Esters. III. Influences of Solvents, Diffusion Rates, Temperature, and Other Factors on Yield and Molecular Weight in the Alkaline Interfacial Polycondensation of Hydroquinone and Phenylphosphonic Dichloride, *Macromolecules*, 1970, **3**(2), 253–256.
107. F. Millich and C. E. Carraher, Interfacial Synthesis of Polyphosphonate and Polyphosphate Esters .5. Poly(Phosphonate Esters) Containing Thymidine, 2-Deoxy-D-Ribose, or Xanthine, *J. Polym. Sci. A1*, 1971, **9**(6), 1715–1721.
108. L. Chen and Y. Z. Wang, Aryl Polyphosphonates: Useful Halogen-Free Flame Retardants for Polymers, *Materials*, 2010, **3**(10), 4746–4760.
109. V. N. Sharov and A. L. Klebanskii, Polymers based on cyclic polyfluoralkylene alkyl (aryl) phosphonates, *Polym. Sci. USSR*, 1973, 2777–2782.
110. K. Kałużynski, J. Libiszowski and S. Penczek, Poly(2-hydro-2-oxo-l,3,2-dioxaphosphorinane)*). Preparation and NMR Spectra, *Die Makromol. Chem.*, 1977, **178**(10), 2943–2947.
111. T. Steinbach, S. Ritz and F. R. Wurm, Water-Soluble Poly(phosphonate)s via Living Ring-Opening Polymerization, *ACS Macro Lett.*, 2014, **3**(3), 244–248.
112. T. Steinbach and F. R. Wurm, Degradable Polyphosphoester-Protein Conjugates: "PPEylation" of Proteins, *Biomacromolecules*, 2016, **17**(10), 3338–3346.
113. D. Russo, M. Plazanet, J. Teixeira, M. Moulin, M. Hartlein, F. R. Wurm and T. Steinbach, Investigation into the Relaxation Dynamics of Polymer-Protein Conjugates Reveals Surprising Role of Polymer Solvation on Inherent Protein Flexibility, *Biomacromolecules*, 2016, **17**(1), 141–147.
114. T. Wolf, T. Steinbach and F. R. Wurm, A Library of Well-Defined and Water-Soluble Poly(alkyl phosphonate)s with Adjustable Hydrolysis, *Macromolecules*, 2015, **48**(12), 3853–3863.
115. E. K. Macdonald and M. P. Shaver, An expanded range of catalysts for synthesising biodegradable polyphosphonates, *Green Mater.*, 2016, **4**(2), 81–88.
116. T. Wolf, J. Nass and F. R. Wurm, Cyclohexyl-substituted poly(phosphonate)-copolymers with adjustable glass transition temperatures, *Polym. Chem.*, 2016, **7**(17), 2934–2937.
117. B. Obermeier and H. Frey, Poly(ethylene glycol-co-allyl glycidyl ether)s: a PEG-based modular synthetic platform for multiple bioconjugation, *Bioconjugatr Chem.*, 2011, **22**(3), 436–444.
118. T. Wolf, T. Rheinberger and F. R. Wurm, Thermoresponsive coacervate formation of random poly(phosphonate) terpolymers, *Eur. Polym. J.*, 2017.
119. T. Wolf, T. Rheinberger, J. Simon and F. R. Wurm, Reversible Self-Assembly of Degradable Polymersomes with Upper Critical Solution Temperature in Water, *J. Am. Chem. Soc.*, 2017, **139**(32), 11064–11072.

120. H. Lin, T. Wolf, F. R. Wurm and M. A. Kelland, Poly(alkyl ethylene phosphonate)s: A New Class of Non-amide Kinetic Hydrate Inhibitor Polymers, *Energy Fuels*, 2017, **31**(4), 3843–3848.
121. H. R. Kricheldorf and G. Schwarz, Poly(thioester)s, *J. Macromol. Sci., Part A*, 2007, **44**(6), 625–649.
122. C. G. Overberger and J. K. Weise, Anionic ring-opening polymerization of thiolactones, *J. Am. Chem. Soc.*, 1968, **90**(13), 3533–3537.
123. C. G. Overberger and J. K. Weise, Optical rotatory dispersion studies of asymmetric poly(thiol esters), *J. Am. Chem. Soc.*, 1968, **90**(13), 3538–3543.
124. C. G. Seefried and J. V. Koleske, Lactone polymers. V. Viscoelastic properties of poly-ε-caprolactone and poly-ε-thiocaprolactone, *J. Macromol. Sci., Part B*, 1974, **10**(4), 579–589.
125. F. Sanda, D. Jirakanjana, M. Hitomi and T. Endo, Anionic Ring-Opening Polymerization of ε-Thionocaprolactone, *Macromolecules*, 1999, **32**(24), 8010–8014.
126. F. Sanda, D. Jirakanjana, M. Hitomi and T. Endo, Cationic ring-opening polymerization of ε-thionocaprolactone: Selective formation of polythioester, *J. Polym. Sci., Part A: Polym. Chem.*, 2000, **38**(22), 4057–4061.
127. K. Shimokawa, M. Kato and S. Matsumura, Enzymatic Synthesis and Chemical Recycling of Polythiocaprolactone, *Macromol. Chem. Phys.*, 2011, **212**(2), 150–158.
128. T. J. Bannin and M. K. Kiesewetter, Poly(thioester) by Organocatalytic Ring-Opening Polymerization, *Macromolecules*, 2015, **48**(16), 5481–5486.
129. P. P. Datta and M. K. Kiesewetter, Controlled Organocatalytic Ring-Opening Polymerization of epsilon-Thionocaprolactone, *Macromolecules*, 2016, **49**(3), 774–780.
130. N. S. Isaacs, The Chemistry of Episulfides. 1. The reactivities of propylene sulfide and propylene oxide towards aniline, *Can. J. Chem.*, 1966, **44**, 395–402.
131. S. Boileau, G. Champetier and P. Sigwalt, Polymérisation anionique du sulfure de propylène, *Die Makromol. Chem.*, 1963, **69**(1), 180–192.
132. C. Bonnans-Plaisance and G. Levesque, Homo- and copolymerization of unprotected 2-(hydroxymethyl)thiirane initiated by quaternary ammonium salts of dithiocarboxylic acids, *Macromolecules*, 1989, **22**(4), 2020–2023.
133. C. Bonnans-Plaisance, S. Courric and G. Levesque, Functional polythiiranes, *Polym. Bull.*, 1992, **28**(5), 489–495.
134. E. Nol, C. Bonnans-Plaisance and G. Levesque, A New Initiator System for the Living Thiiranes Ring-Opening Polymerization: A Way toward Star-Shaped Polythiiranes, *Macromolecules*, 1999, **32**(13), 4485–4487.
135. A. Rehor, N. Tirelli and J. A. Hubbell, A New Living Emulsion Polymerization Mechanism: Episulfide Anionic Polymerization, *Macromolecules*, 2002, **35**(23), 8688–8693.
136. A. Napoli, N. Tirelli, G. Kilcher and A. Hubbell, New Synthetic Methodologies for Amphiphilic Multiblock Copolymers of Ethylene Glycol and Propylene Sulfide, *Macromolecules*, 2001, **34**(26), 8913–8917.

137. A. Napoli, M. Valentini, N. Tirelli, M. Muller and J. A. Hubbell, Oxidation-responsive polymeric vesicles, *Nat. Mater.*, 2004, **3**(3), 183–189.
138. S. Cerritelli, D. Velluto and J. A. Hubbell, PEG-SS-PPS: reduction-sensitive disulfide block copolymer vesicles for intracellular drug delivery, *Biomacromolecules*, 2007, **8**(6), 1966–1972.
139. S. S. Yu, R. L. Scherer, R. A. Ortega, C. S. Bell, C. P. O'Neil, J. A. Hubbell and T. D. Giorgio, Enzymatic- and temperature-sensitive controlled release of ultrasmall superparamagnetic iron oxides (USPIOs), *J. Nanobiotechnol.*, 2011, **9**, 7.

CHAPTER 11

Organopolymerization of Acrylic Monomers

WUCHAO ZHAO AND YUETAO ZHANG*

State Key Laboratory of Supramolecular Structure and Materials, College of Chemistry, Jilin University, Changchun, Jilin, 130012, China
*Email: ytzhang2009@jlu.edu.cn

11.1 Introduction

The previous chapters discussed the organocatalytic ring-opening polymerization (ROP) of cyclic monomers, such as cyclic esters and epoxides. We will focus on the recent advancements that have been made in the (co)polymerization of acrylic monomers by using organic catalysts, with the polymerization mechanism as the main clue throughout this chapter. Polymerization activities can be significantly affected by the differences in electronic properties and steric hindrance due to the structural diversities of monomers.[1,2] Moreover, there is no single type of polymerization or catalyst system that can meet all the demanding and challenging requirements of polymerization. To achieve controlled/living polymerization, it might need different combinations of catalysts, or even a co-catalyst for initiating the polymerization or enhancing the polymerization activity.[3] In addition to the above, the choice of solvent also has a drastic impact on the polymerization performance. These factors will be reflected in our discussion of organic polymerization of polar vinyl monomers in detail. It should be noted that organoaluminum-catalyzed polymerization will also be included.

According to the polymerization mechanism, organopolymerization can be divided into four types of polymerization: group transfer polymerization

Polymer Chemistry Series No. 31
Organic Catalysis for Polymerisation
Edited by Andrew Dove, Haritz Sardon and Stefan Naumann

Published by the Royal Society of Chemistry, www.rsc.org

(GTP), anionic, zwitterionic and radical polymerization. It should be noted that radical polymerization relying on free radicals, such as nitroxide-mediated radical polymerization,[4] atom transfer radical polymerization,[5] and reversible addition–fragmentation chain transfer polymerization[6] will not be included in this chapter. First, the GTP (Section 11.2) can be catalyzed either by organic bases or organic acids. Then, Section 11.3 will be devoted to the chronological development of Lewis pair polymerization (LPP) due to its unique polymerization characteristic. Finally, anionic and zwitterionic polymerizations will be described together in Section 11.4. Anionic polymerization is a form of chain-growth polymerization that encompasses the polymerization of vinyl monomers with strong electron-withdrawing groups. In zwitterionic polymerization, the key intermediate is a zwitterion, which can be formed either from polymerization catalyzed by Lewis pairs composed of a Lewis acid and a Lewis base (so called "LPP") or from the nucleophilic attack of monomers by organic catalysts.

11.2 Organocatalytic Group Transfer Polymerization

As a major breakthrough disclosed in polymer chemistry by Webster *et al.* in 1983, group transfer polymerization (GTP) is employed for the living polymerization of (meth)acrylic monomers at ambient temperature,[7] involving the repetitive Mukaiyama–Michael reaction catalyzed by a Lewis base or a Lewis acid.[7,8] Initially, GTP was discovered and developed to overcome the need for low temperatures in anionic polymerization of acrylic monomers.[8] During GTP, since the end of the polymer chain was capped by silyl group transferred from the silyl ketene acetal (SKA), and thus furnishing the less active chain ends than that of living anionic polymerization, this is one of the reasons why GTP generally can be carried out at room temperature (RT) and is more tolerant to some functional groups.[9] However, specific catalysts are required for the activation of SKA or various monomers investigated in the GTP. A series of Lewis basic catalysts have been developed, such as tris(dialkylamino)sulfonium salts and tetraalkylammonium salts of $SiMe_3F^{2-}$, HF^{2-}, F^-, CN^-, N_3^-, oxyanions, and hydrogen bioxyanions, exhibiting better performance for the polymerization of methacrylate than acrylates due to their high catalytic activity.[10–12] On the other hand, Lewis acids, such as zinc halides, organoaluminums, and mercury(II) iodide, are also found to be effective for polymerization of acrylates.[10,13] After the first organocatalytic living ROP of lactide using 4-(dimethylamino)pyridine (DMAP) as catalyst,[14] N-heterocyclic carbenes (NHCs), have been found to be capable of efficiently catalyzing the GTP of both methyl methacrylate (MMA) and *tert*-butyl acrylate (*t*BA) in a living/controlled manner.[15,16] It turned out that the application of organic catalysts in GTP is very effective for the polymerization of vinyl monomers, which could be extended beyond the limits of conventional GTP in terms of monomer scope, catalytic activity, molecular weight and polymer structure control. Both basic catalyst, such as NHCs, organic strong bases, phosphines, polar donor solvents

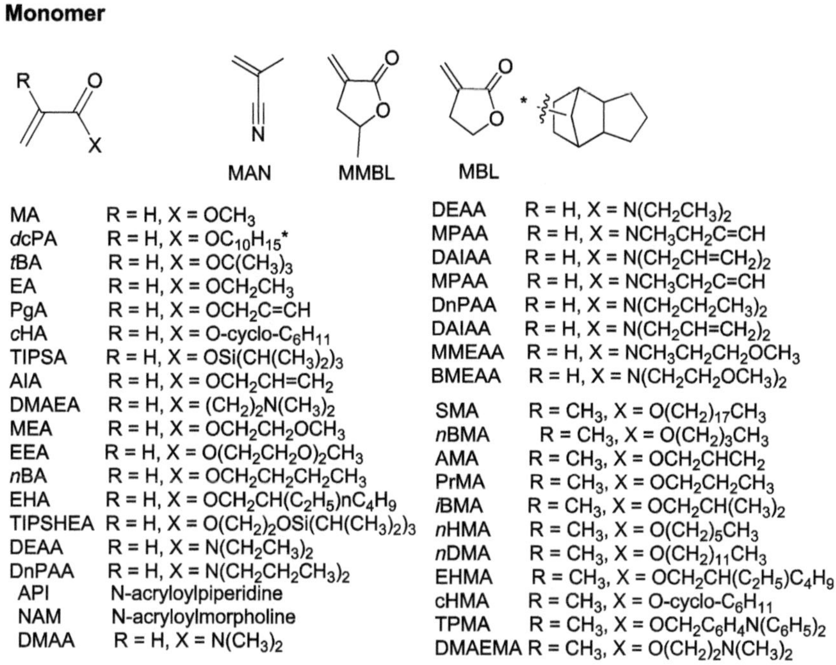

Figure 11.1 Monomers employed in organocatalytic GTP.

and acidic catalyst, such as, tris(pentafluorophenyl)aluminum ($Al(C_6F_5)_3$), tris(pentafluorophenyl)borane ($B(C_6F_5)_3$) with a silylating agent, triphenylmethyl salts, and organic strong Brønsted acids have been utilized as organocatalysts for GTP.

We aim to summarize the recent developments of metal-free GTP since 2000 as follows: including the features and scopes of Lewis basic and Lewis acidic catalysts, monomers (Figure 11.1), SKA and hydrosilane (Figure 11.2) employed in the organocatalytic GTP as well as the precise control of diverse polymer architectures, such as block copolymers and star-shaped polymers.

11.2.1 Organic-base-catalyzed GTP

Lewis bases (LBs), especially strongly basic but weakly nucleophilic ones, are commonly used in organic catalysis to suppress side effects. There are two different mechanisms proposed for Lewis base catalyzed GTP: one is the *associative mechanism*, in which the silyl group coordinated with a nucleophilic catalyst is transferred intramolecularly to the incoming monomer through pentacoordinate anionic silicon species (path a, Scheme 11.1); the other is the *dissociative mechanism*, which involves the ester enolate anion as propagating species and a rapid, reversible complexation or termination of small concentrations of enolate anion with SKA or its polymer homologue (path b, Scheme 11.1.). In the associative mechanism, the activation and

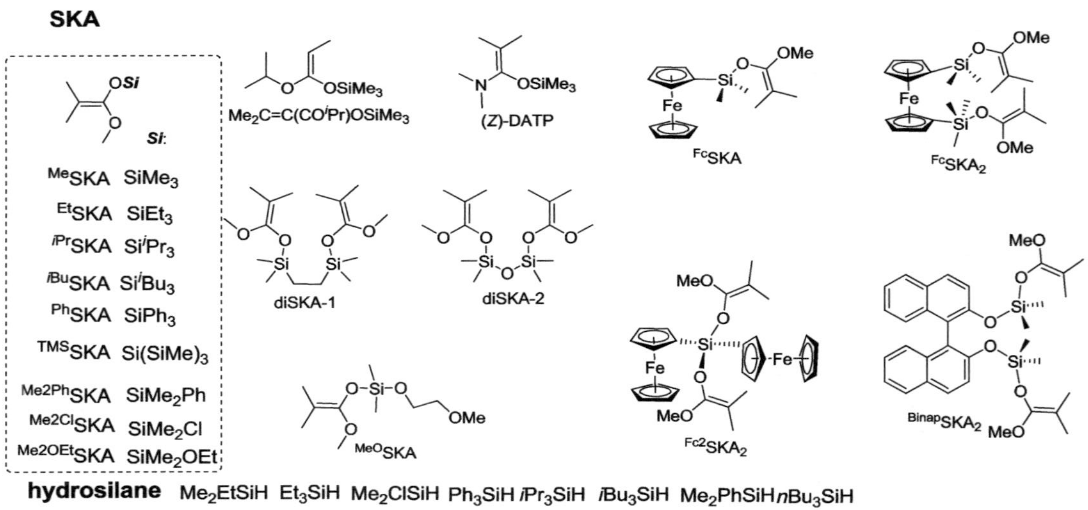

Figure 11.2 SKA and hydrosilane utilized in organocatalytic GTP.

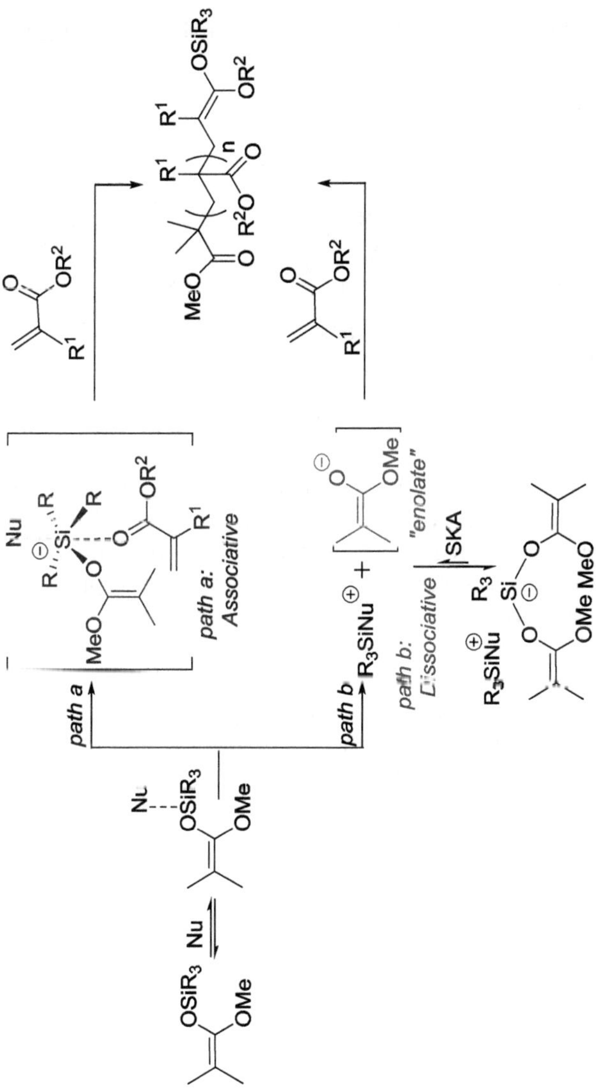

Scheme 11.1 Proposed mechanism of LB-catalyzed GTP.

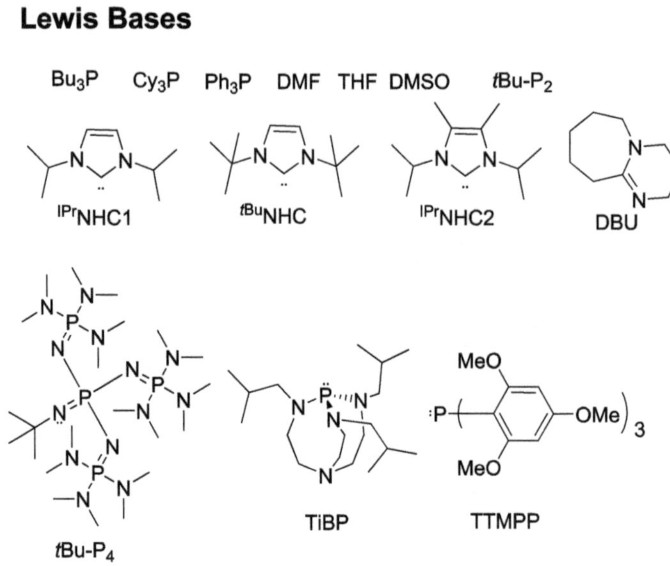

Figure 11.3 Lewis bases employed in organocatalytic GTP.

deactivation steps of the silyl enolate must be in a rapid equilibrium process to ensure that the system has a good control over the polymerization to produce desired polymer products. Meanwhile, in the dissociative mechanism the rate of reversible reaction between the production of enol anions and the formation of hypervalent compounds should be faster than the rate of chain propagation so that controlled polymerization can be accomplished.

Various Lewis basic catalysts, such as NHCs, proazaphosphatranes, phosphazene bases, phosphines and polar donor solvents (DMF, THF, and DMSO), have been widely applied in GTP as organocatalysts since 2008 (Figure 11.3). GTPs catalyzed by organic bases developed in this book are listed in Table 11.1. In general, a 0.05–20 mol% catalyst loading of these organocatalysts are used relative to the initiator, producing polymers with a predictable number-average molecular weight (M_n) and a narrow molecular weight distribution (MWD). It should be noted that such Lewis basic organocatalytic GTP generally exhibits features of rapid initiation and fast growth but is not well-controlled on the tacticity of polymer. Moreover, some polar donor solvents can promote the polymerization even in the absence of a catalyst.

11.2.1.1 NHCs

In 2008, Taton and Gnanou *et al.* first employed NHCs, such as 1,3-diisopropyl-imidazol-2-ylidene (^{IPr}NHC1) and 1,3-di-*tert*-butyl-imidazol-2-ylidene (^{tBu}NHC), for GTP of MMA and *t*BA at RT with methyl trimethylsilyl

Table 11.1 GTP catalyzed by organic bases.

Entry	Catalyst	Initiator	Solvent	Polymer	M_n (kg mol^{-1})	$Đ$ (M_w/M_n)	Ref.
1	^{tBu}NHC	^{Me}SKA	THF	PMMA	2.6–110	1.10–1.30	15
2	^{tBu}NHC	^{Me}SKA	THF	P*t*BuA	4.8	1.30	15
3	^{tBu}NHC	^{Me}SKA	TOL	P*t*BuA	11.5	1.60	15
4	^{tBu}NHC	^{Me}SKA	THF	PDMAEMA	18.0	1.19	18
5	^{tBu}NHC	^{Me}SKA	THF	PDMA	37.0	1.17	18
6	^{tBu}NHC	^{Me}SKA	THF	P*n*BA	33.0	1.31	18
7	^{IPr}NHC2	^{Me}SKA	THF	PMMA	4.1–18.9	1.14–1.72	16
8	^{IPr}NHC2	^{Me}SKA	THF	P*t*BA	6.1–205	1.14–1.47	16
9	^{IPr}NHC1	^{Me}SKA	THF	PMMA	19.0	1.07	18
10	^{IPr}NHC1	^{Me}SKA	THF	P*n*BA	31.0	1.39	18
11	^{IPr}NHC1	^{Me}SKA	TOL	P*t*BuA	4.9–22.2	1.19–1.40	15
12	^{IPr}NHC1	^{Me}SKA	THF	P*t*BA	5.0–40.0	1.09–1.19	18
13	^{IPr}NHC1	^{Me}SKA	TOL	P*t*BA	5.1–14.0	1.25–1.45	18
14	^{IPr}NHC1	^{Me}SKA	THF	PDMAEMA	6.5–34.0	1.10–1.15	18
15	^{IPr}NHC1	^{Me}SKA	THF	PDMAEA	23.0	1.46	18
16	^{IPr}NHC1	^{Me}SKA	THF	PDMA	24.0–42.0	1.15	18
17	^{IPr}NHC1	^{Me}SKA	DMF	PMAN	5.0–18.0	1.37–1.51	18
18	^{IPr}NHC1	^{Me}SKA	THF	PMMA	10.8–22.5	1.08–1.09	15
19	^{IPr}NHC1	^{Me}SKA	TOL	PMMA	10.9	1.30	15
20	^{IPr}NHC1	^{Me}SKA	THF	P*t*BuA	5.8–33.0	1.15–1.25	15
21	TTMPP	^{Me}SKA	THF	PMMA	3.0–10.6	1.13–1.15	25
22	TTMPP	^{Me}SKA	Bulk	PMMA	3.6–57.0	1.30–1.37	25
23	TTMPP	^{Me}SKA	THF	P*t*BA	2.1–7.8	1.12–1.45	25
24	TiBP	^{Me}SKA	THF	PMMA	6.5–55.9	1.05–1.14	20
25	*t*Bu-P_4	^{Me}SKA	TOL	PMMA	6.5–110	1.15–1.32	20

dimethylketene acetal (^{Me}SKA) as initiator, producing polymers with molar masses matching the [monomer]$_0$/[SKA]$_0$ ratio and narrow MWD ($Đ < 1.30$). It is noted that the obtained PMMA with a 110 kg mol^{-1} molar mass has not been achieved using regular anionic catalysts for GTP.[15] Such an NHC-based catalyst system was also applied to the sequential GTP of (meth)acrylic monomers in THF for the synthesis of block copolymers, including alkyl (meth)acrylate monomer units, blocks deriving from *N*,*N*-dimethylacrylamide (DMAA) and methacrylonitrile as well, regardless of the order of addition of the two monomers.[17] Subsequently, Taton *et al.* used ^{IPr}NHC1 and ^{tBu}NHC to further expand the scope of polymerized monomers, and they found that 2-(dimethylamino)ethyl methacrylate (DMAEMA), *n*-butyl acrylate (*n*BA), 2-(dimethylamino)ethylacrylate (DMAEA), DMAA, and methacrylonitrile (MAN) can also be polymerized in a controlled fashion.[18] Among the five screened monomers, the polymerization of DMAEMA had the lowest polymerization rate, but well-defined PDMAEMA could be achieved with $M_n = 6.5–34.0$ kg mol^{-1} and $Đ = 1.10–1.09$. ^{IPr}NHC1 exhibited better control over the polymerization of DMAEMA than ^{tBu}NHC. The GTP of the other monomers, such *n*BA and DMAEA, by ^{IPr}NHC1 is less controlled and yielded polymers with broad MWD up to 1.46.

Taton and Gnanou *et al.* advocated that the GTP catalyzed by ^{IPr}NHC1 and ^{tBu}NHC was promoted *via* an associative mechanism based on the following evidence: (1) the polymerization rate dramatically increased with the increase in the concentration of initiator; (2) the reaction of ^{Me}SKA and ^{IPr}NHC1 in 1/1 molar ratio did not form enolate-type species as revealed by ^{29}Si and ^{13}C NMR spectroscopy; (3) well-defined PMMA could be obtained with a ^{IPr}NHC1/^{Me}SKA (1 : 1) catalyst system; (4) termination reactions such as backbiting or internal isomerization could be drastically minimized during NHC-catalyzed GTP of acrylates; (5) regardless of the addition order of monomer, the sequential GTP of MMA and acrylates afforded block copolymers.[19]

Meanwhile, Waymouth and Hedrick *et al.* also demonstrated that in the presence of ^{Me}SKA, 1,3-diisopropyl-4,5-dimethylimidazol-2-ylidene (^{IPr}NHC2) serves as an effective neutral nucleophilic catalyst for the organocatalytic GTP of MMA and *t*BA at RT, yielding PMMA and P*t*BA with predictable M_n and narrow MWD.[16] Waymouth and Hedrick *et al.* assumed that the GTP catalyzed by ^{IPr}NHC2 proceeds *via* a dissociative mechanism. Because they found in their research that: first, the rate of polymerization was inversely related to the amount of the catalyst, consistent with the theoretical polymerization rate equation of the GTP proceeding *via* a dissociative mechanism, and secondly, the tacticity of the resulting PMMA (*mm*/*mr*/*rr* = 0.09/0.46/0.45) was similar to that obtained from the tris(dimethy1amino)sulfonium bifluoride ($TASHF_2$) catalyzed GTP, in which an enolate anion was generated,[10] it is believed that the dissociative mechanism tends to afford a polymer with a lower stereoregularity than the associative mechanism.

11.2.1.2 Organic Superbase

In 2011, Kakuchi and co-workers employed various organic superbases as efficient catalysts for GTP of MMA.[20] Among these investigated superbases, both proazaphosphatrane (P(*i*-BuNCH$_2$CH$_2$)$_3$-N, TiBP) and phosphazene base (1-*tert*-butyl-4,4,4-tris(dimethylamino)-2,2-bis[tris(dimethylamino)-phosphoranylidenamino]-2Λ^5,4Λ^5-catenadi(phosphazene), *t*Bu-P$_4$) exhibited silicon activation ability for precise synthesis of PMMA with controlled M_n (up to 109.6 kg mol^{-1}) and narrow MWD. The living nature of both polymerizations could be confirmed by MALDI-TOF MS analysis and post-polymerization experiments. A dissociative mechanism was proposed for TiBP and *t*Bu-P$_4$-catalyzed GTP based on the following evidence: (1) the stereoregularity of the resultant PMMA was quite similar to that obtained from the $TASHF_2$-catalyzed GTP; (2) the Mukaiyama aldol reaction catalyzed by proazaphosphatranes generated an enolate anion from ^{Me}SKA[21] and all the reactions involved organosilicon compounds catalyzed by *t*Bu-P$_4$ proceeded through the generation of anionic species.[22] Later, the *t*Bu-P$_4$ catalyst was also combined with a multivalent initiator for controlled GTP of different monomers, such as DMAEMA, stearyl methacrylate (SMA), allyl methacrylate (AMA), and 4-(*N*,*N*-diphenylamino) (TPMA), producing star-shaped polymers with low $Đ$ value (<1.17).[23,24]

11.2.1.3 Phosphines

In 2012, Taton and co-workers reported the first example of phosphine, tris(2,4,6-trimethoxyphenyl)phosphine (TTMPP), as an efficient catalyst for GTP of MMA and *t*BA with ^{Me}SKA as initiator in THF at RT.[25] With 10 mol% catalyst loading relative to ^{Me}SKA initiator, TTMPP/^{Me}SKA exhibited a better control over GTP of MMA than that of *t*BA. The PMMA produced in THF had predictable M_n and narrow MWD ($Đ < 1.37$ in bulk and $Đ < 1.45$ in THF). Sequential GTP of TTMPP was utilized to synthesize block copolymer PMMA-*b*-P*t*BA with controlled M_n and final $Đ$ value <1.30. As shown in the first order kinetic plot for TTMPP-catalyzed GTP of MMA in THF, an induction period of about 2 h might be ascribed to the occurrence of a dissociative GTP mechanism, involving the formation of dormant bisenolate-type species, which was further confirmed by the similar tacticity of the obtained PMMA (*mm*/*mr*/*rr* = 0.06/0.42/0.52) to that observed from a PMMA synthesized by the tris(dimethylamino)sulfonium difluorotrimethylsilicate ($TASHF_2SiMe_3$) catalyst system at RT,[10] since $TASF_2SiMe_3$ was thought to catalyze GTP *via* a dissociation mechanism. However, the formation of enolate-type species was not observed in ^{13}C or ^{29}Si NMR experiments.

11.2.1.4 Polar Donor Solvents

In 2016, Chen and co-workers employed neutral SKA alone to initiate either controlled or extremely rapid anionic polymerization in polar donor solvents such as DMF, without additional catalysts employed for the typical GTP systems.[26] More specifically, the GTP of MMA in DMF with a mono-SKA such as ^{Me}SKA is relatively slow (several hours to the completion of polymerization of 100 eq. MMA) but controlled and remarkably efficient, producing PMMA with predictable M_n values, low $Đ$ values (≤1.20), and high initiation efficiencies ($I^*\% \geq 80$). In contrast, the GTP of MMA by di-SKAs linked by the oxo, ferrocenyl, or binaphthyl bridge exhibited several orders of magnitude rate enhancement (a few seconds to completion) than the polymerization by the mono-SKA, but uncontrolled, producing PMMA with high $Đ$ value (2.00–4.00) and much higher than the predicted M_n, and thus yielding a low $I^*\%$ of 36. It is also noted that the donor ability of the solvent plays a critical role in promoting the activity of the GTP by neutral SKA alone.

Chen and co-workers proposed the dissociative mechanism (Scheme 11.2) for GTP using di-SKAs in DMF based on the following evidence: first, the system was relatively active but uncontrolled, which indicated the formation of reactive enolate anion. Second, the highly reactive and unstable enolate anion was also responsible for the indefinite kinetics of this di-SKA system. Third, the polymerization rate of the di-SKA system was more than three orders of magnitude higher than the mono-SKA system, as the current mono-SKA cannot access this pathway without chelation provided by the second SKA moiety present in the di-SKA structure. Fourth, the low initiation efficiency was associated with the disfavored equilibrium to form the highly

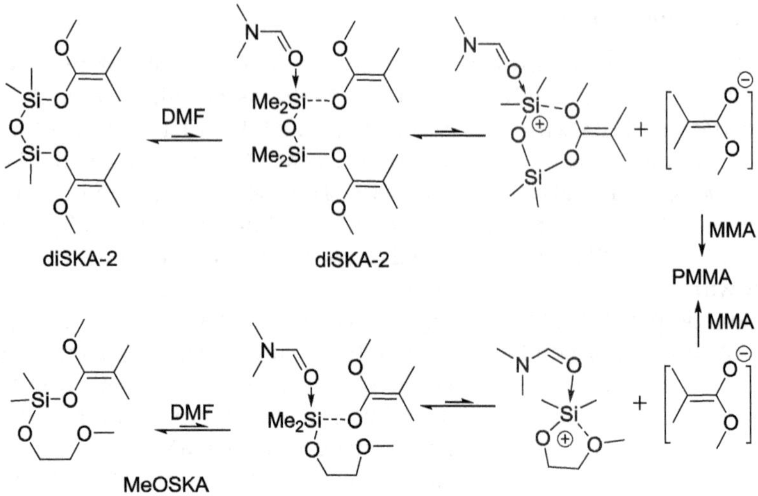

Scheme 11.2 Proposed mechanism of GTP in DMF.
Adapted with permission from *Macromolecules* **2016**, *49*, 8075–8087. Copyright (2016) American Chemical Society.

reactive silylium cation and enolate anion and some decomposition of the enolate anion. Fifth, all the investigated di-SKA showed essentially identical polymerization activity and stereoselectivity, regardless of the structure and chirality, because the active species involved was the same enolate anion. It is worth noting that the stereoselectivity ($rr = 58 \pm 2\%$) of the PMMA by these SKAs at 25 °C in DMF is similar to that obtained from the $TASHF_2$-catalyzed GTP at 20 °C in THF (56% *rr*).[10] Sixth, control reactions using mono-SKA initiator, ^{MeO}SKA, with a flexible and electron donating sidearm on the silicon, also showed rapid polymerization of MMA (TOF up to 4.8×10^4 h^{-1}) and had a similar performance to those di-SKA initiators. This result further confirmed that the proposed coordination was essential but not limited to the di-SKAs, as long as the intermediate (Scheme 11.2) can be stabilized. And as for the mono-SKA system, the polymerization rate decreased with increase in the initiator concentration, which is also indicative of a dissociative mechanism.

11.2.2 Organic-acid-catalyzed GTP

In contrast to the requirement of a high catalyst loading (typically 10–20 mol% relative to monomer) for the conventional Lewis acid (LA)-catalyzed polymerizations of acrylic monomers represented by zinc halides and alkyl aluminums,[13] organic-acid-catalyzed GTP could be applied to more monomers and organic acids. For instance, $B(C_6F_5)_3$ with a silylating agent and strong organic Brønsted acids (Figure 11.4) could be extended to GTP of various monomers (as shown in Figure 11.1).[27] In general, the amount of LA is 1–50 mol% relative to the initiator, and non-polar donor solvents such as TOL, cyclohexane and dichloromethane are beneficial to

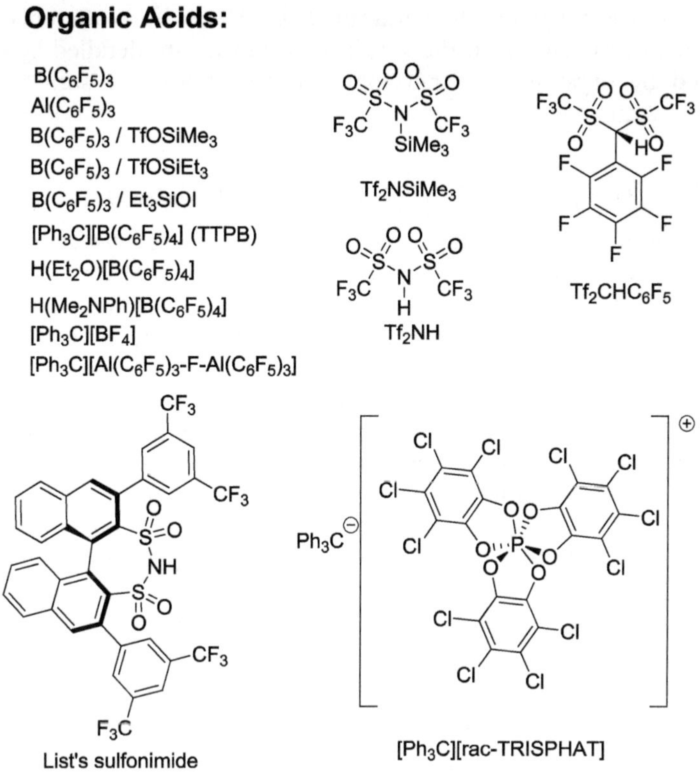

Figure 11.4 Organic acids employed in organocatalytic GTP.

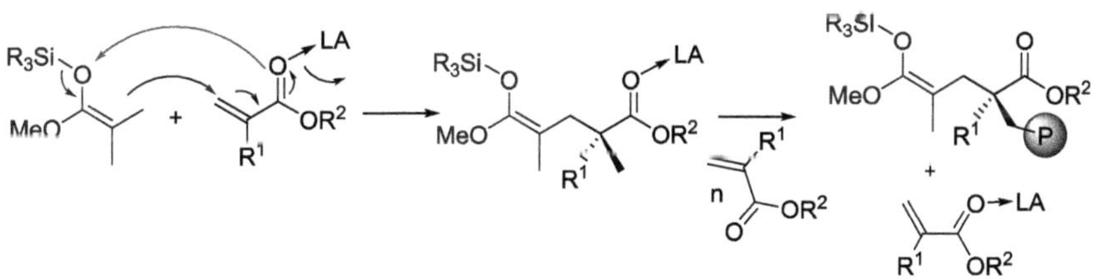

Scheme 11.3 Mechanism of LA-catalyzed GTP.

polymerization but polar donor solvents such as DMF and THF inhibit GTP. Organic-acid-catalyzed GTP proceeds with the activation of monomer mechanism, as shown in Scheme 11.3, in which the polymerization was initiated with the intermolecular Michael-addition of the ester enolate group of the SKA to the vinyl group of the LA-activated monomer; meanwhile, the Si–O bond of SKA was cleaved and the silyl group was transferred to the carbonyl group of the LA-activated monomer to generate the single-monomer addition species or the active propagating species. In the propagation cycle, the catalyst was released from the propagating chain to the incoming monomer, followed by the intermolecular Michael-addition to generate the polymeric SKA intermediate. The subtle differences in the mechanism resulting from the employed catalysts are detailed below. GTP catalyzed by organic acids developed in this book are summarized in Tables 11.2 and 11.3.

11.2.2.1 *Tris(pentafluorophenyl)borane ($B(C_6F_5)_3$)*

In the presence of silylating agent, the metal-free Lewis acid, $B(C_6F_5)_3$-catalyzed controlled GTP of ethyl acrylate (EA) and MMA was reported in 2000.[27] The $B(C_6F_5)_3$/silylating agent catalyst system had better control over GTP of EA than that by silylating agent $(C_2H_5)_3SiOTf)$ or $B(C_6F_5)_3$ alone, producing predictable M_n and a low $Đ$ value. In 2015, Kakuchi and co-workers developed various combinations of organic acid and $B(C_6F_5)_3$ with a silyl ketene acetal (SKA) for GTP of *N*,*N*-diethylacrylamide (DEAA).[28] Among four different SKA initiators investigated, the combination of ^{Et}SKA with $B(C_6F_5)_3$ exhibited better control over the MWD. Such catalyst system was also extended to various *N*,*N*-disubstituted acrylamides (DAAs) (Figure 11.1), including DMAA, DEAA, *N*,*N*-di-*n*-propylacrylamide (D*n*PAA), *N*-acryloylpiperidine (API), *N*-acryloylmorpholine (NAM), *N*-(2-methoxyethyl)-*N*-methylacrylamide (MMEAA), *N*,*N*-bis(2-methoxyethyl)acrylamide (BMEAA), *N*,*N*-diallylacrylamide (DAlAA), and *N*-methyl-*N*-propargylacrylamide (MPAA). Both kinetic studies and chain extension experiment indicated that the GTP catalyzed by the ^{Et}SKA/$B(C_6F_5)_3$ system exhibited living characteristics, which enabled the synthesis of homo-block copolymers composed of different acrylamide blocks and hetero-block copolymers of PMMA-*b*-PDAA by the sequential GTP method. An activated monomer mechanism was also proposed for the polymerization (Scheme 11.3).

More recently, Zhang *et al.* employed a SKA/$B(C_6F_5)_3$ Lewis pair (LP) to catalyze GTP of polar acrylic monomers at RT, including MMA and bio-renewable cyclic monomers γ-methyl-α-methylene-γ-butyrolactone (MMBL) and α-methylene-γ-butyrolactone (MBL).[29] The *in situ* monitored reaction of SKA with $B(C_6F_5)_3$ indicated the formation of frustrated Lewis pairs (FLPs, see Section 11.3 for a detailed description of this term), although it is sluggish for MMA polymerization, such a FLP system exhibits high activity and living GTP of MMBL and MBL, producing well-defined polymers with predictable M_n and narrow MWD. Systematic mechanistic investigation, including the characterization of key reaction intermediates, polymerization kinetics and polymer structures, has led to a bimolecular, activated monomer propagation mechanism. Moreover, well-defined PMMBL-*b*-PMBL, PMMBL-*b*-PMBL-*b*-PMMBL, and random copolymers were successfully prepared with predictable M_n and narrow MWD using this method.

Although the SKAs are good initiators for polar vinyl monomers, it is difficult to control the molar mass of the obtained polymer due to its instability toward moisture and impurities. On the other hand, $B(C_6F_5)_3$ is an efficient dual catalyst, which not only catalyzes the 1,4-hydrosilation of an α, β-unsaturated ketone with hydrosilane to produce SKA,[30] but also promotes

Table 11.2 GTP catalyzed by organic Brønsted acids.

Entry	Catalyst	Initiator	Solvent	Polymer	M_n (kg mol^{-1})	$Đ$ (M_W/M_n)	Ref.
1	$HB(C_6F_5)_4$	^{Me}SKA	DCM	PMMBL	40.2	1.27	37
2	$HB(C_6F_5)_4$	^{iBu}SKA	DCM	PMMBL	40.0	1.07	37
3	$[H(Et_2O)_2][B(C_6F_5)_4]$	^{Me}SKA	DCM	PMMA	1.69–74.7	1.04–1.12	42
4	$[HN(Me_2)Ph][B(C_6H_5)_4]$	^{Me}SKA	DCM	PMMA	18.5	1.07	42
5	List's sulfonimide	^{Me}SKA	DCM	PDMMA	30.2	1.13	42
6	$HNTf_2$	(*Z*)-DATP	TOL	PDMMA	3.2–53.9	1.06–1.20	43
7	$HNTf_2$	^{Me}SKA	DCM	PMMA	3.9–17.0	1.04–1.08	41
8	$C_6F_5CHTf_2$	^{iPr}SKA	TOL	PMA	2.9–108	1.03–1.07	47
9	Me_3SiNTf_2	^{iPr}SKA	TOL	PMEA	14.4–140	1.03–1.05	46
10	Me_3SiNTf_2	^{iPr}SKA	TOL	PEEA	18.9	1.06	46
11	Me_3SiNTf_2	^{iPr}SKA	TOL	PTIPSHEA	21.2	1.02	46
12	Me_3SiNTf_2	^{iPr}SKA	TOL	AIA	11.0	1.04	46
13	Me_3SiNTf_2	^{iPr}SKA	TOL	PgA	18.6	1.10	46
14	Me_3SiNTf_2	^{iPr}SKA	TOL	TIPSA	10.7	1.15	46
15	$B(C_6F_5)_3/Me_3SiNTf_2$	Me_2EtSiH	DCM	PMAM	3.1–27.5	1.04–1.13	33

Table 11.3 GTP catalyzed by organic Lewis acids.

Entry	Catalyst	Initiator	Solvent	Polymer	M_n (kg mol^{-1})	$Đ$ (M_W/M_n)	Ref.
1	TTPB	^{iBu}SKA	TOL	P*n*BA	24.3	1.07	36
2	TTPB	^{iBu}SKA	TOL	PMMA	21.6	1.07	36
3	TTPB	^{iBu}SKA	TOL	PBMA	49.1	1.08	36
4	TTPB	di-SKA1	DCM	PMMA	11.3–49.9	1.28–1.71	38
5	TTPB	di-SKA2	DCM	PMMA	47.4–209	1.29–1.84	38
6	TTPB	di-SKA1	DCM	PMMBL	21.4–23.2	1.41–1.53	38
7	TTPB	di-SKA2	DCM	PMMBL	26.1–50.2	1.36–2.07	38
8	TTPB	di-SKA2	TOL	PMMA	152	1.46	38
9	TTPB	^{iBu}SKA	DCM	PMMBL	18.8–548	1.01–1.06	37
10	$B(C_6F_5)_3$	Et_3SiH	DCM	P*n*BA	6.5	1.08	31
11	$B(C_6F_5)_3$	Et_3SiH	DCM	PMMA	5.6	1.06	32
12	$B(C_6F_5)_3$	nBu_3SiH	DCM	PMMA	3.4	1.11	32
13	$B(C_6F_5)_3$	Me_2PhSiH	DCM	PMMA	3.2	1.09	32
14	$B(C_6F_5)_3$	Ph_3SiH	DCM	PMMA	1.1	1.23	32
15	$B(C_6F_5)_3$	iPr_3SiH	DCM	PMMA	10.0	1.17	32
16	$B(C_6F_5)_3$	Me_2PhSiH	DCM	PPrMA	4.6	1.11	32
17	$B(C_6F_5)_3$	Me_2PhSiH	DCM	P*n*HMA	6.4	1.09	32
18	$B(C_6F_5)_3$	Me_2PhSiH	DCM	P*n*DMA	6.4	1.05	32
19	$B(C_6F_5)_3$	Me_2PhSiH	DCM	PEHMA	5.6	1.06	32
20	$B(C_6F_5)_3$	Me_2PhSiH	DCM	P*i*BMA	4.4	1.13	32
21	$B(C_6F_5)_3$	Me_2PhSiH	DCM	PcHMA	4.7	1.15	32
22	$B(C_6F_5)_3$	Me_2PhSiH	DCM	PDEAA	2.7–87.3	1.15–3.89	33
23	$B(C_6F_5)_3$	$MePb_2SiH$	DCM	PDEAA	29.7–181	1.26–1.90	33
24	$B(C_6F_5)_3$	Et_3SiH	DCM	PDEAA	18.6	1.28	33
25	$B(C_6F_5)_3$	Me_2EtSiH	DCM	PDEAA	3.1–54.3	1.16–1.22	33
26	$SiEt_3^+$	Et_3SiH	TOL	PMMA	3.5–40.1	1.05–1.09	34
27	$Al(C_6F_5)_3$	^{iBu}SKA	DCM	PMMBL	11.7–194	1.02–1.08	39

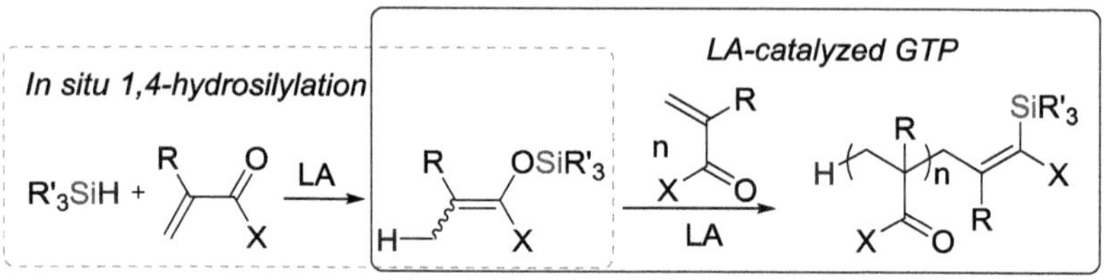

Scheme 11.4 Proposed mechanism of GTP by using LA and hydrosilane.

the GTP for polymer synthesis. Therefore, Kakuchi *et al.* designed a new GTP method to utilize the $B(C_6F_5)_3$-catalyzed the 1,4-hydrosilylation of *n*BA by using moisture-tolerant hydrosilane to generate SKA initiator for GTP of *n*BA, producing well-defined polymers with predictable M_n and narrow MWD.[31] As shown in Scheme 11.4, SKA was *in situ* generated from the 1,4-hydrosilylation of an α, β-unsaturated ester (the monomer) with a hydrosilane using LA as catalyst prior to the polymerization; the following steps were similar to that for the SKA/LA system as described in Scheme 11.3. Six hydrosilanes were used to clarify the effect of the hydrosilane structures on the GTP of *n*BA. It turned out that the alkylsilyl structures improved the polymerization control in the order $Me_2PhSiH > MePh_2SiH > Ph_3SiH$. It appeared that Me_2PhSiH was the most suitable hydrosilane for the GTP of *n*BA and exhibited the fastest polymerization rate, taking 0.2 h for complete monomer conversion.

Such a hydrosilane/$B(C_6F_5)_3$ system was also used to *in situ* generate an SKA initiator for the GTP of alkyl methacrylates.[32] The *in situ* formation of SKA from the 1,4-hydrosilylation of MMA was confirmed by the MALDI-TOF MS and ^1H NMR measurements of the obtained polymers. Among the screened hydrosilanes, Me_2PhSiH was the most suitable hydrosilane for the polymerization, producing well-defined polymers with predictable M_n and extremely marrow MWD ($Đ < 1.10$). The living nature of GTP polymerization of MMA catalyzed by $Me_2PhSiH/B(C_6F_5)_3$ was further verified by kinetic studies, chain-extension experiments and MALDI-TOF MS analysis of the resulting polymers. In addition to MMA, this polymerization method was also applied to the other methacrylate monomers. It turned out that the chemical structures of the monomers significantly affected the process of polymerization, since this GTP method could be applied to primary alkyl methacrylate monomers, while the polymerization rate of the secondary alkyl methacrylates were obviously slower than those of the primary alkyl methacrylate and no polymerization proceeded for the tertiary butyl methacrylate.

In 2016, the hydrosilane/$B(C_6F_5)_3$ system was further extended to GTP of *N*,*N*-disubstituted acrylamide (DAAm) through the *in situ* formation of SKA initiator *via* the 1,4-hydrosilylation of DAAm with hydrosilane.[33] Screen experiments indicated that both the chemical structures of hydrosilane and monomers showed significant effects on the control of the polymerization. More specifically, Me_2EtSiH with the least steric bulkiness was the most suitable hydrosilane; monomers showing a strong coordination with $B(C_6F_5)_3$ such as DEtAAm and DMeAAm only had better control over the M_n and MWD, while monomers exhibiting a relatively weak coordination with $B(C_6F_5)_3$ such as MorAAm, DAlAAm and BMEAAm furnished well-defined polymer products. The kinetic and mechanistic studies indicated that the structure of the monomer also affected the rate-determining step of polymerization, such as the abstraction of the hydride from hydrosilane by $B(C_6F_5)_3$ during 1,4-hydrosilation and propagation. The employment of a small portion of Me_3SiNTf_2 as an additional catalyst could significantly enhance the polymerization rate, thus inhibiting unnecessary thermally induced self-polymerization. By using $B(C_6F_5)_3$ and Me_3SiNTf_2 as a double catalytic system, α-end-functionalization of poly(*N*,*N*-disubstituted acrylamide) (PDAAm) was synthesized for the first time by the *in situ* generation of functional SKA through 1,4-hydrosilylation of functional methacrylamides, which shows no polymerization reactivity in the Lewis acid-catalyzed GTP, followed by the Me_3SiNTf_2-catalyzed GTP of DAAms.

In 2015, Chen and co-workers developed the first silylium (R_3Si^+)-catalyzed living polymerization of MMA by the *in situ* generated SKA initiator through the "R_3Si^+"-catalyzed 1,4-hydrosilylation of monomer with hydrosilane, producing polymers with well-controlled M_n values (close to the calculated ones), narrow MWD ($Đ = 1.05–1.09$), and well-defined chain structures.[34] The R_3Si^+ was formed from the reaction of R_3SiH (Et_3SiH or Me_2PhSiH) with $[Et_3Si(L)]^+[B(C_6F_5)_4]^-$ (L = TOL), which catalyzes the 1,4-hydrosilylation of MMA *via* FLP-type activation of the hydrosilane in the form of an isolable silylium–silane complex, $[Et_3Si–H–SiR_3]^+[B(C_6F_5)_4]^-$. Both kinetic studies and MALDI-TOF analyses of the polymer end groups are consistent with the previously established "R_3Si^+"-catalyzed propagation mechanism in that the C–C bond forming step *via* intermolecular Mukaiyama–Michael addition of the propagating species to the R_3Si^+-activated monomer is the rate-limiting step and the release of the "R_3Si^+" to the incoming monomer is relatively fast.

11.2.2.2 Triphenylmethyl Salts

It has been stated that it is difficult to synthesize PMMA with $M_n \geq 60$ kg mol^{-1} through traditional GTP of MMA using a SKA and a nucleophilic catalyst such as a bifluoride or oxyanion (0.1–1.0 mol% relative to initiator). On the other hand, Lewis acidic catalysts such as zinc halides or alkylaluminum chloride and a much higher catalyst loading (typically 10 mol% based on monomer) is required to achieve a reasonable degree of control over the GTP of acrylates. To overcome these limitations, Chen and co-workers developed a new catalyst system based on the *oxidative activation* of SKA with catalyst triphenylmethyl tetrakis(pentafluorophenyl)borate $Ph_3CB(C_6F_5)_4$ for living GTP of alkyl (meth)acrylates as well as biorenewable monomers MMBL and MBL at RT.[35–37] As shown in the *oxidative activation* mechanism (Scheme 11.5), the vinylogous hydride abstraction of SKA with triphenylmethyl cation (Ph_3C^+) generated a Ph_3CH and a R_3Si^+activated monomer. The Michael addition of SKA to the R_3Si^+activated monomer yielded the first MMA addition product of the polymerization, the highly active ambiphilic silicon propagation species containing both nucleophilic (SKA, which attacks the activated monomer) and electrophilic (Me_3Si^+, which activates the incoming monomer) catalyst sites, thereby promoting controlled polymerization *via* cooperative catalysis. Kinetic studies indicated

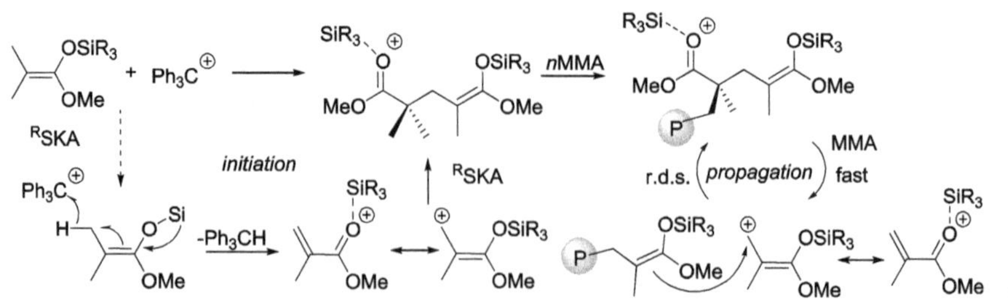

Scheme 11.5 Proposed mechanism of GTP catalyzed by triphenylmethyl salt. Adapted from ref. 42 with permission from John Wiley and Sons, Copyright © 2010 WILEY-VCH Verlag GmbH & Co. KGaA, Weinheim.

that the step of C–C bond formation *via* intermolecular Michael addition of the polymeric SKA to the silyl cation-activated MMA is the rate-determining step, and the release of the catalyst from the ester group of the growing polymer chain to the incoming MMA is relatively fast. This $Ph_3CB(C_6F_5)_4/^{Me}$SKA catalyst system could produce PMMA with low to high M_n values (>100 kg mol^{-1}) and narrow MWD ($Đ = 1.04–1.12$), and the catalyst loading can be as low as 0.025 mol% (based on the monomer).

Later, studies of structure–reactivity relationships for the above mentioned $Ph_3CB(C_6F_5)_4/^{Me}$SKA catalyst system have included effects on the polymerization activity and the degree of control by (1) structures of initiator (alkyl, silyl, germyl, and stannyl acetals); (2) variations in alkyl and silyl groups of silyl acetals in catalyst; (3) different anion structure of trityl salt in activators; and (4) chemical structure of monomer (methacrylates *vs.* acrylates).[36] One of the most interesting and significant result from this study is that remarkable selectivity of the silyl group structure of the SKA initiator was noticed for different monomer structure: initiators with small silyl groups, such as ^{Me}SKA, promoted highly active and efficient polymerization of methacrylates, but they are poor initiators for polymerization of less sterically hindered, active α-H acrylate monomers. On the other hand, initiators with bulky silyl groups, such as ^{iBu}SKA exhibited low activity toward methacrylate polymerization but exceptionally high activity (reaching completion in 1 min), efficiency (quantitative initiation efficiency) and degree of control (the obtained P*n*BA with narrow MWD ($Đ = 1.07–1.17$) for M_n up to 75.1 kg mol^{-1} or higher M_n (>100 kg mol^{-1}) at the expense of broader MWD) for acrylate polymerization at ambient temperature in polar non-coordinating (CH_2Cl_2), aromatic (toluene), or aliphatic (cyclohexane) solvents.

Living GTP of two naturally biorenewable butyrolactone-based vinylidene monomers MBL and MMBL could also be achieved by using the above-mentioned, ambiphilic silicon propagation species derived from the SKA initiator upon *in situ* oxidative activation with a catalytic amount of the $Ph_3CB(C_6F_5)_4$ activator.[37] Through investigations into effects of SKA (thus the resulting R_3Si^+ catalyst) and activator (thus the resulting counteranion) structures, the combination $Ph_3CB(C_6F_5)_4/^{iBu}$SKA is found to be the most active and controlled system for (M)MBL polymerization, producing polymers with controlled low to high M_n values (up to 548 kg mol^{-1}) and narrow MWD ($Đ = 1.01–1.06$). Kinetic studies indicated that this GTP of (M)MBL proceeds with the same mechanism with that for PMMA, as shown in Scheme 11.5. It is noted that the obtained atactic polymers exhibited a high T_g of 194 °C (PMBL) and 225 °C (PMMBL), which is much higher than that of the typical atactic PMMA (105 °C). The living nature of the $Ph_3CB(C_6F_5)_4/^{iBu}$SKA catalyst system enabled the synthesis of well-defined block copolymer PMBL-*b*-PMMA and PMMBL-*b*-PMMA and corresponding random copolymers exhibiting unimodal and narrow MWD ≤ 1.03.

However, there is limitation in the polymerizations under highly dilute initiator or catalyst conditions and in the stereochemical control of

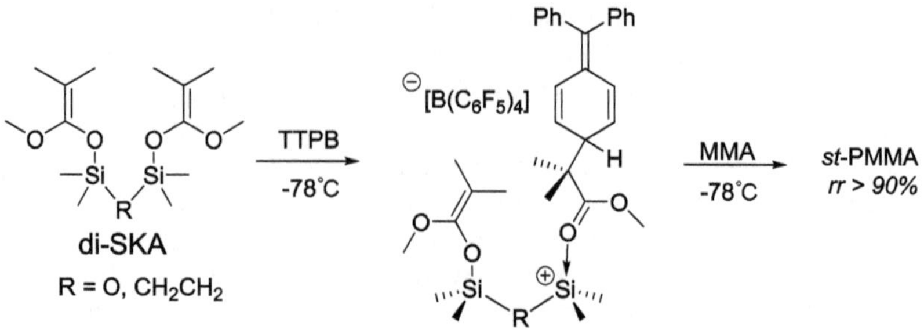

Scheme 11.6 Generation of dinuclear silylium-enolate active species.

polymerization catalyzed by the mono-SKA catalyst systems. To overcome the limitation, Chen and co-workers developed a novel dinuclear silylium-enolate active species derived from activation of ethyl- or oxo-bridged disilicon enolate (*i.e.* disily ketene acetal, di-SKA) compounds with $Ph_3CB(C_6F_5)_4$ for GTP of MMA and renewable methylene butyrolacone monomers.[38] Compared with a mononuclear SKA-base catalyst system, this unimolecular bifunctional catalyst system consisting of both an electrophilic silylium catalyst site and a nucleophilic silicon enolate initiating site are much more active for the polymerization of MMA and exhibited unique polymerization and kinetic characteristics as well as a rate enhancement by a factor of >40 and high syndiotacticity of PMMA ($rr = 92\%$ at -78 °C) (Scheme 11.6). The oxo-bridged silylium-enolate species is controlled and is about 3.7 times more active for polymerization of MMA than the ethyl-bridged one, probably due to the higher reactivity of oxo-bridged active species rendered by the inductive electron-withdrawing effect of the oxygen atom and to their relative proximity between the silylium catalyst and enolate initiating site. The activity difference is even more striking for polymerization of MMBL. Kinetic study coupled with the characterization of the silylium enolate active species suggested a unimolecular propagation mechanism, involving an intramolecular delivery of the polymeric enolate nucleophile to the activated monomer by the silylium ion electrophile being placed in proximity in the same catalyst molecule (Scheme 11.7). Effects of activator, temperature and solvent on the tacticity of the resulting polymer were also investigated and it was found that $Ph_3CB(C_6F_5)_4$ is the most effective with the highest activity and degree of control over the polymerization. It is noted that the dinuclear active species are not only quite active in low temperature but also produce highly stereoregular polymers, compared with the inactivity of mono-SKA based active species at temperatures below -20 °C.

11.2.2.3 Tris(pentafluorophenyl) Aluminum ($Al(C_6F_5)_3$)

In 2018, Zhang and co-workers reported GTP of conjugated polar alkenes, including MMA, MBL and MMBL by an SKA/$Al(C_6F_5)_3$ catalyst system.[39]

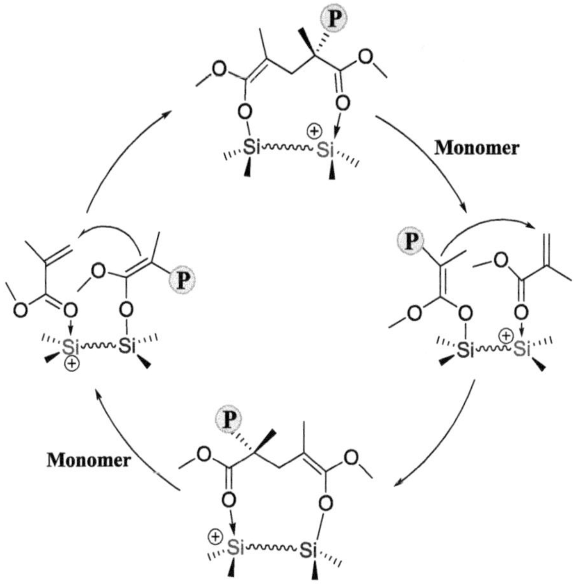

Scheme 11.7 Intramolecular Michael addition pathway for dinuclear silylium-enolate active species.

In comparison with the poor control over the polymerization of MMA by both $Al(C_6F_5)_3$/hydrosilane[40] and $Al(C_6F_5)_3$/SKA catalyst system, the polymerization of MMBL by bulky SKA (^{iBu}SKA)/$Al(C_6F_5)_3$ is living and thus produces well-defined PMMBL with predictable M_n (up to 179 kg mol^{-1}), narrow MWD (*Đ* as low as 1.02) and high initiation efficiency ($I^*\% \geq 97$), by varying the [MMBL]/[^{iBu}SKA] ratio from 100 to 1600. Moreover, the living/controlled nature of $Al(C_6F_5)_3$/SKA also enabled the synthesis of di/tri-block and random copolymerization of MMBL and MBL with controlled M_n and narrow MWD. The combined mechanistic studies involving isolation and characterization of single-monomer-addition intermediates that simulate the active propagating species (Scheme 11.8), polymerization kinetics, and characterization of polymer chain ends have led to a polymerization mechanism. The polymerization is initiated *via* intermolecular Michael addition of the SKA enolate group to the vinyl group of the $Al(C_6F_5)_3$-activated monomer, while the silyl group is transferred to the carbonyl group of the monomer and $Al(C_6F_5)_3$ to the oxygen atom of SKA; the coordinated $Al(C_6F_5)_3$ is released to the incoming monomer, followed by repeated intermolecular Michael additions in the subsequent propagation cycle.

11.2.2.4 Trifluoromethanesulfonimide (Tf_2NH)

Apart from strong Lewis acidic catalysts, a strong organic Brønsted acid, trifluoromethanesulfonimide (Tf_2NH) could also serve as an efficient activator to combine with initiator ^{Me}SKA for GTP of MMA in CH_2Cl_2 at ambient temperature (Figure 11.4), producing PMMA with controlled M_n values from

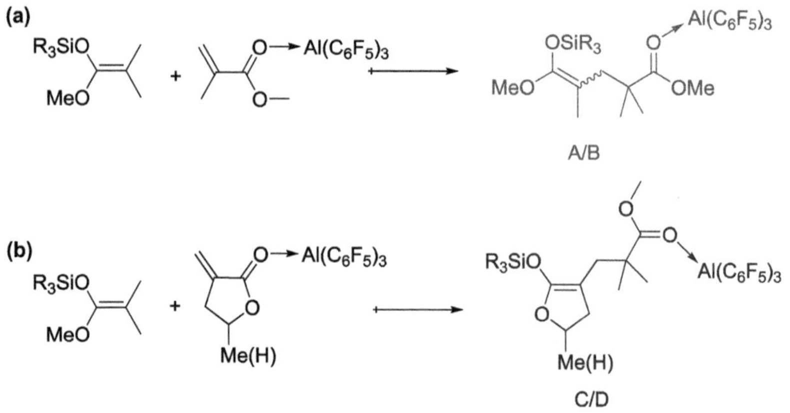

Scheme 11.8 Formation of single-monomer addition intermediates in GTP catalyzed by $Al(C_6F_5)_3$. (a) A or B denotes *Z*/*E* isomers of MMA derivatives; and (b) C or D denotes the corresponding MMBL or MBL derivatives. Adapted with permission from *Macromolecules* **2018**, *51*, 1296 – 1307. Copyright (2018) American Chemical Society.

3.86 to 17.0 kg mol^{-1} and $Đ$ = 1.04–1.08 by varying the initial ratio of $[MMA]_0/[^{Me}SKA]_0$.[41] The living nature of such a catalyst system could be verified by the following three lines of evidence: (1) success achieved in a chain-extension experiment; (2) the linear increase of M_n of PMMA with an increase in monomer conversion; (3) identification of the chain-end group of the resulting PMMA only showing one set of peaks, corresponding to the molecular weight of PMMA possessing the initiator ^{Me}SKA residue and the desilylated chain end. On the other hand, unlike the GTP of MMA promoted by $H(Et_2O)_2[B(C_6F_5)_4]$ (*vide infra*),[42] the kinetics of the current polymerization showed a first-order dependence on the monomer concentration, which may be attributed the difference in the coordination ability of Tf_2N^- and $B(C_6F_5)_4^-$. The syndiotacticity (*rr*) of the obtained PMMA from the polymerization with a 100 : 1 : 0.02 $[MMA]_0 : [^{Me}SKA]_0 : [Tf_2NH]_0$ ratio was increased from 72% to 90% when the polymerization temperature was decreased from 27 °C to −55 °C.

As shown in the proposed mechanism (Scheme 11.9A), the *N*-(trimethylsilyl)bis(trifluoromethanesulfonyl)imide (Me_3SiNTf_2), derived from the reaction of ^{Me}SKA with $HNTf_2$, is the actual catalyst to activate the monomer. The reaction of Me_3SiNTf_2 with an extremely low amount of impurities led to the formation of Tf_2NH, which would further react with ^{Me}SKA, and thus regenerating Me_3SiNTf_2. Based on this, the system also possesses a feature of "self-repair" (*vide infra*). The propagation steps were similar to those proposed for GTP catalyzed by the $Ph_3CB(C_6F_5)_4$/SKA catalyst system (Scheme 11.5). The proposed mechanism of syndiotactic control on the polymerization (Scheme 11.9B) indicated that the less steric hindrance between the chain end of PMMA and the incoming monomer in the structure was considered more favorable for the highly selective formation of the *r* diad in the GTP of MMA.

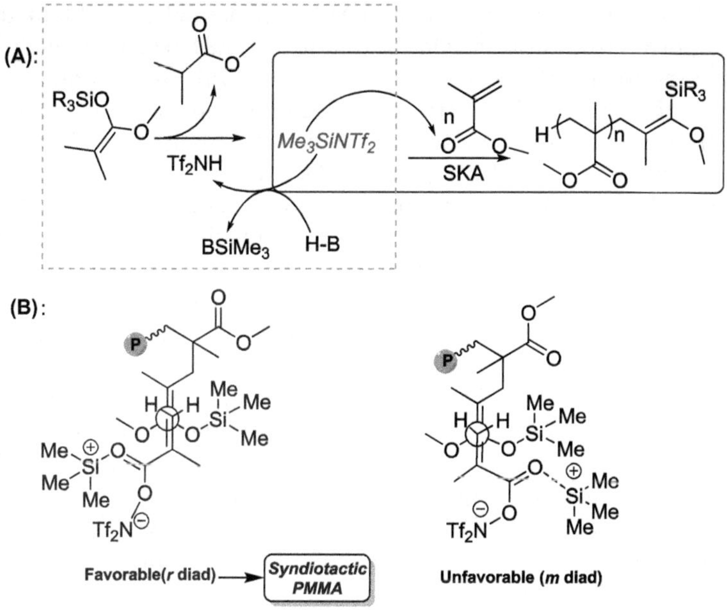

Scheme 11.9 Mechanisms of (A) $HNTf_2$-catalyzed GTP and (B) the formation of syndiotactic PMMA.

In 2010, the combination of Tf_2NH with (*Z*)-1-(dimethylamino)-1-trimethylsiloxy-1-propene ((*Z*)-DATP, Figure 11.2) was also found to be effective for GTP of DMAA at 0 °C in toluene, producing PDMAA with controlled M_n values ranging between 3.24–53.9 kg mol^{-1} and narrow MWD (≤1.17).[43] The living nature of the polymerization was confirmed by the kinetic studies, a post-polymerization experiment and the chain end analysis of the obtained PDMAA. As shown in both ^{1}H and ^{13}C NMR spectra, the GTP of DMAA catalyzed by (*Z*)-DATP/Tf_2NH system produced the *r* dyad-rich PDMAA in a polar solvent or at a lower temperature, which is different from that for the conventional radical polymerization.[44]

11.2.2.5 N-(Trimethylsilyl)bis(trifluoromethanesulfonyl)imide (Me_3SiNTf_2)

To avoid the partial consumption of SKA that reacted with Tf_2NH, Kakuchi *et al.* directly employed Me_3SiNTf_2 to combine with ^{Me}SKA for GTP of MMA to synthesize stereospecific star-shaped PMMA.[45] Compared with Tf_2NH/^{Me}SKA, polymerizations catalyzed by Me_3SiNTf_2/^{Me}SKA not only had the same first-order kinetic and living polymerization feature but also higher initiation efficiency values of 0.94–1.00. Using initiators possessing three, four and six ^{Me}SKA groups, 3-, 4-, and 6-armed star-shaped PMMAs with predictable M_n and unimodal, narrow MWD could be obtained by the Tf_2NH/SKA system at −55 °C.

It is difficult to control the M_n and MWD of the polymer obtained from anionic polymerization of acrylates, due to the side reactions of the ester carbonyl group and the susceptible α-hydrogen with the anionic initiators and the active chain ends. Therefore, it is a remaining and challenging task to synthesize multiblock acrylate polymers. Kakuchi and coworkers utilized the combination of Me_3SiNTf_2 and ^{iPr}SKA for GTP of various acrylate monomers (Figure 11.1), including *n*BA, *t*BA, methyl acrylate (MA), ethyl acrylate (EA), 2-methoxyethyl acrylate (MEA), 2-(2-ethoxyethoxy)ethyl acrylate (EEA), allyl acrylate (AlA), 2-ethylhexyl acrylate (EHA), triisopropylsilyl acrylate (TIPSA), 2-(triisopropylsiloxy)ethyl acrylate (TIPSHEA), (dimethylamino)ethyl acrylate (DMAEA), propargyl acrylate (PgA), dicyclopentanyl acrylate (*d*cPA), and cyclohexyl acrylate (*c*HA).[46] GTP of all acrylate monomers rapidly proceeds in a living/controlled fashion, producing well-defined homo polyacrylates, except for *t*BA and DMAEA. The living nature of polymerization enabled the post-polymerizations of polymers synthesized with MA, EA, *n*BA, and MEA and the sequential synthesis of di- and multiblock polymeracrylates, such as the AB and BA diblock copolymers, $(ABC)_4$ dodecablock terpolymer, $(ABCD)_3$ dodecablock quaterpolymer, and ABCDEF hexablock sestopolymer.

11.2.2.6 Pentafluorophenylbis(triflyl)methane ($Tf_2CHC_6F_5$)

In 2012, $Tf_2CHC_6F_5$ was utilized as an organocatalyst in combination with different initiators, such as ^{Me}SKA, ^{Et}SKA and ^{iPr}SKA, for GTP of MA in toluene at RT.[47] Among three SKA investigated, ^{iPr}SKA was found to be the most effective initiator for the polymerization. The living nature of the polymerization catalyzed by the ^{iPr}SKA/Tf_2NSiMe_3 system was further verified by the identification of a chain-end group of the obtained PMA with MALDI-TOF MS spectrometry, kinetic studies, and chain extension experiments, which allowed for the synthesis of high molecular weight PMA with M_n of up to 108 kg mol^{-1} and narrow MWD (*Đ* as low as 1.07). Moreover, the well-defined block copolymers PMA-*b*-P*n*BA and P*n*BA-*b*-PMA were readily synthesized by the sequential addition of MA and *n*BA due to the living nature of the ^{iPr}SKA/$Tf_2CHC_6F_5$-catalyzed GTP of the alkyl acrylates. Later, both Tf_2NH and $Tf_2CHC_6F_5$ were screened for the effects of their counter anions on the GTP of DMAA initiated with (*Z*)-DATP.[48] Both the equilibrium constant for the monomer activation and the rate constant for the propagation reaction for $Tf_2CHC_6F_5$ are higher than those for Tf_2NH at all the examined temperatures, indicating that a less nucleophilic counter anion was more favorable for both the monomer activation and the propagation reaction.

11.2.2.7 List's Sulfonimide and the Other Organic Brønsted Acids

Three strong organic Brønsted acids, such as $H(Et_2O)_2[B(C_6F_5)_4]$, $H(Me_2NPh)[B(C_6F_5)_4]$, and (*R*)-3,3′-bis[3,5-bis(trifluoromethyl)phenyl]-1,10-binaphthyl-2,20-disulfonimide (List's sulfonimide), were employed to activate SKA to produce silylium ions for high-speed living GTP of polar vinyl monomers, such as methacrylates, acrylates and acrylamides.[42] Among these, the combination of $H(Et_2O)_2[B(C_6F_5)_4]$ with SKA is the most effective for the polymerization, achieving a high catalyst turn-over frequency (TOF) of 6.0×10^3 h^{-1} for polymerization of MMA with ^{Me}SKA as initiator or an exceptionally high TOF of 2.4×10^5 h^{-1} for polymerization of *n*BA with ^{iPr}SKA as initiator, with a low silylium catalyst loading (≤0.05 mol% relative to monomer). The living characteristic of the polymerization system has been demonstrated by kinetic studies, success achieved in chain extension and synthesis of well-defined block copolymers. As highly reactive and moisture-sensitive species, silylium catalysts R_3Si^+ can be consumed by even an extremely low level of protic impurities. Therefore, it is striking to see that ^{R}SKA/$H(Et_2O)[B(C_6F_5)_4]$ exhibited a unique catalyst self-healing feature (Scheme 11.10), as demonstrated by the catalyst "self-repair" studies, thereby offering more attractive polymerization catalysis features. All PMMAs produced by this catalyst system at ambient temperature are syndio-rich polymers with a syndiotacticity of ~72% *rr*. Decreasing the temperature from 25 °C to −40 °C, the syndiotacticity of the obtained PMMA can be increased from 72% to 87% *rr*, with the polymerization rate drastically diminished by more than two orders of magnitude. It should be noted that $H(Et_2O)_2[B(C_6F_5)_4]$ was also an effective catalyst for GTP of BMA, *n*BA and MMBL, when combined with different SKA.[36,37]

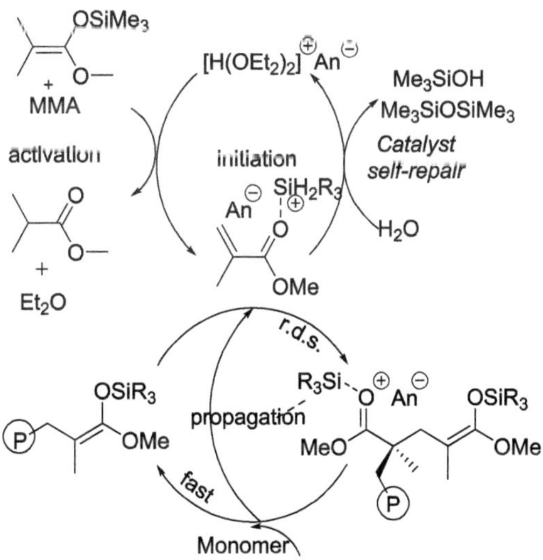

Scheme 11.10 Mechanisms of $[H(OEt_2)_2][An]$-catalyzed GTP. Adapted from ref. 42 with permission from John Wiley and Sons, Copyright © 2010 WILEY-VCH Verlag GmbH & Co. KGaA, Weinheim.

11.2.3 Copolymerization of Acrylic Monomers Using Organocatalyzed GTP

In comparison with techniques as ionic polymerizations, free-radical polymerizations, or conventional GTP, organocatalyzed GTP produces more stable chain ends during the polymerization process. Moreover, the fact that occurrence of side reactions is reduced or even inhibited by these organic basic catalysts while the catalytic activity can be greatly enhanced by organic acidic catalysts, significantly improved the living characteristics of organocatalytic GTP. These advantages provide the possibility to control the architecture of the polymer, thus achieving the successful synthesis of block, star-shaped and end-functionalized polymers. The above-described copolymerization will not be discussed in more detail. Well-defined block polymers produced by organocatalyzed GTP of acrylic monomers developed in this book are summarized in Tables 11.4 and 11.5. In addition, investigations towards preparation of end-functionalized[33,49,50] and star-shaped[23,24,45,51–55] polymers not involving special catalytic process are not presented here, and the interested reader may refer to related reviews.[56,57]

Table 11.4 GTP catalyzed by organic bases.

Entry	Catalyst	Initiator	Solvent	Polymer	M_n (kg mol^{-1})	$Đ$ (M_W/M_n)	Ref.
1	^{tBu}NHC	^{Me}SKA	TOL	PMMA-*b*-P*t*BA	21.9	1.3	19
2	^{tBu}NHC	^{Me}SKA	THF	PMMA-*b*-P*t*BA-*b*-PMMA	12.6	1.4	19
3	^{IPr}NHC1	^{Me}SKA	THF	PMMA-*b*-PDMAEA	11.0	1.3	17
4	^{IPr}NHC1	^{Me}SKA	THF	PMMA-*b*-PDMAEMA	22.0	1.1	17
5	^{IPr}NHC1	^{Me}SKA	THF	PMMA-*b*-PDMAA	16.9	1.09	17
6	^{IPr}NHC1	^{Me}SKA	THF	PMMA-*b*-PMAN	71.0	1.18	17
7	^{IPr}NHC1	^{Me}SKA	THF	PMMA-*b*-P*t*BA-*b*-PDMAA	5.3	1.09	17
8	^{IPr}NHC1	^{Me}SKA	THF	PDMAEMA-*b*-PDMAA	13.0	1.29	17
9	^{IPr}NHC1	^{Me}SKA	THF	PDMAEMA-*b*-PDMAEA	4.8	1.21	17
10	^{IPr}NHC1	^{Me}SKA	THF	P*t*BA-*b*-PMMA	12.0	1.27	17
11	^{IPr}NHC1	^{Me}SKA	THF	P*t*BA-*b*-PDMAA	21.3	1.2	17
12	^{IPr}NHC1	^{Me}SKA	THF	P*t*BA-*b*-PDMAEMA	36.7	1.23	17
13	^{IPr}NHC1	^{Me}SKA	THF	PDMAA-*b*-PDMAEMA	17.6	1.4	17
14	^{IPr}NHC1	^{Me}SKA	THF	PDMAA-*b*-PMMA	23.9	1.3	17
15	^{IPr}NHC1	^{Me}SKA	THF	PMMA-*b*-P*n*BA	16.0	1.5	19
16	^{IPr}NHC1	^{Me}SKA	TOL	PMMA-*b*-P*n*BA	18.2	1.6	19
17	^{IPr}NHC2	^{Me}SKA	THF	PMMA-*b*-P*t*BA	11.8	1.14	16
18	TTMPP	^{Me}SKA	THF	PMMA-*b*-P*t*BA	22.0	1.19	25
19	*t*Bu-P_4	^{Me}SKA	TOL	PTPMA-*b*-PMMA	15.0–85.0	1.06–1.14	20

Table 11.5 GTP catalyzed by organic acids.

Entry	Catalyst	Initiator	Solvent	Polymer	M_n ($kg\,mol^{-1}$)	$Đ$ (M_w/M_n)	Ref.
1	TTPB	^{iBu}SKA	DCM	PMMA-*b*-P*n*BA	44.8	1.09	36
2	TTPB	^{iBu}SKA	TOL	PMMA-*b*-P*n*BA	45.3	1.13	36
3	TTPB	^{iBu}SKA	DCM	PMMA-*b*-PMBL	68.4	1.01	37
4	TTPB	^{iBu}SKA	DCM	PMMA-*b*-PMMBL	68.1	1.03	37
5	TTPB	^{iBu}SKA	DCM	PMMBL-*b*-PMBL	117	1.02	37
6	$H(Et_2O)_2[B(C_6F_5)_4]$	^{iBu}SKA	TOL	PMMA-*b*-P*n*BA	42.5	1.08	42
7	Tf_2NSiMe_3	^{iPr}SKA	TOL	P*n*BA-*b*-PMEA	6.64	1.04	46
8	Tf_2NSiMe_3	^{iPr}SKA	TOL	PMEA-*b*-P*n*BA	6.86	1.05	46
9	Tf_2NSiMe_3	^{iPr}SKA	TOL	P*n*BA-*b*-PALA	6.24	1.07	46
10	Tf_2NSiMe_3	^{iPr}SKA	TOL	PALA-*b*-P*n*BA	6.58	1.13	46
11	Tf_2NSiMe_3	^{iPr}SKA	TOL	P*n*BA-*b*-PPgA	6.32	1.07	46
12	Tf_2NSiMe_3	^{iPr}SKA	TOL	PPgA-*b*-P*n*BA	6.12	1.05	46
13	Tf_2NSiMe_3	^{iPr}SKA	TOL	(PEHA-*b*-P*n*BA-*b*-PEA)$_4$	16.89	1.03	46
14	Tf_2NSiMe_3	^{iPr}SKA	TOL	(P*n*BA-*b*-PEA-*b*-PMEA-*b*-PMA)$_3$	13.52	1.02	46
15	Tf_2NSiMe_3	^{iPr}SKA	TOL	P*dc*PA-*b*-P*n*BA-*b*-PEHA-	9.24	1.05	46
				b-PEA-*b*-PMA-*b*-P*c*HA			46
16	Tf_2NSiMe_3	^{Me}SKA	TOL	PMMA-*b*-PNAM	11.4	1.11	46
17	$B(C_6F_5)_3$	^{Et}SKA	DCM	PDEAA-*b*-PDMAA	5.69	1.28	28
18	$B(C_6F_5)_4$	^{Et}SKA	DCM	PDMAA-*b*-PDEAA	6.64	1.16	28
19	$B(C_6F_5)_5$	^{Et}SKA	DCM	PDEAA-*b*-PDALAA	8.69	1.17	28
20	$B(C_6F_5)_6$	^{Et}SKA	DCM	PDALAA-*b*-PDEAA	7.67	1.15	28
21	$B(C_6F_5)_7$	^{iBu}SKA	DCM	PMMBL-*b*-PMBL	28.3	1.03	29
22	$B(C_6F_5)_8$	^{iBu}SKA	DCM	PMMBL-*b*-PMBL-*b*-PMMBL	37.1	1.06	29
23	$Al(C_6F_5)_3$	^{iBu}SKA	DCM	PMMBL-*b*-PMBL	21.4	1.02	39
24	$Al(C_6F_5)_4$	^{iBu}SKA	DCM	PMMBL-*b*-PMBL-*b*-PMMBL	31.9	1.03	39

11.3 Polymerization of Acrylic Monomers by Organic Lewis Pairs

The frustrated Lewis pairs (FLPs) chemistry has caught chemists' imagination and attracted much intense interests ever since the seminal work of FLP chemistry was reported by Stephan and Erker.[58,59] The application of FLPs has been well established in the small molecule chemistry, such as activation of small molecules,[60–63] catalytic hydrogenation reactions,[64–66] and new reactivity/reaction developments.[67–70] In the area of macromolecular synthesis, the FLP or a classical Lewis adduct (CLA)-catalyzed polymerization has also attracted increasing attention since the first report on polymerization by FLPs based on a bulky NHC or phosphine LB and a bulky, strong LA, such as $Al(C_6F_5)_3$.[71] The employed monomers and organic LAs, such as $HB(C_6F_5)_2$, $B(C_6F_5)_3$, AlR_3 (R = Me, Et, Ph, *i*Bu, *etc.*), $MeAl(BHT)_2$ (BHT = 2,6-di-*tert*-butyl-4-methylphenoxide), $Al(C_6F_5)_3$ and $ClAl(C_6F_5)_2$ are shown in Figure 11.5 while organic LBs, such as, NHCs, NHOs, phosphines, amines are shown in Figure 11.6, respectively. Such Lewis pair polymerizations (LPPs) generally proceed with a bimolecular mechanism (Scheme 11.11) as follows: in the initiation step, after the activation of the monomer by the LA, the LB nucleophilicly attacks the activated monomer to form a zwitterionic intermediate. In the propagation step, the intermediates and activated monomers undergo continuous 1,4-additions to produce polymer chains. To prevent the formation of adducts with Lewis acids during the LPP process, the employment of polar donor solvents (such as THF and DMF) should be avoided. In the following section, we will detail the development of organocatalytic LPP of polar vinyl monomers in chronological order. Polymerizations catalyzed by LPP developed in this book are summarized in Table 11.6.

In 2010, Chen and co-workers first reported the Al-based LPs-catalyzed polymerization of polar vinyl monomers, including linear MMA and its biorenewable, cyclic analogue MBL as well as MMBL.[72] LAs, such as $Al(C_6F_5)_3$, $AlMe_3$, $B(C_6F_5)_3$ and $MeAl(BHT)_2$ were used as a catalyst for the activation of monomers, while several organic Lewis bases (LBs) were employed as the nucleophile, including phosphines (tBu_3P, Mes_3P (Mes = 2,4,6-$Me_3C_6H_2$), and Ph_3P) as well as NHCs (^{tBu}NHC and 1,3-di-mesitylimidazolin-2-ylidene (^{Mes}NHC)). Due to the inactivity or low activity of the combination of $AlMe_3$, $B(C_6F_5)_3$ or $MeAl(BHT)_2$ with phosphines, this contribution was mainly focused on investigations carried out with $Al(C_6F_5)_3$ as LA. It is noted that the different activation procedures can significantly affect the polymerization activity. For example, the premixing of either $Al(C_6F_5)_3$·TOL adduct or unsolvated $Al(C_6F_5)_3$ with tBu_3P followed by addition of MMA caused no monomer consumption for up to 24 h. However, the polymerization rate was drastically increased by premixing the preformed $Al(C_6F_5)_3$·MMA with tBu_3P (which cleanly generates the zwitterionic phosphonium enoaluminate active propagating species) followed by the addition of MMA (800 eq.), quantitative monomer conversion was achieved in 1 h and produced high molecular weight polymer. Therefore, a more convenient but highly active polymerization was developed by premixing $Al(C_6F_5)_3$·TOL adduct with MMA, followed by addition of the base to start the polymerization. With identification of such efficient $tBu_3P/Al(C_6F_5)_3$ pairs with appropriate procedure, the scope of LBs for the polymerization of MMA was expanded. It turned out that the cooperativity of the LA and LB sites of Lewis pairs is essential to achieve an effective polymerization system. The less nucleophilic Mes_3P is unable to generate the zwitterionic active species through the attack of the $Al(C_6F_5)_3$-activated monomer, and thus $Mes_3P/Al(C_6F_5)_3$ is completely ineffective for MMA polymerization. On the other hand, $Ph_3P/Al(C_6F_5)_3$ (TOF = 4.8×10^4 h^{-1}), ^{Mes}NHC·$Al(C_6F_5)_3$ (TOF = 4.8×10^4 h^{-1}) and ^{tBu}NHC/$Al(C_6F_5)_3$ (TOF = 3.2×10^3 h^{-1}) pairs are highly active for MMA polymerization. The formation of classic acid/base adducts did not quench

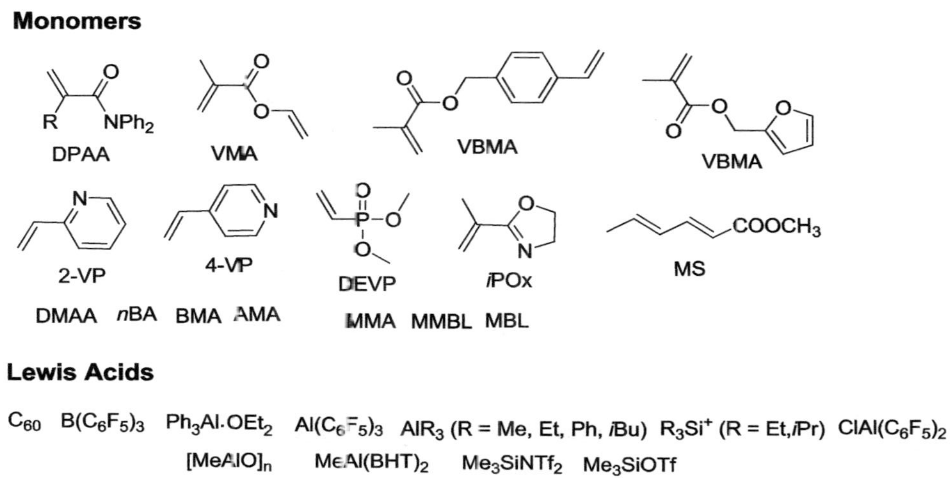

Figure 11.5 Monomers and LAs employed in LPP.

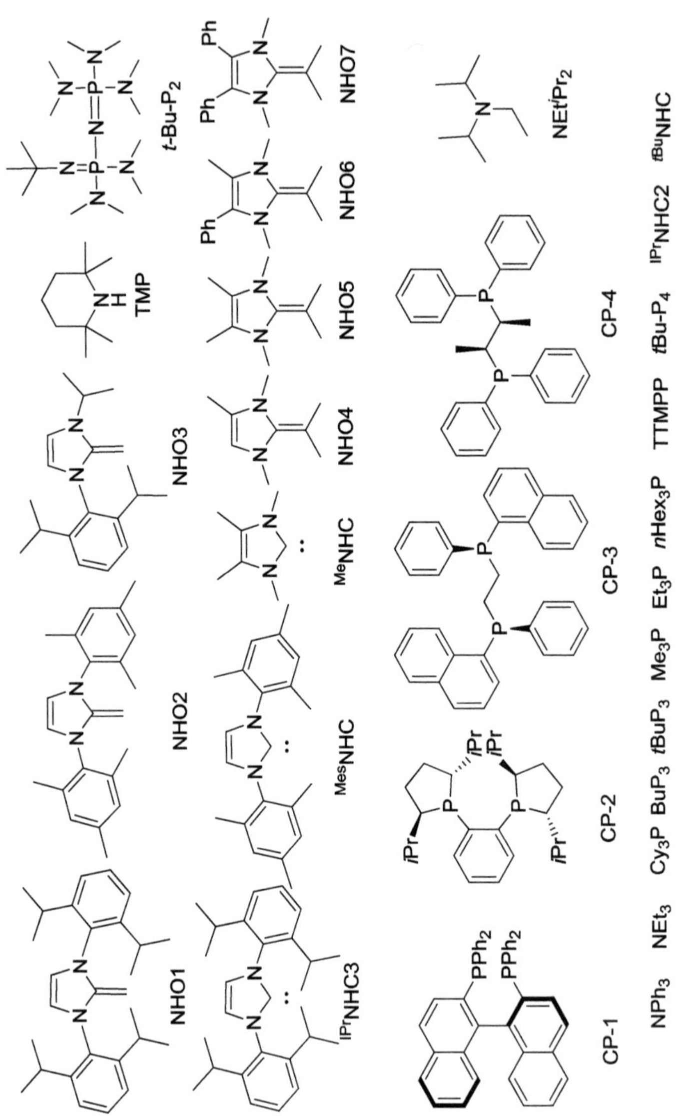

Figure 11.6 LBs employed in LPP.

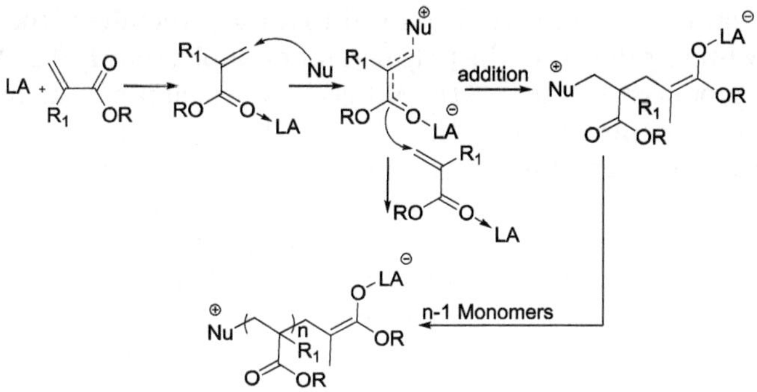

Scheme 11.11 Molecular, activated monomer propagation mechanism in LPP.

Table 11.6 Polymerizations catalyzed by LPP.

Entry	LA	LB	Solvent	Polymer	M_n (kg mol^{-1})	$Đ$ (M_W/M_n)	Ref.
1	$Al(C_6F_5)_3$·MMA	tBu$_3$P	TOL	PMMA	283–397	1.42–1.72	72
2	$Al(C_6F_5)_3$·TOL	^{tBu}NHC	DCM	PMBL	163	1.28	72
3	$Al(C_6F_5)_3$·TOL	^{tBu}NHC	DCM	PMMBL	139	1.15	72
4	$Al(C_6F_5)_3$·MMA	tBu$_3$P	DCM	PMMA	369–380	1.41–1.47	72
5	$Al(C_6F_5)_3$·MMA	^{tBu}NHC	DCM	PMMA	600	1.34	72
6	$Al(C_6F_5)_3$·MMA	TPT	TOL	PMMA	159	1.17	73
7	$Al(C_6F_5)_3$·MMA	TPT	DCM	PMMA	125	1.23	73
8	$Al(C_6F_5)_3$	^{Mes}NHC	DCM	PMMBL	62.8	1.42	73
9	$Al(C_6F_5)_3$	tBu$_3$P	TOL	PDPAA	357	1.31	73
10	$Al(C_6F_5)_3$	CP-1	TOL	PMMA	370	1.13	73
11	$Al(C_6F_5)_3$	tBu-P$_4$	TOL	PMMA	212	1.34	73
12	$AlPh_3$	Et_3P	TOL	PFMA	64.0	1.16	87
13	$AlPh_3$	Me_3P	TOL	PnBMA	42.0	1.00	87
14	$AlMe_3$	Me_3P	TOL	PDMMA	127	1.08	87
15	$AlPh_3$	Me_3P	TOL	PDEVP	55.0	1.42	87
16	$AlEt_3$	Cy_3P	TOL	P(4-VP)	46.0	1.20	87
17	Me_3SiNTf_2	TTMPP	TOL	PMMA	7.5–23.8	1.03–1.08	98
18	$B(C_6F_5)_3$	TMP	DCM	PMMBL	67.5–255	1.54–1.92	97
19	$Al(C_6F_5)_3$	^{tBu}NHC	TOL	P(2-VP)	10.3–80.2	1.55–4.71	82
20	$B(C_6F_5)_3$	^{iPr}NHC2	TOL	PMMA	11.1–18.4	1.38–1.65	96
21	$B(C_6F_5)_3$	^{iPr}NHC2	DCM	PMMBL	33.7–111	1.10–1.15	96
22	MeAl(BHT)$_2$	^{tBu}NHC	THF	poly(MS)	23.0	1.10	99
23	$Al(C_6F_5)_3$	NHO3	TOL	PVBMA	58.0	1.22	81
24	$Al(C_6F_5)_3$	NHO3	TOL	PVMA	138.0	1.28	81
25	$Al(C_6F_5)_3$	NHO3	TOL	PAMA	130–640	1.21–1.28	81
26	$Al(C_6F_5)_3$	NHO3	TOL	PMMA-*co*-PBMA	480.0	1.31	81
27	$Al(C_6F_5)_3$	tBu$_3$P	TOL	PAMA	190.0	1.31	81
28	MeAl(BHT)$_2$	NHO4	TOL	PMMA	30.4–347	1.05–1.09	94
29	MeAl(BHT)$_3$	NHO4	TOL	PMMA-*b*-PBnMA-*b*-PMMA	65.9	1.1	94

reactivity towards MMA. It is also noted that the structure of the base can drastically impact the M_n of the polymer, for example, the M_n (525 kg mol^{-1}) of PMMA produced by the ^{tBu}NHC/Al(C_6F_5)$_3$ pair is approximately 20 times higher than that produced by the ^{Mes}NHC/Al(C_6F_5)$_3$ pair under the same conditions. Moreover, this Al(C_6F_5)$_3$ based catalyst system has been also applied to the polymerization of biorenewable monomer MBL and MMBL. Incomplete monomer conversion was obtained for polymerization of MBL catalyzed by both phosphine/Al(C_6F_5)$_3$ and NHC/Al(C_6F_5)$_3$. Owing to the good solubility of PMMBL in dichloromethane, the LPP of MMBL is homogeneous and highly effective and achieved quantitative monomer conversion within 10 min, thus giving a high molecular weight polymer.

Later in 2012, inspired by the exceptional polymerization activity of Al(C_6F_5)$_3$-based CLA and FLP exhibiting for the polymerization of conjugated polar alkenes, Chen and co-workers further expanded the scope of the LPP with large number of LPs, consisting of eleven LAs as well as ten achiral and four chiral LBs.[73] Among the 11 LAs used, Al(C_6F_5)$_3$-based LPs are far more active and effective than other LA-based LPs. Several types of LBs can also be paired with Al(C_6F_5)$_3$ for highly efficient LPP, such as phosphine, NHC as well as phosphazene superbase (*i.e.* *t*Bu-P$_4$). Among the 14 LBs investigated, the combination of *t*Bu-P$_4$ with Al(C_6F_5)$_3$ exhibits the highest activity of LPs, achieving a remarkably high TOF of 9.6×10^3 h^{-1} and producing PMMA with $M_n = 212$ kg mol^{-1} and $Đ = 1.34$. The polymers produced by LPs, including chiral LPs at RT, are typically atactic (PMMBL with ~47% *mr*) or syndio-rich (PMMA with 67.8–77.9% *rr*), but highly syndiotactic PMMA with *rr* ≈ 91% could be produced by chiral or achiral LPs at −78 °C. Among the twelve kinds of monomers used, neither the Al(C_6F_5)$_3$/P^{t}Bu$_3$ pair nor the Al(C_6F_5)$_3$/^{Mes}NHC pair exhibited any reactivity towards polymerization of bulky furfuryl methacrylate (FMA) and five-membered lactones, including γ-butylrolactone (γ-BL), γ-valerolactone (γ-VL), and α-angelica lactone (α-AL), but only incomplete monomer conversion was obtained for the polymerization of ε-caporalactone (ε-CL) and *n*BA using the Al(C_6F_5)$_3$/P^{t}Bu$_3$ pair. In comparison with much slower polymerization of *N,N*-diphenylacrylamide (DPAA) by Al(C_6F_5)$_3$/P^{t}Bu$_3$, both Al(C_6F_5)$_3$/P^{t}Bu$_3$ and Al(C_6F_5)$_3$/^{tBu}NHC are highly active for the polymerization of DMAA. For polymerization of diethyl vinylphosphonate (DEVP), Al(C_6F_5)$_3$/P^{t}Bu$_3$ is active but sluggish while Al(C_6F_5)$_3$/^{Mes}NHC exhibited much enhanced reactivity. For polymerization of MMA, all Al(C_6F_5)$_3$/LB pairs are highly effective except Al(C_6F_5)$_3$/Mes$_3$P. Both Al(C_6F_5)$_3$/P^{t}Bu$_3$ and Al(C_6F_5)$_3$/^{tBu}NHC are quite effective for MBL polymerization, despite being a heterogeneous process, while the polymerization of MMBL by these Al(C_6F_5)$_3$/LB pairs is homogeneous and highly effective, producing polymers with high molecular weight. It is noted that M_n of the PMMBL produced by Al(C_6F_5)$_3$/^{tBu}NHC is about twice that produced by Al(C_6F_5)$_3$/^{Mes}NHC. The combination of systematic mechanistic studies, characterization of zwitterionic phophonium and imidazolium enolaluminate, the active species of the current LPP system, and kinetic studies led to a bimolecular, activated monomer propagation mechanism (Scheme 11.11), which was also confirmed by computational studies.

As a newly emerging class of organopolymerization catalyst, N-heterocyclic olefins (NHOs),[74–76] possessing a strongly polarized C=C double bond, have attracted attention in recent years due to their strong nucleophilicity and easily tunable stereochemistry property. When combined with LAs, such as $Al(C_6F_5)_3$ or $AlCl_3$, NHOs can rapidly initiate polymerization of various polar monomers such as MMA, BMA, DMAA and DPAA, producing high molecular weight polymers with relatively narrow MWD.[77] Activation of these monomers by LA is the prerequisite for realizing high-speed polymerization. Otherwise, the stable NHO·$Al(C_6F_5)_3$ adducts will be formed by the reaction of NHO and LA. The polymers produced by such NHO-based LPs at RT are syndio-rich, $rr = 70–79\%$. Although random copolymer PMMA-*co*-PBMA with $M_n = 480.0\ \text{kg mol}^{-1}$ and $Đ = 1.31$ could be obtained from such LPs-catalyzed polymerization with a 400/400/1 $[MMA]_0/[BMA]_0/[\textbf{NHO3}]_0$ ratio at RT in toluene, the attempt to synthesize block copolymers proved to be unsuccessful. The formation of six-membered lactone chain ends could be confirmed by both electrospray ionization time-of-flight mass spectrometry (ESI-TOF MS) and NMR spectroscopy, which is ascribed to the nucleophilic backbiting of the polymeric anion to the carboxyl carbon of the adjacent unit, and the backbiting is responsible for the non-living polymerization. The low initiation efficiency of NHO-based LPs is attributed to formation of NHO·$Al(C_6F_5)_3$ adducts during the polymerization, while the production of a lactone chain end leads to a complete termination of chain propagation (Scheme 11.12, termination b).

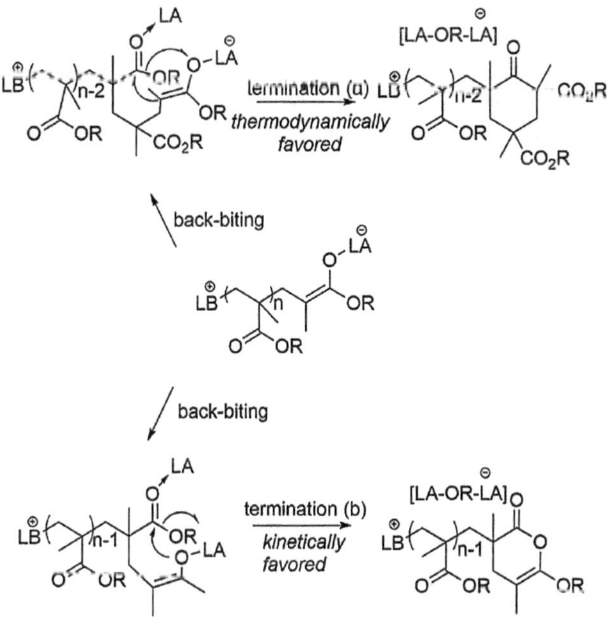

Scheme 11.12 The proposed two possible backbiting chain-termination pathways (a) thermodynamically favored and (b) kinetically favored in LPP of methacrylates.

The selective polymerization of one vinyl group of divinyl monomers has attracted much more attention due to the wide application of polymers bearing pendant vinyl groups on the backbone in material chemistry.[78–80] Lu and co-workers employed $Al(C_6F_5)_3$-based LPs for highly regioselective and active polymerization of various asymmetric divinyl polar monomers at the methylacrylic C=C bond under mild conditions, including 4-vinyl-benzyl methacrylate (VBMA), vinyl methacrylate (VMA), and ally methacrylate (AMA).[81] In comparison with the obtained insoluble or partly soluble polymers through anion and radical polymerization of divinyl polar monomers, the polymers with high M_n (up to 640 kg mol^{-1}) and narrow MWD ($Đ<1.5$) produced by LPs are soluble in various organic solvents. ESI-TOF MS studies demonstrated that the polymerization process of VBMA mediated by LPs only consumes the methyl acrylic C=C bond and selectively retained the pendant styrene C=C bond.

In 2014, Chen and co-workers reported the first example of $Al(C_6F_5)_3$/NHC FLPs catalyzed polymerization of polar vinyl monomers bearing the C=C–C=N functionality, such as 2-vinyl pyridine (2-VP) and 2-isopropenyl-2-oxazoline (*i*POx), into medium to high molecular weight, *N*-functionalized vinyl polymers.[82] For example, P(2-VP) could be obtained with $M_n = 38.4$–315 kg mol^{-1} and $Đ = 1.55$–2.04 by ^{tBu}NHC/$Al(C_6F_5)_3$ LPs. When switching from ^{tBu}NHC to ^{Mes}NHC, the obtained P(2-VP) had a small $M_n = 10.3$ kg mol^{-1} and high $Đ = 4.71$. Compared with polymerization of 2-VP, the LPP of *i*POx had a lower polymerization rate and produced P*i*POx with generally low M_n and broad MWD. Neither the combination of NHCs with $B(C_6F_5)_3$ or $AlEt_3$ nor $Al(C_6F_5)_3$/phosphine (tBu_3P, Mes_3P, or Ph_3P) pairs were effective for polymerization of 2-VP and *i*POx. The zwitterionic intermediate derived from the activation of 2-VP by ^{tBu}NHC/$Al(C_6F_5)_3$ showed no activity towards the polymerization of 2-VP unless the additional LA was added, which was in line with the bimolecular, activated monomer propagation mechanism.

Although significant progress has been achieved in the polymerization catalyzed by FLPs or CLAs as well as in the expansion of scope of monomer and LPs, polymerization kinetics and chain-termination mechanisms for the polymerization of conjugated polar alkenes by FLPs is scarce. Chen and co-workers investigated the mechanistic aspects of polymerization of conjugated polar alkenes by FLPs based on NHC/$Al(C_6F_5)_3$, by focusing on the characterization of active propagating intermediates, propagation kinetics, and chain termination pathways.[83] It is noted that the oligomer of MMBL produced by ^{Mes}NHC/$Al(C_6F_5)_3$ showed one major series of mass ions, indicating a living, linear chain, accompanied by similar chain ends with PMMBL produced by NHC alone[84] (*vide infra*). Kinetic studies revealed that the polymerization of 2-VP by ^{tBu}NHC/$Al(C_6F_5)_3$ proceeds with a bimolecular, activated monomer propagation mechanism in that the C–C bond forming step *via* intermolecular Michael addition of the propagating species to the LA-activated monomer is the rate-limiting step, and the release of the LA catalyst from its coordination to the last inserted monomer unit in the growing polymer chain to the incoming monomer is relatively fast. The LPP of conjugated polar alkenes such as methacrylates is accompanied by competing chain-termination side reactions (Scheme 11.12): (a) the *C*-ester enolate active chain end nucleophilic attacks the activated *antepenultimate* ester group of the growing polymer chain to generate a cyclic *β*-ketoester-terminated polymer chain. This type of intramolecular backbiting cyclization was previously observed in the anionic polymerization of acrylates[85] and polymerization of acrylates by metallocenium catalyst.[86] (b) Another termination pathway was proposed to proceed *via* nucleophilic attack of the activated adjacent ester group of the growing polymer chain by the *O*-ester enolate active chain end to generate a six-membered lactone (*δ*-valerolactone)-terminated polymer chain, which was consistent with the Lu's report (*vide supra*). Analysis of the chain-end group of the obtained low molecular weight polymer sample produced by ^{Mes}NHC/Al(C_6F_5)$_3$ with MALDI-TOF MS spectroscopy provided evidence for such chain termination side reactions but cannot conclusively convey which pathway is adopted by such LP-promoted polymerizations. Finally, computational studies revealed that the pathway leading to a *δ*-valerolactone is kinetically favored but thermodynamically disfavored over the pathway leading to the *β*-ketoester chain end.

The ill-controlled LPPs led to several disadvantages such as high M_n, broad MWD and thus yielding low initiation efficiency and the incapability of synthesizing block copolymers due to the chain termination side reaction. In 2016, Rieger and co-workers utilized simple but highly active combinations of LB such as phosphines and LA such as organoaluminum compounds to achieve catalytic polymerization of diverse Michael-type monomers with high precision, producing polymers with controlled M_n and microstructure.[87] They evaluated effects of different parameters influencing the LPPs by using the well-established fluoride ion affinity (FIA) index for quantification of the Lewis acidity[88–90] and Tolman angle (Θ in degrees) as a scale for the steric demand of LB.[91] For instance, as mentioned above, polymerization of electron-rich acrylates by highly acidic Al(C_6F_5)$_3$/LB pairs showed incomplete monomer conversions and broad MWD,[73] probably attributed to the deactivation of the catalyst resulting from the strong interaction of acrylate and highly acidic LA (FIA: 552.1 kg mol^{-1}).[90] Switching to the combination of less acidic Ph_3Al (FIA: 442.7 kJ mol^{-1}) [92] with PMe_3, *n*BMA was quantitatively converted to P*n*BMA with $M_n = 42.0$ kg mol^{-1} and narrow $Đ = 1.00$. The polymerization of *n*BMA by $AlPh_3$/*n*Hex_3P exhibited a decreased initiation efficiency. On the other hand, the trend that an increase in steric hindrance (Θ: PMe_3 (117°) < PEt_3 (132°) < PCy_3 (170°))[91] led to lower initiation efficiencies (I*%: PMe_3 (93) > PEt_3 (56) > PCy_3 (12)) clearly revealed the influence of the steric hindrance of the LB on the initiation efficiency. Moreover, they found the steric demand of the LPs also influenced the tacticity of the polymer product, such as steric encumbered LPs yield a polymer with higher syndiotacticity (*rr* = 78%) and broad MWD ($Đ = 1.30$), compared with that produced by steric unhindered LPs (*rr* = 61% and $Đ = 1.11$). For *t*BMA, with higher electron-donating and more steric hindrance, LPs composed of less acidic LA ($AlMe_3$, (FIA: 343.0 kJ mol^{-1}))[93] with sterically unhindered PMe_3 ($\Theta = 117°$),[91] exhibited an unparalleled control over the polymerization of *t*BMA in the field of LPP. All monomers were polymerized within 60 min, producing P*t*BMA with $M_n = 61.0$ kg mol^{-1} (initiation efficiency up to 93%) and extremely low $Đ$ value of 1.01. DMAA and DEVP, and even 4-vinylpyridine (4-VP) were also polymerized in a controlled way, regardless of polymerization temperature. In addition, ESI-MS analysis clearly identified phosphine as the initiation group and kinetic studies revealed this polymerization follows a bimolecular, activated monomer propagation mechanism (Scheme 11.13). Although the polymerization of *t*BMA by a CLA of $AlMe_3/PMe_3$ is controlled (predictable M_n, narrow MWD, and high I^* value), it was not interrogated using the living polymerization protocol. Therefore, no chain extension and block copolymer were reported for this strategy.

Although FLPs and CLAs exhibited high activity for polymerization of conjugated polar alkenes, their application is hampered by both low initiation efficiencies and chain-termination side reactions,[77,83] evidenced by the much higher observed M_n than the calculated M_n and broad MWD (or large $Đ$ values) of the resulting polymers, thus giving rise to low initiation efficiencies (I^*) and the inability to produce well-defined block copolymers. Zhang and co-workers employed the combination of a strongly nucleophilic NHO with a sterically encumbered but modestly strong LA $MeAl(4\text{-}Me\text{-}2,6\text{-}^tBu_2\text{-}C_6H_2O)_2$ ($MeAl(BHT)_2$) to achieve highly effective, living polymerization of alkyl methacrylates.[94] At first, $NHO/Al(C_6F_5)_3$ showed high reactivity for MMA polymerization, producing polymers with predictable M_n, narrow MWD, and thus yielding high initiation efficiencies. However, both the partially successful chain extension experiments and the formation of cyclic chain ends derived from the backbiting chain termination process as indicated by the chain end analysis revealed that the polymerization by $NHO/Al(C_6F_5)_3$ is rapid and controlled (over M_n and MWD) but not a living process. To achieve a highly effective and living polymerization by LPs, it is crucial to finely balance the acidity and steric hindrance of the Lewis acid and the steric and electronic interplay between the LA and LB counterpart. Among a series of organoaluminum LAs screened, $MeAl(BHT)_2$ appeared to be the best LA to combine with the screened NHOs to achieve a controlled and living polymerization of MMA. The corresponding analysis of the obtained low molecular weight PMMA with MALDI-TOF MS spectroscopy also showed no evidence for a cyclic backbiting chain end. The stoichiometric reaction of NHO and $MeAl(BHT)_2{\cdot}MMA$ generated zwitterionic intermediates as two isomers (*Z*/*E*), which promoted the first example of non-interacting, true FLP-promoted living polymerization of less bulky methacrylate,

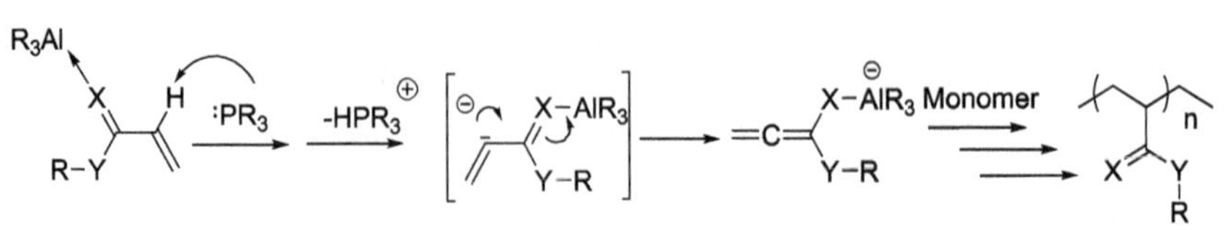

Scheme 11.13 Proposed deprotonation pathway of acrylate by P/Al pairs.

particularly MMA, producing predictable M_n (up to 351 kg mol^{-1}), and narrow MWD ($Đ = 1.05$). It is worth noting that the formation of backbiting cyclic chain ends was observed for polymerization by the combination of NHO with the other LAs such as $AlEt_3$, $AlMe_3$, and even $Ph_3Al{\cdot}OEt_2$, possessing similar acidity to that of $MeAl(BHT)_2$. The living nature of this FLP-promoted polymerization has been further verified by kinetic studies, which also enable the synthesis of well-defined diblock and ABA triblock copolymers with narrow MWD ($Đ = 1.09$–1.13), regardless of the comonomer addition order. These results should stimulate future efforts in developing highly active and living LP catalyst systems for the synthesis of well-defined polymers, thereby further expanding the utility of the FLP chemistry in polymer synthesis.

In the FLP chemistry, the bulky $B(C_6F_5)_3$ LA is commonly used to combine with a bulky NHC or phosphine LBs to display FLP-induced or enhanced reactivity in the activation of small molecules. Interestingly, in the macromolecular synthesis, $Al(C_6F_5)_3$-based FLPs with LB could rapidly polymerize acrylics, including MMA and biorenewable MBL and MMBL to high molecular weight polymers, while it is a challenging task to achieve highly effective FLPs comprising the borane congener $B(C_6F_5)_3$ and the bulky LBs since it is inactive for such polymerization. In 2014, Chen and co-workers developed the first highly active LPP of MMBL by phosphine (P)/borane (B) LPs.[95] Six **P/B** intra- or intermolecular LPs (Figure 11.7) with different degrees of "frustration" (from FLPs to interacting FLPs to CLAs) were employed as catalysts to investigate the mechanism of the relationship between the degree of LP "frustration" and the polymerization activity and observed an

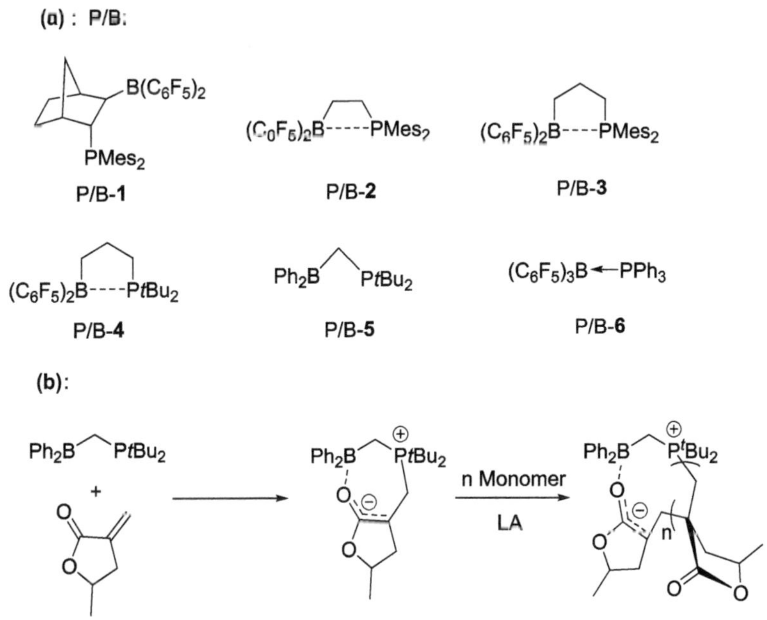

Figure 11.7 P/B pairs and adduct of (a) P/B and (b) MMBL.

obvious inverse relationship: the more frustrated LP had a lower polymerization activity, and the CLA adduct had the highest polymerization rate, which indicated that **P** and **B** were *highly interacting*. Besides the strength between **P** and **B**, there are two other factors also having an effect on polymerization: (1) the stronger basicity of the **P** site gave higher polymerization activity; (2) the stronger acidity of the **B** site also led to a faster polymerization rate. Hence, the most active $Ph_3B{\cdot}B(C_6F_5)_3$ CLA system is brought about by a good compromise between **B** site acidity, **P** site basicity, steric crowding around **P**, and the strength of the **P–B** association in solution, which further highlighted the significant importance of the cooperativity of the **P** and **B** sites of LPs on achieving high polymerization activity. It should be noted that the initiation efficiency of these LP series was low (generally below 61%); at the ratio of $[MMBL]_0/[\mathbf{P/B\text{-}6}]_0 = 800/1$ the initiation efficiency increased up to 88%, but side reactions occurred, indicated by trimodal MWD. Mechanism studies revealed that the zwitterionic intermediate (Figure 11.7) derived from the reaction of MMBL and LP was the active species but cannot initiate the polymerization of MMBL by itself. An additional amount of LA was needed to activate the monomer, which was consistent with previous reports (*vide supra*).

In 2015, Chen and co-workers further developed the first highly active and efficient $NHC/B(C_6F_5)_3$ LPs for polymerization of MMA, MMBL and AMA.[96] Among the three investigated LPs, $^{IPr}NHC_2/B(C_6F_5)_3$ is the most active (~3× and ~120× more active than $^{Mes}NHC/B(C_6F_5)_3$ and $^{tBu}NHC/B(C_6F_5)_3$, respectively) and an effective catalyst system for the polymerization of MMA. PMMA was obtained with $M_n = 10.8–25.6$ kg mol^{-1}, $Đ = 1.36–2.13$, and $rr = 74–80\%$ by using $B(C_6F_5)_3$/NHC LPs in toluene at RT. Kinetic studies revealed the polymerization follows the bimolecular, activated monomer propagation mechanism. In addition, MMBL could be polymerized at high-speed by $NHC/B(C_6F_5)_3$ LPs (TOF ≥ 48 000 h^{-1}), producing PMMBL with medium to high molecular weight (M_n up to 489 kg mol^{-1}) and narrow MWD ($Đ < 1.17$). Moreover, such $NHC/B(C_6F_5)_3$ LPs could chemoselectively polymerize the conjugated vinyl group of AMA while leaving the non-conjugated vinyl group in the allyl moiety intact. The obtained PMMA is syndiotactic ($rr = 83\%$), non-cross-linked and soluble in common solvents, and is thus suitable for further functionalization.

Chen and co-workers also discovered that, when paired with $E(C_6F_5)_3$ (E = Al, B), 2,2,6,6-tetramethylpiperidine (TMP) and Et_3N exhibited a contrasting reactivity toward linear MMA and cyclic MMBL.[97] More specifically, neither $TMP/E(C_6F_5)_3$ nor $Et_3N/E(C_6F_5)_3$ LPs polymerize MMA. In contrast, both $TMP/E(C_6F_5)_3$ and $Et_3N/E(C_6F_5)_3$ promoted high-speed polymerization of MMBL, achieving quantitative monomer conversion in less than 3 min even at a low catalyst loading of 0.0625 mol%. It is noted that TMP-based LPs with either $Al(C_6F_5)_3$ or $B(C_6F_5)_3$ are more effective for the polymerization of MMBL than that by Et_3N-based LPs, achieving PMMBL with a high M_n value of 129 kg mol^{-1}, relatively broad MWD ($Đ = 2.21$) and up to 96 000 h^{-1} TOF at 0.125 mol% catalyst loading.

More recently, Taton and co-workers developed a metal-free LP composed of a weak LA such as Me_3SiNTf_2 and a simple phosphine for the highly active and chemoselective polymerization of MMA.[98] To avoid deactivation of the catalyst resulting from the strong interaction in NHC-based LPs,[81] the phosphine-based LP with the relative weaker interaction is more favorable for the polymerization. Any of the three commercially available phosphines (P^tBu_3, P^nBu_3 or tris(2,4,6-trimethoxyphenyl)phosphine (TTMPP)) paired with Me_3SiNTf_2 achieved complete monomer conversion within 3–15 h in toluene at 25 °C while no polymerization was observed when switching to less electrophilic Me_3SiOTf_2. The TTMP/Me_3SiNTf_2 pair produced well-defined PMMA with predictable M_n and narrow MWD ($Đ<1.10$). Similar to a "bimetallic mechanism", two eq. of Me_3SiNTf_2 were also required to activate a first eq. of MMA in the initiation step and a second eq. in the propagation step. The conjugated addition of phosphine onto the Me_3-$SiNTf_2$-activated MMA yielded a α-phosphonium silyl ketene acetal (α-PSKA) ion pair, rather than a zwitterionic intermediate formed in the LP system (*vide supra*). Chain propagation proceeds with the addition of the α-PSKA to an incoming activated MMA, accompanied by the release of one eq. of Me_3SiNTf_2 (Scheme 11.14). Successful chain extension experiments revealed that PMMA produced by TTMP/Me_3SiNTf_2 could be reactivated. The syndiotacticy of PMMA obtained by Me_3SiNTf_2/TTMPP LP was comparable to that by Me_3SiNTf_2-catalyzed GTP of MMA (*vide supra*).[45] Furthermore, two possible chain termination pathways were proposed for polymerization of MMA by Me_3SiNTf_2/TTMPP LP based on the analysis of ^{31}P NMR and MALDI-TOF MS. Both computational and experimental data support the existence of a cooperative mechanism during the Me_3SiNTf_2/TTMPP LP promoted polymerization of MMA.

As industrially important thermoplastic resins, intense attention has been focused on achieving control over M_n, the regioselectivity, and stereochemistry of the polymerization of conjugated dienes, especially in the synthesis of diene-based cyclic polymers with special characteristics due to

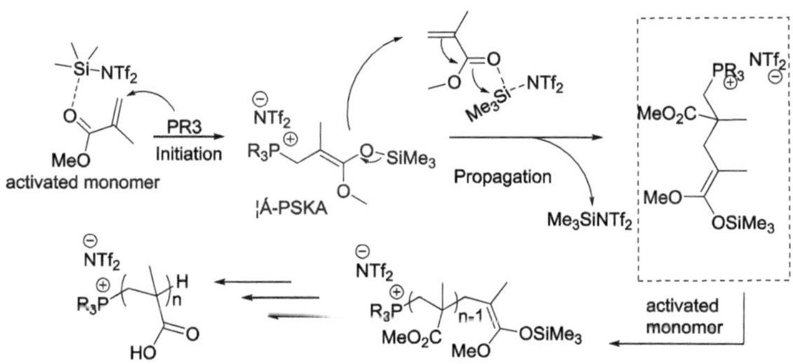

Scheme 11.14 Proposed mechanism of polymerization of MMA catalyzed by Me_3SiNTf_2/PR_3.

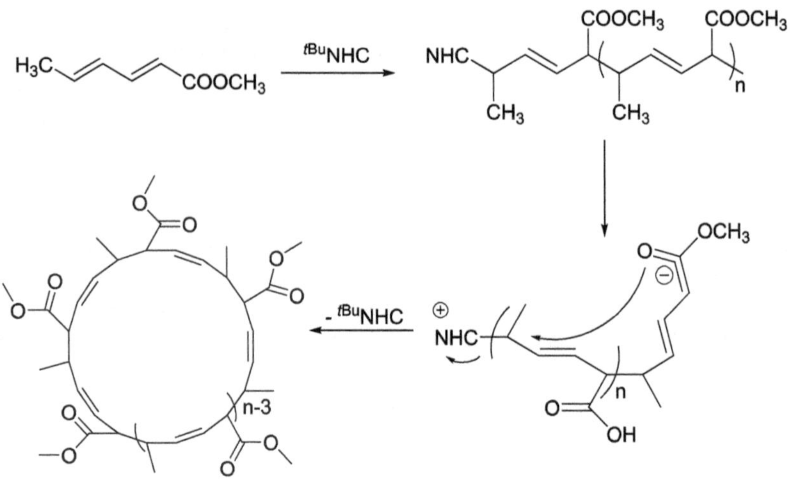

Scheme 11.15 Ring-closing pathway in LPP of MS.

the absence of polymer termini. So far there are only a few reports dealing with synthetic strategies for vinyl monomers as well as diene-based cyclic polymers. In 2017, Takusu and co-workers reported ^{tBu}NHC/MeAl(BHT)$_2$ LPs for polymerization of methyl sorbate (MS) to achieve well-defined cyclic poly(methyl sorbate)s (poly(MS)).[99] Although (*E*,*E*)-methyl sorbate (MS) could be polymerized by ^{tBu}NHC alone in DMF or THF, the obtained polymer with relatively broad MWD ($Đ = 1.2–2.1$) might be attributed to the chain transfer side reaction *via* nucleophilic substitution of the propagating ester enolate to the electrophilic carbon bonded to the NHC to form the cyclic polymer accompanied by regeneration of ^{tBu}NHC. The addition of bulky MeAl(BHT)$_2$ into the NHC-catalyzed anionic polymerization of MS not only accelerated quantitative monomer consumption but also inhibited the proton transfer by coordination to the propagating anion as well as monomer molecules, which shifted the balance from linear polymer to cyclic polymer. The analysis of the chain ends by MALDI-TOF MS and the DSC and viscosity measurements of the resulting polymer and its hydrogenated analogue provided evidence for the formation of cyclic poly(MS)s. Just like a chain-expansion polymerization, the α-terminal imidazolinium group acted as a countercation neighboring the propagating anion in this anionic polymerization, in which the cyclic propagating chain preferred ring closing rather than H-transfer, as shown in Scheme 11.15.

11.4 Other Types of Organopolymerization of Polar Vinyl Monomers

In addition to the above-mentioned GTP and LPP, the organopolymerization using organic compounds, such as NHC, NHO, phosphazene base and so on,

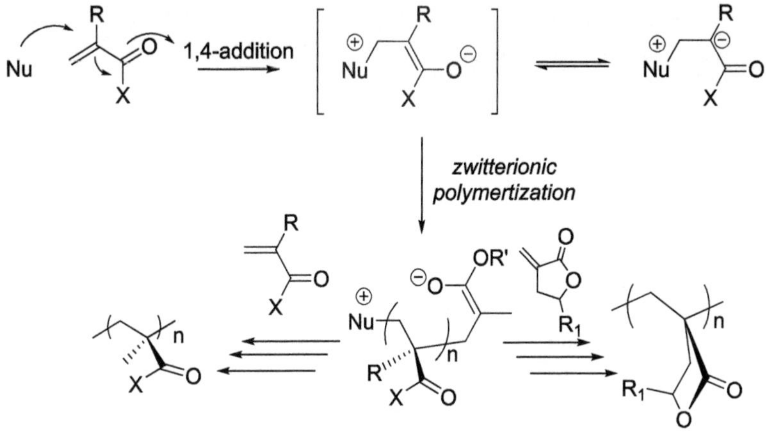

Scheme 11.16 Nucleophilic-induced vinyl polymerization in the absence of any co-initiator.

as catalysts or initiators for polymerization of polar vinyl monomers became a powerful strategy in polymer synthesis, especially when metal-free products or processes are of primary concern. Such organopolymerization adopts either a zwitterionic or anionic polymerization mechanism. As shown in Scheme 11.16, in the polymerization initiated by a zwitterionic active species derived from the attack of monomer by a nucleophile (Nu, *e.g.* NHC, NHO), organic LBs could serve as a direct initiator, while in anionic polymerization. The LB (*i.e.* phosphazene base) extracts a proton from the monomer to generate counterion pairs and the anionic, active species to continuously initiate polymerization. Moreover, various types of organopolymerization have also been developed, such as organic electron donors promoted polymerization, NHC-catalyzed oxa-Michael addition polymerization, proton-transfer polymerization (HTP) and the polymerization of MMA catalyzed by CO_2-protected NHC. According to the classification of catalysts, these polymerizations will be described as follows. The structures of polar vinyl monomers and catalysts employed for organopolymerization are shown in Figures 11.8 and 11.9. The related polymerization results developed in this book are summarized in Table 11.7.

11.4.1 NHC and CO_2-protected NHC

Due to its unique reactivity and selectivity, NHCs have not only been used in many different types of organic reactions, but also polymer synthesis through ring-opening polymerization, group transfer polymerization, Lewis pair polymerization (LPP) and step-growth polymerization.[100–102] On the other hand, several reactions of NHCs and acrylic substrates have also been discovered, such as tail-to-tail dimerization,[103–105] cyclotetramerization[106] and umpolung of Michael acceptors.[107] Chen and co-workers developed the

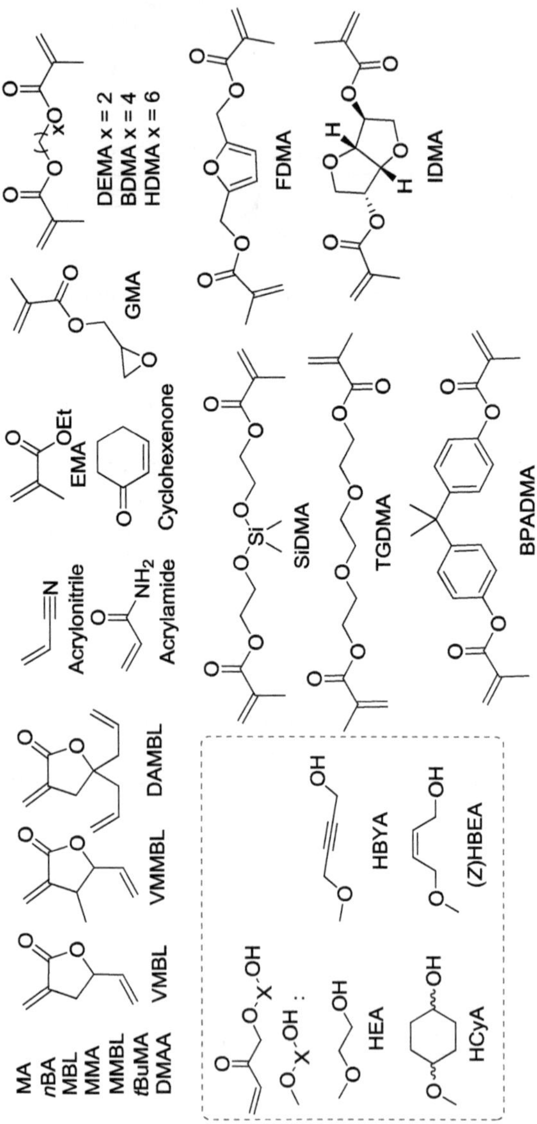

Figure 11.8 Monomers employed in organopolymerization.

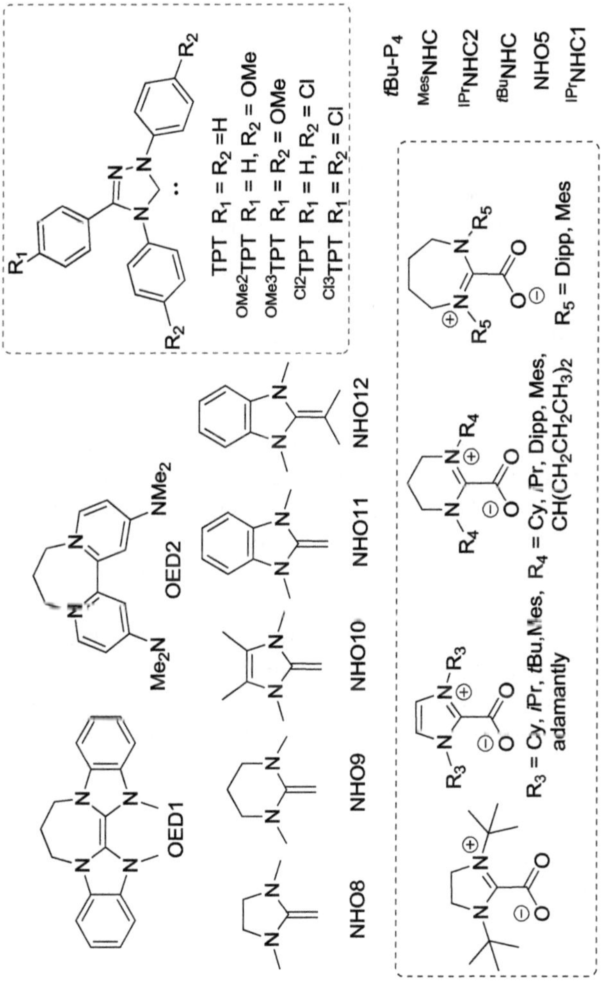

Figure 11.9 Organocatalysts employed in organopolymerization of polar vinyl monomers.

Table 11.7 Other type organopolymerizations.

Entry	Catalyst	Solvent	Polymer	M_n (kg mol^{-1})	$Đ$ (M_W/M_n)	Ref.
1	^{tBu}NHC	DMF	PMMA	33.2	1.99	108
2	^{tBu}NHC	DMF	PFMA	14.7	1.94	108
3	^{tBu}NHC	DMF	PMMBL	38.5–88.7	1.45–2.33	84
4	*t*Bu-P_4	DMF	PMMBL	23.2–111	1.89–3.98	120
5	TPT	THF	PHEA	1.2–1.5	1.6–1.7	112
6	^{tBu}NHC/BnOH	DMF	PMMA	0.9–8.0	1.1–1.9	114
7	^{tBu}NHC/BnOH	DMF	PMA	0.7–3.9	1.2–2.0	114
8	^{tBu}NHC	DMF	PVMBL	65.9–73.8	2.17–2.99	113
9	^{tBu}NHC	DMF	PMVMBL	8.18–10.3	1.23–1.36	113
10	OED2	Bulk	PEMA	9.9–28	1.33–5.76	121
11	OED2	Bulk	PMMA	12–13.2	1.30–1.67	121
12	OED2	Bulk	P*n*BA	1.7	3.59	121
13	OED2	DMF	P*n*BA	5.6	1.52	121
14	TPT	TOL	PBDMA	9.42–15.8	1.70–1.90	110
15	TPT	TOL	PTGDMA	11.2–15.3	1.68–2.02	111
16	TPT	TOL	PFDMA	9.52–14.4	2.27–3.02	111
17	TPT	TOL	PIDMA	5.29–10.9	1.81–2.30	111
18	NHO5	THF	PDMAA	254–654	1.32–1.59	122
19	*t*Bu-P_4/MeOH	THF	PBMA-*co*-PCL	13.2–25.4	1.28–1.72	118
20	*t*Bu-P_4/MeOH	THF	PMMA-*co*-PCL	16.1–28.2	1.45–2.32	117

first example of NHC-mediated rapid organopolymerization of polar vinyl monomers, without the requirement of any other initiating or catalyzing components.[84,108] For MMA, there exists a remarkable selectivity of the NHC structure for the three different types of reactions it promotes: enamine formation (single-monomer addition) by ^{Mes}NHCs; (tail-to-tail) dimerization by TPT, and polymerization by ^{tBu}NHC. For MMBL, all three NHCs promoted rapid polymerization. The rate of the polymerization is strongly affected by the relative nucleophilicity of the NHC catalysts, with the most nucleophilic ^{tBu}NHC in the series exhibiting the highest activity, and the less nucleophilic ^{Mes}NHC showing noticeably lower activity, and the least nucleophilic TPT often showing no activity at all. The polymerization of MMBL by ^{tBu}NHC produced medium- or high-molecular weight polymers in less than one minute, thus giving a high TOF of greater than 4.8×10^4 h^{-1}. The M_n of the obtained PMMBL could be adjusted by varying the $[\text{MMBL}]_0/[^{tBu}\text{NHC}]_0$ ratio (<800), although there was a gradual broadening MWD ($Đ = 1.68–2.11$). For polymerizations carried out in DMF with a $[\text{MMBL}]_0/[^{tBu}\text{NHC}]_0$ ratio of over 800, chain transfer occurred. For instance, 1000–12000 eq. of MMBL could be quantitatively converted into polymer with a narrow range of M_n (70.0–85.0 kg mol^{-1}) and initiator efficiency up to 1600%. It is noted that MMBL polymerization by ^{tBu}NHC in toluene or THF is heterogeneous and slower while it is homogenous and extremely rapid in more polar solvent DMF. Kinetics studies indicated that the MMBL polymerization catalyzed by ^{tBu}NHC in DMF is first-order with respect to monomer concentration and second-order in [^{tBu}NHC] concentration. Computational studies showed the slowest step of the polymerization was the formation of ^{tBu}NHC-MMBL adducts, *i.e.*, generation of zwitterionic active propagating species, which was in agreement with the observed first-order dependence on MMBL concentration. However, the reason why there was a second-order in [^{tBu}NHC] was still elusive. The combined studies by MALDI-TOF MS, GPC, NMR, and computational methods concluded that the PMMBL produced by NHCs at RT is a linear polymer with different chain-end groups, depending on polymerization conditions. The computational studies provided mechanistic insights into reactivity and selectivity between two competing pathways for each NHC-monomer zwitterionic adduct: enamine formation/ dimerization through proton transfer *vs.* polymerization through conjugated addition. The polymerization adopted a zwitterionic pathway, as illustrated in Scheme 11.16, and formed cyclic polymer chains.

In contrast, Taton and co-workers disclosed that ^{tBu}NHC could polymerize MMA into PMMA in DMF at RT, whereas the reaction with a 1 : 2 ^{IPr}NHC1/MMA ratio furnished an unusual imidazolium-enolate cyclodimer, which could be characterized by NMR spectroscopy, X-ray diffraction, and MS. Density functional theory calculations revealed that the occurrence of two competitive low-energy pathways, involving either a basic activation of the alcohol or a nucleophilic activation led to the unexpected difference between ^{tBu}NHC (polymerization) and ^{IPr}NHC1 (cyclodimerization) in the reaction with MMA.[109]

More recently, Chen and co-workers developed a proton (H)-transfer polymerization (HTP) to convert dimethacrylates to unsaturated polyesters by an NHC capable of promoting intermolecular umpolung condensation through proton transfer.[110,111] This HTP proceeds through the step-growth propagation cycle *via* enamine intermediates, consisting of the proposed conjugate addition-proton transfer-NHC release fundamental steps (as shown in Scheme 11.17). The added radical inhibitors (4-methoxyphenol or catechol) not only suppressed the radically induced vinyl-addition polymerization under HTP conditions (typically at 80–120 °C), but also facilitated effective proton transfer after each monomer enchainment, thus achieving enhanced polymerization activity and producing unsaturated polyesters PEDMA with M_n up to 16.1 kg mol^{-1} and $Đ = 1.1$–1.9. Thermal measurements indicated that the glass transition temperature (T_g) decreased with an increase of the methylene unit (x) in the monomer structure: −51 °C for PEDMA ($x = 2$), −55 °C for PBDMA ($x = 4$), and −60 °C for PHDMA ($x = 6$), whereas T_g increased with an increase in the rigidity of the main chain, following the order: PSiDMA (−63.3 °C) < PBDMA (−54.7 °C) < PTGDMA (−47.7 °C) < PFDMA (2.76 °C) < PIDMA (40.4 °C) < PBPADMA (60.0 °C). The structure/reactivity relationship study revealed that ^{OMe2}TPT was both a good leaving group and a strong nucleophile, and thus showed the highest HTP activity and producing unsaturated polyester PBDMA with the highest $M_n = 16.7$ kg mol^{-1} and $Đ = 1.64$, while Cl-substituted TPT derivatives were the least active and efficient. On the one hand, the formation of bis(enamine) intermediates was investigated by monitoring ^{1}H NMR. The reactivity was found to decrease in the following the order: BPADMA > IDMA > FDMA > TGDMA > SiDMA > BDMA, which was in line with the decreasing electron deficiency of the methacrylate double bond

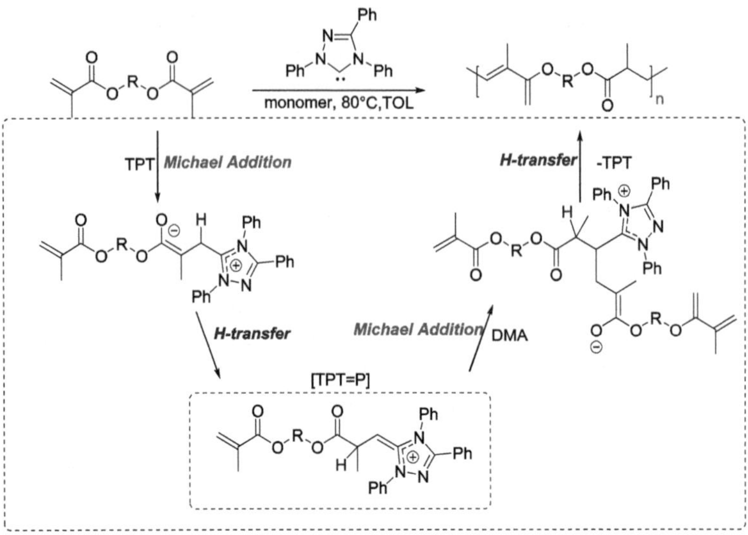

Scheme 11.17 Proposed mechanism for the HTP of dimethacrylates forming unsaturated polyesters.

and the rigidity of the DMA structure. On the other hand, the generation of polyesters was investigated by monitoring the polymerization with a 5/1/0.1 DMA/TPT/MP ratio at 80 °C by ^{1}H NMR, all DMA monomers were readily polymerized with decreasing polymerization activity in the order of: IDMA > BPADMA ≈ TGDMA > FDMA ≈ BDMA > SiDM, which was slightly different from the reactivity of the bis(enamine) intermediate formation. Computational studies have provided mechanistic insights into the tail-to-tail dimerization coupling step as a suitable model for the propagation cycle of the HTP.

Aza- or oxa-Michael addition polymerizations or hydrogen transfer polymerization of NH or OH functionalized vinyl monomers are useful procedures for the synthesis of polyamide and poly (ether-ester)s. Recently, Matsuoka and co-workers reported the first example of NHC-catalyzed oxa-Michael addition polymerization to produce poly(ester-ether)s with alicyclic, alkene, and alkyne groups in the main chain at RT, in contrast to the requirement of high temperatures for the polymerizations by conventional catalysts.[112] The analysis of polymer structure by NMR, ESI-MS analyses and methanolysis enabled the estimation of the frequency of the transesterification. A Lewis base mechanism was proposed on the identification of the terminal structure with ESI-MS: NHC first reacted with monomer to generate the zwitterionic intermediate followed by H-transfer to form a hydroxyl functional triazolium and an alkoxide living chain end, which could be added onto the monomer *via* oxa-Michael addition. The subsequent proton transfer produced the poly(ether-ester) and generated the active alkoxide again (Scheme 11.18). The attack of the ester carbonyls in the main chain or monomer by alkoxide resulted in the transesterification.

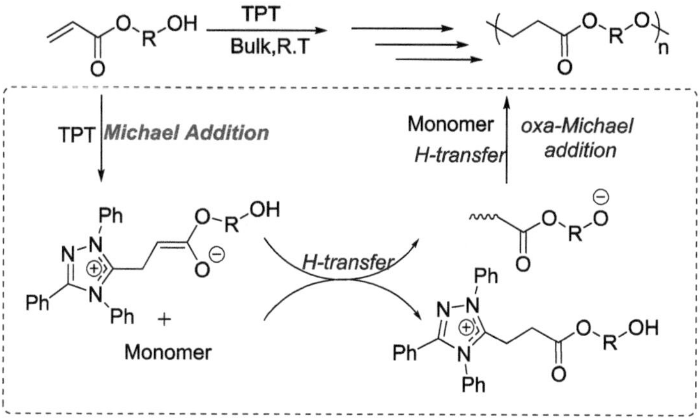

Scheme 11.18 Proposed mechanism of the oxa-Michael addition polymerization of acrylates catalyzed by TPT.

It has been a challenging task to achieve complete chemoselectivity in the polymerization of multivinyl polar monomers due to their derivation potential to form diverse advanced functional materials for various applications, currently achievable only by a metal- or metalloid-mediated polymerization process but in a non-catalytic fashion. Chen and co-workers reported the first example of NHC promoted organic and chemoselective polymerization of multivinyl-functionalized γ-butyrolactones only at the conjugated α-methylene double bond, leaving the γ-vinyl double bond intact.[113] In particular, the polymerization of γ-vinyl-α-methylene-γ-butyrolactone (VMBL) by ^{tBu}NHC in DMF is highly active and efficient, achieving the highest TOF of 80 000 h^{-1} even with an exceptionally low catalyst loading of 50 ppm. The resulting PVBML can be either thermally cured into cross-linked materials or post-functionalized with the thiol-ene "click" reaction to achieve complete conversion of all pendant vinyl groups into the corresponding thioether. Miqueu and Taton also employed ^{tBu}NHC as an organocatalyst to catalyze the selective Michael addition of simple alcohols onto MA or MMA in DMF at RT.[114] The M_n of the obtained polymers could be adjusted by varying the ratio of $[\text{monomer}]_0/[\text{ROH}]_0$, producing PMMA ($M_n = 0.9$–$8.0$ kg mol^{-1}, $Đ = 1.1$–1.9) and PMA ($M_n = 0.7$–3.9 kg mol^{-1}, $Đ = 1.2$–2.0). The use of hydroxyl-terminated poly(ethylene oxide) (PEO-OH) as macro-initiator led to the preparation of amphiphilic block copolymer PEO-*b*-PMMA. The combined analysis of UV, NMR and MALDI-TOF MS indicated that some of the polymer chain-ends were capped by ROH. DFT calculations revealed that there exist two competitive concerted pathways: one is the activated initiator/chain-end mechanism, while the other is the activated monomer mechanism (Scheme 11.19).

Compared to free NHCs, carbon dioxide (CO_2)-protected NHCs (NHC-carboxylates) provide enormous advantages, such as easy preparation and storage. Their latency allows them to serve as thermally latent initiators

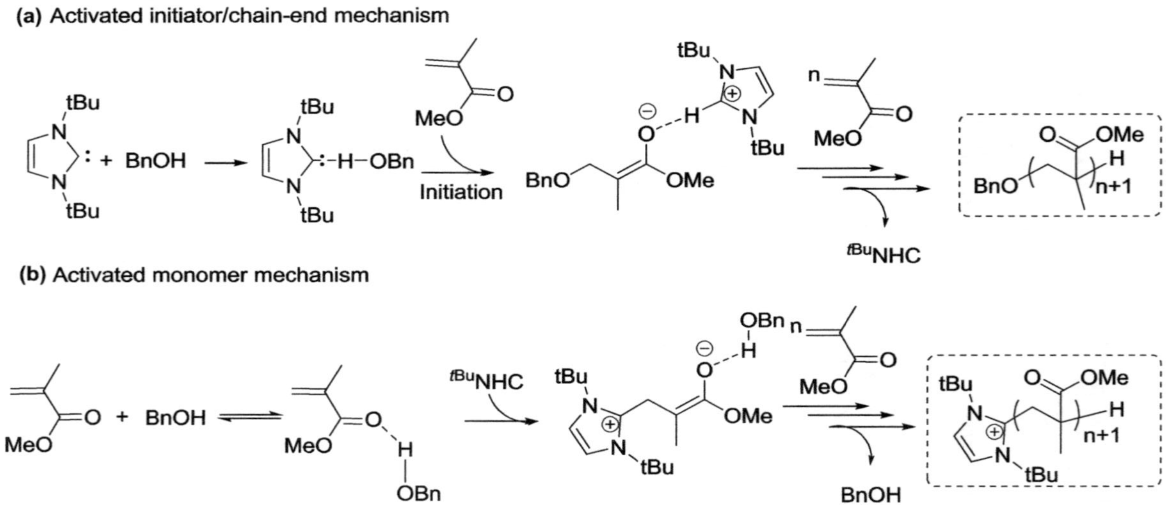

Scheme 11.19 Proposed mechanism of polymerization of MMA catalyzed by ^{tBu}NHC/BnOH.

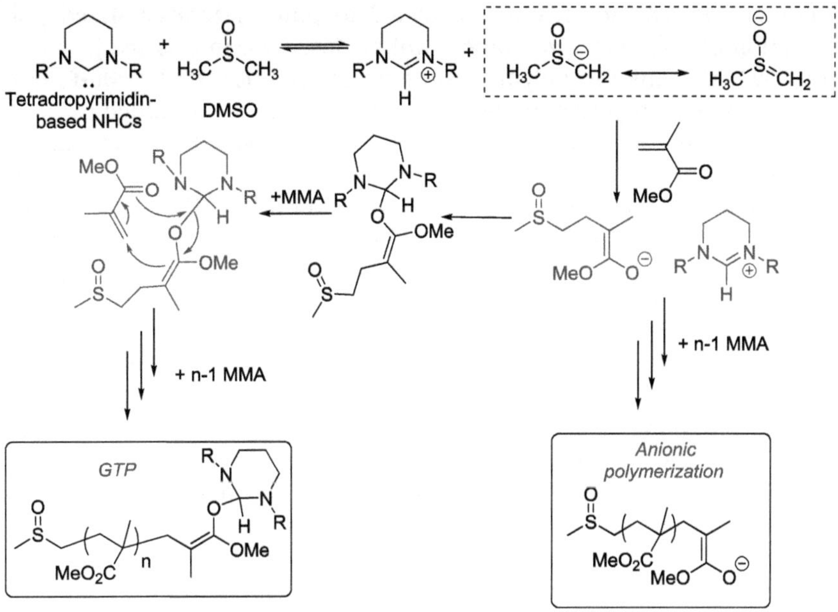

Scheme 11.20 Proposed mechanism of polymerization of MMA catalyzed by NHC in DMSO.
Adapted from ref. 123 with permission from Elsevier, Copyright 2016.

through the decarboxylation to release free NHC for the reaction. Buchmeiser and co-workers investigated the influence and activity of various NHC-carboxylates on the polymerization of MMA.[115] Such NHC-carboxylates are truly latent pre-catalysts, delivering a high yield of PMMA upon heating but are completely inactive at RT for MMA polymerization. ^{tBu}NHC-CO_2 can polymerize MMA in both polar and non-polar solvents as well as in bulk, producing PMMA with $M_n = 8.0–25.0\ \mathrm{kg\,mol^{-1}}$ and relatively broad MWD ($Đ = 2.6–2.7$), comparable to data reported previously.[108] The direct polymerization of MMA was also enabled by tetrahydropyrimidinium-2-carboxylates in DMSO, producing PMMA with higher $M_n = 13.0–80.0\ \mathrm{kg\,mol^{-1}}$ and broader MWD ($Đ = 1.74–4.4$). NHC-CO_2 probably acted as Brønsted bases rather than nucleophiles. DMSO was deprotonated by free NHC to produce the "dimsyl" anion, which could initiate the anionic polymerization of MMA (Scheme 11.20). Alternatively, a GTP-style operation mode was also proposed through transferring the NHC-enolate initiator to the living chain end.

11.4.2 Phosphazene Base

Different from NHCs, phosphazenes are strong Brønsted bases but weak nucleophiles. In the field of macromolecular synthesis, they have been widely used catalysts for ROP and anionic polymerization of polar vinyl monomers. Recent advancements related to phosphazene-catalyzed polymerization of polar vinyl monomers will be described in this section.

In 1993, the anionic polymerization of MMA initiated by *t*Bu-P_4/ethyl acetate (EA) was reported by Seebach and co-workers, producing PMMA having $M_n > 10.0$ kg mol^{-1} and relatively broad MWD ($Đ = 1.11$–2.29).[116] Recently, Zhang *et al.* employed *t*Bu-P_4 for copolymerization of ε-CL and various methacrylate monomers, including MMA,[117] *t*BMA[118] and glycidylmethacrylate (GMA)[119] in the presence of initiator (*i.e.* MeOH, 1-naphthalenemethanol (NMA), ethylene glycol (EG) or ethyl acetate (EA)) (Scheme 11.21). NMR and thermal measurements indicated that the cyclic ester and vinyl monomer form random copolymers. A hybrid copolymerization mechanism was proposed as follows: firstly, the alcohol was activated by *t*Bu-P_4 through intermolecular hydrogen bonding. The nucleophilic attack of the monomer (MMA or ε-CL) by activated alcohol produces alkoxide and enolate active centers, which would continuously react with incoming monomers to form a long active chain. In such a copolymerization, each monomer can form propagating species, which can further combine with either similar or dissimilar monomers to form a random copolymer. This strategy can be applied to produce a number of new polymers with interesting properties.

This organic phosphazene superbase, *t*Bu-P_4, is also utilized to initiate high-speed polymerization of biorenewable monomers MBL and MMBL, without requirement of an organic acid or a nucleohile as a co-initiating component.[120] Quantitative conversion of MMBL could be achieved within 1 min in DMF at RT even with a low *t*Bu-P_4 loading of 0.1 mol% or 0.02 mol%, furnishing polymers with medium to high molecular weight and relatively broad MWD. Kinetics studies of the polymerization by *t*Bu-P_4 showed that a zero-order dependence on [MBL] with an induction time of 7 h in the absence of a chain transfer agent (CTA), but a first-order dependence on [MBL] without an induction period in the presence of CTA. These results indicated the chain initiation was slow relative to the subsequent monomer additions in *t*Bu-P_4 promoted MBL polymerization. Combined experimental and computational studies led to the proposed mechanism to elucidate the process of chain initiation, chain propagation and chain termination in more detail (Scheme 11.22).

11.4.3 Organic Electron Donors

Due to their neutral structures and exceptional redox potentials, strong organic electron donors (OEDs) demonstrated efficiency in the reduction of challenging substrates, for which metallic reducing agents are usually used. In the presence of small amounts of organic reducing agents, Broggi and Vanelle achieved the first metal-free, chain-growth polymerization of various activated alkene and cyclic ester monomers in short reaction times under mild conditions, thus producing polymers with good polydispersities (generally $Đ < 2.00$) and without the requirement of co-initiators or activation

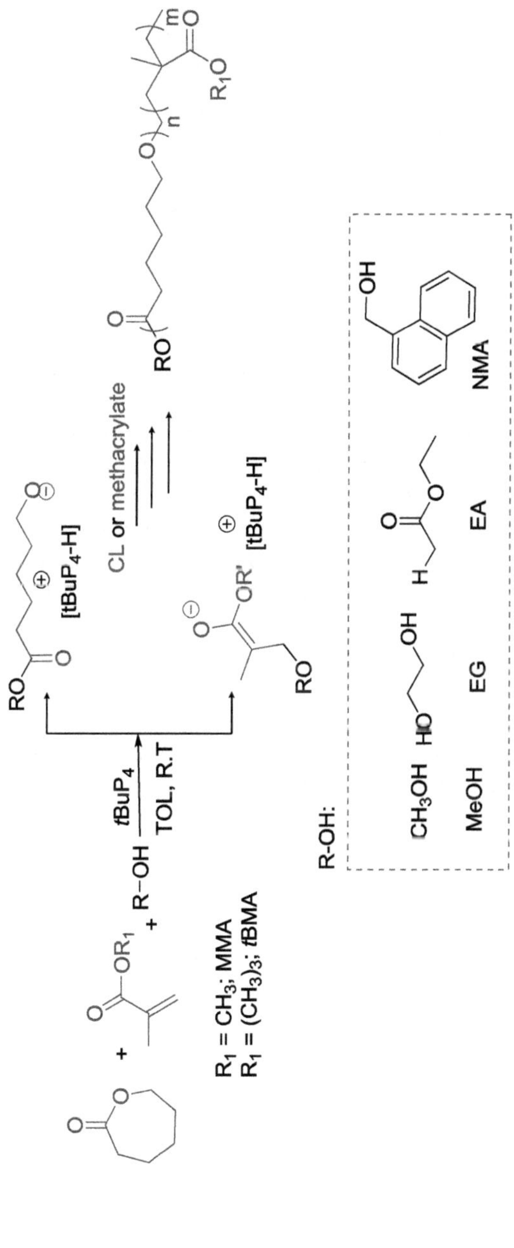

Scheme 11.21 Hybrid copolymerization of ε-CL with methacrylic monomers

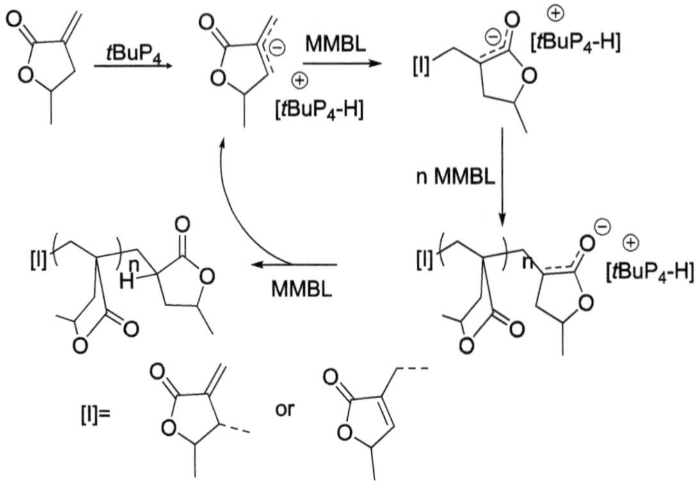

Scheme 11.22 Polymerization of MMBL directly initiated by tBu-P_4.

by photochemical, electrochemical, or other methods.[121] For instance, PEMA could be obtained with $M_n = 9.9–28.0\ kg\,mol^{-1}$ and $Đ = 1.33–5.76$ catalyzed by **OED-2** in DMF or bulk at RT. In addition, the block polymer PMMA-*b*-PEMA was also obtained by sequential addition of monomers. The rate of the chain-growth polymerization was strongly affected by the solvent. The reaction only took place in highly polar aprotic solvents and the conversion rate increased with the polarity of the solvent. It is noted that this OED system could initiate the polymerization of a lactone and various alkenes bearing activating substituents but did not succeed with styrene or vinyl acetate. Polymerization was proposed to proceed *via* electrons transferred from OED to monomer, leading to monomer reduction, and producing anionic intermediates, which could initiate the polymerization, as illustrated in Scheme 11.23.

11.4.4 N-heterocyclic olefins (NHOs)

The charge separation resulting from the presence of a strongly polarized double bond enable NHOs to serve as powerful nucleophiles, organobase or ligands. Naumann and Falivene developed NHO-based zwitterionic organopolymerization of acrylic monomers, including MA, MMA, *t*BMA and DMAA.[122] In contrast to saturated five- and six-membered compounds as well as benzimidazolium-derivatives, only imidazolium-derivatives could polymerize these monomers. It is noted that the substituents at the exocyclic carbon played an important role in polymerization performance. For instance, the employment of NHO without substituent at the exocyclic carbon would lead to the formation of spirocycles (Scheme 11.24) or inter-/intramolecular deprotonation, thus resulting in deactivation. In contrast, NHOs bearing exocyclic methyl groups could suppress such deactivation and

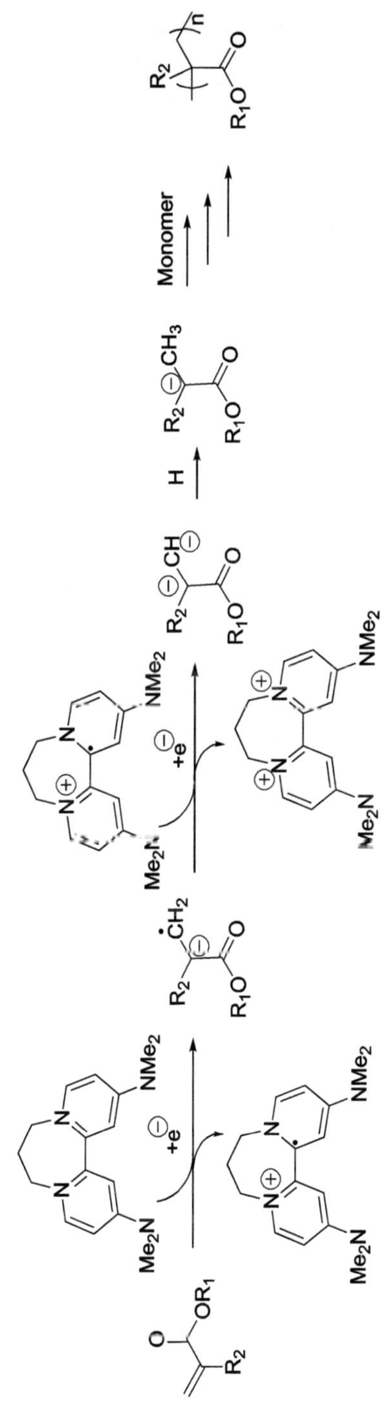

Scheme 11.23 Proposed mechanism of acrylate catalyzed by OED2.

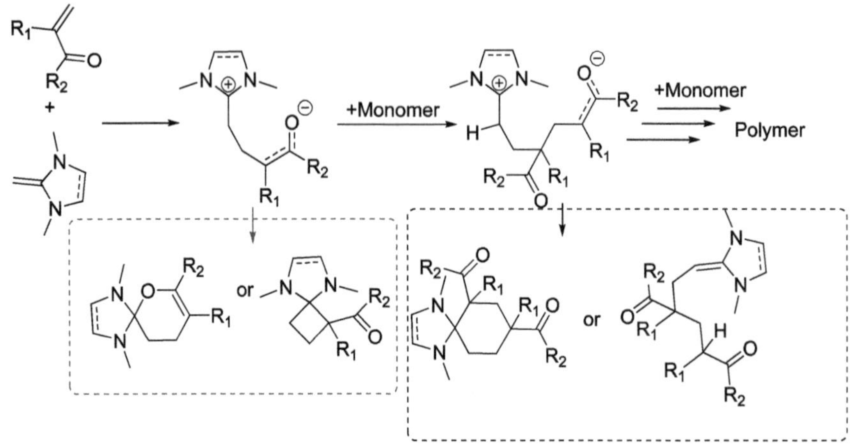

Scheme 11.24 Proposed polymerization pathways of acrylic monomers catalyzed by NHO.

achieved better control over polymerization. Compared with the uncontrolled polymerization by an additive-free NHO system, the addition of LiCl as a μ-type ligand for the polymerization of DMAA results in a rapid and quantitative monomer conversion, furnishing highly isotactic PDMAA with high molecular weight ($M_n = 254–654\ \mathrm{kg\,mol^{-1}}$, $Đ = 1.32–1.59$).

11.5 Summary and Outlook

This chapter has encompassed contributions regarding organopolymerization of polar vinyl monomers over recent years. As discussed above, it is recognized that no single type of polymerization or catalyst system can meet all the demanding requirements and challenges. GTP is the most powerful strategy for the polymerization of polar vinyl monomers to produce various types of polymers, multi-block copolymers, and also star-shape and functionalized polymers, involving over twenty initiators of SKAs and hydrosilanes, and more than 40 different monomers. Most GTPs are well-controlled/living polymerizations. However, there are some disadvantages that cannot be ignored for GTP, such as low polymerization rate (taking several or even tens of hours to reach completion) and difficulty in synthesizing polymers with high molecular weight. In contrast, LPP is typically rapid and efficient for the polymerization of conjugated polar alkenes. However, the application of LPP is hampered by both low initiation efficiencies and chain-termination side reactions, as revealed by the much higher experimental M_n than the calculated M_n and broad MWD of the resulting polymers, thus giving rise to low initiation efficiencies (I^*). Therefore, it is difficult to prepare well-defined block copolymers. To achieve living/controlled LPP, the acidity and steric hindrance of the Lewis acid and the steric and electronic interplay between the LA and LB counterpart must be finely balanced. Regarding zwitterionic polymerization or anionic polymerization catalyzed by small organic molecules, despite many catalysts being employed, the resulting polymers were obtained with relative broad MWD and uncontrolled M_n, in contrast to that by GTP or LPP.

In addition to the aforementioned, organopolymerization of polar vinyl monomers faces a few additional challenges: (1) intolerance for special functional groups places obstacles in synthesizing advanced functional materials; (2) the stereoregularity and topology of polymer cannot be well-controlled; (3) difficulty in achieving well-defined high molecular weight polymers due to the poor control over the polymerization; (4) the new type of polymer materials need to be enriched. However, opportunities and challenges exist side by side, with the understanding of organocatalytic polymerization and development of organic synthesis, organopolymerization of polar vinyl monomers is expected to be a prosperous branch of polymer chemistry.

References

1. D. Baskaran and A. H. E. Müller, *Prog. Polym. Sci.*, 2007, **32**, 173–219.
2. B. S. Soller, S. Salzinger and B. Rieger, *Chem. Rev.*, 2016, **116**, 1993–2022.
3. E. Y.-X. Chen, *Chem. Rev.*, 2009, **109**, 5157–5214.
4. J. Nicolas, Y. Guillaneuf, C. Lefay, D. Bertin, D. Gigmes and B. Charleux, *Prog. Polym. Sci.*, 2013, **38**, 63–235.
5. K. Matyjaszewski and J. Xia, *Chem. Rev.*, 2001, **101**, 2921–2990.
6. S. Perrier and P. Takolpuckdee, *J. Polym. Sci., Part A: Polym. Chem.*, 2005, **43**, 5347–5393.
7. O. W. Webster, W. R. Hertler, D. Y. Sogah, W. B. Farnham and T. V. RajanBabu, *J. Am. Chem. Soc.*, 1983, 5706–5708.
8. O. W. Webster, *J. Polym. Sci., Part A: Polym. Chem.*, 2000, **38**, 2855–2860.
9. R. Zhuang and A. H. E. Müller, *Macromolecules*, 1995, **28**, 8043–8050.
10. D. Y. Sogah, W. R. Hertler, O. W. Webster and G. M. Cohen, *Macromolecules*, 1987, **20**, 1473–1488.
11. I. B. Dicker, G. M. Cohen, W. B. Farnham, W. R. Hertler, E. D. Laganis and D. Y. Sogah, *Macromolecules*, 1990, **23**, 4034–4041.
12. C. S. Patrickios, W. R. Hertler, N. L. Abbott and T. A. Hatton, *Macromolecules*, 1994, **27**, 930–937.
13. W. R. Hertler, D. Y. Sogah and O. W. Webster, *Macromolecules*, 1984, **17**, 1417–1419.
14. F. Nederberg, E. F. Connor, M. Möller, T. Glauser and J. L. Hedrick, *Angew. Chem., Int. Ed.*, 2001, **40**, 2712–2715.
15. J. Raynaud, A. Ciolino, A. Baceiredo, M. Destarac, F. Bonnette, T. Kato, Y. Gnanou and D. Taton, *Angew. Chem., Int. Ed.*, 2008, **47**, 5390–5393.
16. M. D. Scholten, J. L. Hedrick and R. M. Waymouth, *Macromolecules*, 2008, **41**, 7399–7404.
17. J. Raynaud, N. Liu, M. Fèvre, Y. Gnanou and D. Taton, *Polym. Chem.*, 2011, **2**, 1706–1712.

18. J. Raynaud, N. Liu, Y. Gnanou and D. Taton, *Macromolecules*, 2010, **43**, 8853–8861.
19. J. Raynaud, Y. Gnanou and D. Taton, *Macromolecules*, 2009, **42**, 5996–6005.
20. T. Kakuchi, Y. Chen, J. Kitakado, K. Mori, K. Fuchise and T. Satoh, *Macromolecules*, 2011, **44**, 4641–4647.
21. V. R. Chintareddy, K. Wadhwa and J. G. Verkade, *J. Org. Chem.*, 2009, **74**, 8118–8132.
22. K. Kobayashi, M. Ueno and Y. Kondo, *Chem. Commun.*, 2006, 3128–3130.
23. J.-C. Hsu, Y. Chen, T. Kakuchi and W.-C. Chen, *Macromolecules*, 2011, **44**, 5168–5177.
24. Y. Chen, K. Fuchise, A. Narumi, S. Kawaguchi, T. Satoh and T. Kakuchi, *Macromolecules*, 2011, **44**, 9091–9098.
25. M. Fevre, J. Vignolle, V. Heroguez and D. Taton, *Macromolecules*, 2012, **45**, 7711–7718.
26. J. Chen, R. R. Gowda, J. He, Y. Zhang and E. Y.-X. Chen, *Macromolecules*, 2016, **49**, 8075–8087.
27. K. Ute, H. Ohnuma and T. Kitayama, *Polym. J.*, 2000, **32**, 1060–1062.
28. S. Kikuchi, Y. Chen, K. Kitano, K. Takada, T. Satoh and T. Kakuchi, *Polym. Chem.*, 2015, **6**, 6845–6856.
29. L. Hu, W. Zhao, J. He and Y. Zhang, *Molecules*, 2018, **23**, 665–684.
30. N. Asao, T. Ohishi, K. Sato and Y. Yamamoto, *J. Am. Chem. Soc.*, 2001, **123**, 6931–6932.
31. K. Fuchise, S. Tsuchida, K. Takada, Y. Chen, T. Satoh and T. Kakuchi, *ACS Macro Lett.*, 2014, **3**, 1015–1019.
32. Y. Chen, K. Kitano, S. Tsuchida, S. Kikuchi, K. Takada, T. Satoh and T. Kakuchi, *Polym. Chem.*, 2015, **6**, 3502–3511.
33. S. Kikuchi, Y. Chen, K. Kitano, S. Sato, T. Satoh and T. Kakuchi, *Macromolecules*, 2016, **49**, 3049–3060.
34. T. Xu and E. Y.-X. Chen, *J. Polym. Sci., Part A: Polym. Chem.*, 2015, **53**, 1895–1903.
35. Y. Zhang and E. Y.-X. Chen, *Macromolecules*, 2008, **41**, 36–42.
36. Y. Zhang and E. Y.-X. Chen, *Macromolecules*, 2008, **41**, 6353–6360.
37. G. M. Miyake, Y. Zhang and E. Y. X. Chen, *Macromolecules*, 2010, **43**, 4902–4908.
38. Y. Zhang, L. O. Gustafson and E. Y. Chen, *J. Am. Chem. Soc.*, 2011, **133**, 13674–13684.
39. L. Hu, J. He, Y. Zhang and E. Y. X. Chen, *Macromolecules*, 2018, **51**, 1296–1307.
40. J. Chen and E. Y. Chen, *Angew. Chem., Int. Ed.*, 2015, **54**, 6842–6846.
41. R. Kakuchi, K. Chiba, K. Fuchise, R. Sakai, T. Satoh and T. Kakuchi, *Macromolecules*, 2009, **42**, 8747–8750.
42. Y. Zhang, F. Lay, P. García-García, B. List and E. Y.-X. Chen, *Chem. – Eur. J.*, 2010, **16**, 10462–10473.

43. K. Fuchise, R. Sakai, T. Satoh, S. Sato, A. Narumi, S. Kawaguchi and T. Kakuchi, *Macromolecules*, 2010, **43**, 5589–5594.
44. W. Liu, T. Nakano and Y. Okamoto, *Polym. J.*, 2000, **32**, 771–777.
45. Y. Chen, K. Takada, K. Fuchise, T. Satoh and T. Kakuchi, *J. Polym. Sci., Part A: Polym. Chem.*, 2012, **50**, 3277–3285.
46. K. Takada, T. Ito, K. Kitano, S. Tsuchida, Y. Takagi, Y. Chen, T. Satoh and T. Kakuchi, *Macromolecules*, 2015, **48**, 511–519.
47. K. Takada, K. Fuchise, Y. Chen, T. Satoh and T. Kakuchi, *J. Polym. Sci., Part A: Polym. Chem.*, 2012, **50**, 3560–3566.
48. K. Fuchise, Y. Chen, K. Takada, T. Satoh and T. Kakuchi, *Macromol. Chem. Phys.*, 2012, **213**, 1604–1611.
49. Y. Chen, K. Takada, N. Kubota, O.-T. Eric, T. Ito, T. Isono, T. Satoh and T. Kakuchi, *Polym. Chem.*, 2015, **6**, 1830–1837.
50. K. Takada, K. Fuchise, N. Kubota, T. Ito, Y. Chen, T. Satoh and T. Kakuchi, *Macromolecules*, 2014, **47**, 5514–5525.
51. S. Kikuchi, Y. Chen, K. Fuchise, K. Takada, J. Kitakado, S. Sato, T. Satoh and T. Kakuchi, *Polym. Chem.*, 2014, **5**, 4701–4709.
52. O.-T. Eric, Y. Chen, K. Takada, S.-I. Sato, T. Satoh and T. Kakuchi, *Polym. Chem.*, 2015, **6**, 7841–7850.
53. S. Kikuchi, Y. Chen, E. Ichinohe, K. Kitano, S. Sato, Q. Duan, X. Shen and T. Kakuchi, *Macromolecules*, 2016, **49**, 4828–4838.
54. T. K. Georgiou and C. S. Patrickios, *Macromolecules*, 2007, **40**, 2335–2343.
55. D. Kafouris, E. Themistou and C. S. Patrickios, *Chem. Mater.*, 2006, **18**, 85–93.
56. Y. Chen and T. Kakuchi, *Chem. Rec.*, 2016, **16**, 2161–2183.
57. K. Fuchise, Y. Chen, T. Satoh and T. Kakuchi, *Polym. Chem.*, 2013, **4**, 4278–4291.
58. D. W. Stephan, *Science*, 2016, **354**, 1248.
59. D. W. Stephan and G. Erker, *Angew. Chem., Int. Ed.*, 2015, **54**, 6400–6441.
60. D. W. Stephan, *Acc. Chem. Res.*, 2015, **48**, 306–316.
61. A. Simonneau, R. Turrel, L. Vendier and M. Etienne, *Angew. Chem., Int. Ed.*, 2017, **56**, 12268–12272.
62. Z. Jian, G. Kehr, C. G. Daniliuc, B. Wibbeling, T. Wiegand, M. Siedow, H. Eckert, M. Bursch, S. Grimme and G. Erker, *J. Am. Chem. Soc.*, 2017, **139**, 6474–6483.
63. M. Trunk, J. F. Teichert and A. Thomas, *J. Am. Chem. Soc.*, 2017, **139**, 3615–3618.
64. V. Fasano, L. D. Curless, J. E. Radcliffe and M. J. Ingleson, *Angew. Chem., Int. Ed.*, 2017, **56**, 9202–9206.
65. D. J. Scott, M. J. Fuchter and A. E. Ashley, *Chem. Soc.Rev.*, 2017, **46**, 5689–5700.
66. S. Li, G. Li, W. Meng and H. Du, *J. Am. Chem. Soc.*, 2016, **138**, 12956–12962.

67. G. Q. Chen, G. Kehr, C. Mück-Lichtenfeld, C. G. Daniliuc and G. Erker, *J. Am. Chem. Soc.*, 2016, **138**, 8554–8559.
68. K. Liu, R. A. Lalancette and F. Jäkle, *J. Am. Chem. Soc.*, 2017, **139**, 18170–18173.
69. K. Chernichenko, M. Lindqvist, B. Kótai, M. Nieger, K. Sorochkina, I. Pápai and T. Repo, *J. Am. Chem. Soc.*, 2016, **138**, 4860–4868.
70. J. B. Zhu and E. Y.-X. Chen, *J. Am. Chem. Soc.*, 2015, **137**, 12506–12509.
71. Q. Wang, W. Zhao, J. He, Y. Zhang and E. Y.-X. Chen, *Macromolecules*, 2017, **50**, 123–136.
72. Y. Zhang, G. M. Miyake and E. Y.-X. Chen, *Angew. Chem., Int. Ed.*, 2010, **49**, 10158–10162.
73. Y. Zhang, G. M. Miyake, M. G. John, L. Falivene, L. Caporaso, L. Cavallo and E. Y.-X. Chen, *Dalton Trans.*, 2012, **41**, 9119–9134.
74. A. Fürstner, M. Alcarazo, R. Goddard and C. W. Lehmann, *Angew. Chem., Int. Ed.*, 2008, **47**, 3210–3214.
75. B. Maji, M. Horn and H. Mayr, *Angew. Chem., Int. Ed.*, 2012, **51**, 6231–6235.
76. S. M. Al-Rafia, A. C. Malcolm, S. K. Liew, M. J. Ferguson, R. McDonald and E. Rivard, *Chem. Commun.*, 2011, **47**, 6987–6989.
77. Y.-B. Jia, Y.-B. Wang, W.-M. Ren, T. Xu, J. Wang and X.-B. Lu, *Macromolecules*, 2014, **47**, 1966–1972.
78. X. Sui, M. A. Hempenius and G. J. Vancso, *J. Am. Chem. Soc.*, 2012, **134**, 4023–4025.
79. A. E. Cherian, F. C. Sun, S. S. Sheiko and G. W. Coates, *J. Am. Chem. Soc.*, 2007, **129**, 11350–11351.
80. R. Wang, W. Chen, F. Meng, R. Cheng, C. Deng, J. Feijen and Z. Zhong, *Macromolecules*, 2011, **44**, 6009–6016.
81. Y.-B. Jia, W.-M. Ren, S.-J. Liu, T. Xu, Y.-B. Wang and X.-B. Lu, *ACS Macro Lett.*, 2014, **3**, 896–899.
82. J. He, Y. Zhang and E. Y.-X. Chen, *Synlett*, 2014, **25**, 1534–1538.
83. J. He, Y. Zhang, L. Falivene, L. Caporaso, L. Cavallo and E. Y.-X. Chen, *Macromolecules*, 2014, **47**, 7765–7774.
84. Y. Zhang, M. Schmitt, L. Falivene, L. Caporaso, L. Cavallo and E. Y. Chen, *J. Am. Chem. Soc.*, 2013, **135**, 17925–17942.
85. D. Baskaran, *Prog. Polym. Sci.*, 2003, **28**, 521–581.
86. W. R. Mariott, A. Rodriguez-Delgado and E. Y.-X. Chen, *Macromolecules*, 2006, **39**, 1318–1327.
87. M. G. Knaus, M. M. Giuman, A. Pöthig and B. Rieger, *J. Am. Chem. Soc.*, 2016, **138**, 7776–7781.
88. J. C. Haartz and D. H. McDaniel, *J. Am. Chem. Soc.*, 1973, **95**, 8562–8565.
89. J. W. Larson and T. B. McMahon, *J. Am. Chem. Soc.*, 1985, **107**, 766–773.
90. L. O. Müller, D. Himmel, J. Stauffer, G. Steinfeld, J. Slattery, G. Santiso-Quiňones, V. Brecht and I. Krossing, *Angew. Chem., Int. Ed.*, 2008, **47**, 7659–7663.
91. T. E. Müller, D. Michael and P. Mingos, *Transition Met. Chem.*, 1995, **20**, 533–539.

92. A. Y. Timoshkin and G. Frenking, *Organometallics*, 2008, **27**, 371–380.
93. R. Damrauer, M. Krempp, N. H. Damrauer, M. W. Schmidt and M. S. Gordon, *J. Am. Chem. Soc.*, 1993, **115**, 5218–5226.
94. Q. Wang, W. Zhao, S. Zhang, J. He, Y. Zhang and E. Y. X. Chen, *ACS Catal.*, 2018, 3571–3578.
95. T. Xu and E. Y. Chen, *J. Am. Chem. Soc.*, 2014, **136**, 1774–1777.
96. J. Chen and E. Y. X. Chen, *I. J. Chem.*, 2015, **55**, 216–225.
97. J. Chen and E. X. Chen, *Molecules*, 2015, **20**, 9575–9590.
98. W. N. Ottou, E. Conde-Mendizabal, A. Pascual, A.-L. Wirotius, D. Bourichon, J. Vignolle, F. Robert, Y. Landais, J.-M. Sotiropoulos, K. Miqueu and D. Taton, *Macromolecules*, 2017, **50**, 762–774.
99. Y. Hosoi, A. Takasu, S. I. Matsuoka and M. Hayashi, *J. Am. Chem. Soc.*, 2017, **139**, 15005–15012.
100. S. Naumann and A. P. Dove, *Polym. Int.*, 2016, **65**, 16–27.
101. M. Fèvre, J. Pinaud, Y. Gnanou, J. Vignolle and D. Taton, *Chem. Soc. Rev.*, 2013, **42**, 2142–2172.
102. S. Naumann and A. P. Dove, *Polym. Chem.*, 2015, **6**, 3185–3200.
103. S.-I. Matsuoka, Y. Ota, A. Washio, A. Katada, K. Ichioka, K. Takagi and M. Suzuki, *Org. Lett.*, 2011, **13**, 3722–3725.
104. T. Kato, Y. Ota, S.-I. Matsuoka, K. Takagi and M. Suzuki, *J. Org. Chem.*, 2013, **78**, 8739–8747.
105. S.-I. Matsuoka, N. Awano, M. Nakazawa and M. Suzuki, *Tetrahedron Lett.*, 2016, **57**, 5707–5711.
106. S.-I. Matsuoka, S. Namera, A. Washio, K. Takagi and M. Suzuki, *Org. Lett.*, 2013, **15**, 5916–5919.
107. A. T. Biju, M. Padmanaban, N. E. Wurz and F. Glorius, *Angew. Chem., Int. Ed.*, 2011, **50**, 8412–8415.
108. Y. Zhang and E. Y. Chen, *Angew. Chem., Int. Ed.*, 2012, **51**, 2465–2469.
109. W. N. Ottou, D. Bourichon, J. Vignolle, A. L. Wirotius, F. Robert, Y. Landais, J.-M. Sotiropoulos, K. Miqueu and D. Taton, *Chem. – Eur. J.*, 2014, **20**, 3989–3997.
110. M. Hong and E. Y. Chen, *Angew. Chem., Int. Ed.*, 2014, **53**, 11900–11906.
111. M. Hong, X. Tang, L. Falivene, L. Caporaso, L. Cavallo and E. Y.-X. Chen, *J. Am. Chem. Soc.*, 2016, **138**, 2021–2035.
112. S.-I. Matsuoka, S. Namera and M. Suzuki, *Polym. Chem.*, 2015, **6**, 294–301.
113. R. R. Gowda and E. Y.-X. Chen, *ACS Macro Lett.*, 2016, **5**, 772–776.
114. W. N. Ottou, D. Bourichon, J. Vignolle, A. L. Wirotius, F. Robert, Y. Landais, J. M. Sotiropoulos, K. Miqueu and D. Taton, *Chem. – Eur. J.*, 2015, **21**, 9447–9453.
115. S. Naumann, F. G. Schmidt, R. Schowner, W. Frey and M. R. Buchmeiser, *Polym. Chem.*, 2013, **4**, 2731–2740.
116. T. Pietzonka and D. Seebach, *Angew. Chem., Int. Ed.*, 1993, **105**, 716–717.
117. H. Yang, J. Xu, S. Pispas and G. Zhang, *Macromolecules*, 2012, **45**, 3312–3317.

118. J. Xu, H. Yang and G. Zhang, *Macromol. Chem.Phys.*, 2013, **214**, 378–385.
119. H. Yang, J. Xu, S. Pispas and G. Zhang, *RSC Adv.*, 2013, **3**, 6853–6858.
120. M. Schmitt, L. Falivene, L. Caporaso, L. Cavallo and E. Y.-X. Chen, *Polym. Chem.*, 2014, **5**, 3261–3270.
121. J. Broggi, M. Rollet, J. L. Clément, G. Canard, T. Terme, D. Gigmes and P. Vanelle, *Angew. Chem., Int. Ed.*, 2016, **128**, 6098–6103.
122. S. Naumann, K. Mundsinger, L. Cavallo and L. Falivene, *Polym. Chem.*, 2017, **8**, 5803–5812.
123. W. N. Ottou, H. Sardon, D. Mecerreyes, J. Vignolle and D. Taton, *Prog. Polym. Sci.*, 2016, **56**, 64–115.

CHAPTER 12

Organocatalyzed Step-growth Polymerization

AMAURY BOSSION,[a,b] KATHERINE V. HEIFFERON,[c] NICOLAS ZIVIC,[a] TIMOTHY E. LONG[c] AND HARITZ SARDON*[a]

[a] POLYMAT and Polymer Science and Technology Department, Faculty of Chemistry, University of the Basque Country UPV/EHU, Paseo Manuel de Lardizabal 3, 20018 Donostia-San Sebastián, Spain; [b] Laboratoire de Chimie des Polymères Organiques (LCPO), UMR 5629-CNRS-Université de Bordeaux – Institut National Polytechnique de Bordeaux, 16 Avenue Pey Berland, 33607 Pessac, France; [c] Department of Chemistry, Macromolecules Innovation Institute, Virginia Tech, Blacksburg, VA 24061, USA
*Email: haritz.sardon@ehu.eus

12.1 Introduction

Since the pioneering work of Carothers on polycondensation and the commercialization of Nylon-6,6 by DuPont, step-growth polymerization techniques have gained increased attention over the past 50 years.[1,2] Although polymers obtained *via* chain-growth methods dominate the overall polymer production, accounting for nearly 90% of the commodity polymers, step-growth polymers play a key role, mostly as engineering plastics and high-performance polymeric materials (Scheme 12.1).

The step-growth polymerization can proceed either *via* polymerization between two monomers whose functionalities are equal to 2, *i.e.* AA and BB where A functional group reacts with B functional group, or *via* homopolymerization of AB type monomers. In comparison with chain growth

Polymer Chemistry Series No. 31
Organic Catalysis for Polymerisation
Edited by Andrew Dove, Haritz Sardon and Stefan Naumann

Published by the Royal Society of Chemistry, www.rsc.org

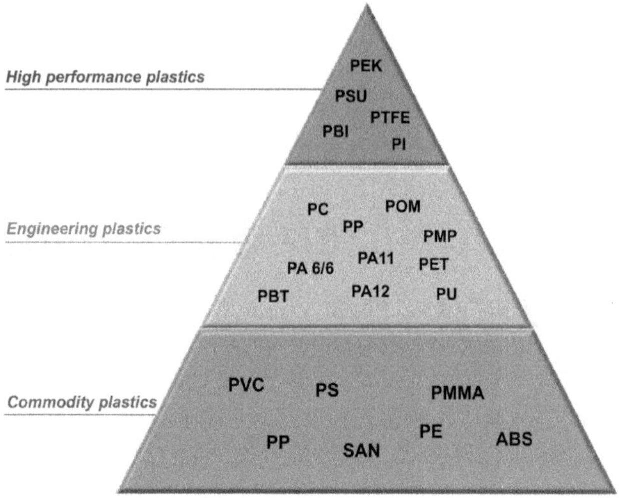

Scheme 12.1 Schematic representation of the plastics classification.

polymerizations, which proceed *via* propagation steps, step-growth polymerizations proceed in a systematic fashion building from monomers to dimers, trimers, tetramers, and so forth (Figure 12.1). As such, high molar mass polymers are only achieved at high monomer conversions.[3]

Unfortunately, most step-growth polymerizations suffer from decomposition of monomer(s) during the polymerization due to the harsh conditions sometimes implemented (high temperature, high pressure), unwanted side reactions, and unfavorable equilibrium (reversible polycondensation). In all cases, these occurrences limit the polymerization from obtaining high conversions and consequently high molar mass. Thus, a catalyst is most of the time utilized to mediate the polymerization and drive the conversion above 99%. Catalysts in polymer synthesis play an important role in increasing the rate of reactions and reducing side reactions that occur with improved selectivity.

For years, the catalysts of choice have been based on metal systems encompassing a variety of acids, bases, and metal complexes, such as organotin compounds. Due to the detrimental effect of catalyst residues on final polymer properties and high cost of removal, in the last 15 years, organocatalysts have emerged as an interesting tool for different polymerizations including the step-growth polymerization processes.[4] Several types of organocatalysts, including Brønsted/Lewis acids or bases, and mono- or bi-component bifunctional catalytic systems have been applied to step-growth polymerization. It has been established that they operate through several different activating mechanisms, which can enhance the polymerization rate. These organocatalysts have been extensively used in ring-opening polymerization (ROP) processes and in anionic, zwitterionic and group-transfer polymerizations. However, their utilization in step-growth methods remains underexplored.

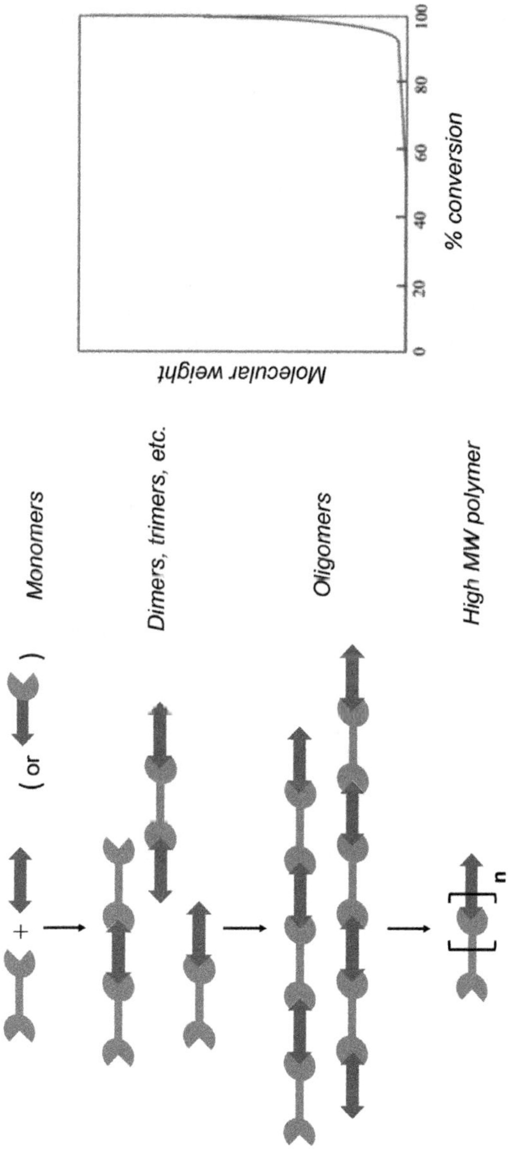

Figure 12.1 Representative mechanism and molecular weight dependence on extent of conversion of step growth polymerization.

The present chapter focuses on the open literature dealing with the organocatalyzed step-growth polymerizations of common polymers, such as polyurethanes, polyureas, polyesters, polycarbonates, and polyethers. In particular, this chapter will highlight recent advances in organocatalyzed step-growth polymerization with the aim of comparing the activity and selectivity of these catalysts with the more commonly used metal catalysts.

12.2 Step-growth Polymerization Catalyzed by Brønsted and Lewis Bases

12.2.1 Alkyl Amines and Pyridine Derivatives

Due to their powerful nucleophilicity, alkyl amines and pyridine derivatives have been extensively employed as organocatalysts for a variety of organic reactions, as well as in polymerization processes. In particular, triethylamine (TEA), 1,4-diazabicyclo [2,2,2]octane (DABCO), and 4-dimethylaminopyridine (DMAP) have been shown to activate the step-growth polymerization of isocyanate and non-isocyanate based polyurethanes (PUs), polycarbonates and polyaldols.

In PU synthesis, tertiary amines are historically the most employed organic bases, DABCO and 2,2′-bis-(dimethylaminoethyl ether) (BDMAEE) being the most industrially employed ones.[5] The use of DABCO became popular due to its ability to activate both isocyanate-hydroxyl and isocyanate-water reactions. However, the mechanism of the amine-catalyzed urethane formation is still uncertain. The first kinetic studies to understand the urethane formation under basic conditions were carried out back in the 1940s by Baker *et al.* According to the authors, this reaction involved first the nucleophilic activation of the isocyanate with the base, prior to the alcohol addition.[6,7] However, this mechanism raised some controversies in the academic field. Schwetlick *et al.* proposed another mechanism that was later confirmed by density functional theory (DFT) calculations as the most dominant pathway. This mechanism involved first the protonation of the catalyst, followed by the nucleophilic addition of the alcohol onto the isocyanate, and formation of the urethane moiety *via* proton transfer from the catalyst to the complex (Scheme 12.2).[8,9]

However, despite the wide use of tertiary amines, such as DABCO, to catalyze the PU synthesis, Cramail *et al.* showed that when using this catalyst

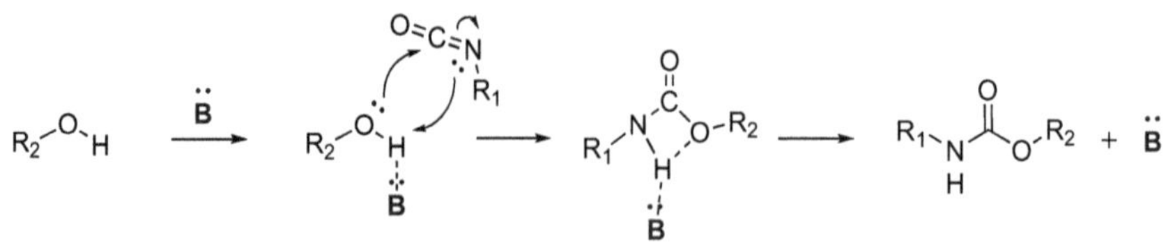

Scheme 12.2 General mechanism of isocyanate/alcohol reaction in the presence of a base as catalyst.

to activate the polymerization between aliphatic isocyanates, such as isophorone diisocyanate (IPDI) and diols, the reaction rate was not significantly enhanced compared to non-catalyzed reaction.[10] Moreover, the fact that tin-based catalysts still surpass these systems in terms of reactivity has shown that isocyanate-hydroxyl reactions require organocatalysts which possess stronger basicity, such as guanidines and amidines.

Besides PUs obtained by conventional isocyanates chemistry, non-isocyanate polyurethanes (NIPUs) have also been synthesized in the presence of tertiary amines. In 2004, Diakoumakos and Kotzev investigated the catalytic activity of TEA and piperazine on the bulk polyaddition of a cyclocarbonate resin at 25 and 60 °C with aliphatic and aromatic diamines.[11] When TEA (1 wt%) was employed as catalyst, the activation energy was lowered in comparison to the non-catalyzed reaction (5.23 kJ mol^{-1} and 6.33 kJ mol^{-1} respectively), and full conversions were achieved faster at both temperatures. The mechanism was hypothesized to involve first the nucleophilic activation of the carbonate moiety by the tertiary amine, followed by the nucleophilic addition of the amine (Scheme 12.3).

Similar findings were described later in 2014 by D'Mello *et al.* while studying the polyaddition of novel cyclic carbonates based on cashew nut shell liquid with aliphatic and cyclic diamines, namely 1,6-hexamethylenediamine and isophorone diamine. This polymerization occurred in bulk at 120 and 150 °C using TEA as catalyst.[12] The polymerization proceeded faster at high temperature and the attained molar masses of the non-crosslinked polymers were in the range $2700 \leq M_n \leq 4600$ Da ($2.7 \leq Đ \leq 6.8$).

In 2013, an efficient method to prepare PEG-based NIPUs in water was reported (Scheme 12.4).[13] The authors performed the polycondensation of a highly reactive linear pentafluorophenyl dicarbonate and Jeffamine in the presence of TEA catalyst and achieved NIPUs with a M_n range of 15–16.5 kDa ($Đ = 1.9$) within 1 h.

Following the same concept, Sardon *et al.* prepared NIPU nanoparticles using surfactant-assisted interfacial polymerization between pentafluorophenyl dicarbonates and poly(oxyethylene) (bis)amine in the presence of TEA catalyst.[14] NIPU nanoparticles with sizes in the range of 200–300 nm and molar masses up to $M_n = 27$ kDa were obtained. In this case, TEA deprotonated the diamine, thus facilitating its nucleophilic attack on the dicarbonate. Although this process represents a strategy to perform polymerization in water, the use of pricey pentafluorophenol derivatives and the lengthy purification process limit its industrial implementation.

Simple organic molecules, such as 4-dimethylaminopyridine (DMAP) and 4-pyrrolidinopyridine (PPY), acting as nucleophilic activators, are effective catalysts for the ROP of cyclic monomers in the presence of a suitable nucleophilic initiator.[15,16]

The potential of DMAP as an organocatalyst for the step-growth polymerization has only been exploited for the polycondensation of dimethyl carbonate and diols (Scheme 12.5).[17–19] Sun *et al.* found that DMAP outperformed other catalyst families, such as guanidines, amidines, and

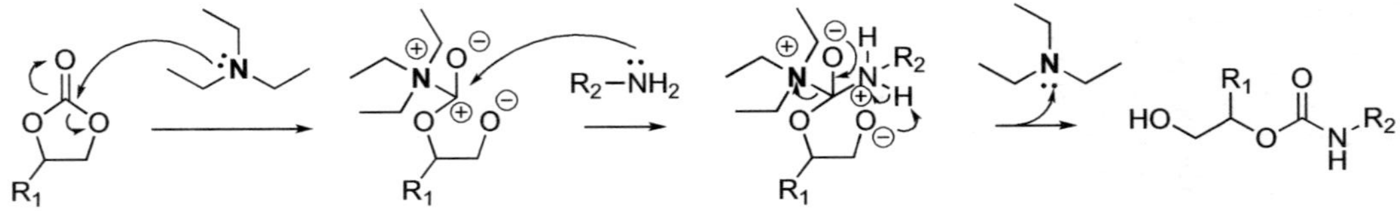

Scheme 12.3 Proposed mechanism of the activation of cyclic carbonates with triethylamine. Adapted from ref. 11 with permission from John Wiley and Sons, Copyright © 2004 WILEY-VCH Verlag GmbH & Co. KGaA, Weinheim.

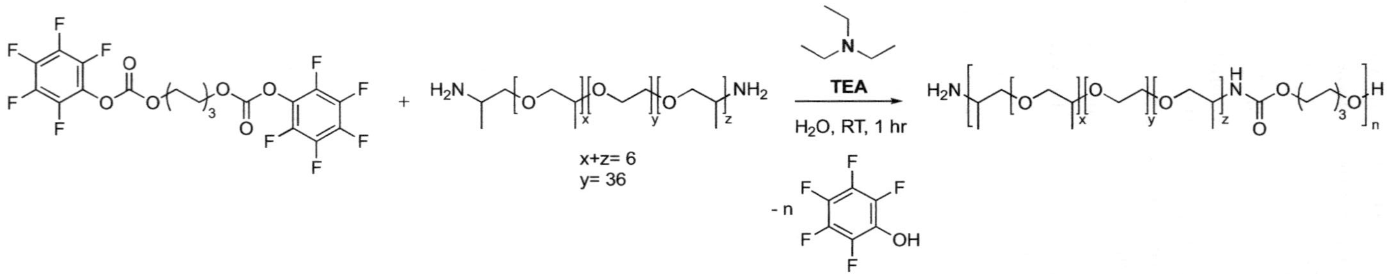

Scheme 12.4 Step-growth polycondensation of linear pentafluorophenyl dicarbonate and Jeffamine at room temperature using TEA as catalyst.[13]

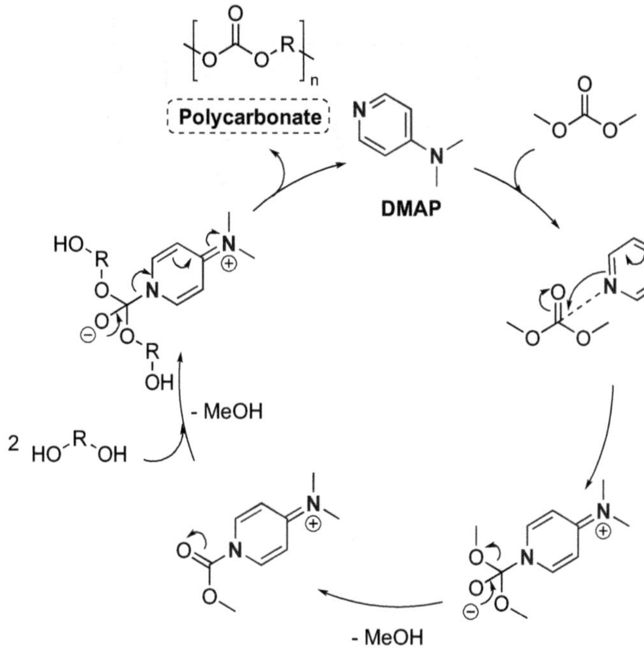

Scheme 12.5 Possible mechanism for the step-growth polymerization of dimethyl carbonate and diols using DMAP as nucleophilic activator.[15]

thioureas for the synthesis of polycarbonates.[17] After optimizing the polymerization conditions (molar ratio of 2 : 1 : 0.01 DMC:diol:DMAP), and performing the polymerization at 170 °C, polycarbonates were obtained with a molar mass up to 52 kDa. Others have shown that higher amounts of dimethylcarbonate (DMC) were necessary to achieve high molar mass when the molar mass of the diol was high.[20]

While DMAP promotes the reaction between aliphatic diols and carbonates it appears less effective in the presence of aromatic diols. Haba *et al.* performed the polymerization of bisphenol-A and diphenyl carbonate using 1 mol% DMAP at high temperatures (up to 215 °C) due to the reduced nucleophilicity of aromatic diols. Unfortunately, the monomer conversion remained low.[21] A similar behavior was observed when using other nucleophilic activators, such as PPY.[17]

Recently, Taton *et al.* screened various cyclic and acyclic secondary and tertiary amine compounds as potential catalysts for the synthesis of polyaldols through the repetition of direct intermolecular aldolization reactions between bis(aldehyde)s and bis(ketone)s (Scheme 12.6).[22]

These step-growth polymerizations were performed at room temperature with a catalyst loading of 30 mol%. Among the tested catalysts, only pyrrolidine gave significant results as monomer conversion reached 55–60%. Molar masses up to $M_n = 17$ kDa ($Đ = 8.5$) were obtained after 48 h of reaction in THF. Interestingly, the authors found that the use of pyrrolidine

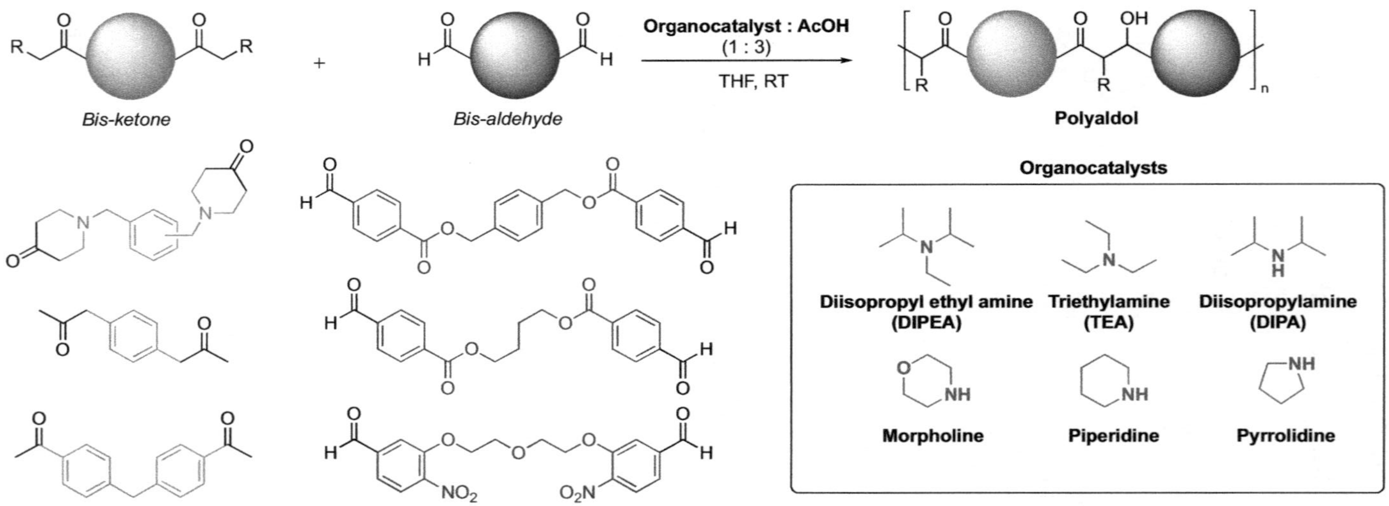

Scheme 12.6 Synthesis of polyaldols from direct step-growth polymerization of bis(aldehyde)s with bis(ketone)s using amines as catalysts.[22]

in conjunction with acetic acid in a 1 to 3 molar ratio gave higher molar masses. The mechanism involves the electrophilic activation of the carbonyl bis(ketone) group *via* protonation or hydrogen bonding between the oxygen atom and the acetic acid, which subsequently facilitates the nucleophilic attack of the pyrrolidine, forming iminium and enamine intermediates (Scheme 12.7). Addition of an aldehyde leads to the formation of the aldol, which can dehydrate to form its conjugated ketone, as confirmed by NMR spectroscopy analysis.

Others have reported the aldol polymerization of acetaldehyde catalyzed by tertiary amines, such as trimethylamine, TEA, tri-*n*-propylamine, and tri-*n*-amylamine.[23]

12.2.2 Amidines and Guanidines

In comparison with tertiary amines, tetra-alkylated guanidines, such as 1,5,7-triazabicyclo[4.4.0]dec-5-ene (TBD, MeCN pK_a TBDH$^+$ = 26), penta-alkylated guanidines like *N*-methyl 7-methyl-1,5,7-triazabicyclo[4.4.0]dec-5-ene (MTBD, MeCN pK_a MTBDH$^+$ = 25.5), and amidines, such as 1,8-diazabicyclo[5.4.0]undec-7-ene (DBU, MeCN pK_a DBUH$^+$ = 24.3), are stronger organic bases due to the great stability of their conjugated acid. These organo-bases have been extensively used in ROP processes, and have shown to be extremely

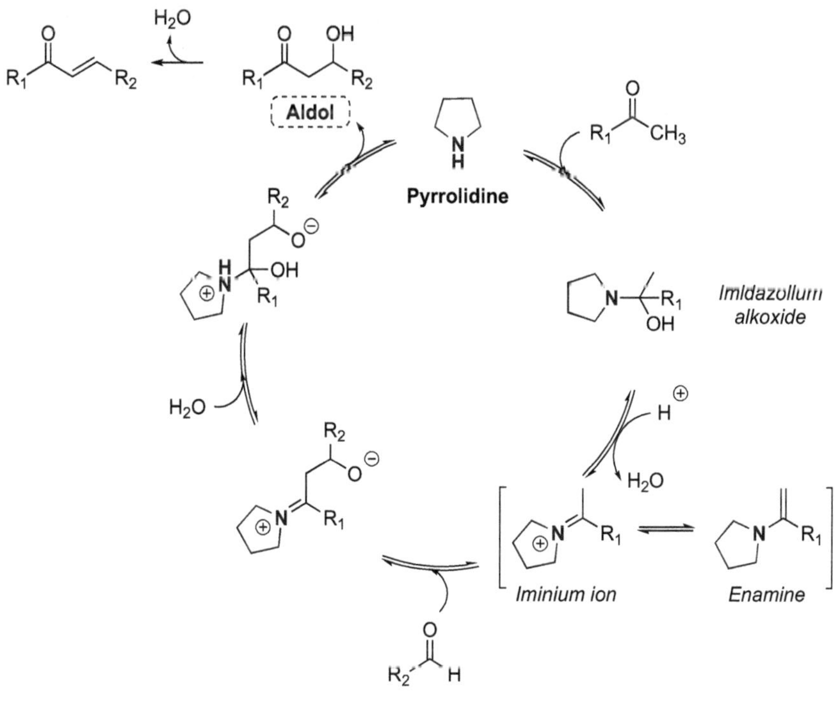

Scheme 12.7 General mechanism for the aldolization of ketone with aldehyde using secondary amines as catalysts.[22]

powerful catalysts for the polymerization of cyclic esters and carbonates.[24] Recent studies revealed potential in the area of step-growth polymerizations. In particular, these Brønsted bases have been applied extensively to the step-growth polymerization of isocyanate and alcohol. In a pioneering work, Landais *et al.* screened different cyclic and acyclic guanidines, as well as amidines, for the synthesis of PUs. The reaction was performed in bulk at 60 °C by mixing equimolar amounts of IPDI and poly(tetramethylene oxide)650 in the presence of 1 mol% of catalyst (Scheme 12.8).[10]

In comparison with dibutyltin dilaurate (DBTDL), which achieved complete conversion after 1 h, the reactions employing DBU or MTBD as catalyst reached full PU conversion within 15 min. Using 1 mol% of DBU, molar masses M_w up to 74 kDa ($Đ = 1.9$) were obtained. In 2016, Sardon *et al.* completed this study by preparing polyether-based PUs in solution at room temperature with DBU as the catalyst. Using trans-1,4-cyclohexylene diisocyanate, PUs with molar mass up to 188 kDa were obtained.[25] Interestingly, the authors found that cyclic guanidines were much more reactive than their acyclic counterparts although they have similar pK_a values in acetonitrile (25.4 and 25.43, respectively).[26,27] Moreover, even though TBD is a stronger base than MTBD, its reactivity toward the polymerization was much lower. A general base mechanism in which the base catalyst acts as a deprotonating agent should imply that the higher the basicity of the catalyst the higher the reaction rate. In regard to the results obtained, the authors claimed that another mechanism must be implied. As shown in Scheme 12.9 below, the mechanism involves the formation of adducts *via* the reaction of 1 equiv. of guanidine with 2 equiv. of isocyanate in a so-called nucleophilic mechanism, which can further trigger the polymerization leading to PUs.

Amidines and guanidines have also been extensively employed for the synthesis of NIPUs. In particular, TBD which represents the most studied bicyclic guanidine, showed to be an efficient catalyst for the aminolysis reaction of cyclic carbonates with amines.[28] Applied to difunctional material, the TBD-catalyzed polymerization achieved NIPUs with molar mass up to $M_n = 53.4$ kDa ($Đ = 1.38$) (Scheme 12.10). In comparison, the non-catalyzed polyaddition yielded NIPU with lower molar mass ($M_n = 5.43$ kDa, $Đ = 1.39$).

Chen *et al.* further investigated the use of TBD as catalyst for the step-growth polymerization of lignin-based dicyclic carbonates and diamines.[29] For example, using hexamethylenediamine and 5 mol% TBD, full conversion was achieved within 24 h and NIPUs with molar mass of $M_n = 30$ kDa ($Đ = 1.7$) were obtained. In comparison, the non-catalyzed reaction exhibited conversion of 93% and lower molar masses ($M_n = 23$ kDa, $Đ = 1.9$). Using a similar catalytic amount of TBD, Caillol *et al.* prepared NIPU foams from step-growth polymerization of five-membered cyclic carbonates (trimethylolpropane tris-carbonate or polypropyleneoxide bis-carbonate) with diamines (Priamine1073 or Jeffamine EDR148).[30]

Similarly to the ROP of cyclic carbonates, it is hypothesized that the activation mechanism of TBD occurs *via* H-bonding (Scheme 12.11). However, as shown by Landais *et al.*, TBD may also act as a nucleophilic activator.[10]

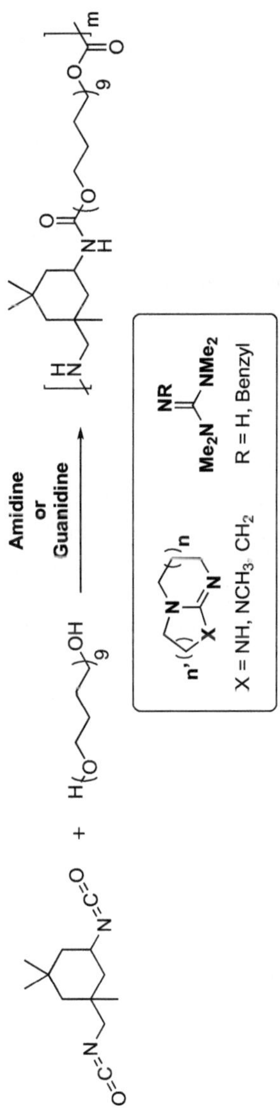

Scheme 12.8 Step-growth polymerization of PU from IPDI and PTMO-650 with cyclic and acyclic guanidines or amidines as catalyst.[10]

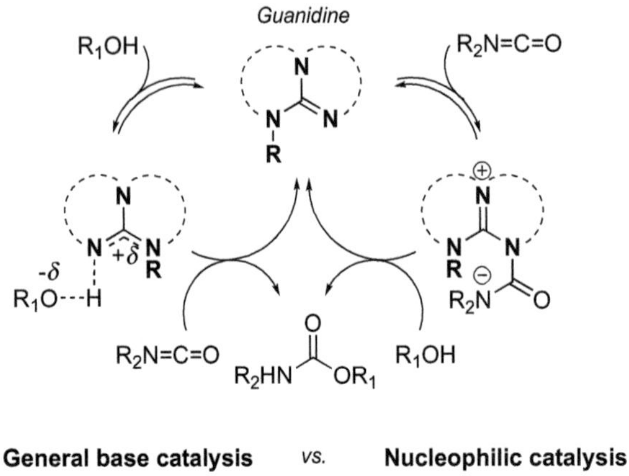

Scheme 12.9 Possible mechanisms for the guanidine-catalyzed reaction of isocyanate and alcohol.
Adapted from ref. 10 with permission from American Chemical Society, Copyright 2012.

As another alternative to conventional PUs, some authors have reported the polycondensation reactions of activated dicarbamates with diols using TBD as catalyst.[31–34] For example, Unverferth *et al.* reported the polycondensation reactions of dimethyl dicarbamates derived from castor oil and diols using TBD catalyst (Scheme 12.12).[31] While low temperatures (110 °C) only led to low molar mass NIPUs, high temperatures (160 °C) resulted in the degradation of both the dicarbamates and TBD. As a compromise to prevent degradation from happening, the authors added 0.1 equiv. of TBD in three increments while increasing the temperature gradually from 120 to 160 °C. Under these optimized conditions, NIPUs with molar masses of up to $M_n = 24.6$ kDa ($Đ = 1.95$) were achieved.

As with conventional PUs, polyureas comprise an area in which researchers have tried to find greener alternatives to commonly used isocyanates.[35] TBD has also been employed as a catalyst in the synthesis of isocyanate-free polyureas. For example, polymerization of dicarbamates with poly(propylene glycol)-based diamines in the presence of 10 mol% of TBD afforded segmented polyureas with molar masses in the range of $M_n = 27–36$ kDa ($Đ = 1.41–1.85$) (Scheme 12.13).[36]

Recently, the same group has shown than alkoxy-alkyl molecules, such as KOMe or KO-*t*-Bu, were much more reactive than TBD for the formation of urea starting from hexylamine and dimethylcarbonate.[37]

Despite the wide use of weak organic bases, such as trimethylamine, to accelerate the polycondensation reaction of phosgene or phosgene derivatives with bisphenol-A, their low basicity does not allow high molar mass polycarbonates to be obtained in the presence of dimethyl carbonate and diphenyl carbonate. As a result, stronger organic bases, such as guanidines

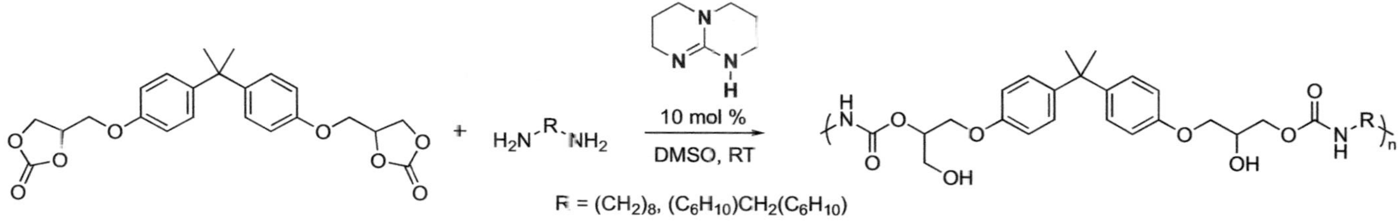

Scheme 12.10 Step-growth polymerization of five-membered cyclic dicarbonate and diaminooctane or bis-(4-aminocyclohexyl) methane at room temperature in DMSO using 10 mol% of TBD as catalyst.
Adapted from ref. 28 with permission from Elsevier, Copyright 2013.

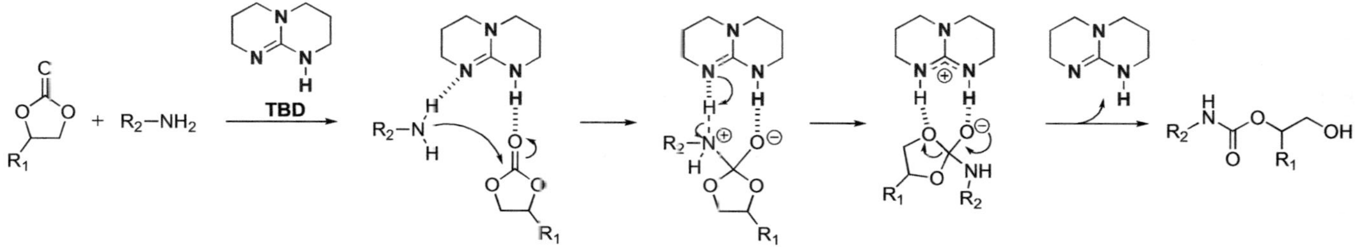

Scheme 12.11 Activation mechanism of TBD-catalyzed step-growth polyaddition of cyclic carbonates with diamines.

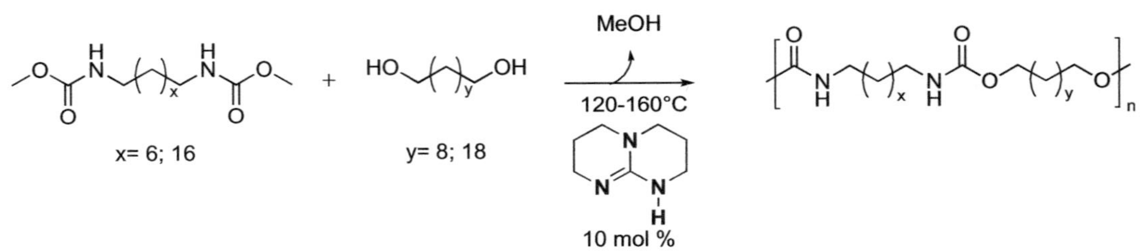

Scheme 12.12 Step-growth polycondensation of linear dicarbamate derived from ricinoleic acid and diols at 130 °C using 10 mol% of TBD as catalyst.
Adapted from ref. 31 with permission from John Wiley and Sons, © 2013 WILEY-VCH Verlag GmbH & Co. KGaA, Weinheim.

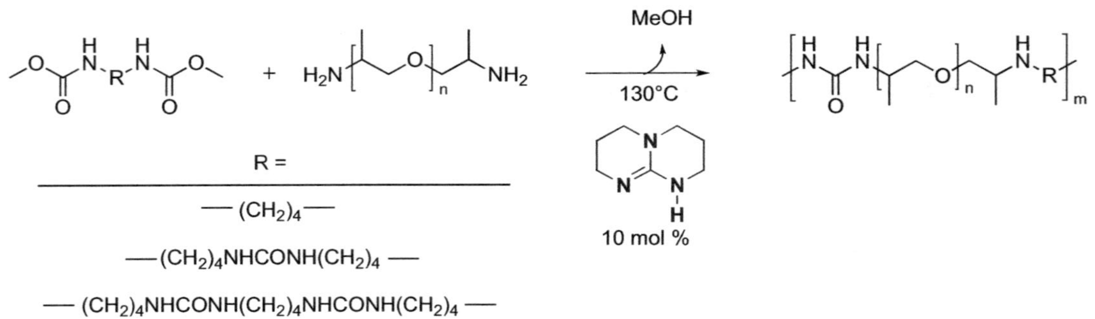

Scheme 12.13 Synthesis of segmented non-isocyanate polyureas from dicarbamates and poly(propylene glycol)-based diamines at 130 °C using 10 mol% of TBD as catalyst.[36]

and amidines, have been deemed necessary. Among them, TBD, MTBD, and DBU have been extensively employed.[17,38,39] For example, using 0.5 mol% of TBD, Meier *et al.* prepared polycarbonates with M_n up to 33 kDa starting from 1,6-hexanediol and dimethyl carbonate (Scheme 12.14).[40] Varying the catalyst concentration, they observed that higher amounts of TBD lowered the molar masses as a result of the cleavage of the terminal methyl carbonates which afforded the formation of terminal allyl carbonate groups.

Recently, Malkoch *et al.* prepared activated carbonates with carbonylimidazolide moieties, which allowed the reaction to be performed at low temperature and in solution in the presence of DBU.[41] The authors reported a library of rigid to flexible polycarbonates having molar masses ranging from 5 to 20 kDa ($Đ = 1.3$–2.9).[39]

Cramail *et al.* compared the reactivity of different organocatalyst families, such as the guanidines TBD and MTBD, and the amidine base DBU with sodium methoxide, a widely used transesterification catalyst in oleochemistry for the synthesis of polyesters.[42] Among them, only TBD could compete with the metal catalyst due to its dual activation behavior. Similarly, dimethyl adipate or dimethyl sebacate with an excess of 1,3-propanediol, 2,3-butanediol, 1,4-butanediol or 1,12-dodecanediol were polymerized at about 120 °C in the presence of TBD to yield polyesters with M_n of 2.1 kDa ($Đ = 1.6$) (Scheme 12.15).[43] Polycondensation of α,ω-bifunctional fatty acids using 5 mol% of TBD were also reported.[44]

Although TBD has been shown to be an effective organocatalyst for the synthesis of polyesters, Flores *et al.* recently demonstrated the great potential of DBU and DMAP to promote the step-growth polymerization of PET from dimethyl therephatlate and ethylene glycol at 200 °C.[45] Specifically, the kinetic study revealed that the catalytic activity of DBU was less than twice as reactive as TBD and the rate constant of DMAP was four times higher than TBD (Figure 12.2).

The potential of amidines and guanidines has also been investigated in novel organocatalyzed polymerization processes. For example, Hedrick *et al.* studied the organocatalyzed polymerization reaction between fluoroarene electrophiles and silyl thioethers (Scheme 12.16).[46] The authors compared the reactivity of different catalysts and kinetic studies performed in DMF at room temperature revealed that DBU, TBD and *N*,*N*′-dicyclohexyl-4-morpholineformamide were all effective catalysts. Polymerization performed between TMS-protected 1,6-hexanedithiol and hexafluorobenzene with a catalyst loading as low as 0.5 mol% attained full conversion in 15 min and yielded to polymers with molar masses of $M_n = 6$–7 kDa ($Đ = 2.5$–4). Based on DFT calculations, the authors stated that the mechanism followed a nucleophilic aromatic substitution between the silyl protected thioethers and perfluoroarenes in which the organocatalyst plays the role of a dual-activator.

In a second example, Long *et al.* reported in 2016 the synthesis of bio-based and semi-crystalline non-isocyanate poly(amide-hydroxyurethane)s (Scheme 12.17).[47] The authors performed the polymerization between a

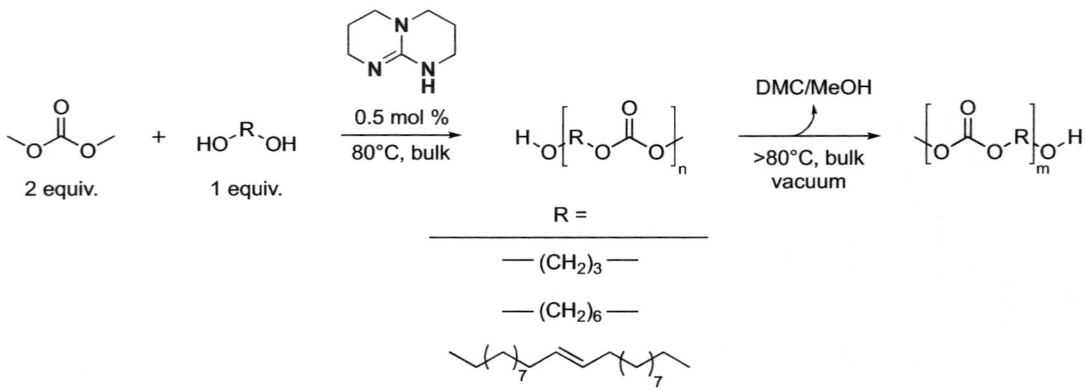

Scheme 12.14 Step-growth polymerization of dimethyl carbonate and 1,6-hexanediol using TBD as catalyst.[40]

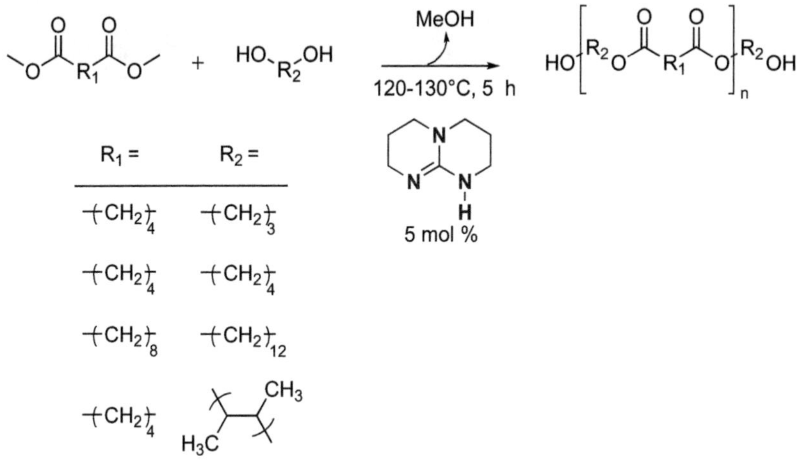

Scheme 12.15 Step-growth polymerization of diesters with diols using TBD as catalyst.
Adapted from ref. 43 with permission from John Wiley and Sons, Copyright © 2011 Wiley Periodicals, Inc.

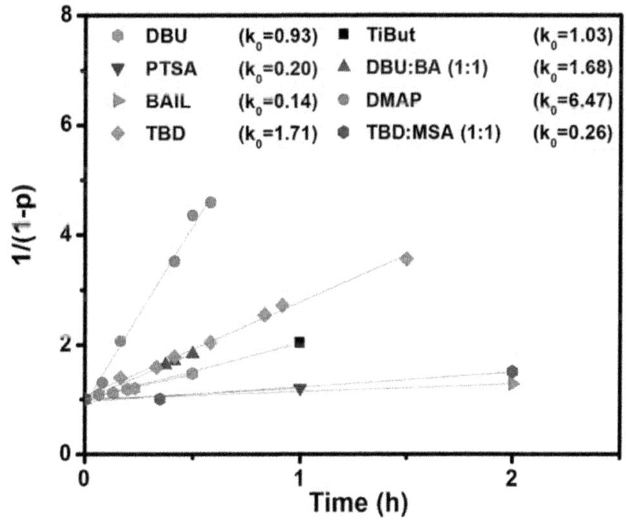

Figure 12.2 Second order linear plot of $1/(1-p)$ as a function of time for the catalyzed polymerization (at 200 °C). Rate constants k_0 were determined from the slopes. p is the fraction of monomer converted to polymer. Adapted from ref. 45 with permission from Elsevier, Copyright 2018.

monomer containing two functionalities, cyclic carbonate and methyl ester respectively, which have the cooperative ability to react with primary diamines in the presence of TBD to yield to poly(amide-hydroxyurethane)s.

Truong and Dove also demonstrated that in the presence of the appropriate catalyst/solvent system, the regioselectivity of a thiol-yne polyaddition

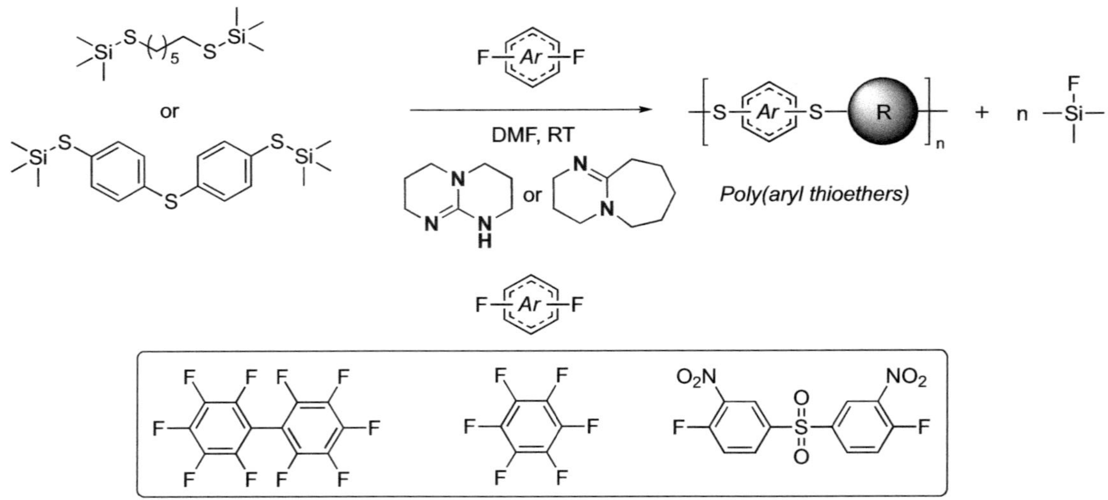

Scheme 12.16 Preparation of poly(aryl thioethers) *via* the reaction of trimethylsilane-protected thioethers with perfluoroarenes using TBD or DBU as catalysts.

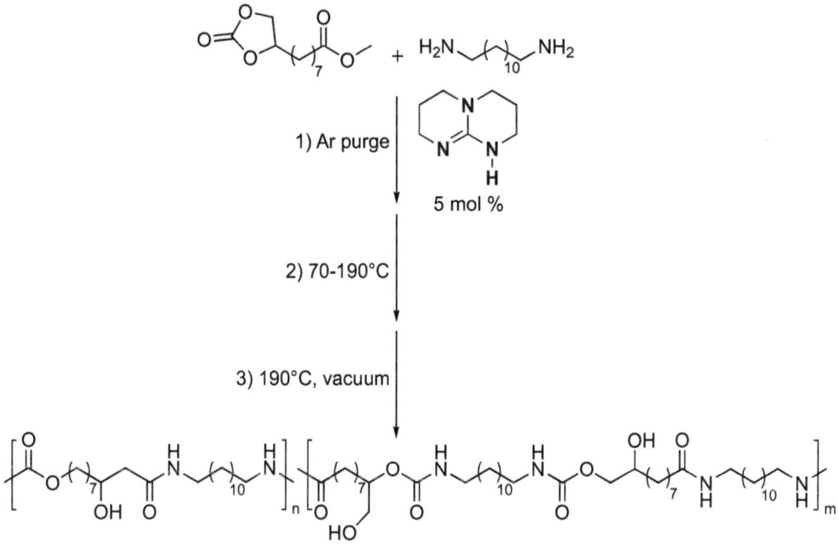

Scheme 12.17 Synthesis of non-segmented poly(amide-hydroxyurethane)s from AB-type monomer and diamine using TBD as catalyst.[47]

reaction can easily be controlled using amidines and guanidines.[48] Applied to difunctional materials, high molar mass *cis* and *trans* elastomers could be isolated *via* step-growth polymerization of 1,6-hexanedithiol and propane-1,3-diyl dipropiolate (Scheme 12.18).[49]

12.2.3 Phosphazenes

Organic phosphazene superbases have been employed as organocatalysts in organic chemistry and polymerization processes because they feature good solubility in most organic solvents, are easy to handle, and can easily generate reactive anionic species due to their high basicity. These strong non-nucleophilic Brønsted bases, first introduced in 1987 by Schwesinger *et al.*, have been used as effective catalysts for the ROPs of heterocycles and vinyl monomers.[50,51]

Recently in 2016, Zhang *et al.* investigated the use of *t*-BuP$_4$ (MeCN pK_a BH$^+$ = 42.7)[51] in step-growth polymerization. PUs were generated based on poly(ethylene oxide) (PEO) and IPDI.[52] The authors discovered that when adding IPDI to a PEO containing solution, a gel was formed almost instantaneously. The authors suggested that under such strong basic conditions, chain-growth homopolymerization of the isocyanate moiety occurred, leading to a cross-linked polyisocyanate structure. Incorporation of different thioureas, such as 1,3-diphenylthiourea, was performed to attenuate the basicity of the phosphazene catalyst and produced PUs

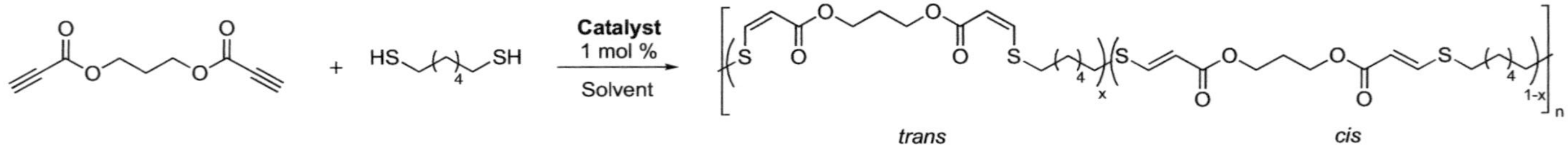

Catalyst	Solvent	x
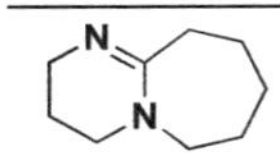	$CHCl_3$	0.8
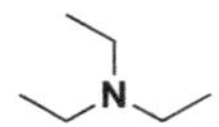	DMF/$CHCl_3$ (1/1)	0.7
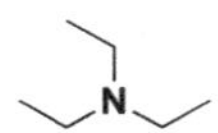	DMF/$CHCl_3$ (3/7)	0.52
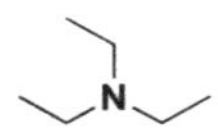	DMF/$CHCl_3$ (3/17)	0.32

Scheme 12.18 Organocatalyzed synthesis of thiol-yne elastomers from dialkyne and dithiol. Adapted from ref. 49, https://doi.org/10.1002/anie.201606750, under the terms of the CC BY 4.0 license, https://creativecommons.org/licenses/by/4.0/.

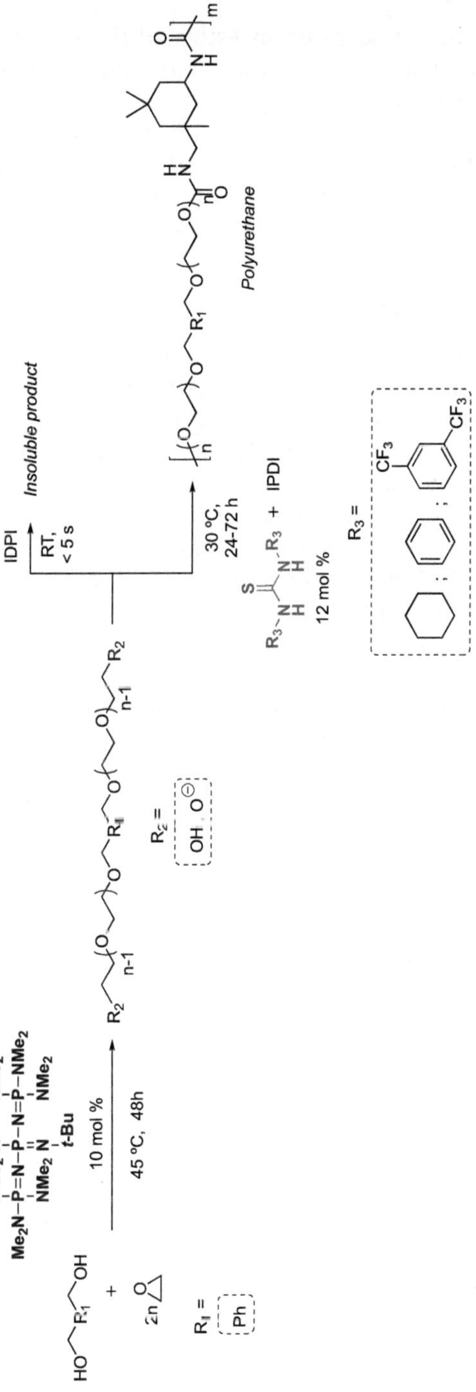

Scheme 12.19 One-pot sequential synthesis of PEO-based PU using a phosphazene and a thiourea derivative as catalysts.[52]

with molar mass up to $M_n = 67.8$ kDa and a dispersity of 1.59 after 72 h of reaction (Scheme 12.19).

Later in 2017, the same group investigated the use of a less basic phosphazene catalyst, namely *tert*-butylimino-tris(dimethylamino)phosphorane (*t*-BuP_1, MeCN pK_a $BH^+ = 26.9$), for the synthesis of PUs.[53,54] After synthesizing poly(3,4-dihydrocoumarin-*alt*-ethylene oxide) macro-diol in toluene using *t*-BuP_1 as catalyst, hexamethylene diisocyanate (HMDI) was added into the reaction mixture and stirred for 24 h at room temperature to achieve PUs with a molar mass of $M_n = 43.9$ kDa ($Đ = 2.1$).

12.2.4 N-heterocyclic Carbenes

N-heterocyclic carbenes (NHCs) were among the first organocatalysts employed for polymer synthesis, providing a straightforward and metal-free synthetic strategy to a wide range of polymers.[55–57] This relates to the near unlimited structural diversity of NHCs, which allows fine tuning of their steric and electronic properties enabling their use as either Brønsted or Lewis bases.[58–60] Extensively employed to catalyze the ROP of cyclic esters, such as D,L-lactide and lactones, the utilization of NHCs has been largely expanded to the ROP of other heterocycles, including *O*-carboxyanhydrides, *N*-substituted carboxyanhydrides, phosphoesters, carbonates, siloxanes and carbosiloxanes, lactams, epoxides and *N*-activated aziridines, as well as the polymerization of alkyl (meth)acrylates.[55–57] NHCs have also proven particularly efficient in step-growth polymerization, for instance, in the synthesis of PUs, aliphatic and aromatic polyesters.[24,35,55–57]

In a recent review by Taton *et al.*, NHCs were used to catalyze the step-growth polymerization of aliphatic diisocyanates and diols.[55] Linear PUs were obtained under relatively mild conditions (30–50 °C) using 1 mol% of 1,3-bis(ditertiobutyl)imidazol-2-ylidene relative to the monomer.[61] In this article, the authors suggested a basic mechanism involving activation of the alcohol by the NHC catalyst for the organocatalyzed step-growth polymerization of aliphatic diisocyanates and diols, at 30–50 °C, using a catalyst loading <1 mol%.[61] Linear PUs with $M_n = 3–5$ kDa have been achieved by adding the NHC organocatalyst to a solution of both monomers. The authors observed that deactivation of the NHC occurred when the diisocyanate monomer was added prior to the diol, and rapid polymerization was observed by adding the diol first, which suggested that the polymerization followed a hydrogen-bonding mechanism.

Terephtaldehyde has also been self-polymerized *via* the NHC-catalyzed benzoin condensation mechanism (Scheme 12.20).[62] Among NHCs tested, the typical "umpolung" triazolydene catalyst has been found to be the most efficient, leading to oligobenzoin-type polymers. According to Breslow *et al.*, the reaction follows a mechanism in which the active moiety is a thiazolydine or carbene formed upon deprotonation of the thiazolium salt by the strong base.[63] Activation of the carbonyl of the benzaldehyde by the carbene, which operates as a nucleophilic catalyst, leads to the formation of an imidazolium alkoxide that rearranges to eneaminol forming the so-called

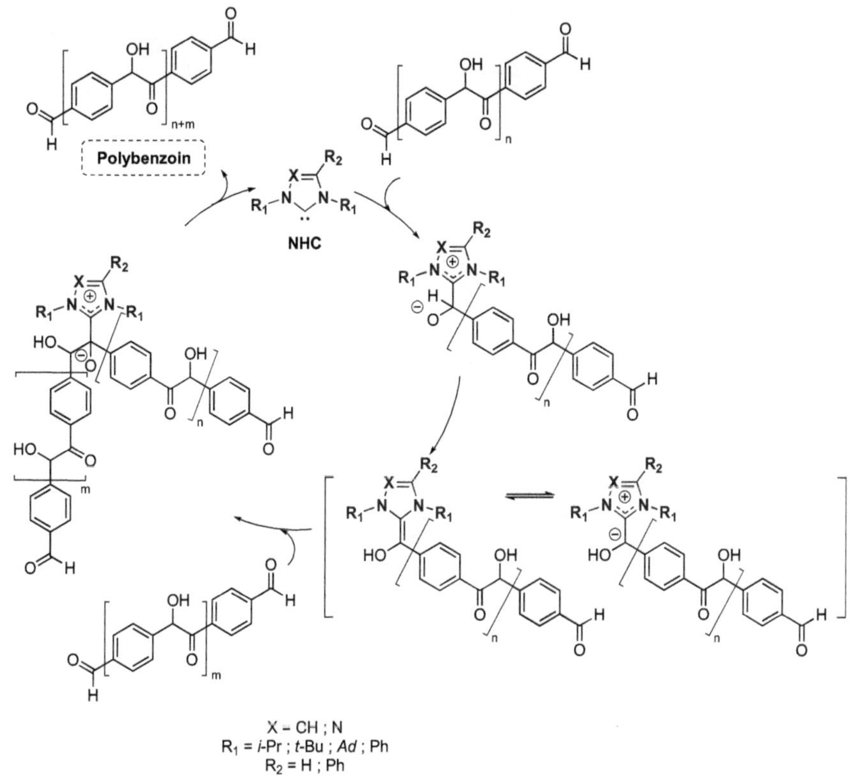

Scheme 12.20 Mechanism of the step-growth polymerization of terephthalaldehyde using N-heterocyclic carbenes as catalysts.[62]

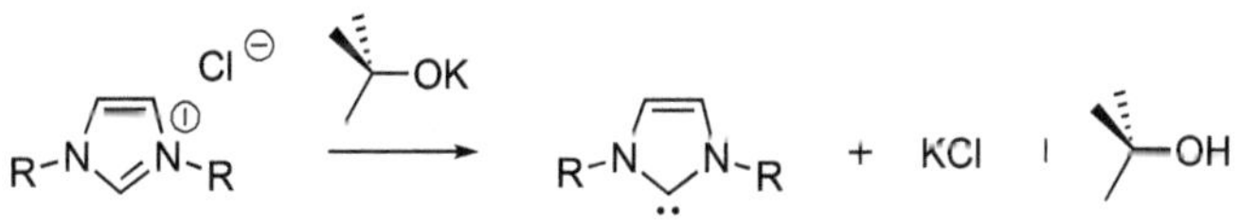

Scheme 12.21 *In situ* generation of N-heterocyclic carbenes from their corresponding imidazolium chloride salts.[64]

"Breslow intermediate". Addition of a second benzaldehyde molecule regenerates the carbene (Scheme 12.20).

The ability of NHCs to catalyze transesterification reactions from the preparation of polyesters *via* step-growth polymerization has been described through one example.[64] The catalysts were prepared *in situ* from their respective imidazolium chloride salts using potassium *tert*-butoxide due to the high moisture sensitivity of NHCs (Scheme 12.21).

For example, Hedrick *et al.* reported the synthesis of PET by NHC organocatalysis.[64] For that purpose, a two-step synthetic procedure involving the preparation of bis(2-hydroxyethyl)terephtalate (BHET), followed by

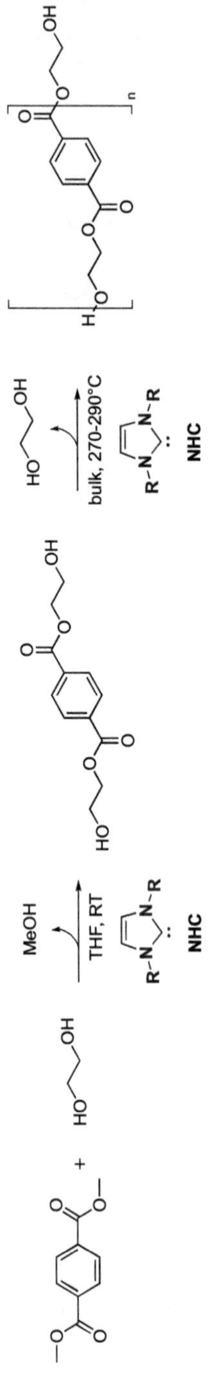

Scheme 12.22 Step-growth polymerization of bis(2-hydroxyethyl) terephthalate using NHCs as catalyst.[64]

self-condensation of this precursor within 1 h at 250 °C, has been implemented (Scheme 12.22).

Aliphatic polycarbonates have also been synthesized by NHC-catalyzed polycondensation between DMC and various diols.[65] To this end, 1 mol% NHC-CO_2 adducts serving as pre-catalysts have been employed in bulk conditions at 150 °C under reduced pressure for 1 h. Homopolymers and random copolymers with average molar masses (M_n) in the range 1.7–7.4 kDa and a dispersity between 1.4 and 2.8 have thus been obtained. The authors proposed a mechanism where first the carbene is generated *in situ* by decarboxylation then the activation of the alcohol by the NHC *via* a hydrogen-bonding mechanism occurs before its addition on the dicarbonate monomer (Scheme 12.23).

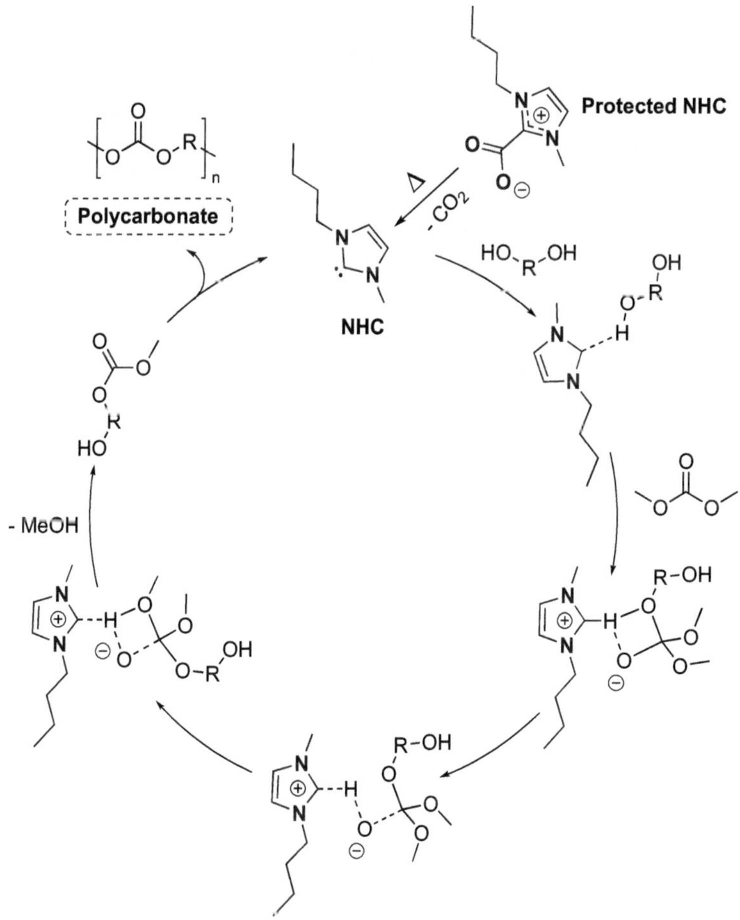

Scheme 12.23 Proposed mechanism for the step-growth polymerization of dimethyl carbonate and diols using 1-n-butyl-3-methylimidazol-2-carboxylate protected carbenes as catalyst.[65]

In summary, among all organo-bases investigated in organocatalyzed step-growth polymerization, and the nitrogen-containing bases tested, DBU and TBD proved to be efficient and versatile catalysts not only for the well-studied PU synthesis, but also for the preparation of NIPUs, polycarbonates, polyesters and more recently novel types of step-growth polymers. Thanks to the wide variety of commercially available guanidine and amidine-based catalysts, such molecules are becoming attractive organocatalysts for polymer synthesis. On the other hand, even if phosphazene derivatives or NHCs exhibit an adapted selectivity for polymer synthesis, they are still affected by their relatively high cost, toxicity and moisture sensitivity.

12.3 Step-growth Polymerization Catalyzed by Brønsted and Lewis Acids

12.3.1 Sulfonic and Sulfonamide Acids

As recently reviewed by Taton *et al.*, strong and "super strong" sulfonic and sulfonamides acids have been widely investigated in the ROP of cyclic esters and carbonates.[24] In contrast, only few examples of acid-catalyzed step-growth polymerization have been reported in the literature.

One of the most successful application of Lewis acids in polymerizations relates to the step-growth polymerization of aliphatic diisocyanates with diols.[66–68] Compared to organic bases, very few studies about the employment of organic acids for PU polymerizations have been published, and those reported exhibited poor results in comparison with tin-catalyzed reactions.[68] It was only in 2013 that Sardon *et al.* conducted a rational study using several different organic acids as potential catalysts for the polymerization of HMDI and poly(ethylene glycol)1500 (PEG_{1500}) in dichloromethane at room temperature (Scheme 12.24).[66] The results were compared between the DBTDL-catalyzed, non-catalyzed, and acid-catalyzed reactions.

Using 5 mol% of "super strong" acids, such as triflic acid (TFA, $pK_a = -13$[69]) and bis(trifluoromethanesulfonyl)imide ($pK_a = -18$[70]) PUs with molar masses around $M_w = 24$ kDa ($Đ = 1.5$) were afforded within 6 h. Surprisingly, the reaction proceeded faster than the tin-catalyzed and DBU-catalyzed reactions, which required 24 h to reach full conversion. Sulfonic acids, such as methanesufonic acid (MSA) and *p*-toluenesulfonic acid (PTSA) or the phosphoric derivative, diphenyl phosphate (DPP, $pK_a = 1.1$), also exhibited catalysis towards urethane reactions with a slower rate compared to the previous cases. On the other hand, weaker carboxylic acids, such as acetic acid (AcOH) and trifluoroacetic acid, were found to be totally ineffective. In comparison, the non-catalyzed reaction resulted in negligible conversions. According to DFT calculations, the mechanism follows a dual hydrogen-bonding route *via* electrophilic activation of the isocyanate, followed by the nucleophilic activation of the alcohol (Scheme 12.25). Although protonation of the isocyanate could either occur on the nitrogen or oxygen

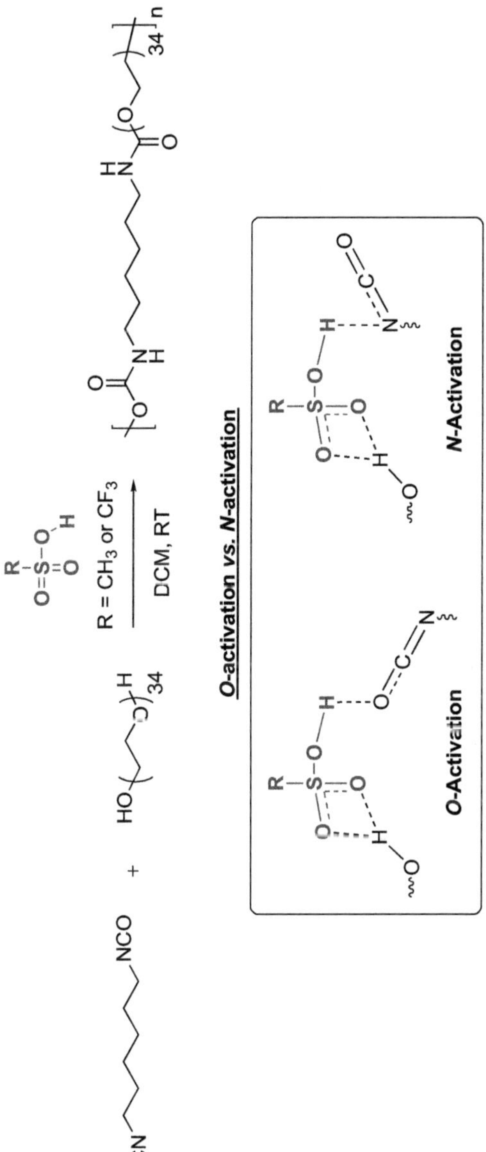

Scheme 12.24 *O*-activation *vs.* *N*-activation mechanism for the step-growth polymerization of hexamethylene diisocyanate and poly(ethylene glycol)1500 in dichloromethane at room temperature.
Adapted from ref. 66 with permission from Elsevier, Copyright 2016.

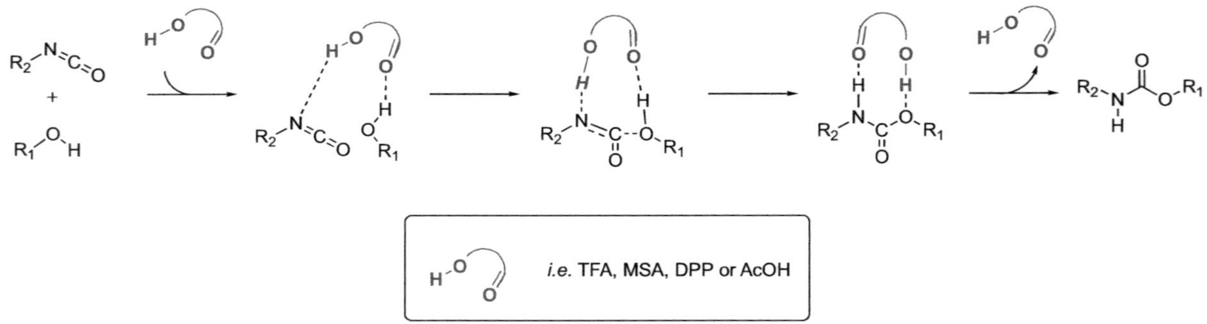

Scheme 12.25 *N*-activation catalytic mechanism for urethane formation using organic acids as catalysts as determined by DFT calculations.
Adapted from ref. 66 with permission from American Chemical Society, Copyright 2013.

atoms, the reaction followed a preference for *N*-activation over *O*-activation pathways. This implies that the strength of a given acid and the nucleophilicity of its conjugate base are major criteria of the urethane formation.

Thermally responsive PEG-based PUs synthesized from IPDI, different molar mass PEG-diols ($M_n = 600$, 1000 and 1500 Da) soft segments, and 2,2-bis(hydroxymethyl)-propionic acid (bis-MPA) were recently prepared.[71] Adequate choice of catalyst is key when employing monomers featuring functional groups. In order to prevent deactivation of the carboxylic functional group with the base catalyst, the authors performed the reaction in the presence of 5 mol% MSA as catalyst. The poly(ether urethanes) exhibited thermoresponsive properties in aqueous media and manifested a tunable LCST that ranged from 30 °C to 70 °C by increasing the poly(ethylene glycol) content.

Another successful polymerization dominated by organic acids deals with the preparation of polyacetals (Figure 12.3).

The formation of aliphatic polyacetals can be directly achieved by the acid-catalyzed polycondensation of a diol with an aldehyde-containing molecule whose functionality is equal to 2 (*i.e.* ketone, aldehyde) (Figure 12.3a). Electrophile activation of the carbonyl moiety prior to the addition of the alcohol leads to a formation of a hemiacetal. Due to the instability of the hemiacetal under acidic condition, the reaction can be driven toward the formation of an acetal *via* addition of a second alcohol if the water by-product is removed (Scheme 12.26).[72–75] Among all the Brønsted acids available, PTSA has been used extensively to catalyze the polyacetal formation.

For example, Takeshi Endo *et al.* prepared amphiphilic polyacetal by the polycondensation of poly(ethylene glycol) PEG ($M_n = 400$ Da) and lilial, an aldehyde widely used in fragrance applications (Scheme 12.27).[72] Performing the polymerization in cyclohexane under reflux in the presence of 1 mol% of PTSA monohydrate (*p*-TsOH) as catalyst, the authors reported polyacetal having a molar mass of $M_n = 7.6$ kDa ($Đ = 2.0$).

Similarly, Miller *et al.* synthesized bioaromatic polyacetals from the PTSA–catalyzed polycondensation of polyvinyl alcohol (PVA) ($M_n = 23$ kDa) or different tetraols, such as erythritol, pentaerythritol and ditrimethylolpropane with aromatic aldehydes (Scheme 12.28).[73–75]

Polycondensation of α,ω-diacetals yielding polyacetals was also studied (Figure 12.3c).[76–79] Linear α,ω-diols were first converted to the corresponding α,ω-diacetals, which were then subjected to the action of an acid catalyst under vacuum. A low boiling acetal was removed during the polymerization, thus driving the step-growth polymerization to completion. For example, Miller's group performed the step-growth polycondensation of linear α,ω-diacetals derived from α,ω-diols with dimethoxymethane.[77] The results showed that short diols (ethylene glycol, 1,3 propanediol, and 1,4-butanediol) led to low polymeric yields due to the potential formation of the corresponding five-, six-, and seven-membered acetal rings. Longer diols, however, were successfully polymerized and high molar masses were

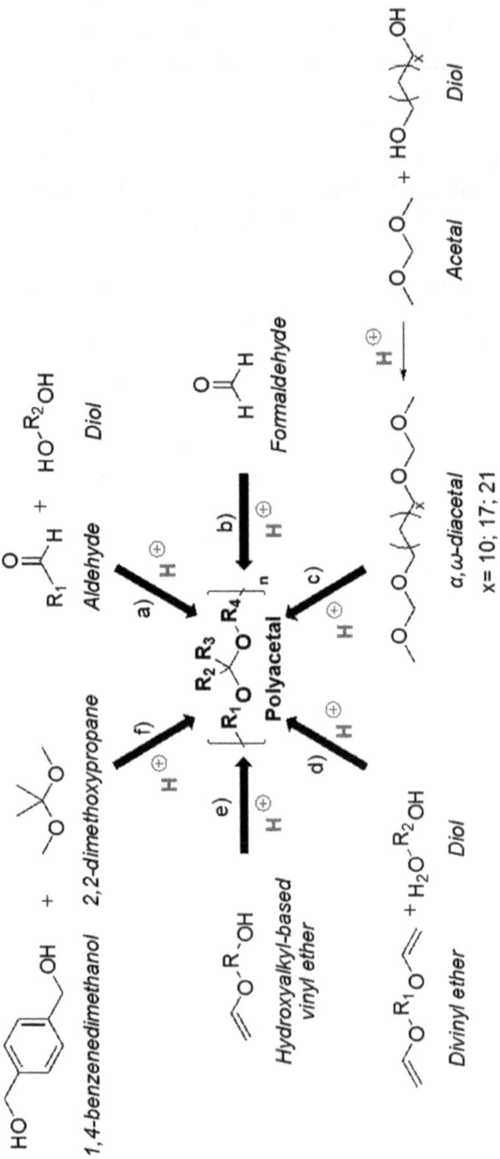

Figure 12.3 Acid-catalyzed routes toward polyacetals.

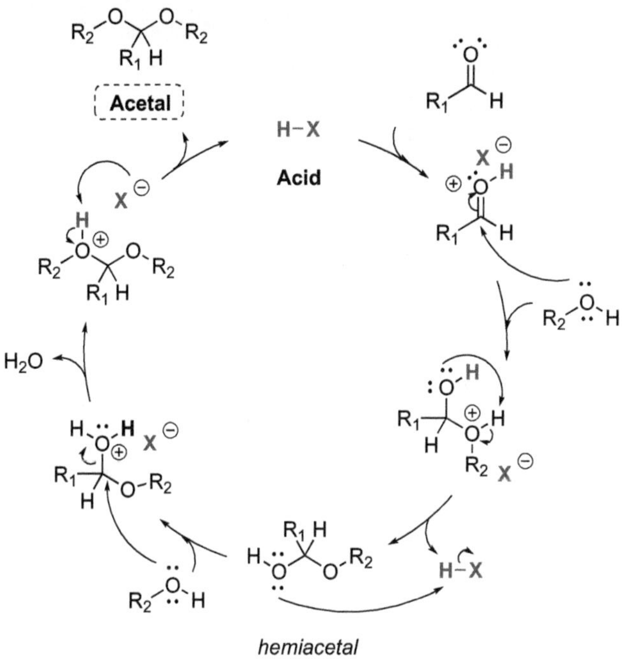

Scheme 12.26 General mechanism for the acetalization of aldehyde with alcohol using acid as catalyst.

afforded ($M_n = 20.7$ to 38.5 kDa, $Đ = 1.83$ to 2.48) in relatively good yields (>64%).

As an entirely different route, polyacetals can be obtained from the acid catalyzed polyaddition of vinyl ethers and alcohols. Such strategies have the advantage of leading to the desired product without formation of by-products (Figures 12.3d and e).[80–90] The polyaddition of divinyl ethers and diols was first developed by Heller *et al.* in 1980.[80] These polyacetals were prepared from the condensation reaction between 1,4-divinylbutane or di-ethyleneglycol divinyl ether with *trans*-1,4-cyclohexanedimethanol, bisphenol-A or *cis/trans*-1,4-cyclohexanediol in THF in the presence of *p*-TsOH catalyst. This straightforward strategy has been extensively studied by a variety of groups who incorporated different functional groups into the polyacetal structure.[82–85,87]

Analogous to this system, hydroxyalkyl-based vinyl ethers can undergo self-polyaddition. For example, Zhang and Ruckenstein described in 2000 the self-polyaddition of 2-hydroxyethyl vinyl ether, 4-hydroxybutyl vinyl ether, and 6-hydroxyhexyl vinyl ether using THF as solvent and pyridinium *p*-toluenesulfonate (PTS) as catalyst.[81] Later, Hashimoto *et al.* conducted a comparative analysis of the effect of different catalysts, such as *p*-toluene sulfonic anhydride (PTSAA), *p*-TsOH, PTS, hydrochloric acid (HCl) and boron trifluoride diethyl etherate (BF_3OEt_2), on the polyaddition of

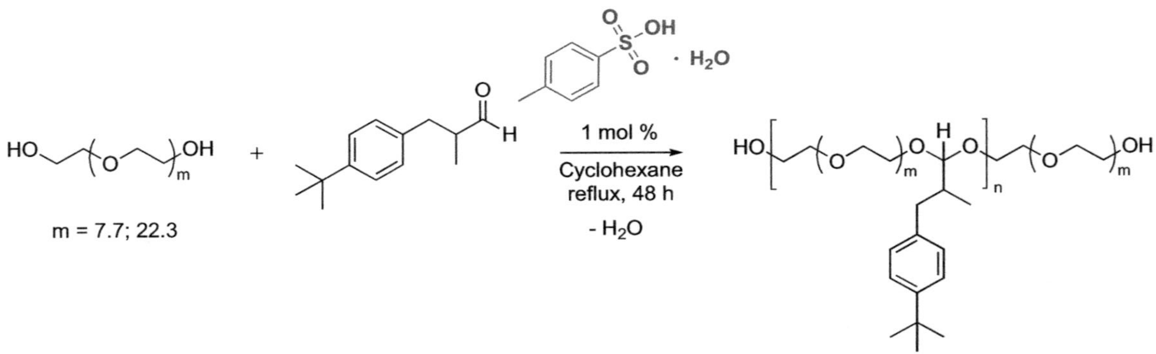

Scheme 12.27 Step-growth polymerization of poly(ethylene glycol) and lilial using *p*-toluenesulfonic acid monohydrate as catalyst. Adapted from ref. 72 with permission from John Wiley and Sons, Copyright © 2010 Wiley Periodicals, Inc.

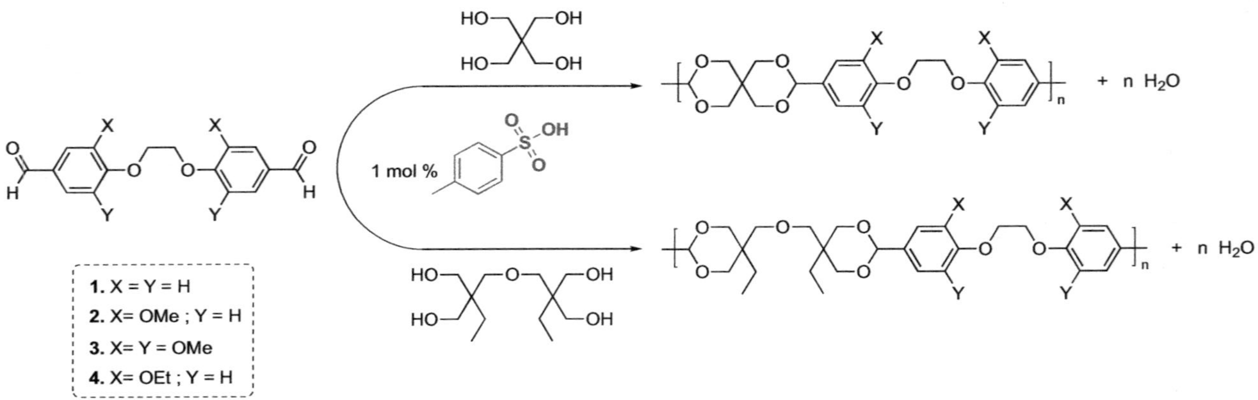

Scheme 12.28 Step-growth polycondensation of bioaromatic polyacetals from dialdehydes and tetraols using acid as catalyst.[73]

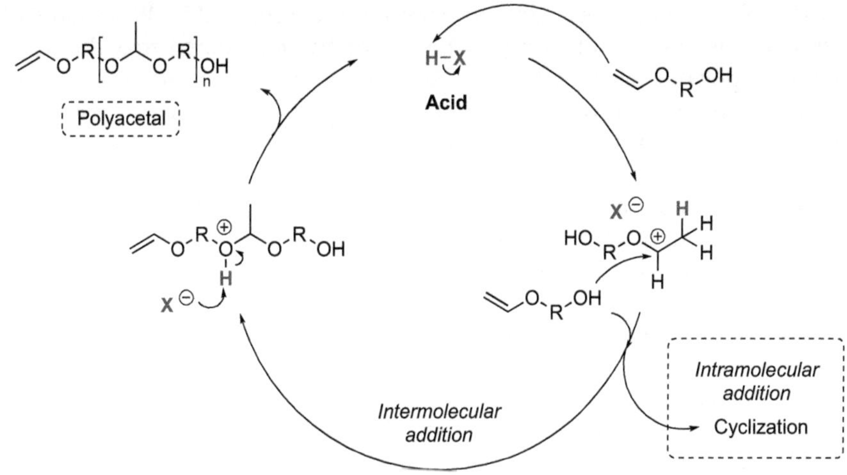

Scheme 12.29 Proposed mechanism for the step-growth polymerization of hydroxyalkyl vinyl ether using acids as catalysts.[83]

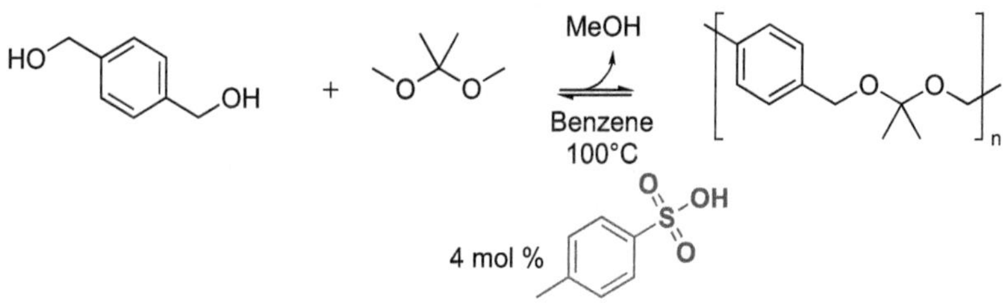

Scheme 12.30 Step-growth polymerization of 1,4-benzenedimethanol and 2,2-dimethoxypropane using PTSA as catalyst.[92]

4-hydroxybutyl vinyl ether in THF at 0 °C for 72 h.[90] It was discovered that no polymerization occurred without catalyst due to the high stability of the monomer; however, the polymerization was successful in the presence of acid-catalysts. In particular, *p*-TsOH and PTSAA led to high molar mass polyacetals $M_n = 66.5–110$ kDa ($Đ = 1.59–1.64$). A mechanism was proposed as follows: the acid catalyst first generates a carbocation that further reacts with the hydroxyl group of another monomer to form an acetal linkage after deprotonation. Studies found that a small amount of cyclic acetal was produced by intramolecular addition (Scheme 12.29). 2-hydroxyethyl vinyl ether, diethylene glycol monovinyl ether, cyclohexane dimethanol monovinyl ether, and tricyclodecane dimethanol monovinyl ether were also polymerized under the optimal conditions demonstrated in these previous examples.

Finally, the formation of acetal linkages during polymerization can also be achieved by the poly(transacetalization) reaction between a diol and a ketal-containing monomer (Figure 12.3f).[91–93] Murthy *et al.* thus prepared a linear polyketal by polycondensation of 1,4-benzenedimethanol and 2,2-dimethoxypropane in benzene at 100 °C with PTSA as catalyst (Scheme 12.30).[92]

Poly(1,4-phenyleneacetone dimethyleneketal) was obtained in 48% yield with a molar mass of $M_w = 4$ kDa ($Đ = 1.54$). The authors stated that the acetal exchange polymerization was a two-step procedure beginning with the formation of dimers and trimers. The second step encompasses the distillation of the methanol by-product to drive the reaction toward the formation of the polymer.

Chatterjee and Ramakrishnan recently reported the synthesis of hyperbranched polyacetals *via* melt poly(transacetalization) of AB_2-type monomers, where A was a hydroxyl group and B groups were dialkyl acetals, such as dibutylacetal or dihexylacetal.[93] The polymerization was performed at 100 °C using mildly acidic pyridinium camphorsulfonate (PCS) as catalyst; $M_w = 21$ kDa and $Đ = 3.0$ were reported; the high dispersity is typical for hyperbranched polymers prepared *via* self-condensation of AB_2 monomers. In a similar fashion, Taton *et al.* studied the polytransacetalization and polyacetalization of AB_2-type monomers namely, *p*-hydroxymethylbenzaldehyde, dimethylacetal, and *p*-hydroxymethylbenzaldehyde, under inert atmosphere in toluene or THF. This system used Brønsted acids as catalyst, such as PTSA (20 mol%), camphor-10-sulfonic acid (CSA; 20 mol%), or PCS (2 mol%) (Scheme 12.31).[91] In the case of the polyacetalization of *p*-hydroxymethylbenzaldehyde, the authors used molecular sieves as a water scavenger in THF and observed that PTSA, which possesses a higher acidity than CSA, provided higher conversions and molar masses in both cases. In the case of the polytransacetalization of *p*-hydroxymethylbenzaldehyde dimethylacetal, PCS (20 mol%) was found to be a suitable catalyst for the reaction (85% conversion) with molar masses reaching $M_w = 20.4$ kDa ($Đ = 1.7$) in 2 h. Overall, the authors found that the polymerization of *p*-hydroxymethylbenzaldehyde dimethylacetal led to hyperbranched polyacetals with a degree of branching around 50% in contrast to *p*-hydroxymethylbenzaldehyde which provided quantitative branching. This arises from the different reactivities of the intermediates formed upon polymerization.

In the late 1980s, Kuo and Chen, among others, studied the acid catalyzed polymerization of esters. They followed the kinetics of polymerization between adipic acid and different diols, namely 1,4-butanediol, 1,6-hexanediol and ethylene glycol at 140–160 °C in the presence of PTSA as catalyst (Scheme 12.32).[94–96] The authors discovered that in all cases the activation energy of the acid-catalyzed reaction was lower than the self-catalyzed method and decreased with the increase of the diol chain length. Also it was shown that acid catalysts were more effective towards the esterification reaction between carboxylic acids and diol than between diesters and diols.[45]

Moyori *et al.* used strong Brønsted acids, such as bis(1,1,2,2,3,3,4,4,4-nonafluoro-1-butanesulfonyl)imide (Nf_2NH) for the bulk polycondensation of adipic acid with 3-methyl-1,5-pentanediol at 60 °C under reduced pressure (Scheme 12.33).[97] The use of Nf_2NH at a concentration of 0.1 mol% achieved molar masses up to $M_n = 19.1$ kDa ($Đ = 2.2$). Interestingly, it was

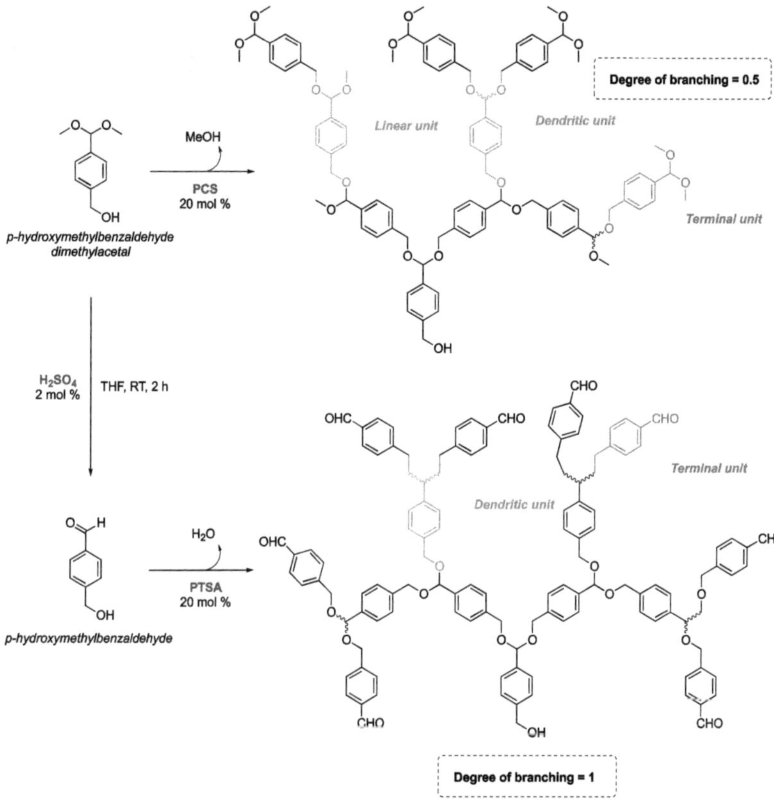

Scheme 12.31 Synthesis of hyperbranched polyacetals with different degrees of branching from the step-growth polymerization of *p*-hydroxymethylbenzaldehyde dimethylacetal and *p*-hydroxymethylbenzaldehyde using acids as catalyst.[91]

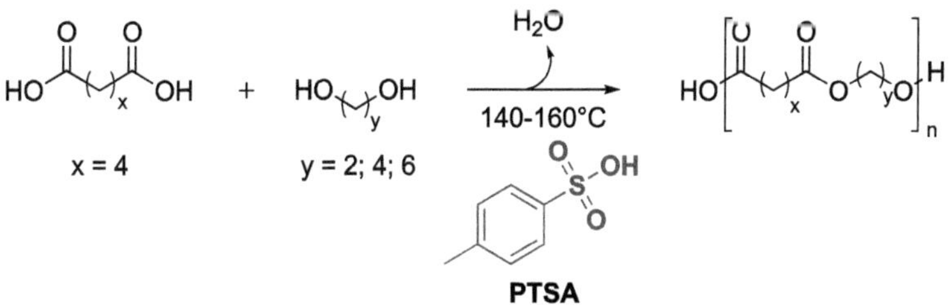

Scheme 12.32 Step-growth polymerization of adipic acid with diols using PTSA as catalyst.[96]

found that Nf_2NH could be recovered by sublimation and recycled without loss of catalytic activity.

One of the main drawbacks of these processes is the necessity of removing the water or alcohol by-product under reduced pressure due to the

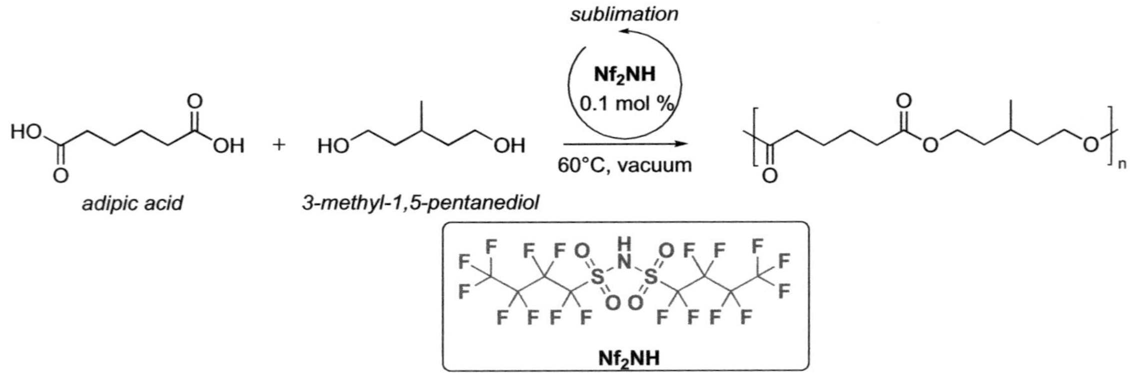

Scheme 12.33 Step-growth polymerization of adipic acid with 3-methyl-1,5-pentanediol using Nf_2NH strong Brønsted acid as catalyst.[97]

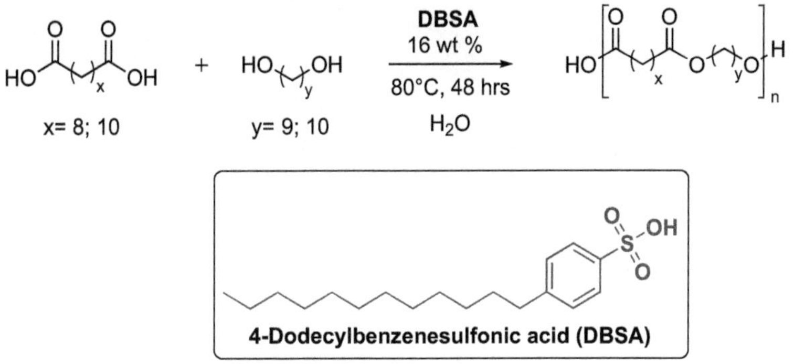

Scheme 12.34 Step-growth polymerization of diacids with diols by emulsion in water using dodecylbenzenesulfonic acid as surfactant type catalyst.[100]

equilibrium between the condensation reaction and hydrolysis. Using a surfactant-type Brønsted acid like dodecylbenzenesulfonic acid (DBSA), Kobayashi *et al.* were able to catalyze the direct esterification of carboxylic acids with alcohols by emulsion in water. Since the reaction occurs at the interface of the micelles, the water by-product stays in the aqueous phase, thus the reaction does not require the removal of water from the media and can proceed at relatively low temperature and atmospheric pressure.[98,99] Applied to difunctional materials, Takasu *et al.* synthesized different polyesters at 80 °C using 16 wt% of DBSA as catalyst with molar mass M_w up to 10 kDa ($Đ=2.0$) (Scheme 12.34).[100]

The acid-catalyzed aldol polymerization has also been described by Zhang *et al.*[101] They recently synthesized two conjugated isoindigo-based polymers *via* aldol polycondensation in the presence of 60 mmol% of PTSA or HCl (Scheme 12.35). Both PTSA and HCl afforded similar results in terms of monomer conversion and molar masses (M_n up to 59.5 kDa within 16 h). However, HCl required a longer reaction time (96 h) than PTSA (16 h) to achieve these results.

12.3.2 Phosphoric Acid and Derivatives

Phosphoric acid and its derivatives are relatively mild Brønsted acids (pK_a around 2–4 in DMSO). These catalysts have shown to be some of the best catalysts to perform the controlled ROP of cyclic esters and cyclic carbonates due to their mild character. In spite of their success in ROP polymerizations, only two examples can be found in the literature about the use of phosphoric-based acids in step-growth polymerizations. In the first one, Sardon *et al.* demonstrated in their catalytic study for the synthesis of PUs that diphenyl phosphate can mediate the polymerization of diisocyanates and diols in a modest fashion compared to stronger acids.[66] In the second example, Sokolsky-Papkov *et al.* studied the effect of phosphoric acid (H_3PO_4) and sulfuric acid (H_2SO_4) as catalyst on the polycondensation in

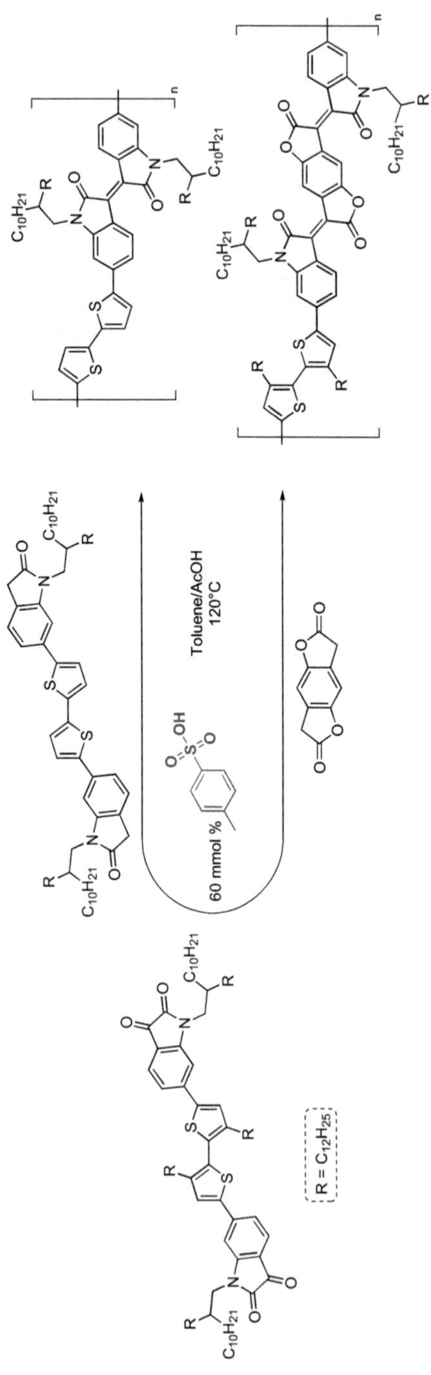

Scheme 12.35 Synthesis of two conjugated isoindigo-based polymers *via* aldol polycondensation using PTSA as catalyst.[101]

bulk at 190 °C under vacuum of adipic acid and sebacic acid with diols.[102] Overall, the optimal conditions were obtained using 0.5 wt% of H_3PO_4 and a small excess of diol. For example, poly(butylene adipate) was obtained at a molar mass of $M_w = 23$ kDa ($Đ = 1.4$). In contrast, the molar mass obtained in the non-catalyzed reaction was approximately three times lower.

12.3.3 (Thio)ureas

Urea and thiourea derivatives can be easily prepared from the reaction between iso(thio)cyanate derivatives and amines. These Brønsted acids are popular thanks to their ability to form hydrogen-bonded complexes with carbonyl groups *via* the so-called bidentate binding interactions and therefore have been utilized successfully in a large variety of organocatalytic transformations as well as in organocatalyzed ROP processes.[103,104] Owing to their strong σ-electron withdrawing effect that increases the acidity of

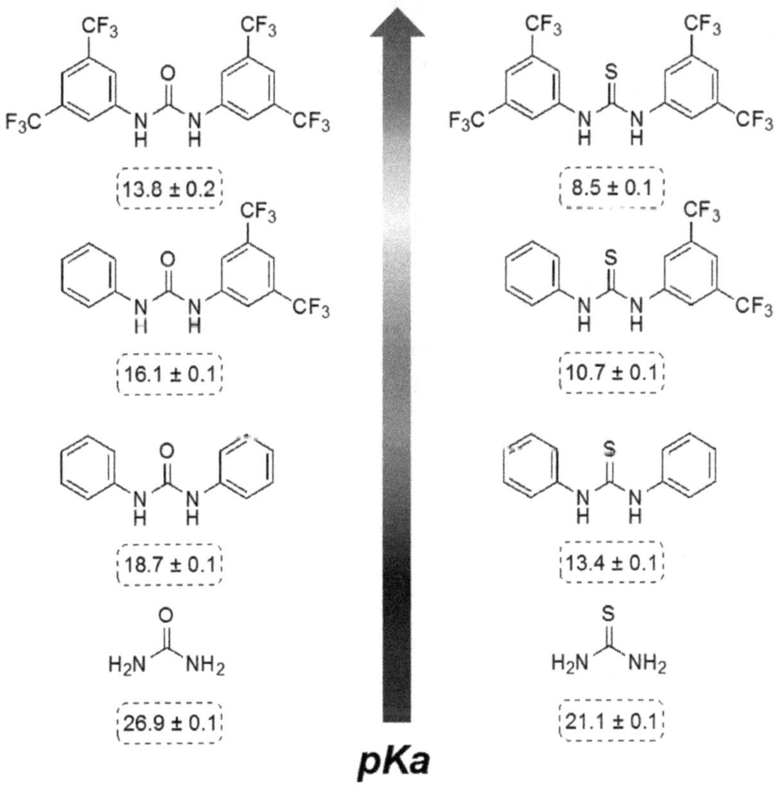

Figure 12.4 Structure and pK_a value in DMSO of different ureas and thioureas showing the influence of the 3,5-bis(trifluoromethyl)phenyl rings on their acidity.
Adapted from ref. 105 with permission from American Chemical Society, Copyright 2012.

double H-bonding, 3,5-bis(trifluoromethyl)phenyl ring-containing ureas and thioureas have been extensively studied (Figure 12.4).[105] As evidenced by the pK_a values of these compounds, urea derivatives are less acidic than their thiourea counterparts.

One of the most impressive results related to the use of (thio)ureas is associated with their ability to catalyze the aminolysis of cyclic carbonates. In 2014, Andrioletti *et al.* conducted a rational study of aminolysis reactions between five-membered cyclic carbonates and various amines. They reported the use of (1-(3,5-bis(trifluoromethyl)phenyl)-3-butylthiourea and 1-(3,5-bis(trifluoromethyl)phenyl)-3-cyclohexylthiourea) as potential organocatalysts for the model reaction between propylene carbonate and cyclohexylamine.[106] At a catalyst concentration of 5 mol%, both thioureas showed similar reactivities to TBD with conversion that reached more than 60% within 1 h of reaction at room temperature. Interestingly, at low catalyst loading, the cyclohexylphenyl-based thiourea surpassed TBD. Applied to a wide range of (thio)ureas, the same group discovered that the catalytic activity of the different thio(urea)s were mainly governed by their acidity.[107] Limited monomer conversions were observed in the case of the most acidic thio(urea)s because of catalyst deprotonation by the amines preventing activation of the carbonate and in the case of the most basic catalysts, which did not activate the carbonyl strongly enough to allow nucleophilic attack. In general, the tested aromatic urea derivatives afforded higher monomer conversion than their corresponding thioureas and were less likely to be deactivated through deprotonation due to lower acidity. In 2016, Caillol *et al.* described the preparation at room temperature of flexible isocyanate-free PU foams by copolymerization of trimethylolpropane tris-carbonate, polypropylene oxide bis-carbonate, and Jeffamine EDR148 as diamine with 1 mol% of cyclohexylphenyl-based thiourea as catalyst.[108]

Considering the success of thioureas in the ROP process of cyclic carbonates, Sun *et al.* screened different thioureas as catalysts for the synthesis of polycarbonates using DMC and 1,4-butanediol.[17] These reactions led to polycarbonates with lower molar masses (4–11 kDa) than when using DMAP or MTBD (16 and 17 kDa, respectively). Similarly to ROP processes where bifunctional activation of monomer and initiator lead to better polymerization kinetics, the authors envisioned the use of a base, such as MTBD, with thiourea as co-catalyst. Thioureas were introduced as co-catalysts in the reaction with MTBD, but the combination was unsuccessful and the polycarbonates molar mass decreased significantly. The authors believe that the carbonyl groups were activated by thiourea through double hydrogen bonding, giving rise to more stable intermediates, lowering the reaction rate, and consequently reducing the molar masses (Scheme 12.36).

12.3.4 Brønsted Acid Ionic Liquids (BAILs)

In recent years, Brønsted acid ionic liquids (BAILs) have attracted special attention as efficient organocatalysts for the synthesis of organic molecules

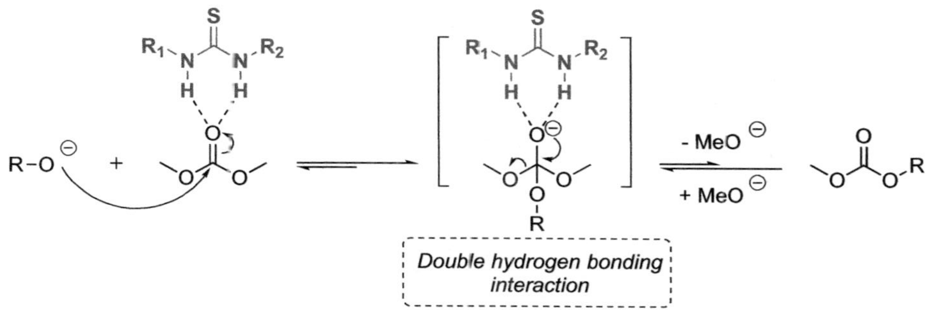

Scheme 12.36 Proposed mechanism for the step-growth polymerization of dimethyl carbonate and diols using thiourea as catalyst.[17]

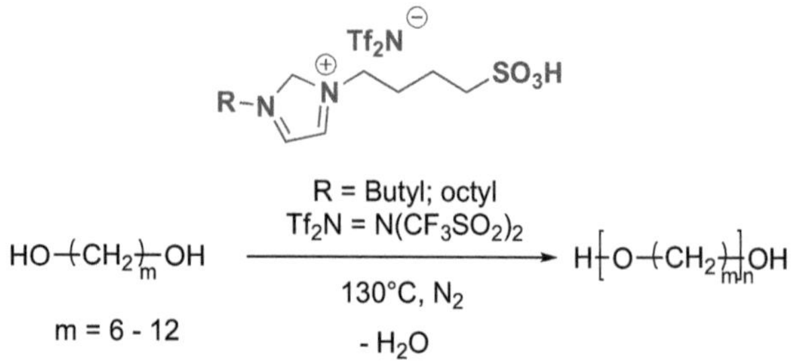

Scheme 12.37 Synthesis of poly(ether)s from the polycondensation of diols using Brønsted acid ionic liquids (BAILs) as reaction media and catalysts.[110]

as well as in polymerization processes. BAILs combine the catalytic activity of Brønsted acid with the high thermal stability and low vapor pressure of ionic liquids. They also present the advantages of being easy to recover from the reaction medium *via* extraction, for example, and are reusable several times.[109] For example, Fradet *et al.* used BAILs, namely 4-(3′-butyl-1′-imidazolio)-1-butanesulfonic acid bis(trifluoromethylsulfonyl)imidide ([BBSIm]Tf_2N) and 4-(3′-octyl-1′-imidazolio)-1-butane-sulfonic acid bis(trifluoromethylsulfonyl)imidide ([OBSIm]Tf_2N), as reaction media and catalysts for the synthesis of poly(ether)s from aliphatic diols (Scheme 12.37).[110] One of the main drawbacks of polyether synthesis is that water by-product must be removed to achieve high molar masses, and harsh polymerization conditions such as high temperatures must be employed. These conditions are usually incompatible with the use of Brønsted acids because of their low boiling points and low degradation temperatures. To minimize these drawbacks, aliphatic diol HO–$(CH_2)_m$–OH ($m = 6$–12) and BAILs were mixed together in a 2 : 1 molar ratio and the reaction performed at 130 °C under N_2 flow. In all cases, good yields and relatively high molar masses were reported ($M_w \geq 20$ kDa). 1,8-octanediol and [BBSIm]Tf_2N, polyether exhibited a 77% yield with a molar mass of $M_w = 42$ kDa ($Đ = 2.2$).

In 2010, Fradet and colleagues performed the self-polyaddition of hydroxy acid to obtain high molar weight poly(12-hydroxydodecanoic acid) in BAILs (Scheme 12.38).[111] Performing the reaction at 90 °C, using 3-alkyl-1-(butyl-4-sulfony-l)imidazolium hydrogen sulfate as both medium and catalyst in a 1 : 1 molar ratio with the hydroxy acid, high molar mass polyesters were obtained within 2 h ($M_n = 40$ kDa, $Đ = 2.2$). Increasing the temperature to 110 °C and 130 °C led to higher molar mass ($M_n = 43.5$ kDa, $Đ = 2.11$; $M_n = 49.9$ kDa, $Đ = 2.04$, respectively). In comparison, at 110 °C without any catalyst or with a catalytic amount of PTSA, only oligomers were obtained ($M_n < 700$ Da). While investigating the effect of BAIL-to-hydroxyl acid molar ratio from 1 : 3 to 3 : 1, the authors found that higher molar masses were obtained for the 1 : 1 molar ratio ($M_n \approx 43$ kDa). Higher monomer concentrations provoked highly viscous reaction mixtures, while higher BAIL concentration led to dilute media, which in both cases slowed the reaction rate.

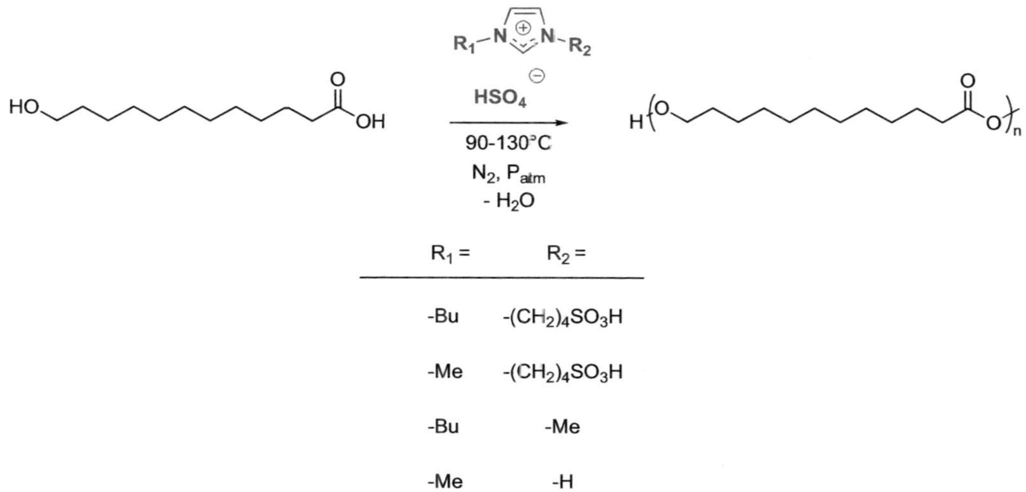

Scheme 12.38 Step-growth polymerization of 12-hydroxydodecanoic acid using BAIL as both medium and catalyst.[111]

BAILs not containing sulfonic groups only led to oligomers under the same conditions due to lower acidities.

In summary, weak to strong organic acids have been investigated as potential catalysts in step-growth polymerization. Among them, sulfonic-based acids, such as PTSA or MSA, have emerged as efficient promoters leading to relatively high molar mass polymers. Their potential has been extensively exploited in PUs, polyesters and notably in polyacetals formation. However, these catalysts suffer from low boiling points and degradation at high temperatures, thus preventing polymerization from occurring under harsh conditions. Lately, BAILs have emerged as an interesting alternative thanks to their higher thermal stability and great recyclability.

12.4 Step-growth Polymerization Catalyzed by Organic Ionic Salts

Lack of thermal stability of common organocatalysts prevent these systems from obtaining similar polymerization properties to the metal-based catalysts at high temperatures. Recent developments have focused on the preparation of novel organic ionic salts catalysts based on DBU:BA or TBD:MSA mixtures. For example, these protic ionic salts have shown to combine the excellent catalytic ability of organocatalysts with the thermal stability of metal catalysts, making them suitable candidates for both polymerization and depolymerization of polyesters in a more sustainable cycle. Sardon *et al.* recently screened some ionic salts of 1,5,7-triazabicyclo[4.4.0]dec-5-ene:methane sulfonic acid (TBD:MSA) and 1,8-diazabicyclo[5.4.0]undec-7-ene:benzoic acid (DBU:BA) for the bulk polymerization of PET from dimethyl terephthalate (DMT) and ethylene glycol (EG) at high temperatures (Scheme 12.39).[45] Thermal analysis revealed that the protic ionic salts could resist thermal degradation up to >250 °C. Kinetic studies revealed that the catalytic activity of DBU:BA (1 : 1) salt was competitive in comparison to the conventional organometallic-based catalysts employed in PET synthesis. PET with molar mass up to $M_n = 10.7$ kDa ($Đ = 2.9$) was obtained using 5 mol% of DBU:BA ionic salt. However, the molar masses remain relatively low in comparison with commercial PET.

Similarly, TBD:MSA ionic salt appeared to be an ideal organocatalyst for the chemical recycling of PET (Scheme 12.40).[112] In less than 2 h at 180 °C, the salt was found to fully depolymerize commercial PET into highly pure BHET (90% conversion). Interestingly, the catalyst could be reused at least five times without loss of activity. In addition, the obtained BHET could be used as a monomer source for the preparation of new PET with similar thermal properties to that produced using a conventional polycondensation process with $M_n = 12$ kDa, using 5 mol% of the same catalyst.

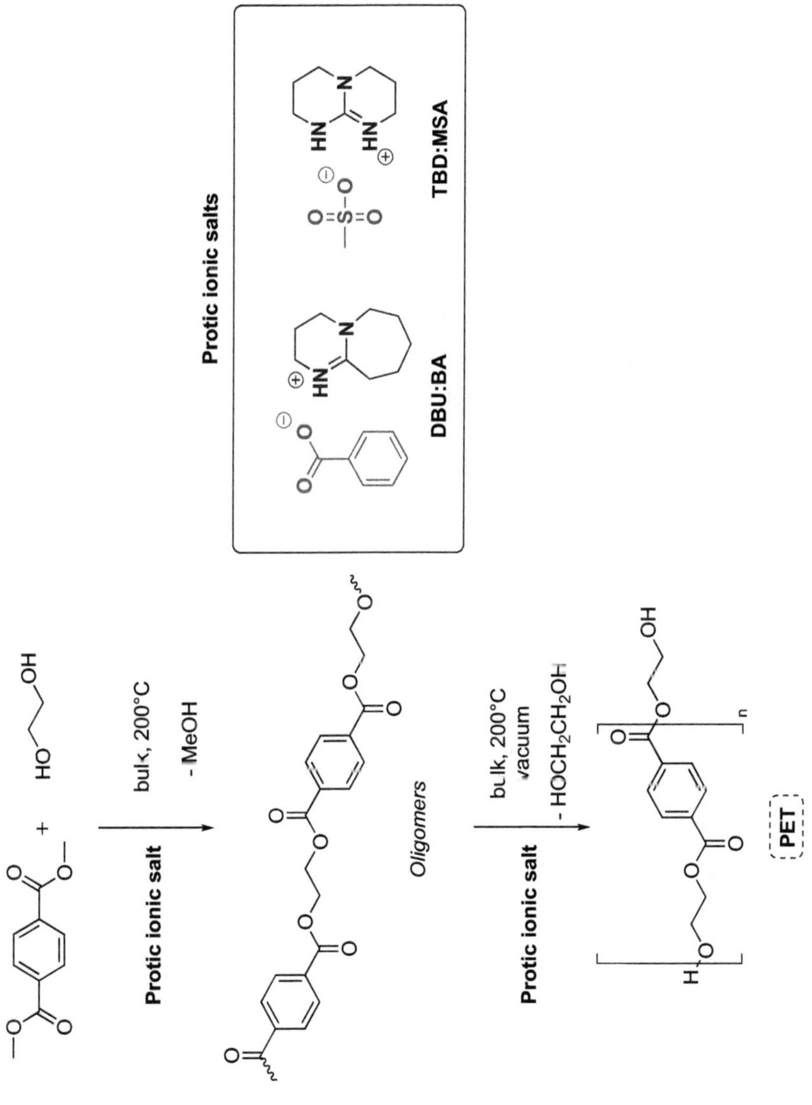

Scheme 12.39 Step-growth bulk polymerization of PET from dimethyl terephthalate (DMT) and ethylene glycol using ionic salts as organocatalysts.[45]

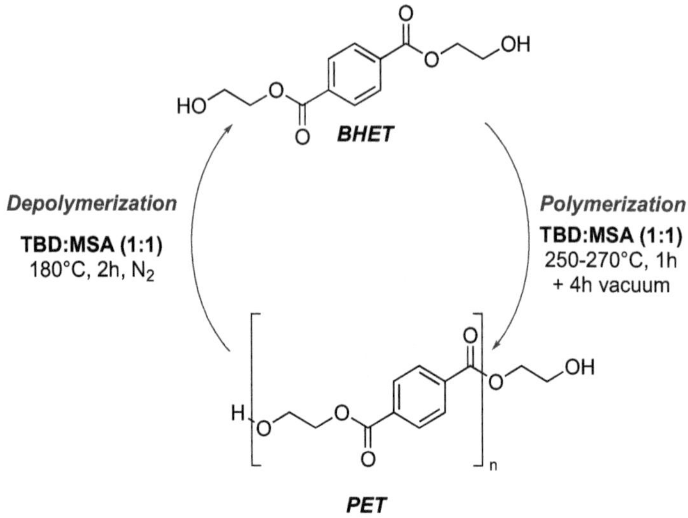

Scheme 12.40 PET-to-PET depolymerization–repolymerization cycle using TBD:MSA ionic salt as organocatalyst.[112]

In summary, similarly to BAILs, protic ionic salt catalysts combine excellent thermal stability, good reactivity and facile recyclability, which make them suitable catalysts for the synthesis of polyesters notably.

12.5 Summary and Outlook

As discussed in depth in this chapter, organocatalysts encompass a large variety of compounds from Brønsted/Lewis acids or bases to mono- or bi-component bifunctional catalytic systems, which have shown to actively catalyze not only chain-growth ROP, but also recently step-growth polymerizations. Among them, guanidine such as TBD and phosphoric-based Brønsted acid such as PTSA have been extensively applied to the synthesis of high molar mass polymers such as PUs, polycarbonates, polyesters and polyacetals. These commercially available organocompounds demonstrated in some cases comparable or better performances in comparison with the commonly used organometallics.

Strong organic acids such as TFA or PTSA appeared to be ideal candidates for the synthesis of PUs and polyacetals. On the other hand, polycarbonates NIPUs and polyesters have been successfully obtained in good yield and relatively high molar masses using strong bases such as the tertiary amine DMAP, the amidine base DBU and the guanidine base TBD.

The removal of trace metal catalysts from final polymers still plague industrial processes due to the high cost, difficult/time consuming methods, and toxicity when unremoved. The synthesis of metal-free polymers, however, benefits from their lower toxicity and thus, are more convenient to use, especially when designing polymers for bioelectronic and biomedical domains. More importantly, as described in this chapter, some organocatalysts have shown to not only possess comparable or better activities to metal catalysts but can also be easily recycled and reused several times. It is however, important to note that the utilization of some organic activators, such as phosphazene derivatives, or the relative high catalyst concentration used in most of the organocatalyzed step-growth polymerization, still represent major drawbacks in terms of safety, cost and polymer properties.

Overall, one can thus reasonably expect that organocatalysis in step-growth polymerization processes will continue to revolutionize the field, ranging from more thermally stable polymers that are more amenable to recycling to novel sequences where regioselectivity will enable novel thermomechanical performance. Organocatalysis will also expand the current scope of reactions that are amenable to the formation of high molar mass, ensuring high degrees of conversions at reaction conditions that were previously unattainable.

Abbreviations

AcOH	acetic acid
BAIL	[BBSIm]Tf_2N, 4-(3′-butyl-1′-imidazolio)-1-butanesulfonic acid bis(trifluoromethylsulfonyl)imidide; Brønsted acid ionic liquid
BDMAEE	2,2′-bis-(dimethylaminoethyl ether)
BF_3OEt_2	boron trifluoride diethyl etherate
BHET	bis(2-hydroxyethyl)terephtalate
bis-MPA	2,2-bis(hydroxymethyl)-propionic acid
CSA	camphor-10-sulfonic acid
Đ	dispersity
DABCO	1,4-diazabicyclo[2,2,2]octane
DBSA	dodecylbenzenesulfonic acid
DBTDL	dibutyltin dilaurate
DBU	1,8-diazabicyclo[5.4.0]undec-7-ene
DCM	dichloromethane
DFT	density functional theory
DMAP	4-dimethylaminopyridine
DMC	dimethylcarbonate
DMF	*N,N*-dimethylformamide
DMSO	dimethyl sulfoxide
DMT	dimethyl terephthalate
DPP	diphenyl phosphate
EG	ethylene glycol
HCl	hydrochloric acid
HMDI	hexamethylene diisocyanate
H_3PO_4	phosphoric acid
H_2SO_4	sulfuric acid

IPDI	isophorone diisocyanate
KOMe	potassium methoxide
KO-*t*-Bu	potassium *tert*-butoxide
LCST	lower critical solution temperature
MeCN	acetonitrile
M_n	number-average molar mass
MSA	methanesulfonic acid
MTBD	*N*-methyl 7-methyl-1,5,7-triazabicyclo[4.4.0]dec-5-ene
M_w	weight-average molar mass
Nf_2NH	bis(1,1,2,2,3,3,4,4,4-nonafluoro-1-butanesulfonyl)imide
NHC	N-heterocyclic carbene
NHC-CO_2	azolium-2-carboxylate
NIPU	non-isocyanate polyurethane
NMR	nuclear magnetic resonance
[OBSIm]Tf_2N	4-(3′-octyl-1′-imidazolio)-1-butane-sulfonic acid bis-(trifluoromethylsulfonyl) imidide
PCS	pyridinium camphorsulfonate
PEG	polyethylene glycol
PEO	polyethylene oxide
PET	poly(ethyleneterephtalate)
PPY	4-pyrrolidinopyridine
PTSA	*p*-toluene sulfonic acid
PTSAA	*p*-toluene sulfonic anhydride
PTS	pyridinium *p*-toluenesulfonate
p-TsOH	*p*-toluene sulfonic acid monohydrate
PU	polyurethane
PVA	polyvinyl alcohol
ROP	ring-opening polymerization
RT	room temperature (23 ± 5 °C)
TBD	1,5,7-triazabicyclo[4.4.0]dec-5-ene
t-BuP_1	*tert*-butylimino-tris(dimethylamino)phosphorene
t-BuP_4	1-*tert*-butyl-4,4,4-tris(dimethylamino)-2,2-bis[tris(dimethylamino)-phosphoranylidenamino]-$2\lambda^5,4\lambda^5$-catenadi(phosphazene)
TEA	triethylamine
TFA	triflic acid
THF	tetrahydrofuran
TMS	trimethylsilyl

Acknowledgements

This work has been supported by the EU Horizon 2020 Research and Innovation Framework Program – European Joint Doctorate – SUSPOL-EJD [grant number 642671].

References

1. H. F. Mark and J. I. Kroschwitz, in *Encyclopedia of Polymer Science and Engineering*, Wiley-VCH Verlag GmbH, New York, 2nd edn, 1984.
2. G. Allen and J. C. Bevington, *Comprehensive Polymer Science: The Synthesis, Characterization, Reactions & Applications of Polymers*, Pergamon Press, 1989.
3. G. G. Odian, *Principles of Polymerization*, Wiley, 1981.
4. M. E. Rogers, T. E. Long and S. R. Turner, in *Synthetic Methods in Step-growth Polymers*, John Wiley & Sons, Inc., 2003, pp. 1–16.
5. A. L. Silva and J. C. Bordado, *Catal. Rev.*, 2004, **46**, 31–51.
6. J. W. Baker and J. Gaunt, *J. Chem. Soc.*, 1949, **0**, 9–18.
7. J. W. Baker and J. B. Holdsworth, *J. Chem. Soc.*, 1947, **0**, 713–726.
8. K. Schwetlick, R. Noack and F. Stebner, *J. Chem. Soc., Perkin Trans. 2*, 1994, **0**, 599–608.
9. M. Hatanaka, *Bull. Chem. Soc. Jpn.*, 2011, **84**, 933–935.
10. J. Alsarraf, Y. A. Ammar, F. Robert, E. Cloutet, H. Cramail and Y. Landais, *Macromolecules*, 2012, **45**, 2249–2256.
11. C. D. Diakoumakos and D. L. Kotzev, *Macromol. Symp.*, 2004, **216**, 37–46.
12. M. Kathalewar, A. Sabnis and D. D'Mello, *Eur. Polym. J.*, 2014, **57**, 99–108.
13. H. Sardon, A. C. Engler, J. M. W. Chan, D. J. Coady, J. M. O'Brien, D. Mecerreyes, Y. Y. Yang and J. L. Hedrick, *Green Chem.*, 2013, **15**, 1121–1126.
14. A. Bossion, G. O. Jones, D. Taton, D. Mecerreyes, J. L. Hedrick, Z. Y. Ong, Y. Y. Yang and H. Sardon, *Langmuir*, 2017, **33**, 1959–1968.
15. F. Nederberg, E. F. Connor, M. Möller, T. Glauser and J. L. Hedrick, *Angew. Chem., Int. Ed.*, 2001, **40**, 2712–2715.
16. A. P. Dove, *ACS Macro Lett.*, 2012, **1**, 1409–1412.
17. J. Sun and D. Kuckling, *Polym. Chem.*, 2016, **7**, 1642–1649.
18. L. Meabe, N. Lago, L. Rubatat, C. Li, A. J. Müller, H. Sardon, M. Armand and D. Mecerreyes, *Electrochim. Acta*, 2017, **237**, 259–266.
19. L. Meabe, T. V. Huynh, N. Lago, H. Sardon, C. Li, L. A. O'Dell, M. Armand, M. Forsyth and D. Mecerreyes, *Electrochim. Acta*, 2018, **264**, 367–375.
20. L. Meabe, H. Sardon and D. Mecerreyes, *Eur. Polym. J.*, 2017, **95**, 737–745.
21. O. Haba, I. Itakura, M. Ueda and S. Kuze, *J. Polym. Sci., Part A: Polym. Chem.*, 1999, **37**, 2087–2093.
22. N. Liu, A. Martin, F. Robert, J.-M. Vincent, Y. Landais, J. Vignolle, H. Cramail and D. Taton, *Macromolecules*, 2014, **47**, 525–533.
23. E. F. Degering and T. Stoudt, *J. Polym. Sci.*, 1951, **7**, 653–656.
24. W. N. Ottou, H. Sardon, D. Mecerreyes, J. Vignolle and D. Taton, *Prog. Polym. Sci.*, 2016, **56**, 64–115.

25. A. Basterretxea, Y. Haga, A. Sanchez-Sanchez, M. Isik, L. Irusta, M. Tanaka, K. Fukushima and H. Sardon, *Eur. Polym. J.*, 2016, **84**, 750–758.
26. I. Kaljurand, T. Rodima, I. Leito, I. A. Koppel and R. Schwesinger, *J. Org. Chem.*, 2000, **65**, 6202–6208.
27. I. Kaljurand, A. Kütt, L. Sooväli, T. Rodima, V. Mäemets, I. Leito and I. A. Koppel, *J. Org. Chem.*, 2005, **70**, 1019–1028.
28. R. H. Lambeth and T. J. Henderson, *Polymer*, 2013, **54**, 5568–5573.
29. Q. Chen, K. Gao, C. Peng, H. Xie, Z. K. Zhao and M. Bao, *Green Chem.*, 2015, **17**, 4546–4551.
30. A. Cornille, S. Dworakowska, D. Bogdal, B. Boutevin and S. Caillol, *Eur. Polym. J.*, 2015, **66**, 129–138.
31. M. Unverferth, O. Kreye, A. Prohammer and M. A. R. Meier, *Macromol. Rapid Commun.*, 2013, **34**, 1569–1574.
32. M. Firdaus and M. A. R. Meier, *Green Chem.*, 2013, **15**, 370–380.
33. C. Duval, N. Kébir, A. Charvet, A. Martin and F. Burel, *J. Polym. Sci., Part A: Polym. Chem.*, 2015, **53**, 1351–1359.
34. A. Martin, L. Lecamp, H. Labib, F. Aloui, N. Kébir and F. Burel, *Eur. Polym. J.*, 2016, **84**, 828–836.
35. H. Sardon, A. Pascual, D. Mecerreyes, D. Taton, H. Cramail and J. L. Hedrick, *Macromolecules*, 2015, **48**, 3153–3165.
36. D. Tang, D.-J. Mulder, B. A. J. Noordover and C. E. Koning, *Macromol. Rapid Commun.*, 2011, **32**, 1379–1385.
37. S. Ma, C. Liu, R. J. Sablong, B. A. J. Noordover, E. J. M. Hensen, R. A. T. M. van Benthem and C. E. Koning, *ACS Catal.*, 2016, **6**, 6883–6891.
38. M. K. Kiesewetter, M. D. Scholten, N. Kirn, R. L. Weber, J. L. Hedrick and R. M. Waymouth, *J. Org. Chem.*, 2009, **74**, 9490–9496.
39. D. Hult, S. García-Gallego, T. Ingverud, O. C. J. Andrén and M. Malkoch, *Polym. Chem.*, 2018, **9**, 2238–2246.
40. H. Mutlu, J. Ruiz, S. C. Solleder and M. A. R. Meier, *Green Chem.*, 2012, **14**, 1728–1735.
41. J. V. Olsson, D. Hult, S. García-Gallego and M. Malkoch, *Chem. Sci.*, 2017, **8**, 4853–4857.
42. B. Testud, D. Pintori, E. Grau, D. Taton and H. Cramail, *Green Chem.*, 2017, **19**, 259–269.
43. D. Tang, B. A. J. Noordover, R. J. Sablong and C. E. Koning, *J. Polym. Sci., Part A: Polym. Chem.*, 2011, **49**, 2959–2968.
44. O. Türünç and M. A. R. Meier, *Macromol. Rapid Commun.*, 2010, **31**, 1822–1826.
45. I. Flores, J. Demarteau, A. J. Müller, A. Etxeberria, L. Irusta, F. Bergman, C. Koning and H. Sardon, *Eur. Polym. J.*, 2018, **104**, 170–176.
46. N. H. Park, G. dos, P. Gomes, M. Fevre, G. O. Jones, I. V. Alabugin and J. L. Hedrick, *Nat. Commun.*, 2017, **8**, 166.
47. K. Zhang, A. M. Nelson, S. J. Talley, M. Chen, E. Margaretta, A. G. Hudson, R. B. Moore and T. E. Long, *Green Chem.*, 2016, **18**, 4667–4681.
48. V. X. Truong and A. P. Dove, *Angew. Chem., Int. Ed.*, 2013, **52**, 4132–4136.

49. C. A. Bell, J. Yu, I. A. Barker, V. X. Truong, Z. Cao, A. V. Dobrinyin, M. L. Becker and A. P. Dove, *Angew. Chem., Int. Ed.*, 2016, **55**, 13076–13080.
50. R. Schwesinger and H. Schlemper, *Angew. Chem., Int. Ed. Engl.*, 1987, **26**, 1167–1169.
51. J. Zhao, N. Hadjichristidis and H. Schlaad, in *Anionic Polymerization*, Springer, Tokyo, 2015, pp. 429–449.
52. Y. Xia, Y. Chen, Q. Song, S. Hu, J. Zhao and G. Zhang, *Macromolecules*, 2016, **49**, 6817–6825.
53. L. Zhang, F. Nederberg, R. C. Pratt, R. M. Waymouth, J. L. Hedrick and C. G. Wade, *Macromolecules*, 2007, **40**, 4154–4158.
54. H. Zhang, S. Hu, J. Zhao and G. Zhang, *Macromolecules*, 2017, **50**, 4198–4205.
55. M. Fèvre, J. Pinaud, Y. Gnanou, J. Vignolle and D. Taton, *Chem. Soc. Rev.*, 2013, **42**, 2142–2172.
56. S. Naumann and A. P. Dove, *Polym. Chem.*, 2015, **6**, 3185–3200.
57. S. Naumann and A. P. Dove, *Polym. Int.*, 2016, **65**, 16–27.
58. D. Bourissou, O. Guerret, F. P. Gabbaï and G. Bertrand, *Chem. Rev.*, 2000, **100**, 39–92.
59. D. Enders, O. Niemeier and A. Henseler, *Chem. Rev.*, 2007, **107**, 5606–5655.
60. M. N. Hopkinson, C. Richter, M. Schedler and F. Glorius, *Nature*, 2014, **510**, 485.
61. O. Coutelier, M. E. Ezzi, M. Destarac, F. Bonnette, T. Kato, A. Baceiredo, G. Sivasankarapillai, Y. Gnanou and D. Taton, *Polym. Chem.*, 2012, **3**, 605–608.
62. J. Pinaud, K. Vijayakrishna, D. Taton and Y. Gnanou, *Macromolecules*, 2009, **42**, 4932–4936.
63. R. Breslow, *J. Am. Chem. Soc.*, 1958, **80**, 3719–3726.
64. G. W. Nyce, J. A. Lamboy, E. F. Connor, R. M. Waymouth and J. L. Hedrick, *Org. Lett.*, 2002, **4**, 3587–3590.
65. P. U. Naik, K. Refes, F. Sadaka, C.-H. Brachais, G. Boni, J.-P. Couvercelle, M. Picquet and L. Plasseraud, *Polym. Chem.*, 2012, **3**, 1475–1480.
66. H. Sardon, A. C. Engler, J. M. W. Chan, J. M. García, D. J. Coady, A. Pascual, D. Mecerreyes, G. O. Jones, J. E. Rice, H. W. Horn and J. L. Hedrick, *J. Am. Chem. Soc.*, 2013, **135**, 16235–16241.
67. H. Sardon, J. M. W. Chan, R. J. Ono, D. Mecerreyes and J. L. Hedrick, *Polym. Chem.*, 2014, **5**, 3547–3550.
68. R. P. Subrayan, S. Zhang, F. N. Jones, V. Swarup and A. I. Yezrielev, *J. Appl. Polym. Sci.*, 2000, **77**, 2212–2228.
69. D. J. Coady, H. W. Horn, G. O. Jones, H. Sardon, A. C. Engler, R. M. Waymouth, J. E. Rice, Y. Y. Yang and J. L. Hedrick, *ACS Macro Lett.*, 2013, **2**, 306–312.
70. A. Kütt, T. Rodima, J. Saame, E. Raamat, V. Mäemets, I. Kaljurand, I. A. Koppel, R. Y. Garlyauskayte, Y. L. Yagupolskii, L. M. Yagupolskii, E. Bernhardt, H. Willner and I. Leito, *J. Org. Chem.*, 2011, **76**, 391–395.

71. L. I. Ronco, A. Basterretxea, D. Mantione, R. H. Aguirresarobe, R. J. Minari, L. M. Gugliotta, D. Mecerreyes and H. Sardon, *Polymer*, 2017, **122**, 117–124.
72. Y. Wang, H. Morinaga, A. Sudo and T. Endo, *J. Polym. Sci., Part A: Polym. Chem.*, 2011, **49**, 596–602.
73. A. G. Pemba, M. Rostagno, T. A. Lee and S. A. Miller, *Polym. Chem.*, 2014, **5**, 3214–3221.
74. M. Rostagno, E. J. Price, A. G. Pemba, I. Ghiriviga, K. A. Abboud and S. A. Miller, *J. Appl. Polym. Sci.*, 2016, **133**(45), 44089–44100.
75. M. Rostagno, S. Shen, I. Ghiviriga and S. A. Miller, *Polym. Chem.*, 2017, **8**, 5049–5059.
76. S. Chikkali, F. Stempfle and S. Mecking, *Macromol. Rapid Commun.*, 2012, **33**, 1126–1129.
77. A. G. Pemba, J. A. Flores and S. A. Miller, *Green Chem.*, 2013, **15**, 325–329.
78. B. S. Rajput, S. R. Gaikwad, S. K. Menon and S. H. Chikkali, *Green Chem.*, 2014, **16**, 3810–3818.
79. B. S. Rajput, U. Chander, K. Arole, F. Stempfle, S. Menon, S. Mecking and S. H. Chikkali, *Macromol. Chem. Phys.*, 2016, **217**, 1396–1410.
80. J. Heller, D. W. H. Penhale and R. F. Helwing, *J. Polym. Sci., Polym. Lett. Ed.*, 1980, **18**, 293–297.
81. H. Zhang and E. Ruckenstein, *J. Polym. Sci., Part A: Polym. Chem.*, 2000, **38**, 3751–3760.
82. R. Tomlinson, M. Klee, S. Garrett, J. Heller, R. Duncan and S. Brocchini, *Macromolecules*, 2002, **35**, 473–480.
83. T. Hashimoto, K. Ishizuka, A. Umehara and T. Kodaira, *J. Polym. Sci., Part A: Polym. Chem.*, 2002, **40**, 4053–4064.
84. R. Tomlinson, J. Heller, S. Brocchini and R. Duncan, *Bioconjugate Chem.*, 2003, **14**, 1096–1106.
85. M. J. Vicent, R. Tomlinson, S. Brocchini and R. Duncan, *J. Drug Targeting*, 2004, **12**, 491–501.
86. T. Hashimoto, A. Umehara, M. Urushisaki and T. Kodaira, *J. Polym. Sci., Part A: Polym. Chem.*, 2004, **42**, 2766–2773.
87. J. Rickerby, R. Prabhakar, M. Ali, J. Knowles and S. Brocchini, *J. Mater. Chem.*, 2005, **15**, 1849–1856.
88. X.-C. Sui, Y. Shi and Z.-F. Fu, *Aust. J. Chem.*, 2010, **63**, 1497–1501.
89. V. K. Garripelli, J.-K. Kim, R. Namgung, W. J. Kim, M. A. Repka and S. Jo, *Acta Biomater.*, 2010, **6**, 477–485.
90. A. Iinuma, T. Hashimoto, M. Urushisaki and T. Sakaguchi, *J. Appl. Polym. Sci.*, 2016, **133**(41), 44088–44098.
91. N. Liu, J. Vignolle, J.-M. Vincent, F. Robert, Y. Landais, H. Cramail and D. Taton, *Macromolecules*, 2014, **47**, 1532–1542.
92. M. J. Heffernan and N. Murthy, *Bioconjugate Chem.*, 2005, **16**, 1340–1342.
93. S. Chatterjee and S. Ramakrishnan, *Macromolecules*, 2011, **44**, 4658–4664.

94. S. D. Hamann, D. H. Solomon and J. D. Swift, *J. Macromol. Sci., Chem.*, 1968, **2**, 153–174.
95. S. Chen and K.-C. Wu, *J. Polym. Sci., Polym. Chem. Ed.*, 1982, **20**, 1819–1831.
96. C.-T. Kuo and S.-A. Chen, *J. Polym. Sci., Part A: Polym. Chem.*, 1989, **27**, 2793–2803.
97. T. Moyori, T. Tang and A. Takasu, *Biomacromolecules*, 2012, **13**, 1240–1243.
98. K. Manabe, X.-M. Sun and S. Kobayashi, *J. Am. Chem. Soc.*, 2001, **123**, 10101–10102.
99. K. Manabe, S. Iimura, X.-M. Sun and S. Kobayashi, *J. Am. Chem. Soc.*, 2002, **124**, 11971–11978.
100. A. Takasu, A. Takemoto and T. Hirabayashi, *Biomacromolecules*, 2006, **7**, 6–9.
101. G. Zhang, Y. Dai, Y. Liu, J. Liu, H. Lu, L. Qiu and K. Cho, *Polym. Chem.*, 2017, **8**, 3448–3456.
102. M. Sokolsky-Papkov, R. Langer and A. J. Domb, *Polym. Adv. Technol.*, 2011, **22**, 502–511.
103. P. R. Schreiner, *Chem. Soc. Rev.*, 2003, **32**, 289–296.
104. N. E. Kamber, W. Jeong, R. M. Waymouth, R. C. Pratt, B. G. G. Lohmeijer and J. L. Hedrick, *Chem. Rev.*, 2007, **107**, 5813–5840.
105. G. Jakab, C. Tancon, Z. Zhang, K. M. Lippert and P. R. Schreiner, *Org. Lett.*, 2012, **14**, 1724–1727.
106. M. Blain, L. Jean-Gérard, R. Auvergne, D. Benazet, S. Caillol and B. Andrioletti, *Green Chem.*, 2014, **16**, 4286–4291.
107. M. Blain, H. Yau, L. Jean-Gérard, R. Auvergne, D. Benazet, P. R. Schreiner, S. Caillol and B. Andrioletti, *ChemSusChem*, 2016, **9**, 2269–2272.
108. A. Cornille, C. Guillet, S. Benyahya, C. Negrell, B. Boutevin and S. Caillol, *Eur. Polym. J.*, 2016, **84**, 873–888.
109. M. Vafaeezadeh and H. Alinezhad, *J. Mol. Liq.*, 2016, **218**, 95–105.
110. S. Zhang, A. Féret, H. Lefebvre, M. Tessier and A. Fradet, *Chem. Commun.*, 2011, **47**, 11092–11094.
111. E.-M. Dukuzeyezu, H. Lefebvre, M. Tessier and A. Fradet, *Polymer*, 2010, **51**, 1218–1221.
112. C. Jehanno, I. Flores, A. P. Dove, A. J. Müller, F. Ruipérez and H. Sardon, *Green Chem.*, 2018, **20**, 1205–1212.

CHAPTER 13

Organocatalyzed Controlled Radical Polymerizations

MATTHEW D. RYAN, RYAN M. PEARSON AND GARRET M. MIYAKE*

Department of Chemistry, Colorado State University, Fort Collins, Colorado 80523, USA
*Email: Garret.Miyake@colostate.edu

13.1 Fundamentals of Organocatalyzed Controlled Radical Polymerizations

13.1.1 Free-radical Polymerizations

Over half of all polymers produced on an industrial scale are synthesized using free radical approaches.[1] Free-radical polymerizations exhibit excellent functional group tolerance and can be performed under relatively moderate reaction conditions, making them desirable for industrial implementation. Furthermore, the rate of curing in a radical polymerization can be beneficial for applications including coatings[2] and dentistry.[3] Free-radical polymerizations are commonly initiated by decomposition of an initiator to generate a radical that can proceed in polymerization propagation.[4,5] However, the inherent reactivity of radical and intrinsic bimolecular termination pathways challenge the ability to produce polymers with controlled compositions or architectures (Figure 13.1).[6] For free-radical polymerization, the distribution of polymer molecular weights in a given sample, defined as dispersity ($Đ$), is largely dependent on the dominant

Polymer Chemistry Series No. 31
Organic Catalysis for Polymerisation
Edited by Andrew Dove, Haritz Sardon and Stefan Naumann

Published by the Royal Society of Chemistry, www.rsc.org

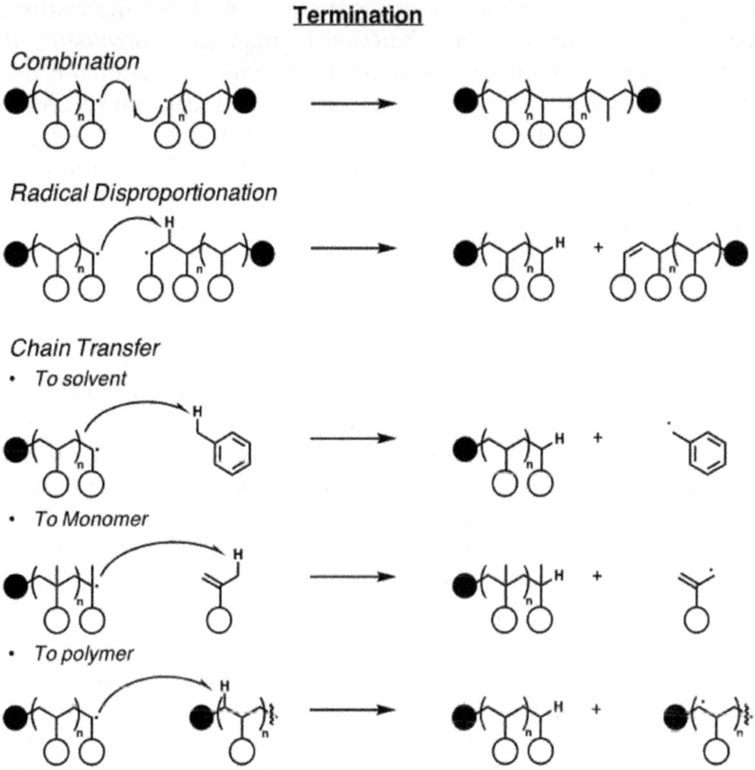

Figure 13.1 Reaction schemes for various potential termination processes in a radical polymerization.

mechanism of termination, and theoretically approaches $Đ = 1.5$ for combination in absence of chain transfer and $Đ = 2.0$ for disproportionation.

These characteristics directly control the properties of a polymer and enable expanded application potential as well as enhanced fundamental understanding of structure–property relationships of new polymers. As such, although free-radical polymerizations are excellent at producing commodity polymers, polymerization methodologies that can synthesize polymers with controlled molecular weights, compositions, and architectures are needed for the synthesis of advanced materials.

13.1.2 Controlled Radical Polymerizations

Over the last several decades the field of controlled radical polymerizations (CRPs) has evolved to deliver a plethora of approaches to harness reactive radicals for the synthesis of precision macromolecules.[7] CRPs are methods which allow for the synthesis of polymers with predictable molecular weights (MWs), narrow molecular weight distributions (*i.e.* low $Đ$), controlled chemical compositions, tailored architectures, and defined chain-end group functionality.[8] During the course of polymerization, the polymer MW grows linearly as a function of monomer conversion, the *Đ* remains low, and the chain-end group retains chemical functionality derived from the initiator. CRP techniques are based on a process called reversible-deactivation, mediated by activity of a catalyst[1] or operation of a degenerative chain-transfer process[9] (Figure 13.2). The reversible-deactivation process is an equilibrium between an active species capable of growing the polymer MW and a dormant state in which reactivity is temporarily quenched.

The iterative cycling between active "on" and dormant "off" states is modulated by catalysts, initiators, or additives[10–12] influenced by thermal,[7] photochemical,[13] or electrochemical stimuli.[14] Successful mediation of this equilibrium controls the concentration of active radicals in solution and minimizes undesirable bimolecular termination events. Due to this mechanistic feature, the observable characteristics of CRPs more resemble those of a living polymerization than those of free-radical polymerizations. A living polymerization proceeds in the absence of termination or chain-transfer reactions. Certain experimental criteria have been established in determining if a polymerization can be classified as living or displaying living characteristics: (1) the polymerization will continue until all monomer is consumed and if more monomer is added polymerization will resume, (2) the polymer MW will grow linearly as a function of monomer conversion, (3) the number of actively growing polymer chains is constant throughout the polymerization, (4) the polymer MW is controlled by stoichiometry, (5) the polymer *Đ* is relatively low (<1.30), (6) block polymers are prepared through consecutive monomer addition, and (7) chain-end functionalized polymers are prepared in high efficiency (Figure 13.3).[15]

Free-radical polymerization deviates from these conditions on many fronts, including the presence of significant termination pathways, which typically result in polymers with high *Đ* (1.5–2.0) and the inability to produce polymers with absolute chain-end group functionality. Although CRPs are not technically a living polymerization, due to the presence of the reversible-deactivation pathway, functionally they display the characteristics of a living polymerization. The likelihood of termination through disproportionation

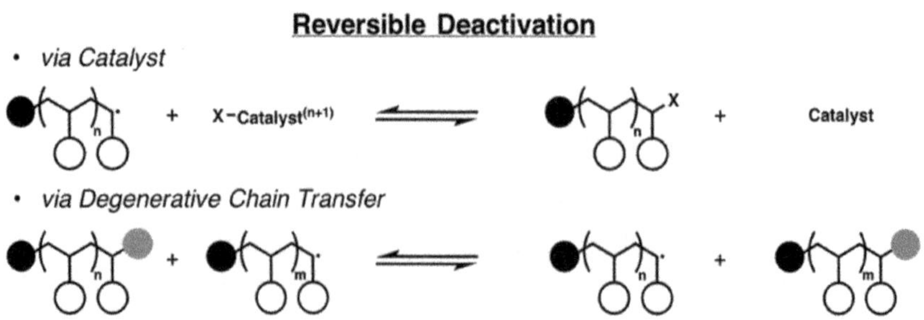

Figure 13.2 Reaction schemes for reversible-deactivation processes mediated by catalyst (top) or degenerative chain transfer (bottom).

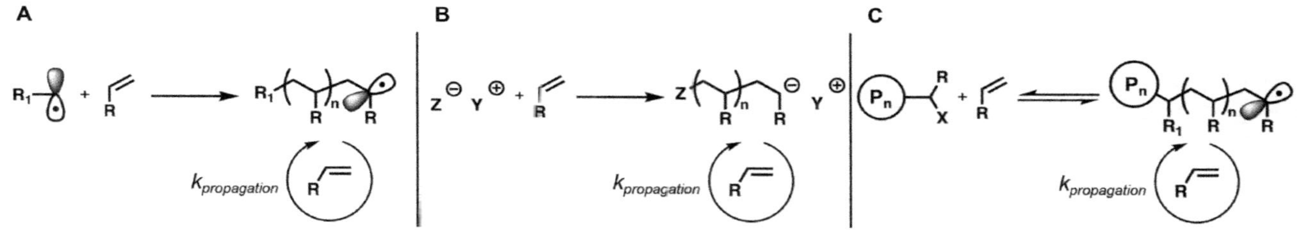

Free Radical Polymerization	Vs.	**Living Polymerization**	Vs.	**Controlled Radical Polymerization**
• Polymer chains rapidly terminate. • Polymer MW does not grow linearly as a function of monomer conversion. • The polymer MW can be controlled by stoichiometry. • The polymer *Đ* is typically between 1.5-2.0, depending on the dominant termination mechanism. • Block polymers cannot be prepared by consecutive monomer addition. • Chain-end functionalized polymers are not prepared in high efficiency.		• Polymerization will continue until all monomer is consumed. • Polymer MW grows linearly as a function of monomer conversion. • The number of actively growing polymer chains is constant throughout the polymerization. • The polymer MW is controlled by stoichiometry. • The polymer *Đ* is relatively low (< 1.3), and typically approaches 1.0. • Block copolymers can be prepared by sequential monomer addition. • Chain-end functionalized polymers are prepared in high efficiency.		• Polymerization will continue until all monomer is consumed. • Polymer MW grows linearly as a function of monomer conversion. • The number of actively growing polymer chains is constant. • The polymer MW is controlled by stoichiometry. • The polymer *Đ* is relatively low (< 1.3). • Block polymers are prepared through consecutive monomer addition. • Chain-end functionalized polymers can be prepared in high efficiency.

Figure 13.3 General reaction schemes for a traditional free-radical polymerization (A), living-anionic polymerization (B), and controlled-radical polymerization (C), evaluated against the seven characteristics for a living polymerization (bottom).

or combination is dramatically decreased through dictating a low instantaneous concentration of active radical, polymer MW grows linearly as a function of monomer conversion, Đ is relatively low (<1.30), and polymers with active chain-end group functionality can be isolated in high efficiency.

The general advantages of CRPs over many living polymerization methodologies include accessible reaction conditions without necessitating the complete absence of water, wide monomer scope, and the ability of the polymer to be isolated with a functionalized chain-end group, which can be utilized for downstream syntheses. The isolated chain-end group functionalized polymer, commonly referred to as a "macro-initiator" in subsequent reactions, allows for production of a myriad of higher order polymeric architectures. The macro-initiator can be broadly functionalized, including direct chemical modification of the chain-end group or the use of the macro-initiator in the synthesis of higher-order polymeric architectures. (Figure 13.4).[16]

CRP methods have been developed based on the concept of reversible-deactivation, all operating some type of equilibrium between the "On" and "Off" states of the polymerization. The most common CRPs include nitroxide-mediated polymerization (NMP),[17–19] reversible-addition fragmentation chain-transfer (RAFT) polymerization,[20–22] and atom transfer radical polymerization (ATRP).[23–25] To achieve control over the polymerization, NMP employs stable nitroxyl radicals to deactivate activated carbon-centered radicals, RAFT operates a degenerative chain-transfer mechanism, and ATRP affects the transfer of a halogen atom between the polymer chain-end group and a catalyst (*e.g.* copper). ATRP has emerged as the most-studied CRP due to ease of operation and mild reaction conditions, wide availability of stable and inexpensive reaction components, and wide

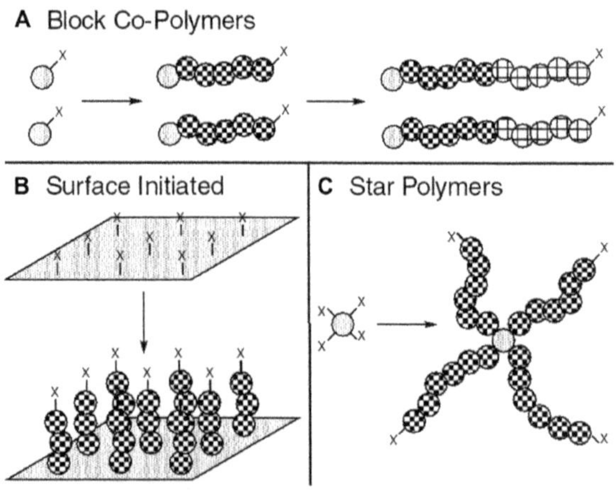

Figure 13.4 Examples of various polymeric architectures accessed by way of active chain-end group functionalized macro-initiators, such as block co-polymers (A), patterned surfaces (B), and star polymers (C), with active chain-end groups denoted X.

monomer scope.[26] Although RAFT also benefits from a wide monomer scope and ease of operation, the high cost of chain-transfer agents presents a barrier. NMP differs from ATRP and RAFT by having a somewhat limited monomer scope. Each CRP strategy possesses advantages and disadvantages in comparison to other CRPs,[27] dictated by the specific polymerization mechanism for reversible-deactivation.

The development of organocatalytic CRP systems has become a major field of research and presents a key advantage by avoiding potential for metal contamination in the final polymer product, facilitating the use of the materials in biomedical and electronic applications.[28] NMP and RAFT do not traditionally require a metal catalyst, while more recently organocatalyzed ATRP (O-ATRP) has been developed.

For the scope of this chapter, the focus is organocatalyzed CRPs (O-CRPs), restricted to methods wherein an organic catalyst provides the primary control over the polymerization and no other mechanisms such as degenerative chain transfer in the absence of catalyst. Additionally, discussion is restricted to methods that have shown strong evidence of satisfying the following determining characteristics of an O-CRP: use of an organic catalyst, production of polymers with predictable MWs *via* linear plots of polymer MW as a function of monomer conversion, low (<1.10) or relatively low $Đ$ (<1.30), the ability to isolate polymers with an active chain-end group functionality, and in an efficient polymerization as evidenced by a high (>80%) initiator efficiency (I^*) (the ratio between the theoretical number average molecular weight ($M_{n,theo}$) and the experimentally measured number-average molecular weight ($M_{n,exp}$)).

A key distinction in the focus of this chapter is that between "organocatalytic" *vs.* "organo-initiated", the former of which requires a deeper analysis. The defining characteristic in differentiating the two scenarios is if the catalyst controls the reversible-deactivation process or if it solely generates the initiating radical species to begin propagation. As the degree of control in an O-CRP is dictated by the balance between activation and deactivation, catalyst involvement in both processes is an important factor to be considered. Because of this distinction, most examples of NMP and RAFT are not discussed in this chapter, as the mechanism for reversible-deactivation is one of degenerative chain transfer (for RAFT) or not involving activity of a catalyst (for NMP), although exceptions do exist and are discussed.

13.1.3 Photo-mediated Controlled Radical Polymerizations

To date, the majority of organocatalyzed CRPs are photo-driven processes. To expose a carbon-centered radical during initiation, an organic catalyst must activate an initiator and affect a homolytic bond cleavage. This cleavage is typically mediated through a reduction event *via* a catalyst to create the active radical and grow the polymer chain, and the complementary oxidized state of the catalyst to deactivate the growing polymer chain, respectively.

The most common species participating in this reactivity is an alkyl halide (for ATRP) or a trithiocarbonate/xanthate (for RAFT). The alkyl halide ATRP initiators are commonly designed so that the generated radical closely mimics the structure of the monomers to be polymerized, and the trithiocarbonate/xanthate RAFT initiators employ a variety of structural features to modulate reactivity.[29] After adding some number of monomer units to the growing polymer chain, the active radical is deactivated by the complementary deactivating form of the catalyst, which is formed during the activation step, returning the polymer chain-end group to a dormant state. Because of the inherent nature of radical polymerizations, some degree of termination is inevitable, and traditional biomolecular radical quenching mechanisms still exist in organocatalyzed CRPs.

In order for organocatalysts to activate ATRP or RAFT initiators *via* reductive or oxidative processes, the use of light as a driving force is commonplace. Ground state organic molecules do not typically participate in single-electron redox processes while a photoexcited molecule is generally both a stronger reductant and oxidant. Examining the specific case of an initiator to be reduced, there are two available reaction mechanisms for these organic photocatalysts (PCs): reductive or oxidative quenching pathways (Figure 13.5).[30] In the reductive quenching cycle, a sacrificial electron donor (denoted D for electron donor) must be present for the excited state photocatalyst (PC*) to oxidize, thus generating a highly reactive, reduced

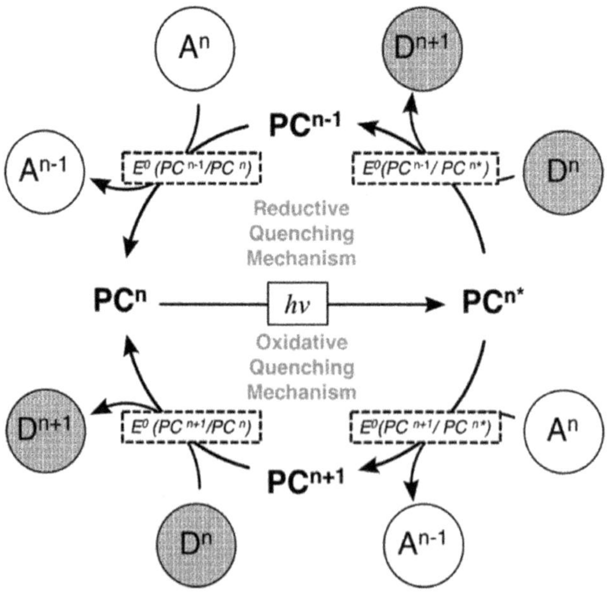

Figure 13.5 Photoredox catalysis mediated by organocatalysts can proceed through reductive quenching cycles by way of sacrificial electron donors (D) (top), or oxidative quenching cycles by way of electron acceptors (A) (bottom).

photocatalyst (PC^{n-1}) species. This strong reductant is then capable of transferring an electron to an acceptor species (denoted A for electron acceptor), which in the case of an O-CRP is the initiating or dormant polymer chain. In the oxidative quenching cycle, the PC* can directly reduce the acceptor species, *i.e.* initiating polymerization, and thus convert to the oxidized photocatalyst (PC^{n+1}). PC^{n+1} can then act as an electron acceptor and deactivate the polymerization while re-generating the PC. In general, processes not requiring additives to generate the reactive state of the photocatalyst are preferred (*i.e.* oxidative quenching for a strong reductant, and reductive quenching for a strong oxidant) because the additives can participate in undesirable reactivity forming byproducts. If the photocatalyst can perform electron transfer to substrate directly from its excited state, there are fewer pathways for side-reactivity *via* the need for extraneous electron donors or acceptors.

Additionally, photo-mediated reversible-deactivation mechanisms provide the opportunity for spatial and temporal control over the polymerization through irradiation procedures.[31] A powerful deployment of this affect may involve a mask or template placed over a surface, which allows the stimuli to control the polymerization in specific patterns on the surface (Figure 13.6A). Furthermore, polymerization can be started, stopped, and resumed by irradiation sequences. Experimentally, conversion with respect to time can be plotted, and polymerization only proceeds during irradiation, while no monomer conversion or polymer growth is observed during the "off" periods (Figure 13.6B). Spatial and temporal control over CRPs has been utilized in various advanced applications such as gradient surface functionalization,[32] representing a promising future avenue for this methodology.

13.1.4 Photocatalysis: Photophysical and Electrochemical Considerations

To efficiently harness light energy as a driving force for reactions, it is of the utmost importance to understand the photophysical and electrochemical characteristics of the PCs employed in polymerization. Some of the important quantifiable metrics of an organic photocatalyst necessary to operate a successful CRP have been identified as wavelength of maximum absorption (λ_{max}, nm), molar absorptivity (ε, $M^{-1}\,cm^{-1}$), excited state lifetime (τ, (ns/μs/s)), excited state reduction potential ($E^0[PC^{\bullet+}/PC^*]$), excited state oxidation potential ($E^0[PC^{\bullet}/PC^{+*}]$), ground state oxidation potential ($E^0[PC^{\bullet+}/PC]$), and ground state reduction potential ($E^0[PC^{\bullet-}/PC]$). Together, these features represent key properties that need to work in concert to allow for efficient electron transfer between the catalyst and polymer chain-end group during both initiation and deactivation processes.

These characteristics can be considered as preliminary design principles for organic photocatalysts. It is ideal that the PCs utilize lower-energy visible light as the driving force for reaction in order to minimize side reactions that

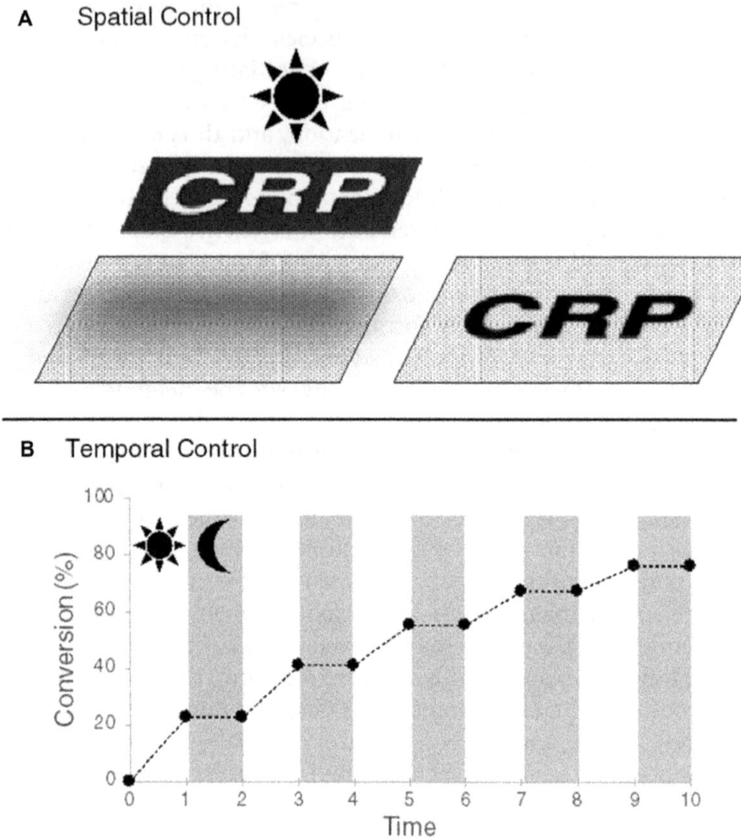

Figure 13.6 Schematic example of spatial control *via* a surface patterning approach, where the photo-mask directs polymerization on the substrate (A), and temporal control exemplified by a characteristic pulsed-irradiation sequence, with monomer conversion occurring only in periods of time exposed to light (B).

can occur from direct interaction between high-energy light and substrate.[33] The overlap of absorption of the PC with the emission of the light source in conjunction with the light flux will, in part, determine the concentration of excited state molecule at any given time. In concert with the overlap between PC absorption and light source emission, the efficiency at which the PC is able to reach the reactive state, described *via* the singlet quantum yield ($\Phi_{Singlet}$) or triplet quantum yield ($\Phi_{Triplet}$), must be considered. There are exceptions to the idea of decreased electron transfer efficiency from the singlet state[34] but it is generally accepted that the longer-lived triplet excited state is the predominant driving force in photoredox catalysis.[35,36] If the photoexcitation of the PC to PC* is not efficient and intersystem crossing from the singlet to the triplet manifold is low, decay to the ground state is

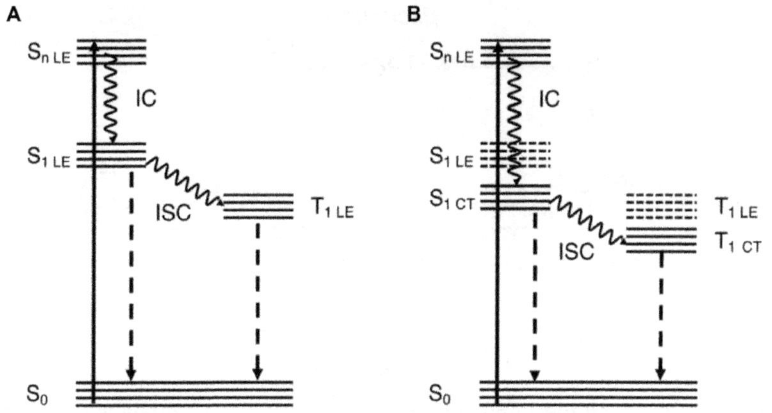

Figure 13.7 Jablonski diagrams for the photophysical process leading to the population of excited states of LE character (A, left) and CT character (B, right). The solid arrows represent photoexcitation from S_0 to S_n, and the dashed arrows represent radiative and non-radiative decay.

likely to out-compete a successful bimolecular encounter with the substrate, and ultimately prevents efficient electron transfer in photo-CRPs.

Understanding the photoexcitation process of a photocatalyst is necessary to understand the polymerization mechanism. Jablonski diagrams are typically used to visualize the photophysical processes of which a compound can undergo: absorption of a photon from the ground to the excited state, internal conversion (IC) to a lower energy, intersystem crossing (ISC), as well as non-productive (for catalysis) radiative or non-radiative decay processes from both the singlet or triplet state (Figure 13.6). Additionally, the nature of an excited compound can differ dramatically. Upon photoexcitation, a species can retain electronic charge in the same region from where the electron was excited (denoted LE for local excitation, Figure 13.7A) or can access a charge-transfer (CT) excited state, where electronic charge is transferred intramolecularly (Figure 13.7B). CT is proposed to be beneficial in catalysis due to longer lived excited state lifetimes and a lower probability of back electron transfer to the ground state.

13.2 Organocatalyzed Atom Transfer Radical Polymerization

13.2.1 Mechanistic Cycle

13.2.1.1 Introduction

The oxidative quenching mechanistic cycle for O-ATRP is superior to the reductive cycle because of the avoidance of sacrificial electron donors,[37] and is composed of the three general mechanistic elements of a CRP (activation, propagation, deactivation) (Figure 13.8). In the first step, activation requires

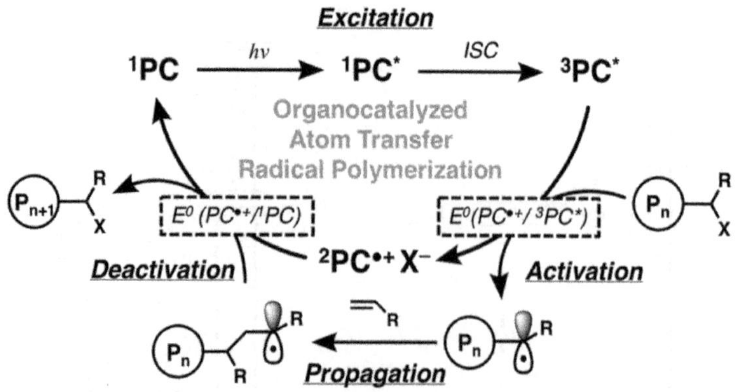

Figure 13.8 Proposed mechanism for O-ATRP, where: (1) *excitation* is composed of initial population of a singlet excited state (S_1) after absorption of a photon, followed by intersystem crossing to the triplet state (T_1); (2) *activation* involves electron transfer from $^3PC^*$ with reduction potential $E^0(PC^{\bullet +}/PC^*)$ to initiator or dormant polymer chain-end group generating active radical and $^2PC^{\bullet +}/X^-$; (3) *propagation* for polymer MW growth; and (4) *deactivation* by $^2PC^{\bullet +}/X^-$ with oxidation potential $E^0(PC^{\bullet +}/PC)$, returning the polymer chain-end group to a dormant state and re-generating PC.

an excited state photocatalyst, and the excitation process can be divided into two different processes: (1) initial excitation from light then internal conversion to the first singlet excited state ($^1PC^*$), followed by (2) intersystem crossing (ISC) to a long-lived triplet state. From here, $^3PC^*$ is assumed to be the primary species from which electron transfer to the initiator or dormant polymer chain-end group occurs, although reactivity from the $^1PC^*$ is also possible. For the activation step, PCs with <-0.7 V *vs.* saturated calomel electrode (SCE) are desirable in order to reduce the carbon halide bond typically found in common ATRP initiators. In a successful process, a minimal number of monomer units are added to the growing polymer chain, the oxidized ground state of PC, ($^2PC^{\bullet +}/X^-$), is responsible for deactivation of the active radical back to the halide-capped dormant state. The balance between the activity of PC* and $^2PC^{\bullet +}/X^-$ is what comprises the reversible-deactivation mechanism and influences the degree of control exercised over an O-ATRP.

13.2.1.2 *Activation*

Fast initiation is beneficial in O-ATRP to establish a constant concentration of growing polymer chains equal to the initial initiator concentration[22] and, more importantly, establish the necessary concentration of deactivating species for control.[38] The combination of efficient initiation and deactivation, and thus accessing a high I^*, allows the ability to target specific polymer MWs by stoichiometry and monomer conversion. For an efficient initiation in O-ATRP, one must consider the photophysical and electrochemical characteristics of the employed PC. First, the $E^0(^2PC^{\bullet+}/^3PC^*)$ must be sufficiently reducing to activate an alkyl halide initiator, and the subsequent halide-capped polymer. According to Marcus theory, the outer sphere electron transfer mechanism, wherein the species donating and receiving the transferred electron remain separated as opposed to connected *via* a chemical bond, requires some degree of over-potential to accomplish this process at an acceptable rate.[39] Furthermore, the τ of the PC* should be long to ensure diffusion for productive bimolecular collision with the alkyl halide initiator, typically necessitating $\tau > 10$ ns.[40,41]

13.2.1.3 Propagation

O-ATRP has been well developed for methacrylate monomers with a wide range of functionality, but remains relatively under explored or uncontrolled in the case of styrene[27] and acrylonitrile.[42] An impactful challenge in this field is to extend the methodology to acrylates, vinyl acetate and other less-activated olefinic monomers. Realizing this advancement will necessitate the design of PCs that have the tailored reduction and oxidation potentials to efficiently activate the appropriate initiators, and deactivate radicals of varied reactivity and rates of propagation.

13.2.1.4 Deactivation

Similar to traditional metal-catalyzed ATRP, to maintain control over an O-ATRP the rate of deactivation must be faster than the rate of polymerization propagation, ensuring a low concentration of active radicals in solution. Because of this fact, the $E^0(^2PC^{\bullet+}/^1PC)$ must be sufficiently more positive than -0.7 V *vs.* SCE to oxidize the active radical. The stability of $PC^{\bullet+}$ is also a necessary consideration for deactivation, as the catalyst must cycle between the reduced and oxidized states through many cycles.

13.2.1.5 Role of Light Overall

When considering the mechanistic cycle of O-ATRP, only activation is directly involved in the photochemical process; however, all downstream processes are controlled by the efficiency of this photochemical step and the preceding photophysical processes that led to it.[43] Specifically, the efficiency of activation (*i.e.* the reduction event), directly corresponds to the formation and overall concentration of $^2PC^{\bullet+}$ in solution, which imparts control in the O-ATRP.

13.2.2 Catalyst Families

13.2.2.1 Introduction

To date, there are five major classes of organic photocatalysts that have been shown to operate O-ATRP: polycyclic aromatic hydrocarbons (PAHs), phenothiazines, phenazines, phenoxazines, and carbazoles (Figure 13.9). In

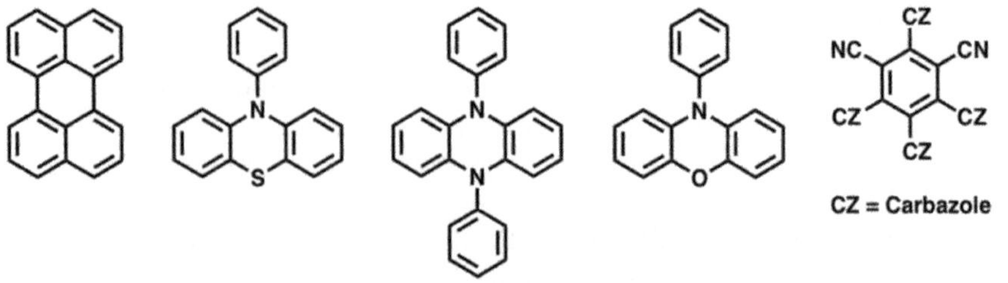

1 2 3 4 5

Figure 13.9 Representative species for the five different classes of organic catalysts shown to operate O-ATRP: PAHs (1), phenothiazine (2), phenazine (3), phenoxazine (4), carbazole (5).

particular, phenothiazines, phenazines, and phenoxazines have been shown to provide a linear growth in polymer MW as a function of monomer conversion, polymers with predictable MWs and low dispersity, while retaining the active chain-end group functionality. PAH catalyzed O-ATRP currently suffers from a lack of predictable polymer MW due to low I^*, and carbazole catalyzed O-ATRP produces polymers with high $Đ$ (>1.50).

13.2.2.2 PAHs

Perylene, a polyaromatic hydrocarbon (PAH) and the simplest rylene dye, has potential as an organic photocatalyst for O-ATRP built upon its well-characterized excited state properties,[44] and use as a photoinitiator for other polymerization mechanisms.[45–47] Specifically, perylene possesses a $E^0(^2PC^{\bullet +}/^1PC^*) = -1.87$ V *vs.* SCE from the S_1 state, and $E^0(^2PC^{\bullet +}/^3PC^*) = -0.58$ V *vs.* SCE from the T_1 state.[48–51] The first report of perylene as an organic photocatalyst in a radical mediated polymerization showed that it was able to polymerize methacrylates, acrylates, and styrene,[52–54] albeit uncontrolled for acrylates and styrene, through an oxidative quenching cycle, and provided polymers achieving relatively low $Đ$ (~1.30) under irradiation from white light. However, linear growth in polymer MW as a function of monomer conversion under pulsed irradiation conditions was not observed.

Further investigations with the perylene catalyst, such as the use of increased catalyst loading and optimized irradiation conditions, have shown this catalyst system to operate a more controlled O-ATRP in both flow[55] and batch[17] conditions. Under these experimental conditions, both addressing the issue of irradiation intensity as an essential experimental variable, perylene mediated O-ATRP, displaying linear growth in polymer MW as a function of monomer conversion. Additionally, perylene has shown success as a photocatalyst in the production of higher order polymeric architectures *via* an O-ATRP method.[56] Like perylene, other PAHs such as pyrene and anthracene have shown promise as catalysts for UV-light driven O-ATRP, but current methods suffer from low conversions (<30%), relatively high $Đ$ (1.4–1.5), and low I^*,[57] similar to the original perylene report.[48]

13.2.2.3 Phenothiazines

Phenothiazines are another class of photocatalysts wherein initial use was inspired by previous broad application in the field of organic electronics,[58–61] and identified to find success in O-ATRP due to its photophysical characteristics; in particular the strong excited state reduction potential of 10-phenyl phenothiazine $E^0(^2PC^{\bullet+}/^1PC^*) = -1.97$ V *vs.* SCE.[62] Unlike perylene, its λ_{max} falls at 343 nm, well below the visible regime, and its ε is ten times lower at 3200 $M^{-1}\,cm^{-1}$. Due to these photophysical traits, higher energy blue LEDs are used to mediate these polymerizations.

13.2.2.4 Phenazines

The class of 5,10-diaryl dihydrophenazines were the first reported PCs able to operate a visible-light driven O-ATRP[63] and produce polymers with low $Đ$ (<1.10).[64] Phenazines benefit from long-lived excited states,[65] stable radical cations,[66–68] and a two-step synthesis to access a variety of catalyst structures.[69] DFT calculations aided in the discovery of the first phenazines applied in O-ATRP, and through chemical modification these species spanned a range of predicted $E^0(^2PC^{\bullet+}/^3PC^*) = -2.06$ V to -2.36 V *vs.* SCE, with absorption profiles tailing into the visible range, enabling polymerization irradiation with white LEDs. Early polymerization data indicated the phenazine PCs operating the most controlled O-ATRPs (as evidenced by linear growth in polymer MW as a function of monomer conversion) had electron-withdrawing groups as *N*-aryl substituents.

Computational investigation compared the excited state singly occupied molecular orbitals of phenazine PCs containing *N*-aryl substituents with electron-withdrawing groups *vs.* electron-donating groups on which revealed that the former class accesses triplet excited states of charge transfer character, where electron density in T_1 is localized to one *N*-aryl substituent. The CT nature of these excited states are analogous to the metal to ligand CT observed in commonly used iridium and ruthenium PCs,[70] and are proposed to benefit photoredox catalysis (see Section 13.1.4). Using this feature as a potential design principle, two additional phenazine PCs containing 1- and 2-naphthyl *N*-aryl substituents were synthesized, and the 1-naphthyl derivative polymerized MMA to relatively low PDI (<1.30) while be able to achieve quantitative I^*.

Due to the nature of the CT excited states, phenazines accessing a CT state exhibit solvatochromism in their emission, where the energy of emission varies with the polarity of the solvent.[71–73] This solvatochromic effect was observed through emission across the entire visible spectrum, with PCs emitting blue in low polarity solution, and red in high polarity.[74] Speculating the importance of this unique property on performance in O-ATRP, a collection of phenazines computationally predicted to access charge transfer excited states were used in the polymerization of MMA in solvents of varying polarity, from hexanes to DMF, and compared to PCs predicted to access excited states of local excitation (LE) character.[43] It was found that the CT phenazines were capable of operating a highly controlled O-ATRP in a wide range of solvent polarities, whereas the LE phenazines only performed reasonably well in the high polarity solvents.

13.2.2.5 Phenoxazines

Phenoxazines are a class of molecules that are isoelectronic to phenothiazines, wherein the sulfur atom is replaced with an oxygen atom. Similar to the phenazines, their photophysical properties and charge transfer characteristics have been well-characterized, and thus this class of molecules has found use in a variety of fields, including biological imaging[75] and OLEDs.[76,77] A series of highly reducing *N*-aryl phenoxazines ($E^0(^2PC^{\bullet+}/^3PC^*)$ of -1.9 V to -2.1 V *vs.* SCE) were shown to operate a controlled O-ATRP, but necessitated the use of UV light (365 nm).[78]

To improve this catalytic system, the phenoxazine PC can be chemically modified at the 3 and 7 positions *via* a two-step synthesis, which shifted the absorption profile into the visible range. Specifically, addition of biphenyl substituents to the 3 and 7 positions of the molecule shifted the λ_{max} from ~320 nm for the original PCs, to ~390 nm, and increased the molar absorptivity from $\varepsilon = 7848\ M^{-1}\,cm^{-1}$ to $\varepsilon = 26\,635\ M^{-1}\,cm^{-1}$, providing substantial tailing into the visible region.[52] More recently, this phenoxazine PC was found to have extremely long-lived triplet excited states, with $\tau = 480 \pm 50\ \mu s$, orders of magnitude longer than their iridium and ruthenium counterparts.[39]

13.2.2.6 Carbazoles

Carbazoles currently represent the least explored family of photocatalysts in CRPs, although the initial report is promising.[79] It is well known that carbazole-containing molecules, such as the carbazolyl dicyanobenzene donor–acceptor complex, can access CT excited states, and have thus have been exploited in organic electronic applications.[80]

The carbazole catalyzed O-ATRP of MMA displays characteristics of an O-CRP, such as linear growth in polymer MW as a function of monomer conversion, and the synthesis of block polymers. However, the *Đ* is high (>1.50), possibly due to the presence of side reactivity from the catalyst. Impressively, the carbazole catalyst operates at ppm concentration, offering a method for precision polymer synthesis using an exceptionally low amount of catalyst.

13.2.2.7 Summary

Overall, O-ATRP has significantly evolved since the first examples of perylene and 10-phenyl phenothiazine in 2014. Through mechanistic investigation, structure-property relationships have guided the synthesis of new PCs in the phenothiazine, phenazine, and phenoxazine classes,[81] and opened doors to their applications in small molecule catalysis[82,83] and the production of higher order polymeric architectures.[84] Further efforts in this field aim to understand how subtle factors such as the degree of charge transfer in the excited state and structural re-organization during electron transfer affect the overall performance in O-ATRP. Additionally, leveraging existing mechanistic understanding, new families of potential PCs continue to be developed.[85]

13.3 Organocatalyzed Reversible Addition–Fragmentation Chain-transfer Polymerization

13.3.1 Mechanism

The mechanism of organocatalyzed RAFT is proposed to consist of the same fundamental CRP steps, but with the added component of a degenerative chain-transfer mechanism (Figure 13.10). For this section, mechanistic processes are discussed if the activity of a catalyst is used to exert control over the polymerization, although it should be noted that the chain-transfer agent in a degenerative transfer pathway can mediate a controlled process on its own. First, a PC absorbs a photon and accesses PC*. Second, electron transfer to an initiator causes homolytic cleavage and production of a radical to undergo propagation. This step simultaneously produces the oxidized form of the PC and a proposed ion pair with the anion fragment of the iniferter. Differing RAFT from O-ATRP, at this stage the propagating polymer chain can undergo a degenerative chain-transfer process with a second iniferter molecule, thus controlling the polymerization. Third, the oxidized PC can deactivate the growing polymer chain by back electron transfer, thus completing the reversible-deactivation cycle.

A number of organic dyes have been successfully used in the RAFT polymerization of various methacrylates, demonstrating the characteristics of a highly controlled CRP.[86] Interestingly, the use of Eosin Y as the catalyst was explored in both oxidative and reductive quenching cycles, the latter of which employed triethylamine as an electron source and allowed for an efficient polymerization to proceed in the presence of oxygen. Eosin Y was also shown to mediate the RAFT polymerization of vinyl ketone monomers, exhibiting the same tolerance to oxygen as for the methacrylates.[87]

In 2015, 10-phenyl phenothiazine was shown to operate a highly controlled RAFT polymerization of acrylates and acrylamides, producing polymer with $Đ < 1.20$.[88] This methodology displayed the characteristics of a CRP, with confirmed chain-end group functionality *via* MALDI-TOF, linear growth in polymer MW as a function of monomer conversion, and displaying the traditional on/off plot in a pulsed irradiation experiment. This polymerization system has also found use in advanced materials applications, including the functionalization of proteins *via* a grafting from strategy,[89] and as a method to enable "living additive manufacturing".[90] Catalytic systems for RAFT are continuously being developed, such as

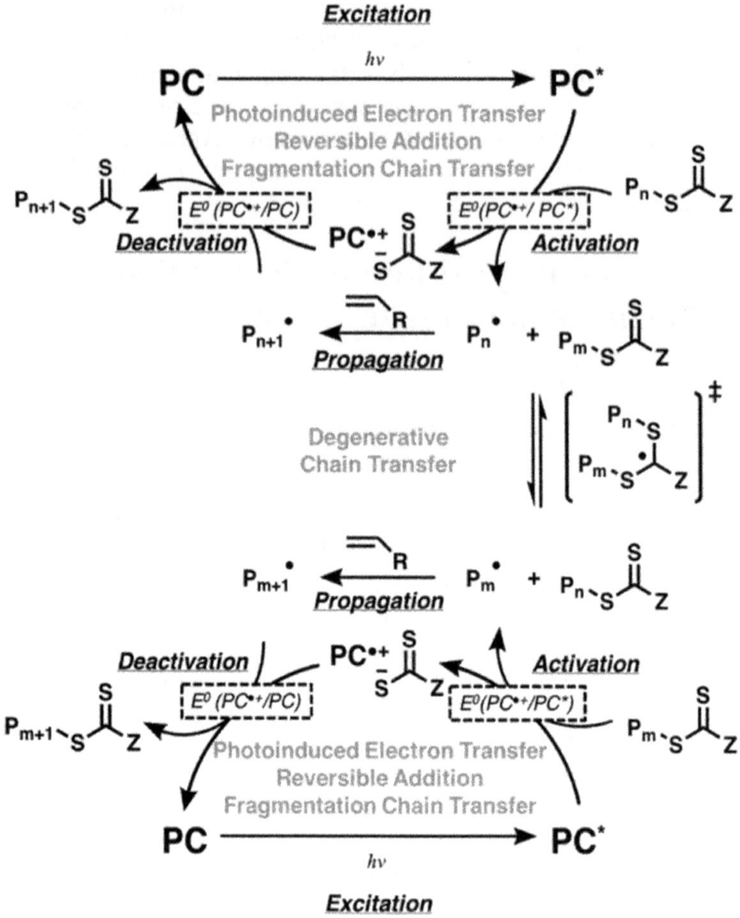

Figure 13.10 Proposed mechanism of organocatalyzed RAFT polymerization where: (1) *excitation* generates PC*; (2) *activation* involves electron transfer from PC* with reduction potential $E^0(PC^{\bullet+}/PC^*)$ to initiator or dormant polymer chain-end group generating active radical and $PC^{\bullet+}/X^-$; (3) *propagation* for polymer MW growth *or* degenerative chain transfer; and (4) *deactivation* by $^2PC^{\bullet+}/X^-$ with oxidation potential $E^0(PC^{\bullet+}/PC)$, returning the polymer chain-end group to a dormant state and regenerating PC.

organic electron donor–acceptor photoredox catalysts, where the organic catalyst and iniferter are covalently bound.[91]

13.4 Reversible Complexation Mediated Radical Polymerization

Iodine transfer polymerization (ITP) is a CRP that derives its control from an iodo-compound mediated degenerative transfer process.[92] The first example

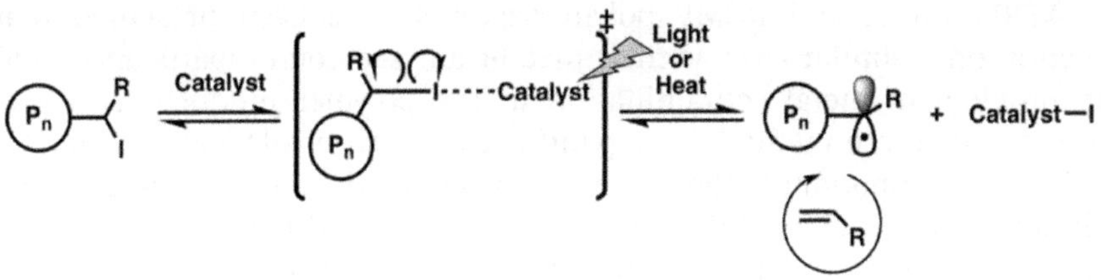

Figure 13.11 Reaction scheme for the proposed mechanism of reversible complexation mediated radical polymerization (RCMP) driven by light or thermal energy.

of this polymerization used reactivity between a conventional radical initiator and an iodofluoro-compound to enter the DT process, although the mechanism of control was not enabled by the activity of a catalyst species. It was then found that an organic catalyst, *N*-iodosuccinimide, could increase the efficiency of the DT process, and this polymerization was thus named reversible chain-transfer catalyzed polymerization (RTCP).[93] Even with this advancement, primary control over the polymerization was not due solely to activity of the catalyst, as it functioned as a DT promoter.

Following the use of *N*-iodosuccinimide as the first organic catalyst to mediate an ITP, use of amine catalysts enabled the polymerization to derive primary control from activity of the catalyst, and was termed reversible complexation mediated radical polymerization (RCMP) (Figure 13.11).[94,95] A photo-mediated version of RCMP has also been presented, which relies on a visible-light absorbing complex between the iodo-compound and an organic chromophore.[96]

This methodology is similar to ATRP and RAFT in the way that a catalyst mediates the reversible-deactivation equilibrium, but differs in the nature of the controlling mechanism. Uniquely, this mechanism does not require a photocatalyst to homolytically break a bond *via* a formal redox process, but the formation of a complex between the catalyst and iodide chain-end group. In the amine-catalyzed systems, C–I bond cleavage is caused by the amine abstracting an iodine from the polymer (polymer-I), generating the radical for propagation, and a complex between the amine and iodine radical (simplified to catalyst-I). Specifically, it is proposed that two of the iodine radicals combine to generate an I_2/amine complex,[97] which acts as the deactivator for the active radical. This polymerization can also operate using various organic iodide salts[98] and pseudo-halide anion catalysts such as azide, thiocyanate, and cyanate. Interestingly, in the azide case the polymer-I bond cleavage is proposed to operate specifically *via* halogen bonding.[99]

13.5 Future Outlook

Although a rather new family of methodologies, the development of organocatalyzed CRPs is an exciting field of research that has provided more sustainable methods of precision polymer synthesis. New catalytic systems for ATRP, RAFT, and RCMP polymerizations have been presented that function on a similar level as their metal-catalyzed counterparts and avoid the problem of metal contamination in the polymer product. As a prerequisite for these methods to expand their utility, challenges remain. Arguably most important is the gaining of mechanistic understanding, which will allow for the rational design of new catalysts and polymerization optimizations. The use of light as a driving force for the reaction is simultaneously an additional handle for polymerization optimization and capabilities while also presenting an added complication. This research will facilitate the expansion of available monomer scope to other less-activated monomers, and the use of these methods in applications that can benefit from metal-free preparation.

Specifically, expansion of the O-ATRP monomer scope to acrylates and vinyl acetate is an existing challenge, and one that will likely be solved through increased understanding of catalyst properties necessary to control the polymerization of highly reactive monomers. The use of O-ATRP in the production of higher order polymeric architectures is underway and will expand from linear block polymers and star polymers to more complex structures. The development of catalysts for RAFT that allow for a wider scope of chain-transfer reagents will simultaneously increase the accessibility of this method and enable further application of this method in biological applications.[100] It is expected that the fundamental properties of CRPs will continue to be benefited by under-explored intermolecular interactions, such as the use of halogen bonding in RCMP, to develop the next generation of methodologies.

References

1. K. Matyjaszewski and T. P. Davis, *Handbook of Radical Polymerization*, Wiley-Interscience, Hoboken, 2002.
2. J. P. Fouassier, X. Allonas and D. Burget, *Prog. Org. Coat.*, 2003, **47**, 16–36.
3. L. G. Lovell, H. Lu, J. E. Elliott, J. W. Stansbury and C. N. Bowman, *Dent. Mater.*, 2001, **17**, 504–511.
4. J.-F. Fouassier and J. Lalevee, *Photoinitiators for Polymer Synthesis*, Wiley-VCH, Weinheim, Germany, 2012.
5. Y. Yagci, S. Jockusch and N. J. Turro, *Macromolecules*, 2010, **43**, 6245–6260.
6. A. Goto and T. Fukuda, *Prog. Polym. Sci.*, 2004, **29**, 329–385.
7. W. A. Braunecker and K. Matyjaszewski, *Prog. Polym. Sci.*, 2007, **32**, 93–146.
8. B. M. Rosen and V. Percec, *Chem. Rev.*, 2009, **109**, 5069–5119.
9. M. Destarac, D. Charmot, X. Franck and S. Z. Zard, *Macromol. Rapid Commun.*, 2000, **21**, 1035–1039.
10. K. Min, H. Gao and K. Matyjaszewski, *J. Am. Chem. Soc.*, 2005, **127**, 3825–3830.

11. K. Matyjaszewski, W. Jakubowski, K. Min, W. Tang, J. Huang, W. A. Braunecker and N. V. Tsarevsky, *Proc. Natl. Acad. Sci. U. S. A.*, 2006, **103**, 15309–15314.
12. W. Jakubowski, K. Min and K. Matyjaszewski, *Macromolecules*, 2006, **39**, 39–45.
13. M. Chen, M. Zhong and J. J. Johnson, *Chem. Rev.*, 2016, **116**, 10167–10211.
14. A. J. D. Magenau, N. C. Strandwitz, A. Gennaro and K. Matyjaszewski, *Science*, 2011, **332**, 81–84.
15. R. P. Quirk and B. Lee, *Polym. Int.*, 1992, **27**, 359–367.
16. K. Matyjaszewski and N. V. Tsarevsky, *J. Am. Chem. Soc.*, 2014, **136**, 6513–6533.
17. D. H. Solomon, E. Rizzardo and P. Cacioli, *U. S. Pat.*, 4,581,429, CSIRO, 1986.
18. C. J. Hawker, A. W. Bosman and E. Harth, *Chem. Rev.*, 2001, **101**, 3661–3688.
19. J. Nicolas, Y. Guillaneuf, C. Lefay, D. Bertin, D. Gigmes and B. Charleux, *Prog. Polym. Sci.*, 2013, **38**, 63–235.
20. J. Chiefari, Y. K. Chong, F. Ercole, J. Krstina, J. Jeffery, T. P. T. Le, R. T. A. Mayadunne, G. F. Meijs, C. L. Moad, G. Moad, E. Rizzardo and S. H. Thang, *Macromolecules*, 1998, **31**, 5559–5562.
21. G. Moad, J. Chiefari, Y. K. Chong, J. Krstina, R. T. A. Mayadunne, A. Postma, E. Rizzardo and S. H. Thang, *Polym. Int.*, 2000, **49**, 993–1001.
22. S. Perrier and P. Takolpuckdee, *J. Polym. Sci., Part A: Polym. Chem.*, 2005, **43**, 5347–5393.
23. M. Kato, M. Kamigaito, M. Sawamoto and T. Higashimura, *Macromolecules*, 1995, **28**, 1721–1723.
24. J. S. Wang and K. Matyjaszewski, *J. Am. Chem. Soc.*, 1995, **117**, 5614–5615.
25. K. Matyjaszewski and J. Xia, *Chem. Rev.*, 2001, **101**, 2921–2990.
26. K. Matyjaszewski and N. V. Tsarevsky, *Nat. Chem.*, 2009, **1**, 276–288.
27. R. B. Grubbs, *Polym. Rev.*, 2011, **51**, 104–137.
28. N. V. Tsarevsky and K. Matyjaszewski, *Chem. Rev.*, 2007, **107**, 2270–2299.
29. D. J. Keddie, G. Moad, E. Rizzardo and S. H. Thang, *Macromolecules*, 2012, **45**, 5321–5342.
30. C.-J. Wallentin, J. D. Nguyen, P. Finkbeiner and C. R. J. Stephenson, *J. Am. Chem. Soc.*, 2012, **134**, 8875–8884.
31. F. A. Liebfarth, K. M. Mattson, B. P. Fors, H. A. Collins and C. J. Hawker, *Angew. Chem., Int. Ed.*, 2013, **52**, 199–210.
32. B. P. Fors, J. E. Poelma, M. S. Menyo, M. J. Robb, D. M. Spokoyny, J. W. Kramer, J. H. Waite and C. J. Hawker, *J. Am. Chem. Soc.*, 2013, **135**, 14106.
33. *Handbook of Synthetic Photochemistry*, ed. A. Albini and M. Fagnoni, Wiley-VCH, Weinhein, 2010.
34. S. Nath, H. Pal, D. K. Palit, A. V. Sapre and J. P. Mittal, *J. Phys. Chem. A*, 1998, **102**, 5822.

35. N. J. Turro, *Modern Molecular Photochemistry*, University Science Books, 1991.
36. G. J. Kavarnos and N. J. Turro, *Chem. Rev.*, 1986, **86**, 401–449.
37. C. Kutahya, F. S. Aykac, G. Yilmaz and Y. Yagci, *Polym. Chem.*, 2016, **7**, 6094–6098.
38. K. Matyjaszewski, T. E. Patten and J. Xia, *J. Am. Chem. Soc.*, 1997, **119**, 674–680.
39. R. A. Marcus, *Angew. Chem., Int. Ed. Engl.*, 1993, **32**, 1111–1121.
40. D. M. Arias-Rotondo and J. K. McCusker, *Chem. Soc. Rev.*, 2016, **45**, 5803–5820.
41. G. J. Kavarnos and N. J. Turro, *Chem. Rev.*, 1986, **86**, 401–449.
42. X. Pan, M. Lamson, J. Yan and K. Matyjaszewski, *ACS Macro Lett.*, 2015, **4**, 192–196.
43. M. D. Ryan, R. M. Pearson, T. A. French and G. M. Miyake, *Macromolecules*, 2017, **50**, 4616–4622.
44. F. Wurthner, *Chem. Commun.*, 2004, 1564–1579.
45. P. Xiao, F. Dumur, M. Frigoli, B. Graff, F. Morlet-Savary, G. Wantz, H. Bock, J. P. Fouassier, D. Gigmes and J. Lalevee, *Eur. Polym. J.*, 2014, **53**, 215–222.
46. P. Xiao, F. Dumur, B. Graff, D. Gigmes, J. P. Fouassier and J. Lalevee, *Macromol. Rapid Commun.*, 2013, **34**, 1452–1458.
47. M.-A. Tehfe, F. Dumur, B. Graff, D. Gigmes, J. P. Fouassier and J. Lalevee, *Macromol. Chem. Phys.*, 2013, **214**, 1052–1060.
48. J. C. Bachman, R. Kavian, D. J. Graham, D. Y. Kim, S. Noda, D. G. Nocera, Y. Shao-Horn and S. W. Lee, *Nat. Commun.*, 2015, **6**, 7040.
49. T. N. Singh-Ratchford and F. N. Castellano, *J. Phys. Chem. Lett.*, 2010, **1**, 195.
50. M. Parac and S. Grimme, *Chem. Phys.*, 2003, **292**, 11.
51. K. Kikuchi, T. Niwa, Y. Takahashi, H. Ikeda and T. Miyashi, *J. Phys. Chem.*, 1993, **97**, 5070.
52. G. M. Miyake and J. C. Theriot, *Macromolecules*, 2014, **47**, 8255–8261.
53. G. M. Miyake, J. C. Theriot, M. D. Ryan, R. M. Pearson, T. A. French, A. Lockwood, C. Musgrave and C.-H. Lim, *U. S. Pat. Appl.*, 62/245,804, 2015.
54. J. C. Theriot, M. D. Ryan, T. A. French, R. M. Pearson and G. M. Miyake, *J. Visualized Exp.*, 2016, **110**, e53571.
55. B. L. Ramsey, R. M. Pearson, L. R. Beck and G. M. Miyake, *Macromolecules*, 2017, **50**, 2668–2674.
56. C. Aydogan, G. Yilmaz and Y. Yagci, *Macromolecules*, 2017, DOI: 10.1021/acs.macromol.7b02240.
57. A. Allushi, S. Jockusch, G. Yilmaz and Y. Yagci, *Macromolecules*, 2016, **49**, 7785.
58. H. N. Tian, X. C. Yang, R. K. Chen, Y. Z. Pan, L. Li, A. Hagfeldt and L. C. Sun, *Chem. Commun.*, 2007, 3741.

59. W. J. Wu, J. B. Yang, J. L. Hua, J. Tang, L. Zhang, Y. T. Long and H. Tian, *J. Mater. Chem.*, 2010, **20**, 1772.
60. L. Yao, Y. Pan, X. Tang, Q. Bai, F. Shen, F. Li, P. Lu, B. Yang and Y. J. Ma, *Phys. Chem. C*, 2015, **119**, 17800.
61. T. Godet-Bar, J.-C. Lepretre, O. Le Bacq, J.-Y. Sanchez, A. Deronzier and A. Pasturel, *Phys. Chem. Chem. Phys.*, 2015, **17**, 25283.
62. X. Pan, C. Fang, M. Fantin, N. Malhotra, W. Y. So, L. A. Peteanu, A. A. Isse, A. Gennaro, P. Liu and K. Matyjaszewski, *J. Am. Chem. Soc.*, 2016, **138**, 2411.
63. J. C. Theriot, C.-H. Lim, H. Yang, M. D. Ryan, C. B. Musgrave and G. M. Miyake, *Science*, 2016, **352**, 1082–1086.
64. T. E. Patten, J. Xia, T. Abernathy and K. Matyjaszewski, *Science*, 1996, **272**, 866–868.
65. Y. Du, R. M. Pearson, C.-H. Lim, S. M. Sartor, M. D. Ryan, H. Yang, N. H. Damrauer and G. M. Miyake, *Chem. – Eur. J.*, 2017, **23**, 10962–10968.
66. S. Hiraoka, T. Okamoto, M. Kozaki, D. Shiomi, K. Sato, T. Takui and K. Okada, *J. Am. Chem. Soc.*, 2004, **126**, 58.
67. Y. Masuda, M. Kuratsu, S. Suzuki, M. Kozaki, D. Shiomi, K. Sato, T. Takui and K. Okada, *Polyhedron*, 2009, **28**, 1950.
68. L. L. Gordienko and A. G. Chukhlantseva, *Theor. Exp. Chem.*, 1972, **5**, 616.
69. M. D. Ryan, J. C. Theriot, C.-H. Lim, H. Yang, A. G. Lockwood, N. G. Garrison, S. R. Lincoln, C. B. Musgrave and G. M. Miyake, *J. Polym. Sci., Part A: Polym. Chem.*, 2017, **55**, 3017–3027.
70. C. K. Prier, D. A. Rankic and D. W. C. MacMillan, *Chem. Rev.*, 2013, **113**, 5322–5363.
71. H. A. Frank, J. A. Bautista, J. Josue, Z. Pendon, R. G. Hiller, F. P. Sharples, D. Gosztola and M. R. Wasielewski, *J. Phys. Chem. B*, 2000, **104**, 4569–4577.
72. E. Ragnoni, M. Di Donato, A. Lagatti, A. Lapini and R. Righini, *J. Phys. Chem. B*, 2015, **119**, 420–432.
73. M. L. Horng, K. Dahl, G. Jones and M. Maroncelli, *Chem. Phys. Lett.*, 1999, **315**, 363–370.
74. C.-H. Lim, M. D. Ryan, B. G. McCarthy, J. C. Theriot, S. M. Sartor, N. H. Damrauer, C. B. Musgrave and G. M. Miyake, *J. Am. Chem. Soc.*, 2017, **139**, 348–355.
75. R. Sun, W. Liu, Y. Xu, J. Lu, J. Ge and M. Ihara, *Chem. Commun.*, 2013, **49**, 10709.
76. T. Okamoto, M. Kozaki, M. Doe, M. Uchida, G. Wang and K. Okada, *Chem. Mater.*, 2005, **17**, 5504.
77. Y. Zhu, A. P. Kulkarni, P. Wu and S. A. Jenekhe, *Chem. Mater.*, 2008, **20**, 4200.
78. R. P. Pearson, C.-H. Lim, B. G. McCarthy, C. B. Musgrave and G. M. Miyake, *J. Am. Chem. Soc.*, 2016, **138**, 11399–11407.
79. Z. Huang, Y. Gu, X. Liu, L. Zhang, Z. Cheng and X. Zhu, *Macromol. Rapid Commun.*, 2017, **38**, 600461.

80. H. Uoyama, K. Goushi, K. Shizu, H. Nomura and C. Adachi, *Nature*, 2012, **492**, 234–238.
81. J. C. Theriot, B. G. McCarthy, C.-H. Lim and G. M. Miyake, *Macromol. Rapid Commun.*, 2017, **38**, 1700040.
82. Y. Du, R. M. Pearson, C.-H. Lim, S. M. Sartor, M. D. Ryan, H. Yang, N. H. Damrauer and G. M. Miyake, *Chem. – Eur. J.*, 2017, **23**, 10962–10968.
83. S. O. Poelma, G. L. Burnett, E. H. Discekici, K. M. Mattson, N. J. Treat, Y. Luo, Z. M. Hudson, S. L. Shankel, P. G. Clark, J. W. Kramer, C. J. Hawker and J. Read de Alaniz, *J. Org. Chem.*, 2016, **81**, 7155–7160.
84. B. L. Buss, L. R. Beck and G. M. Miyake, *Polym. Chem.*, 2018, **9**, 1658–1665.
85. C. Kutahya, A. Allushi, R. Isci, J. Kreutzer, T. Ozturk, G. Yilmaz and Y. Yagci, *Macromolecules*, 2017, **50**, 6903–6910.
86. J. Xu, S. Shanmugam, H. T. Duong and C. Boyer, *Polym. Chem.*, 2015, **6**, 5615–5624.
87. I.-H. Lee, E. H. Discekici, A. Anastasaki, J. Read de Alaniz and C. Hawker, *J. Polym. Chem.*, 2017, **8**, 3351–3356.
88. M. Chen, M. J. MacLeod and J. A. Johnon, *ACS Macro Lett.*, 2015, **4**, 566–569.
89. B. S. Tucker, M. L. Coughlin, C. A. Figg and B. S. Sumerlin, *ACS Macro Lett.*, 2017, **6**, 452–457.
90. M. Chen, Y. Gu, A. Singh, M. Zhong, A. M. Jordan, S. Biswas, L. T. J. Korley, A. C. Balazs and J. A. Johnson, *ACS Cent. Sci.*, 2017, **3**, 124–134.
91. J. Xu, S. Shanmugam and C. Boyer, *ACS Macro Lett.*, 2015, **4**, 926–932.
92. M. Tatemoto, Y. Yutani, K. Fujiwara, *Eur. Pat.*, 272698, 1988.
93. A. Goto, N. Hirai, T. Wakada, K. Nagasawa, Y. Tsujii and T. Fukuda, *Macromolecules*, 2008, **41**, 6261–6264.
94. A. Goto, T. Suzuki, H. Ohfuji, M. Tanishima, T. Fukuda, Y. Tsujii and H. Kaji, *Macromolecules*, 2011, **44**, 8709–8715.
95. A. Goto, Y. Tsujii and H. Kaji, in *Fundamentals of Controlled/Living Radical Polymerization*, ed. N. V. Tsarevsky and B. S. Sumerlin, Royal Society of Chemistry, UK, 2013, pp. 250–286.
96. A. Ohtsuki, L. Lei, M. Tanishima, A. Goto and H. Kaji, *J. Am. Chem. Soc.*, 2015, **137**, 5610–5617.
97. P. Lacroix-Desmazes, J. Tonnar and B. Boutevin, *Macromol. Symp.*, 2007, **248**, 150–157.
98. A. Goto, A. Ohtsuki, H. Ohfuji, M. Tanishima and H. Kaji, *J. Am. Chem. Soc.*, 2013, **135**, 11131–11139.
99. C.-G. Wang and A. Goto, *J. Am. Chem. Soc.*, 2017, **139**, 10551–10560.
100. C. Boyer, V. Bulmus, T. P. Davis, V. Ladmiral, J. Liu and S. Perrier, *Chem. Rev.*, 2009, **109**, 5402–5436.

CHAPTER 14

Organocatalysis for Depolymerisation

CORALIE JEHANNO,[a,b] JEREMY DEMARTEAU[b] AND ANDREW P. DOVE*[a]

[a] School of Chemistry, University of Birmingham, Edgbaston, Birmingham, UK B15 2TT; [b] POLYMAT, University of the Basque Country UPV/EHU, Joxe Mari Korta Center, Avda. Tolosa 72, 20018 Donostia-San Sebastián, Spain
*Email: a.dove@bham.ac.uk

14.1 Introduction

While the major focus of catalyst design and advancement in the field of polymer science to date has been in the construction of polymers from a range of different monomers, as a consequence of the societal and environmental impact they have made, significant focus is now being directed at how catalysts can be used and developed to depolymerise polymers either back to monomers or to other useful small molecules or materials. While current polymer recycling technology mainly focusses on either pyrolysis into combustible fuels or "downcycling" of the resulting product by secondary/mechanical recycling, this leads to a loss of the materials from the value chain and can, in the case of mechanical recycling, entail a loss of strength, contamination, discolouration or decrease in polymer properties that affects the usability of the end products.[1]

As a consequence of their high stability (as designed to be suitable for their application), forcing conditions are usually required to transform polymers into small molecules with high yield and purity. Thus, different techniques have been applied such as microwave assistance,[2–4] supercritical

Polymer Chemistry Series No. 31
Organic Catalysis for Polymerisation
Edited by Andrew Dove, Haritz Sardon and Stefan Naumann

Published by the Royal Society of Chemistry, www.rsc.org

conditions[5,6] or catalysis[7–9] to enhance the efficiency of depolymerisation reactions. In particular, the application of the known catalytic tools of organic chemistry to the depolymerisation processes using well-known stable and highly active organometallic catalysts, such as zinc or lead acetates, sodium/potassium sulphate or titanium phosphate, have been largely studied. Despite their advantages, metal-based catalysts have several drawbacks that include: challenging separation from the crude product, which leads to a low-quality final material; poor selectivity in the degradation process leads to a mixture of oligomers that are difficult to re-process. Organic catalysts have therefore a key role to play, as with their application for polymerisation methodologies, their lower environmental and ecological impact than many metal-based alternatives make them highly attractive. Mechanistic studies and experimental data have demonstrated the importance of the hydrogen-bond network in the control of activity, selectivity and architecture of the resulting polymers.[10,11] Applied to polymer degradation and, in particular, to depolymerisation involving transesterification, mechanisms promoted by organocatalysts may lead to the formation of highly pure small molecules that are in turn suitable for subsequent polymerisations. The aim of this chapter is to cover the contribution of organocatalysts to the depolymerisation field, outlining the advances for organic base, organic acid and ionic liquid catalysts as well as comparing them with classical organometallic catalysts for these processes.

14.2 Recycling of Commodity Polymers

An important technological challenge involves the design of suitable pathways for the degradation of commonly used commodity polymers, mainly polycarbonates, polyesters and polyurethanes. These materials are extensively used across a range of sectors that include packaging, building, automotive and electronics with the majority of those produced never being recycled. Several catalytic approaches have been investigated for the depolymerisation of these materials.

14.2.1 Organic Bases

Organic bases are efficient catalysts for a large variety of base-mediated transformations both in organic and polymer synthesis. They are a powerful tool that has found particular use in a range of transesterification reactions.[11–13] The so-called "superbases", such as amidines, guanidines or phosphazenes have been found to be extremely active catalysts for the ring-opening polymerisation (ROP) of cyclic esters,[14–17] and in particular lactide.[18,19] As therefore may be expected, their ability to catalyse depolymerisation *via* similar "general base" mechanisms have driven interest in depolymerisation as it demands precise scission of the polymer backbone to obtain unique building blocks.

In 2011 Hedrick and co-workers reported the glycolysis of PET using a guanidine base, 1,5,7-triazabiciclo [4,4,0] dec-ene (TBD, Scheme 14.1).[20] In a large excess of ethylene glycol (EG) (16 eq.), at 190 °C, pellets of waste PET beverage bottles were degraded in 3.5 h. The major product (78% after crystallisation from water) of this reaction was bis(2-hydroxyethyl) terephthalate (BHET), a convenient monomer for a subsequent polymerisation back to PET (Scheme 14.2-I). Insoluble impurities were identified as short oligomers of PET and additives to the polymer (isophtalic acid, diethylene glycol, and cyclohexane dimethanol). Additional experiments using coloured PET bottles led to slower glycolysis with a lower BHET yield (64%). These results are comparable to ones, encountered in the literature, obtained with usual organometallic catalysts employed for this reaction such as acetate (zinc, manganese).[21,22] A complementary density functional theory (DFT) computational study has demonstrated that both TBD and EG played a role in the depolymerisation mechanism, activating transesterification *via* H-bonding.[23]

In a subsequent work, the efficiency of a range of other nitrogen bases was investigated to establish a correlation between their basicity (pK_a) and catalytic activity.[24] Glycolysis appeared to be more efficient (more rapid with lower undesirable oligomers content) in the case of strong bases such as TBD, 1,8-diazabicyclo[5.4.0]undec-7-ene (DBU) or 1,5-diazabicyclo [4.3.0]non-5-ene (DBN) compared to bases with a lower pK_a such as 1-methylimidazole (NMI) or dimethylaniline (DMA) (Scheme 14.1). Surprisingly, DBU demonstrated a higher efficiency than TBD for depolymerisation despite its lower pK_a. Using different chain-length alcohols, from EG to 1-octanol (Scheme 14.2-II), differences in reactivity were found. For alcohols having a 4-carbon or longer chains, TBD exhibited higher catalytic activity than DBU. Computational investigation revealed that, for longer chains, the acid/base character of TBD provides a simultaneous activation of the carbonyl group of the (poly)ester and the nucleophilic group of the alcohol, leading to a faster reaction. By contrast, in the case of short-chain nucleophilic reagents, a large excess of the alcohol leads to the activation of the carbonyl of the polymer by the hydroxyl group, increasing the reaction rate, thus annihilating the advantage of TBD bifunctionality in this case.

The ability of TBD to catalyse the aminolytic depolymerisation of waste PET was also investigated to produce a large range of crystalline terephthalamide compounds (Scheme 14.2-III).[25] The thermal and mechanical

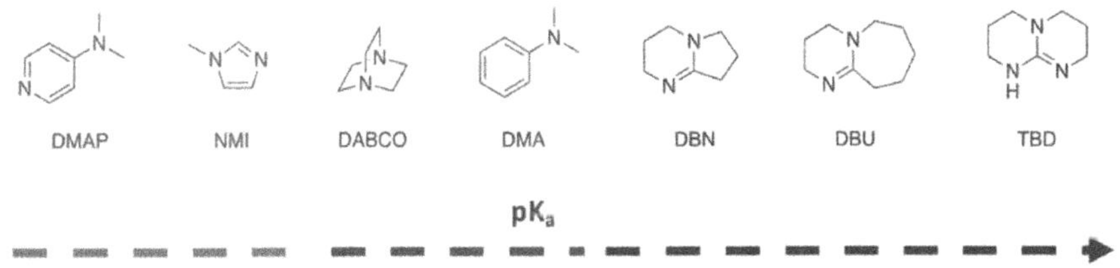

Scheme 14.1 Organic bases described in this chapter, arranged according to their basicity.

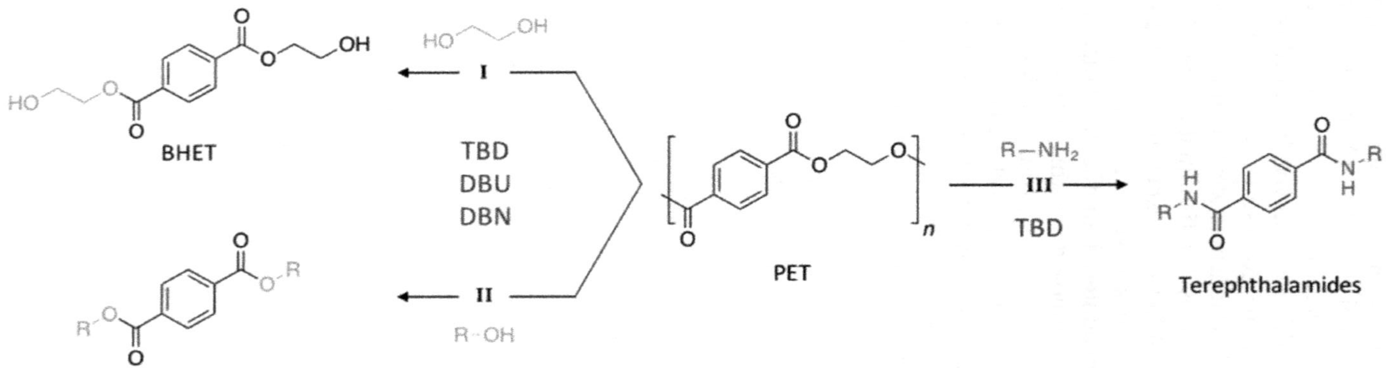

Scheme 14.2 Depolymerisation of PET through glycolysis (I), alcoholysis (II) and aminolysis (III).

properties were dependant on the amine used as the reagent and provided building blocks for high added-value materials. Typical depolymerisations were performed at temperatures between 110 and 190 °C for 1 to 18 h, synthesising terephthalamides in reasonable yields. Once again, computational support studies suggested that the bifunctionality of TBD plays an important role in the mechanism, especially in the activation of the carbonyl group *via* H-bonding, which makes it more efficient than other organic bases for aminolysis.

More recently, organic bases as catalysts have gained attention for the depolymerisation of other polymers. Indeed, Leibfarth *et al.* reported the efficiency of TBD to degrade polylactic acid (PLA), at room temperature, to obtain valuable building blocks.[26] Similarly to PET depolymerisation with a range of alcohols, the degradation of PLA led to a mixture of products, in this case lactate esters and their dimers. However, with ethanol as reagent and solvent, using 2.5 mol% of catalyst, the complete degradation of PLA reached to the recovery of more than 95% of ethyl lactate (Scheme 14.3a). Additionally, the depolymerisation of poly(L-lactide) with benzyl alcohol was shown to retain the stereochemistry of the product with chiral HPLC analysis showing that >95% of the product was *S*-benzyl lactate. Moreover, the tolerance of the reaction to the incorporation of various polymerisable groups to the ester products presented the opportunity to produce new polymers by step growth methods (Scheme 14.3b). The same process was also applied to polyglycolide (PG).

The depolymerisation of bisphenol A-based polycarbonate (BPA-PC) has been also investigated using organic bases as catalysts. Using DBU (10 mol%) at 100 °C in an excess of ethanol or methanol, BPA-PC was degraded within 30 min to provide a reasonable yield of bisphenol A (BPA) and the respective organic carbonates (Scheme 14.4a).[27] Reutilisation of the catalyst was explored, *via* subsequent addition of BPA-PC *in situ*, demonstrating the ability of DBU to catalyse the degradation for several cycles. Notably, however, the reaction time increased with successive loads of fresh polymer, from 30 min for the first feed to 4 h for the fifth one. Further investigation into the nature of the catalytic species revealed that a DBU-BPA adduct was formed in the crude reaction product. While still catalytically active, it is less active than DBU itself, hence explaining the sequential loss in activity. The investigation of other bases has shown that weaker bases, 1,4-diazabicyclo[2.2.2]octane (DABCO, Scheme 14.1) and 4-dimethylaminopyridine (DMAP, Scheme 14.1) are less active than DBU requiring 4 to 6× longer reaction times to reach equilibrium. These observations support the tendency previously reported that catalytic performances of organic bases for transesterification increases with their basicity.

In another very recent publication, the use of TBD as catalyst was reported for the methanolysis of BPA-PC into BPA and dimethyl carbonate (DMC).[28] Investigations into solvents and catalysts have demonstrated that the best result was obtained using DMC as solvent, at 75 °C, with 2 mol% of TBD and 10 eq. of methanol. The use of DMC, one of the products of the

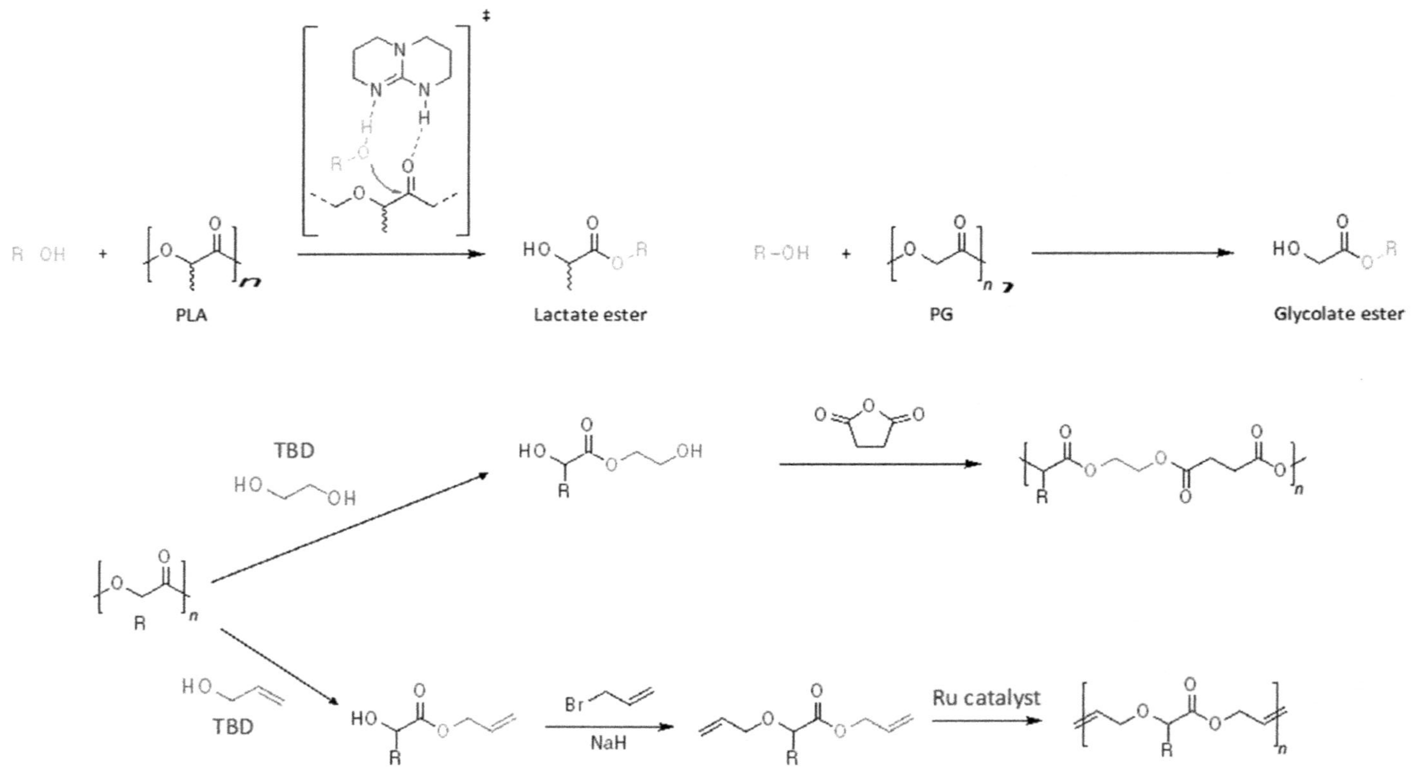

Scheme 14.3 (Top) Depolymerisation of PLA and transition state proposed for the TBD-mediated reaction and (bottom) depolymerisation and subsequent polymerisation of PG and PLA for the synthesis of new materials.

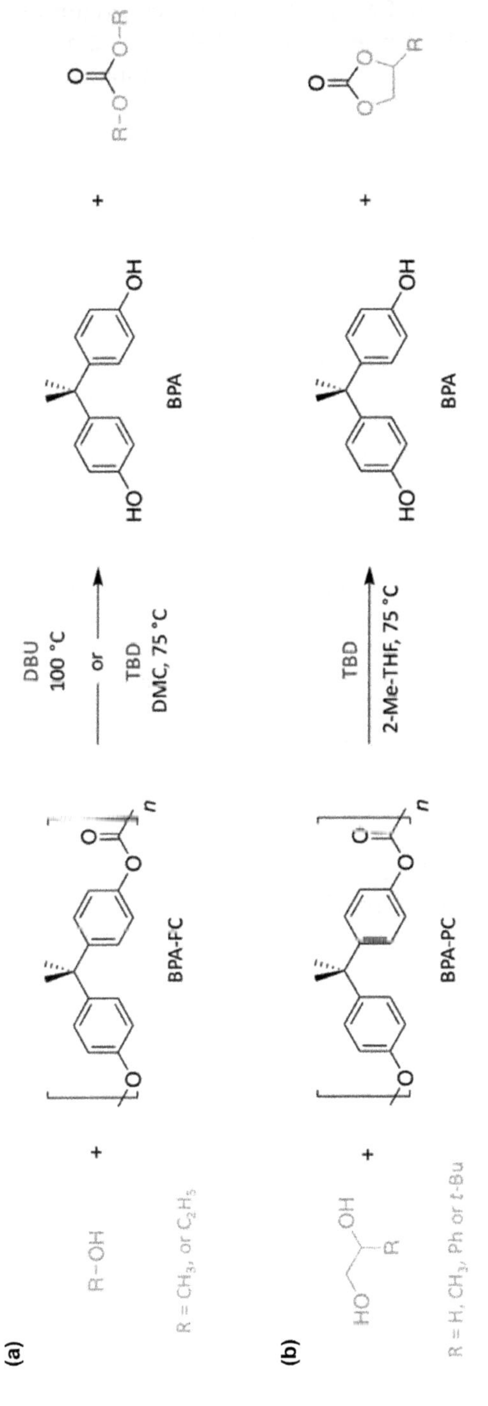

Scheme 14.4 Organocatalysed BPA-PC depolymerisation using (a) alcohols as reagent, (b) diols as reagent.

depolymerisation, as solvent for the reaction enables an easier and faster separation, which simplifies the overall process. Additionally, the authors explored the depolymerisation of BPA-PC in 2-methyltetrahydrofuran (2-Me-THF), using small diols, to obtain 5-membered cyclic carbonates in reasonable yields (Scheme 14.4b).

14.2.2 Organic Acids

Organic bases as catalyst reactions have been widely investigated for both polymerisation and depolymerisation. By contrast, although organic acids are attractive building blocks for polymerisation, few examples of organic acids as catalyst are reported in the literature. Studies involving organic acids for catalysing depolymerisation are also rare but oxidative methods involving acidic catalysts have been reported to be very efficient for the depolymerisation of polyolefins.[29–31] In the first report of using acid catalysts for polymer degradation, typical aqueous acidic solutions were used as catalyst for the degradation of discarded polyamide-6 (PA-6) waste fibres ($M_n = 12$ kg mol^{-1}).[32] After dissolution in concentrated solution of formic acid, degradation products of different molecular weights were recovered through fractional precipitation, raising a minimal M_n of 580 g mol^{-1} in 20 h. Interestingly, with hydrochloric acid and sulfuric acid, aminocaproic acid was identified as the major component (94% and 78%, respectively) and was isolated in high purity, eventually also allowing for subsequent polymerisation.

More recently, Kamimura and coworkers reported the degradation of polyamide-66 (PA-66) into valuable chemicals using the combination of supercritical methanol as solvent and organic acids as catalyst.[33] Depolymerisation with 8 eq. of glycolic acid in methanol, at 270 °C for 6 h, yielded 75% of dimethyl adipate and 50% of 1,6-hexanediol. Similar yields were obtained with other acids such as lactic or benzoic acid while use of esters or weaker acids provided comparable yields of dimethyl adipate, but lower yields of 1,6-hexanediol. The authors proposed that the scission of the amide bond was promoted by the acid catalyst to yield dimethyl adipate and hexamethylenediamine before the subsequent degradation of the diamine into various compounds, including 1,6-hexanediol, was promoted by supercritical methanol (Scheme 14.5). This diol being a relevant monomer for subsequent preparations of polyesters or polyurethanes, its synthesis from commercial PA-6 constitutes a step towards economically viable depolymerisation processes.

One of the major challenges for polymer recycling is the degradation of polyolefins. Their C–C backbone is extremely stable and harsh conditions are necessary to afford and control their degradation. Several methods involving pyrolysis, supercritical conditions or organometallic catalysts have already been employed to degrade polyolefins into light alkanes as well as different oxidative methods. In particular, the depolymerisation of polyisoprene-based materials, made more accessible by the reactivity of the C=C double bond that remains in the polymer structure, has been investigated with the use of *m*-chlorobenzoic acid,[29] phenylhydrazine[30] or

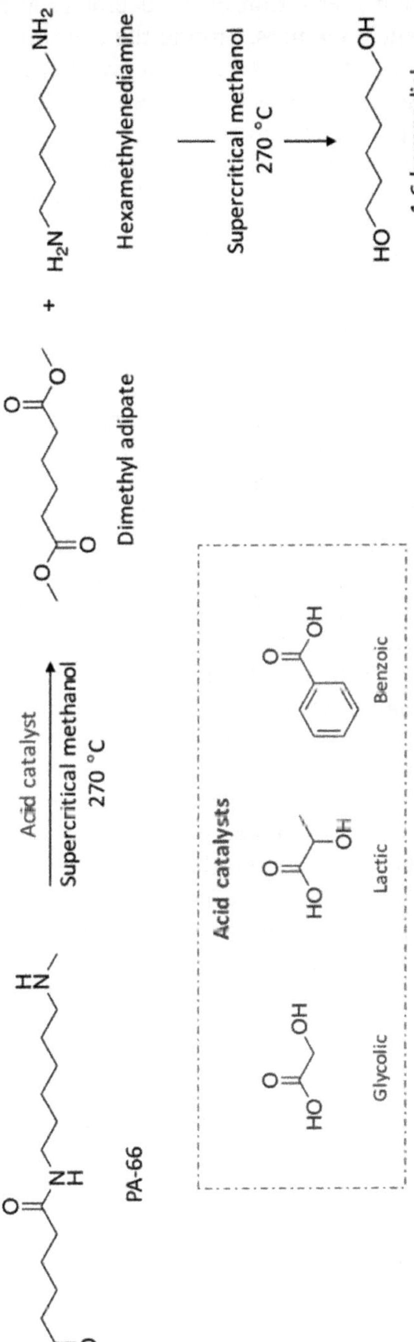

Scheme 14.5 Depolymerisation of PA-66 catalysed by organic acids in supercritical methanol.

potassium persulfate.[31] In all cases, a drastic reduction of the molecular weight was reported, characterisation of the degradation products revealed the presence of telechelic oligomers. Among the acid catalysts explored for the depolymerisation of both natural and synthetic rubber (poly(*cis*-isoprene)), periodic acid is the most widely studied.

In 2005, Phinyocheep *et al.* reported the use of periodic acid (H_5IO_6 or HIO_4, $2H_2O$) as catalyst for the oxidative degradation of epoxidised natural rubber (ENR).[34] Periodic acid is used for its ability to release periodate anions able to cleave 1,2-difunctional compounds. At low temperature (30 °C) after 8 h the molecular weight of ENR was reduced from 15 $kg\,mol^{-1}$ to 0.4 $kg\,mol^{-1}$. Applying the same process to natural rubber (NR) also resulted in degradation of the polymer with the molecular weight decreasing from 20 $kg\,mol^{-1}$ to 1 $kg\,mol^{-1}$ in 30 h. It was postulated that for NR, the first step of the reaction involved the epoxidation of NR into ENR by the periodic acid, which is why the longer reaction times were required (Scheme 14.6).

In a subsequent study, Sadaka *et al.* adapted the method to directly depolymerise ground waste tyres.[36] Experiments confirmed that the oxidative mechanism first involves the epoxidation of the double bond of the polyisoprene before the cleavage of the *in situ*-formed oxirane (Scheme 14.6). Controlling the quantity of periodic acid, low molecular weight telechelic polymers, with aldehyde end-groups, were synthesised. These telechelic oligomers were then used as precursors for subsequent synthesis of other relevant materials for the rubber industry.

The degradation of polyethylene (PE) presents a more significant challenge as it necessitates breaking the more energetically stable C–C single bond. PE is one of the simplest and cheapest polymers to synthesise, rendering its annual production one of the most important ones worldwide. An innovative initiative has been taken very recently by Bäckström *et al.* to organocatalytically depolymerise polyethylene waste. Merging microwave assistance and acidic catalysed oxidation, PE has been shown to degrade into dicarboxylic acids using diluted nitric acid solution in 1 h at 180 °C. With the aim of recycling low density polyethylene (LDPE) freezer bags, diluted nitric acid solution was used to obtain complete degradation. After 3 h, 71% of the degradation product were dicarboxylic acids of different chain-length, from malonic (3 carbons) to pimelic acid (7 carbons).[37] However, the mechanisms of the oxidative reactions are not fully understood and the apparent random oxidations of the polymer chain, leading to side-reactions, complicate subsequent use of the degradation products for the synthesis of new materials.

14.2.3 Alcohols and Amines

Alcohols and amines are respectively weak acids and weak bases that are primarily used as nucleophilic agents in depolymerisation chemistry. Given their reactivity and ability to co-catalyse polymerisations,[20,23] it is not surprising that they also catalyse the degradation on their own. In particular, low weight glycols and amines have been used to depolymerise different

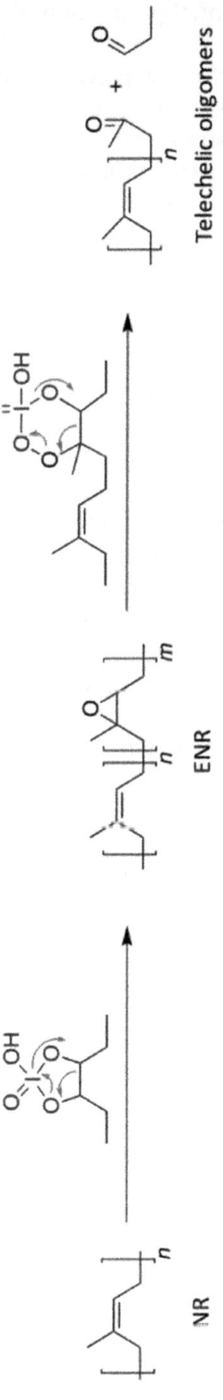

Scheme 14.6 Depolymerisation of polyisoprene using periodic acid.[35]

kinds of polyurethanes, targeting the isolation of the initial polyol. As a consequence of their large annual production, the recycling of polyurethane (PU) is particularly relevant.

In 1993, Xue *et al.* successfully depolymerised discarded PU rigid foam through glycolysis using diethylene glycol (DEG) as reagent and ethanolamine (EA) as catalyst. The resulting degradation product was identified as a glycolysate, which could be used as additive for the synthesis of epoxy resins.[38] However, no recent studies have been reported without the use of alkaline salts or metal complexes as catalyst. The glycolysis of PU rigid foam was reported using organometallic catalysts, demonstrating that low molecular-weight glycols and alkanolamines are not sufficiently catalytic to depolymerise the polymer by themselves.[39,40]

Depolymerisation of PU flexible foams using alkanolamines was first reported in 1994 by Kanaya and Takahashi.[41] A diphenyl isocyanate-based PU was depolymerised using EA to obtain pure polyether polyol, without the need of additional catalyst. Authors suggested that the mechanism involved alcoholysis rather than aminolysis (Scheme 14.7a). A few years later, Borda *et al.* studied the use of diethanolamine (DEA) to catalyse the glycolysis of PU flexible foams, with EG, 1,2-propylene glycol, triethylene glycol or low molecular weight poly(ethylene glycol) as reagent and solvent.[42] The resulting polyol phase was directly employed for synthesising an industrial adhesive mixture. Using an equimolar quantity of EG and DEA at 180 °C, PU flexible foams depolymerised in less than 2 h (Scheme 14.7b). In this study, an alcoholysis mechanism was the primary degradation route, however, analysis of released gases during the reaction demonstrated the presence of CO_2, which suggested that formation of water led to elimination of carbon dioxide *via* competing hydrolysis (Scheme 14.7c). More recently, the efficiency of DEA and EA were compared with metal catalysts such as acetate metallic salts, tin oxides and tin laurates, for the degradation of PU flexible foams.[43] At 200 °C, the presence of alkanolamines demonstrated a significant enhancement in the depolymerisation rate compared to the non-catalysed reaction; the best results were obtained with zinc acetate as catalyst. However, DEA showed similar performance to barium acetate while EA performed as well as potassium salts. Dibutyl tin dilaurate, butyl tin oxide and hydroxy butyl tin oxide demonstrated poor to very poor degradation over the same reaction duration.

Finally, the glycolysis of thermoplastic polyurethane elastomer (TPU), based on 4,4′-diphenylmethane diisocyanate and polyether polyol, was also conducted using low-weight glycols such as DEG or EG, and EA as catalyst.[44] Treated with the following mixture, DEG:EG:EA (9 : 9 : 2) and lithium acetate as co-catalyst, at 160–190 °C, the TPU depolymerised dephasing into two distinct layers. After 3 h of reaction, GPC, TGA as well as FTIR and NMR spectroscopic characterisations of the less dense liquid phase revealed a polyether polyol that corresponded to the starting polyol. Indeed, comparison with the industrial polyethylene glycol used to prepare TPU demonstrated that the same molecular weight and similar polydispersity is obtained.

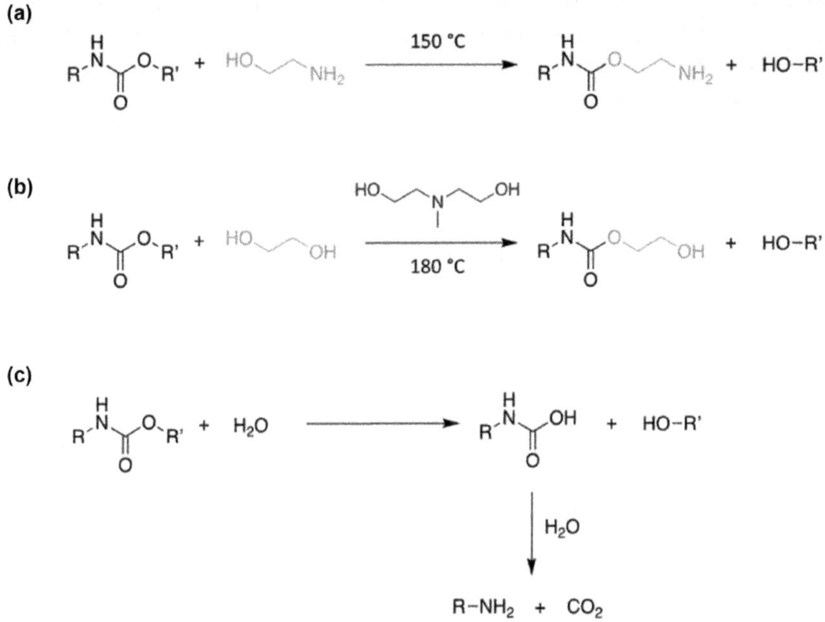

Scheme 14.7 Depolymerisation of PU (a) using EA, (b) using EG as reagent and DEA as catalyst and (c) side reaction.

14.2.4 Ionic Liquids and Acid–Base Salts

The use of ionic liquids (ILs) – defined as a salt in the liquid state with melting point lower than 100 °C – or other acid–base salts is appealing on account of their high thermal stability, which prevents their degradation under the demanding, high temperature conditions required for many depolymerisation processes. As well as retaining activity for multiple cycles of depolymerisation (satisfying the need for low cost catalysts), this property prevents the highly undesirable colouration of the final product. ILs have the added advantage of being liquid and hence the use of organic solvents can also be prevented.

In an early report, the exploration of the performance of DBU-based salts, including complexes formed by different ratios of DBU and benzoic acid (DBU-BA) were studied for depolymerisation of PET.[24] Glycolysis was slightly slower compared to DBU alone but the organic salt, DBU:BA (1 : 1) molar ratio, had the advantage of not aging when exposed to air, facilitating the depolymerisation procedure. Most recently, some of us reported the use of an innovative acid:base catalyst for the glycolysis of PET and the subsequent polymerisation of the resulting degradation product into a new polymer.[45] An equimolar mixture of TBD and methane sulfonic acid (MSA), led to a very stable protic ionic salt, which resists thermal degradation up to 400 °C. Organic molecules typically exhibit poor thermal stability, making their persistence at high temperature one of the main challenges for organocatalysis. Unfortunately, depolymerisation usually requires elevated temperature to overcome the biphasic nature of the depolymerisation mixtures. At 180 °C, in the presence of EG, the TBD:MSA (1 : 1) salt efficiently depolymerised post-consumer PET bottle waste in less than 2 h. Under optimised conditions, over 90% of high purity monomeric BHET was recovered through a simple crystallisation in water. Additionally, this catalyst has also been shown to be highly efficient for the re-polymerisation of the BHET monomer to yield PET with similar properties than the commercial material. Finally, the catalyst has been recycled up to six times without any loss in the catalytic activity. These highly stable catalysts present an interesting platform for industrially viable depolymerisation/polymerisation of PET.

The use of ILs as an efficient media has been primarily developed for the depolymerisation of PA-6 (Scheme 14.8). Quaternary ammonium salts such as *N*-methyl-*N*-propylpiperidinium (PP13) and *N*,*N*,*N*-trimethyl-*N*-propylammonium (TMPA), together with bis(trifluoromethane sulphonyl)-imide (TFSI) as counter anion, successfully depolymerised PA-6 in 5 to 6 h, at 300 °C.[46] The polymer was converted into ε-caprolactam in satisfactory yield (43–55%), without requiring addition of a further catalyst. Adding 5 wt% of DMAP as added catalyst, improved the depolymerisation efficiency such that 86% of monomer was recovered. Further investigations into optimisation of the reaction conditions demonstrated the importance of the temperature with 7% monomer isolated at 270 °C, 55% at 330 °C and 6% at 360 °C. For temperatures below 300 °C, a large portion of oligomeric polyamides in the crude product explained the low quantity of ε-caprolactam collected. For temperatures above 300 °C, formation of by-products such as *N*-methyl- and *N*-propyl lactams indicated that ILs decomposed at these temperatures, which hinders the depolymerisation to reach good yield. Extension of this methodology to polyamide-12 (PA-12) also led to isolation of the corresponding laurolactam; however the yield did not exceed 17%, most likely a consequence of the closing of a 12-member ring being energetically disfavoured.[47] The ILs were able to be recycled, with the process being repeated up to five times without observing any loss of depolymerisation efficiency. In a subsequent work, the same group demonstrated the possibility of depolymerising PA-6 using DMAP salts in the same ionic liquids. DMAP salts prepared with iodine and imidazolium as counter anion were able to catalyse the reaction, several times, achieving up to 79% of monomer recovery. However, this performance being under the yield obtained with DMAP alone, this catalyst was not investigated further for depolymerisation reactions.[48]

Urea and urea-based ionic liquids were also applied to the catalysis of polymer degradation. Musale and Shukla showed that choline chloride ([Ch][Cl]) urea salts are highly efficient catalysts for the aminolysis of PET.[49] Compared to glycolysis, depolymerisation of PET using amines is faster and generally more selective. Using EA or DEA, degradation of PET under reflux is complete in less than 30 min, providing the corresponding amides in reasonable to good yields. [Ch][Cl] urea mediated depolymerisation demonstrated an improved performance compared to urea, 69 and 80% for EA

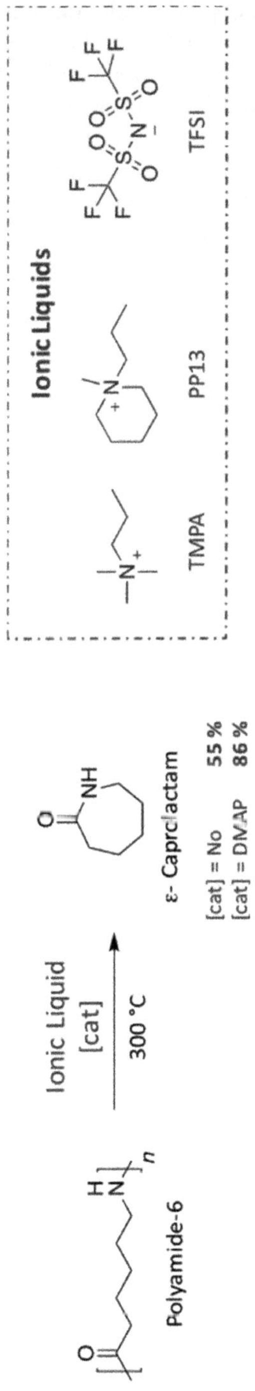

Scheme 14.8 Depolymerisation of PA-6 into ε-caprolactam using ILs as catalyst.

and DEA, respectively, against 55 and 66% for urea alone. However, the reaction using the corresponding zinc salt, namely [Ch][Cl] $ZnCl_2$, afforded higher monomer yields, up to 82 and 95% respectively (Figure 14.1A). Similar results were obtained for the urea-catalysed glycolysis of PET wastes. At 180 °C, using EG as nucleophile and solvent, monomer was collected with a yield of 78%.[50] Moreover, residual EG and urea were recycled, up to ten times, and every recycling step yielded similar final ratio of BHET, demonstrating that no loss of activity was experienced. *In situ* IR and DFT calculations emphasised the predominant role of the H-bonds formed between urea and EG. As they are known to form H-bond with alcohol groups, tetraalkylammonium-based amino acid-functionalised ILs were used as

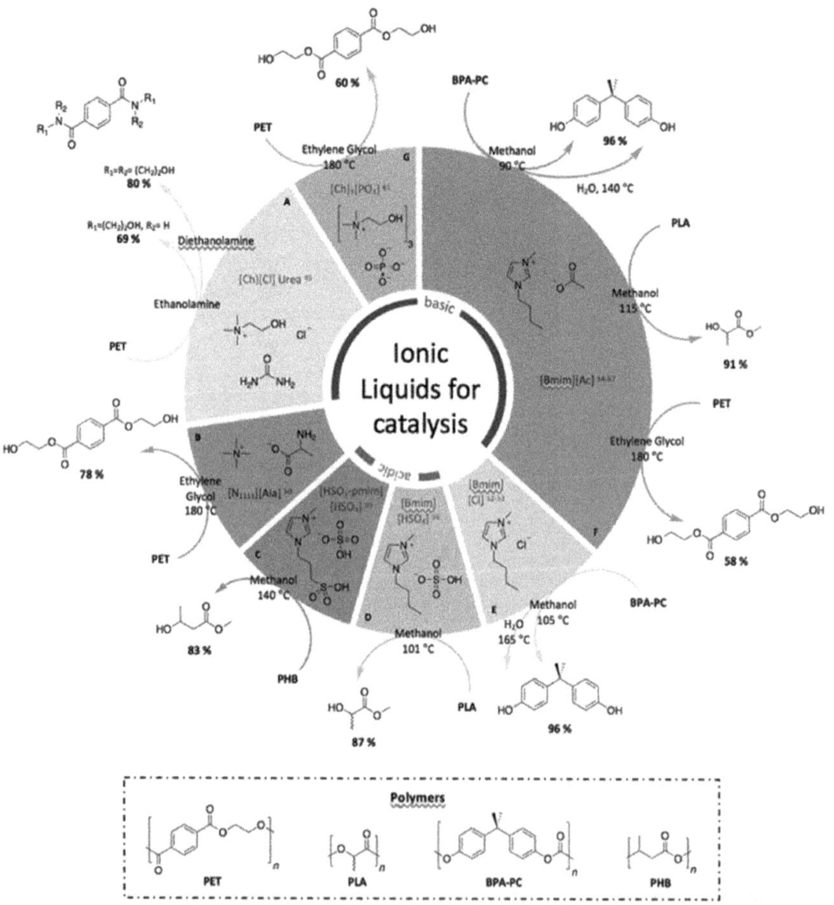

Figure 14.1 Ionic liquids as catalyst for depolymerisation. (A) [Ch][Cl] Urea, (B) [N_{1111}][Ala], (C) [HSO_3-pmim][HSO_4], (D) [Bmim][HSO_4], (E) [Bmim][Cl], (F) [Bmim][Ac], (G) $[Ch]_3[PO_4]$ where Bmim = 1-butyl-3-methylimidazolium, Ch = choline, N_{1111} = tetramethylammonium.

catalyst to further enhance the efficiency of the depolymerisation. Using tetramethylammonium alaninate [N_{1111}][Ala], a similar yield of monomer was obtained in a reduced time, 50 min at 170 °C against 190 min for urea alone. This last result consolidates the key role of H-bonding in the depolymerisation mechanism (Figure 14.1B).

In an environmental-friendly perspective, Sun *et al.* studied the application of low-cost and biocompatible ionic liquids to the solubilisation and depolymerisation of PET.[51] The authors demonstrated the ability for cholinium phosphate ([$Ch]_3[PO_4$]) to solubilise PET at relatively low temperature (120 °C) and subsequent glycolysis of PET led to BHET being obtained in 60% yield before recycling the IL for subsequent depolymerisations. NMR and IR spectroscopic analysis suggested a bifunctional activation of the system with the cation activating the carbonyl group of PET while the anion simultaneously activated one hydroxyl group of EG, similar to the previous example with TBD:MSA (1 : 1) ionic salt (Figure 14.1G).

Imidazolium-based ILs have been increasingly studied for a wide range of applications for a couple of decades, as green solvent, catalyst or for separation technologies. At the end of the 2000s, the capability of imidazolium ionic liquids to degrade PET was also investigated.[52] These ILs had the advantage of solubilising the material but depolymerisation required additional catalysts such as zinc acetate, tetrabutyltitanate or solid superacid to be completed. The authors did not specify the nature or yield of product obtained but characterisation revealed an average molecular weight of 790 g mol^{-1} for the isolated species. Although depolymerisation did not afford degradation to the monomeric unit, this first work was an encouraging proof of concept for this family of IL.

In 2010, the same imidazolium-based ILs as reported above for PET degradation were investigated by Liu *at al.* for the depolymerisation of BPA-PC. ILs with different inorganic anions (Cl, Br, BF_4, PF_6) and various chain natures for the *N*-alkyl imidazolium moieties have been synthesised to catalyse the methanolysis of the polymer.[53] Although most of the ILs did not display any catalytic activity, 1-butyl-3-methylimidazolium chloride ([Bmim][Cl]) provided complete depolymerisation of BPA-PC after 2.5 h at 105 °C. High yields (up to 96%) of both monomers, BPA and DMC, were isolated and the ionic liquid was reused up to eight times with no significant loss of the catalytic activity being observed (Figure 14.1E). Subsequently, Al Sabagh *et al.* showed that [Bmim][Cl] and [Bmim][Br] did not display any catalytic activity for the glycolysis of PET, however the more basic [Bmim][Ac] fully degraded the polymer in 3 h.[54] Once again, the IL was, then, recycled for subsequent polymerisation with no loss in the catalytic activity (Figure 14.1F). In another approach, Nunes *et al.* sought to improve depolymerisation with ILs by using neutral [Bmim][BF_4] for the depolymerisation of PET in supercritical ethanol.[55] A slight but significant improvement was observed compared to the use of supercritical conditions without ILs. Under the best conditions, the depolymerisation led to 98% yield of diethyl terephthalate, the monomer resulting from the ethanolysis of PET, in 45 min.

In successive studies, the same conditions were applied to the hydrolysis and methanolysis of BPA-PC testing both [Bmim][Cl] and 1-butyl-3-methylimidazolium acetate ([Bmim][Ac]) as catalyst, obtaining BPA in very good yields for each reaction (Table 14.1). Using [Bmim][Cl] as catalyst with water at 165 °C , the polymer fully depolymerised in 3 h, to afford up to 95% yield of BPA (Figure 14.1E).[56] Similarly, with [Bmim][Ac], in the same amount of time, 0.5 g of methanol for 1 g of BPA-PC was needed to provide high yield of BPA (96%), at 90 °C.[57] The same monomeric yield was afforded through hydrolysis of the BPA-PC using 1.5 g of the same IL and 0.35 g of water, at 140 °C (Figure 14.1F).[58] Notably, in hydrolysis, the catalyst requires higher temperatures and higher loading to afford the same performance as in methanolysis. Indeed, kinetic studies for reactions with [Bmim][Ac] indicate that depolymerisations are first-order with hydrolysis exhibiting a higher activation energy (228 kJ mol^{-1}) than methanolysis (168 kJ mol^{-1}), supporting the harsher conditions required while water was the reagent. Furthermore, [Bmim][Ac] exhibited higher catalytic activity than [Bmim][Cl], depolymerising BPA-PC under milder conditions, using lower temperatures and a smaller amount of catalyst (Table 14.1, columns 3 and 5). The authors suggested that the enhanced performance of [Bmim][Ac] in depolymerising BPA-PC is related to the better solubility of the polymer in this IL. Indeed, at 105 °C, [Bmim][Ac] is able to solubilise 1 g of BPA-PC in 5 g of IL while in [Bmim][Cl] and at the same temperature the material barely swells. Finally, the methanolysis of PLA into methyl lactide has also been investigated using different imidazolium ionic liquids.[59] Comparison of the catalytic activity of [Bmim][Cl], [Bmim][PF_6], [Bmim][Ac] and [Bmim][HSO_4] revealed that the neutral ILs ([Bmim][Cl] and [Bmim][PF_6]) were inactive for depolymerisation whereas the basic [Bmim][Ac] and acidic [Bmim][HSO_4] completed degradation of the polymer in 3 h at 115 °C. The efficacy of acid and base catalysts for PLA depolymerisation by transesterification is well known and hence this result is not surprising. [Bmim][Ac] appeared to be slightly more active than its acidic homologue, reaching 96% conversion of PLA for an isolated methyl lactide yield of 91% (Figures 14.1D and F).

As described earlier, basic ILs have been widely investigated for alcoholysis of oxygen-containing polymers, in particular, PET, BPA-PC or PLA. In contrast, acidic ILs have received much less attention. One notable application has explored the methanolysis of the polyester, poly(3-hydroxybutyrate),

Table 14.1 Hydrolysis and methanolysis of BPA-PC using imidazolium ionic liquids.

Ref.	Reagent (R)	Catalyst (IL)	Temperature (°C)	Reaction time	R:IL:PC[a]	BPA yield (%)
52	Methanol	[Bmim][Cl]	105	2 h 30 min	1.5 : 1 : 1	96
53	Water	[Bmim][Cl]	165	3 h	10 : 1.5 : 1	95
54	Methanol	[Bmim][Ac]	90	3 h	0.5 : 0.5 : 1	96
55	Water	[Bmim][Ac]	140	3 h	0.35 : 1.5 : 1	96

[a]Mass ratio.

(PHB), a bio-sourced and biodegradable polymer, using 1-methyl-3-(3-sulfopropyl) imidazolium hydrogen sulfate [HSO_3-pmim][HSO_4] as catalyst.[60] Efficient depolymerisation to methyl 3-hydroxybutyrate monomer in 83% yield after 3 h occurred at 140 °C. Interestingly, the ionic liquid prepared with the same cation but with dihydrogen phosphate ($H_2PO_4^-$) anion was not active for depolymerisation, most likely a consequence of the less acidic nature of $H_2PO_4^-$, ($pK_a = 7.21$)[61] compared to HSO_4^- ($pK_a = 1.99$)[61] (Figure 14.1C).

An interesting example of polymer recycling applied to industrial waste has been recently reported for the microwave-assisted degradation of fibres-reinforced plastics (FRP), using ionic liquids.[62] Different ILs were explored as solvent and catalyst for the transformation of constituent polyesters of FRP into phthalic anhydride. At 340 °C, using [PP13][TFSI], depolymerisation, gave up to 80% yield of monomeric material within minutes (Scheme 14.9). Phthalic anhydride, pure glass fibres and the ionic liquid were recovered through direct distillation of the crude product. The IL has then been used for several iterations of the depolymerisation, obtaining comparable monomer yields with no degradation observed.

ILs have also been used as solvent and catalyst for depolymerisation of polyolefins. In 2000, Adams *et al.* described first the use of IL for the cracking of both high and low density polyethylene (HDPE and LDPE) using chloroaluminate(III) ionic liquids.[63] At temperatures below 200 °C, the reaction led to low-volatility alkanes. Although this IL is metal based, this first example opened the way to other initiatives using IL as catalyst for polymer degradation. More recently, the depolymerisation of waste tyres in a hydrophobic ionic liquid was reported. Using trihexyl-(tetradecyl)-phosphonium chloride (Cyphos101) and *N*,*N*-dioctylimidazolium bromide (C_8 C_8 ImBr), metathesis of waste tyres led to the production of low molecular weight telechelic polymers of controlled lengths.[64] The resulting oligomers are interesting intermediates for synthesis of innovative polymers by methathesis polymerisation for the rubber industry.

14.3 Innovative Polymers and Their End-of-life Option

Although the recycling of commonly used polymers is of great interest for both sustainable and economical purposes, the ultimate aim for polymer chemists is to conceive polymers together with their degradation route. As a battery can be charged and discharged theoretically infinitely, a material should be depolymerised and re-polymerised with minimal loss of quantity or change in the final properties.

Although the necessity of discarding and recycling polymers is a contemporary challenge, some early attempts in designing sustainable cycle including polymers degradation are of particular importance. In one early example, pyridinium *p*-toluenesulfonate (PPTS), was considered as catalyst for the condensation of common dicarboxylic acid with divinyl ether.[65]

Scheme 14.9 Depolymerisation of FRP in ionic liquids.

The catalyst enabled the insertion of different functional groups, resulting in materials with tuneable thermolysis temperature (Scheme 14.10a). Indeed, at the corresponding temperature, polymers degrade, presumably into di-acid and CO_2. This noteworthy example emphasises the advantage of some organocatalysts to tolerate functional groups, allowing the synthesis of materials with a broader range of properties.

While poly(hydroxyl butyrate), PHB, can be considered to be a well known polymer, it is not used widely and until recently has typically been produced by a microbial method to yield high molecular weight material. Melchiors *et al.* investigated the depolymerisation of PHB using *p*-toluenesulfonic acid (PTSA) as catalyst.[66] Rather than resulting in the hydroxyl butyric acid starting material that may be expected, the resulting degradation product was characterised as the corresponding cyclic trimer (TBL), obtained in a yield of 80%, in THF and at 100 °C (Scheme 14.10b). Thermodynamic values in full agreement with the experimental data were determined using bomb calorimetry experiments. However, the polymerisation of TBL back to PHB was determined to be thermodynamically non-feasible although the authors succeeded in demonstrating that it was practically possible suggesting an error in the calorimetry measurement that was attributed to polymer crystallinity as well as the polymer–monomer interactions.

More recently, Hong and Chen achieved the polymerisation of poly-(γ-butyrolactone) (PγBL) with fairly high molecular weight *via* ROP of γ-butyrolactone (γ-BL).[67] Considered until then as a "non-polymerisable" monomer as a consequence of its very low ring strain, γ-BL was polymerised at −40 °C using a powerful lanthanide-based catalyst. The polymerisation

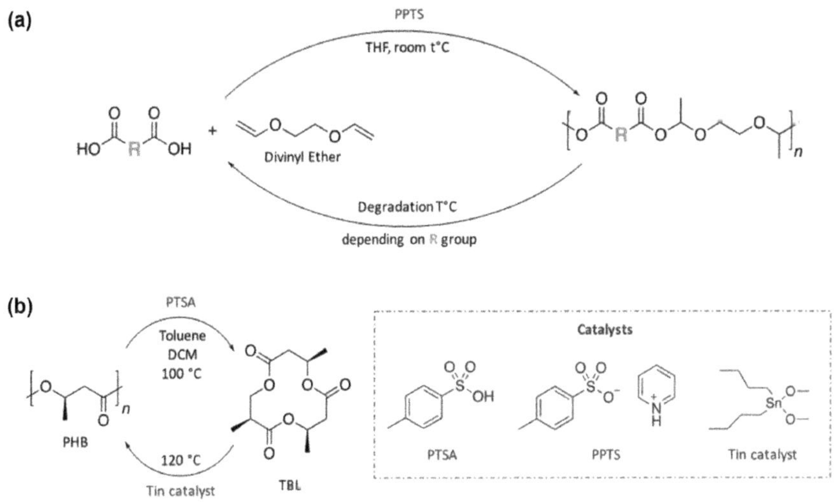

Scheme 14.10 (a) Depolymerisation of PHB using acidic catalysis and (b) tuneable polymerisation of poly(hemiacetal ester)s and their thermolysis degradation.

reached high monomer conversion, up to 90%, and led to both linear and cyclic structures. In a second time, depolymerisation of the polymer using either the same catalyst or TBD, allowed full conversion back into γ-BL in minutes, at room temperature (Scheme 14.11a). Subsequently, this circular polymerisation/depolymerisation was shown to be possible with the phospazene superbase *tert*-Bu-P_4 as catalyst.[68] Although ^{t}Bu-P_4 can initiate the polymerisation reaction alone, the resulting polymer was achieved with only 30% yield with $M_n = 26.4$ kg mol^{-1}, but the use of benzyl alcohol as initiator increased PγBL conversion, up to 90%, with comparable molecular weight (Scheme 14.11b). Heating the polymer at 260 °C, the starting monomer and the initiator were completely recovered. As a result, a complete sustainable cycle has been designed involving the ROP of γ-BL using an organocatalyst and depolymerisation of the resulting polymer *via* thermolysis.

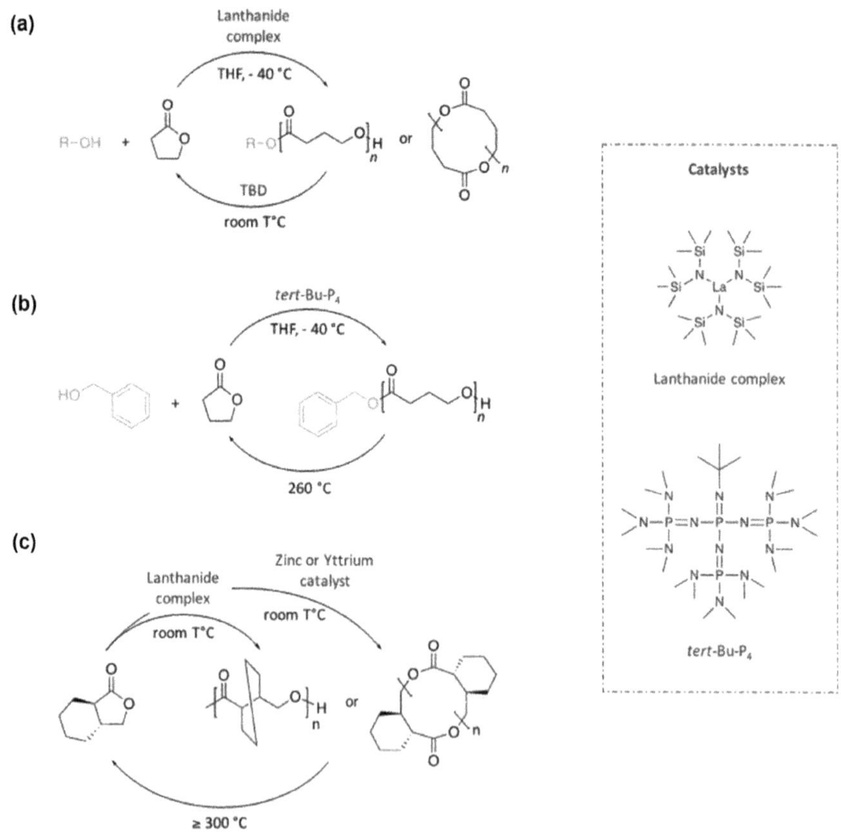

Scheme 14.11 Polymerisation of γ-BL into PγBL and its depolymerisation pathway using (a) metal catalyst and TBD or (b) *tert*-Bu-P_4 and thermolysis, and (c) polymerisation of substituted γ-BL and its thermal degradation.

More recently, Chen and coworkers have extended this concept to a substituted γ-butyrolactone monomer that was designed such that the polymer would have better thermal and mechanical properties than the unsubstituted derivative.[69] The monomer incorporates a cyclohexyl ring fused to the α- and β-positions of the butyrolactone ring but, importantly leaves the γ-position unsubstituted (Scheme 14.11c). With this design, not only is the ring strain of the cyclic structure enhanced, thus increasing the thermodynamic polymerisability, but the rate of polymerisation also rises due to the release of steric pressure at the ester side. Their results suggest that maintaining the γ-position unsubstituted is essential to obtain a monomer with sufficient ring strain to be readily polymerised. Steric hindrance and stereoregularity result after polymerisation in a semi-crystalline polymer, however retention of the five-membered γ-lactone core ensures that the thermal recyclability is preserved. The authors found that by blending two enantiomerically pure polymers of opposite stereochemistry in a 1 : 1 stoichiometric ratio, stereocomplexed materials could be obtained that displayed melting temperatures that are more than 75 °C higher than either constituent polymer. Importantly, using either lanthanide-based or organic catalysts, the polymers could be readily depolymerised directly back to the starting monomer. While to date only reported with metal-based catalysts (*e.g.* $ZnCl_2$ salts), it demonstrates a future opportunity for organocatalysis and innovative polymer design.

In a similar vein, Hillmyer and coworkers recently designed a complete sustainable cycle for innovative recyclable polyurethane foams[70] and thermoset elastomers.[71] A limonene-based polycarbonate was produced using TBD as organocatalyst for the polymerisation while stannous octoate was catalysing the depolymerisation. Even though organocatalysis was not involved in the depolymerisation step, it is interesting to note that the switch in the nature of the catalyst led to polymerisation or depolymerisation, allowing reversible transformation of the material, mediating the reaction with one catalyst or the other.

Interestingly, Fahnhorst *et al.* recently described what they called the "divergent chemical recycling" of a malic acid-based polyvalerlactone.[72] In a two-step synthesis, 4-carbomethoxyvalerolactone **1** was obtained to create the innovative polyester **P1**. This polymer was depolymerised using two different catalysts, which resulted in the generation of two different monomers (Scheme 14.12). When degraded with tin octanoate, **P1** returned to the initial monomer **1** *via* a transesterification reaction however, degradation employing DBU as catalyst, afforded the methacrylate-like monomer **2** through a water-elimination mechanism. The subsequent polymerisation of this distinct monomer provided the polymethacrylate analogue **P2**. This "divergent" degradation illustrates previous remarks about the inherent differences between the mechanisms promoted by organometallics and organocatalysts. The choice of the catalysts plays an important role as it impacts the nature of the final products, more importantly in the case of depolymerisation reactions, as possibilities to cleave the polymer backbone are multiple.

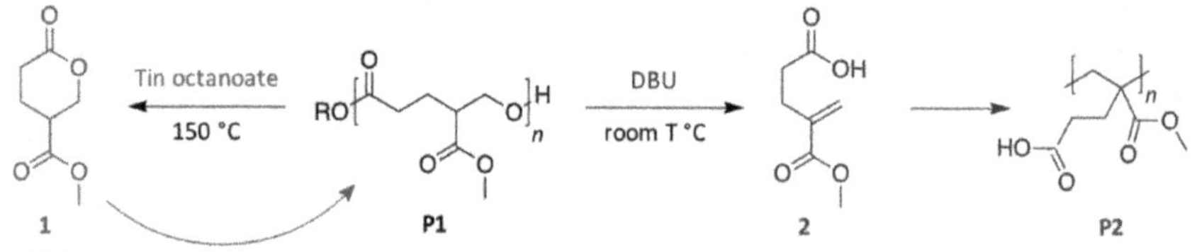

Scheme 14.12 Two different pathways for the chemical recycling of a carbomethoxylated polyvalerolactone.

14.4 Conclusion

Depolymerisation of plastic waste is complicated by the many different chemical structures of the polymers. In turn, this means that there is no "one size fits all" solution to depolymerisation and each polymer family requires specific studies to find the best depolymerisation method. The majority of studies have focussed on transesterification to depolymerise PET or for oxidation of isoprene in a controlled manner. Clearly, expanding the scope of the organocatalytic families used here, as well as developing routes to recycle a wider range of polymeric materials, is a critical target for the field. As in the case of polymerisation chemistry, acidic and basic organocatalysts have different advantages although in the specific case of depolymerisation, salts and ionic liquids may provide an inherent advantage over either on account of their greater thermal stability and their ability to solubilise otherwise insoluble polymers.

Beyond working with existing commodity polymer materials, a focus of the field moving forward must be the consideration of how new materials can be depolymerised at the same time as they are created/developed. As described in the last part of this chapter, the creation of sustainable cycles involving the depolymerisation/polymerisation of materials *via* organocatalysis is a promising tool for the design of innovative polymers jointly with their degradation pathways. Improvement in organocatalyst design will be highly beneficial to this field, leading to more economically and environmentally sustainable methods for polymer production.

References

1. W. McDonough and M. Braungart, *Cradle to Cradle – Remaking the Way We Make Things*, North Point Press, 2002.
2. P. Sharma, B. Lochab, D. Kumar and P. K. Roy, *ACS Sustainable Chem. Eng.*, 2016, **4**, 1085–1093.
3. R. V. Shah, V. S. Borude and S. R. Shukla, *J. Appl. Polym. Sci.*, 2013, **127**, 323–328.
4. Y. Hongmei, M. Yongshuai, Z. Weilu and Z. Dong, *Adv. Mater. Res.*, 2012, **550**, 280–283.

5. Y. Yang, Y. Lu, H. Xiang, Y. Xu and Y. Li, *Polym. Degrad. Stab.*, 2002, **75**, 185–191.
6. A. Kamimura, Y. Oishi, K. Kaiso, T. Sugimoto and K. Kashiwagi, *ChemSusChem*, 2008, **1**, 82–84.
7. Y. Chujo, H. Kobayashi and Y. Yamashita, *J. Polym. Sci., Part A: Polym. Chem.*, 1989, **27**, 2007–2014.
8. C.-H. Chen, C.-Y. Chen, Y.-W. Lo, C.-F. Mao and W.-T. Liao, *J. Appl. Polym. Sci.*, 2001, **80**, 943–948.
9. M. Ghaemy and K. Mossaddegh, *Polym. Degrad. Stab.*, 2005, **90**, 570–576.
10. A. P. Dove, *ACS Macro Lett.*, 2012, **1**, 1409–1412.
11. M. K. Kiesewetter, E. J. Shin, J. L. Hedrick and R. M. Waymouth, *Macromolecules*, 2010, **43**, 2093–2107.
12. N. E. Kamber, W. Jeong, R. M. Waymouth, R. C. Pratt, B. G. G. Lohmeijer and J. L. Hedrick, *Chem. Rev.*, 2007, **107**, 5813–5840.
13. K. Fukushima, *Polym. J.*, 2016, **48**, 1103–1114.
14. R. C. Pratt, B. G. G. Lohmeijer, D. A. Long, R. M. Waymouth and J. L. Hedrick, *J. Am. Chem. Soc.*, 2006, **128**, 4556–4557.
15. S. Venkataraman, V. W. L. Ng, D. J. Coady, H. W. Horn, G. O. Jones, T. S. Fung, H. Sardon, R. M. Waymouth, J. L. Hedrick and Y. Y. Yang, *J. Am. Chem. Soc.*, 2015, **137**, 13851–13860.
16. C. Tan, S. Xiong and C. Chen, *Macromolecules*, 2018, **51**, 2048–2053.
17. S. Naumann, P. B. V. Scholten, J. A. Wilson and A. P. Dove, *J. Am. Chem. Soc.*, 2015, **137**, 14439–14445.
18. A. Chuma, H. W. Horn, W. C. Swope, R. C. Pratt, L. Zhang, B. G. G. Lohmeijer, C. G. Wade, R. M. Waymouth, J. L. Hedrick and J. E. Rice, *J. Am. Chem. Soc.*, 2008, **130**, 6749–6754.
19. T. S. Stukenbroeker, J. S. Bandar, X. Zhang, T. H. Lambert and R. M. Waymouth, *ACS Macro Lett.*, 2015, **4**, 853–856.
20. K. Fukushima, O. Coulembier, J. M. Lecuyer, H. A. Almegren, A. M. Alabdulrahman, F. D. Alsewailem, M. A. Mcneil, P. Dubois, R. M. Waymouth, H. W. Horn, J. E. Rice and J. L. Hedrick, *J. Polym. Sci., Part A: Polym. Chem.*, 2011, **49**, 1273–1281.
21. R. López-Fonseca, I. Duque-Ingunza, B. de Rivas, S. Arnaiz and J. I. Gutiérrez-Ortiz, *Polym. Degrad. Stab.*, 2010, **95**, 1022–1028.
22. D. E. Nikles and M. S. Farahat, *Macromol. Mater. Eng.*, 2005, **290**, 13–30.
23. H. W. Horn, G. O. Jones, D. S. Wei, K. Fukushima, J. M. Lecuyer, D. J. Coady, J. L. Hedrick and J. E. Rice, *J. Phys. Chem.*, 2012, **116**, 12389–12398.
24. K. Fukushima, D. J. Coady, G. O. Jones, H. A. Almegren, A. M. Alabdulrahman, F. D. Alsewailem, H. W. Horn, J. E. Rice and J. L. Hedrick, *J. Polym. Sci., Part A: Polym. Chem.*, 2013, **51**, 1606–1611.
25. K. Fukushima, J. M. Lecuyer, D. S. Wei, H. W. Horn, G. O. Jones, H. A. Al-Megren, A. M. Alabdulrahman, F. D. Alsewailem, M. A. McNeil, J. E. Rice and J. L. Hedrick, *Polym. Chem.*, 2013, **4**, 1610–1616.
26. F. A. Leibfarth, N. Moreno, A. P. Hawker and J. D. Shand, *J. Polym. Sci., Part A: Polym. Chem.*, 2012, **50**, 4814–4822.

27. E. Quaranta, D. Sgherza and G. Tartaro, *Green Chem.*, 2017, **19**, 5422–5434.
28. T. Do, E. R. Baral and J. G. Kim, *Polymer*, 2018, **143**, 106–114.
29. S. Gillier-Ritoit, D. Reyx, I. Campistron, A. Laguerre and R. Pal Singh, *J. Appl. Polym. Sci.*, 2003, **87**, 42–46.
30. A. El Hamdaoui, D. Reyx, I. Campistron and S. F. Tétouani, *Eur. Polym. J.*, 1999, **35**, 2165–2183.
31. C. W. Phetphaisit and P. Phinyocheep, *J. Appl. Polym. Sci.*, 2003, **90**, 3546–3555.
32. S. R. Shukla, A. M. Harad and D. Mahato, *J. Appl. Polym. Sci.*, 2006, **100**, 186–190.
33. H. Matsumoto, Y. Akinari, K. Kaiso and A. Kamimura, *J. Mater. Cycles Waste Manage.*, 2017, **19**, 326–331.
34. P. Phinyocheep, C. W. Phetphaisit, D. Derouet, I. Campistron and J. C. Brosse, *J. Appl. Polym. Sci.*, 2005, **95**, 6–15.
35. F. Sadaka, *Etude de la dégradation contrôlée de polydiènes: Application au recyclage des déchets pneumatiques*, Le Mans, 2010.
36. F. Sadaka, I. Campistron, A. Laguerre and J.-F. Pilard, *Polym. Degrad. Stab.*, 2012, **97**, 816–828.
37. E. Bäckström, K. Odelius and M. Hakkarainen, *Ind. Eng. Chem. Res.*, 2017, **56**, 14814–14821.
38. S. Xue, F. He, M. Omoto, T. Hidai and Y. Imai, *Kobunshi Ronbunshu*, 1993, **50**, 847–853.
39. M. Murai, M. Sanou, T. Fujimoto and F. Baba, *J. Cell. Plast.*, 2003, **39**, 15–27.
40. M. M. A. Nikje and M. Nikrah, *J. Macromol. Sci., Part A*, 2007, **44**, 613–617.
41. K. Kanaya and S. Takahashi, *J. Appl. Polym. Sci.*, 1994, **51**, 675–682.
42. J. Borda, G. Pásztor and M. Zsuga, *Polym. Degrad. Stab.*, 2000, **68**, 419–422.
43. L. M. Dos Santos, C. L. P. Carone, J. Dullius, R. Ligabue and S. Einloft, *Polím.: Cienc. Tecnol.*, 2013, **23**, 608–613.
44. X. Wang, H. Chen, C. Chen and H. Li, *Fibers Polym.*, 2011, **12**, 857.
45. C. Jehanno, I. Flores, A. P. Dove, A. J. Müller, F. Ruipérez and H. Sardon, *Green Chem.*, 2018, **20**, 1205–1212.
46. A. Kamimura and S. Yamamoto, *Org. Lett.*, 2007, **9**, 2533–2535.
47. A. Kamimura and S. Yamamoto, *Polym. Adv. Technol.*, 2008, **19**, 1391–1395.
48. S. Yamamoto and A. Kamimura, *Chem. Lett.*, 2009, **38**, 1016–1017.
49. R. M. Musale and S. R. Shukla, *Int. J. Plast. Technol.*, 2016, 1–15.
50. Q. Wang, X. Yao, S. Tang, X. Lu, X. Zhang and S. Zhang, *Green Chem.*, 2012, **14**, 2559–2566.
51. S. Jian, L. Dajiang, P. Young Robert, G. Cruz Alejandro, G. Isern Nancy, S. Timo, R. Cort John, A. Simmons Blake and S. Seema, *ChemSusChem*, 2018, **11**, 781–792.
52. H. Wang, Z. Li, Y. Liu, X. Zhang and S. Zhang, *Green Chem.*, 2009, **11**, 1568–1575.

53. F. Liu, Z. Li, S. Yu, X. Cui and X. Ge, *J. Hazard. Mater.*, 2010, **174**, 872–875.
54. A. M. Al-Sabagh, F. Z. Yehia, A.-M. M. F. Eissa, M. E. Moustafa, G. Eshaq, A.-R. M. Rabie and A. E. ElMetwally, *Ind. Eng. Chem. Res.*, 2014, **53**, 18443–18451.
55. C. S. Nunes, M. J. V. da Silva, D. C. da Silva, A. dos, R. Freitas, F. A. Rosa, A. F. Rubira and E. C. Muniz, *RSC Adv.*, 2014, **4**, 20308–20316.
56. L. Li, F. Liu, Z. Li, X. Song, S. Yu and S. Liu, *Fibers Polym.*, 2013, **14**, 365–368.
57. F. Liu, L. Li, S. Yu, Z. Lv and X. Ge, *J. Hazard. Mater.*, 2011, **189**, 249–254.
58. X. Song, F. Liu, L. Li, X. Yang, S. Yu and X. Ge, *J. Hazard. Mater.*, 2013, **244–245**, 204–208.
59. X. Song, X. Zhang, H. Wang, F. Liu, S. Yu and S. Liu, *Polym. Degrad. Stab.*, 2013, **98**, 2760–2764.
60. X. Song, H. Wang, F. Liu and S. Yu, *Polym. Degrad. Stab.*, 2016, **130**, 22–29.
61. J. Bjerrum, *Stability Constants of Metal-ion Complexes, with Solubility Products of Inorganic Substances: Inorganic Ligands*, Chemical Society, 1958.
62. A. Kamimura, S. Yamamoto and K. Yamada, *ChemSusChem*, 2011, **4**, 644–649.
63. C. J. Adams, M. J. Earle and K. R. Seddon, *Green Chem.*, 2000, **2**, 21–24.
64. A. Mouawia, A. Nourry, A.-C. Gaumont, J.-F. Pilard and I. Dez, *ACS Sustainable Chem. Eng.*, 2017, **5**, 696–700.
65. H. Otsuka and T. Endo, *Macromolecules*, 1999, **32**, 9059–9061.
66. M. Melchiors, H. Keul and H. Höcker, *Macromolecules*, 1996, **29**, 6442–6451.
67. M. Hong and E. Y.-X. Chen, *Nat. Chem.*, 2016, **8**, 42–49.
68. M. Hong and E. Y.-X. Chen, *Angew. Chem.*, 2016, **128**, 4260–4265.
69. J.-B. Zhu, E. M. Watson, J. Tang and E. Y.-X. Chen, *Science*, 2018, **360**, 398–403.
70. D. K. Schneiderman, M. E. Vanderlaan, A. M. Mannion, T. R. Panthani, D. C. Batiste, J. Z. Wang, F. S. Bates, C. W. Macosko and M. A. Hillmyer, *ACS Macro Lett.*, 2016, **5**, 515–518.
71. J. P. Brutman, G. X. De Hoe, D. K. Schneiderman, T. N. Le and M. A. Hillmyer, *Ind. Eng. Chem. Res.*, 2016, **55**, 11097–11106.
72. G. W. Fahnhorst and T. R. Hoye, *ACS Macro Lett.*, 2018, **7**, 143–147.

CHAPTER 15

Organic Catalysis Outlook: Roadmap for the Future

ANDREW P. DOVE

School of Chemistry, University of Birmingham, Edgbaston, Birmingham, UK B15 2TT
Email: A.Dove@bham.ac.uk

Organic catalysis has undoubtedly made a significant impact in the field of polymerisation. From the initial report of nucleophilic transesterification ring-opening polymerisation,[1] the field has come a long way in a short time. Mirroring the more broad organocatalysis field, there have been many discoveries that have enabled more complex polymer structures to be accessed and a greater range of materials through expansion of the types of polymerisation that organocatalysis can now mediate. Organocatalysts can be generally split into five categories (Figure 15.1): nucleophiles, bases, acids, supramolecular catalysts and redox mediators that fundamentally operate by different mechanisms. This gives the field the ability to access materials *via* a wide range of techniques that has extended from (trans)esterification chemistries in ring-opening polymerisation of cyclic esters and polycondensation into a diverse range of polymer backbones through other ring-opening and step growth polymerisations as well as group-transfer polymerisation (GTP) and controlled radical techniques such as atom-transfer radical polymerisation (ATRP) (Figure 15.1). While significant progress has been made across this landscape there still exists potential for translating catalytic discoveries towards the synthesis of polymers by new routes and for further innovation and catalyst development.

Polymer Chemistry Series No. 31
Organic Catalysis for Polymerisation
Edited by Andrew Dove, Haritz Sardon and Stefan Naumann

Published by the Royal Society of Chemistry, www.rsc.org

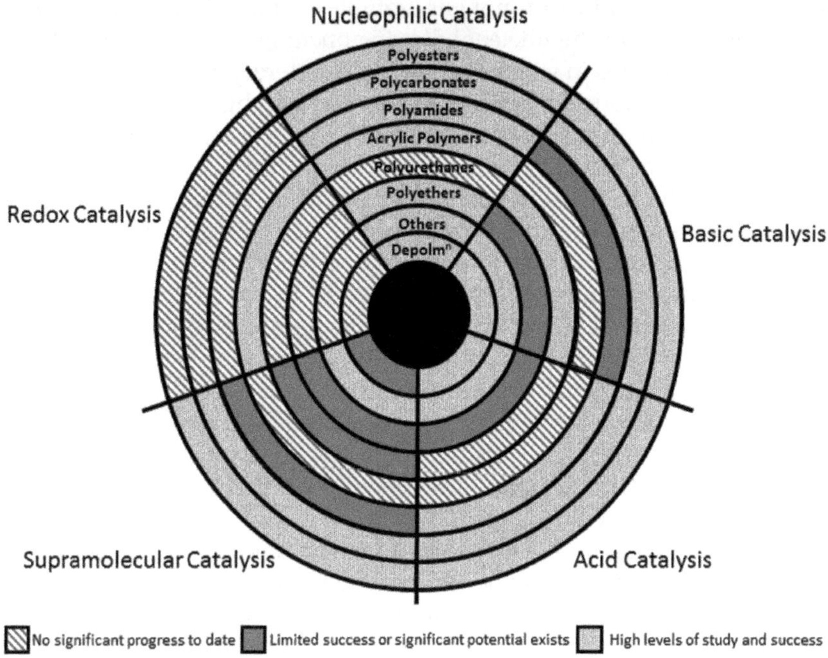

Figure 15.1 Summary of organic catalysis application in polymerisation methods.

Of the many advantages of organic catalysts, their 'green' or environmentally friendly aspect could present them as attractive alternatives to metal-based analogues. While clearly not applicable to all organocatalysts applied, the reduction in the need for heavy metal mining and refinement that could be afforded by a switch to organocatalytically mediate processes could present a major ecological advantage. Furthermore, the low environmental impact and biological toxicity of some leading organic catalysts such as 4-dimethylaminopyridine (DMAP) or amino-acid based catalysts could provide a big advantage in the application of polymers that are intended to be released into the environment, such as in agricultural applications, and they can be applied in products that are compostable in which the absence of heavy metals negates the need to remove them from polymers to prevent further ecological effect. Furthermore, this advantage makes organic catalysts ideal for the production of polymers that are intended for biological or microelectronics applications in which metal contaminants are highly undesirable and costly to remove. The growing field of catalytic depolymerisation of plastic waste for circular or added-value plastic economies will also benefit from these critical aspects of organic catalysts moving forward.

One of the more striking aspects of the field is that the use of organic catalysis is starting to allow materials to be synthesised that have not been, and perhaps could not easily be, accessed through metal-catalysed routes.

Perhaps one of the most prominent examples is the selective generation of cyclic polymers from the nucleophilic ring-opening polymerisation of cyclic esters and carbonates in which typically an N-heterocyclic carbene acts as an initiator for a zwitterionic polymerisation that backbites to selectively form cyclic species.[2] In another example, the recently reported nucleophilic addition of thiols to activated alkynes has only been reported with organic catalysts.[3] While it is likely possible to activate the thiols to the thiolate anion using basic metal salts, only through the versatility of organic catalysis and the precise control over the basicity of the system can the resultant stereochemistry of the resultant double bond be controlled. This in turn, controls the mechanical properties of the resultant elastomeric materials in a similar way to how the stereocontrol in polyisoprene dictates the mechanical properties of natural rubber and gutta percha. While other examples exist in the areas of switchable catalysts, monomer selectivity and new materials, this work highlights the potential to apply organocatalysis to create new materials that offer advantages over existing state-of-the-art or commonly used materials.

Despite the many successes in using organic catalysis in all of these fields, there remains significant scope for future development. Generally, organocatalysts have a lower thermal stability than their organo-metallic counterparts, which is potentially limiting in many industrial processing methodologies where high temperatures are commonly used to overcome factors such as melt processing of semi-crystalline polymers or polycondensations to drive off residual water. Ironically, these are most prevalent in the areas where organocatalysis was first used for polymerisation, *i.e.* ring-opening polymerisation of lactide in which temperatures of >180 °C are required to control the melt viscosity of the polymer as it is formed.[4] While a range of catalysts have shown significant potential at lower temperatures, they typically degrade at high temperatures, which leads to both deactivation and undesirable discolouration of the resulting material. To overcome this challenge, thermally stable and highly active organocatalysts will be required. The solution may be found in the exploitation of the remarkable thermal stability of acid:base salts and ionic liquids, which are stable at these temperatures and of which some have been found to have activities for polymerisation and depolymerisation reactions.[5] Delivery of such technologies may be important to realise the industrial translation of the fundamental work; however, this approach may not be applicable across all polymerisation mechanisms and alternative processing methods may be required to adapt organocatalytic solutions to industrial scales.

Perhaps one of the major limitations to the furthering of organocatalytic processes for wider application is their activity when compared directly to metal-based species. While significant progress in the ring-opening of lactide as presented catalysts such as N-heterocyclic carbenes,[6,7] phosphazene bases[8] and 1,5,7-triazabicyclo[4.4.0]dec-5-ene (TBD)[9] as amongst the most active reported, challenging or even exceeding the reactivity of the most active metal-based species,[10] in many other polymerisation processes, their activity significantly lags behind metallo-organic catalysts. For example, while N-heterocyclic olefin catalysts were reported to be the most active organic catalysts ever reported for the ring-opening polymerisation of epoxide monomers such as propylene oxide,[11] they still do not compare with the rates of polymerisation that can be achieved by metal-based species for the same transformation. New catalyst design and approaches will be required if this class of catalyst are to rival metal-based species across the spectrum of techniques. It is also an important challenge to answer as we seek to transition from fossil-based and non-recyclable materials into ecologically degradable or sustainably sourced polymers where high productivities and turnover frequencies, comparable to those seen in olefin polymerisation, may be required to justify scaling of processes to make them economically viable.

Beyond activity, polymerisation control remains an important challenge in some organocatalyzed polymerisations that leaves scope for further innovation in organic catalyst and process design. While supramolecular thiourea-based catalysts and their subsequently reported analogies have enabled extreme precision in the ring-opening polymerisation of cyclic esters and carbonates to present selectivity for ring-opening against side-chain transesterification and hence loss of control, this success has not been repeated across the whole range of organocatalytic polymerisation family. For example in the anionic/nucleophilic polymerisation or GTP of acrylic monomers, the control of the polymerisation is often lost, most likely a result of competing side reactions at the side chain. Despite this, obtaining control over these processes presents a significant opportunity to create materials that would be challenging to produce by other means or by offering opportunities to keep catalysts 'alive' for the creation of advanced crosslinked materials such as vitrimers.[12] Extending the concept of polymerisation control beyond control of reactivity, there is significant scope for further advances in the field of stereoselective organocatalyzed polymerisation. While some organocatalysts are able to offer high levels of stereoselectivity, they are typically observed at low temperatures, thus making them difficult to translate to an industrial environment. The discovery of catalyst species that are able to mediate highly stereoselective polymerisations at ambient temperature, ideally under melt conditions, is a priority for this area. While most activity in this area has focussed on the development of stereoselective catalysts for lactide ring-opening polymerisation, there also exists significant scope to extend the use of existing and new, 'tailor-made' catalysts to a wider range of monomers such as epoxides, acrylates and many other monomers that contain chiral or prochiral centres. The development of polymers with defined stereochemistry is important to their ultimate application and is likely to become a more important area of study as the rich stereochemistry of naturally sourced monomers is further explored and exploited.

Another emerging area in organocatalysis is the use of light and other external stimuli to activate catalysis. Perhaps the initial report in this area focussed on the thermal 'on demand' activation of triazolium carbene catalysts for the ring-opening polymerisation of lactide, however the rise of light activated chemistries has seen interest in this area rise on account of the ability to enable spatio-temporal control in polymerisation. While the area of organocatalytic ATRP has widely used light and redox triggers to activate/deactivate polymerisation, the use of light to activate/deactivate ring-opening polymerisation is in its infancy with only a handful of examples reported.[13–18] Most only consider the single activation of acids or bases for mediating polymerisation; however, the examples of switchable N-heterocyclic carbene[16] or phenol/ketone tautomer[18] organocatalysts provides the ability to switch activity on and off. Most recently, this concept was extended to moderate activity rather than be applied as an empirical switch.[18] This approach allowed the control over the monomer incorporation in the copolymerisation of δ-valerolactone and trimethylene carbonate, which presents interesting opportunities to create materials with precisely defined degradation profiles and behaviour. No doubt extension of these concepts to other polymerisations and techniques will yield further advances in materials design that could not be achieved in another manner.

Further progress in the field of organic catalysis for polymerisation will no doubt be made by product diversification using existing species or those that are closely related. Some significant challenges have only been partially addressed but there are many opportunities to advance work in these areas or in areas of polymerisation that organic catalysis has not yet made an impact provide a challenge in which progress could be game changing. Further development in ATRP or other controlled radical polymerisation methodologies is inevitable and needed to advance this fledging field to compete with the more commonly used Cu-mediated processes. This area has come a long way in a short time and has much promise to create a wide range of acrylic polymers by organocatalytic methods. Some initial reports of the organocatalytic copolymerisation of oxirane and oxetane monomers with carbon dioxide have been made. At present, the resultant materials have not been able to access significant molecular weights, typically they are limited to *ca.* 5 $kg\,mol^{-1}$ although it is noted that these materials and molecular weights are well suited to application as polyols for the synthesis of polyurethanes. The importance of CO_2 as a feedstock make this an obvious area for further organocatalytic innovation. While some organo-mediated click reactions have been applied in step-growth processes, there remains more scope to exploit the efficiency of these reactions to create high molecular weight polymers that are new to the World and may have interesting and useful properties.

Beyond the seemingly possible, the synthesis of electroactive polymers remains dominated by metal-mediated reactions such as Stille couplings and Grignard metathesis (GRIM) polymerisation methodologies. These processes inevitably leave behind metal residues in the polymer that may affect their ultimate application and organocatalytic methods to create these materials could transform the area and make the chemistry even more accessible to a wider range of researchers in the same way that use of 1,8-diazabicyclo[5.4.0]undec-7-ene (DBU) has opened up the ring-opening polymerisation of lactide by overcoming the limitations and lack of ease of use of tin(II) octanoate.[19] Perhaps the ultimate challenge for the field will be the synthesis of polyolefins from unsaturated monomers. Some initial steps have been taken in the realisation of organocatalytic ring-opening metathesis polymerisation (ROMP).[20] Here, the ring strain of the olefin can be used to drive the polymerisation but publications are to date limited to the ROMP of norbornene as monomer but significant space exists to broaden the monomer scope as well as improve both the efficiency and speed of these methodologies. Ultimately it may lead to the discovery of organocatalytic routes for the polymerisation of unstrained olefins such as ethylene, propylene or isoprene.

While clearly innovation in organic catalyst design is coming from the polymer community, the broader field of organic catalyst design is one that is highly active in the mediation of organic chemistry transformations across a dizzying range of reactions. These catalyst species and reactions may present new opportunities for improved organocatalytic polymerisation or for the further diversification of the polymers that can be accessed *via* these routes. Some key catalytic innovations that might be brought across from the organic chemistry community include a greater focus on asymmetric synthesis, use in aqueous media or in multicomponent reactions among many other themes.

The future remains bright for organocatalyzed polymerisations. There has been a huge amount of progress in the last 15 years. With a continued drive to improve access to existing materials and access materials that are hard or impossible to access by more traditional metal-catalysed processes, new discoveries will continue to be made and drive the field forward.

References

1. F. Nederberg, E. F. Connor, M. Moller, T. Glauser and J. L. Hedrick, *Angew. Chem., Int. Ed.*, 2001, **40**, 2712–2715.
2. H. A. Brown and R. M. Waymouth, *Acc. Chem. Res.*, 2013, **46**, 2585–2596.
3. C. A. Bell, J. Y. Yu, I. A. Barker, V. X. Truong, Z. Cao, A. V. Dobrinyin, M. L. Becker and A. P. Dove, *Angew. Chem., Int. Ed.*, 2016, **55**, 13076–13080.
4. L. Mezzasalma, A. P. Dove and O. Coulembier, *Eur. Polym. J.*, 2017, **95**, 628–634.
5. C. Jehanno, I. Flores, A. P. Dove, A. J. Muller, F. Ruiperez and H. Sardon, *Green Chem.*, 2018, **20**, 1205–1212.
6. G. W. Nyce, T. Glauser, E. F. Connor, A. Mock, R. M. Waymouth and J. L. Hedrick, *J. Am. Chem. Soc.*, 2003, **125**, 3046–3056.
7. A. P. Dove, R. C. Pratt, B. G. G. Lohmeijer, D. A. Culkin, E. C. Hagberg, G. W. Nyce, R. M. Waymouth and J. L. Hedrick, *Polymer*, 2006, **47**, 4018–4025.
8. L. Zhang, F. Nederberg, R. C. Pratt, R. M. Waymouth, J. L. Hedrick and C. G. Wade, *Macromolecules*, 2007, **40**, 4154–4158.

9. R. C. Pratt, B. G. G. Lohmeijer, D. A. Long, R. M. Waymouth and J. L. Hedrick, *J. Am. Chem. Soc.*, 2006, **128**, 4556–4557.
10. O. Dechy-Cabaret, B. Martin-Vaca and D. Bourissou, *Chem. Rev.*, 2004, **104**, 6147–6176.
11. S. Naumann, A. W. Thomas and A. P. Dove, *Angew. Chem., Int. Ed.*, 2015, **54**, 9550–9554.
12. D. Montarnal, M. Capelot, F. Tournilhac and L. Leibler, *Science*, 2011, **334**, 965–968.
13. I. A. Barker and A. P. Dove, *Chem. Commun.*, 2013, **49**, 1205–1207.
14. C. K. Fu, J. T. Xu and C. Boyer, *Chem. Commun.*, 2016, **52**, 7126–7129.
15. P. K. Kuroishi and A. P. Dove, *Chem. Commun.*, 2018, **54**, 6264–6267.
16. B. M. Neilson and C. W. Bielawski, *Chem. Commun.*, 2013, **49**, 5453–5455.
17. X. Sun, J. P. Gao and Z. Y. Wang, *J. Am. Chem. Soc.*, 2008, **130**, 8130–8131.
18. F. Eisenreich, M. Kathan, A. Dallmann, S. P. Ihrig, T. Schwaar, B. M. Schmidt and S. Hecht, *Nature Catalysis*, 2018, DOI: 10.1038/s41929-018-0091-8.
19. B. G. G. Lohmeijer, R. C. Pratt, F. Leibfarth, J. W. Logan, D. A. Long, A. P. Dove, F. Nederberg, J. Choi, C. Wade, R. M. Waymouth and J. L. Hedrick, *Macromolecules*, 2006, **39**, 8574–8583.
20. K. A. Ogawa, A. E. Goetz and A. J. Boydston, *J. Am. Chem. Soc.*, 2015, **137**, 1400–1403.

Subject Index